T0401198

Bioimaging in Neurodegeneration

BIOIMAGING IN NEURODEGENERATION

EDITED BY

PATRICIA A. BRODERICK, PhD

Department of Physiology and Pharmacology,
City University of New York Medical School;
Department of Neurology, New York University
School of Medicine; NYU Comprehensive
Epilepsy Center, New York, NY

DAVID N. RAHNI, PhD

Department of Chemistry and Physical Sciences
Pace University, Pleasantville, NY

EDWIN H. KOLODNY, MD

Department of Neurology
New York University School of Medicine
New York, NY

999 Riverview Drive, Suite 208
Totowa, New Jersey 07512

humanapress.com

For additional copies, pricing for bulk purchases, and/or information about other Humana titles, contact Humana at the above address or at any of the following numbers: Tel.: 973-256-1699; Fax: 973-256-8341; E-mail:humana@humanapr.com; Website: humanapress.com

Cover design by Patricia F. Cleary.

Cover illustrations: FOREGROUND, TOP: (*left*) Transaxial slice at the level of the striatum showing uptake of ^{99}m Tc-TRODAT-1 in dopamine transporters in a normal healthy volunteer; (*right*) a patient with hemi-PD exhibits a unilateral decrease in the uptake of ^{99}m Tc-TRODAT-1 in the side contralateral to clinical symptoms, most severely in the putamen (Chapter 2, Figs. 1 and 3; *see* full captions and discussion on p. 15.) FOREGROUND, MIDDLE: Regional NAA/Cr decrease in AD (Chapter 9, Fig. 2; *see* complete caption on p. 98 and discussion on p. 96). FOREGROUND, BOTTOM: Proton MRS (TE = 144 ms) in Canavan's disease demonstrating marked elevation in NAA caused by aspartoacylase deficiency (Chapter 21, Fig. 11; *see* full caption on p. 253 and discussion on p. 251). BACKGROUND: Hippocampal and entorhinal cortex boundary definition (Chapter 9, Fig. 1; *see* full caption on p. 97 and discussion on p. 96).

This publication is printed on acid-free paper. ∞
ANSI Z39.48-1984 (American National Standards Institute) Permanence of Paper for Printed Library Materials.

eISBN 1-59259-888-9

Printed in the United States of America. 10 9 8 7 6 5 4 3 2 1

Library of Congress Cataloging-in-Publication Data

Bioimaging in neurodegeneration / edited by Patricia A. Broderick, David N.
Rahni, Edwin H. Kolodny.
p. cm.
Includes bibliographical references and index.

Additional material to this book can be downloaded from http://extras.springer.com.

ISBN 1-58829-391-2 (alk. paper)
1. Brain--Degeneration--Imaging. I. Broderick, Patricia A., 1949- II.
Rahni, David N. III. Kolodny, Edwin H.
RC394.D35B55 2005
616.8'04757--dc22

2004026624

Preface

Bioimaging is in the forefront of medicine for the diagnosis and treatment of neurodegenerative disease. Conventional magnetic resonance imaging (MRI) uses interactive external magnetic fields and resonant frequencies of protons from water molecules. However, newer sequences, such as magnetization-prepared rapid acquisition gradient echo (MPRAGE), are able to seek higher levels of anatomic resolution by allowing more rapid temporal imaging. Magnetic resonance spectroscopy (MRS) images metabolic changes, enabling underlying pathophysiologic dysfunction in neurodegeneration to be deciphered. Neurochemicals visible with proton ^1H MRS include *N*-acetyl aspartate (NAA), creatine/phosphocreatine (Cr), and choline (Cho); NAA is considered to act as an in vivo marker for neuronal loss and/or neuronal dysfunction. By extending imaging to the study of elements such as iron—elevated in several neurodegenerative diseases—laser microprobe studies have become extremely useful, followed by X-ray absorption fine-structure experiments.

Positron emission tomography (PET) and single-photon emission tomography (SPECT) have become important tools in the differential diagnosis of neurodegenerative diseases by allowing imaging of metabolism and cerebral blood flow. PET studies of cerebral glucose metabolism use the glucose analog [^{18}F] fluorodeoxyglucose analog ([^{18}F]FDG) and radioactive water ($H_2{}^{15}O$) and SPECT tracers use ^{99m}Tc-hexamethylpropylene amine oxime, (^{99m}Tc-HMPAO), and ^{99m}Tc-ethylcysteinate dimer (^{99m}Tc-ECD). Moreover, direct imaging of the nigrostriatal pathway with 6-[^{18}F]-fluoro-1-3,4-dihydroxyphenylalanine (FDOPA) in combination with PET technology, may be more effective at differentiating neurodegenerative diseases than PET or SPECT alone. Radioactive cocaine and the tropane analogs directly measure dopamine (DA) transporter binding sites and ^{99m}Tc-TRODAT-1 is a new tracer that could move imaging of the DA neuronal circuitry from the research environment to the clinic. [^{123}I] altropane SPECT may equal and further advance FDOPA PET.

Surgical treatments of neurodegenerative diseases are gaining attention as craniotomies become more routine, and as patients opt for surgery because they experience intractable responses to pharmacotherapy for neurodegeneration. These treatments fall into three categories: lesion ablation, deep brain stimulation (DBS), and restorative therapies such as nerve growth factor infusion or DA cell transplantation along the nigrostriatal pathway, particularly in Parkinson's disease. Also, electron micrographics image amyloid β aggregation in Alzheimer's disease (AD) and MRI (gadolinium enhanced) has been successfully exploited to image neuroinflammation in AD. MR-based volumetric imaging helps to predict the progression of AD via mild cognitive impairment (MCI) studies.

Novel neuroimaging technologies, such as neuromolecular imaging (NMI) with a series of newly developed BRODERICK PROBE® sensors, directly image neurotransmitters, precursors, and metabolites in vivo, in real time and within seconds, at separate and selective waveform potentials. NMI, which uses an electrochemical basis for detection, enables the differentiation of neurodegenerative diseases in patients who present with mesial versus neocortical temporal lobe epilepsy. In fact, NMI has some remarkable similarities to MRI insofar as there is technological dependence on electron and proton transfer, respectively, and further dependence is seen in both NMI and MRI on tissue composition such as lipids. NMI has already been joined with electrophysiological (EEG) and electromyographic (EMG) studies to enhance detection capabilities; the integration of NMI with MRI, PET, and SPECT can be envisioned as the next advance.

The tracer molecule, [^{11}C] α-methyl-L-tryptophan (AMT) is already used with PET to study serotonin (5-HT) deficiencies, presumably attributable to kynurenine enhancement in neocortical epilepsy patients. Moreover, AMT PET, in addition to FDG PET, provides reliable diagnosis for pediatric epilepsy syndromes such as West's syndrome. Important in children with cortical dysplasia (CD), FDG PET delineates areas of altered glucose, which can be missed by MRI. The new tracer, [^{11}C] flumazenil used with PET (FMZ PET), has found utility in the detection of epileptic foci in CD patients with partial epilepsies, and yet normal structural imaging is observed. Another new 5-HT1A tracer for PET imaging in abnormal dysplastic tissue is a carboxamide compound called [^{18}F]FCWAY.

Diagnosis of neocortical epilepsy has been significantly advanced by IOS or intrinsic optical signal imaging. IOS has its basis in the light absorption properties of electrophysiologically active neural tissue, activity caused by focal alterations in blood flow, oxygenation of hemoglobin, and scattering of light. IOS can map interictal spikes, onsets and offsets, and horizontal propagation lines. Thus, IOS is useful for diagnosing "spreading epileptiform depression." As with NMI, IOS holds promise for intraoperative cortical mapping wherein ictal and interictal margins can be more clearly defined. As does intraoperative MRI (iMRI) with neuronavigation, these technologies provide what is called "guided neurosurgery." Correlative imaging of general inhalational anesthetics such as nitrous oxide (N_2O) during intraoperative surgery is made possible by NMI technologies with nano- and microsensors.

NMI and MRI also enable the differential detection of white matter versus gray matter in discrete neuroanatomic substrates in brain, detection which is critical to both the epilepsies and the leukodystrophies. Although NMI is in its early stage in this arena, the immediate and distinct waveforms that distinguish white from gray matter are impressive. Moreover, the early finding of a leukodystrophy by MRI, particularly relevant for metachromatic leukodystrophy (MLD), Krabbe's disease (KD), and X-linked adrenoleukodystophy (ALD), allows clinicians therapeutic interventions before overt symptoms are exhibited. Imaging technologies, pathologies, clinical features, and treatments for these and other leukodystrophies, including peroxisomal disorders and leukodystrophies with macrocrania (Canavan's disease and Alexander's disease), are presented here in precise detail. The van der Knaap syndrome is a recently described leukodystrophy in vacuolating megaloencephalic leukoencephalopathy (VMI). This vanishing white matter disease highlights the potential of MRS imaging, which was used in its identification.

Bioimaging in Neurodegeneration provides extensive detail on pediatric mitochondrial disease, including imaging, pathologies, clinical features, and treatment or lack of treatment. It is extremely important to note that in pediatric mitochondrial cytopathies, a frequent finding on MRI is abnormal myelination, and infants with leukoencephalopathies, especially leukodystrophies, should be evaluated for mitochondrial cytopathy. Infarct-like, often transient lesions not confined to vascular territories are the imaging hallmark of mitochondrial myopathy, encephalopathy, lactic acidosis, and stroke-like episodes (MELAS). [^{31}P] MRS, which can measure transient changes in nonoxidative adenosine triphosphate (ATP) synthesis, and [^{1}H] MRS, which can measure lactate, are included in the mitochondrial imaging technologies.

Thus, *Bioimaging in Neurodegeneration* fulfills the current need to bring together neurodegeneration with bio- and neuroimaging technologies that actually enable diagnosis and treatment. Professionals in neurology, psychiatry, pharmacology, radiology, and surgery are among many who will greatly benefit. Neurodegenerative disease is divided into four areas, i.e., Parkinson's disease, Alzheimer's disease, the epilepsies, and the leukodystrophies. Chapter authors were selected for their formidable expertise in each field of medicine, their expertise in imaging technologies, and their scholarly contributions to medicine and science. Our appreciation is extended to them, and their staffs, for their fine research. We thank the editors and staff at Humana Press for their excellent assistance and support.

Patricia A. Broderick, PhD
David N. Rahni, PhD
Edwin H. Kolodny, MD

Contents

Preface v

Contributors ix

Companion CD *(Inside Back Cover)* xi

Prologue: *Nano- and Microimaging Surgical Anesthesia in Epilepsy Patients* xiii
Patricia A. Broderick, David N. Rahni, and Steven V. Pacia

I. PARKINSON'S DISEASE

1 Magnetic Resonance Imaging and Magnetic Resonance Spectroscopy in Parkinson's Disease: *Structural vs Functional Changes* 3
W. R. Wayne Martin

2 Positron Emission Tomography and Single-Photon Emission Tomography in the Diagnosis of Parkinson's Disease: *Differential Diagnosis From Parkinson-Like Degenerative Diseases* 13
Paul D. Acton

3 Positron Emission Tomography in Parkinson's Disease: *Cerebral Activation Studies and Neurochemical and Receptor Research* 25
André R. Troiano and A. Jon Stoessl

4 [^{123}I]-Altropane SPECT: *How It Compares to Other Positron Emission Tomography and Single-Photon Emission Tomography Dopamine Transporters in Early Parkinson's Disease* 37
Hubert H. Fernandez, Paula D. Ravin, and Dylan P. Wint

5 Positron Emission Tomography and Embryonic Dopamine Cell Transplantation in Parkinson's Disease 45
Yilong Ma, Vijay Dhawan, Curt Freed, Stanley Fahn, and David Eidelberg

II. ALZHEIMER'S DISEASE

6 Neurotoxicity of the Alzheimer's β-Amyloid Peptide: *Spectroscopic and Microscopic Studies* 61
David R. Howlett

7 Functional Imaging and Psychopathological Consequences of Inflammation in Alzheimer's Dementia 75
Jan Versijpt, Rudi A. Dierckx, and Jakob Korf

8 Neurotoxic Oxidative Metabolite of Serotonin: *Possible Role in Alzheimer's Disease* 85
Ladislav Volicer, Monika Z. Wrona, Wayne Matson, and Glenn Dryhurst

9 Predicting Progression of Alzheimer's Disease With Magnetic Resonance 95
Kejal Kantarci and Clifford R. Jack, Jr.

10 Stages of Brain Functional Failure in Alzheimer's Disease: *In Vivo Positron Emission Tomography and Postmortem Studies Suggest Potential Initial Reversibility and Later Irreversibility* 107
Stanley I. Rapoport

III. EPILEPSY

11 Neocortical Epilepsy: *α-Methyl-L-Tryptophan and Positron Emission Tomography Studies* 123
Jun Natsume, Andrea Bernasconi, and Mirko Diksic

12 Pediatric Cortical Dysplasia: *Positron Emission Tomography Studies* 131
Bharathi Dasan Jagadeesan, Csaba Juhász, Diane C. Chugani, and Harry T. Chugani

13 Bioimaging L-Tryptophan in Human Hippocampus and Neocortex: *Subtyping Temporal Lobe Epilepsy* 141
Steven V. Pacia and Patricia A. Broderick

14 In Vivo Intrinsic Optical Signal Imaging of Neocortical Epilepsy 149
Sonya Bahar, Minah Suh, Ashesh Mehta, and Theodore H. Schwartz

15 Intraoperative Magnetic Resonance Imaging in the Surgical Treatment of Epilepsy 177
Theodore H. Schwartz

16 Periodic Epileptiform Discharges Associated With Increased Cerebral Blood Flow: *Role of Single-Photon Emission Tomography Imaging* 193
Imran I. Ali and Noor A. Pirzada

17 Imaging White Matter Signals in Epilepsy Patients: *A Unique Sensor Technology* 199
Patricia A. Broderick and Steven V. Pacia

IV. LEUKODYSTROPHY (WHITE MATTER) DISEASES

18 Overview of the Leukoencephalopathies: *An MRI Point of View* 209
Edwin H. Kolodny

19 Pyramidal Tract Involvement in Adult Krabbe's Disease: *Magnetic Resonance Imaging and Proton Magnetic Resonance Spectroscopy Abnormalities* 215
Laura Farina, Alberto Bizzi, and Mario Savoiardo

20 Imaging Leukodystrophies: *Focus on Lysosomal, Peroxisomal, and Non-Organelle Pathology* 225
Annette O. Nusbaum

21 Advanced Magnetic Resonance Imaging in Leukodystrophies 239
Edwin Y. Wang and Meng Law

22 Childhood Mitochondrial Disorders and Other Inborn Errors of Metabolism Presenting With White Matter Disease 261
Adeline Vanderver and Andrea L. Gropman

23 Mitochondrial Disease: *Brain Oxidative Metabolism Studied by* ^{31}P, ^{1}H, *and* ^{13}C *Magnetic Resonance Spectroscopy, Functional Magnetic Resonance Imaging, and Positron Emission Tomography* 297
Graham J. Kemp

Index 309

Contributors

PAUL D. ACTON, PhD • *Department of Radiology, University of Pennsylvania, Philadelphia, PA*

IMRAN I. ALI, MD • *Comprehensive Epilepsy Program, Medical College of Ohio, Toledo, OH*

SONYA BAHAR, PhD • *Department of Neurological Surgery, Weill Cornell Medical College, New York Presbyterian Hospital, New York, NY*

ANDREA BERNASCONI, MD • *Department of Neurology and Neurosurgery, McGill University, Montreal, QC, Canada*

ALBERTO BIZZI, MD • *Department of Neuroradiology, Instituto Nazionale Neurologico "C. Besta," Milan, Italy*

PATRICIA A. BRODERICK, PhD • *Department of Physiology and Pharmacology, City University of New York Medical School; Department of Neurology, New York University School of Medicine; NYU Comprehensive Epilepsy Center, New York, NY*

DIANE C. CHUGANI, PhD • *Department of Pediatrics, Wayne State University; PET Center, Children's Hospital of Michigan, Detroit, MI*

HARRY T. CHUGANI, MD • *Department of Pediatrics, Wayne State University; PET Center, Children's Hospital of Michigan, Detroit, MI*

VIJAY DHAWAN, PhD • *Center for Neurosciences, North Shore-Long Island Jewish Research Institute, New York University School of Medicine, Manhasset, NY*

RUDI A. DIERCKX, MD, PhD • *Department of Nuclear Medicine, University Hospital of Gent, Gent, Belgium*

MIRKO DIKSIC, PhD • *Department of Neurology and Neurosurgery, McGill University, Montreal, QC, Canada*

GLENN DRYHURST, PhD • *Department of Chemistry and Biochemistry, University of Oklahoma, Norman, OK*

DAVID EIDELBERG, MD • *Center for Neurosciences, North Shore-Long Island Jewish Research Institute, New York University School of Medicine, Manhasset, NY*

STANLEY FAHN, MD • *Department of Neurology, Columbia College of Physicians and Surgeons, New York, NY*

LAURA FARINA, MD • *Department of Neuroradiology, Instituto Nazionale Neurologico "C. Besta," Milan, Italy*

HUBERT H. FERNANDEZ, MD • *Department of Neurology, McKnight Brain Institute/University of Florida, Gainesville, FL*

CURT FREED, MD • *Neuroscience Center and Division of Clinical Pharmacology and Toxicology, University of Colorado Health Sciences Center, Denver, CO*

ANDREA L. GROPMAN, MD, FAAP, FACMG • *Division of Genetics and Metabolism, Departments of Pediatrics and Neurology, Georgetown University, Washington, DC*

DAVID R. HOWLETT, PhD • *Neurology & G.I. Centre of Excellence for Drug Discovery, GlaxoSmithKline, Harlow, Essex, UK*

CLIFFORD R. JACK, JR., MD • *Department of Radiology, Mayo Clinic, Rochester, MN*

BHARATHI DASAN JAGADEESAN, MD • *Department of Pediatrics, Wayne State University; PET Center, Children's Hospital of Michigan, Detroit, MI*

CSABA JUHÁSZ, MD, PhD • *Department of Pediatrics, Wayne State University; PET Center, Children's Hospital of Michigan, Detroit, MI*

KEJAL KANTARCI, MD • *Department of Radiology, Mayo Clinic, Rochester, MN*

GRAHAM J. KEMP, MA, DM, FRCPath, ILTM • *Division of Metabolic and Cellular Medicine, University of Liverpool, Liverpool, UK*

EDWIN H. KOLODNY, MD • *Department of Neurology, NYU School of Medicine, New York, NY*

JAKOB KORF, PhD • *Department of Biological Psychiatry, University Hospital of Groningen, Groningen, The Netherlands*

MENG LAW, MD, FRACR • *Department of Radiology, NYU Medical Center, New York, NY*

YILONG MA, PhD • *Center for Neurosciences, North Shore–Long Island Jewish Research Institute, New York University School of Medicine, Manhasset, NY*

W. R. WAYNE MARTIN, MD, FRCPC • *Movement Disorder Clinic, University of Alberta/Glenrose Rehabilitation Hospital, Edmonton, Alberta, Canada*

WAYNE MATSON PhD • *Systems Biology Section, Edith Nourse Rogers Memorial Veterans Hospital, VA New England Health Care System, Bedford, MA*

ASHESH MEHTA, MD, PhD • *Department of Neurological Surgery, Weill Cornell Medical College, New York Presbyterian Hospital, New York, NY*

JUN NATSUME, MD, PhD • *Department of Pediatrics, Japanese Red Cross Nagoya First Hospital, Nagoya, Aichi, Japan*

ANNETTE O. NUSBAUM, MD • *Department of Radiology, NYU School of Medicine, New York, NY*

STEVEN V. PACIA, MD • *Department of Neurology, NYU School of Medicine; NYU Comprehensive Epilepsy Center, New York, NY*

NOOR A. PIRZADA, MD • *Comprehensive Epilepsy Program, Medical College of Ohio, Toledo, OH*

DAVID N. RAHNI, PhD • *Department of Chemistry and Physical Sciences, Pace University, Pleasantville, NY*

STANLEY I. RAPOPORT, MD • *Brain Physiology and Metabolism Section, National Institute on Aging, National Institute on Health, Bethesda, MD*

PAULA D. RAVIN, MD • *Department of Neurology, University of Massachusetts Medical Center, Worcester, MA*

MARIO SAVOIARDO, MD • *Department of Neuroradiology, Instituto Nazionale Neurologico "C. Besta," Milan, Italy*

THEODORE H. SCHWARTZ, MD • *Department of Neurological Surgery, Weill Cornell Medical College, New York Presbyterian Hospital, New York, NY*

A. JON STOESSL, MD, FRCPC • *Pacific Parkinson's Research Centre, University of British Columbia, Vancouver, BC, Canada*

MINAH SUH, PhD • *Department of Neurological Surgery, Weill Cornell Medical College, New York Presbyterian Hospital, New York, NY*

ANDRÉ R. TROIANO, MD • *Pacific Parkinson's Research Centre, University of British Columbia, Vancouver, BC, Canada*

ADELINE VANDERVER, MD • *Child Neurology, Children's National Medical Center, Washington, DC*

JAN VERSIJPT, MD, PHD • *Department of Nuclear Medicine, University Hospital of Gent, Belgium; Department of Biological Psychiatry, University Hospital of Groningen, Groningen, The Netherlands*

LADISLAV VOLICER, MD, PhD • *Boston University School of Medicine and Edith Nourse Rogers Memorial Veterans Hospital, VA New England Health Care System, Bedford, MA*

EDWIN Y. WANG, MD • *Department of Radiology, NYU Medical Center, New York, NY*

DYLAN P. WINT, MD • *Department of Psychiatry, University of Florida/McKnight Brain Institute, Gainesville, FL*

MONIKA Z. WRONA, PhD • *Department of Chemistry and Biochemistry, University of Oklahoma, Norman, OK*

Companion CD

for *Bioimaging in Neurodegeneration*

Color versions of illustrations listed here may be found on the Companion CD attached to the inside back cover. The image files are organized into folders by chapter number and are viewable in most Web browsers. The number following "f" at the end of the file name identifies the corresponding figure in the text. The CD is compatible with both Mac and PC operating systems.

Chapter 2 Figs. 1–5

Chapter 3 Fig. 2

Chapter 5 Figs. 1, 4

Chapter 7 Fig. 1

Chapter 10 Figs. 2, 3, and 6

Chapter 11 Fig. 1

Chapter 14 Figs. 1, 3, 6, 8, 9, 11–17

Chapter 15 Figs. 1–5, 11

Chapter 16 Figs. 2–5

Prologue

Nano- and Microimaging Surgical Anesthesia in Epilepsy Patients

Patricia A. Broderick, PhD, David N. Rahni, PhD, and Steven V. Pacia, MD

Nitrous oxide (N_2O) is a simple and small molecule, consisting of two nitrogen atoms and one oxygen atom (Fig. 1). Yet, its anesthetic, analgesic, and psychotropic properties are indisputable *(1–3)*. Nitrous oxide is reported to act via opiate mechanisms because it induces met-enkephalin and β-endorphin release in rat and human, and the antinociceptic properties of nitrous oxide are reversible by naloxone *(4,5)*. Also, but likely not exclusively, nitrous oxide may exert its effects via glutamate receptors, that is, administration of (80%) nitrous oxide to rat hippocampus depresses excitatory currents evoked by *N*-methyl-D-aspartate *(6,7)*.

The combination of nitrous oxide and oxygen has found its way into prehospital emergency treatment of pain *(2)*. Under the proprietary names, Entonos® and Dolonox®, this combination in a 40–60% ratio is used by paramedics when treating acute myocardial infarction *(8)*. In some areas of the world, it is used in emergency medicine in lieu of opioid analgesics for the management of painful injuries *(9)*.

In the hospital setting, intraoperatively nitrous oxide is used adjunctly with other general anesthetics for its well-known "second gas effect," a phenomenon that is caused by its ability to diffuse quickly from alveoli. However, nitrous oxide, even in combination with oxygen, rarely is used alone in surgery because it is a relatively weak general anesthetic (low blood/gas solubility partition coefficient).

Interestingly, in studies used to map the effects of analgesics on pain, cerebral substrates for the nociceptive effects of nitrous oxide have been identified. Using low concentrations (20%) nitrous oxide was imaged using positron emission tomography and cerebral blood flow (rCBF). Inhalation of 20% nitrous oxide was found to be associated with enhanced rCBF in the anterior cingulate cortex (area 24), decreased rCBF in the hippocampus, posterior cingulate (areas 23,24), and decreased rCBF in the secondary visual cortices (areas 18,19; ref. *10*).

Despite the importance of this small and simple molecule in surgery, emergency medicine, and dentistry alone, there are virtually little or no direct techniques available to detect nitrous oxide unchanged in living tissue. Our purpose here is to present such a technique using neuromolecular imaging (NMI) and carbon based nano- and microsensors.

We describe the experimental design for and the results from in vitro assays, i.e., studies of nitrous oxide as N_2O is diffused into an electrochemical cell as well as those from in vivo assays, i.e., studies of nitrous oxide which has stabilized in living tissue from N_2O infusion.

This is the first report of the experimental assay for the gaseous solution, nitrous oxide and the results from such, in vitro. High purity (99.9%) commercially available nitrous oxide (T.W. Smith, Brooklyn, NY) was diffused using a flowmeter, calibrated at 10 psi, into an electrochemical cell containing saline/phosphate buffer for 5 min to allow the gas to reach saturation at room temperature. The flowmeter was purchased from Fisher Scientific (Bridgewater, NJ). Nitrous oxide concentrations in the approximate range of 10–100 μ*M* were achieved. Figure 2 shows a representative recording of nitrous oxide detection in vitro. Nitrous oxide detection occurred at the oxidation (half-wave) potential of 0.53 ± 0.02 V. In addition, DA and 5-HT signals are shown because increasing concentrations of DA and 5-HT were aliquoted into the electrochemical cell for use as standards. Thus, studies with the monoamines were conducted, which show the selective detection of nitrous oxide in the presence of the monoamine neurotransmitter.

Procedures for the detection of neurotransmitters and neurochemicals by BRODERICK PROBE® sensors are described *(11–20)*.

This is also the first report of the experimental assay for the gaseous solution, nitrous oxide, and the results from such, in vivo. Resected living tissue (hippocampal and neocortical) from temporal lobe epilepsy (TLE) patients was studied. Methods for patient classification and methods for delineating neurochemical profiles are previously published. Patients were administered nitrous oxide-oxygen anesthesia in a 40/60% concentration during intraoperative surgery. Figure 3 shows a representative recording from an NTLE patient, in vivo. These images from TLE patients show reliable nitrous oxide signals that occurred at the oxidation (half-wave) potential of 0.53 ±

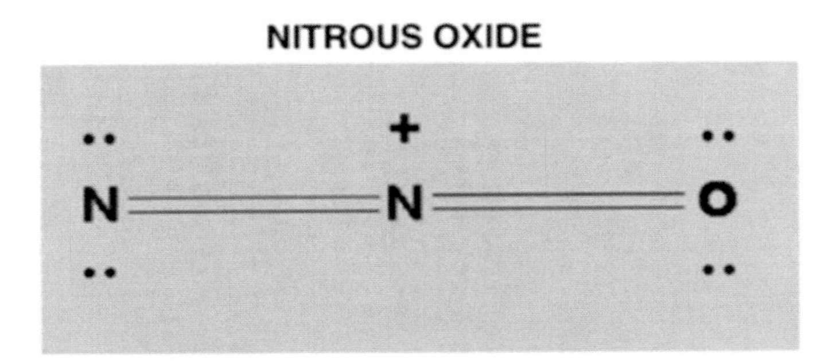

Fig. 1. The nitrous oxide molecule.

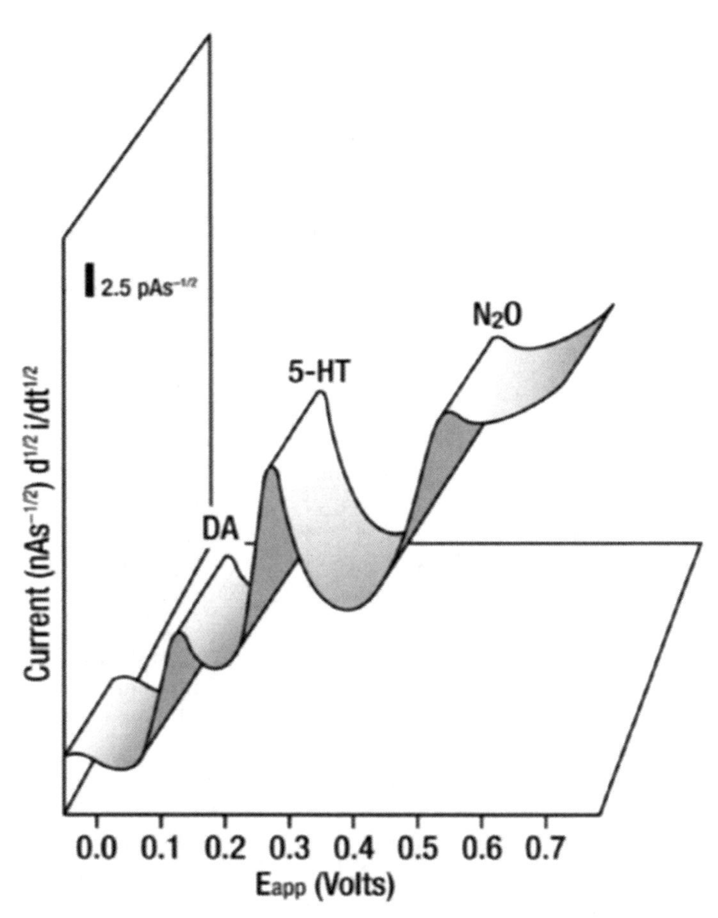

Fig. 2. A representative recording showing the detection of DA, 5-HT, and nitrous oxide (N_2O) in saline/phosphate buffer by the BRODERICK PROBE® stearate sensor in vitro. The oxidation (half-wave) potential for N_2O is 0.53 V ± 0.02 V. Oxidation potentials for monoamines confirm previous reports *(11–20)*. Note that the sensitivity of the sensor for the monoamine, 5-HT, is approximately two- to threefold greater than that for the monoamine DA as previously reported *(19)*.

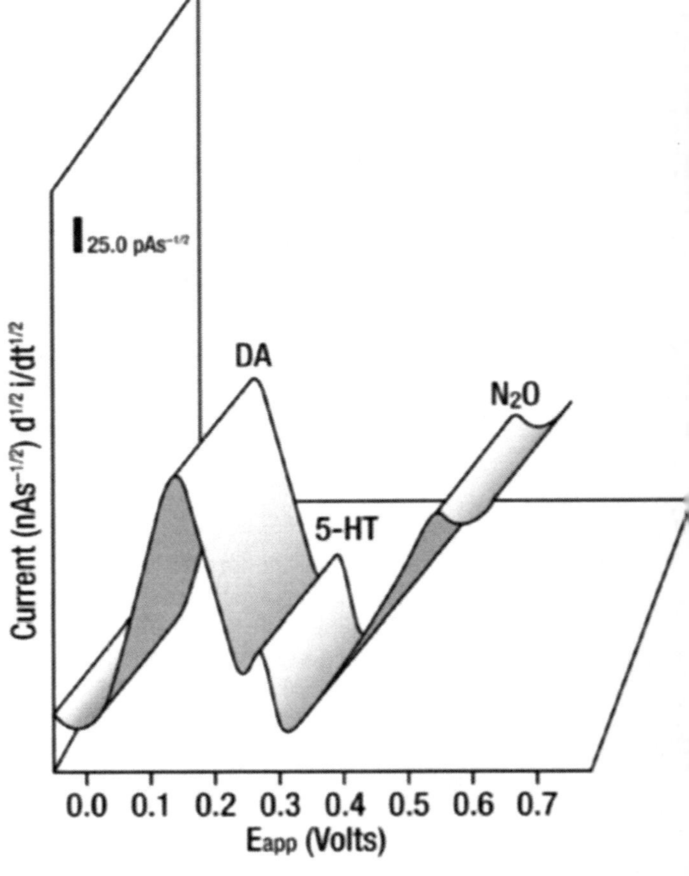

Fig. 3. A representative recording showing the detection of DA, 5-HT, and nitrous oxide (N_2O) in vivo in a TLE patient, specifically an NTLE patient, a subtype of TLE that typically exhibits DA and 5-HT; the study was performed with a BRODERICK PROBE® stearate sensor. The oxidation (half-wave) potential for N_2O is 0.53 V ± 0.02 V. Oxidation potentials for monoamines confirm previous reports *(11–18)*. Note that in vivo data exhibit greater concentrations for catecholamines than for indoleamines with the noted exception of white matter imaging *(20)*.

0.02 V, which is consistent with the detection of nitrous oxide at the same oxidation potential in vitro. Further evidence for the reliable detection of nitrous oxide comes from studies in this laboratory which has shown the separate detection of nitric oxide (NO) at an approximate oxidation potential of 0.75V *(21)*. These data are in general agreement with detection of NO using the carbon fiber electrode *(22)*. Moreover, the reliable detection of nitrous oxide comes from the known predictive ability to detect oxygen at early negative potentials, again separating the detection of oxygen from that of nitrous oxide. Finally, the stability of nitrous oxide at physiological temperatures is known (url: www.chm.bris.ac.uk/motm/n2o/n2oh.htm [*23;* retrieved on or about June 3, 2004]). Thus, these studies confirm the selective detection of nitrous oxide in the absence and presence of the monoamine neurotransmitters.

Table 1 shows the neuroanatomic location of nitrous oxide signals imaged in distinct neocortical neuroanatomic structures and hippocampal subparcellations. As expected, there was no

Table 1
Nitrous Oxide (N_2O) in Resected Temporal Tissues From Human Epilepsy Patients

Patient Number	*Epilepsy Type*	*N_2o Signals Imaged in:*	
2	MTLE	Neocortex (G)	
3	NTLE	Neocortex (G,W);	HPC (Granular cells)
4	NTLE	Neocortex (G,W);	HPC (Polymorphic layer)
5	MTLE	Neocortex (G)	
6	MTLE	HPC (Subiculum)	
8	MTLE	HPC (Pyramidal layer)	
9	NTLE	HPC (Pyramidal layer)	
10	NTLE	Neocortex (G)	
13	MTLE	Neocortex (G,W);	HPC (Polymorphic, Pyramidal, Mol. layers)
14	MTLE	Neocortex (G);	HPC (Subiculum)
15	MTLE	Neocortex (G,W);	HPC (Polymorphic layer)
16	MTLE	Neocortex (G,W)	

G is gray matter; W is white matter; Patients 7, 11, and 12: MTLE did not exhibit N_2O; NTLE, neocortical temporal lobe epilepsy; MTLE, mesial temporal lobe epilepsy.

apparent difference in the degree of nitrous oxide imaging between NTLE and MTLE patients, given the caveat that there were more MTLE patients than NTLE patients.

In summary, the significance of imaging nitrous oxide anesthesia in living tissue is paramount in neurology, neurosurgery, emergency medicine, toxicology, substance abuse (*see* ref. *24* for a recent review), and dentistry. Moreover, the BRODERICK PROBE® sensors image nitrous oxide signals on line yet separately from monoamines, metabolites, precursors, These data provide promise for optimizing such a technology for selective nano- and micro-monitoring and measuring nitrous oxide intraoperatively.

ACKNOWLEDGMENTS

Protocol for human studies was approved by CUNY and NYU Investigational Review Boards. We thank the FACES Campaign, Parents Against Childhood Epilepsy (PACE Foundation), and the National Institute of Health, NIH/NIGMS SCORE AWARD # SO 6 GM 08168 for partial financial support. The authors gratefully acknowledge Bridget T. O'Sullivan, O.P., M.A., Msgr. Scanlan High School, and Karen Schulz, Humana Press, for secretarial and artistic assistance, respectively. We also thank Ratna Medicherla, CUNY medical student, for assistance with formula and table format.

REFERENCES

1. Archer WH. Life and letters of Horace Wells, discoverer of anesthesia. J Am Coll Dent 1944;11:81.
2. Thompson PL, Lown B. Nitrous oxide as an analgesic in acute myocardial infarction. J Am Med Assoc 1976;235:924.
3. Gillman MA. Nitrous oxide abuse in perspective. Clin Neuropharmacol 1992;15:297.
4. Zuniga JR, Joseph SA, Knigge KM. The effects of nitrous oxide on the central endogenous pro-opiomelanocortin system in the rat. Brain Res 1987;420:57–65.
5. Chapman CR, Benidetti C. Nitrous oxide effects on cerebral evoked potential to pain: Partial reversal with a narcotic antagonist. Anesthesiology 1979;51:135–138.
6. Mennerick S, Jevtovic-Todorovic V, Todorovic SM, Shen W, Olney JW, Zorumski CF. Effect of nitrous oxide on excitatory and inhibitory synaptic transmission in hippocampal cultures. J Neurosci 1998;18:9716–9726.
7. Jevtovic-Todorovic V, Todorovic SM, et al. Nitrous oxide (laughing gas) is an NMDA antagonist, neuroprotectant and neurotoxin. Nat Med 1998;4:460–463.
8. O'Leary U, Puglia C, Friching T D, Kowey PR. Nitrous oxide anesthesia in patients with ischemic chest discomfort: effect on beta-endorphins. J Clin Pharmacol 1987;27:957.
9. Malamed SF, Clark MS. Nitrous oxide-oxygen: a new look at a very old technique. J Calif Dent Assoc 2003;31:397–403.
10. Gyulai F, Firestone LL, Mintum M, Winter P. In Vivo Imaging of human limbic responses to nitrous oxide inhalation. Anesth Analg 1996;83:291–298.
11. Broderick PA, Pacia SV, Doyle WK, Devinsky O. Monoamine neurotransmitters in resected hippocampal subparcellations from neocortical and mesial temporal lobe epilepsy patients: in situ microvoltammetric studies. Brain Res 2000;878:49–63.
12. Pacia SV, Doyle WK, Broderick PA. Biogenic amines in the human neocortex in patients with neocortical and mesial temporal lobe epilepsy: identification with in situ micovoltammetry. Brain Res 2001;899:106–111.
13. Pacia SV, Broderick PA. Bioimaging L-tryptophan in human hippocampus and neocortex: subtyping temporal lobe epilepsy. In: Broderick PA, Rahni DN, Kolodny EH, editors. Bioimaging in Neurodegeneration, Totowa, NJ: Humana Press, 2005, pp. 141–147.
14. Broderick PA. Distinguishing in vitro electrochemical signatures for norepinephrine and dopamine. Neurosci Lett 1988;95:275–280.
15. Broderick PA. Characterizing stearate probes in vitro for the electrochemical detection of dopamine and serotonin. Brain Res 1989;495:115–121.
16. Broderick PA. Microelectrodes and their use in cathodic electrochemical arrangement with telemetric application. 1995; US Patent #5,433,710.1996; European Patent #90914306.7.

17. Broderick PA. Microelectrodes and their use in an electrochemical arrangement with telemetric application. 1999; US Patent #5,938,903.
18. Broderick, PA, Pacia SV. Identification, diagnosis and treatment of neuropathologies, neurotoxicities, tumors and brain and spinal cord injuries using microelectrodes with microvoltammetry. 2002;#PCT/USO2/11244. Pending 2002; US 10/118,571. Pending.
19. Broderick PA. Striatal neurochemistry of dynorphin-(1–13): In vivo electrochemical semidifferential analyses. Neuropeptides 1987;10: 369–386.
20. Broderick PA, Pacia SV. Imaging white matter signals in epilepsy patients: A unique sensor technology. In: Broderick PA, Rahni DN, Kolodny EH, editors. Bioimaging in Neurodegeneration, Totowa, NJ: Humana Press, 2005, pp. 199–206.
21. Rahni DN, Pacia SV, Broderick PA, A novel microvoltammetric approach for the determination of nitrous and nitric oxides: human epilepsy. Soc. Neurosci. Abstr. 2001, Orlando, Fl.
22. Crespi F, Campagnola M, Neudeck A, et al. Can voltammetry measure nitrogen monoxide (NO) and/or nitrites? J Neurosci Methods 2001;109(1):59–70.
23. Url: www.chm.bris.ac.uk/motm/n$_2$o/n$_2$oh.htm [retrieved on or about June 3rd, 2004]).
24. Bonson KR, Baggott M. Emerging drugs of abuse: use patterns and clinical toxicity. In: Massao EJ, Broderick PA, Mattson JL, Schardein JL, Schlaepfer TE, editors. Handbook of Neurotoxicology, Vol. 2, Totowa, NJ: Humana Press, 2002, pp. 223–257.

PARKINSON'S DISEASE I

1 Magnetic Resonance Imaging and Magnetic Resonance Spectroscopy in Parkinson's Disease

Structural vs Functional Changes

W. R. Wayne Martin, MD, FRCP

SUMMARY

At present, conventional magnetic resonance imaging (MRI) shows no convincing structural changes in Parkinson's disease (PD) itself, but it may be useful in helping to distinguish PD from other neurodegenerative parkinsonian syndromes. Magnetic resonance spectroscopy (MRS) also may provide useful information in distinguishing PD from disorders such as multiple system atrophy. The general field of MRI and MRS is evolving rapidly, and a number of new developments may provide relevant information. Novel pulse sequences, for instance, may provide more information regarding substantia nigra pathology in PD. The use of MR technologies to measure regional concentrations of brain iron should provide more information regarding the relationship between iron accumulation and parkinsonian symptoms. MRS provides a sensitive tool to investigate the possible contribution of abnormal brain energy metabolism to the pathogenesis of PD. MRS also allows the assessment of other metabolite changes in PD, for example, providing for the evaluation of associated changes in regional brain glutamate content. Last, functional MRI provides the potential to evaluate, in a noninvasive fashion, the role played by the basal ganglia in motor control and cognition in normal individuals as well as in PD.

Key Words: Parkinson's disease; T_2-weighted imaging; T2* effect; brain iron; substantia nigra imaging; progressive supranuclear palsy; hummingbird sign; multiple system atrophy; corticobasal degeneration; voxel-based morphometry; diffusion-weighted imaging; magnetic resonance spectroscopy; brain energy metabolism; mitochondrial function; ^{31}P-magnetic resonance spectroscopy; functional MRI; event-related fMRI.

From: *Bioimaging in Neurodegeneration*
Edited by P. A. Broderick, D. N. Rahni, and E. H. Kolodny

1. INTRODUCTION

The continuing evolution of new techniques for imaging the central nervous system has produced significant advances in the investigation of patients with neurodegenerative disorders and in our understanding of basal ganglia function. Although magnetic resonance imaging (MRI) has made possible the correlation of structural abnormalities identified in vivo with specific neurologic syndromes such as parkinsonism, changes in cerebral function do not always parallel changes in structure. Magnetic resonance spectroscopy (MRS) has provided insights into some of the underlying metabolic abnormalities, thereby providing further insights relating to the underlying pathophysiology of these neurodegenerative syndromes. Brain activation studies with functional MR imaging (fMRI) have provided additional information regarding the abnormalities in brain function associated with these disorders.

2. MRI IN PARKINSONISM

Conventional MRI is based primarily on the interplay between external magnetic fields and the resonant frequency of water protons in tissue. Image contrast is related to the specific imaging parameters used but typically represents a complex function of proton density and the longitudinal (spin-lattice) relaxation time (T_1) and transverse (spin–spin) relaxation time (T_2) of protons in tissue. Inhomogeneities in the magnetic field induced by tissue attributes also have an important effect on image contrast, termed T_2*. T_1-weighted images tend to display superior gray–white matter differentiation compared with T_2-weighted sequences, allowing a clear delineation, for example, of the head of the caudate nucleus from the lenticular nucleus and of the cerebral cortex from adjacent white matter. Newer sequences, such as magnetization-prepared rapid acquisition gradient echo (MPRAGE), allow for very short imaging times and high anatomical resolution. Volumetric studies using MPRAGE sequences with multiple thin slices are useful in studies that quantify tissue atrophy. On T_2-weighted images from high field strength MRI systems (1.5 T and greater), the lenticular nucleus is readily subdivided into the globus pallidus and putamen, with the former structure displaying reduced signal intensity as compared with the latter. This differentiation is not present at birth but becomes evident within the first year or two of life, gradually increasing through the first three decades. The pallidal signal then remains relatively constant until the sixth or seventh decade, after which the signal attenuation becomes more prominent. Similar areas of reduced signal are also seen in the midbrain (red nucleus and substantia nigra pars reticulata), the dentate nucleus and, to a lesser extent, the putamen.

The prominent low-signal regions on T_2-weighted images correlate with sites of ferric iron accumulation as determined in vitro by Perls' Prussian blue stain *(1)*. Tissue iron produces a local inhomogeneity in the magnetic field that dephases proton spins, resulting in signal loss and decreased T_2 relaxation times (the T_2* effect). This iron susceptibility effect is best observed on heavily T_2-weighted images *(2,3)*. The T_2 changes are related to the strength of the static magnetic field and are not observed in normal individuals who are imaged with low field strength systems. It is important to note that gradient echo sequences are much more sensitive to iron-induced susceptibility changes than are the turbo spin echo (or fast spin echo) sequences that have become the routine method for producing T_2-weighted images on many clinical magnets.

In the early days of clinical MRI, the anatomical detail evident in images of the midbrain led investigators to evaluate this technology in patients with Parkinson's disease (PD), in whom the major neuropathological changes relate to neuronal loss from the midbrain substantia nigra pars compacta. As noted previously, T_2-weighted images show a prominent low signal area in the red nucleus and the substantia nigra pars reticulata, structures which are separated by the substantia nigra pars compacta. A narrowing, or smudging of this high signal zone separating the red nucleus and the pars reticulata has been reported in PD, consistent with the well-established pathological involvement in this area *(4,5)*. However, conventional MRI is not sufficiently sensitive at present to detect these changes in routine clinical applications. Although nigral changes in PD may be detected in population studies, technical factors, such as slice thickness, partial volume averaging, and head positioning, make it difficult to define reproducible abnormalities in individual patients in a structure as small as the substantia nigra. An alternative MRI approach to the study of midbrain pathology has been reported recently. This approach uses two inversion recovery pulse sequences, based on the hypothesis that contrast in T_1-weighted imaging depends mainly on the intracellular space and that T_1-weighted sequences are sensitive to the changes in intracellular volume that occur with cell death *(6)*. One sequence is designed to suppress peduncular white matter and the other to suppress nigral gray matter. These investigators reported structural changes in the nigra with this technique, even in the earliest cases of symptomatic disease. Hu et al.reported that structural changes in the nigra of patients with PD detected with inversion recovery sequences correlate with measures of striatal dopaminergic function using fluorodopa/positron emission tomography (PET; *7*).

Conventional MRI is of value in helping to differentiate PD from other neurodegenerative parkinsonian disorders. In progressive supranuclear palsy (PSP), MRI has been reported to delineate atrophy of the midbrain with a dilated cerebral aqueduct and enlarged perimesencephalic cisterns *(8)*. Patients with PSP have significantly decreased midbrain diameter than patients with PD and control subjects *(9)*. Atrophy of the rostral midbrain tegmentum produces the "hummingbird sign" on T_1-weighted midsagittal images, possibly corresponding to involvement of the rostral interstitial nucleus of the medial longitudinal fasciculus, a structure involved in the control of vertical eye movements *(10)*. Changes in the superior colliculus also may correlate with the eye findings, which are prominent in this condition *(11)*. Increased signal in periaqueductal regions also may be seen, coincident with the neuropathologic finding of gliosis in this region *(12)*. This technology is particularly valuable in the differential diagnosis of PSP, with up to 33% of patients with the typical clinical presentation having evidence of multiple cerebral infarcts on MRI *(13)*. Typical changes in a patient with PSP are illustrated in Fig. 1. In corticobasal degeneration, decreased signal intensity in the lenticular nucleus on T_2-weighted images, ventricular enlargement, and asymmetrical cortical atrophy has been reported *(14)*.

Imaging changes in multiple system atrophy (MSA), which help to differentiate this condition from PD, have been reported *(15–18)*. MSA typically is classified as a parkinsonian subtype (MSA-P), previously known as striatonigral degeneration (SND), and a cerebellar subtype (MSA-C), previously called olivopontocerebellar atrophy. The most widely reported change in MSA has been the presence, particularly on high field strength instruments, of a low signal in the putamen on T_2-weighted images. A "slit-like void" in the putamen on T_2-weighted images (and to a lesser degree on T_1-weighted images) in patients with SND has been reported, which Lang et al. *(19)* suggest is characteristic, if not pathognomonic, of this disorder. In this study, the MRI change correlated with the extent of neuronal loss and gliosis and with the pattern of iron deposition evident at autopsy. These authors and others *(20)* also reported a high signal rim on the lateral border of the putamen in SND on T_2-weighted images. Wakai et al. *(18)* reported in addition, atrophy in the putamen correlating with the severity of parkinsonian symptoms. Kraft et al. *(21)* have suggested that the combination of dorsolateral putamenal low-signal with a high-signal lateral rim is highly specific for MSA since it was found in 9 of their 15 MSA patients but in none of their 65 patients with PD and none of their 10 patients with PSP. In contrast, these authors found that putamenal low signal alone did not exclude a diagnosis of PD. A distinctive pontine high-signal abnormality, the "hot cross bun" sign, caused by a loss of pontine neurons and myelinated transverse pontocerebellar fibers with the preservation of the corticospinal tracts running craniocaudally, has been reported in patients with MSA *(22)*, although it is now clear that this appearance is not specific to this disorder *(23)*. Typical changes in a patient with suspected MSA-P are illustrated in Fig. 2. In olivopontocerebellar atrophy, MRI may show substantial atrophy of the cerebellar cortex and pons, accompanied by marked enlargement of the fourth ventricle *(24)*.

Schrag et al. *(25)* studied the specificity and sensitivity of routine MRI in differentiating atypical parkinsonian syn-

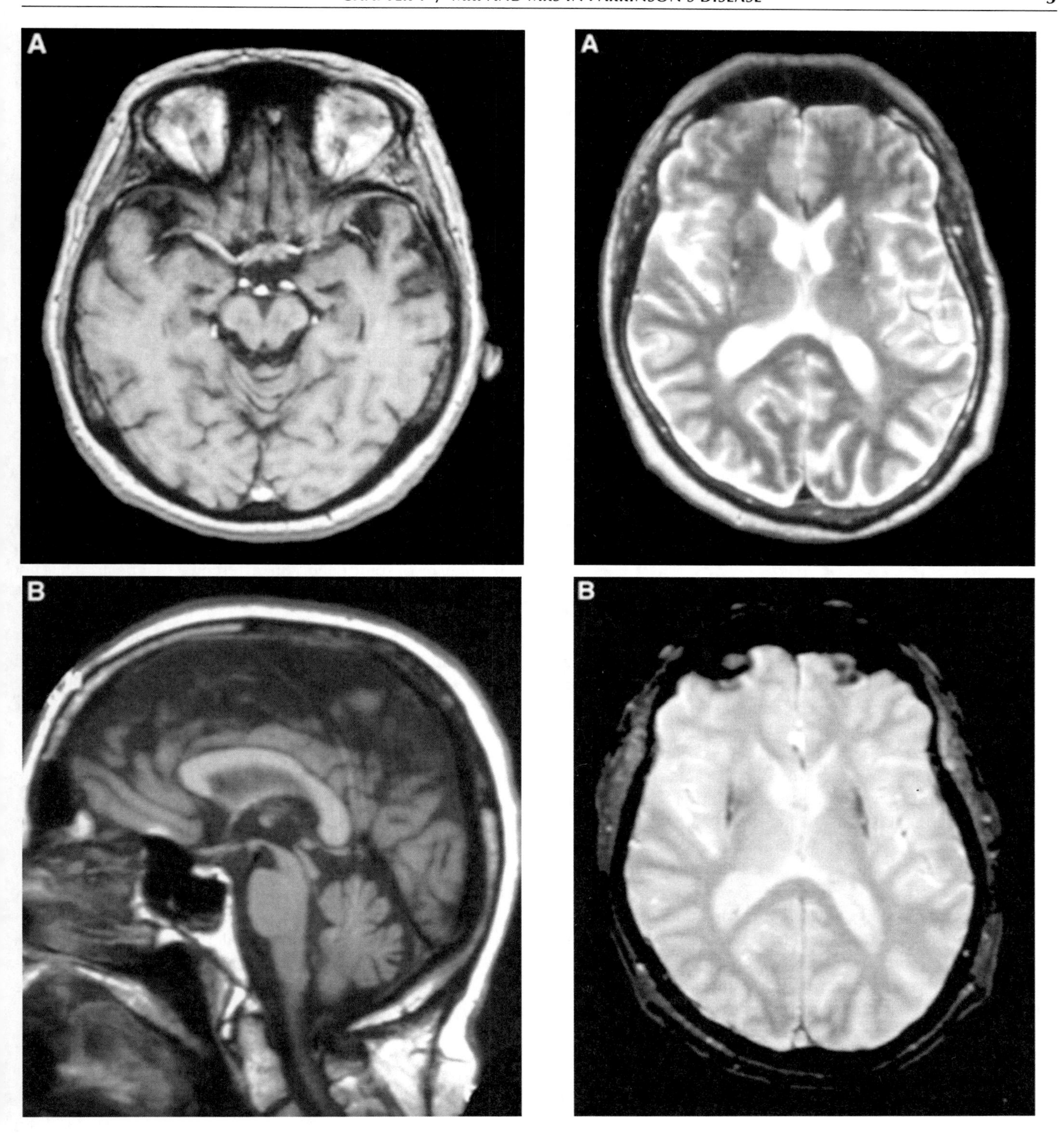

Fig. 1. Axial (**A**) and sagittal (**B**) images showing midbrain atrophy in PSP. Atrophy of the rostral midbrain tegmentum produces the "hummingbird sign" evident on the sagittal image.

Fig. 2. Axial images of the striatum and midbrain in the parkinsonian subtype of MSA. T_2-weighted images (**A**) suggest a "slit-like" void with a high signal rim on the lateral border of the putamen, although the changes are subtle. Gradient echo images (**B**) show a more definite low-signal abnormality in the putamen. Pontine images (**C**, next page) show the "hot cross bun" sign.

dromes. In this report, more than 70% of patients with PSP and more than 80% of those with MSA-C could be classified correctly. In contrast, only approx 50% of patients with MSA-P could be classified correctly.

Three-dimensional volumetric measurements in PD, PSP, and MSA have demonstrated normal striatal, cerebellar, and brainstem volumes in PD but reduced mean striatal and

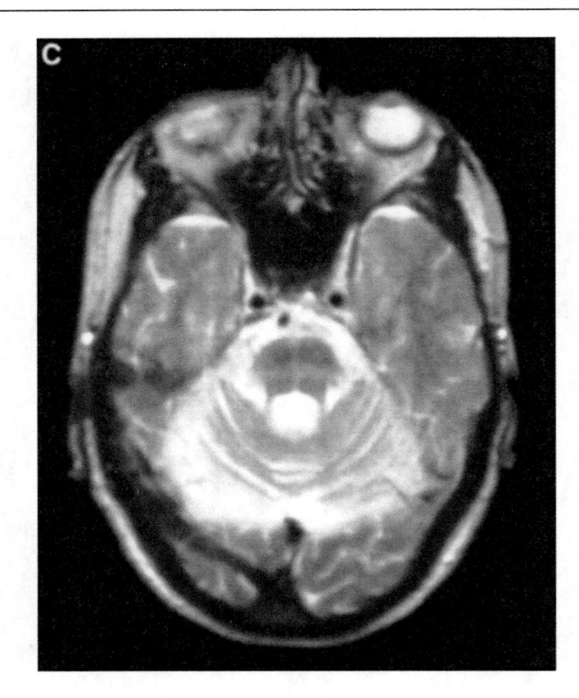

Fig. 2. (continued)

brainstem volumes in patients with MSA and PSP *(26)*. Those with MSA also showed a reduction in cerebellar volume. Although there was substantial overlap in volumetric measurements between groups, a discriminant analysis demonstrated effective discrimination of most of the MSA and PSP patients from the normal and PD groups. This analysis, however, did not separate PD patients from controls. The authors concluded that MRI-based volumetry might provide a marker to discriminate typical and atypical parkinsonism.

Voxel-based morphometry (VBM) recently has been applied to avoid the biases often associated with region of interest-guided volumetric measurements. VBM allows an objective, unbiased, comprehensive assessment of anatomical differences of gray and white matter throughout the brain, unconstrained by arbitrary region of interest selection. In MSA-P, VBM revealed selective cortical atrophy affecting primary and higher order motor areas, such as the supplementary motor area and anterior cingulate cortex *(27)*. Not surprisingly, VBM revealed a significant loss of cerebellar and brainstem volume in MSA-C *(28)*.

The analysis of serial MRI studies has been suggested *(29)* as a technique that is capable of demonstrating a characteristic pattern and progression of atrophy in a single patient with MSA in whom pathological confirmation of diagnosis was subsequently available. In this study, T_1-weighted volumetric scans acquired at two points 14 mo apart were analyzed using a previously validated nonlinear matching algorithm to obtain a voxel-by-voxel measure of volume change. The greatest rates of atrophy were reported in the pons, middle cerebellar peduncles, and the immediately adjacent midbrain and medulla.

Diffusion weighted imaging is a technique used to study the random movement of water molecules in the brain. Diffusion can be quantified by applying field gradients with varying degrees of diffusion sensitization, allowing the calculation of the apparent diffusion coefficient in tissue. Because the brain is organized in bundles of fiber tracts, water molecules move mainly along these structures, whereas diffusion perpendicular to the fiber tracts is restricted. Neuronal loss may alter the barriers restricting diffusion and increase the mobility of water molecules within the tissue architecture. In comparison with PD patients and control subjects, patients with MSA-P have been shown to have significantly higher apparent diffusion coefficient values in the putamen, with complete discrimination of MSA-P from PD based on these values *(30)*. Similar observations in comparison with controls have been reported in patients with PSP, although the results indicate that diffusion-weighted imaging does not differentiate PSP from MSA-P *(31)*.

As the above summary indicates, several different MRI techniques have been applied to these studies. Because of the variability in technique and the small number of patients included in most of these reports, further investigation is required to determine which MRI methodologies provide the highest sensitivity and specificity for the structural changes associated with PD and related neurodegenerative disorders. Although MRI may be of some benefit in differential diagnosis, it has not yet surpassed the role of the clinical neurologist in identifying these disorders.

MRI does play a major role, however, in the investigation of parkinsonism that may occur secondary to various structural brain lesions. Brain tumors may infiltrate or compress the basal ganglia and brainstem or, by vascular compression, may produce relative basal ganglia ischemia, thereby causing parkinsonism. Involvement of premotor frontal cortex by tumor may produce similar symptoms. Parkinsonism has been reported to occur in association with normal pressure or obstructive hydrocephalus, subdural hematoma, or multiple infarcts. MRI has a well-established role in the diagnosis of these disorders.

Focal basal ganglia lesions may also be demonstrated with MRI in some metabolic or toxic disorders associated with parkinsonism. Low-signal changes on all pulse sequences typically are present in basal ganglia calcification associated with idiopathic hypoparathyroidism or pseudohypoparathyroidism. Bilateral symmetrical necrosis of the globus pallidus may be evident in carbon monoxide *(32)* or cyanide poisoning *(33)* or multiple other toxic/metabolic disorders *(34)*.

3. QUANTITATIVE ESTIMATION OF REGIONAL BRAIN IRON WITH MRI

The adult brain has a very high iron content, particularly in the basal ganglia. Direct postmortem measurements have shown nonheme brain iron to be very low throughout the brain at birth but to increase gradually in most parts of the brain during the first two decades of life *(35)*. Brain iron concentration is maximal in the globus pallidus, substantia nigra, red nucleus, caudate, and putamen. Abnormally elevated iron levels are evident in various neurodegenerative disorders, including PD, in which increased iron in the substantia nigra has been reported *(36,37)*. Laser microprobe studies indicate that iron normally accumulates within neuromelanin granules of nigral neurons and that iron levels within these granules are significantly increased in PD *(38)*. Extended X-ray absorption fine-structure experiments have shown that ferritin is the only storage protein detectable in both control and parkinsonian brain, with increased loading of ferritin with iron in PD *(39)*.

Although ferritin in aqueous solution has a strong effect on transverse relaxation times, these changes are much less prominent in tissue. Estimation of transverse relaxation times in patients with PD, using a 1.5-T whole-body imaging system, showed reduced T_2 values in substantia nigra, caudate, and putamen in PD patients as compared with healthy controls *(40)*. The decrease was small, however, and because of substantial overlap between groups, the investigators were unable to differentiate individual patients from controls with T_2 measurements. Vymazal et al. *(41)* reported nonsignificant T_2 shortening in the substantia nigra in PD consistent with iron accumulation. A more complex relationship between brain iron changes and disease state in PD, however, was suggested by Ryvlin et al. *(42)*. These authors reported decreased T_2 in the pars compacta of PD patients, regardless of disease duration, but increased T_2 values in the putamen and pallidum in those with duration of illness greater than 10 yr. In this study, putamen transverse relaxation time correlated positively with disease duration. Others have observed an imperfect correlation between T_2 values and quantitative assays of iron and ferritin *(43)*. The lack of a more substantial T_2 difference between PD and controls is not surprising because regional iron content is only one of several determinants of transverse relaxation times in tissue.

Several investigators have reported MR methods designed to estimate iron content directly. Bartzokis et al. used the influence of the strength of the external magnetic field on ferritin-induced T_2 changes to derive an index of regional tissue ferritin levels *(44)*. This method involves the measurement of transverse relaxation rates in the same patient with two different field strength instruments. Using this method, patients with earlier-onset PD (onset before the age of 60) were suggested to have increased ferritin in the substantia nigra, putamen, and globus pallidus, whereas later-onset patients had decreased ferritin in the substantia nigra reticulata *(45)*. Others have exploited the fact that paramagnetic substances, such as iron, create local magnetic field inhomogeneities that alter transverse relaxation times in the brain *(46,47)*. We have developed a method that quantifies the effects of paramagnetic centers sequestered inside cell membranes, based on the interecho time dependence of the decay of transverse magnetization caused by local field inhomogeneities that are the result of intracellular paramagnetic ions *(47)*. Because the concentration of brain iron is much greater than that of other paramagnetic ions, such as manganese and copper, this method enables the estimation of regional indices of brain iron content.

We used this technique to show a strong direct relationship between age and both putamen and caudate iron content *(48)*. This age-related increase may increase the probability of free-radical formation in the striatum, thereby representing a risk factor for the development of disorders such as PD in which nigrostriatal neurons may be affected by increased oxidant stress, although it should be noted that iron bound to ferritin might be relatively nonreactive and therefore unlikely to induce tissue damage. We also have reported a significant increase in iron content in the putamen and pallidum in PD and a correlation with the severity of clinical symptomatology with more severely affected patients having a higher iron content in these structures *(49)*. We have applied this methodology in striatonigral degeneration and have shown an increase in putamen iron content in this disorder, beyond the 95% confidence limit for inclusion in the PD group, even when considering severity of clinical symptomatology *(50)*. Our observations are summarized in Fig. 3. Using an alternate method developed by Ordidge and colleagues *(46)*, Gorell et al. reported an increase in iron-related MR contrast in the substantia nigra in PD, with a correlation between the increase and disease severity as indicated by simple reaction time *(51)*. Changes compatible with increased nigral iron content also have been reported by Graham et al. using a partially refocused interleaved multiple echo pulse sequence at 1.5 T, although these investigators suggested reduced iron content in the putamen *(52)*.

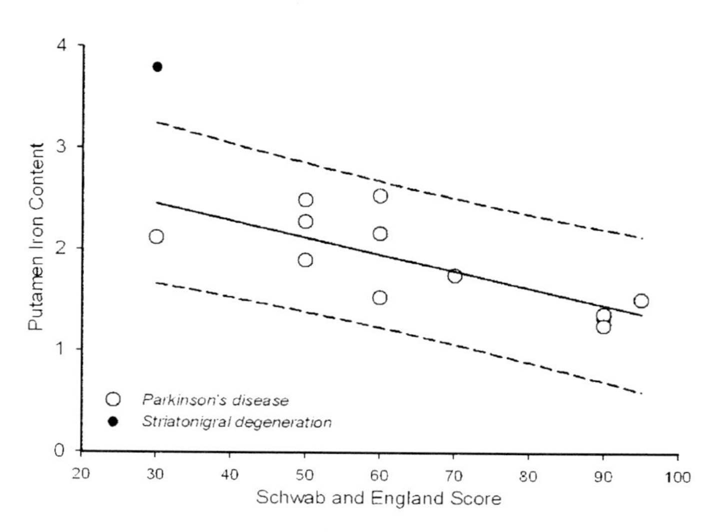

Fig. 3. Putamen iron content in PD and a patient with clinical diagnosis of the parkinsonian subtype of multiple system atrophy, expressed as a function of disease severity based on the Schwab and England activities of daily living score. The regression line and 95% confidence limits for inclusion in the PD group are indicated.

In summary, the application of MRI methods to study regional brain iron content is in its infancy. Much further investigation is required to determine which techniques are best able to provide quantifiable images that correspond to independent measures of brain iron and to determine the sensitivity and specificity of changes in basal ganglia iron content that might be associated with PD and other neurodegenerative disorders.

4. MRS IN PARKINSONISM

MRS provides a noninvasive method of quantifying the concentration of MR-visible metabolites in the brain. The technique is based on the general principle that the resonant frequency of a specific metabolite depends on its chemical environment. Most clinical MRS studies have concentrated on the metabolites visible with proton (^{1}H) spectroscopy and measured in single, localized tissue volumes in the brain. The metabolites of interest that can be most readily studied with ^{1}H-MRS at long echo times include *N*-acetyl aspartate (NAA), creatine/phosphocreatine (Cr), and choline (Cho). Altered neuronal membrane synthesis and degradation can produce changes in Cho *(53)*. NAA is contained almost exclusively within neurons *(54)* and is therefore considered to act as an in vivo marker of neuronal loss or dysfunction. Reduced regional NAA concentration has been reported in conditions characterized by neuronal or axonal loss *(55–59)*. Most frequently, NAA measurements have been based on the regional NAA/Cr ratio. The rationale for the use of the Cr resonance as the denominator is based on the concept that creatine and phosphocreatine are in chemical equilibrium and that the total concentration of both compounds is expected to remain unchanged by neurodegenerative disease processes. Alternate methods are now available for the measurement of metabolite concentrations, such as those using water as an internal standard for calibration *(60)* and those involving calibration to an external standard *(61)*, which should allow for a significant improvement of the quantitative accuracy of ^{1}H-MRS.

Several studies that used ^{1}H-MRS in PD have been reported, reviewed by Davie *(62)* and systematically by Clarke and Lowry *(63)*. A large multicenter study of 151 patients with PD showed no significant difference in either NAA/Cr or NAA/Cho ratios between controls and patients who were not taking levodopa *(64)*. Other groups *(65–67)* have reported similar results. Similarly, in a small study using absolute metabolite quantitation, NAA, Cr, and Cho concentrations in PD did not differ significantly from those in controls *(68)*. Although no significant difference was observed in the NAA/Cr ratio in levodopa-treated patients with PD, Ellis et al. reported reduced NAA/Cr in drug-naïve PD patients compared with both the treated PD group and with the control group *(69)*. These authors suggested that the reduced ratio in PD might reflect a functional abnormality of neurons in the putamen that can be reversed with levodopa treatment. Clarke and Lowry have reported a significant decrease in the NAA/Cho ratio in PD because of an increase in the absolute concentration of Cho unassociated with a change in NAA concentration *(70)*.

O'Neill et al. have recently reported a small study using quantitative ^{1}H-MRS in which multiple tissue volumes were assessed *(71)*. This study applied more rigorous MR methodology than previous studies, including tissue segmentation to correct for varying gray and white matter content within the voxel of interest in different patients and the use of spectroscopic imaging to provide more widespread sampling of multiple brain regions. Observations from this study included decreased Cr concentration in the substantia nigra but normal NAA and Cho levels, suggesting that the use of simple ratios such as NAA/Cr may be misleading in PD. No differences in NAA, Cr, or Cho content were observed between PD patients and controls in either basal ganglia or multiple cortical volumes. This study also reported that the volumes of putamen, globus pallidus, and prefrontal cortical gray matter were significantly reduced in PD vs. age-matched controls. A negative correlation was observed between the volume of the substantia nigra pars compacta and that of the other basal ganglia nuclei in controls but not PD, although the volume measurements were based on MRI sequences yielding a relatively low spatial resolution. Cortical changes in PD have also been reported with a reduced NAA/Cr ratio in motor cortex *(72)* and in temporoparietal cortex *(73)*, possibly related to impaired neuronal function resulting from a loss of thalamocortical excitatory input.

In contrast to the apparent lack of changes in PD with MR spectroscopy, there does appear to be a significant reduction in basal ganglia NAA concentration in MSA. Davie reported ^{1}H-MRS studies in the lentiform nucleus in controls and in patients with PD and with clinically probable MSA *(65)*. In MSA, there was a significant reduction in absolute levels of NAA, particularly in patients with the SND subtype of MSA, suggesting that the spectroscopic measurement of NAA levels in the lentiform nucleus may provide a clinically useful technique to help differentiate MSA from PD. Federico et al. showed similar changes in MSA *(74)*. However, in a study using absolute quantitation of metabolite concentrations, NAA was reported to be unchanged in MSA *(70)*.

Axelson et al. have reported an alternate approach to the analysis of spectroscopic data based on pattern recognition utilizing an artificial neural network *(75)*. Conventional data analysis in this study showed no significant abnormalities in metabolite ratios in PD, whereas trained neural networks could distinguish control from PD spectra with considerable accuracy.

Several additional issues contribute to the variability of the results observed in these spectroscopic studies. Most of the studies reported only a small number of patients, often with fewer than 10 patients in each group. The patient groups themselves have varied somewhat from study to study, with some patients having early, untreated PD, and others having more advanced disease that requires treatment. Variabilities in the MR technique itself, for example, in the choice of echo time, may lead to heterogeneous results. Lastly, the high iron content in the basal ganglia has a significant impact on the ability to obtain reproducible high-resolution spectra from this region and may impact on the accuracy of quantitative results extracted from the spectra.

5. ENERGY METABOLISM IN PD

Although the etiology of PD is unknown, the possibility of an underlying defect in mitochondrial metabolism has been addressed in several biochemical studies *(76)*. There is evidence of reduced complex I activity in the substantia nigra in PD, and Gu et al. have suggested that a mitochondrial DNA abnormality may underlie this complex I defect in at least a subgroup of PD patients *(77)*. Studies in other tissues, however, have produced conflicting results, perhaps in part because biochemical studies involve removal of mitochondria from their natural milieu, with consequent mechanical disruption and a loss of normal control mechanisms. In contrast, MRS provides the potential to study mitochondrial metabolism in vivo.

The rate of intracellular energy metabolism is reflected by the ratio of inorganic phosphate (Pi) to phosphocreatine (PCr), readily measured with ^{31}P-MRS. The measurement of this ratio in resting muscle has been shown to be a useful diagnostic test for mitochondrial disease *(78)*. Penn et al. have used ^{31}P-MRS to investigate energy metabolism in muscle in patients with PD. The Pi/PCr ratio was significantly increased in PD, suggesting a small, generalized mitochondrial defect *(79)*. Further studies are needed to determine whether these changes are limited to a clinically definable subset of parkinsonian individuals. ^{31}P-MRS studies of brain have recently been reported in MSA and PD *(80)*. In these studies, patients with MSA showed significantly increased Pi content and reduced PCr content, whereas those with PD showed significantly increased Pi but unchanged PCr, suggesting abnormal energy metabolism in both disorders.

The combination of ^{31}P-MRS and fluorodeoxyglucose/PET has been used to suggest that temporoparietal cortical glycolytic and oxidative metabolism are both impaired in nondemented PD patients *(81)*. These observations are consistent with a previous report of temporoparietal cortical reduction in NAA/Cr ratio in nondemented PD patients, which correlated with measures of global cognitive decline independently of motor impairment *(73)*.

An alternate approach to study energy metabolism is with ^{1}H-MRS. Normal brain energy production is derived from the oxidative metabolism of glucose by way of the Krebs cycle and, ultimately, the electron transport chain. A defect at the level of either of the two latter processes will result in decreased metabolism of pyruvate through these pathways, and increased production of lactate. Regional brain lactate concentrations can be readily assessed with ^{1}H-MRS. For example, this methodology has been used to demonstrate increased occipital lactate levels, thereby suggesting impaired energy metabolism in Huntington's disease *(82)*. We have found similar changes in some but not all patients with PD (W. R. W. Martin, unpublished observations), providing further evidence for the presence of a mitochondrial defect in this disorder. In contrast, however, Hoang et al. *(83)* have reported normal energy metabolism in the putamen and in occipital and parietal lobes when using both ^{31}P- and ^{1}H-MRS in patients with PD.

6. FUNCTIONAL MRI IN PD

Motor activation studies provide a means to investigate the regional cerebral mechanisms involved in motor control in normal subjects and in patients with disorders affecting these control systems. Typical fMRI experiments involve measurement of regional blood oxygen level-dependent signal increases associated with specific activation paradigms. These signal changes occur as a result of the increased local cerebral blood flow and altered oxyhemoglobin concentration associated with neuronal activation. Functional MRI experiments have extended our knowledge of disordered motor control systems, based on the extensive previous experience obtained with PET/motor activation studies in control subjects and in patients with PD. PET studies have suggested that cortical motor areas, such as the supplementary motor area (SMA), seem to be underactive in akinetic parkinsonian patients *(84,85)*, whereas other motor areas, such as the parietal and lateral premotor cortex and the cerebellum, appear to be overactive *(86)*. In comparison with PET, the application of MR-based methodology has allowed for improvements in both spatial and temporal resolution in activation studies.

Sabatini et al. *(87)* compared the changes induced by a complex sequential motor task performed with the right hand in six PD patients in the "off" state to six normal subjects. In control subjects, significant activation was seen in the left primary sensorimotor cortex, the left lateral premotor cortex, bilateral parietal cortex, the anterior cingulate cortex, and in the rostral and caudal parts of the supplementary motor area SMA. In PD patients, significant activation was seen bilaterally in primary sensorimotor cortex (left more than right), in bilateral parietal cortex, in cingulate cortex, and in the caudal but not the rostral SMA. Between-group comparisons showed increased activation of the rostral SMA and the right dorsolateral prefrontal cortex in controls, and increased activation of primary sensorimotor cortex, premotor cortex, and parietal cortex bilaterally, as well as the cingulate cortex, and the caudal SMA in PD. The decreased rostral SMA activation in PD is consistent with previous PET studies *(84,85)* and is thought to reflect a decrease in the positive efferent feedback from the basal ganglia-thalamocortical motor loop due to striatal dopamine depletion. The widespread increased activation in other motor areas also is consistent with previous PET studies *(86)* and suggests an attempt to recruit parallel motor circuits to overcome the functional deficit of the striatocortical motor loops. The high resolution of fMRI allowed the SMA to be subdivided into two functionally distinct areas in this study with the rostral component corresponding best to the decreased activation noted on previous PET studies *(87)*.

Event-related fMRI directly reflects signal changes associated with single movements, thereby avoiding the problem of a prolonged acquisition time, which may confound data by invoking cerebral processes unrelated to movement. Haslinger et al. *(88)* used this technique to study cerebral activation associated with single joystick movements in controls and in PD patients, both in the akinetic "off" state, and again in the "on" state after taking levodopa. Control subjects activated primary sensorimotor and adjoining cortex, as well as the rostral SMA. Patients in the "off" state showed significant underactivity in rostral SMA, as well as increased activation in primary motor and lateral premotor cortex bilaterally. These results are similar to those reported in the previous block design fMRI study described by Sabatini et al. *(87)*. In the "on" state, there was relative normalization of the impaired activation in the rostral SMA and of the increased activation in primary motor and premotor cortices. This event-related study provides an example of the exquisite sensitivity that can be achieved with fMRI, sufficient to demonstrate the metabolic/hemodynamic changes associated with the neuronal activity involved in generating a single voluntary movement.

7. SUMMARY AND CONCLUSIONS

At present, conventional MRI shows no convincing structural changes in PD itself, but it may be useful in helping to distinguish PD from other neurodegenerative parkinsonian syndromes and from the occasional case of parkinsonism secondary to a focal brain lesion. MRS also may provide useful information in distinguishing PD from disorders such as MSA.

The general field of MRI and MRS is evolving rapidly, and there are a number of areas in which we can expect new developments to provide relevant information. Novel pulse sequences may provide more information regarding substantia nigra pathology in PD. The use of MR as a tool to measure regional iron concentrations should provide more information regarding the relationship between iron accumulation and parkinsonian symptoms. MRS provides a sensitive tool for the researcher to investigate in vivo the possible contribution of abnormalities in brain energy metabolism to the pathogenesis of PD. MRS also allows the assessment of other metabolite changes in PD, for example, providing for the evaluation of the potential importance of changes in regional brain glutamate content. Lastly, fMRI provides the potential to evaluate, in a noninvasive fashion, the role played by the basal ganglia in motor control and in cognition in normal individuals as well as in PD.

REFERENCES

1. Rutledge JN, Hilal SK, Schallert T, Silver AJ, Defendini RD, Fahn S. Magnetic resonance imaging of Parkinsonisms. In: Fahn S, Marsden CD, Calne D, Goldstein M, editors. Recent Developments in Parkinson's Disease. Florham Park, NJ: Macmillan; 1987:123–134.
2. Olanow CW. Magnetic resonance imaging in parkinsonism. Neurol Clin 1992;10::405–420.
3. Rutledge JN, Hilal SK, Silver AJ, et al. Study of movement disorders brain iron by MR. AJNR Am J Neuroradiol 1987;8:397–411.
4. Duguid JR, De la Paz R, DeGroot J. Magnetic resonance imaging of the midbrain in Parkinson's disease. Ann Neurol 1986;20:744–747.
5. Braffman BH, Grossman RI, Goldberg HI, et al. MR imaging of Parkinson disease with spin-echo gradient-echo sequences. AJR Am J Roentgenol 1989;152:159–165.
6. Hutchinson M, Raff U. Structural changes of the subsantia nigra in Parkinson's disease as revealed by MR imaging. AJNR Am J Neurorad 2000;21:697–701.
7. Hu M.T.M., White SJ, Herlihy AH, et al. A comparison of 18F-dopa PET inversion recovery MRI in the diagnosis of Parkinson's disease. Neurology 2001;56:1195–1200.
8. Savoiardo M, Strada L, Girotti F, et al. MR imaging in progressive supranuclear palsy Shy-Drager syndrome. J Comput Assist Tomogr 1989;13:555–560.
9. Warmuth-Metz M, Naumann M, Csoti I, Solymosi L. Measurement of the midbrain diameter on routine magnetic resonance imaging. Arch Neurol 2001;58:1076–1079.
10. Kato N, Arai K, Hattori T. Study of the rostral midbrain atrophy in progressive supranuclear palsy. J Neurol Sci 2003;210:57–60.
11. Savoiardo M, Girotti F, Strada L, Ciceri E. Magnetic resonance imaging in progressive supranuclear palsy other parkinsonian disorders. J Neural Transm Suppl 1994;42:93–110.
12. Yagishita A, Oda M. Progressive supranuclear palsy: MRI pathological findings. Neuroradiology 1996;38:(Suppl 1):S60–S66.
13. Dubinsky RM, Jankovic J. Progressive supranuclear palsy a multi-infarct state. Neurology 1987;37:570–576.
14. Hauser RA, Olanow CW, Gold M, et al. Magnetic resonance imaging of corticobasal degeneration. J Neuroimaging 1996;6:222–226.
15. Drayer BP, Olanow W, Burger P, et al. Parkinson plus syndrome: diagnosis using high field MR imaging of brain iron. Radiology 1986;159:493–498.
16. Stern MB, Braffman BH, Skolnick BE, Hurtig HI, Grossman RI. Magnetic resonance imaging in Parkinson's disease parkinsonian syndromes. Neurology 1989;39:1524–1526.
17. Pastakia B, Polinsky R, Di Chiro G, Simmons JT, Brown R, Wener L. Multiple system atrophy (Shy-Drager syndrome): MR imaging. Radiology 1986;159:499–502.
18. Wakai M, Kume A, Takahashi A, Ando T, Hashizume Y. A study of parkinsonism in multiple system atrophy: clinical MRI correlation. Acta Neurol Scand 1994;90:225–231.
19. Lang AE, Curran T, Provias J, Bergeron C. Striatonigral degeneration: iron deposition in putamen correlates with the slit-like void signal of magnetic resonance imaging. Can J Neurol Sci 1994;21:311–318.
20. Konagaya M, Konagaya Y, Iida M. Clinical magnetic resonance imaging study of extrapyramidal symptoms in multiple system atrophy. J Neurol Neurosurg Psychiatry 1994;57:1528–1531.
21. Kraft E, Schwarz J, Trenkwalder C, Vogl T, Pfluger T, Oertel, WH. The combination of hypointense hyperintense signal changes on T2-weighted magnetic resonance imaging sequences: a specific marker of multiple system atrophy? Arch Neurol 1999;56:225–228.
22. Schrag A, Kingsley D, Phatouros C, et al. Clinical usefulness of magnetic resonance imaging in multiple system atrophy. J Neurol Neurosurg Psychiatry 1998;65:65–71.
23. Muqit MMK, Mort D, Miszkiel KA, Shakir RA. "Hot cross bun" sign in a patient with parkinsonism secondary to presumed vasculitis. J Neurol Neurosurg Psychiatry 2001;71:565–566.
24. Savoiardo M, Strada L, Girotti F, et al. Olivopontocerebellar atrophy: MR diagnosis relationship to multiple system atrophy. Radiology 1990;174:693–696.
25. Schrag A, Good CD, Miszkiel K, et al. Differentiation of atypical parkinsonian syndromes with routine MRI. Neurology 2000;54:697–702.
26. Schulz JB, Skalej M, Wedekind D, et al. Magnetic resonance imaging-based volumetry differentiates idiopathic Parkinson's syndrome from multiple system atrophy progressive supranuclear palsy. Ann Neurol 1999;45:65–74.
27. Brenneis C, Seppi K, Schocke MF, et al. Voxel-based morphometry detects cortical atrophy in the Parkinson variant of multiple system atrophy. Mov Disord 2003;18:1132–1138.
28. Specht K, Minnerop M, Abele M, Reul J, Wullner U, Klockgether T. In vivo voxel-based morphometry in multiple system atrophy of the cerebellar type. Arch Neurol 2003;60:1431–1435.
29. Schott JMM, Simon JE, Fox NC, et al. Delineating the sites progression of in vivo atrophy in multiple system atrophy using fluid-registered MRI. Mov Disord 2003;18:955–958.
30. Schocke MFH, Seppi K, Esterhammer R, et al. Diffusion-weighted MRI differentiates the Parkinson variant of multiple system atrophy from PD. Neurology 2002;58:575–580.
31. Seppi K, Schocke MFH, Esterhammer R, et al. Diffusion-weighted imaging discriminates progressive supranuclear palsy from PD, but not from the Parkinson variant of multiple system atrophy. Neurology 2003;60:922–927.
32. Davis PL. The magnetic resonance imaging appearances of basal ganglia lesions in carbon monoxide poisoning. Magn Reson Imaging 1986;4:489–490.
33. Rosenberg NL, Myers JA, Martin WRW. Cyanide-induced parkinsonism: clinical, MRI 6-fluorodopa positron emission tomography studies. Neurology 1989;39:142-144.
34. Marsden CD, Lang AE, Quinn NP, et al. Familial dystonia visual failure with striatal CT lucencies. J Neurol Neurosurg Psychiatry 1986;49:500–509.
35. Hallgren B, Sourander P. The effect of age on the nonhaemin iron in the human brain. J Neurochem 1958;3:41–51.
36. Dexter DT, Wells FR, Lees AJ, et al. Increased nigral iron content alterations in other metal ions occurring in brain in Parkinson's disease. J Neurochem 1989;52:1830–1836.
37. Sofic E, Riederer P, Heinsen H, et al. Increased iron (III) total iron content in post mortem substantia niga of parkinsonian brain. J Neural Transm 1988;74:199–205.
38. Good PF, Olanow CW, Perl DP. Neuromelanin-containing neurons of the substantia nigra accumulate iron aluminum in Parkinson's disease: a LAMMA study. Brain Res 1992;593:343–364
39. Griffiths PD, Dobson BR, Jones GR, Clarke DT. Iron in the basal ganglia in Parkinson's disease: an in vitro study using extended X-ray absorption fine structure cryo-electron microscopy. Brain 1999;122:667–673.
40. Antinoni A, Leenders KL, Meier D, Oertel WH, Boesiger P, Anliker M. T2 relaxation time in patients with Parkinson's disease. Neurology 1993;43:697–700.
41. Vymazal J, Righini A, Brooks RA, et al. T1 and T2 in the brain of healthy subjects, patients with Parkinson disease, patients with

multiple system atrophy: relation to iron content. Radiology 1999;211:489–495.
42. Ryvlin P, Broussolle E, Piollet H, Viallet F, Khalfallah Y, Chazot G. Magnetic resonance imaging evidence of decreased putamenal iron content in idiopathic Parkinson's disease. Arch Neurol 1995;52: 583–588
43. Chen JC, Hardy PA, Kucharczyk W, et al. MR of human postmortem brain tissue: correlative study between T2 assays of iron ferritin in Parkinson Huntington disease. AJNR Am J Neuroradiol 1993;14:275–281.
44. Bartzokis G, Aravagiri M, Oldendorf WH, Mintz J, Marder SR. Field dependent transverse relaxation rate increase may be a specific measure of tissue iron stores. Magn Reson Med 1993;29:459–464.
45. Bartzokis G, Cummings JL, Markham CH, et al. MRI evaluation of brain iron in earlier- later-onset Parkinson's disease normal subjects. Magn Reson Imaging 1999;17:213–222
46. Ordidge RJ, Gorell JM, Deniau JC, Knight RA, Helpern JA. Assessment of relative brain iron concentrations using T2-weighted T2*-weighted MRI at 3 Tesla. Magn Reson Med 1994;32:335–341.
47. Ye FQ, Martin WRW, Allen PS. Estimation of the brain iron in vivo by means of the interecho time dependence of image contrast. Magn Reson Med 1996;36:153–158.
48. Martin WRW, Ye FQ, Allen PS. Increasing striatal iron content associated with normal aging. Mov Disord 1998;13:281–286.
49. Ye FQ, Allen PS, Martin WRW. Basal ganglia iron content in Parkinson's disease measured with magnetic resonance. Mov Disord 1996;11:243–249.
50. Martin WRW, Roberts TE, Ye FQ, Allen PS. Increased basal ganglia iron in striatonigral degeneration: in vivo estimation with magnetic resonance. Can J Neurol Sci 1998;25:44–47.
51. Gorell JM, Ordidge RJ, Brown GG, Deniau, J-C, Buderer NM, Helpern JA. Increased iron-related MRI contrast in the substantia nigra in Parkinson's disease. Neurology 1995;45:1138–1143.
52. Graham JM, Paley MNJ, Grunewald RA, Hoggard N, Griffiths PD. Brain iron deposition in Parkinson's disease imaged using the PRIME magnetic resonance sequence. Brain 2000;123:2423–2431.
53. Vion-Dury J, Meyerhoff DJ, Cozzone PJ, Weiner MW. What might be the impact on neurology of the analysis of brain metabolism by in vivo magnetic resonance spectroscopy? J Neurol 1994;241:354–371.
54. Unrejak J, Williams SR, Gadian DG, Noble M. Proton nuclear magnetic resonance spectroscopy unambiguously identifies different neural cell types. J Neurosci 1993;13:981–989.
55. Matthews PM, Francis G, Antel J, Arnold DL. Proton magnetic resonance spectroscopy for metabolic characterisation of plaques in multiple sclerosis. Neurology 1991;41:1251–1256.
56. Chong WK, Sweeney B, Wilkinson ID, et al. Proton spectroscopy of the brain in HIV infection: correlation with clinical, immunologic MR imaging findings. Radiology 1993;188:119–124.
57. Shino A, Matsuda M, Morikawa S, Inubushi T, Akiguchi I, Handa J. Proton magnetic resonance spectroscopy with dementia. Surg Neurol 1993;39:143–147.
58. Gideon P, Henriksen O, Sperling B, et al. Early time course of N-acetylaspartate, creatine phosphocreatine, compounds containing choline in the brain after acute stroke. A proton magnetic resonance spectroscopy study. Stroke 1992;23:1566–1572.
59. Cwik V, Hanstock C, Allen PS, Martin WRW. Estimation of brainstem neuronal loss in amyotrophic lateral sclerosis with in vivo proton magnetic resonance spectroscopy. Neurology 1998;50: 72–77.
60. Christiansen P, Henriksen O, Stubgaard M, Gideon P, Larsson HBW. In vivo quantification of brain meabolites by 1H MRS using water as an internal standard. Magn Reson Imaging 1993;11:107–108.
61. Michaelis T, Merboldt KD, Bruhn H, Hanicke W, Frahm J. Absolute concentrations of metabolites in the adult human brain in vivo: quantification of localized proton MR spectra. Radiology 1993;187:219–227
62. Davie C. The role of spectroscopy in parkinsonism. Mov Disord 1998;13:2–4.
63. Clarke CE, Lowry M. Systematic review of proton magnetic resonance spectroscopy of the striatum in parkinsonian syndromes. Eur J Neurol 2001;8:573–577.
64. Holshauser BA, Komu M, Moller HE, et al. Localised proton NMR spectroscopy in the striatum of patients with idiopathic Parkinson's disease: a multicenter pilot study. Magn Reson Med 1995;33:589–594.
65. Davie CA, Wenning GK, Barker GJ, et al. Differentiation of multiple system atrophy from idiopathic Parkinson's disease using proton magnetic resonance spectroscopy. Ann Neurol 1995;37: 204–210.
66. Cruz CJ, Aminoff MJ, Meyerhoff DJ, Graham SH, Weiner MW. Proton MR spectroscopic imaging of the striatum in Parkinson's disease. Magn Reson Imaging 1997;15:619–624.
67. Tedeschi G, Litvan I, Bonavita S,. et al. Proton magnetic resonance spectroscopic imaging in progressive supranuclear palsy, Parkinson's disease corticobasal degeneration. Brain 1997;120: 1541–1552.
68. Clarke CE, Lowry M, Horsman A. Unchanged basal ganglia N-acetylaspartate glutamate in idiopathic Parkinson's disease measured by proton magnetic resonance spectroscopy. Mov Disord 1997;12:297–301.
69. Ellis CM, Lemmens G, Williams SCR, et al. Changes in putamen N-acetylaspartate choline ratios in untreated levodopa treated Parkinson's disease: a proton magnetic resonance spectroscopy study. Neurology 1997;49:438–444.
70. Clarke CE, Lowry M. Basal ganglia metabolite concentrations in idiopathic Parkinson's disease multiple system atrophy measured by proton magnetic resonance spectroscopy. Eur J Neurol 2000;7: 661–665.
71. O'Neill J, Schuff N, Marks WJ, Feiwell R, Aminoff MJ, Weiner MW. Quantitative 1H magnetic resonance spectroscopy MRI of Parkinson's disease. Mov Disord 2002;17:917–927.
72. Lucetti C, del Dotto P, Gambaccini G, et al. Proton magnetic resonance spectroscopy (1H-MRS) of motor cortex basal ganglia in de novo Parkinson's disease patients. Neurol Sci 2001;22:69–70.
73. Hu MT.M., Taylor-Robinson SD, Chaudhuri KR, et al. Evidence for cortical dysfunction in clinically nondemented patients with Parkinson's disease: a proton MR spectroscopy study. J Neurol Neurosurg Psychiatry 1999;67:20–26.
74. Federico F, Simone IL, Lucivero V, et al. Usefulness of proton magnetic resonance spectroscopy in differentiating parkinsonian syndromes. Italian J Neurol Sci 1999;20:223–229.
75. Axelson D, Bakken, IJ., Gribbestad IS, Ehrnholm B, Nilsen G, Aasly J. Applications of neural network analyses to in vivo 1H magnetic resonance spectroscopy of Parkinson disease patients. J Magn Reson Imaging 2002;16:13–20.
76. DiMauro S. Mitochondrial involvement in Parkinson's disease: the controversy continues. Neurology 1993;43:2170–2171.
77. Gu M, Cooper JM, Taanman JW, Schapira AHV. Mitochondrial DNA transmission of the mitochondrial defect in Parkinson's disease. Ann Neurol 1998;44:177–186.
78. Matthews PM, Allaire C, Shoubridge EA, Karpati G, Carpenter S, Arnold DL. In vivo muscle magnetic resonance spectroscopy in the clinical investigation of mitochondrial disease. Neurology 1991;41:114–120.
79. Penn AMW, Roberts T, Hodder J, Allen PS, Zhu G, Martin WRW. Generalized mitochondrial dysfunction in Parkinson's disease detected by magnetic resonance spectroscopy of muscle. Neurology 1995;45:2097–2099.
80. Barbiroli B, Martinelli P, Patuelli A, et al. Phosphorus magnetic resonance spectroscopy in multiple system atrophy Parkinson's disease. Mov Disord 1999;14:430–435.
81. Hu MTM, Taylor-Robinson SD, Chaudhuri KR, et al. Cortical dysfunction in non-demented Parkinson's disease patients. A combined 31P-MRS 18FDG-PET study. Brain 2000;123:340–352.
82. Jenkins BG, Koroshetz WJ, Beal MF, Rosen BR. Evidence for impairment of energy metabolism in vivo in Huntington's disease using localized 1H NMR spectroscopy. Neurology 1993;43:2689–2695.

83. Hoang TQ, Bluml S, Dubowitz DJ, et al. Quantitative proton-decoupled 31P MRS 1H MRS in the evaluation of Huntington's Parkinson's diseases. Neurology 1998;50:1033–1040.
84. Samuel M, Ceballos-Baumann AO, Blin J, et al. Evidence for lateral premotor parietal overactivity in Parkinson's disease during sequential biamanual movements. A PET study. Brain 1997;120:963–976.
85. Jahanshahi M, Jenkins, IH., Brown RG, Marsden CD, Passingham RE, Brooks DJ. Self-initiated versus externally triggered movements. I. An investigation using measurement of regional cerebral blood flow with PET movement-related potentials in normal Parkinson's disease subjects. Brain 1995;118:913–933.
86. Playford ED, Jenkins IH, Passingham RE, Nutt J, Frackowiak RS, Brooks DJ. Impaired mesial frontal putamen activation in Parkinson's disease: a positron emission tomography study. Ann Neurol 1992;32:151–161.
87. Sabatini U, Boulanouar K, Fabre N, et al. Cortical motor reorganization in akinetic patients with Parkinson's disease: a functional MRI study. Brain 2000;123:394–403.
88. Haslinger B, Erhard P, Kampfe N, et al. Event related functional magnetic resonance imaging in Parkinson's disease before after levodopa. Brain 2001;124:558–570.

2 Positron Emission Tomography and Single-Photon Emission Tomography in the Diagnosis of Parkinson's Disease

Differential Diagnosis From Parkinson-Like Degenerative Diseases

PAUL D. ACTON, PhD

SUMMARY

Parkinsonian symptoms are associated with a number of neurodegenerative disorders, such as Parkinson's disease, multiple system atrophy, and progressive supranuclear palsy. Positron emission tomography (PET) and single-photon emission tomography (SPECT) now are able to visualize and quantify changes in cerebral blood flow, glucose metabolism, and neurotransmitter function produced by parkinsonian disorders. Both PET and SPECT have become important tools in the differential diagnosis of these diseases and may have sufficient sensitivity to detect neuronal changes before the onset of clinical symptoms. Imaging is now being used to elucidate the genetic contribution to Parkinson's disease and in longitudinal studies to assess the efficacy and mode of action of neuroprotective drug and surgical treatments.

Key Words: Imaging; Parkinson's disease; multiple system atrophy; progressive supranuclear palsy; essential tremor; differential diagnosis; positron emission tomography (PET); single-photon emission tomography (SPECT); dopamine transporter; dopamine receptor; cerebral blood flow; cerebral glucose metabolism.

1. INTRODUCTION

The differential diagnosis of the various parkinsonian disorders based on clinical symptoms alone is difficult *(1–3)*. Clinical criteria for the diagnosis of Parkinson's disease (PD) provide high sensitivity for detecting parkinsonism but show poor specificity for identifying brainstem Lewy body disease or for differentiating atypical and typical PD *(4)*. Tremor is a classic feature of PD, although this can also be found in patients with progressive supranuclear palsy (PSP) and multiple system atrophy (MSA). Similarly, a general criterion for diagnosing PD is a good, sustained response to levodopa (L-dopa) therapy, although, again, this also is found in some patients with MSA and dopa-responsive dystonia. Indeed, some post mortem histopathological studies have shown that as many as 25% of all patients who were diagnosed with PD before death had been misdiagnosed *(1,2)*. Detecting preclinical disease by using biochemical markers for neurodegeneration has not been successful. Familial PD sometimes exhibits a mutation of the α-synuclein gene, but this cannot be used as a genetic marker for the majority of cases because the pathogenesis is rarely related to genetic mutation. These observations have contributed to the motivation for developing objective neuroimaging techniques that can differentiate between these disorders.

Structural changes induced by parkinsonian diseases are generally small and often only evident when the disease is in an advanced stage. Consequently, the diagnostic accuracy of anatomical imaging modalities (e.g., magnetic resonance imaging [MRI]) in neurodegenerative disorders is poor *(5)*. Preceding changes in brain morphology, alterations in the way the brain consumes glucose, or disruptions in regional cerebral blood flow (rCBF) may provide useful indicators of neurodegeneration. However, it is likely that changes in neurotransmitter function, most notably in the dopaminergic system, will become evident long before structural, metabolic, or blood flow variations.

In general, positron emission tomography (PET) and single-photon emission tomography (SPECT) imaging have provided a better platform for the diagnosis of parkinsonian disorders than MRI. Functional imaging of neurodegenerative disease with PET and SPECT has followed two main paths; studies of blood flow and cerebral metabolism to detect abnormal tissue functioning or imaging of the dopaminergic neurotransmitter system to study the loss of dopamine neurons.

2. IMAGING BLOOD FLOW AND METABOLISM

PET studies of cerebral glucose metabolism have used the glucose analog [^{18}F]fluorodeoxyglucose ([^{18}F]FDG), whereas radioactive water ($H_2{}^{15}O$), and the SPECT tracers ^{99m}Tc-hexamethylpropylene amine oxime (^{99m}Tc-HMPAO) and ^{99m}Tc-ethylcysteinate dimer (^{99m}Tc-ECD) are markers of cere-

From: *Bioimaging in Neurodegeneration*
Edited by P. A. Broderick, D. N. Rahni, and E. H. Kolodny

bral blood flow and perfusion. Striatal glucose metabolism and perfusion are generally found to be normal in PD *(6–10)*, although some studies have demonstrated an asymmetry of striatal metabolism *(11)*. Interestingly, atypical parkinsonian disorder has been differentiated from idiopathic PD by the appearance of striatal metabolic abnormalities in the atypical group *(12)*, which may provide a useful adjunct to routine clinical examination. Many studies have shown more global cortical hypometabolism or hypoperfusion or a loss of posterior parietal metabolism with a pattern similar to that observed in Alzheimer's, and other neurodegenerative diseases *(8,9,13–18)*. Others have used the differences in regional metabolism or rCBF to discriminate between PD and MSA *(10,19)* or PSP *(20)*. Studies of blood flow and glucose metabolism in patients with pure Lewy body disease with no features of Alzheimer's disease have consistently shown biparietal, bitemporal hypometabolism, a pattern that was once thought to represent the signature of Alzheimer's.

Imaging studies of glucose metabolism and CBF have shown important changes concomitant with degeneration in cognitive performance or autonomic failure *(21–25)*. Although blood flow studies have shown a poor correlation with laterality in hemi-Parkinson's patients *(26)*, there are clearly dramatic changes in CBF and glucose metabolism resulting from cognitive impairment *(18,27–29)*. This is similar in detail to patients with other neurodegenerative disorders, such as Alzheimer's disease and dementia with Lewy bodies *(30)*, but may be useful to distinguish vascular parkinsonism *(31)*. Interestingly, patients with gait disorders generally exhibit an internal verbal cue to compensate for the loss of control in the motor cortex *(32)*. Blood flow PET imaging also has been used to study the effects of novel therapies, such as Voice Treatment, on the reorganization of brain function to compensate for motor dysfunction *(33)*. Studies of blood flow and metabolism have indicated conflicting results when the on- and off-dopamine replacement therapy conditions are compared. Both reduced *(11,18,34)* and normal *(35,36)* regional glucose metabolism and rCBF have been reported after L-dopa treatment.

Recent advances in image analysis, using the voxel-based statistical techniques, such as statistical parametric mapping *(37,38)*, may provide greater accuracy in detecting focal changes in rCBF. These techniques compare changes in rCBF, voxel-by-voxel, or in glucose metabolism to identify regions of statistically significant differences. Although statistical parametric mapping has found important applications in studies of blood flow and metabolic changes in neurodegenerative disease *(39,40)*, it is limited to the comparison of groups of subjects, rather than the diagnosis of individuals. Other statistical methodologies have been developed to attempt to automate the diagnosis of patients with PD and other parkinsonian disorders, based on scans of individual subjects *(41,42)*.

3. IMAGING THE DOPAMINERGIC SYSTEM

In general, the diagnostic accuracy of CBF and glucose metabolism in differentiating neurodegenerative disorders is relatively poor in comparison with direct imaging of the dopaminergic nigrostriatal pathway *(20)*. Early PET studies of the nigrostriatal pathway used the uptake of 6-[^{18}F]fluoro-l-3,4-dihydroxyphenylalanine ([^{18}F]fluorodopa) as a measure of the integrity of dopamine neurons *(43,44)*. [^{18}F]fluorodopa measures changes in aromatic L-amino decarboxylase activity, which is dependent on the availability of striatal dopaminergic nerve terminals and is proportional to the number of dopamine neurons in the substantia nigra *(45)*.

Quantitative parameters associated with [^{18}F]fluorodopa uptake, such as the striatal-to-background uptake ratio, and the influx rate constant, have been shown to be useful indicators of dopaminergic degeneration in PD and other syndromes *(46–67)*. Indeed, [^{18}F]fluorodopa and PET are often regarded as the "gold standard" in the detection of dopamine neuronal loss *(68)*, although the contributions from SPECT imaging, and other direct measures of the dopaminergic binding sites, both pre- and postsynaptic, are increasing *(55,56,69–71)*. The analysis of [^{18}F]fluorodopa PET studies is known to have a number of serious potential problems. [^{18}F]fluorodopa is metabolized into a number of diffusible and nondiffusible labeled metabolites ([^{18}F]3-*O*-methyl-fluorodopa (3OMFD) in peripheral and brain tissue, and [^{18}F]dopamine (FDA), [^{18}F]3-4-dihydroxyphenylacetic acid (FdopaC), and [^{18}F]homovanillic acid in brain tissue). A further issue with the distribution of [^{18}F]fluorodopa in PET scans is the kinetic rate constants tend to disagree with in vitro measurements by a large factor (up to 10 times lower) *(72–76)*. Despite the fact that in vivo measurements of the decarboxylation rate, k_3, gave values considerably lower than in vitro measurements, it has been concluded that k_3 accurately reflects striatal aromatic L-amino decarboxylase activity in vivo with [^{18}F]fluorodopa PET *(75,76)*. Other technical considerations, which are common to all PET and SPECT imaging techniques, include partial volume effects *(62)*, which decrease the apparent striatal uptake of these tracers due to the limited resolution of the scanner.

Direct measurements of dopamine transporter binding sites are possible with [^{11}C]cocaine *(77)*, or the cocaine analogs 2β-carbomethoxy-3β-[4-iodophenyl] tropane (β-CIT) and *N*-ω-fluoropropyl-2β-carbomethoxy-3β-[4-iodophenyl] tropane (FP-CIT), labeled with either ^{18}F or ^{11}C for PET or ^{123}I for SPECT *(78–80)*. Other dopamine transporter ligands include *N*-[3-iodopropen-2-yl]-2β-carbomethoxy-3β-[4-chlorophenyl] tropane ([^{123}I]IPT) *(81)*, its 4-fluorophenyl analog [^{123}I]altropane *(82)*, 2β-carbomethoxy-3β-[4-fluorophenyl] tropane ([^{11}C]CFT) *(83)*, and [^{11}C]d-threo-methylphenidate *(84)*. Of particular importance is the recent development of the first successful ^{99m}Tc-labeled dopamine transporter ligand, ^{99m}Tc-Technetium[2-[[2-[[[3-(4-chlorophenyl)-8-methyl-8-azabicyclo[3.2.1] oct-2-yl]-methyl](2-mercaptoethyl) amino]-ethyl] amino] ethane-thiolato-*N*2,*N*2',*S*2,*S*2'] oxo-[1*R*-(exo-exo)] (^{99m}Tc-TRODAT-1) *(85, 86)*. Because ^{99m}Tc is so much more widely available and less expensive than ^{123}I, this new tracer could move imaging of the dopaminergic system from a research environment into routine clinical practice, particularly with simplified imaging protocols *(87)*.

Several tracers exist for imaging postsynaptic dopamine D2 receptors, using radioactively labeled dopamine receptor antagonists. The most widely used for SPECT include *S*-(-)-3-iodo-2-hydroxy-6-methoxy-*N*-[(1-ethyl-2-pyrrolidinyl) methyl] benzamide ([^{123}I]IBZM) *(88–90)*, *S*-5-iodo-7-*N*-[(1-ethyl-2-

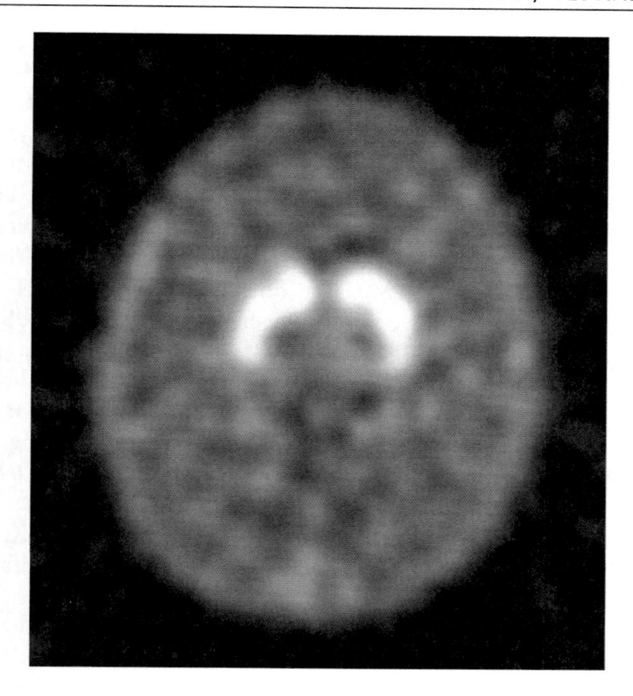

Fig. 1. Transaxial slice at the level of the striatum showing uptake of ^{99m}Tc-TRODAT-1 in dopamine transporters in a normal healthy volunteer. Good uptake is seen in the caudate nucleus and putamen, with background activity throughout the rest of the brain. Image courtesy of Dr. Andrew Newberg, University of Pennsylvania. *See* color version on Companion CD.

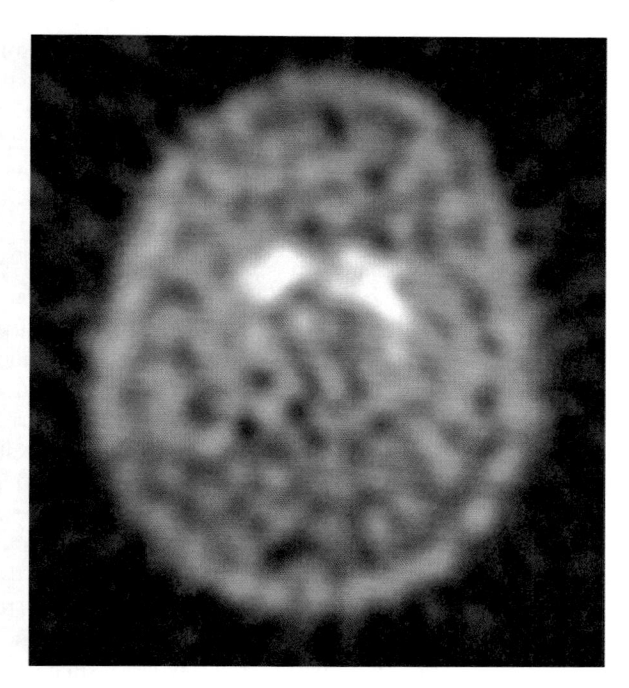

Fig. 2. Uptake of ^{99m}Tc-TRODAT-1 in the striatum of a patient with mild Parkinson's disease shows bilateral decrease in tracer concentration, particularly in the putamen, indicating a loss of dopaminergic neurons in these brain regions. Image courtesy of Dr. Andrew Newberg, University of Pennsylvania. *See* color version on Companion CD.

pyrrolidinyl) methyl] carboxamido-2,3-dihydrobenzofuran ([^{123}I]IBF) *(91,92)*, *S*-*N*-[(1-ethyl-2-pyrrolidinyl) methyl]-5-

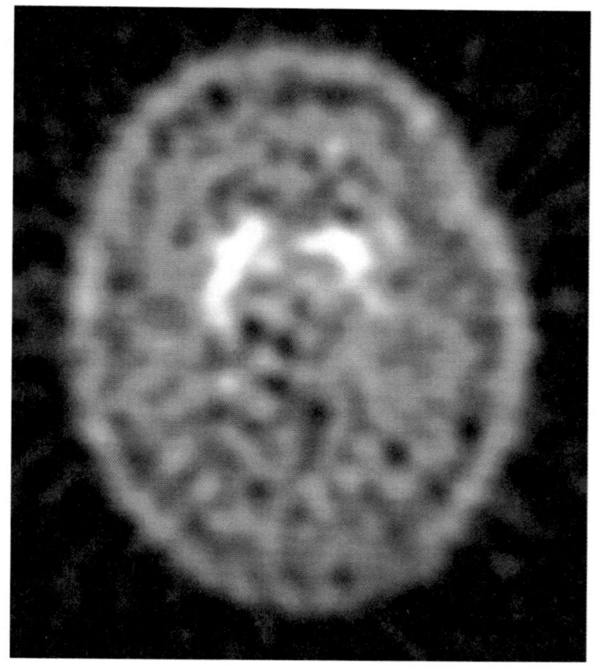

Fig. 3. A patient with hemi-PD exhibits a unilateral decrease in the uptake of ^{99m}Tc-TRODAT-1 in the side contralateral to clinical symptoms, most severely in the putamen. Although the clinical symptoms are confined to one side of the body, there also is reduced tracer uptake in the ipsilateral side, indicating a preclinical reduction in dopaminergic neurons, and demonstrating the sensitivity of imaging techniques for measuring dopaminergic dysfunction before clinical symptoms become apparent. Image courtesy of Dr. Andrew Newberg, University of Pennsylvania. *See* color version on Companion CD.

iodo-2,3-dimethoxybenzamide ([^{123}I]epidepride) *(93,94)* and for PET include *S*-(-)-3,5-dichloro-*N*-[(1-ethyl-2-pyrrolidinyl)] methyl-2-hydroxy-6-methoxybenzamide ([^{11}C]raclopride) *(95)* and [^{11}C] or [^{18}F]*N*-methylspiroperidol *(96,97)*.

PET and SPECT studies of radiotracer binding to postsynaptic dopamine receptors and presynaptic dopamine transporters have proved to be powerful techniques for quantifying the loss of dopaminergic neurons in normal aging *(98–107)*, PD *(67,108–162)* and other neurodegenerative disorders *(48, 143,163–184)*. Studies of neuronal degeneration associated with the effects of normal aging have indicated that, whereas dopamine transporter concentrations decrease as a natural consequence of aging, the changes are small compared with the effects of disease *(106)* (Fig. 1). PET and SPECT studies have indicated a consistent pattern of dopaminergic neuronal loss in PD, usually with more pronounced depletion in the putamen rather than in the caudate (Fig. 2). In addition, there is frequently a marked asymmetry, particularly in the early stages of the disease (Fig. 3), and a good correlation with symptom severity *(114,161)* and illness duration *(152)*. Most importantly, imaging studies may be sensitive enough to detect very early PD *(4,61,115,123,130,141,185–189)*, perhaps even before clinical symptoms become apparent.

Characteristically, PD begins with unilateral symptoms of motor deficit, which gradually progress bilaterally over time. Studies of patients with early hemi-PD have shown that, despite the subject exhibiting only one-sided clinical symptoms, the

PET and SPECT findings demonstrated bilateral decreases in tracer binding, with a greater reduction in the side contralateral to the clinical signs *(41,115,123,190,191)*. The ability of PET and SPECT to detect presymptomatic PD may have important consequences for the screening of familial PD *(187,188,192)*. PET and SPECT studies of parkinsonian kindreds have implicated a genetic foundation for familial PD, including mutations in the *parkin* gene. Hereditary parkinsonism has been detected in asymptomatic relatives with heterozygous *parkin* mutations, using imaging to determine the extent of neuronal damage *(187,193–196)*. Indeed, PET and SPECT imaging of the dopaminergic system is able to demonstrate presynaptic dysfunction in asymptomatic relatives, which is fully compatible with early parkinsonism *(187)*. Even subjects with apparently normal alleles exhibited reduced dopaminergic function on imaging, indicating a preclinical disease in these subjects that is likely to progress to full PD *(187)*. The same features were observed in asymptomatic twins, both monozygotic and dizygotic, of a sibling with parkinsonism *(188)*.

Although most of the PET and SPECT imaging studies have shown highly significant differences between groups of Parkinson's patients and age-matched normal controls, the statistically significant differential diagnosis of an individual subject is more problematic. Patients with severe PD are easily separated from healthy controls even by simple visual inspection of striatal images, quantified using some form of discriminant analysis *(122,126,150,152,164,186,197)*possessing a sensitivity and specificity close to 100% in the proper clinical setting. The differentiation between PD and vascular parkinsonism *(173,182)* and between PD and drug-induced parkinsonism *(133)* also appears possible using imaging of the dopaminergic system. However, patients presenting much earlier in the course of the disease are more difficult to detect, with potentially significant overlap with an age-matched control group *(185,198)* and consequential loss of diagnostic accuracy. The situation may be further complicated if the early differential diagnosis between several neurodegenerative disorders is required. Many of the symptoms associated with parkinsonian disorders are nonspecific, which is why the accurate clinical diagnosis of these diseases is difficult. Studies have shown little difference between radiotracer binding to dopamine transporters in patients with PD, MSA, or PSP *(164,166)* (Fig. 4). Based on current methods of analysis, it appears that the detection of early PD, or the differential diagnosis between various neurodegenerative disorders, may not be possible in individual cases based on imaging of a single neurotransmitter system alone *(121)*. Interestingly, progress on the differential diagnosis of PD and other parkinsonian disorders may come from PET and SPECT imaging outside the brain. Recent studies of the functional integrity of postganglionic cardiac sympathetic neurons, using [^{123}I]MIBG or [^{11}C]HED, have indicated a distinct difference between cardiac autonomic dysfunction in patients with PD and those with MSA *(199,200)*.

4. MULTIMODALITY AND MULTITRACER STUDIES

The relative merits of anatomical and functional imaging have been combined in some studies utilizing either several different radiotracers or data from both MRI/PET or SPECT. Regional glucose metabolism has been studied in parkinsonian disorders with [^{18}F]FDG and PET, and the data combined with striatal [^{18}F]fluorodopa uptake measurements to give an improved diagnostic indicator, and a better understanding of the underlying disease processes *(19,51,59,201)*. However, it should be noted that the improvement was relatively small over the good predictive capabilities of [^{18}F]fluorodopa by itself in these patient groups. Some studies have used the complementary information coming from structural MRI and functional [^{18}F]FDG PET in distinguishing between control subjects and patients with MSA *(53,202–205)*, in whom both focal MRI hypointensities, changes in striatal and midbrain size, and reduced glucose metabolism occurred on the side contralateral to clinical symptoms. Magnetic resonance spectroscopy adds an important new probe to complement functional PET and SPECT imaging studies *(204)*. Other studies have combined data from MRI and postsynaptic dopamine receptor concentrations using [^{123}I]IBZM and SPECT, giving useful information on the involvement of multiple brain regions in PSP *(206)* and MSA *(207)*.

However, the greatest discrimination between various neurodegenerative disorders may be found using PET or SPECT imaging of both pre- and postsynaptic dopaminergic function *(116,121,208)*. A study of [^{123}I]β-CIT and [^{123}I]IBZM binding in patients with early PD showed marked unilateral reductions in dopamine transporters measured by [^{123}I]β-CIT concomitant with elevated dopamine D2 receptor binding of [^{123}I]IBZM *(209)*. Recent SPECT studies investigating pre- and postsynaptic dopamine binding sites in the differential diagnosis of PD, MSA, and PSP have shown promising results, with a reduction in dopamine transporter availability in all diseases, and some discrimination between disorders in the pattern of dopamine D2 receptor concentrations *(142,210,211)* (Fig. 5). Similar results were observed in PET studies of early Parkinson's patients, where striatal [^{18}F]fluorodopa uptake was reduced and [^{11}C]raclopride binding was upregulated, with the degree of increase in dopamine receptor binding inversely proportional to disease severity *(47,53)*. These studies also used [^{18}F]FDG imaging of the same patients to determine the optimum combination of neuroreceptor function and glucose metabolism to differentiate between healthy controls and patients with PD *(47)* or MSA *(48,53)*. The results suggest that striatal [^{18}F]FDG and particularly [^{11}C]raclopride are sensitive to striatal function and may help with the characterization of patients with MSA, whereas [^{18}F]fluorodopa can accurately detect nigrostriatal dopaminergic abnormalities consistent with parkinsonian disorders. Other parkinsonian syndromes, such as Wilson disease, a disorder related to copper deposition, have been studied using imaging and demonstrate a significant decline in dopaminergic function, both pre- and postsynaptic, that can be differentiated from idiopathic PD *(170)*.

SPECT imaging of both pre- and postsynaptic dopamine binding sites simultaneously has now been performed in nonhuman primates, using ^{99m}Tc-TRODAT-1 and [^{123}I]IBZM or [^{123}I]IBF, separating the two radiotracers based on their different energy spectra *(212,213)*. The possibility of simultaneously imaging both dopamine transporters and D2 receptors in neurodegenerative disorders is an exciting prospect, providing a unique probe in the investigation and diagnosis of these diseases.

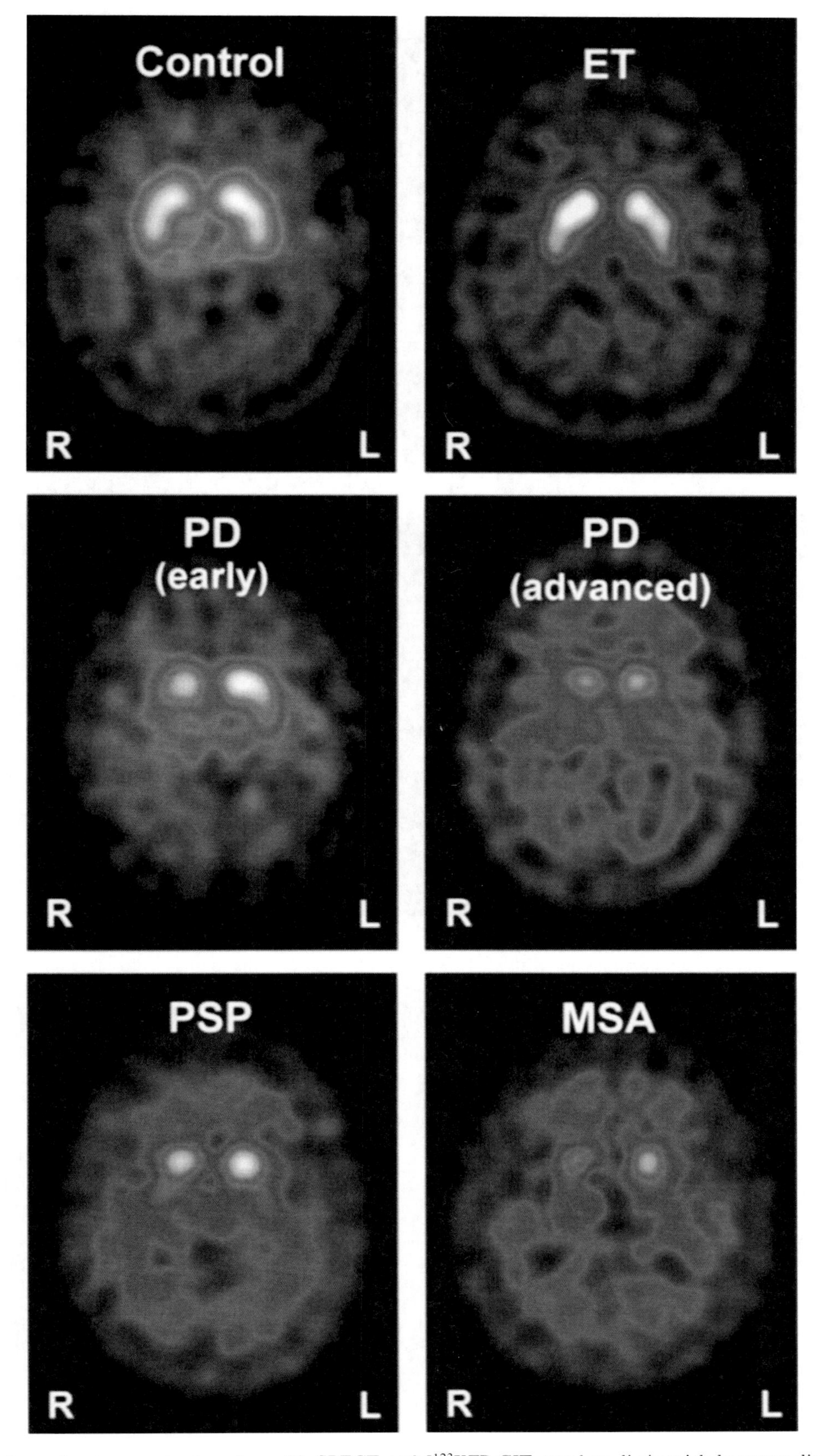

Fig. 4. Presynaptic dopamine transporter imaging with SPECT and [^{123}I]FP-CIT, used to distinguish between disease with and without nigrostriatal deficit. Whereas neurodegenerative parkinsonian syndromes such as PD, MSA, and PSP present with compromised dopamine terminal function, illnesses without involvement of those terminals (e.g., essential tremor [ET]) present with normal findings. Images courtesy of Prof. Klaus Tatsch, University of Munich. *See* color version on Companion CD.

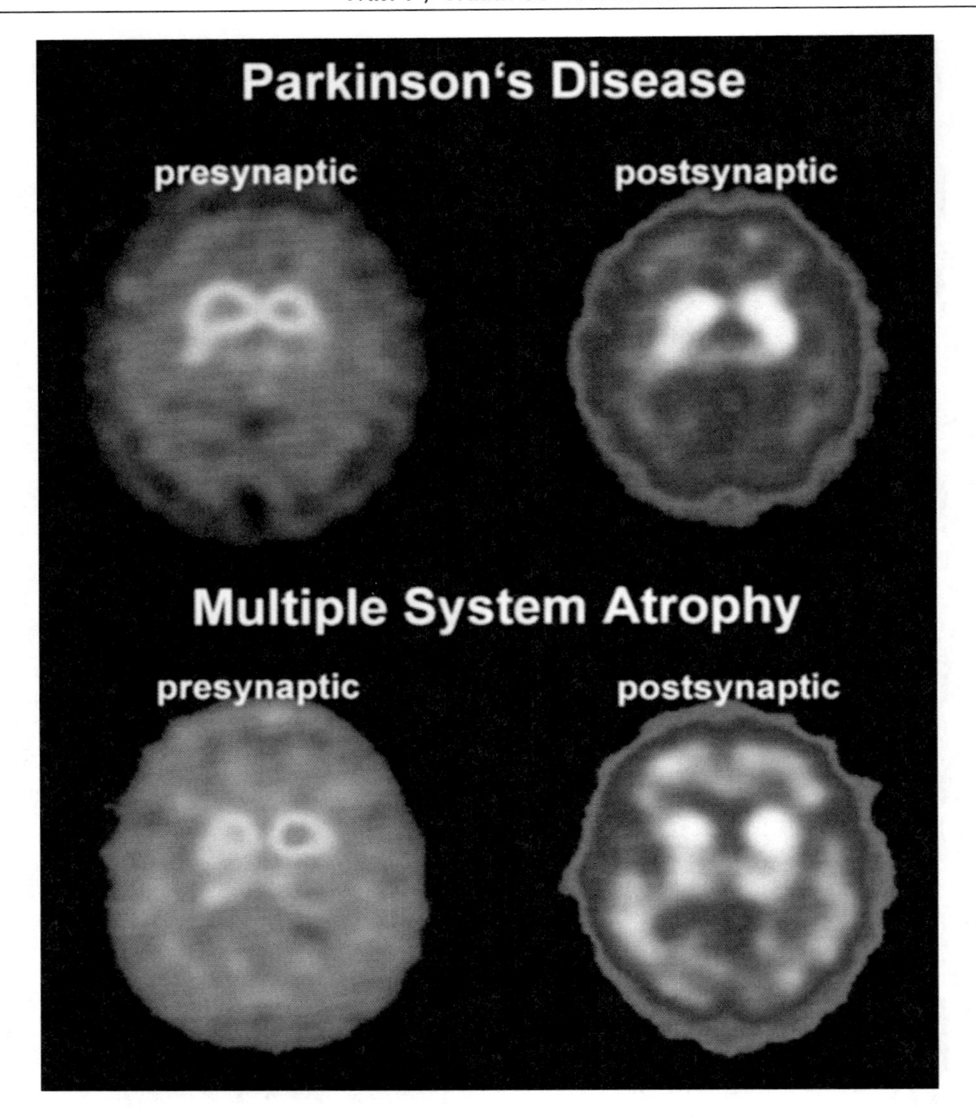

Fig. 5. Postsynaptic dopamine D2 receptor imaging with [^{123}I]IBZM SPECT used to distinguish between PD and atypical parkinsonian syndromes. PD patients present with normal or increased D2 receptor binding suggesting preserved or upregulated postsynaptic receptors, whereas in atypical parkinsonian syndromes (e.g., MSA) D2 receptor binding is reduced, reflecting degeneration of postynaptic receptor binding sites. Presynaptic terminal function, measured with [^{123}I]FP-CIT SPECT, is compromised in both PD and atypical parkinsonian syndromes. Images courtesy of Prof. Klaus Tatsch, University of Munich. *See* color version on Companion CD.

5. IMAGING OTHER NEUROTRANSMITTER SYSTEMS

Although most imaging studies have investigated the effects of parkinsonian disorders on the dopaminergic system, neuropathologic and biochemical studies suggest that serotonin neurons also are affected by the disease process *(214)*. The integrity of serotonin neurons in the midbrain region can be studied using the SPECT tracer [^{123}I]β-CIT *(215,216)*. Although this radioligand is more commonly associated with measurements of the dopamine transporter, it also binds with high affinity to the serotonin transporter. However, owing to the high concentration of dopamine transporters in the striatum, imaging of the serotonergic system with this tracer is limited to the midbrain *(217)*. Despite these technical difficulties, studies suggest that dopamine and serotonin transporters are differentially affected in PD, and serotonin transporters in the midbrain region may not be affected in relatively early stages of PD *(218,219)*. In later stages of the disease, serotonin transporters are reduced in thalamic and frontal areas of the brain and correlate with scores of disease severity *(139)*. Indeed, others have demonstrated that dysfunction in the serotonin system is closely related to the neuropsychiatric symptoms of PD, whereas the dopaminergic system correlates more with motor deficits *(220)*. This is not unexpected, because the serotonin transporter is a well-known target for antidepressant drugs *(221)*. However, the role of imaging the serotonin system in the differential diagnosis of parkinsonian disorders is still unclear and may obtain a more prominent position using the more selective serotonin transporter PET and SPECT ligands that have been developed recently *(222–230)*.

6. CONCLUSION

There are a large number of imaging techniques that can be used to attempt to differentiate between the various neurodegenerative disorders. Taken in isolation, many of them can diagnose parkinsonian disorders with some success. However, the diagnosis at an early stage in the progression of each disease, possibly even before clinical symptoms have become apparent, is much more difficult and may require multiple imaging modalities or combinations of tracers. The widespread availability of SPECT imaging, perhaps combined with newer and less expensive tracers, may lead to the routine implementation of SPECT scanning in the diagnosis of parkinsonian disorders.

ACKNOWLEDGMENTS

The author is indebted to Dr. Andrew Newberg, University of Pennsylvania, and Prof. Klaus Tatsch, University of Munich, for providing the images to accompany this chapter. This work was supported in part by grants from the National Institutes of Health (R01-EB002774 and R01-AG017524).

REFERENCES

1. Raiput AH, Rozdilsky B, Raiput A. Accuracy of clinical diagnosis in parkinsonism—a prospective study. Can J Neurol Sci 1991;12:219–228.
2. Hughes AJ, Daniel SE, Blankson S, Lees AJ. A clinicopathalogic study of 100 cases of Parkinson's disease. Arch Neurol 1993;50: 140–148.
3. Testa D, Filippini G, Farinotti M, Palazzini E, Caraceni T. Survival in multiple system atrophy: a study of prognostic factors in 59 cases. J Neurol 1996;243:401–404.
4. Brooks DJ. The early diagnosis of Parkinson's disease. Ann Neurol 1998;44:S10–S18.
5. Schrag A, Kingsley D, Phatouros C, et al. Clinical usefulness of magnetic resonance imaging in multiple system atrophy. J Neurol Neurosurg Psych 1998;65:65–71.
6. Kuhl DE, Metter EJ, Riege WH, Markham CH. Patterns of cerebral glucose utilisation in Parkinson's disease and Huntingdon's disease. Ann Neurol 1984;15:119–125.
7. Smith FW, Gemmell HG, Sharp PF, Besson JA. Technetium-99m HMPAO imaging in patients with basal ganglia disease. Br J Radiol 1988;61:914–920.
8. Wang SJ, Liu RS, Liu HC, et al. Technetium-99m hexamethylpropylene amine oxime single photon emission tomography of the brain in early Parkinson's disease: correlation with dementia and lateralization. Eur J Nucl Med 1993;20:339–344.
9. Markus HS, Lees AJ, Lennox G, Marsden CD, Costa DC. Patterns of regional cerebral blood flow in corticobasal degenration studied using HMPAO SPECT; comparison with Parkinson's disease and normal controls. Mov Disord 1995;10:179–187.
10. Otsuka M, Ichiya Y, Kuwabara Y, et al K. Glucose metabolism in the cortical and subcortical brain structures in multiple system atrophy and Parkinson's disease: a positron emission tomography study. J Neurol Sci 1996;144:77–83.
11. Dethy S, Van Blercom N, Damhaut P, Wikler D, Hilderbrand J, Goldman S. Asymmetry of basal ganglia glucose metabolism and dopa responsiveness in parkinsonism. Mov Disord 1998;13:275–280.
12. Antonini A, Kazumata K, Feigin A, et al. Differential diagnosis of parkinsonism with [^{18}F]fluorodeoxyglucose and PET. Mov Disord 1998;13:268–274.
13. Pizzolato G, Dam M, Borsato N, et al. [99mTc]-HM-PAO SPECT in Parkinson's disease. J Cerebral Blood Flow Metab 1988;8:S101–S108.
14. Liu RS, Lin KN, Wang SJ, et al. Cognition and 99mTc-HMPAO SPECT in Parkinson's disease. Nucl Med Commun 1992;13:744–748.
15. Eberling JL, Richardson BC, Reed BR, Wolfe N, Jagust WJ. Cortical glucose metabolism in Parkinson's disease without dementia. Neurobiol Aging 1994;15:329–335.
16. Eidelberg D, Moeller JR, Ishikawa T, et al. Early differential diagnosis of Parkinson's disease with ^{18}F-fluorodeoxyglucose and positron emission tomography. Neurology 1995;45:1995–2004.
17. Piert M, Koeppe RA, Giordani B, Minoshima S, Kuhl DE. Determination of regional rate constants from dynamic FDG-PET studies in Parkinson's disease. J Nucl Med 1996;37:1115–1122.
18. Berding G, Odin P, Brooks DJ, et al. Resting regional cerebral glucose metabolism in advanced Parkinson's disease studied in the off and on conditions with [(18)F]FDG-PET. Movement Disorders 2001;16:1014–1022.
19. Otsuka M, Kuwabara Y, Ichiya Y, et al. Differentiating between multiple system atrophy and Parkinson's disease by positron emission tomography with 18F-dopa and 18F-FDG. Ann Nucl Med 1997;11:251–257.
20. Defebvre L, Lecouffe P, Destee A, Houdart P, Steinling M. Tomographic measurements of regional cerebral blood flow in progressive supranuclear palsy and Parkinson's disease. Acta Neurol Scand 1995;92:235–241.
21. Jagust WJ, Reed BR, Martin EM, Eberling JL, Nelson-Abbott RA. Cognitive function and regional cerebral blood flow in Parkinson's disease. Brain 1992;115:521–537.
22. Samuel M, Ceballos-Baumann AO, Boecker H, Brooks DJ. Motor imagery in normal subjects and Parkinson's disease patients: an H215O PET study. Neuroreport 2001;12:821–828.
23. Sawada H, Udaka F, Kameyama M, et al. SPECT findings in Parkinson's disease associated with dementia. J Neurol Neurosurg Psychiatry 1992;55:960–963.
24. Wu JC, Iacono R, Ayman M, et al. Correlation of intellectual impairment in Parkinson's disease with FDG PET scan. Neuroreport 2000;11:2139–2144.
25. Antonini A, De Notaris R, Benti R, De Gaspari D, Pezzoli G. Perfusion ECD/SPECT in the characterization of cognitive deficits in Parkinson's disease. Neuroll Sci 2001;22:45–46.
26. Agniel A, Celsis P, Viallard G, et al. Cognition and cerebral blood flow in lateralised parkinsonism: lack of functional lateral asymmetries. J Neurol Neurosurg Psychiatry 1991;54:783–786.
27. Arahata Y, Hirayama M, Ieda T, et al. Parieto-occipital glucose hypometabolism in Parkinson's disease with autonomic failure. J Neurol Sci 1999;163:119–126.
28. Bissessur S, Tissingh G, Wolters EC, Scheltens P. rCBF SPECT in Parkinson's disease patients with mental dysfunction. J Neural Transmission Suppl 1997;50:25–30.
29. Ceballos-Baumann AO. Functional imaging in Parkinson's disease: activation studies with PET, fMRI and SPECT. J Neurol 2003;250 Suppl 1:15–23.
30. Costa DC, Ell PJ, Burns A, Philpot M, Levy R. CBF tomograms with [99mTc-HM-PAO in patients with dementia (Alzheimer type and HIV) and Parkinson's disease—initial results. J Cerebral Blood Flow Metab 1988;8:S109–S115.
31. De Reuck J, Siau B, Decoo D, et al. Parkinsonism in patients with vascular dementia: clinical, computed- and positron emission-tomographic findings. Cerebrovascr Dis 2001;11:51–58.
32. Albani G, Kunig G, Soelch CM, et al. The role of language areas in motor control dysfunction in Parkinson's disease. Neurol Sci 2001;22:43–44.
33. Liotti M, Ramig LO, Vogel D, et al. Hypophonia in Parkinson's disease: neural correlates of voice treatment revealed by PET. Neurology 2003;60:432–440.
34. Feigin A, Fukuda M, Dhawan V, et al. Metabolic correlates of levodopa response in Parkinson's disease. Neurology 2001;57: 2083–2088.
35. Rougemont D, Baron JC, Collard P, Bustany P, Comar D, Agid Y. Local cerebral glucose utilisation in treated and untreated patients with Parkinson's disease. J Neurol Neurosurg Psychiatry 1984;47: 824–830.
36. Markus HS, Costa DC, Lees AJ. HMPAO SPECT in Parkinson's disease before and after levodopa: correlation with dopaminer-

gic responsiveness. J Neurol Neurosurg Psychiatry 1994;57:180–185.
37. Friston KJ, Holmes AP, Worsley KJ, Poline JP, Frith CD, Frackowiak RSJ. Statistical parametric maps in functional imaging: a general linear approach. Hum Brain Map 1995;2:189–210.
38. Acton PD, Friston KJ. Statistical parametric mapping in functional neuroimaging: beyond PET and fMRI activation studies. Eur J Nucl Med 1998;25:663–667.
39. Imon Y, Matsuda H, Ogawa M, Kogure D, Sunohara N. SPECT image analysis using statistical parametric mapping in patients with Parkinson's disease. J Nuclear Med 1999;40:1583–1589.
40. Lozza C, Marie RM, Baron JC. The metabolic substrates of bradykinesia and tremor in uncomplicated Parkinson's disease. Neuroimage 2002;17:688–699.
41. Acton PD, Mozley PD, Kung HF. Logistic discriminant parametric mapping: a novel method for the pixel-based differential diagnosis of Parkinson's disease. Eur J Nucl Med 1999;26:1413–1423.
42. Slomka PJ, Radau PE, Hurwitz GA, Dey D. Automated three-dimensional quantification of myocardial perfusion and brain SPECT. Comput Med Imaging Graph 2001;25:153–164.
43. Garnett ES, Firnau G, Chan PKH, Sood S, Belbeck LW. [18F]-Fluoro-dopa, an analogue of dopa, and its use in direct external measurements of storage, degradation, and turnover of intracerebral dopamine. Proc Natl Acad Sci USA 1978;75:464.
44. Garnett ES, Firnau G, Nahmias C. Dopamine visualised in the basal ganglia of living man. Nature 1983;305:137.
45. Snow BJ, Tooyama I, McGeer EG. Human positron emission tomographic (fluorine-18)fluorodopa studies correlate with dopamine cell counts and levels. Ann Neurol 1993;34:324–330.
46. Dhawan V, Ma Y, Pillai V, et al. Comparative analysis of striatal FDOPA uptake in Parkinson's disease: ratio method versus graphical approach. J Nucl Med 2002;43:1324–1330.
47. Antonini A, Vontobel P, Psylla M, et al. Complementary positron emission tomographic studies of the striatal dopaminergic system in Parkinson's disease. Arch Neurol 1995;52:1183–1190.
48. Antonini A, Leenders KL, Vontobel P, Maguire RP, Missimer J, Psylla M, Gunther I. Complementary PET studies of striatal neuronal function in the differential diagnosis between multiple system atrophy and Parkinson's disease. Brain 1997;120:2187–2195.
49. Dhawan V, Ishikawa T, Patlak C, et al. Combined FDOPA and 3OMFD PET studies in Parkinson's disease. J Nucl Med 1996;37:209–216.
50. Doudet DJ, Chan GL, Holden JE, et al. 6-(18F)Fluoro-L-DOPA PET studies of the turnover of dopamine in MPTP-induced parkinsonism in monkeys. Synapse 1998;29:225–232.
51. Eidelberg D, Moeller JR, Dhawan V, et al. The metabolic anatomy of Parkinson's disease: complementary (18F)fluordeoxyglucose and (18F)fluorodopa positron emission tomographic studies. Mov Disord 1990;5:203–213.
52. Ghaemi M, Rudolf J, Hilker R, Herholz K, Heiss WD. Increased pineal Fdopa uptake is related to severity of Parkinson's disease—a PET study. J Pineal Res 2001;30:213–219.
53. Ghaemi M, Hilker R, Rudolf J, Sobesky J, Heiss WD. Differentiating multiple system atrophy from Parkinson's disease: contribution of striatal and midbrain MRI volumetry and multi-tracer PET imaging. J Neurol Neurosurg Psychiatry 2002;73:517–523.
54. Hoshi H, Kuwabara H, Leger G, Cumming P, Guttman M, Gjedde A. 6-(18F)fluoro-L-dopa metabolism in living human brain: a comparison of six analytical methods. J Cerebral Blood Flow Metab 1993;13:57–69.
55. Huang WS, Chiang YH, Lin JC, Chou YH, Cheng CY, Liu RS. Crossover study of (99m)Tc-TRODAT-1 SPECT and (18)F-FDOPA PET in Parkinson's disease patients. J Nucl Med 2003;44:999–1005.
56. Huang WS, Ma KH, Chou YH, Chen CY, Liu RS, Liu JC. 99mTc-TRODAT-1 SPECT in healthy and 6-OHDA lesioned parkinsonian monkeys: comparison with 18F-FDOPA PET. Nucl Med Commun 2003;24:77–83.
57. Kawatsu S, Kato T, Nagano-Saito A, Hatano K, Ito K, Ishigaki T. New insight into the analysis of 6-(18F)fluoro-L-DOPA PET dynamic data in brain tissue without an irreversible compartment: comparative study of the Patlak and Logan analyses. Radiat Med 2003;21:47–54.
58. Kuwabara H, Cumming P, Yasuhara Y, et al. Regional striatal DOPA transport and decarboxylase activity in Parkinson's disease. J Nucl Med 1995;36:1226–1231.
59. Laureys S, Salmon E, Garraux G, et al. Fluorodopa uptake and glucose metabolism in early stages of corticobasal degeneration. J Neurol 1999;246:1151–1158.
60. Melega WP, Raleigh MJ, Stout DB, et al. Longitudinal behavioral and 6-(18F)fluoro-L-DOPA-PET assessment in MPTP-hemiparkinsonian monkeys. Exp Neurol 1996;141:318–329.
61. Nurmi E, Ruottinen HM, Bergman J, et al. Rate of progression in Parkinson's disease: a 6-[18F]fluoro-L-dopa PET study. Mov Disord 2001;16:608–615.
62. Rousset OG, Deep P, Kuwabara H, Evans AC, Gjedde AH, Cumming P. Effect of partial volume correction on estimates of the influx and cerebral metabolism of 6-[(18)F]fluoro-L-dopa studied with PET in normal control and Parkinson's disease subjects. Synapse 2000;37:81–89.
63. Ruottinen HM, Partinen M, Hublin C, et al. An FDOPA PET study in patients with periodic limb movement disorder and restless legs syndrome. Neurology 2000;54:502–504.
64. Takikawa S, Dhawan V, Chaly T, et al. Input functions for 6-(fluorine-18)fluorodopa quantitation in parkinsonism: comparative studies and clinical correlations. J Nucl Med 1994;35:955–963.
65. Broussolle E, Dentresangle C, Landais P, et al. The relation of putamen and caudate nucleus 18F-Dopa uptake to motor and cognitive performances in Parkinson's disease. J Neurol Sci 1999;166:141–151.
66. de la Fuente-Fernandez R, Pal PK, Vingerhoets FJ, et al. Evidence for impaired presynaptic dopamine function in parkinsonian patients with motor fluctuations. J Neural Transmission 2000;107:49–57.
67. De La Fuente-Fernandez R, Lim AS, Sossi V, et al. Age and severity of nigrostriatal damage at onset of Parkinson's disease. Synapse 2003;47:152–158.
68. Frey KA. Can SPET imaging of dopamine uptake sites replace PET imaging in Parkinson's disease? Eur J Nucl Med Mol Imaging 2002;29:715–717.
69. Tatsch K. Can SPET imaging of dopamine uptake sites replace PET imaging in Parkinson's disease? For. Eur J Nucl Med Mol Imaging 2002;29:711–714.
70. Brucke T, Djamshidian S, Bencsits G, Pirker W, Asenbaum S, Podreka I. SPECT and PET imaging of the dopaminergic system in Parkinson's disease. J Neurol 2000;247(Suppl 4):2–7.
71. Fernandez HH, Friedman JH, Fischman AJ, Noto RB, Lannon MC. Is altropane SPECT more sensitive to fluoroDOPA PET for detecting early Parkinson's disease? Medical Sci Monitor 2001;7:1339–1343.
72. Cumming P, Boyes BE, Martin WRW, et al. The metabolism of (18F)6-fluoro-L-3,4-dihydroxyphenylalanine in the hooded rat. J Neurochem 1987;48:601.
73. Cumming P, Hausser M, Martin WR, Grierson J, Adam MJ, Ruth TJ. Kinetics of in vitro decarboxylation and the in vivo metabolism of 2–18F- and 6–18F-fluorodopa in the hooded rat. Biochem Pharmacol 1988;37:247–250.
74. Cumming P, Brown E, Damsma G, Fibiger HC. Formation and clearance of interstitial metabolites of dopamine and serotonin in rat striatum: an in vivo microdialysis study. J Neurochem 1992;59:1905–1914.
75. Cumming P, Kuwabara H, Gjedde AH. A kinetic analysis of 6-[18F]fluoro-L-dihydroxyphenylalanine metabolism in the rat. J Neurochem 1994;63:1675–1682.
76. Yee RE, Huang SC, Stout DB, et al. Nigrostriatal reduction of aromatic L-amino acid decarboxylase activity in MPTP-treated squirrel monkeys: in vivo and in vitro investigations. J Neurochem 2000;74:1147–1157.
77. Fowler JS, Volkow ND, Wolf AP, et al. Mapping cocaine binding sites in human and baboon brain in vivo. Synapse 1989;4:371–377.

78. Bergstrom KA, Kuikka JT, Ahonen A, Vanninen E. [123I] β-CIT, a tracer for dopamine and serotonin re-uptake sites: preparation and preliminary SPECT studies in humans. J Nucl Biol Med 1994;38: 128–131.
79. Madras BK, Spealman RD, Fahey MA, Neumeyer JL, Saha JK, Milius RA. Cocaine receptors labeled by [3H]2b-carbomethoxy-3b-(4-fluorophenyl)tropane. Mol Pharmacol 1989;36:518–524.
80. Neumeyer JL, Wang S-Y, Milius RA, et al. [123I]-2b-Carbomethoxy-3b-(4-iodophenyl)tropane: high-affinity SPECT radiotracer of monoamine reuptake sites in brain [letter]. J Med Chem 1991;34:3144–3146.
81. Kung M-P, Essman WD, Frederick D, et al. IPT: a novel iodinated ligand for the CNS dopamine transporter. Synapse 1995;20:316–324.
82. Madras BK, Meltzer PC, Liang AY, Elmaleh DR, Babich J, Fischman AJ. Altropane, a SPECT or PET imaging probe for dopamine neurons: I. Dopamine transporter binding in primate brain. Synapse 1998;29:93–104.
83. Canfield DR, Spealman RD, Kaufman MJ, Madras BK. Autoradiographic localization of cocaine binding sites by [3H]CFT ([3H]WIN 35,428) in the monkey brain. Synapse 1990;6:189–195.
84. Volkow ND, Ding YS, Fowler JS, et al. A new PET ligand for the dopamine transporter: studies in the human brain. J Nucl Med 1995;36:2162–2168.
85. Kung HF, Kim H-J, Kung M-P, Meegalla SK, Plössl K, Lee H-K. Imaging of dopamine transporters in humans with technetium-99m TRODAT-1. Eur J Nucl Med 1996;23:1527–1530.
86. Kung M-P, Stevenson DA, Plössl K, et al. [99mTc]TRODAT-1: a novel technetium-99m complex as a dopamine transporter imaging agent. Eur J Nucl Med 1997;24:372–380.
87. Acton PD, Kushner SA, Kung MP, Mozley PD, Plössl K, Kung HF. Simplified reference region model for the kinetic analysis of [99mTc]TRODAT-1 binding to dopamine transporters in non-human primates using SPET. Eur J Nucl Med 1999;26:518–526.
88. Kung HF, Billings JJ, Guo Y-Z, et al. Preparation and biodistribution of (125I)IBZM: a potential CNS D2 dopamine receptor imaging agent. Nucl Med Biol 1988;15:195–201.
89. Kung HF, Pan S, Kung M-P, Billings JJ, et al. In vitro and in vivo evaluation of (123I)IBZM: a potential CNS D2 dopamine receptor imaging agent. J Nucl Med 1989;30:88–92.
90. Kung HF, Alavi A, Chang W, et al. In vivo SPECT imaging of CNS D2 dopamine receptors: initial studies with iodine-123-IBZM in humans. J Nucl Med 1990;31:573–579.
91. Kung MP, Kung HF, Billings J, Yang Y, Murphy RA, Alavi A. The characterization of IBF as a new selective dopamine D-2 receptor imaging agent. J Nucl Med 1990;31:648–654.
92. Billings JJ, Guo YZ, Kung MP, Kung HF. Localization of IBF as a D-2 dopamine receptor imaging agent in nonhuman primates. Eur J Nucl Med 1993;20:1146–1153.
93. Kessler RM, Ansari MS, Schmidt DE, et al. High affinity dopamine D2 receptor radioligands. 2. [125I]Epidepride, a potent and specific radioligand for the characterization of striatal and extrastriatal dopamine D2 receptors. Life Sci 1991;49:617–628.
94. Kessler RM, Mason NS, Votaw JR, et al. Visualization of extrastriatal dopamine D2 receptors in the human brain. Eur J Pharmacol 1992;223:105–107.
95. Ehrin E, Farde L, de Paulis T. Preparation of 11C-labelled raclopride, a new potent dopamine receptor antagonist: preliminary PET studies of cerebral dopamine receptors in the monkey. Int J Appl Rad Isot 1985;36:269–273.
96. Arnett CD, Wolf AP, Shiue C-Y, et al. Improved delineation of human dopamine receptors using (18F)-N-methylspiroperidol and PET. J Nucl Med 1986;27:1878–1882.
97. Shiue C-Y, Fowler JS, Wolf AP, McPherson DW, Arnett CD, Zecca L. No-carrier-added fluorine-18-labeled N-methylspiroperidol: synthesis and biodistribution in mice. J Nucl Med 1986;27:226–234.
98. Volkow ND, Fowler JS, Wang G-J, et al. Decreased dopamine transporters with age in healthy human subjects. Ann Neurol 1994; 36:237–239.
99. Vandyck CH, Seibyl JP, Malison RT, et al. Age-related decline in striatal dopamine transporter binding with iodine-123-β-CIT SPECT. J Nucl Med 1995;36:1175–1181.
100. Volkow ND, Ding YS, Fowler JS, et al. Dopamine transporters decrease with age. J Nucl Med 1996;37:554–559.
101. Mozley PD, Kim HJ, Gur RC, et al. [I-123]IPT SPECT imaging of CNS dopamine transporters: non-linear effects of normal aging on striatal uptake values. J Nucl Med 1996;37:1965–1970.
102. Martin WRW, Palmer MR, Patlak CS, Calne DB. Nigrostriatal function in man studied with positron emission tomography. Ann Neurol 1989;26:535–542.
103. Sawle GV, Colebatch JG, Shah A, Brooks DJ, Marsden CD, Frackowiak RSJ. Striatal function in normal aging: implications for Parkinson's disease. Ann Neurol 1990;28:799–804.
104. Cordes M, Snow BJ, Cooper S, et al. Age-dependent decline of nigrostriatal dopaminergic function: a positron emission tomographic study of grandparents and their grandchildren. Ann Neurol 1994;36:667–670.
105. Antonini A, Leenders KL, Reist H, Thomann R, Beer HF, Locher J. Effect of age on D2 dopamine receptors in normal human brain measured by positron emission tomography and 11C-raclopride. Arch Neurol 1993;50:474–480.
106. Mozley PD, Acton PD, Barraclough ED, et al. Effects of age on the cerebral distribution of (Tc-99m)TRODAT-1 in healthy humans. J Nucl Med 1999;40:1812–1817.
107. van Dyck CH, Seibyl JP, Malison RT, et al. Age-related decline in dopamine transporters: analysis of striatal subregions, nonlinear effects, and hemispheric asymmetries. Am J Geriatric Psychiatry 2002;10:36–43.
108. Leenders KL. The nigrostriatal dopaminergic system assessed in vivo by positron emission tomography in healthy volunteer subjects and patients with Parkinson's disease. Arch Neurol 1990; 47:1290.
109. Brooks DJ. Detection of preclinical Parkinson's disease with PET. Neurology 1991;41(5 Suppl 2):24-27; discussion 28.
110. Brooks DJ, Ibanez V, Sawle GV, et al. Differing patterns of striatal 18F-dopa uptake in Parkinson's disease, multiple system atrophy, and progressive supranuclear palsy. Ann Neurol 1990;28:547.
111. Brooks DJ. PET and SPECT studies in Parkinson's disease. Baillieres Clin Neurol 1997;6:69–87.
112. Seibyl JP, Marek KL, Quinlan D, et al. Decreased single-photon emission computed tomographic [123I]β-CIT striatal uptake correlates with symptom severity in Parkinson's disease. Ann Neurol 1995;38:589–598.
113. Tatsch K, Schwarz J, Oertel WH, Kirsch C-M. SPECT imaging of dopamine D2 receptors with [123I]IBZM in Parkinsonian syndromes. J Nucl Med 1991;32:1014–1015.
114. Tatsch K, Schwarz J, Mozley PD, et al. Relationship between clinical features of Parkinson's disease and presynaptic dopamine transporter binding assessed with [123I]IPT and single-photon emission tomography. Eur J Nucl Med 1997;24:415–421.
115. Booij J, Tissingh G, Boer GJ, et al. [123I]FP-CIT SPECT shows a pronounced decline of striatal dopamine transporter labelling in early and advanced Parkinson's disease. J Neurol Neurosurg Psychiatry 1997;62:133–140.
116. Tissingh G, Booij J, Winogrodzka A, van Royen EA, Wolters EC. IBZM- and CIT-SPECT of the dopaminergic system in parkinsonism. Neural Transmission Suppl 1997;50:31–37.
117. Asenbaum S, Brücke T, Pirker W, et al. Imaging of dopamine transporters with iodine-123-β-CIT and SPECT in Parkinson's disease. J Nucl Med 1997;38:1–6.
118. Kim HJ, Im JH, Yang SO, et al. Imaging and quantitation of dopamine transporters with iodine-123-IPT in normal and Parkinson's disease subjects. J Nucl Med 1997;38:1703–1711.
119. Fischman AJ, Bonab AA, Babich JW, et al. Rapid detection of Parkinson's disease by SPECT with altropane: a selective ligand for dopamine transporters. Synapse 1998;29:128–141.
120. Muller T, Farahati J, Kuhn W, et al. [123I]β-CIT SPECT visualizes dopamine transporter loss in de novo Parkinsonian patients. Eur Neurol 1998;39:44–48.

121. Booij J, Tissingh G, Winogrodzka A, van Royen EA. Imaging of the dopaminergic neurotransmission system using single-photon emission tomography and positron emission tomography in patients with parkinsonism. Eur J Nucl Med 1999;26:171–182.
122. Prunier C, Bezard E, Montharu J, et al. Presymptomatic diagnosis of experimental Parkinsonism with 123I-PE2I SPECT. Neuroimage 2003;19:810–816.
123. Prunier C, Payoux P, Guilloteau D, et al. Quantification of dopamine transporter by 123I-PE2I SPECT and the noninvasive Logan graphical method in Parkinson's disease. J Nuclear Med 2003;44: 663–670.
124. Radau PE, Linke R, Slomka PJ, Tatsch K. Optimization of automated quantification of 123I-IBZM uptake in the striatum applied to parkinsonism. J Nucl Med 2000;41:220–227.
125. Ransmayrl G, Seppi K, Donnemiller E, et al. Striatal dopamine transporter function in dementia with Lewy bodies and Parkinson's disease. Eur J Nucl Med 2001;28:1523–1528.
126. Anonymous. A multicenter assessment of dopamine transporter imaging with DOPASCAN/SPECT in parkinsonism. Parkinson Study Group. Neurology 2000;55:1540–1547.
127. Antonini A, Moresco RM, Gobbo C, et al. The status of dopamine nerve terminals in Parkinson's disease and essential tremor: a PET study with the tracer (11-C)FE-CIT. Neurol Sci 2001;22:47–48.
128. Bao SY, Wu JC, Luo WF, Fang P, Liu ZL, Tang J. Imaging of dopamine transporters with technetium-99m TRODAT-1 and single photon emission computed tomography. J Neuroimaging 2000;10:200–203.
129. Benamer HT, Patterson J, Wyper DJ, Hadley DM, Macphee GJ, Grosset DG. Correlation of Parkinson's disease severity and duration with 123I-FP-CIT SPECT striatal uptake. Mov Disord 2000; 15:692–698.
130. Berendse HW, Booij J, Francot CM, et al. Subclinical dopaminergic dysfunction in asymptomatic Parkinson's disease patients' relatives with a decreased sense of smell. Ann Neurol 2001;50:34–41.
131. Booij J, Habraken JB, Bergmans P, -et al. Imaging of dopamine transporters with iodine-123-FP-CIT SPECT in healthy controls and patients with Parkinson's disease. J Nucl Med 1998;39:1879–1884.
132. Booij J, Hemelaar TG, Speelman JD, de Bruin K, Janssen AG, van Royen EA. One-day protocol for imaging of the nigrostriatal dopaminergic pathway in Parkinson's disease by (123I)FPCIT SPECT. J Nucl Med1999;40:753–761.
133. Booij J, Speelman JD, Horstink MW, Wolters EC. The clinical benefit of imaging striatal dopamine transporters with (123I)FP-CIT SPET in differentiating patients with presynaptic parkinsonism from those with other forms of parkinsonism. Eur J Nucl Med2001;28:266–272.
134. Chouker M, Tatsch K, Linke R, Pogarell O, Hahn K, Schwarz J. Striatal dopamine transporter binding in early to moderately advanced Parkinson's disease: monitoring of disease progression over 2 years. Nucl Med Commun 2001;22:721–725.
135. Davis MR, Votaw JR, Bremner JD, et al. Initial human PET imaging studies with the dopamine transporter ligand 18F-FECNT. J Nucl Med 2003;44:855–861.
136. Dentresangle C, Veyre L, Le Bars D, et al. Striatal D2 dopamine receptor status in Parkinson's disease: an (18F)dopa and (11C)raclopride PET study. MovDisord 1999;14:1025–1030.
137. Duchesne N, Soucy JP, Masson H, Chouinard S, Bedard MA. Cognitive deficits and striatal dopaminergic denervation in Parkinson's disease: a single photon emission computed tomography study using 123iodine-β-CIT in patients on and off levodopa. Clin Neuropharmacol 2002;25:216–224.
138. Eising EG, Muller TH, Freudenberg L, et al. SPECT imaging with [123I]-β-CIT in Parkinsonism: comparison of SPECT images obtained by a single-headed and a three-headed gamma camera. Nucl Med Commun 2001;22:145–150.
139. Haapaniemi TH, Ahonen A, Torniainen P, Sotaniemi KA, Myllyla VV. [123I]β-CIT SPECT demonstrates decreased brain dopamine and serotonin transporter levels in untreated parkinsonian patients. Mov Disord2001;16:124–130.
140. Happe S, Pirker W, Klosch G, Sauter C, Zeitlhofer J. Periodic leg movements in patients with Parkinson's disease are associated with reduced striatal dopamine transporter binding. J Neurol 2003;250:83–86.
141. Huang WS, Lin SZ, Lin JC, Wey SP, Ting G, Liu RS. Evaluation of early-stage Parkinson's disease with 99mTc-TRODAT-1 imaging. J Nucl Med 2001;42:1303–1308.
142. Ichise M, Kim YJ, Ballinger JR, Vines D, Erami SS, Tanaka F, et al. SPECT imaging of pre- and postsynaptic dopaminergic alterations in L-dopa-untreated PD. Neurology 1999;52:1206–1214.
143. Ilgin N, Zubieta J, Reich SG, Dannals RF, Ravert HT, Frost JJ. PET imaging of the dopamine transporter in progressive supranuclear palsy and Parkinson's disease. Neurology 1999;52:1221–1226.
144. Kaasinen V, Nagren K, Hietala J, et al. Extrastriatal dopamine D2 and D3 receptors in early and advanced Parkinson's disease. Neurology 2000;54:1482–1487.
145. Kaasinen V, Ruottinen HM, Nagren K, Lehikoinen P, Oikonen V, Rinne JO. Upregulation of putaminal dopamine D2 receptors in early Parkinson's disease: a comparative PET study with [11C] raclopride and [11C]N-methylspiperone. J Nucl Med 2000;41:65–70.
146. Kaasinen V, Rinne JO. Functional imaging studies of dopamine system and cognition in normal aging and Parkinson's disease. Neurosci Biobehav Rev 2002;26:785–793.
147. Kaasinen V, Aalto S, K NA, Hietala J, Sonninen P, Rinne JO. Extrastriatal dopamine D(2) receptors in Parkinson's disease: a longitudinal study. J Neural Transmission 2003;110:591–601.
148. Kao PF, Tzen KY, Yen TC, et al. The optimal imaging time for (99Tcm)TRODAT-1/SPET in normal subjects and patients with Parkinson's disease. Nucl Med Commun 2001;22:151–154.
149. Linke R, Gostomzyk J, Hahn K, Tatsch K. [123I]IPT binding to the presynaptic dopamine transporter: variation of intra- and interobserver data evaluation in parkinsonian patients and controls. Eur J Nucl Med 2000;27:1809–1812.
150. Lokkegaard A, Werdelin LM, Friberg L. Clinical impact of diagnostic SPET investigations with a dopamine re-uptake ligand. Eur J Nucl Med Mol Imaging 2002;29:1623–1629.
151. Marek K, Innis R, van Dyck C, et al. [123I]β-CIT SPECT imaging assessment of the rate of Parkinson's disease progression. Neurology 2001;57:2089–2094.
152. Mozley PD, Schneider JS, Acton PD, et al. Binding of (Tc-99m)TRODAT-1 to dopamine transporters in patients with Parkinson's disease and healthy volunteers. J Nucl Med 2000;41: 584–589.
153. Muller U, Wachter T, Barthel H, Reuter M, von Cramon DY. Striatal (123I)β-CIT SPECT and prefrontal cognitive functions in Parkinson's disease. J Neural Transmission 2000;107:303–319.
154. Nakabeppu O, Nakajo M, Mitsuda M, Tsuchimochi S, Tani A, Osame M. Iodine-123 iodobenzofuran (I-123 IBF) SPECT in patients with parkinsonism. Ann Nucl Med 1999;13:447–452.
155. Nurmi E, Ruottinen HM, Kaasinen V, et al. Progression in Parkinson's disease: a positron emission tomography study with a dopamine transporter ligand [18F]CFT. Ann Neurol 2000;47:804–808.
156. Rinne JO, Bergman J, Ruottinen H, et al. Striatal uptake of a novel PET ligand, (18F)β-CFT, is reduced in early Parkinson's disease. Synapse 1999;31:119–124.
157. Rinne JO, Ruottinen H, Bergman J, Haaparanta M, Sonninen P, Solin O. Usefulness of a dopamine transporter PET ligand ((18)F)β-CFT in assessing disability in Parkinson's disease. J Neurol Neurosurg Psychiatry 1999;67:737–741.
158. Sakakibara R, Shinotoh H, Uchiyama T, Yoshiyama M, Hattori T, Yamanishi T. SPECT imaging of the dopamine transporter with ((123)I)-β-CIT reveals marked decline of nigrostriatal dopaminergic function in Parkinson's disease with urinary dysfunction. J Neurol Sci 2001;187:55–59.
159. Shinotoh H, Uchida Y, Ito H, Harrori T. Relationship between striatal [123I]β-CIT binding and four major clinical signs in Parkinson's disease. Ann Nucl Med 2000;14:199–203.
160. Staffen W, Mair A, Unterrainer J, Trinka E, Ladurner G. Measuring the progression of idiopathic Parkinson's disease with (123I) β-CIT SPECT. J Neural Transmission 2000;107:543–552.

161. Staffen W, Mair A, Unterrainer J, Trinka E, Bsteh C, Ladurner G. [123I] β-CIT binding and SPET compared with clinical diagnosis in parkinsonism. Nucl Med Commun 2000;21:417–424.
162. Winogrodzka A, Bergmans P, Booij J, van Royen EA, Janssen AG, Wolters EC. [123I]FP-CIT SPECT is a useful method to monitor the rate of dopaminergic degeneration in early-stage Parkinson's disease. J Neural Transmission 2001;108:1011–1019.
163. Van Royen E, Verhoeff NFLG, Speelman JD, Wolters EC, Kuiper MA, Janssen AGM. Multiple system atrophy and progressive supranuclear palsy. Diminished striatal D2 receptor activity demonstrated by 123IBZM single photon emission computed tomography. Arch Neurol 1993;50:513–516.
164. Burn DJ, Sawle GV, Brooks DJ. Differential diagnosis of Parkinson's disease, multiple system atrophy, and Steele-Richardson-Olszewski syndrome: discriminant analysis of striatal 18F-dopa PET data. J Neurol Neurosurg Psych 1994;57: 278–284.
165. Pirker W, Asenbaum S, Wenger S, et al. Iodine-123-epidepride-SPECT: studies in Parkinson's disease, multiple system atrophy and Huntington's disease. J Nucl Med 1997;38:1711–1717.
166. Brucke T, Asenbaum S, Pirker W, et al. Measurement of the dopaminergic degeneration in Parkinson's disease with [123I]β-CIT and SPECT. Correlation with clinical findings and comparison with multiple system atrophy and progressive supranuclear palsy. J Neural Transm 1997;50:9–24.
167. Messa C, Volonte MA, Fazio F, et al. Differential distribution of striatal [123I]β-CIT in Parkinson's disease and progressive supranuclear palsy, evaluated with single photon emission tomography. Eur J Nucl Med 1998;25:1270–1276.
168. Hierholzer J, Cordes M, Venz S, et al. Loss of dopamine-D2 receptor binding sites in Parkinsonian plus syndromes. J Nucl Med 1998;39:954–960.
169. Barthel H, Sorger D, Kuhn HJ, Wagner A, Kluge R, Hermann W. Differential alteration of the nigrostriatal dopaminergic system in Wilson's disease investigated with [^{123}I]ss-CIT and high-resolution SPET. Eur J Nucl Med 2001;28:1656–1663.
170. Barthel H, Hermann W, Kluge R, et al. Concordant pre- and postsynaptic deficits of dopaminergic neurotransmission in neurologic Wilson disease. Am J Neuroradiol 2003;24:234–238.
171. Brashear A, Mulholland GK, Zheng QH, Farlow MR, Siemers ER, Hutchins GD. PET imaging of the pre-synaptic dopamine uptake sites in rapid-onset dystonia-parkinsonism (RDP). Mov Disord 1999;14:132–137.
172. Eisensehr I, Linke R, Noachtar S, Schwarz J, Gildehaus FJ, Tatsch K. Reduced striatal dopamine transporters in idiopathic rapid eye movement sleep behaviour disorder. Comparison with Parkinson's disease and controls. Brain 2000;123:1155–1160.
173. Gerschlager W, Bencsits G, Pirker W, et al. (123I)β-CIT SPECT distinguishes vascular parkinsonism from Parkinson's disease. Mov Disord 2002;17:518–523.
174. Huang CC, Yen TC, Weng YH, Lu CS. Normal dopamine transporter binding in dopa responsive dystonia. J Neurol 2002;249: 1016–1020.
175. Jeon B, Kim JM, Jeong JM, et al. Dopamine transporter imaging with (123I)-β-CIT demonstrates presynaptic nigrostriatal dopaminergic damage in Wilson's disease. J Neurol Neurosurg Psychiatry 1998;65: 60–64.
176. Jeon BS, Jeong JM, Park SS, et al. Dopamine transporter density measured by (123I)β-CIT single-photon emission computed tomography is normal in dopa-responsive dystonia. Ann Neurol 1998;43:792–800.
177. Katzenschlager R, Costa D, Gerschlager W, et al. [123I]-FP-CIT-SPECT demonstrates dopaminergic deficit in orthostatic tremor. Ann Neurol 2003;53:489–496.
178. Oyanagi C, Katsumi Y, Hanakawa T, et al. Comparison of striatal dopamine D2 receptors in Parkinson's disease and progressive supranuclear palsy patients using [123I] iodobenzofuran single-photon emission computed tomography. J Neuroimaging 2002;12: 316–324.
179. Pirker W, Asenbaum S, Bencsits G, et al. [123I]β-CIT SPECT in multiple system atrophy, progressive supranuclear palsy, and corticobasal degeneration. Mov Disord 2000;15:1158–1167.
180. Pirker W, Djamshidian S, Asenbaum S, et al. Progression of dopaminergic degeneration in Parkinson's disease and atypical parkinsonism: a longitudinal β-CIT SPECT study. Mov Disord 2002;17:45–53.
181. Sjoholm H, Sundsfjord J, Mellgren SI. Beta-CIT-SPECT combined with UPDRS appears to distinguish different parkinsonian conditions. Acta Neurol Scand 2002;105:5–7.
182. Tzen KY, Lu CS, Yen TC, Wey SP, Ting G. Differential diagnosis of Parkinson's disease and vascular parkinsonism by (99m)Tc-TRODAT-1. J Nucl Med 2001;42:408–413.
183. Varrone A, Marek KL, Jennings D, Innis RB, Seibyl JP. [(123)I]β-CIT SPECT imaging demonstrates reduced density of striatal dopamine transporters in Parkinson's disease and multiple system atrophy. Mov Disord 2001;16:1023–1032.
184. Walker Z, Costa DC, Walker RW, et al. Differentiation of dementia with Lewy bodies from Alzheimer's disease using a dopaminergic presynaptic ligand. J Neurol Neurosurg Psychiatry 2002;73: 134–140.
185. Morrish PK, Sawle GV, Brooks DJ. Clinical and [18F]dopa PET findings in early Parkinson's disease. J Neurol Neurosurg Psychiatry 1995;59:597–600.
186. Tissingh G, Bergmans P, Booij J, et al. Drug-naive patients with Parkinson's disease in Hoehn and Yahr stages I and II show a bilateral decrease in striatal dopamine transporters as revealed by [123I]β-CIT SPECT. J Neurol 1998;245:14–20.
187. Hilker R, Klein C, Ghaemi M, et al. Positron emission tomographic analysis of the nigrostriatal dopaminergic system in familial parkinsonism associated with mutations in the parkin gene. Ann Neurol 2001;49:367–376.
188. Laihinen A, Ruottinen H, Rinne JO, et al. Risk for Parkinson's disease: twin studies for the detection of asymptomatic subjects using [18F]6-fluorodopa PET. J Neurol 2000;247(Suppl 2):110–113.
189. Morrish PK, Rakshi JS, Bailey DL, Sawle GV, Brooks DJ. Measuring the rate of progression and estimating the preclinical period of Parkinson's disease with [18F]dopa PET. J Neurol Neurosurg Psychiatry1998;64:314–319.
190. Marek KL, Seibyl JP, Zoghbi SS, et al. [I-123]β-CIT SPECT imaging demonstrates bilateral loss of dopamine transporters in hemi-Parkinson's disease. Neurology 1996;46:231–237.
191. Rakshi JS, Uema T, Ito K. Statistical parametric mapping of three dimensional 18F-dopa PET in early and advanced Parkinson's disease (abstract). Mov Disord 1996;11:147.
192. Maraganore DM, O'Connor MK, et al. Detection of preclinical Parkinson disease in at-risk family members with use of (123I)β-CIT and SPECT: an exploratory study. Mayo Clinic Proc 1999;74: 681–685.
193. Antonini A, Moresco RM, Gobbo C, et al. Striatal dopaminergic denervation in early and late onset Parkinson's disease assessed by PET and the tracer [11C]FECIT: preliminary findings in one patient with autosomal recessive parkinsonism (Park2). Neurol Sci 2002;23(Suppl 2):S51–S52.
194. Khan NL, Brooks DJ, Pavese N, et al. Progression of nigrostriatal dysfunction in a parkin kindred: an (18F)dopa PET and clinical study. Brain 2002;125:2248–2256.
195. Wu RM, Shan DE, Sun CM, et al. Clinical, 18F-dopa PET, and genetic analysis of an ethnic Chinese kindred with early-onset parkinsonism and parkin gene mutations. Mov Disord 2002;17: 670–675.
196. Khan NL, Valente EM, Bentivoglio AR, et al. Clinical and subclinical dopaminergic dysfunction in PARK6-linked parkinsonism: an 18F-dopa PET study. Ann Neurol 2002;52:849–853.
197. Sawle GV, Playford ED, Burn DJ, Cunningham VJ, Brooks DJ. Separating Parkinson's disease from normality: discriminant function analysis of [18F]Dopa PET data. Arch Neurol 1993;51: 237–243.
198. Rinne JO, Laihinen A, Nagren K, Ruottinen H, Ruotsalainen U, Rinne UK. PET examination of the monoamine transporter with

[11C]β-CIT and [11C]β-CFT in early Parkinson's disease. Synapse 1995;21:97–103.
199. Druschky A, Hilz MJ, Platsch G, et al. Differentiation of Parkinson's disease and multiple system atrophy in early disease stages by means of I-123-MIBG-SPECT. J Neurol Sci 2000;175:3–12.
200. Berding G, Schrader CH, Peschel T, et al. [N-methyl 11C]meta-Hydroxyephedrine positron emission tomography in Parkinson's disease and multiple system atrophy. Eur J Nucl Med Mol Imaging 2003;30:127–131.
201. Boecker H, Weindl A, Leenders K, et al. Secondary parkinsonism due to focal substantia nigra lesions: a PET study with [18F]FDG and [18F]fluorodopa. Acta Neurol Scand 1996;93:387–392.
202. Kato T, Kume A, Ito K, Tadokoro M, Takahashi A, Sakuma S. Asymmetrical FDG-PET and MRI findings of striatonigral system in multiple system atrophy. Radiat Med 1992;10:87–93.
203. Kume A, Shiratori M, Takahashi A, et al. Hemi-parkinsonism in multiple system atrophy: a PET and MRI study. J Neurol Sci 1992; 110:37–45.
204. Hu MT, Taylor-Robinson SD, Chaudhuri KR, et al. Cortical dysfunction in non-demented Parkinson's disease patients: a combined (31)P-MRS and (18)FDG-PET study. Brain 2000;123: 340–352.
205. Ghaemi M, Raethjen J, Hilker R, et al. Monosymptomatic resting tremor and Parkinson's disease: a multitracer positron emission tomographic study. Mov Disord 2002;17:782–788.
206. Arnold G, Tatsch K, Oertel WH, et al. Clinical progressive supranuclear palsy: differential diagnosis by IBZM-SPECT and MRI. J Neural Transm 1994;42:111–118.
207. Schulz JB, Klockgether T, Peterson D, et al. Multiple system atrophy: natural history, MRI morphology, and dopamine receptor imaging with 123IBZM-SPECT. J Neurol Neurosurg Psych 1994; 57:1047–1056.
208. Acton PD, Mozley PD. Single photon emission tomography imaging in parkinsonian disorders: a review. Behav Neurol 2000;12:11–27.
209. Wenning GK, Donnemiller E, Granata R, Riccabona G, Poewe W. 123I-β-CIT and 123I-IBZM SPECT scanning in levodopa-naive Parkinson's disease. Mov Disord 1998;13:438–445.
210. Kim YJ, Ichise M, Tatschida T, Ballinger JR, Vines D, Lang AE. Differential diagnosis of parkinsonism using dopamine transporter and D2 receptor SPECT. J Nucl Med 1999;5(Suppl):68P.
211. Kim YJ, Ichise M, Ballinger JR, et al. Combination of dopamine transporter and D2 receptor SPECT in the diagnostic evaluation of PD, MSA, and PSP. Mov Disord 2002;17:303–312.
212. Dresel SHJ, Kung MP, Huang XF, et al. Simultaneous SPECT studies of pre- and post-synaptic dopamine binding sites in baboons. J Nucl Med 1999;40:660–666.
213. Ma KH, Huang WS, Chen CH, et al. Dual SPECT of dopamine system using (99mTc)TRODAT-1 and (123I)IBZM in normal and 6-OHDA-lesioned Formosan rock monkeys. Nucl Med Biol 2002;29:561–567.
214. Miyawaki E, Meah Y, Koller WC. Serotonin, dopamine, and motor effects in Parkinson's disease. Clin Neuropharmacol 1997;20:300–310.
215. Brucke T, Kornhuber J, Angelberger P, Asenbaum S, Frassine H, Podreka I. SPECT imaging of dopamine and serotonin transporters with (123I)β-CIT. Binding kinetics in the human brain. J Neural Transmission 1993;94:137–146.
216. Pirker W, Asenbaum S, Hauk M, et al. Imaging serotonin and dopamine transporters with 123I-β-CIT SPECT: binding kinetics and effects of normal aging. J Nucl Med 2000;41:36–44.
217. Malison RT, Price LH, Berman R, et al. Reduced brain serotonin transporter availability in major depression as measured by (123I)-2b-carbomethoxy-3b-(4-iodophenyl)tropane and single photon emission computed tomography. Biol Psychiatry 1998;11:1090–1098.
218. Kim SE, Lee WY, Choe YS, Kim JH. SPECT measurement of iodine-123-β-CIT binding to dopamine and serotonin transporters in Parkinson's disease: correlation with symptom severity. Neurol Res 1999;21:255–261.
219. Kim SE, Choi JY, Choe YS, Choi Y, Lee WY. Serotonin transporters in the midbrain of Parkinson's disease patients: a study with 123I-β-CIT SPECT. J Nucl Med 2003;44:870–876.
220. Murai T, Muller U, Werheid K, et al. In vivo evidence for differential association of striatal dopamine and midbrain serotonin systems with neuropsychiatric symptoms in Parkinson's disease. J Neuropsychiatry Clin Neurosci 2001;13:222–228.
221. Owens MJ, Nemeroff CB. Role of serotonin in the pathophysiology of depression: focus on the serotonin transporter. Clin Chem 1994;40:288–295.
222. Acton PD, Kung MP, Mu M, et al. Single photon emission tomography imaging of serotonin transporters in the nonhuman primate brain with the selective radioligand [123I]IDAM. Eur J Nucl Med 1999;26:854–861.
223. Acton PD, Mu M, Plössl K, et al. Single photon emission tomography imaging of serotonin transporters in the nonhuman primate brain with [123I]ODAM. Eur J Nucl Med 1999;26:1359–1362.
224. Acton PD, Choi SR, Hou C, Plössl K, Kung HF. Quantification of serotonin transporters in nonhuman primates using [123I]ADAM and single photon emission tomography. J Nucl Med 2001;42: 1556–1562.
225. Choi SR, Hou C, Oya S, et al. Selective in vitro and in vivo binding of [125I]ADAM to serotonin transporters in rat brain. Synapse 2000;38:403–412.
226. Ginovart N, Wilson AA, Meyer JH, Hussey D, Houle S. Positron emission tomography quantification of (11C)-DASB binding to the human serotonin transporter: modeling strategies. J Cereb Blood Flow Metab 2001;21:1342–1353.
227. Kung MP, Hou C, Oya S, Mu M, Acton PD, Kung HF. Characterization of (123I)IDAM as a novel single photon emission tomography tracer for serotonin transporters. Eur J Nucl Med 1999;26: 844–853.
228. Oya S, Kung MP, Acton PD, Hou C, Mu M, Kung HF. A new SPECT imaging agent for serotonin transporters, [I-123]IDAM: 5-iodo-2-[[2-2-[(dimethylamino) methyl] phenyl] thio] benzyl alcohol. J Med Chem 1999;42:333–335.
229. Oya S, Choi SR, Hou C, et al. 2-((2-((dimethylamino)-methyl)phenyl)thio)-5-iodophenylamine (ADAM): an improved serotonin transporter ligand. Nucl Med Biol 2000;27:249–254.
230. Zhuang ZP, Choi SR, Hou C, et al. A novel serotonin transporter ligand: 5-iodo-2-(2-dimethylaminomethylphenoxy)-benzyl alcohol (ODAM). Nucl Med Biol 2000;27:169–175.

3 Positron Emission Tomography in Parkinson's Disease

Cerebral Activation Studies and Neurochemical and Receptor Research

André R. Troiano, MD and A. Jon Stoessl, MD, FRCPC

SUMMARY

The development of positron emission tomography (PET) and the use of radiotracers designed for the study of cerebral blood flow and metabolism have provided new insights into central nervous system function in vivo. Particularly in the case of Parkinson's disease, different approaches have permitted a broader understanding of dopaminergic neurotransmission and the changes that take place during the progressive loss of dopaminergic neurons of the substantia nigra. In this chapter, we review physiological aspects of dopamine metabolism, as well as the patterns of alterations observed in Parkinson's disease and related forms of parkinsonism. Studies of cerebral metabolism during performance of motor behaviors and insights derived from imaging into the complications of long term levodopa treatment (fluctuations and dyskinesias) will also be considered.

Key Words: Positron emission tomography; Parkinson's disease; parkinsonism; dopamine; dopamine receptors.

1. INTRODUCTION

Positron emission tomography (PET) is an invaluable tool for in vivo assessment of central nervous system (CNS) functioning. Based on chemical, pharmacokinetic, and pharmacodynamic properties of radioligands administered before image acquisition, as well as different modes of scanning, PET provides the ability to study regional activation status, cerebral blood flow (CBF; ref. *1*), and neurotransmission. This wide array of functional techniques has disclosed new perspectives in the understanding of neurological disease, particularly in the field of neurodegenerative diseases and movement disorders *(2)*. In Parkinson's disease (PD), such functional studies are establishing the connection between the knowledge of pathology and neurochemistry with recent observations of neurotransmitter metabolism and patterns of regional cerebral activation, as well as their relationship to treatment and evolving disease.

In this chapter, we will highlight the most important applications of PET with respect to the normal control of movement by the basal ganglia (BG) and alterations found in the chronic state of dopamine depletion that characterizes PD and, to a lesser extent, other parkinsonian syndromes. For this purpose, the dopamine system in normal and abnormal conditions, synaptic alterations throughout the course of PD, and the constellation of findings downstream in the cortico-striatal-pallidal-thalamic-cortical (CSPTC) loop will be reviewed.

2. THE DOPAMINE SYSTEM

In the CNS, dopamine (DA) is synthesized by neurons of the pars compacta of the substantia nigra (SNc) and the ventral tegmental area (VTA) in the midbrain. The aromatic amino acid tyrosine is converted to DA by means of two reactions: the first one, involving tyrosine–hydroxylase, the rate-limiting enzyme in this pathway, converts tyrosine into L-3,4-dihydroxyphenylalanine (L-dopa). The second reaction is mediated by aromatic L-amino acid decarboxylase (AADC) and transforms L-dopa into DA, which constitutes approx 80% of all catecholamines (CAs) in the brain *(3)*. Whereas AADC is not thought to be rate limiting for DA synthesis in the normal brain, the situation may be different in PD. DA is packaged via the vesicular monoamine transporter (VMAT2) in synaptic vesicles in concentrations up to 1000 times higher than in cytosol, until release from dopaminergic terminals after an action potential. The membrane dopamine transporter (DAT) is responsible for DA reuptake into presynaptic terminals and is one of the main determinants of DA extracellular concentrations in the normal brain. The remaining DA in the synaptic cleft is converted into homovanillic acid by cathecol-σ-methyltransferase (COMT) and monoamine oxidase *(4)*.

Four main pathways distribute DA throughout the CNS: (1) nigrostriatal; (2) mesolimbic; (3) mesocortical; and (4) tuberoinfundibular. The first three are of neurological interest and the last has endocrine functions. The nigrostriatal pathway

From: *Bioimaging in Neurodegeneration*
Edited by P. A. Broderick, D. N. Rahni, and E. H. Kolodny

projects from the SNc to the dorsal caudate and putamen. The mesolimbic and mesocortical pathways project from the VTA to the telencephalon: the mesolimbic division to the nucleus accumbens (ventral striatum), olfactory tubercle, and limbic system; the mesocortical pathway to the frontal cortex *(3)*. The heterogeneity of DA input to the striatum forms the pathophysiological basis for the anteroposterior pattern of putaminal DA denervation in PD discussed in Section 4.2.

Dopamine receptors are G protein-coupled structures, originally divided into two classes based on their effect on adenylyl cyclase. The D1-like class encompasses D1 and D5 receptors *(3)*, which are positively linked to adenylyl cyclase and are thought to have a predominantly excitatory effect *(5)*. D1 receptors are predominantly postsynaptic and are found in the striatum, substantia nigra pars reticulata (SNr), nucleus accumbens, olfactory tubercle, cerebral cortex, amygdala, and subthalamic nucleus *(6)*. There is, therefore, a dense population of D1 receptors on the terminals of striatonigral projection neurons *(7,8)*. Receptors of the D2-like class exert a negative influence over adenylyl cyclase. Their subtypes—D2, D3, and D4—have inhibitory activity and may be pre- or postsynaptic. D2-like receptors are expressed in the striatum, olfactory tubercle, nucleus accumbens, SNc, and VTA.

Within the striatum, the input nucleus of the CSPTC loop, two subsets of neurons can be distinguished, based on their DA receptor characteristics: the first one has mainly D1 receptors and co-expresses GABA, substance P, and dynorphin, giving rise to the inhibitory direct pathway to the output nuclei internal globus pallidus (GPi) and SNr. The second population is mostly inhibited by DA via D2 receptors, giving rise to the indirect pathway that, in association with enkephalin, ultimately leads to increased GPi/SNr activity *(9)*. It should be noted that, for the purposes of the study of PD and its relationship with different PET patterns, the D1 and D2 receptors will be mainly considered.

3. PET SCAN PATTERNS IN MOVEMENT CONTROL

3.1. ACTIVATION AND REGIONAL CBF PATTERNS IN NORMAL INDIVIDUALS AND IN PD

The classical view of the basal ganglia function under normal circumstances is that of a structure in which five separate parallel circuits, under dual control by the direct and indirect pathways, subserve specific areas of the neocortex. These segregated pathways thus process cortical information from the supplementary motor area (SMA) and premotor cortex (PMC); dorsolateral prefrontal cortex (DLPFC); lateral orbitofrontal cortex; anterior cingulate area; and oculomotor cortex. Their role seems to be the activation of cortical areas mediating motor, oculomotor, or cognitive behaviors, as well as alternating the focus of attention towards novel and rewarding extrapersonal stimuli *(10)*.

Since the introduction of regional CBF (rCBF) and activation studies with PET, it became possible to analyze cortical and subcortical patterns during motor performance. Therefore, either rCBF PET with $H_2{}^{15}O$ to detect changes in local blood delivery during performance of motor or cognitive tasks or glucose metabolism with ^{18}F-fluorodeoxyglucose (^{18}F-FDG) can be used.

Several interrelated aspects of limb movement have been investigated by means of specific paradigms. Jenkins et al. *(11)* studied changes in rCBF related to movement frequency in six right-handed normal subjects moving a joystick with freely selected directions and found focal activation related to increasing speed in posterior SMA, bilateral lateral PMC, contralateral sensorimotor cortex, and ipsilateral cerebellar hemisphere and vermis *(11)*. Also, from the Hammersmith group, a study of $H_2{}^{15}O$ rCBF evaluated patterns of cortical-subcortical activation during performance of a prelearned sequence of unimanual finger movements at a constant pace. The authors observed that progressive complexity of the task was positively correlated with regional increases in rostral SMA, pallidothalamic loop (bilateral pallidal and contralateral thalamus), right precuneus (Brodmann area 7), and ipsilateral PMC (ref. *12*) . Dettmers et al. *(13)* evaluated the neural correlate of strength of finger movement by means of $H_2{}^{15}O$ PET scan. In their study, increasing force of right index finger flexion in six right-handed subjects was related to increases of rCBF in primary sensorimotor cortex, the posterior part of SMA and cingulate sulcus, and the cerebellar vermis.

It is hypothesized that the BG function by activating parts of the cortex into similar frequencies, thereby facilitating the performance of a motor or cognitive action. This pattern of activation, referred to as "focused attention," would be the physiological substrate for the harmonic performance of movement. In PD, the imbalance between direct and indirect pathways as a consequence of dopamine deficiency leads to overactivation of BG output nuclei and subsequent failure to achieve the state of focused attention *(10)*. Several reports have addressed the changes in CBF and ^{18}F-FDG uptake during different tasks in PD patients. In an inhaled $C^{15}O_2$ CBF PET study, Playford et al. *(14)* evaluated six PD patients and six controls at rest during the execution of freely chosen movements and with programmed repetitive forward movement of a joystick. For the free-selection task, both groups had similar increases in blood flow in left sensorimotor and bilateral premotor cortices. However, PD patients failed to activate the SMA and anterior cingulate areas, putamen, thalamus, and cerebellum. In the repetitive task, the control patients had lesser degrees of activation of right DLPFC, whereas for PD patients this was observed for right inferior parietal association cortex and right premotor area *(14)*. Imagination of movement, when examined in normal individuals, resulted in activations in the sensorimotor, inferior parietal, bilateral dorsal premotor (PMC), caudal supplementary motor area, bilateral ventral premotor, right M1 and left superior parietal cortices, left putamen, and right cerebellum. However, a relationship between imagined movement complexity and enhanced activation was observed in contralateral PMC and ipsilateral superior parietal cortex and cerebellar vermis *(15)*. In PD patients, imagination of movements did not translate into greater activation of DLPFC or mesial frontal areas (homolog to SMA in nonhuman primates), and the execution of movement did not activate DLPFC. Other components of motor control evaluated by functional imaging include the differential aspects of activation during unilateral or bilateral movement and the observation of cerebellar overactivation in PD. To categorize the rCBF distribution according to complexity of movement, Goerres et al. *(16)* created a paradigm for $H_2{}^{15}O$ PET, in which six individuals were scanned while performing unimanual, bimanual–symmetric, and bimanual–asymmetric ballistic finger movement. When subjects moved one finger, there was contralateral activation of primary sensorimotor cortex, inferior parietal cortex, and precuneus, whereas bilateral symmetric movement led to bilateral increases in rCBF. Bilateral asymmetric finger movement, in comparison with the other tasks, was associated with augmented CBF in rostral SMA. In a Xenon single-photon emission computed tomography (SPECT) CBF study, Rascol et al. *(17)* compared 12 normal subjects, 12 PD patients not currently on medication, and 16 PD patients on medication (the two latter groups exhibiting mainly the akinetic-dominant form) during performance of unilateral finger-to-thumb movements. Regions of interest were placed over the cerebral and cerebellar hemispheres. Both groups of PD patients on medication and controls had similar values of cerebellar increases in CBF, whereas PD individuals who were studied off medication had significantly enhanced activation of ipsilateral cerebellar hemisphere *(17)*.

Findings of overactivation of extrastriatal areas in PD have been further reproduced and converge with the "preimaging era" description of the paradoxical gait of PD, in which patients with advanced akinetic status and a typical short-stepped gait may experience benefit when helped by visual cues. The possible physiological mechanism underlying this discrepancy was investigated by Hanakawa et al. *(18)* in a study where 10 PD patients on medication and 10 controls underwent Tc-99m exametazime (HMPAO) SPECT imaging after walking on a treadmill. This treadmill was equipped with either parallel or transverse lines. After walking with each of the line orientations, the individuals were scanned. The purpose was to observe changes in rCBF correspondent to changes in cadence (number of steps/minute), more so in PD patients walking on transverse visual cues. While walking on the treadmill equipped with transverse lines, PD patients had marked improvement of cadence as previously reported and showed significantly greater activation of lateral premotor cortex, a region known to have abundant connections to the cerebellum and posterior parietal cortex *(18)*. In the setting of BG dysfunction, leading to a state of impaired activation of SMA and PMC loops responsible for automatic gait, the use of such an alternate pathway may represent an adaptation process.

These reports of distant sites activated during execution of different motor tasks and their underuse in the parkinsonian state led to the proposal of a rather attractive putative imaging network of corticosubcortical structures known to be altered in PD *(19)*. In one of the first articles addressing this issue, ^{18}F-FDG PET was used in three groups: PD patients, presumed multiple system atrophy (MSA), and normal controls. The authors used the scaled subprofile model, a form of principal components analysis, to determine a topographical covariance pattern that would allow differentiation of the groups. Individuals with PD, as a group, had enhanced ^{18}F-FDG uptake in the pallidum, thalamus, and pons, combined with hypometabolism in the SMA and lateral premotor, DLPFC, and parietooccipital association cortices. Most interestingly, this covariance profile correlated with Hoehn and Yahr (H&Y) stage, rigidity, and akinesia scores, but not with tremor scores *(20)*. The same model, when applied to early typical and atypical PD patients, resulted in covariance patterns that were able to distinguish between the groups of patients *(21)*.

This pattern of subcortical hypermetabolism with hypometabolism in related cortical areas was referred to as the Parkinson's disease-related pattern (PRDP) and further showed quantitative potential. The authors compared 23 early PD patients (mean H&Y 2.4 ± 1.3) with 14 patients with more advanced disease (mean H&Y 3.2 ± 1.2) using clinical assessment and ^{18}F-FDG PET. The magnitude of the PDRP was calculated on an individual basis using topographic profile rating. This tool disclosed statistically significant results in the distinction of PDRPs for early and advanced disease and for rigidity and bradykinesia, but once again not for tremor *(22)*. Why were tremor scores not related to PDRP in these two studies? To answer this question, the same group conducted ^{18}F-FDG PET of 16 PD patients, 8 with and 8 without tremor, the former with Unified Parkinson's Disease Rating Scale (UPDRS) tremor score of at least 4, and the latter scoring 0. The groups were otherwise matched for age, duration of disease, H&Y, and composite motor UPDRS (exclusive of tremor). Compared with akinetic patients, the tremulous group had hypermetabolism of the thalamus, pons, and motor association cortex. A cerebellar-midbrain-thalamic-cortical network is thought to be independently active in PD patients featuring resting tremor *(23)*. These results are in keeping with neuropathologic findings in tremor-dominant PD patients, who display more severe neuronal cell loss in medial SNc and retrorubral field A8, which are related to dorsolateral striatum and ventromedial thalamus *(24)*. Patients with essential and writing tremor also were shown to display bilateral cerebellar, red nucleus, and thalamic overactivation *(25)*. Interestingly, although olivary metabolism was not increased in these circumstances, it showed a negative correlation with improvement of tremor and cerebellar activation after alcohol intake in essential tremor individuals *(26)*.

The PDRP was further reproduced in other centers *(27)*, and its imaging substrate was then compared with electrophysiological testing when 42 PD patients underwent ^{18}F-FDG PET and pallidotomy with intraoperative spontaneous single unit activity recording of the internal globus pallidus *(28)*. There was a positive correlation between spontaneous GPi firing rates and ipsilateral anterior thalamic glucose metabolism, but no other significant relationship could beestablished between putamen and caudate. Although GPi activity is expected to inhibit thalamic activity through GABAergic neurotransmission, it should be noted that ^{18}F-FDG measures synaptic rather than cell body activity and that metabolism may be influenced to a greater extent by excitatory activity than by inhibitory input *(29)*. Moreover, other excitatory influences over GPi, such as those from subthalamic nucleus, may also play a role.

In PD patients under treatment with levodopa (L-dopa), PET assessment disclosed attenuation of the covariance profile of PDRP, with decreased metabolism in left putamen, right thalamus, bilateral cerebellum, and left PMC *(30)*. Similarly, these pleiades of CSPTC loop alterations have shown improvement with surgical treatments, such as pallidotomy

(31), pallidal deep brain stimulation *(32)*, and subthalamotomy *(33)*, with varying degrees of increased uptake in cortical areas and diminished activity in the BG after the procedures, and statistically significant correlation with UPDRS motor scores. Nevertheless, medical and surgical treatments may apparently benefit patients by different mechanisms: pallidotomy leads to enhanced ^{18}F-FDG uptake in primary motor cortex, lateral premotor cortex and DLPFC *(31)*, whereas L-dopa treatment is associated with reduced metabolism in prefrontal cortex in nondemented patients, especially in the orbitofrontal cortex *(34)*. It has been suggested that these different patterns of metabolic improvement may explain differences in cognitive outcome following the different interventions *(35)*. Of course, comparisons of this kind should always be viewed with caution because they refer to studies conducted with different populations and using diverse PET protocols.

4. NEUROCHEMICAL AND NEURORECEPTOR IMAGING OF PD USING PET

4.1. RADIOLIGANDS USED IN THE STUDY OF DA METABOLISM

Aside from functional imaging of glucose consumption and rCBF in the CNS, the other cardinal approach to the understanding of PD relies on the ability of radioactive ligands to label specific neurochemical processes and to bind to receptors and transporters in the striatum and other brain regions. The use of such tracers provides an opportunity to anatomically quantify the loss of dopaminergic terminals or function, as well as the synaptic consequences of this loss, and compensatory changes that take place in the course of disease. For practical purposes, we will initially consider tracers according to whether they primarily target presynaptic terminals or postsynaptic receptors. In the former case, it is possible to assess AADC activity with 6-[^{18}F]-fluoro-L-dopa (FDOPA). This ligand reflects the uptake, decarboxylation, and storage of DA (*see* Section 4.2.).

Presynaptic markers directed to DAT are mostly tropane derivatives (cocaine analogs), such as [^{11}C]- or [^{18}F]-2-[β]-carbomethoxy-3[β]-(4-fluorophenyl) tropane (CFT, also known as [^{11}C]- or [^{18}F]-WIN 35,428), carbomethoxy-iodophenyl tropane (CIT), and [^{11}C]-RTI-32 *(36)*. CIT can also be modified with either a fluoroethyl or fluoropropyl group. Because of the delayed time to equilibrium, the ^{18}F-labeled FP-CIT is more suitable than its ^{11}C analog, which has a much shorter half-life *(37)*. [^{11}C]-D-*threo*-methylphenidate (MP) may also be used to target the DAT, with a higher affinity and better pharmacokinetic properties than the majority of the tropanes *(38)*. Although the DAT (but not all of its ligands) is fairly specific for DA, it has the disadvantage of being subject to up- or down-regulation by dopaminergic drugs and in response to extracellular levels of DA *(39–41)*. Thus, findings may be attributable to disease or a consequence of its treatment. Numerous tropanes have been labeled with I-123 for SPECT imaging of the DAT, and there is to date a single ^{99m}Tc labeled tropane *(42)*. SPECT has the advantage of being more widely available than PET. However, its resolution is in general lower, and quantitation can be more difficult *(43)*.

The central VMAT2 ligand [^{11}C]-dihydrotetrabenazine (DTBZ) offers the opportunity to map the protein that packages monoamines into synaptic vesicles. In contrast to the DAT, VMAT2 is thought to be resistant to pharmacological interference, according to the work by Vander Borght et al. *(44)*, in which intact rats treated with dopaminergic medications showed no alteration of DTBZ binding. However, DTBZ labels all monoaminergic neurons, not just those that produce and release DA. This is largely compensated by the minor concentrations of monoamines other than DA in the striatum.

Ligands aimed at postsynaptic DA receptors may bind to D1- or D2-like receptors. For the D1-like class, the tracers used include the antagonists [^{11}C]-SCH-23390, [^{11}C]-SCH-39166, and a number of [^{11}C]-NNC compounds. D2 receptors can be labeled using [^{11}C]- or [^{18}F]- spiperone or its derivatives, [^{18}F]-benperidol *(45)* or [^{11}C]-raclopride (RAC). The latter is a low affinity (low nanomolar) competitive antagonist of D2/D3 receptors and its binding (in contrast to that of spiperone or benperidol) is therefore subject to competitive displacement by DA itself. This property has been exploited to estimate DA release after a variety of interventions—physical and pharmacological—designed to modulate dopaminergic transmission *(46)*. RAC binding reflects a combination of total D2-like receptor density as well as receptor occupancy by DA (or dopaminergic medications). Thus, to determine the significance of changes between different patient groups, one would ideally require multiple scans in order to perform a Scatchard type analysis *(47,48)*. However, changes in receptor density are much less likely to occur over a short time span within an individual, and are therefore more likely to reflect intervention-related changes in receptor occupancy (i.e. changes in DA levels).

4.2. CLINICAL STUDIES

FDOPA was first used to visualize the nigrostriatal system in humans in the early 1980s *(49)*. Studies in parkinsonism have been conducted since that time, when it was first used in MPTP-exposed individuals *(50)* and PD patients *(51)*. After image acquisition, FDOPA uptake is usually calculated using multiple-timepoint graphical analysis *(52–53)*. The typical finding in idiopathic PD (IPD) is an asymmetric FDOPA uptake with relative sparing of the caudate compared to putamen *(39)*, corresponding to the preferential cell loss in the ventrolateral tier of the SNc *(54)*. FDOPA may also identify subclinical deficits of DA production in individuals at genetic *(55)* or environmental *(50)* risk for PD. Of the cardinal signs of PD, the one that has the best correlation with striatal FDOPA is bradykinesia, as rated by the Purdue pegboard and modified Columbia scores *(56)*.

FDOPA uptake is usually quantified using the uptake constant (K_i; based on multiple time graphical analysis) using an arterial plasma-derived input function, or its reference tissue-based analog, the K_{occ}. Alternative metrics include the calculated K_3 (decarboxylation rate constant; ref. *57*) or the striatal/background ratio *(58)*. Although FDOPA uptake is generally taken as "an indirect index of the number of striatal dopamine terminals and nigral neurons" *(59)*, the relationship between FDOPA uptake rate constant, striatal DA levels, and nigral cell

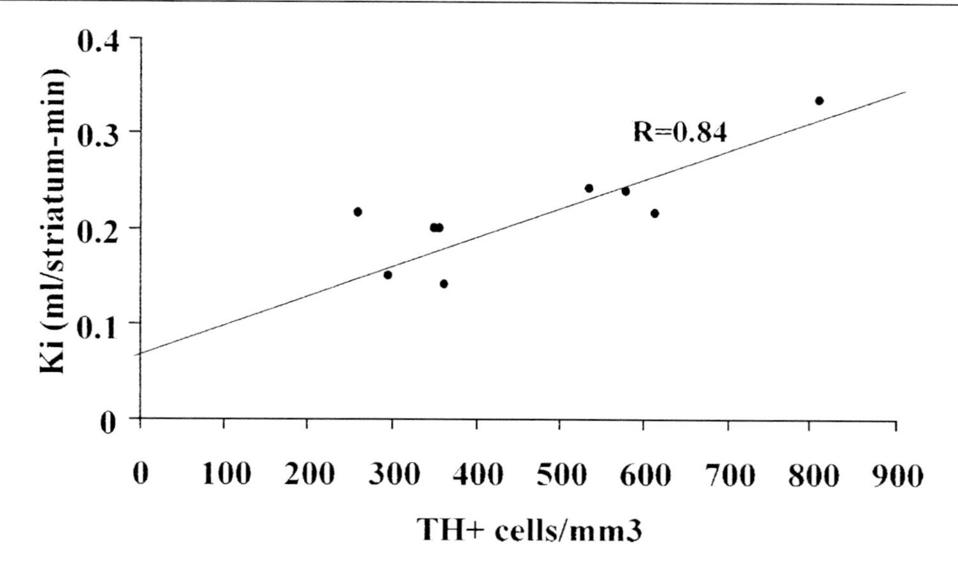

Fig. 1. Fluorodopa uptake measured by PET correlates well with tyrosine hydroxylase cell counts in the substantia nigra (courtesy of E. G. McGeer and UBC-TRIUMF PET group).

counts is imperfect. One investigation found good correlation between the three parameters in monkeys treated with 1-methyl-4-phenyl-1,2,3,6-tetrahydropyridine *(60)*, whereas in a more recent paper, Yee et al. *(61)* described a positive association between FDOPA and striatal DA levels, although no correlation could be established with nigral cell counts. (*See* Fig. 1.)

Methodological differences may contribute to such disparities. The study by Yee et al. was performed in animals with rather mild lesions and the limited dynamic range may have suppressed the correlation. Conversely, in advanced disease FDOPA may not be trapped into synaptic vesicles; thus, K_i (or K_{occ}) cannot be reliably measured *(62)*.

Several studies addressed the issue of FDOPA PET in the diagnosis and, above all, its potential in the differential diagnosis of parkinsonian syndromes in clinical practice. Brooks and colleagues *(63)* used FDOPA and S-[^{11}C]-nomifensine, a DA reuptake blocker, to detect distinctive alterations in IPD compared with parkinsonism-dominant MSA (MSA-P) and pure autonomic failure patients and controls. Direct comparison between IPD and MSA-P disclosed a homogeneous reduction of striatal FDOPA K_i in MSA patients, in contrast with the rostrocaudal gradient in IPD *(63)*. More recently, IPD was compared with MSA-P by means of FDOPA, RAC, and ^{18}F-FDG PET as well as 3D-MRI volumetry. FDOPA had similar K_i values in both conditions and some MSA patients displayed the rostrocaudal gradient typical of IPD. However, unlike IPD, MSA patients had significantly decreased RAC binding in posterior putamen and smaller putamen volumes on MRI *(64)*. These results correspond to the pathological observation that MSA has a more widespread distribution than the presynaptic terminal loss of IPD *(65,66)*. Antonini et al. *(67)* evaluated, using FDG PET, a series of 56 patients with IPD and 48 patients with "red flags" for atypical parkinsonism. They found a combination of altered local metabolism in the caudate, lentiform nucleus, and thalamus that could help to differentiate the two groups.

Posttraumatic parkinsonism was evaluated with FDOPA and compared with IPD. The results showed decreased striatal uptake, with putamen and caudate equally affected, disclosing a probable distinct pathophysiological process *(68)*. Hence, it is not currently possible to rely on PET to consistently and reliably differentiate IPD from other causes of parkinsonism because the rostrocaudal gradient of abnormal FDOPA uptake may be seen in other conditions, especially MSA. Some of the alterations delineating these conditions represent a group effect and are not suitable for clinical practice on an individual basis. The same holds true for progressive supranuclear palsy and corticobasal degeneration. In the latter condition, striatal FDOPA uptake may be reduced in a relatively symmetrical fashion, in contrast to the clinical deficits and abnormalities in glucose metabolism *(69)*.

The rate of disease progression determined with FDOPA uptake was assessed by Nurmi et al. *(70)* on two separate scans after 5 yr follow-up of 10 newly diagnosed PD patients. They found annual relative rates of decline of 5.9 ± 5.1% for caudate, 8.3 ± 6.3% for anterior putamen, and 10.3 ± 4.8% for posterior putamen ($p < 0.001$). Based on these figures and an assumption of linear decline, they estimated the preclinical period of degeneration in the posterior putamen in their cohort to be 6.5 yr *(70)*. In a larger group of patients, the calculated yearly decline in FDOPA K_{occ} was 8.9% for the same region *(71)*. Using data from human studies, other authors have estimated that symptoms onset when FDOPA uptake is between 47% and 62% *(40)* or 75% *(71)* of normal. However, as stated above, the relationship between FDOPA uptake and nigral cell counts should be assessed with caution, taking into account compensatory changes in degenerating DA terminals. Thus, AADC activity may increase in remaining neurons during the course of PD, thereby resulting in increased FDOPA K_i. This initially was observed in animal models and replicated in human investigation. Lee and collaborators *(40)* used FDOPA, DTBZ, and MP in 35 PD patients in various disease

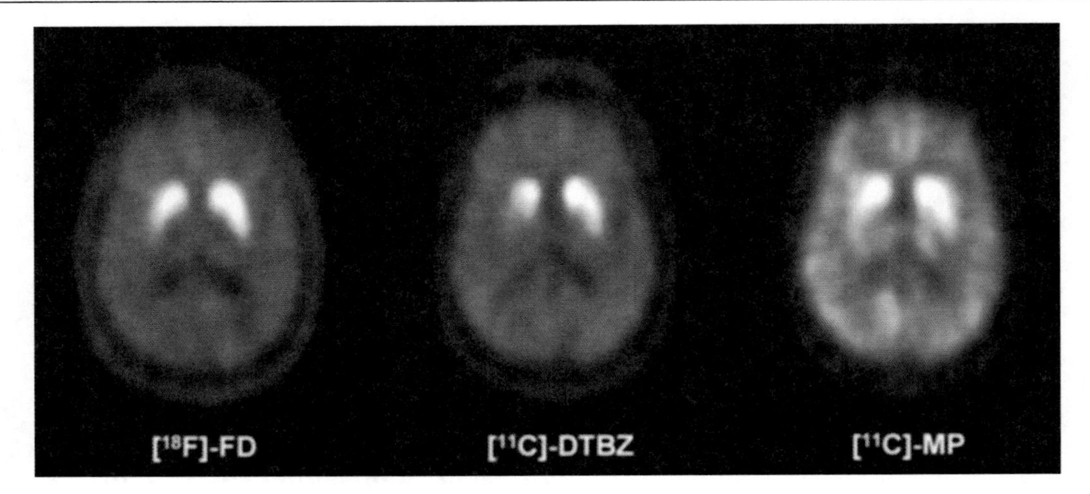

Fig. 2. Axial PET scans in a patient with early PD showing asymmetric reduction of fluorodopa uptake and of dihydrotetrabenazine and D-threo-methylphenidate binding to the vesicular monoamine and membrane dopamine transporter, respectively. There is additionally a rostrocaudal gradient of uptake, with the posterior putamen maximally affected. Courtesy of the UBC-TRIUMF PET group. *See* color version on Companion CD.

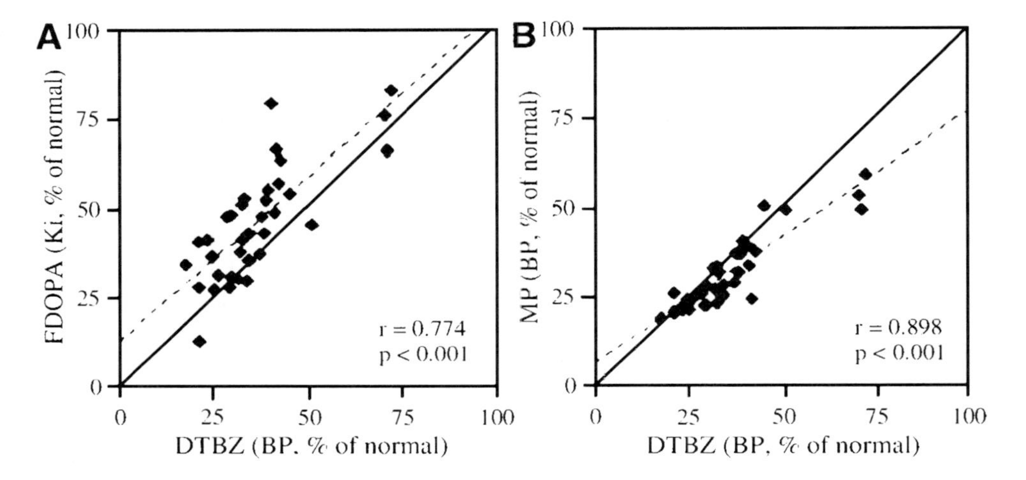

Fig. 3. Relationship between fluorodopa uptake (**A**) and methylphenidate binding to the membrane dopamine transporter (**B**) and dihydrotetrabenazine binding to the vesicular monoamine transporter. All values are expressed as a percentage of age-matched normals. There are statistically significant increases in fluorodopa uptake, in keeping with upregulation of aromatic amino acid decarboxylase, whereas in contrast, dopamine transporter binding is reduced. Both changes may occur in an effort to preserve extracellular levels of dopamine. Preprinted from ref. *40*. © Wiley 2000.

stages and 16 normal subjects. PD individuals had FDOPA uptake values higher than predicted, based on DTBZ binding to the VMAT, whereas MP binding to the dopamine transporter was relatively reduced (Figs. 2 and 3). These discrepancies were more pronounced in the posterior putamen. These data provide a strong argument in favor of a functional imaging correlate of adaptive changes in DA terminals.

In contrast to FDOPA or DTBZ, which map the presynaptic DA terminals involved in PD, the susceptibility of RAC binding to competitive displacement by DA has been used to assess the effect of dopaminergic medications and functional surgery. In normal individuals, repetitive transcranial magnetic stimulation of the motor cortex associated with RAC PET has been shown to selectively release DA in the ipsilateral ventrolateral putamen *(72)* whereas, in keeping with parallel processing in the basal ganglia, transcranial magnetic stimulation of the DLPFC leads to the same effect in the caudate nucleus *(73)*. In early PD, there is increased RAC binding potential (BP), compatible with increased D2/D3 receptor availability secondary to DA denervation *(74)*, and this correlates with ipsilateral FDOPA levels *(75)*. Although the increase in RAC binding could conceivably reflect increased availability of D2 receptors consequent to loss of DA, PET studies with $[^{11}C]N$-methylspiperone also show increased D2 binding. Because this ligand is much less susceptible to displacement by endogenous DA, the increase in D2 binding likely reflects increased receptor density, at least in part *(76)*. In the course of disease, there is a trend for D2 receptors to return towards normal levels *(74,77)*. Dopamine release/RAC displacement have been demonstrated to occur in the striatum of PD patients to a lesser extent than normal subjects during sequential finger movements *(78)* and intravenous methamphetamine challenges *(79)*.

Dopamine agonists may lead to decreased endogenous DA release, presumably via stimulation of presynaptic autoreceptors *(80)*. The administration of L-dopa also reduces RAC binding, which is thought to reflect its conversion to DA that is subsequently released, predominantly in the posterior putamen *(81)*. Six months of treatment with pergolide also resulted in decreased BP compared with pretreatment levels, more so in the posterior striatum than the caudate *(82)*. Two series of six PD patients who had symptomatic improvement after subthalamic nucleus--deep brain stimulation reported that the clinical impact was not related to changes in striatal RAC binding *(83,84)*.

4.3. PET STUDIES IN FAMILIAL PD

A family history of parkinsonism is present in 5–10% of individuals with PD. Currently, at least 10 loci and 5 genes with both autosomal-recessive and autosomal-dominant patterns of inheritance recognized. A genetic component for PD is plausible even for sporadic cases because asymptomatic twins of PD individuals show decreased putaminal FDOPA uptake *(85)*. Of utmost importance in the understanding of sporadic PD is the concept that a number of the recently described mutations result in impaired function of ubiquitin proteasomal function, a mechanism that is related to the clearance of cellular proteins *(86)*.

A French group compared young-onset PD subjects with and without *parkin* mutations (autosomal-recessive *parkin* disease at 6q25.2-27) using FDOPA PET imaging. Both cohorts had similar findings with asymmetrically reduced uptake and caudate-to-putamen rostrocaudal gradient. The authors found no relationship between the type of mutation and PET pattern *(87)*, although another study of 14 individuals with *parkin* mutations found a negative correlation between the number of mutated alleles and FDOPA uptake *(88)*. The FDOPA asymmetry *(89)* and rostrocaudal gradient *(90)* have not been uniformly reproduced in every study of *parkin* kindreds. It is noteworthy that even very young-onset PD patients without *parkin* mutations may present the same abnormalities, as in the case of a 14-yr-old patient with symptoms beginning at age 9 who showed striking reduction of FDOPA uptake in keeping with previous findings regarding asymmetry and preferential putaminal involvement *(91)*. A study of an Italian family with adult-onset pseudodominant *parkin* disease found, aside from the above noted changes in FDOPA uptake, significant reduction in RAC binding in symptomatic patients *(92)*. However, findings of postsynaptic dysfunction in this disorder are not consistent.

Reports of kindreds harboring the α-synuclein mutation at G209A *(93)* and A30P *(94)* disclosed results closely resembling IPD. Familial frontotemporal dementia with parkinsonism linked to chromosome 17 (FTDP17) was evaluated by Pal et al. *(95)* using ^{18}F-FDG, FDOPA, and RAC in three patients of a kindred. They found FDOPA uptake in both caudate and putamen reduced to a similar degree, as well as normal to elevated RAC BP and global reduction of cerebral glucose metabolism, mainly in the frontal lobes *(95)*. Another study mapping DAT with [^{11}C]-CFT in FTDP17 found decreases in striatal uptake (both caudate and putamen) that correlated to the motor UPDRS *(96)*. The extensive changes caused by the tauopathy of FTDP17 explain the lack of striatal rostrocaudal gradient and global cortical hypoperfusion.

4.4. PET AND COMPLICATIONS OF LONG-TERM LEVODOPA THERAPY

After years experiencing a good response to L-dopa, most PD patients develop an unstable state of inconsistent response in which three main complications are usually recognized: 1) motor fluctuations (MF), 2) dyskinesias, and 3) psychiatric disturbances. MFs ultimately encompass dyskinesias. Aberrant responses to L-dopa such as predictable "off" periods are seen in more than 20% of patients after 5 yr of treatment and 70% after 15 yr of treatment *(97)*. The degeneration of presynaptic terminals was formerly raised as the basis of fluctuations, rendering the dopaminergic system unable to store DA. However, investigations of motoric responses to the postsynaptic DA receptor agonist apomorphine have shown that improvement is at least partially dependent upon the disease stage, lending support to the concept that postsynaptic alterations may play a role in the genesis of MF *(98)*.

De la Fuente-Fernandez et al. *(99)*, using FDOPA PET, studied 15 patients who had a stable response to L-dopa and 52 with MFs . Patients with MF had further decreases in striatal FDOPA uptake compared with those who had a stable response to L-dopa. As expected, patients with MF had a more prolonged disease course than the stable subjects, and the fluctuators had an earlier age of disease onset. This decrement in FDOPA K_i persisted even after adjustment for disease course and matching the two groups. However, there was significant FDOPA K_i overlap between the groups. This finding was interpreted as a suggestion of increased DA turnover in the fluctuators *(99)*. This was confirmed by a longitudinal study with RAC PET, in which eight patients with early PD and a good response to L-dopa after an average of 1.5 yr of treatment were evaluated (Fig. 4). The subjects had the first scan washed-out of medications and two additional scans 1 h and 4 h after a single oral dose of L-dopa. After clinical follow-up for a minimum of 3 yr, four patients developed motor fluctuations, and three of those also had dyskinesias, as rated by an examiner blinded to the results of the PET scan. Patients who developed MF had a pattern of DA release consistent with an initial marked increase in DA release, followed by an early return to baseline. This was demonstrated with RAC BP decreases much more pronounced in fluctuators than in stable responders 1 h after L-dopa. Conversely, at 4 h, fluctuators returned to or even surpassed baseline RAC binding values in contrast to the patients with a persistently stable response, who showed a slower but more sustained increase in DA release after treatment with L-dopa. This was the first in vivo evidence that DA turnover is increased preclinically in individuals who will develop transient responses to L-dopa *(100)*. It has not been possible to clearly attribute the development of MF to alterations in postsynaptic DA receptors, except for a negative correlation between time of L-dopa treatment and extent of D1 receptors in putamen *(101–103)*.

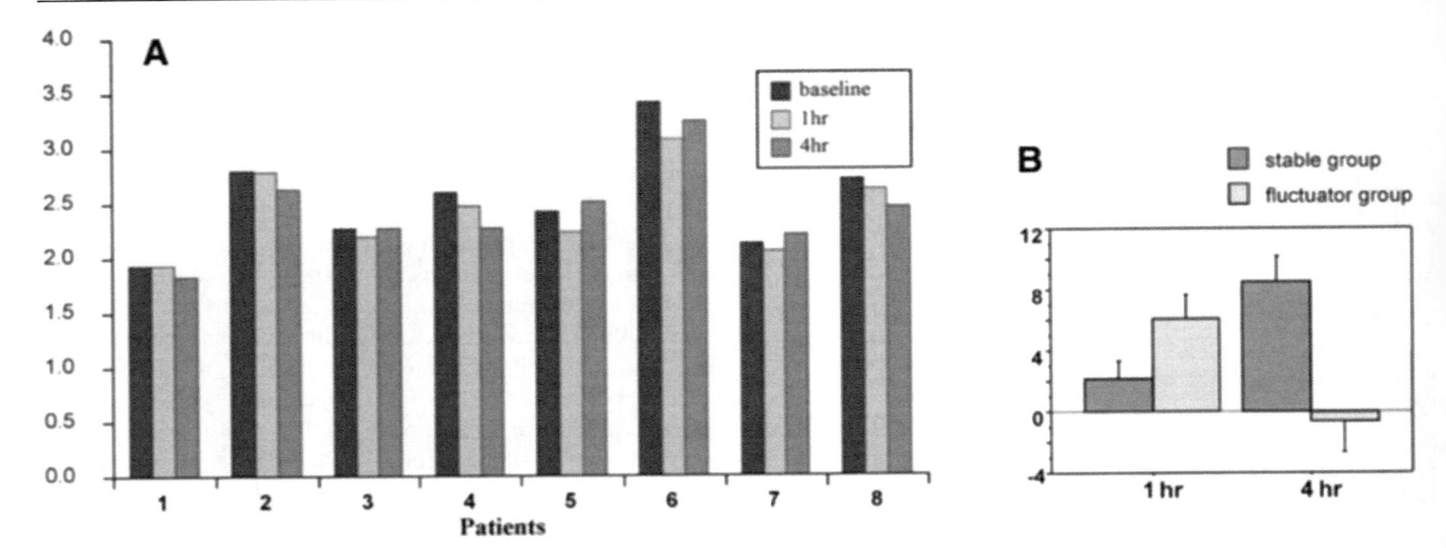

Fig. 4. (A) [^{11}C]Raclopride binding in patients with a stable response to levodopa (L-dopa), measured at baseline, 1 h, and 4 h after a single oral dose of L-dopa. In Patients 3, 5, 6, and 7, there is an early reduction in raclopride binding (representing an increase in extracellular dopamine), followed by an early return to baseline values. These patients went on to develop motor response fluctuations within 3 yr of starting L-dopa treatment. In the other patients, there is a slower and more sustained increase in extracellular dopamine. These patients still had a stable response to levodopa after 3 yr of therapy. **(B)** Mean (+SD) changes in raclopride binding 1 h and 4 h after the oral administration of L-dopa in patients who maintained a stable response to L-dopa vs those who went on to develop motor response fluctuations. Preprinted from ref. *100*. © Wiley 2001. *See* color version on Companion CD.

There is a growing body of evidence that MF and dyskinesias are the result of neuronal changes that occur downstream from striatal dopamine receptors *(104)*. Thus, altering opioid transmission with naloxone or with the kappa opioid antagonist nor-binaltorphimine has been shown to improve dyskinesia without significant negative impact on antiparkinsonian features *(105–107)*. The imaging correlate of these findings was demonstrated by Brooks et al. *(108)* who, in a comparison of dyskinetic and nondyskinetic patients, found no differences in BP for ligands of D1 and D2 receptors. However, binding of the opioid receptor ligand [^{11}C]-diprenorphine was decreased in the striatum and thalamus of dyskinetic patients, compatible with increased levels of endogenous opioids.

4.5. PD AND THE PLACEBO EFFECT

Dopamine is related to signaling of learning and reward. Dopaminergic neurons fire in response to rewarding events or reward-predicting stimuli, regardless of their specific nature, but not for nonnoxious aversive stimuli *(109)*. Functional imaging studies have shown that the dopaminergic system or related cerebral areas play a prominent role in tasks and behaviors related to reward, such as nicotine *(110)* and illicit drug abuse *(111)*, obesity *(112)*, video game playing *(113)*, and the experience of pleasure with music *(114)*. There are data supporting the notion that the symptomatic benefit experienced after placebo intake is a form of reward and, therefore, could be mediated by dopaminergic transmission. Consequently, its relationship with clinical effect in PD should be distinguished on a biochemical basis, especially when it is considered that PD patients may objectively benefit from placebo treatment *(115)*. De la Fuente-Fernandez et al. *(116)* provided insight into the changes that occur in nigrostriatal synapses during exposure to placebo. Using the RAC displacement paradigm, they found substantial DA release in response to placebo in the whole striatum, particularly in the posterolateral putamen. Patients who experienced benefit from placebo released more DA than those who did not. A further study of the same group of subjects addressed the expectation of benefit, which, in contrast to the subjective experience of benefit, led to DA release in the nucleus accumbens (ventral striatum), a region that has been implicated in behavioral responses to rewarding stimuli *(117)*.

5. CONCLUSIONS

With PET, it has become possible to understand the spectrum of changes that take place in cortical and subcortical structures in the course of PD, as well as in related diseases presenting with parkinsonism. The investigation of CBF and metabolism has disclosed patterns of activation in distant sites that previously could be only inferred by means of neuroanatomical and neuropathological studies. Biochemical alterations at the receptor level are now visualized in vivo, and the use of radioligands aimed at different parts of the nigrostriatal synapse makes possible a better understanding of DA metabolism, such as progression of dopaminergic cell loss and adaptive mechanisms in these and other neurotransmitter systems.

Is it hoped that future developments in functional imaging technology, including improved spatial resolution, will permit the quantitative visualization of smaller subcortical structures, such as the brainstem, as well as a wider availability of radioligands targeted to other processes in the CNS.

ACKNOWLEDGMENTS

The work described here was supported by the Canadian Institutes for Health Research, the National Parkinson Foundation (Miami, Inc.), the Pacific Parkinson's Research Institute, and the Canada Research Chairs program (AJS).

REFERENCES

1. Brooks DJ. PET studies on the function of dopamine in health and Parkinson's disease. Ann N Y Acad Sci 2003;991:22–35.
2. de la Fuente-Fernandez R, Stoessl AJ. Parkinson's disease: imaging update. Curr Opin Neurol 2002;15:477–482.
3. Valloni D, Picetti R, Borrelli E. Structure and function of dopamine receptors. Neurosci Behav Rev 2000;24:125–132.
4. Elsworth JD, Roth RH. Dopamine synthesis, uptake, metabolism and receptors: relevance to gene therapy of Parkinson's disease. Exp Neurol 1997;144:4–9.
5. Wooten GF. Anatomy and function of dopamine receptors: understanding the pathophysiology of fluctuations in Parkinson's disease. Parkinsonism Rel Disord 2001;8:79–83.
6. Jackson DM, Westlind-Danielsson A. Dopamine receptors: molecular biology, biochemistry and behavioural aspects. Pharmacol Ther 1994;64:291–370.
7. Harrison MB, Wiley RG, Wooten GF. Selective localization of striatal D1 receptors to striatonigral neurons. Brain Res 1990;528:317–322.
8. Graham WC, Crossman AR. Autoradiographic localization of dopamine D1 binding sites in areas receiving striatal input. Eur J Pharmacol 1987;142:479–481.
9. Alexander GE, Crutcher MD. Functional architecture of basal ganglia circuits: neural substrates of parallel processing. Trends Neurosci 1990;13:266–271.
10. Brown P, Marsden CD. What do the basal ganglia do? Lancet 1998;351:1801–1804.
11. Jenkins IH, Passingham RE, Brooks DJ. The effect of movement frequency on cerebral activation: a positron emission tomography study. J Neurol Sci 1997;151:195–205.
12. Boecker H, Dagher A, Ceballos-Baumann AO, Passingham RE, Samuel M, Friston KJ, Poline J, Dettmers C, Conrad B, Brooks DJ. Role of the human rostral supplementary motor area and the basal ganglia in motor sequence control: investigations with H2 15O PET. J Neurophysiol 1998;79:1070–1080.
13. Dettmers C, Fink GR, Lemon RN, Stephan KM, Passingham RE, Silbersweig D, Holmes A, Ridding MC, Brooks DJ, Frackowiak RS. Relation between cerebral activity and force in the motor areas of the human brain. J Neurophysiol 1995;74:802–815.
14. Playford ED, Jenkins IH, Passingham RE, Nutt J, Frackowiak RS, Brooks DJ. Impaired mesial frontal and putamen activation in Parkinson's disease: a positron emission tomography study. Ann Neurol 1992;32:151–161.
15. Boecker H, Ceballos-Baumann AO, Bartenstein E, Dagher A, Forster K, Haslinger B, Brooks DJ, Schwaiger M, Conrad B. A H(2)(15)O positron emission tomography study on mental imagery of movement sequences – the effect of modulating sequence length and direction. Neuroimage 2002;17:999–1009.
16. Goerres GW, Samuel M, Jenkins IH, Brooks DJ. Cerebral control of unimanual and bimanual movements: an H2(15)O PET study. Neuroreport 1998;9:3631–3638.
17. Rascol O, Sabatini U, Fabre N, Brefel C, Loubinoux I, Celsis P, Senard JM, Montastruc JL, Chollet F. The ipsilateral cerebellar hemisphere is overactive during hand movements in akinetic parkinsonian patients. Brain 1997;120:103–110.
18. Hanakawa T, Fukuyama H, Katsumi Y, Honda M, Shibasaki H. Enhanced lateral premotor activity during paradoxical gait in Parkinson's disease. Ann Neurol 1999;45:329–336.
19. Carbon M, Edwards C, Eidelberg D. Functional brain imaging in Parkinson's disease. Adv Neurol 2003;91:175–181.
20. Eidelberg D, Moeller JR, Dhawan V, Spetsieris P, Takikawa S, Ishikawa T, Chaly T, Robeson W, Margouleff D, Przedborski S. The metabolic topography of parkinsonism. J Cereb Blood Flow Metab 1994;14:783–801.
21. Eidelberg D, Moeller JR, Ishikawa T, Dhawan V, Spetsieris P, Chaly T, Belakhlef A, Mandel F, Przedborski S, Fahn S. Early differential diagnosis of Parkinson's disease with ^{1818}F-fluorodeoxyglucose and positron emission tomography. Neurology 1995;45:1995–2004.
22. Eidelberg D, Moeller JR, Ishikawa T, Dhawan V, Spetsieris P, Chaly T, Robeson W, Dahl JR, Margouleff D. Assessment of disease severity in parkinsonism with fluorine-^{18}18-fluorodeoxy-glucose and PET. J Nucl Med 1995;36:378–383.
23. Antonini A, Moeller JR, Nakamura T, Spetsieris P, Dhawan V, Eidelberg D. The metabolic anatomy of tremor in Parkinson's disease. Neurology 1998;51:803–810.
24. Jellinger KA. Recent developments in the pathology of Parkinson's disease. J Neural Transm Suppl 2002;62:347–376.
25. Wills AJ, Jenkins IH, Thompson PD, Findley LJ, Brooks DJ. A positron emission tomography study of cerebral activation associated with essential and writing tremor. Arch Neurol 1995;52:299–305.
26. Boecker H, Wills AJ, Ceballos-Baumann A, Samuel M, Thompson PD, Findley LJ, Brooks DJ. The effect of ethanol on alcohol-responsive essential tremor: a positron emission tomography study. Ann Neurol 1996;39:650–658.
27. Moeller JR, Nakamura T, Mentis MJ, Dhawan V, Spetsieris P, Antonini A, Missimer J, Leenders KL, Eidelberg D. Reproducibility of regional metabolic covariance patterns: comparison of four populations. J Nucl Med 1999;40:1264–1269.
28. Eidelberg D, Moeller JR, Kazumata K, Antonini A, Sterio D, Dhawan V, Spetsieris P, Alterman R, Kelly PJ, Dogali M, Fazzini E, Beric A. Metabolic correlates of pallidal neuronal activity in Parkinson's disease. Brain 1997;120:1315–1324.
29. Waldvogel D, van Gelderen P, Muellbacher W, Ziemann U, Immisch I, Hallett M. The relative metabolic demand of inhibition and excitation. Nature 2000;406:995–998.
30. Feigin A, Fukuda M, Dhawan V, Przedborski S, Jackson-Lewis V, Mentis MJ, Moeller JR, Eidelberg D. Metabolic correlates of levodopa response in Parkinson's disease. Neurology 2001;57:2083–2088.
31. Eidelberg D, Moeller JR, Ishikawa T, Dhawan V, Spetsieris P, Silbersweig D, Stern E, Woods RP, Fazzini E, Dogali M, Beric A. Regional metabolic correlates of surgical outcome following unilateral pallidotomy for Parkinson's disease. Ann Neurol 1996;39:450–459.
32. Fukuda M, Mentis MJ, Ma Y, Dhawan V, Antonini A, Lang AE, Lozano AM, Hammerstad J, Lyons K, Koller WC, Moeller JR, Eidelberg D. Networks mediating the clinical effects of pallidal brain stimulation for Parkinson's disease: a PET study of resting-state glucose metabolism. Brain 2001;124:1601–1609.
33. Su PC, Ma Y, Fukuda M, Mentis MJ, Tseng HM, Yen RF, Liu HM, Moeller JR, Eidelberg D. Metabolic changes following subthalamotomy for advanced Parkinson's disease. Ann Neurol 2001;50:514–520.
34. Berding G, Odin P, Brooks DJ, Nikkhah G, Matthies C, Peschel T, Shing M, Kolbe H, van den Hoff J, Fricke H, Dengler R, Samii M, Knapp WH. Resting regional cerebral glucose metabolism in advanced Parkinson's disease studied in the off and on conditions with [^{18}F]FDG-PET. Movement Disorders 2001;16:1014–1022.
35. Feigin A, Ghilardi MF, Carbon M, Edwards C, Fukuda M, Dhawan V, Margouleff C, Ghez C, Eidelberg D. Effects of levodopa on motor sequence learning in Parkinson's disease. Neurology 2003;60:1744–1749.
36. Stoessl AJ, de la Fuente-Fernandez R. Dopamine receptors in Parkinson's disease: imaging studies. Adv Neurol 2003;91:65–71.
37. Lundkvist C, Halldin C, Ginovart N, Swahn GC, Farde L. [^{1818}F]-beta-b-CIT-FP is superior to [11C]-beta-b-CIT-FP for quantitation of the dopamine transporter. Nucl Med Biol 1997;24:621–627.
38. Volkow ND, Ding Y-S, Fowler JS, Wang GJ, Logan J, Gatley SJ, Schlyer DJ, Pappas N. A new PET ligand for the dopamine transporter: studies in the human brain. J Nucl Med 1995;36:2162–2168.
39. Stoessl AJ. Neurochemical and neuroreceptor imaging with PET in Parkinson's disease. Adv Neurol 2001;86:215–223.
40. Lee CS, Samii A, Sossi V, Ruth TJ, Schulzer M, Holden JE, Wudel J, Pal PK, de la Fuente-Fernandez R, Calne DB, Stoessl AJ. In vivo positron emission tomographic evidence for compensatory changes in presynaptic dopaminergic nerve terminals in Parkinson's disease. Ann Neurol 2000;47:493–503.
41. Guttman M, Stewart D, Hussey D, Wilson A, Houle S, Kish S. Influence of L-dopa and pramipexole on striatal dopamine transporter in early PD. Neurology 2001;56:1559–1564.

42. Mozley PD, Schneider JS, Acton PD, Plossl K, Stern MB, Siderowf A, Leopold NA, Li PY, Alavi A, Kung HF. Binding of [99mTC] TRODAT-1 to dopamine transporters in patients with Parkinson's disease and in healthy volunteers. J Nucl Med 2000;41:584–589.
43. Marek K, Jennings D, Seibyl J. Single-photon emission tomography and dopamine transporter imaging in Parkinson's disease. Adv Neurol 2003;91:183–191.
44. Vander Borght, T, Kilbourn M, Desmond T, Kuhl D, Frey K. The vesicular monoamine transporter is not regulated by dopaminergic drug treatments. Eur J Pharmacol 1995;294:577–583.
45. Moerlein SM, Perlmutter JS, Markham J, Welch MJ. In vivo kinetics of [^{18}F]-(N-methyl)-benperidol: a novel PET tracer for assessment of dopaminergic D2-like receptor binding. J Cereb Blood Flow Metab 1997;17:833–845.
46. Volkow ND, Wang G-J, Fowler JS, Logan J, Schlyer D, Hitzemann R, Lieberman J, Angrist B, Pappas N, MacGregor R. Imaging endogenous dopamine competition with [11C]raclopride in the human brain. Synapse 1994;16:255–262.
47. Farde L, Hall H, Ehrin E, Sedvall G. Quantitative analysis of D2 dopamine receptor binding in the living human brain by PET. Science 1986;231:258–261.
48. Doudet DJ, Jivan S, Holden JE. In vivo measurement of receptor density and affinity: comparison of the routine sequential method with a nonsequential method in studies of dopamine D2 receptors with [11C] raclopride. J Cereb Blood Flow Metab 2003;23:280–284.
49. Garnett ES, Firnau G, Nahmias C. Dopamine visualized in the basal ganglia of living man. Nature 1983;305:137–138.
50. Calne DB, Langston JW, Martin WR, Stoessl AJ, Ruth TJ, Adam MJ, Pate BD, Schulzer M. Positron emission tomography after MPTP: observations relating to the cause of Parkinson's disease. Nature 1985;317:246–248.
51. Leenders KL, Palmer AJ, Quinn N, Clark JC, Firnau G, Garnett ES, Nahmias C, Jones T, Marsden CD. Brain dopamine metabolism in patients with Parkinson's disease measured with positron emission tomography. J Neurol Neurosurg Psychiatry 1986;49:853–860.
52. Martin WRW, Palmer MR, Patlak CS, Calne DB. Nigrostriatal function in humans studied with positron emission tomography. Ann Neurol 1989;26:535–542.
53. Patlak CS, Blasberg RG. Graphical evaluation of blood-to-brain transfer constants from multiple-time uptake data: generalizations. J Cereb Blood Flow Metab 1985;5:584–590.
54. Fearnley JM, Lees AJ. Ageing and Parkinson's disease: substantia nigra regional selectivity. Brain 1991;114:2283–2301.
55. Piccini P, Burn DJ, Ceravolo R, Maraganore D, Brooks DJ. The role of inheritance in sporadic Parkinson's disease: evidence from a longitudinal study of dopaminergic functions in twins. Ann Neurol 1999;45:577–582.
56. Vingerhoets FJ, Schulzer M, Calne DB, Snow BJ. Which clinical sign of Parkinson's disease best reflects the nigrostriatal lesion? Ann Neurol 1997;41:58–64.
57. Kuwabara H, Cumming P, Reith J, Leger G, Diksic M, Evans AC, Gjedde A. Human striatal L-dopa decarboxylase activity estimated in vivo using 6-[^{1818}F]fluoro-dopa and positron emission tomography: error analysis and application to normal subjects. J Cereb Blood Flow Metab 1993;13:43–56.
58. Dhawan V, Ma Y, Pillai V, Spetsieris P, Chaly T, Belakhlef A, Margouleff C, Eidelberg D. Comparative analysis of striatal FDOPA uptake in Parkinson's disease: ratio method versus graphical approach. J Nucl Med 2002;43:1324–1330.
59. Olanow CW, Watts RL, Koller WC. An algorithm (decision tree) for the management of Parkinson's disease (2001): treatment guidelines. Neurology 2001;56(Suppl 5):S1–S88.
60. Pate BD, Kawamata T, Yamada T, McGeer EG, Hewitt KA, Snow BJ, Ruth TJ, Calne DB. Correlation of striatal fluorodopa uptake in the MPTP monkey with dopaminergic indices. Ann Neurol 1993;34:331–338.
61. Yee RE, Irwin I, Milonas C, Stout DB, Huang SC, Shoghi-Jadid K, Satyamurthy N, Delanney LE, Togasaki DM, Farahani KF, Delfani K, Janson AM, Phelps ME, Langston JW, Barrio JR. Novel observations with FDOPA-PET imaging after early nigrostriatal damage. Mov Disord 2001;16:838–848.
62. Stoessl AJ. Assessing the integrity of the dopamine system in Parkinson's disease: how best to do it? Mov Disord 2001;16:804–806.
63. Brooks DJ, Salmon EP, Mathias CJ, Quinn N, Leenders KL, Bannister R, Marsden CD, Frackowiak RS. The relationship between locomotor disability, autonomic dysfunction and the integrity of the striatal dopaminergic system in patients with multiple system atrophy, pure autonomic failure and Parkinson's disease, studied with PET. Brain 1990;113:1539–1552.
64. Ghaemi M, Hilker R, Rudolf J, Sobesky J, Heiss W-D. Differentiating multiple system atrophy from Parkinson's disease: contribution of striatal and midbrain MRI volumetry and multi-tracer PET imaging. J Neurol Neurosurg Psychiatry 2002;73:517–523.
65. Antonini A, Leenders KL, Vontobel P, Maguire RP, Missimer J, Psylla M, Gunther I. Complimentary PET studies of striatal neuronal function in the differential diagnosis between multiple system atrophy and Parkinson's disease. Brain 1997;120:2187–2195.
66. Shinotoh H, Inoue O, Hirayama K, Aotsuka K, Asahina M, Suhara T, Yamazaki T, Tateno Y. Dopamine D1 receptors in Parkinson's disease and striatonigral degeneration: a positron emission tomography study. J Neurol Neurosurg Psychiatry 1993;56:467–472.
67. Antonini A, Kazumata K, Feigin A, Mandel F, Dhawan V, Margouleff C, Eidelberg D. Differential diagnosis of parkinsonism with [^{1818}F]fluorodeoxyglucose and PET. Mov Disord 1998;13:268–274.
68. Turjanski N, Lees AJ, Brooks DJ. Dopaminergic function in patients with posttraumatic parkinsonism: an ^{1818}FDOPA PET study. Neurology 1997;49:183–189.
69. Laureys S, Salmon E, Garraux G, Peigneux P, Lemaire C, Degueldre C, Franck G. Fluorodopa uptake and glucose metabolism in early stages of corticobasal degeneration. J Neurol 1999;246:1151–1158.
70. Nurmi E, Ruottinen HM, Bergman J, Haaparanta M, Solin O, Sonninen P, Rinne JO. Rate of progression in Parkinson's disease: a 6-[^{1818}F]fluoro-L-dopa PET study. Mov Disord 2001;16:608–615
71. Morrish PK, Rakshi JS, Bailey DL, Sawle GV, Brooks DJ. Measuring the rate of progression and estimating the preclinical period of Parkinson's disease with [^{1818}F]dopa PET. J Neurol Neurosurg Psychiatry 1998;64:314–319.
72. Strafella AP, Paus T, Fraracaccio M, Dagher A. Striatal dopamine release induced by repetitive transcranial magnetic stimulation of the human motor cortex. Brain 2003;126:2609–2615.
73. Strafella AP, Paus T, Barrett J, Dagher A. Repetitive transcranial magnetic stimulation of the human prefrontal cortex induces dopamine release in the caudate nucleus. J Neurosci 2001;21:RC157.
74. Rinne UK, Laihinen A, Rinne JO, Nagren K, Bergman J, Ruotsalainen U. Positron emission tomography demonstrates dopamine D2 receptor supersensivity in the striatum of patients with early Parkinson's disease. Mov Disord 1990;5:55–59.
75. Antonini A, Vontobel P, Psylla M, Gunther I, Maguire PR, Missimer J, Leenders KL. Complementary positron emission tomographic studies of the striatal dopaminergic system in Parkinson's disease. Arch Neurol 1995;52:1183–1190.
76. Kaasinen V, Ruottinen HM, Nagren K, Lehikoinen P, Oikonen V, Rinno JO. Upregulation of putaminal dopamine D2 receptors in early Parkinson's disease: a comparative PET study with [11C] raclopride and [11C]N-methylspiperone. J Nucl Med 2000:41:65–70.
77. Antonini A, Schwarz J, Oertel WH, Pogarell O, Leenders KL. Long-term changes of striatal dopamine D2 receptors in patients with Parkinson's disease: a study with positron emission tomography and [11C]raclopride. Mov Disord 1997;12:33–38.
78. Goerendt IK, Messa C, Lawrence AD, Grasby PM, Piccini P, Brooks DJ; PET study. Dopamine release during sequential finger movements in health and Parkinson's disease: a PET study. Brain 2003;126:312–325.
79. Piccini P, Pavese N, Brooks DJ. Endogenous dopamine release after pharmacological challenges in Parkinson's disease. Ann Neurol 2003;53:647–653.
80. de la Fuente-Fernandez R, Lim AS, Sossi V, Holden JE, Calne DB, Ruth TJ, Stoessl AJ. Apomorphine induced-changes in synaptic dopamine levels: positron emission tomography evidence for presynaptic inhibition. J Cereb Blood Flow Metab 2001;21:1151–1159.

81. Tedroff J, Pedersen M, Aquilonius SM, Hartvig P, Jacobsson G, Langstrom B. Levodopa-induced changes in synaptic dopamine in patients with Parkinson's disease as measured by [11C]-raclopride displacement and PET. Neurology 1996;46:1430–1436.
82. Linazasoro G, Obeso JA, Gomez JC, Martinez M, Antonini A, Leenders KL. Modification of dopamine D2 receptor activity by pergolide in Parkinson's disease: an in vivo study by PET. Clin Neuropharmacol 1999;22:277–280.
83. Strafella AP, Sadikot AF, Dagher A. Subthalamic deep brain stimulation does not induce striatal dopamine release in Parkinson's disease. Neuroreport 2003;14:1287–1289.
84. Hilker R, Voges J, Ghaemi M, Lehrke R, Rudolf J, Koulousakis A, Herholz K, Wienhard K, Sturm V, Heiss WD. Deep brain stimulation of the subthalamic nucleus does not increase the striatal dopamine concentration in parkinsonian humans. Mov Disord 2003;18:41–48.
85. Laihinen A, Ruottinen H, Rinne JO, Haaparanta M, Bergman J, Solin O, Koskenvuo M, Marttila R, Rine UK. Risk for Parkinson's disease: twin studies for the detection of asymptomatic subjects using [^{1818}F]6-fluorodopa PET. J Neurol 2000;247(suppl 2):II110–13.
86. Hardy J, Cookson MR, Singleton A. Genes and parkinsonism. Lancet Neurol 2003;2:221–228.
87. Thobois S, Ribeiro MJ, Lohmann E, Durr A, Pollak P, Rascol O, Guillouet S, Chapoy E, Costes N, Agid Y, Remy P, Brice A, Broussolle E, French Parkinson's Disease Genetics Study Group. Young-onset Parkinson's disease with and without *parkin* gene mutations: a fluorodopa F 18 positron emission tomography study. Arch Neurol 2003;60:713–718.
88. Hilker R, Klein C, Hedrich K, Ozelius LJ, Vieregge P, Herholz K, Pramstaller PP, Heiss WD. The striatal dopaminergic deficit is dependent on the number of mutant alleles in a familiy with mutations in the parkin gene: evidence for enzymatic parkin function in humans. Neurosci Lett 2002;323:50–54.
89. Wu RM, Shan DE, Sun CM, Liu RS, Hwu WL, Tai CH, Hussey J, West A, Gwinn-Hardy K, Hardy J, Chen J, Farrer M, Lincoln S. Clinical, ^{1818}FDOPA PET, and genetic analysis of an ethnic Chinese kindred with early-onset parkinsonism and parkin gene mutations. Mov Disord 2002;17:670–675.
90. Khan NL, Valente EM, Bentivoglio AR, Wood NW, Albanese A, Brooks DJ, Piccini P. Clinical and subclinical dopaminergic dysfunction in PARK6-linked parkinsonism: an ^{1818}FDOPA PET study. Ann Neurol 2002;52:849–853.
91. Pal PK, Leung J, Hedrich K, Samii A, Lieberman A, Nausieda PA, Calne DB, Breakefield XO, Klein C, Stoessl AJ. [^{1818}F]-DOPA positron emission tomography imaging in early-stage, non-parkin juvenile parkinsonism. Mov Disord 2002;17:789–794.
92. Hilker R, Klein C, Ghaemi M, Kis B, Strotmann T, Ozelius LJ, Lenz O, Vieregge P, Herholz K, Heiss WD, Pramstaller PP. Positron emission tomographic analysis of the nigrostriatal dopaminergic system in familial parkinsonism associated with mutations in the parkin gene. Ann Neurol 2001;49:367–376.
93. Samii A, Markopoulou K, Wszolek ZK, Sossi V, Dobko T, Mak E, Calne DB, Stoessl AJ. PET studies of parkinsonism associated with mutation in the alpha-a-synuclein gene. Neurology 1999;53:2097–2102.
94. Kruger R, Kuhn W, Leenders KL, Sprengelmeyer R, Muller T, Woitalla D, Portman AT, Maguire RP, Veenma L, Schroder U, Schols L, Epplen JT, Riess O, Przuntek H. Familial parkinsonism with synuclein pathology: clinical and PET studies of A30P mutation carriers. Neurology 2001;56:1355–1362.
95. Pal PK, Wszolek ZK, Kishore A, de la Fuente-Fernandez R, Sossi V, Uitti RJ, Dobko T, Stoessl AJ. Positron emission tomography in pallido-ponto-nigral degeneration (PPND) family (frontotemporal dementia with parkinsonism linked to chromosome 17 and point mutation in tau gene). Parkinsonism Relat Disord 2001;7:81–88.
96. Rinne JO, Laine M, Kaasinen V, Norvasuo-Heila MK, Nagren K, Helenius H. Striatal dopamine transporter and extrapyramidal symptoms in frontotemporal dementia. Neurology 2002;58:1489–1493.
97. Miyawaki E, Lyons K, Pahwa R, Troster AI, Hubble J, Smith D, Busenbark K, McGuire D, Michalek D, Koller WC. Motor complications of chronic levodopa therapy in Parkinson's disease. Clin Neuropharmacol 1997;20:523–530.
98. Verhagen Metman L, Locatelli ER, Bravi D, Mouradian MM, Chase TN. Apomorphine responses in Parkinson's disease and the pathogenesis of motor complications. Neurology 1997;48:369–372.
99. de la Fuente-Fernandez R, Pal PK, Vingerhoets FJ, Kishore A, Schulzer M, Mak EK, Ruth TJ, Snow BJ, Calne DB, Stoessl AJ. Evidence for impaired presynaptic dopamine function in parkinsonian patients with motor fluctuations. J Neural Transm 2000;107:49–57.
100. de la Fuente-Fernandez R, Lu J-Q, Sossi V, Jivan S, Schulzer M, Holden JE, Lee CS, Ruth TJ, Calne DB, Stoessl AJ. Biochemical variations in the synaptic level of dopamine precede motor fluctuations in Parkinson's disease: PET evidence of increased dopamine turnover. Ann Neurol 2001;49:298–303.
101. Turjanski N, Lees AJ, Brooks DJ. In vivo studies on striatal dopamine D1 and D2 site binding in L-dopa-treated Parkinson's disease patients with and without dyskinesias. Neurology 1997;49:717–723.
102. de la Fuente-Fernandez R, Kishore A, Snow BJ, Schulzer M, Mak E, Huser J, Stoessl AJ, Calne DB. Dopamine D1 and D2 receptors and motor fluctuations in idiopathic Parkinsonism (IP): a simultaneous PET study (abstract). Neurology 1997;48(suppl 2):A208.
103. Kishore A, de la Fuente-Fernandez R, Snow BJ, Schulzer M, Mak E, Huser J, Stoessl AJ, Calne DB. Levodopa-induced dyskinesias in idiopathic parkinsonism (IP): a simultaneous PET study of dopamine D1 and D2 receptors (abstract). Neurology 1997;48(suppl 2):A327.
104. Brotchie JM. Adjuncts to dopamine replacement: a pragmatic approach to reducing the problem of dyskinesia in Parkinson's disease. Mov Disord 1998;13:871–876.
105. Newman DD, Rajakumar N, Flumerfelt BA, Stoessl AJ. A kappa opioid antagonist blocks sensitization in a rodent model of Parkinson's disease. Neuroreport 1997;8:669–672.
106. Klintenberg R, Svenningsson P, Gunne L, Andren PE. Naloxone reduces levodopa-induced dyskinesias and apomorphine-induced rotations in primate models of parkinsonism. J Neural Transm 2002;109:1295–1307.
107. McCormick SE, Stoessl AJ. Blockade of nigral and pallidal opioid receptors suppresses vacuous chewing movements in a rodent model of tardive dyskinesia. Neuroscience 2002;112:851–859.
108. Brooks DJ, Piccini P, Turjanski N, Samuel M. Neuroimaging of dyskinesia. Ann Neurol 2000;47:S154–S158.
109. Schultz W. Predictive reward signal of dopamine neurons. J Neurophysiol 1998;80:1–27.
110. Tsukada H, Miyasato K, Kakiuchi T, Nishiyama S, Harada N, Domino EF. Comparative effects of metamphetamine and nicotine on the striatal [11C]raclopride binding in unanesthetized monkeys. Synapse 2002;45:207–212.
111. Kilts CD, Schweitzer JB, Quinn CK, Gross RE, Faber TL, Muhammad F, Ely TD, Hoffman JM, Drexler KP. Neural activity related to drug craving in cocaine addiction. Arch Gen Psychiatry 2001;58:334–341.
112. Wang GJ, Volkow ND, Logan J, Pappas NR, Wong CT, Zhu W, Netusil N, Fowler JS. Brain dopamine and obesity. Lancet 2001;357:354–357.
113. Koepp MJ, Gunn RN, Lawrence AD, Cunningham VJ, Dagher A, Jones T, Brooks DJ, Bench CJ, Grasby PM. Evidence for striatal dopamine release during a video game. Nature 1998;393:266–268.
114. Blood AJ, Zatorre RJ. Intensely pleasurable responses to music correlate with activity in brain regions implicated in reward and emotion. Proc Natl Acad Sci USA 2001;98:11818–11823.
115. Goetz CG, Leurgans S, Raman R, Stebbins GT. Objective changes in motor function during placebo treatment in PD. Neurology 2000;54:710–714.
116. de la Fuente-Fernandez R, Ruth TJ, Sossi V, Schulzer M, Calne DB, Stoessl AJ. Expectation and dopamine release: mechanism of the placebo effect in Parkinson's disease. Science 2001;293:1164–1166.
117. de la Fuente Fernandez R, Phillips AG, Zamburlini M, Sossi V, Calne DB, Ruth TJ, Stoessl AJ. Dopamine release in human ventral striatum and expectation of reward. Behav Brain Res 2002;136:359–363.

4 [^{123}I]-Altropane SPECT

How It Compares to Other Positron Emission Tomography and Single-Photon Emission Tomography Dopamine Transporters in Early Parkinson's Disease

Hubert H. Fernandez, MD, Paula D. Ravin, MD, and Dylan P. Wint, MD

SUMMARY

Positron emission tomography (PET) scan with [^{18}F]-6-fluoro- dihydroxyphenylalanine (FDOPA) is currently the imaging "gold standard" for diagnosing Parkinson's disease (PD), but this procedure is available at only a limited number of facilities. PET cameras are expensive, they require proximity to a cyclotron, and tests are nonreimbursable. A less costly and more available test, such as a single photon emission computed tomography (SPECT), may thus be helpful in the diagnosis of early or atypical PD, if its sensitivity is comparable with a PET scan. Altropane is an iodinated form of the *N*-allyl analog of WIN 35,428, which acts as a dopamine transport inhibitor. When radiolabeled with the γ-emitting isotope [^{123}I], altropane serves as a SPECT ligand with high affinity and selectivity for the dopamine transporter. It is a good marker for dopamine neurons and is useful in detecting PD. There have now been reported cases that suggest altropane SPECT is comparable, if not possibly superior, to FDOPA PET scans in detecting early cases of PD.

Key Words: Altropane; dopamine transporter ligand; SPECT; PET; Parkinson.

1. MEASURING DOPAMINERGIC ACTIVITY IN EARLY PARKINSON'S DISEASE (PD): ITS CHALLENGES AND SIGNIFICANCE

Early PD can be a difficult diagnosis to be certain of given the spectrum of clinical symptoms and signs with which a patient can present. The diagnostic challenges faced by clinicians are illustrated by estimates that as many as 25% of patients diagnosed with idiopathic PD do not have the characteristic Lewy bodies at autopsy, many years further into the course of disease *(1,2)*. A fraction of these patients have other parkinsonian disorders such as progressive supranuclear palsy (PSP), cortico-basal ganglionic degeneration, and multiple systems atrophy. The clinical presentation of parkinsonism can include symptoms of fatigue, for example, dragging one leg or stiffness in an arm, and can be misconstrued as age-related or, more commonly, as the result of arthritic conditions. A paucity of facial expression and decreased spontaneity of speech or withdrawal from social events, features that often accompany parkinsonism, may be misinterpreted as depression. The occurrence of unilateral "pill-rolling" tremor at rest, although pathognomonic of parkinsonism, is observed in only 40% of patients with idiopathic PD and fewer still with Parkinson's variants such as multiple systems atrophy and PSP *(3)*. Not uncommonly, the action and intention tremor that may appear in parkinsonism can lead to a false diagnosis of benign essential tremor—the most prevalent nonparkinsonian movement disorder. The characteristic pathology all these parkinsonian disorders have in common is the loss of dopaminergic neurons whose presynaptic axons end in the caudate and putamen. A surrogate marker of early disease would ideally measure the loss of dopaminergic neurons and not be influenced by treatment status, age or sex of the patient.

The accurate and early diagnosis of PD and other parkinsonian syndromes have become even more critical as clinical trials in PD have now entered the era of neuroprotection. Soon, it will be critical that PD patients be identified at the earliest possible stage, not only for proper inclusion in neuroprotective clinical trials but also to give the prospective disease-modifying agent the best chance of slowing or stopping PD progression.

At present the diagnosis of parkinsonian disorders during life is based on a careful exclusion of structural disorders in the basal ganglia, clinical observation during a period of time (usually 2 to 5 yr), and a sustained, robust response to dopamine-replacement therapy. Although Rajput et al. *(3)* and Hughes et al. *(4)* have shown that the accuracy of the clinical diagnosis of idiopathic PD improves between the initial diagnosis (65–74%) and the final diagnosis; still, only 76–82% of patients with the final (5–12 yr) diagnosis of PD are found to have substantia nigra loss and Lewy-body type pathology on autopsy. Rajput et al. suggested that response to levodopa may not be

From: *Bioimaging in Neurodegeneration*
Edited by P. A. Broderick, D. N. Rahni, and E. H. Kolodny

specific to the underlying pathology of PD and concluded "there were no clues to distinguish neurofibrillary tangle parkinsonism or profound substantia nigra loss without neuronal inclusions from idiopathic PD" *(3)*. In a community-based retrospective population study of 402 cases of diagnosed parkinsonism published by Meara et al. *(5)*, 74% of cases were "confirmed" as having parkinsonism and 53% as "clinically probable idiopathic PD" when re-evaluated by neurologists following formal diagnostic criteria. This type of inaccuracy has been seen both in movement-disorders specialty clinics and general brain pathology registries.

Several in vivo imaging modalities have been evaluated in patients with parkinsonian disorders. Routine "anatomic" or structural imaging techniques, including magnetic resonance imaging, are generally not useful in distinguishing PD from other neurodegenerative conditions *(6)*. In contrast, "functional" imaging techniques, including positron emission tomography (PET) using a variety of radioligand agents such as ^{18}F-6-fluoro-dihydroxyphenylalanine (FDOPA) and ^{18}F-deoxyglucose (^{18}FDG), have been used with varying degrees of success to differentiate PD from other movement disorders *(7,8)*. Agents evaluated using PET imaging include radiolabeled phenyltropane analogs that have a binding affinity for the dopamine transporter, normally found in high concentrations in the striatal region of the brain. Clinical studies using PET imaging have demonstrated that striatal uptake of several phenyltropane analogs is markedly reduced in patients with PD. However, because of the limited availability and economic costs, the routine use of PET imaging has not been adopted widely.

Problems that arise with using agents that bind to the dopamine transporter protein (DAT) to measure early parkinsonism include uncertainty about the rate of progression of presynaptic dopaminergic cell loss and latency to onset of clinical disease *(9)*. Longitudinal PET and single-photon emission computed tomography (SPECT) studies suggest a relatively rapid or exponential decline of dopaminergic function in early PD, followed by slowing of the degeneration process in the later stages of the disease *(10–12)*. In a SPECT study using [^{123}I]β-CIT, a significant reduction of striatal binding over the course of 2 yr was found in the PD group who had symptoms for fewer than 5 years compared with the PD group who had a longer duration of symptoms *(12)*. However, recent reports using β-CIT show a slower, more linear decline of striatal binding in the first 5 yr of the disease *(13)*. Furthermore, variability on test/retest protocols, such as the one study published using a 3- to 6-wk interval between scans with β-FP-CIT, also demonstrated a 7.4–7.9% fluctuation in measurements within subjects *(14)*. The published multicenter trials for many of the radioligands report poor negative predictive values for both quantitative and qualitative methods of interpreting scans *(15)*. In this regard, altropane SPECT imaging may be more sensitive than FDOPA PET for detecting early PD. A report of two patients with clinically defined early PD, summarized later in this chapter, supports this statement *(15)*. Finally, the influence of dopaminergic therapy on DAT homeostasis is not yet clearly understood and potentially confounds interpretation of both PET and SPECT scans in early parkinsonism. Whether drugs under current study are "neuroprotective" or simply "levodopa sparing" may require imaging of individuals "at risk" for PD during a period of 5–10 yr without drug therapy. However, the feasibility of such study and the ethical implications of withholding putative neuroprotective agents from this control group once abnormal DAT images are perceived makes such a study unlikely to occur.

2. ALTROPANE: ITS NOVEL PROPERTIES

Recently, radiolabeled phenyltropane analogs have been modified for SPECT imaging, a technology that is widely available in nuclear medicine departments and much less costly than PET. The first analog in the United States to be studied extensively in vivo was β-CIT, but its equilibration properties require a 24-h delay between injection and measurement of DAT in the striatum. Furthermore, β-CIT has an equal affinity for the serotonin transporter, which may affect the background counts and tracer distribution *(16)*. Altropane is a new phenyltropane analog (^{123}I-E-2β-carbomethoxy-3β-nortropane or ^{123}I-E-IACT) with pharmacologic properties that make it more practical for clinical application (Fig. 1). It is rapidly and widely distributed after administration and the majority of the drug is cleared in urine during the next 24 to 48 h as iodinated metabolites. It does not significantly bind to human plasma proteins and requires only 5 to 8 mCi to obtain clearly defined uptake in the brain. Image acquisition is optimal at 1 h after injection because of its high binding affinity (K_d, 5.33 nM) preferential of 28:1 for dopamine over serotonin and relatively low nonspecific binding *(17)*. Images are obtained at 10-min intervals for 1 h after injection and radioactivity approaches baseline at less than 2 h. No differences in binding have been noted in healthy subjects older than age 50 of either gender. Adverse reactions in preliminary studies have been frequent but clinically insignificant (e.g., mild headache, slight elevation in blood pressure, bowel frequency).

Phase I trials conducted by Boston Life Sciences, Inc (BLSI) established age-related binding potentials, safety, and radiation dosimetry in a total of 39 healthy subjects (BLSI report to FDA April 9, 1999). The Phase II multi-center trial involved nine study sites, with 12 normal and 25 PD subjects. There was also a post-hoc blinded qualitative assessment of the images to determine the sensitivity and specificity of altropane for diagnosis of PD. Results showed a sensitivity of 95.8%, specificity 100%, positive predictive value 100%, and negative predictive value of 91.7% compared with the standard of a movement disorders specialist (MDS) diagnosis (BLSI report to FDA August 2, 2000). The final study reviewed to date, a Phase III multicenter trial involving 50% parkinsonian and 50% nonparkinsonian movement disorders, was expanded to include 15 sites and a total of 165 patients. The technical variability of different readers and different cameras for data acquisition was reflected in an overall accuracy of blinded interpretation of SPECT images of 79.5% (BLSI report to FDA March 26, 2001). This still represents a meaningful improvement over the non-MDS experience of 30–50% false-positive diagnoses of PD reported in the literature. Another Phase III trial is being launched soon to assess sensitivity and specificity for distinguishing parkinsonian from nonparkinsonian tremor disorders (IND in process). This trial will compare the diagnostic accu-

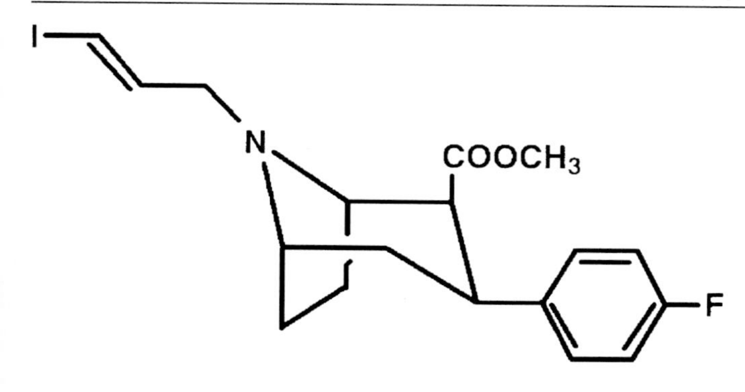

Fig. 1. Altropane ($C_{18}H_{21}NO_2{}^{123}IF$), which has a molecular weight of 425, has the following chemical structure.

racy of altropane SPECT vs general clinicians in a "real-world" setting in distinguishing benign essential tremor, drug-induced tremor and other less well-defined involuntary movements from early parkinsonism, using the MDS diagnosis as the "gold standard."

3. ALTROPANE SPECT VS FDOPA PET IN EARLY PD: LESSONS FROM TWO CASES

We previously reported two cases of typical, early PD showing nondiagnostic FDOPA PET scans but with unilateral striatal reduction of tracer uptake on altropane SPECT obtained within 3 mo of the PET scan *(18,19)*. We summarize the highlights of these cases in this chapter.

3.1. CASE REPORTS

The first case was of a 54-yr-old right-handed man who had noticed a rest tremor of the right hand 1 yr before, along with increased flexion of the elbow. He had no history of central nervous system infections, stroke, use of drugs that might interfere with dopamine effects, recreational drugs, family history of PD or similar disorders, head trauma, or other neurological problems. Magnetic resonance imaging of his brain was normal. His physical examination revealed a mild, persistent rest tremor of the right hand; mild rigidity of the right arm; and mild slowness of finger tapping of the right. He had a mildly diminished right arm swing and right hand tremor when he walked. He had Hoehn-Yahr stage 1 (of 5) PD. Pramipexole, a dopamine agonist, at 2.25 mg/d improved his symptoms.

The second case was of a 38-yr-old right-handed man who had a 1-yr history of intermittent tremor of the right hand aggravated by fatigue and stress. He later noted decreased arm swing on the right when walking, stiffness and joint pain in the right wrist, and general fatigue. His wife had noted decreased facial expression, mild dysarthria, and irritability. Magnetic resonance imaging of his brain prior to his evaluation was normal. There was no history of stroke, central nervous system infection, head trauma, use of dopamine blocking agents, recreational drugs, or family history of neurological disorders. He had mild rigidity in all limbs and mildly decreased amplitude and speed on fine-finger movements, hand opening, and heel tapping on the right. Rest tremor of the right hand was seen primarily with reinforcement maneuvers. Gait was normal except for decreased right arm swing and a right hand tremor. He also had Hoehn-Yahr Stage 1 PD. His SPECT and PET scans were obtained 3 mo later. A significant and sustained response with pergolide, another dopamine agonist, was noted.

3.2. IMAGING METHOD

3.2.1. PET Imaging With 6-[^{18}F]-Fluoro-L-dopa

Both patients were fasted overnight and all anti-PD medications were withheld for 12 h prior to PET imaging. Carbidopa 200 mg was given orally 1.5 h before tracer injection to inhibit decarboxylation. PET studies were performed with a PC-4096 scanner (Scanditronix AB, Sweden) with 15 axial slices and resolution of 6 mm full width at half maximum (fwhm). The performance characteristics of this instrument are well described in the literature *(20)*. Attenuation corrections were performed from transmission data acquired using a rotating pin source containing 68Ge. The PET camera was cross-calibrated with a well scintillation counter. Fluorodopa was prepared following a previously described procedure *(21)*.

After positioning the patient in the gantry of the PET camera, 7.2 mCi (case 1) and 5 mCi (case 2) of FDOPA were injected intravenously over the course of 45 s at the start of imaging. Twelve 10-min sequential emission scans were acquired. Image reconstruction was performed using a conventional filtered back projection algorithm to an in-plane resolution of 7 mm fwhm in a digital matrix of 128 × 128. All projection data were corrected for attenuation, non-uniformity of detector response, dead time, random coincidence, and scattered radiation. Details of these procedures are described elsewhere *(22–25)*.

3.2.2. SPECT Scan Using [^{123}I]-Altropane

To block accumulation of [^{123}I] by the thyroid, both subjects were treated with a 5-d course of saturated solution of potassium iodide (SSKI) starting 24 h before imaging. The brain SPECT scans were performed after intravenous administration of 5 mCi of [^{123}I]-altropane with four sequential 10-min SPECT acquisitions performed from 0 to 40 min. The acquisitions were performed on a MultiSPECT gamma camera (Siemens, Hoffman Estates, IL) equipped with fan-beam collimators with an intrinsic resolution of 4.6 fwhm (for case 1) and a DSI Ceraspect camera with an annular crystal, with resolution of approx 6.4 mm fwhm (for case 2). The SPECT cameras were cross-calibrated by comparing the camera response from a uniform distribution of [^{123}I]-altropane solution in a cylindrical phantom to an aliquot of the same solution measured with a well counter. The images were reconstructed with a standard filtered back projection algorithm using a Butterworth filter. Attenuation corrections were performed using the Chang algorithm *(26)*. Reconstruction and attenuation corrections were performed using dedicated computers interfaced to the gamma cameras. Details of image reconstruction and analysis are described elsewhere *(27)*.

3.3. RESULTS

In the first case, his PET scan showed mild reduction of tracer accumulation throughout the left striatum. However, the reduction in striatal accumulation of the tracer did not reach the quantitative values for PD (*see* Figs. 2 and 3; a normal PET scan is provided for comparison). Quantitative data expressed as striatal/occipital ratios (SORs) are shown in Table 1. The distribution of tracer elsewhere in the brain was unremarkable. The SPECT scan, however, revealed normal activity in the right

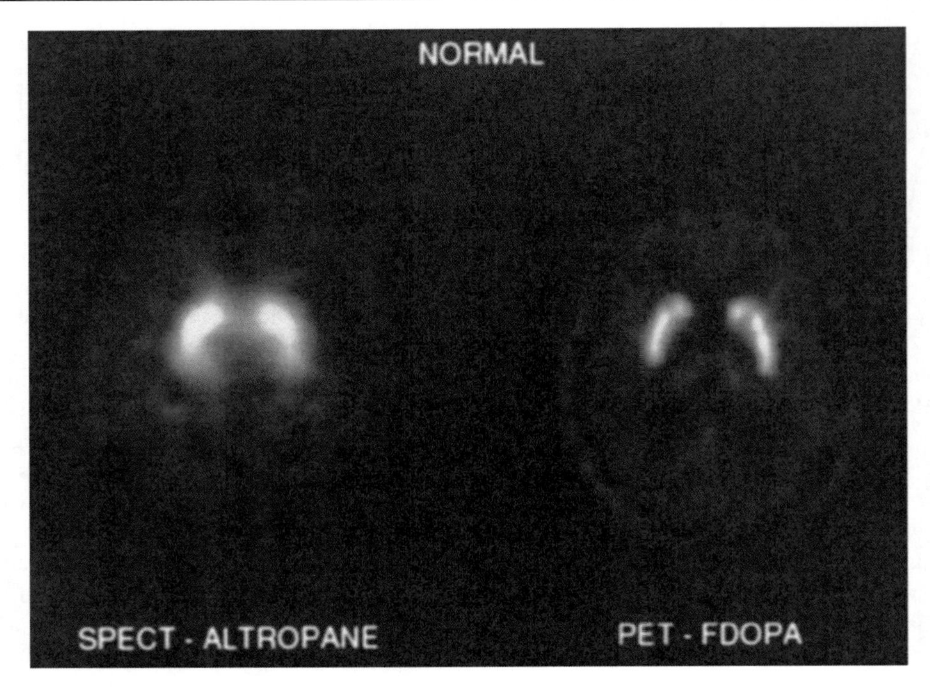

Fig. 2. Left, normal altropane SPECT. Right, normal FDOPA PET. Reprinted with permission from Medical Science Monitor 2001;7:1339–1343.

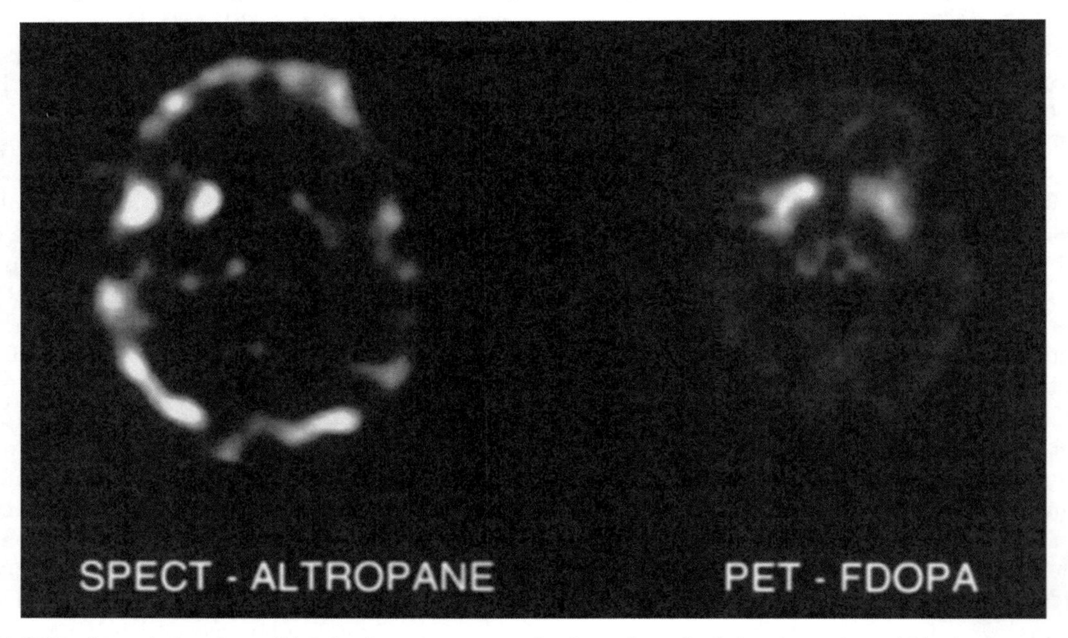

Fig. 3. A FDOPA PET of case 1 showing mild reduction of tracer uptake throughout the left striatum but within the normal range. Altropane SPECT shows markedly reduced activity in the left striatum consistent with PD. Reprinted with permission from Medical Science Monitor 2001;7:1339–1343.

caudate region; mild decreased activity in the right putamen; but markedly decreased activity diffusely in the left striatum consistent with PD (Figs. 2 and 3, Table 1).

In the second case, his PET scan showed mildly reduced uptake in the left striatum (Fig. 4). However, the SORs were not outside the normal range (Table 1). The SPECT scan revealed definite unilateral reduction of uptake in the left posterior putamen consistent with PD (Fig. 4, Table 1).

Both patients carried a clinical diagnosis of probable PD based on the presence of unilateral rest tremor, rigidity, and bradykinesia with a sustained significant response to dopamine agonist monotherapy. Both patients obtained their PET scans prior to SPECT scans but within a 3-mo interval. No medication adjustment or worsening of parkinsonism was noted during this period. Although the SORs of both patients by [^{18}F]-FDOPA PET fell within the normal range, despite mild striatal reduction of tracer uptake, there was a clear reduction of tracer uptake in the striatum contralateral to their parkinsonian side using [^{123}I]-Altropane SPECT. Although parkinsonian ranges of SORs may differ among institutions, making it

Table 1
Quantitative Data on PET Scans (Expressed as Striato-Occipital Rations)

PET Scan	*Normal (mean ± SD)*	*PD (mean ± SD)*	*Case 1*	*Case 2*
Left caudate	2.24 ± 0.3	1.51 ± 0.08	2.68	2.23
Left anterior putamen	2.24 ± 0.3	1.51 ± 0.08	2.64	2.45
Left posterior putamen	2.24 ± 0.3	1.51 ± 0.08	2.13	2.01
Right caudate	2.37 ± 0.15	1.58 ± 0.17	2.77	2.61
Right anterior putamen	2.37 ± 0.15	1.58 ± 0.17	2.73	2.64
Right posterior putamen	2.37 ± 0.15	1.58 ± 0.17	2.23	2.52

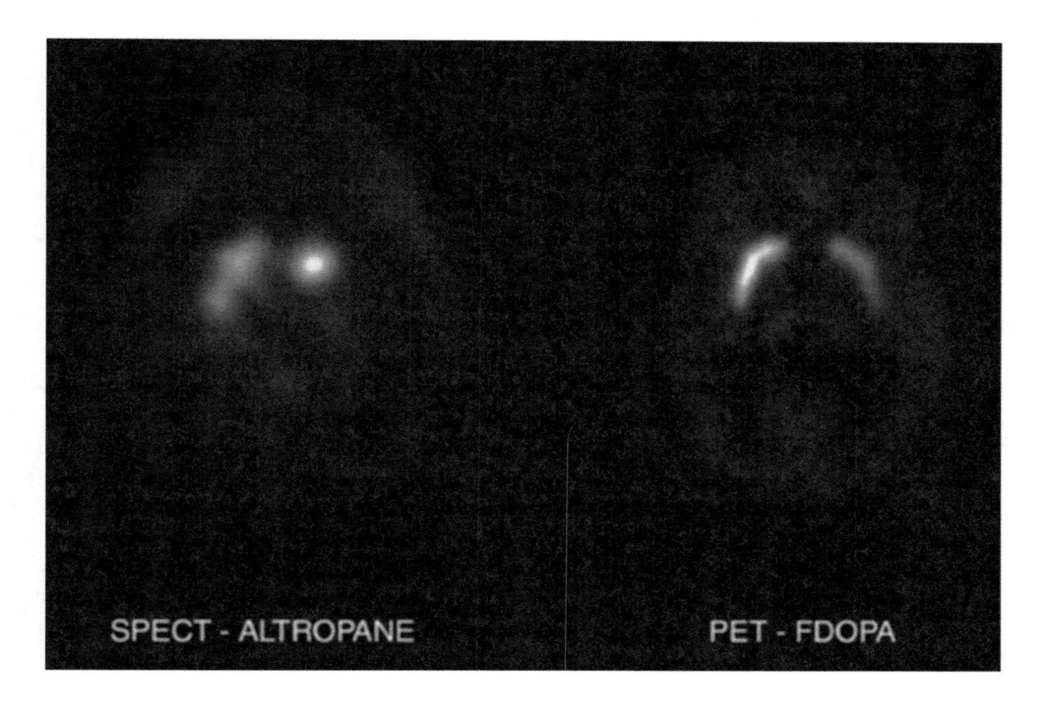

Fig. 4. A FDOPA PET of Case 2 showing mildly reduced uptake in the left striatum within the normal range. Altropane SPECT shows definite unilateral reduction of uptake in the left posterior putamen consistent with PD. Reprinted with permission from Medical Science Monitor 2001;7:1339–1343.

difficult to compare sensitivities of imaging modalities in detecting mild PD, the asymmetry of tracer uptake in the SPECT scans was unmistakable in these two patients. Although further studies are needed to define its potential role in detecting presymptomatic, early, and atypical PD cases, SPECT imaging using [^{123}I]-altropane may be an accessible, rapid, and sensitive imaging modality for detecting early PD.

4. OTHER PROMISING DOPAMINE TRANSPORTER RADIOLIGANDS USED IN EARLY PD

Excitement about novel DAT ligands is increasing as researchers improve synthetic processes and develop creative methods for incorporating radioisotopes. Efforts in this area seek to improve on each of the essential characteristics of a useful DAT ligand: high striatal-to-cerebellar uptake ratio; high selectivity for the DAT; efficient manufacturing process with high purity and specific activity; and striatal localization rate well-matched to the isotope's half-life *(27)*. In addition, academic and industry investigators hope to find DAT ligands that have broader applications. Thus, investigation of these compounds has increasingly turned toward demonstrating and enhancing their clinical utility.

4.1. SPECT RADIOLIGANDS

4.1.1. ^{123}I-β-CIT Analogs

Although ^{123}I-β-CIT is clearly useful in imaging striatal DAT, its pharmacokinetic characteristics delay imaging until the day after radioligand injection *(28)*. It also may be somewhat less specific than desired, having a tendency to accumulate at serotonin transporters *(29,30)*. Thus, N-Ω-fluoralkyl analogs ^{123}I-FP-β-CIT (*N*-3-fluoropropyl-2-β-carboxymethoxy-3-β [4-iodophenyl]-nortropane) and ^{123}I-FE-β-CIT (*N*-(2-fluoroethyl)-2-β-carboxymethoxy-3-β-(4-iodophenyl)-nortropane) were developed. Both of these compounds bind to live and postmortem striatal DAT within 30 min and wash out of the occipital cortex and midbrain within 100 min *(28)*. Their binding specificities for DAT approximate that of ^{123}I-β-CIT (with β-FE-CIT being perhaps less specific; refs. *31,32*). Thus, their pharmacokinetics may be comparable, if not superior, to altropane's (peak striatal binding within 15 min, return to background levels within 120 min; ref. *26*).

In diagnosis and evaluation of PD, multiple studies have demonstrated ^{123}I-FP-β-CIT's ability to reveal reduced DAT density, perhaps more effectively than ^{123}I-β-CIT *(33–35)*. Like its progenitor, ^{123}I-FP-β-CIT is also able to detect bilateral decreases in striatal DAT density in PD patients with minimal or unilateral symptoms *(36)*. However, there is still some disagreement about whether striatal DAT density as measured by ^{123}I-FP-β-CIT correlates with the stage of PD or the severity of PD symptoms *(32,37,38)*.

4.1.2. ^{99m}Tc-TRODAT-1

To improve the clinical utility of DAT SPECT, several groups have investigated radioisotopes with greater stability than ^{123}I. One such radioisotope is ^{99m}technetium (^{99m}Tc). A relatively long half-life makes ^{99m}Tc useful for clinical applications of SPECT because of its improved transportability and ease of manufacturing. In addition, ^{99m}Tc costs approximately $0.30/mCi (^{123}I costs $30/mCi; ref. *39*). ^{99m}Tc-TRODAT-1 ([2-[[2-[[[3-(4-chlorophenyl)-8-methyl-8-azabicyclo [3,2,1] oct-2-yl] methyl] (2-mercaptoethyl) amino] ethyl] amino] ethanethiolato (3-)-*N*2,*N*2',*S*2,*S*2']oxo-[1R-(exo-exo)]), first used in humans in 1996 *(40)*, was the first site-specific ^{99m}Tc-based brain-imaging agent *(39)*. It can be produced from a kit *(41)*, without a cyclotron, which gives it an enormous advantage over all ^{123}I-based radioligands. It takes 4 h to achieve its optimal striatal/cortical binding ratio *(42)*.

^{99m}Technetium-TRODAT-1 has shown some promise in evaluating PD. Uptake of ^{99m}Tc-TRODAT-1 in the posterior putamen was significantly reduced in a controlled study of 42 patients with PD *(43)*. In another study of 34 patients with PD, putaminal uptake of ^{99m}Tc-TRODAT-1 was significantly correlated with Hoehn and Yahr stage; in addition, bilaterally decreased putaminal uptake could be identified in patients with Stage I disease *(44)*. ^{99m}Technetium-TRODAT-1 has also been reported to be able to distinguish idiopathic PD between from vascular parkinsonism *(45)*.

4.2. PET RADIOLIGANDS

Radioligands used in PET imaging are generally taken up into the striatum at a slower rate than altropane. In addition, the expense of PET and the necessity of a nearby cyclotron makes PET less clinically accessible than SPECT. However, the better spatial resolution of PET is important in many research applications, and may increase in importance as knowledge about the neuropathological basis for basal ganglia diseases grows.

4.2.1. β-CFT and Derivatives

β-CFT (2-β-carbomethoxy-3-β-(4-fluorophenyl)tropane, or WIN 35,428) is a cocaine analog that binds selectively to DAT. Radioactive labeling with carbon-11 (^{11}C) produces a specific DAT radioligand, ^{11}C-β-CFT. Practical use of ^{11}C-β-CFT is limited by the mismatch between its relatively slow uptake into striatum (peak levels at 80–90 min; ref. *46*) and the short half-life (20 min) of ^{11}C.

One of the earliest discoveries using ^{11}C-β-CFT was that more than more than 70% destruction of dopaminergic cells was required to produce parkinsonian symptoms in 1-methyl-4phenyl-1,2,3,6-tetrahydropyridine-lesioned monkeys *(47)*. ^{11}C-β-CFT can reliably discriminate between changes in DAT distribution caused by PD and PSP *(48)*. It may also be used to detect dopaminergic cell loss in early PD *(49)*.

β-CFT also has been labeled with fluorine-18 (^{18}F), which has a radioactive half-life of 109.8 min and relatively low-energy positrons; those features allow for longer imaging times and higher-resolution images than with ^{11}C *(50)*. ^{18}F-β-CFT reaches peak striatal binding at approx 225 min. Therefore, although it capably images DAT *(50)* and monitors the progress of PD *(51,52)*, the relationship between its striatal localization and isotope half-life is less-than-ideal.

4.2.2. β-CIT AND DERIVATIVES

Developed in 1993, β-CIT (2-β-carbomethoxy-3-β-(4-iodophenyl)tropane) is another cocaine analog with affinity for DAT. It achieves a striatal/cerebellar radioactivity ratio of 5 within about 90 min after injection *(53)*. It does, however, exhibit some binding nonspecificity and is detectable on serotonin and noradrenaline transporters in the thalamus *(54)*. In addition, its uptake into the striatum can be influenced by cerebral perfusion.

^{11}C-β-CIT-FE may have some advantages over the other cocaine analogs, with more rapid peak binding (60 min; ref. *55*) and similar ability to differentiate between movement disorders and diagnose PD *(56,57)*.

4.2.3. β-CBT and Derivatives

Methyl-2-β-carbomethoxy-3-β-(4-bromophenyl)tropane (β-CBT) was developed to study the role of the halogenated site in DAT selectivity of cocaine analogs *(58)*. Its radioactively brominated form (^{76}Br-β-CBT) reaches peak concentration in the striatum in about 60 min and remains there for about 4 h, with a striatal-to-cerebellar activity ratio between 17 and 22 *(58)*. In a study of 18 patients with PD, ^{76}Br-β-CBT demonstrated reduction of DAT density better than ^{18}fluorodopa but was not able to stratify disease severity *(59)*.

In summary, although PET scan with [^{18}F]-6-flouro-dopa is currently the imaging "gold standard" for diagnosing PD, this procedure is available at only a limited number of facilities. PET cameras are expensive, they require proximity to a cyclotron, and tests are nonreimbursable. Hence, the need for a less expensive and universal test such as a SPECT, especially in detecting early or atypical PD and other movement disorders, still remains. A DAT ligand with ideal pharmacologic properties using SPECT imaging, such as altropane, may prove to be an accessible, rapid and sensitive test in detecting early and/or atypical PD.

REFERENCES

1. Hughes AJ, Daniel SE, Kilford L, et al. Accuracy of clinical diagnosis of Parkinson's disease: a clinico-pathologic study of 100 cases. J Neurol Neurosurg d Psychiatry 1992;55:181–184 .
2. Hughes AJ, Daniel SE, Blankson S, et al. A clinicopathologic study of 100 cases of Parkinson's disease. Arch Neurol 1993;50:140–148.
3. Rajput AH, Rozdilsky B, Rajput A. Accuracy of clinical diagnosis in Parkinson's disease—a prospective study. Can J Neurol Sci 1991; 18:275–278.
4. Hughes AJ, Ben-Schlomo T, Daniel SE, et al. What features improve the accuracy of clinical diagnosis in Parkinson's disease? Neurology 1992;42:1142–1146.
5. Meara J, Bhowmick BK, Hobson P. Accuracy of diagnosis in patients with presumed Parkinson's disease. Age Ageing 1999;28: 99–102.
6. Bhattacharya K, Saadia D, Eisenkraft B, Yahr M, Olanow W, Drayer B, et al. Brain magnetic resonance imaging in multiple-system atrophy parkinson's disease—a diagnostic algorithm. Arch Neurol 2002; 59:835–842.

7. Brooks DJ. Opinion review: functional imaging in relation to Parkinsonian syndromes. J Neurol Sci 1993;115:1–17.
8. Brooks DJ, Ibanez V, Sawle GV, et al. Differing Patterns of striatal F18DOPA uptake in Parkinson's disease, multiple system atrophy, progressive supranuclear palsy. Ann Neurol 1990;28:547–555.
9. Benamer HTS, Patterson J, Wyper DJ, et al. Correlation of Parkinson's disease severity duration with I123-FP-CIT SPECT striatal uptake. Mov Disord 2000;15:4:692–698.
10. Morrish PK, Sawle GV, Brooks DJ. An [18F]dopa-PET clinical study of the rate of progression in Parkinson's disease. Brain 1996;119:585–591.
11. Moorish PK, Rakshi JS, Bailey DL, Sawle GV. Measuring the rate of progression estimating the preclinical period of Parkinson's disease with [18F]dopa-PET. J Neurol Neurosurg Psychiatry 1998;64:314–319.
12. Pirker W, Djamshidian S, Asenbaum S, Gerschlager W, Tribl G, Hoffman M, et al. The progression of degeneration in Parkinson's disease atypical parkinsonism: a longitudinal B-CIT SPECT study. Mov Disord 2002;17: 45–53.
13. Pirker W, Holler I, Gerschlager W, Asenbaum S, Zetting G, Brucke T. Measuring the rate of progression of Parkinson's disease over a 5-year period with B-CIT SPECT. Mov Disord 2003;18:1266–1272.
14. Booij J, Tissingh G, Winogrodzka A, et al. Imaging of the dopaminergic neurotransmission system using single-photon emission tomography positron emission tomography in patients with parkinsonism. Eur J Nucl Med 1999;26:171–182.
15. Marek K, for the Parkinson study group. A multicenter assessment of dopamine transporter imaging with DOPSCAN/SPECT in Parkinsonism. Neurology 2000;55:1540–1547.
16. Innis RB, Seibyl JB, Scanley BE, et al. Single photon computed imaging demonstrates loss of dopamine transporters in Parkinson's disease. Proc Natl Acad Sci USA 1993;90:11965–11969.
17. Madras BK, Meltzer PC, Liang AY, et al. Altropane a SPECT or PET imaging probe for dopamine neurons (parts I and II) Synapse 1998;29:93–104, 105–115.
18. Fernez HH, Friedman JH, Fischman AJ, et al. Is Altropane SPECT more sensitive to fluoroDOPA PET for detecting early Parkinson's disease? Med Sci Monit 2001;7:1339–1343.
19. Fernandez HH, Friedman JH. 18F-Dopa vs dopamine transporter ligands in positron emission tomographic and single-photon emission computed tomographic scans for Parkinson disease. Ann Neurol 2002;59:1973.
20. Rota Kops E, Herzog H, Schmid A, et al. Performance characteristics of eight-ring whole body PET scanner. J Comput Assist Tomogr 1990;14:437–445.
21. Luxen A, Milton P, Bida GT. Remote, semiautomatic production of 6- [18F]fluoro-L-dopa for human studies with PET. Ap Radiat Isot 1990;41:275–281.
22. Martin WRW, Palmer MR, Patlak CS, Calne DB. Nigrostriatal function in man studied with positron emission tomography. Ann Neurol 1989;26:535–542.
23. Patlak CS, Blasberg RG, Fenstermacher JD. Graphical evaluation of blood-to-brain transfer constants from multiple-time uptake data. J Cereb Blood Flow Metab 1983;3:1–7.
24. Patlak CS, Blasberg RG. Graphical evaluation of blood-to-brain transfer constants from multiple-time uptake data. Generalization. J Cereb Blood Flow Metab 1985;5:584–590.
25. Chang L. A method for attenuation correction in computed tomography. IEEE Trans Nucl Sci, 1987;25:638–643.
26. Fischman AJ, Bonab AA, Babich JW, et al. Rapid detection of Parkinson's disease by SPECT with altropane: a selective lig for dopamine transporters. Synapse 1998;29:128–141.
27. Fischman AJ, Babich JW, Elmaleh DR, et al. SPECT imaging of dopamine transporter sites in normal MPTP-treated rhesus monkeys. J Nucl Med 1997;38:144–150.
28. Abi-Dargham A, Gelman MS, DeErausquin GA, et al. SPECT imaging of dopamine transporters in human brain with [^{123}iodine]-fluoroalkyl analogs of β-CIT. J Nucl Med 1996;37:1129–1133.
29. Innis R, Baldwin R, Sybirska E, et al. Single photon emission computed tomography imaging of monoamine reuptake sites in primate brain with [^{123}I]CIT. Eur J Pharmacol 1991;200:369–370.
30. Neumeyer JL, Wang SY, Milius RA, et al. [^{123}I]2-β-carbomethoxy-3 beta-(4-iodophenyl)tropane: high-affinity SPECT radiotracer of monoamine reuptake sites in brain. J Med Chem 1991;34:3144–3146.
31. Gunther I, Hall H, Halldin C, et al. ^{125}I-β-CIT-FE ^{125}I-β-CIT-FP are superior to ^{125}I-β-CIT for dopamine transporter visualization: autoradiographic evaluation in the human brain. Nucl Med Biol 1997;24:629–634.
32. Kuikka JT, Akerman K, Bergstrom KA, et al. ^{123}Iodine labeled N-(2fluoroethyl)-2-β-carboxymethoxy-3-β-(4-iodophenyl) nortropane for dopamine transporter imaging in the living human brain. Eur J Nucl Med 1995;22:682–686.
33. Booij J, Tissingh G, Boer GJ, et al. [^{123}I]FP-CIT SPECT shows a pronounced decline of striatal dopamine transporter labeling in early advanced Parkinson's disease. J Neurol Neurosurg Psychiatry 1997;62:133–140.
34. Booij J, Habraken JB, Bergmans P, et al. Imaging of dopamine transporters with [^{123}I]FP-CIT SPECT in healthy controls patients with Parkinson's disease. J Nucl Med 1998;39:1879–1884.
35. Seibyl JP, Marek K, Sheff K, et al. ^{123}Iodine-β-CIT [^{123}I]FP-CIT SPECT measurement of dopamine transporters in healthy subjects Parkinson's patients. J Nucl Med 1998;39:1500–1508.
36. Katzenschlager R, Costa D, Gacinovic S, et al. [^{123}Iodine]FP-CIT SPECT in the early diagnosis of PD presenting as exercise-induced dystonia. Neurology 2002;59:1974–1976
37. Vermeulen RJ, Wolters EC, Tissingh G, et al. Evaluation of [^{123}I]β-CIT binding with SPECT in controls, early late Parkinson's disease. Nucl Med Biol 1995;22:985–991.
38. Tissingh G, Booij J, Bergmans P, et al. ^{123}Iodine-N-omega fluoropropyl-2-β -carbomethoxy-3-β-(4-iod ophenyl)tropane SPECT in healthy controls early-stage, drug-naive Parkinson's disease. J Nucl Med 1998;39:1143–1148.
39. Kung HF. Development of Tc^{99m} labeled tropanes: TRODAT-1, as a dopamine transporter imaging agent. Nucl Med Biol 2001;28:505–508.
40. Kung HF, Kim HJ, Kung MP, et al. Imaging of dopamine transporters in humans with technetium^{99m}-TRODAT-1. Eur J Nucl Med 1996;23:1527–1530.
41. Choi SR, Kung MP, Plossl K, et al. An improved kit formulation of a dopamine transporter imaging agent: [Tc^{99m}]TRODAT-1. Nucl Med Biol 1999;26:461–466.
42. Kao PF, Tzen KY, Yen TC, et al. The optimal imaging time for [^{99}Tc^{m}]TRODAT-1/SPET in normal subjects patients with Parkinson's disease. Nucl Med Commun 2001;22:151–154.
43. Mozley PD, Schneider JS, Acton PD, et al. Binding of [^{99m}Tc]TRODAT-1 to dopamine transporters in patients with Parkinson's disease in healthy volunteers. J Nucl Med 2000;41:584–589.
44. Huang WS, Lin SZ, Lin JC, et al. Evaluation of early-stage Parkinson's disease with ^{99m}Tc-TRODAT-1 imaging. J Nucl Med 2001;42:1303–1308.
45. Tzen KY, Lu CS, Yen TC, et al. Differential diagnosis of Parkinson's disease vascular parkinsonism by ^{99m}Tc-TRODAT-1. J Nucl Med 2001;42:408–413.
46. Tsukada H, Nishiyama S, Kakiuchi T, et al. Ketamine alters the availability of striatal dopamine transporter as measured by [^{11}C]-β-CFT [^{11}C]-β-CIT-FE in the monkey brain. Synapse 2001;42:273–280.
47. Hantraye P, Brownell AL, Elmaleh D, et al. Dopamine fiber detection by [^{11}C]-CFT PET in a primate model of parkinsonism. Neuroreport 1992;3:265–268.
48. Ilgin N, Zubieta J, Reich SG, et al. PET imaging of the dopamine transporter in progressive supranuclear palsy Parkinson's disease. Neurology 1999;52:1221–1226.
49. Frost JJ, Rosier AJ, Reich SG, et al. Positron emission tomographic imaging of the dopamine transporter with ^{11}C-WIN 35,428 reveals marked declines in mild Parkinson's disease. Ann Neurol 1993;34:423–431.

50. Gu XH, Zong R, Kula NS, et al. Synthesis biological evaluation of a series of novel N- or O-fluoroalkyl derivatives of tropane: potential positron emission tomography (PET) imaging agents for the dopamine transporter. Bioorg Med Chem Lett 2001;11:3049–3053.
51. Nurmi E, Ruottinen HM, Kaasinen V, et al. Progression in Parkinson's disease: a positron emission tomography study with a dopamine transporter lig $[^{18}F]$CFT. Ann Neurol 2000;47:804–808.
52. Nurmi E, Bergman J, Eskola O, et al. Progression of dopaminergic hypofunction in striatal subregions in Parkinson's disease using $[^{18}F]$CFT PET. Synapse 2003;48:109–115.
53. Muller L, Halldin C, Farde L, et al. $[^{11}C]\beta$-CIT, a cocaine analogue. Preparation, autoradiography preliminary PET investigations. Nucl Med Biol 1993;20(3):249–255.
54. Farde L, Halldin C, Muller L, et al. PET study of $[^{11}C]\beta$-CIT binding to monoamine transporters in the monkey human brain. Synapse 1994;16:93–103.
55. Halldin C, Farde L, Lundkvist C, et al. $[^{11}C]$-β-CIT-FE, a radioligand for quantitation of the dopamine transporter in the living brain using positron emission tomography. Synapse 1996;22: 386–390.
56. Lucignani G, Gobbo C, Moresco RM, et al. The feasibility of statistical parametric mapping for the analysis of positron emission tomo-graphy studies using ^{11}C-2-β-carbomethoxy-3-β-(4-fluorophenyl)-tropane in patients with movement disorders. Nucl Med Commun 2002;23:1047–1055.
57. Antonini A, Moresco RM, Gobbo C, et al. The status of dopamine nerve terminals in Parkinson's disease essential tremor: a PET study with the tracer $[^{11}C]$FE-CIT. Neurol Sci 2001;22:47–48.
58. Maziere B, Loc'h C, Muller L, et al. ^{76}Br-β-CBT, a PET tracer for investigating dopamine neuronal uptake. Nucl Med Biol 1995;22: 993–997.
59. Ribeiro MJ, Vidailhet M, Loc'h C, et al. Dopaminergic function dopamine transporter binding asseüŸed with positron emission tomography in Parkinson disease. Arch Neurol 2002;59:580–586.

5 Positron Emission Tomography and Embryonic Dopamine Cell Transplantation in Parkinson's Disease

Yilong Ma, PhD, Vijay Dhawan, PhD, Curt Freed, MD, Stanley Fahn, MD, and David Eidelberg, MD

SUMMARY

Parkinson's disease (PD) is a common movement disorder marked by progressive degeneration of dopamine (DA) neurons in the substantia nigra and striatum. The hallmark of motor symptoms in PD includes the resting tremor, rigidity, bradykinesia, and posture instability. Medical therapy to replace lost DA works well initially but becomes ineffective and less tolerated over time. The chronic use of dopaminergic medications leads to motor fluctuations and dyskinesias in patients at more advanced stages. The transplantation of viable DA tissue into the brain is a promising new treatment to reinnervate neurons along the nigrostriatal pathway. In patients with PD, cell survival and clinical benefit have been observed after fetal nigral grafting. Position emission tomography allows in vivo imaging of neuropathophysiology resulting from dopaminergic dysfunction that is inherent in PD by measuring cerebral blood flow, metabolism, and neuroreceptor binding. It offers a unique window for assessing the functional recovery of the brain and its clinical correlation after medical or surgical interventions. In this chapter, we describe the use of positron emission tomography in providing sensitive biomarkers in PD and its application in evaluating the surgical outcome of embryonic DA cell transplantation.

Key Words: Fetal tissue transplantation; Parkinson disease; neurodegeneration; neurosurgical intervention; emission tomography.

1. INTRODUCTION TO POSITRON EMISSION TOMOGRAPHY (PET) IMAGING

PET is a modern imaging system for visualizing and quantifying cerebral function in the human brain. It is based on the use of a number of short-lived radiotracers that are directly involved in physiological processes. Current scanners have a three-dimensional image resolution of 2–4 mm and are sufficient enough to study small animals and humans respectively. Conventionally, PET data are acquired as dynamic images over a time window of up to 90 min after intravenous administration of a radiotracer and analyzed using a region of interest (ROI) approach. Time activity curves are calculated automatically from a set of anatomical areas determined on PET or coregistered magnetic resonance image (MRI). Functional parameters are calculated by using compartment analysis with an input function from arterial blood sampling or a reference tissue assumed to have a low and nonspecific uptake. At present, PET studies are performed noninvasively without taking blood samples to simplify the protocol. Radiotracer binding in the DA system is measured by a simple activity ratio between the striatum and the reference tissue or a kinetic constant computed from multiple time graphical analysis of dynamic series *(1,2)*.

Brain mapping algorithms such as statistical parametric mapping (SPM) represent a complementary and more accurate way for analyzing PET imaging data. This method allows the objective detection of localized and uneven changes in functional parameter over the whole brain, independent of any previous assumptions implicit in a ROI analysis. Statistical comparison is performed on a voxel basis after transforming all images into the stereotactic Talairach coordinate system *(3,4)*. Areas with significant changes are examined by overlaying the active clusters onto an MRI brain template created in the same standard brain space. This approach has greatly improved our ability to measure gradual alteration of brain function in neurological disorders as well as to investigate disease progression and therapeutical interventions.

2. IMAGING PATHOPHYSIOLOGY OF PD

In vitro and in vivo studies in humans have demonstrated that idiopathic PD is preceded by early degeneration of DA neurons in the ventrolateral substantia nigra pars compacta projecting to the posterior and dorsal putamen *(5–8)*. With disease progression, nigrostriatal projections to more anterior and ventral putamen areas begin to decrease, with late loss of projections to the caudate nucleus. This leads to widespread decline

From: *Bioimaging in Neurodegeneration*
Edited by P. A. Broderick, D. N. Rahni, and E. H. Kolodny

of brain function through impaired dopaminergic projections to and from the cortex.

Motor symptoms in PD patients develop after a preclinical period and evolve from unilateral to bilateral involvement. A clinical diagnosis is made if the patient shows at least two of the four cardinal signs of PD with a good response to levodopa treatment. A variable degree of cognitive impairment in frontal function is also present during advanced stages of the disease. The severity of the overall symptoms is evaluated objectively by standardized ratings such as Unified Parkinson's Disease Rating Scale (UPDRS). PET imaging has provided important insight into the functional anatomy of PD by measuring changes in biochemical, hemodynamic and metabolic systems of the brain.

2.1. PRESYNAPTIC DA FUNCTION

[^{18}F]fluorodopa (FDOPA) is the radioligand used mostly for quantifying the nigrostriatal dopaminergic dysfunction in PD. This tracer measures the rate of FDOPA decarboxylation and subsequent storage in the dopaminergic nerve terminals. FDOPA uptake is estimated by using a striatal/occipital ratio (SOR: striatal/occipital – 1) *(9,10)* or influx constant (K_i) computed from dynamic data *(11–13)*.

It has been established that FDOPA uptake is reduced in the posterior putamen but relatively preserved in the caudate nucleus and anterior putamen in early stages of PD *(13–16)*. FDOPA uptake is decreased in both putamen (60%) and caudate (40%) in patients with advanced PD (Fig. 1). This has been valuable in discriminating PD from normal controls and also in early differential diagnosis of PD from other atypical parkinsonisms. More importantly, FDOPA uptake indices in putamen and caudate have been consistently shown to correlate negatively with the severity of motor symptoms in PD. FDOPA binding is a suitable marker to follow disease progression *(17,18)*.

SPM analysis of K_i maps has detected significant reductions of FDOPA uptake not only in the bilateral striata and substantia nigra but also in the midbrain and pons with increasing severity of PD *(15)*. K_i in the caudate has a negative association with performance in the attention-demanding interference task. K_i in the frontal cortex has a positive correlation with performance in the digit span, verbal fluency, and verbal immediate-recall tests. This suggests that dysfunction of the DA system has an impact on the cognitive impairment of patients with PD.

We have recently demonstrated that the simple uptake index SOR is statistically comparable with influx constant K_i in revealing striatal DA deficiency in PD and predicting clinical correlation with objective measures of disease severity *(10)*. The measurement of K_i takes a longer time and may not be tolerated by patients with severe movement disorders. However, SOR can be measured relatively easily in a short period of time and allows us to shorten and further simplify the protocol for longitudinal clinical trials.

The accuracy of FDOPA uptake measured in PET studies is confounded by two factors: (1) transport of metabolites across the blood–brain barriers may affect quantification of FDOPA binding *(22,23)*. This potentially makes FDOPA binding insensitive to age-related decline in presynaptic dopaminergic function. Therefore, striatal FDOPA uptake indices may overestimate the number of dopaminergic nerve terminals.

Imaging of dopamine transporter (DAT) is another way for probing the impaired nigrostriatal dopaminergic system in PD. DAT is expressed on dopaminergic nigral terminals, and quantification of striatal DAT appears to be directly related to the extent of nigral cell degeneration *(24)*. This has received more attention in recent years as radiotracers that bind to the striatal DAT have been successfully developed for both PET and SPECT imaging. The most common agents are the cocaine analogs, such as (^{123}I)β carbomethoxy-iodophenyl tropane (CIT) and (^{18}F)FP-βCIT *(25, 26)*, as well as [β]-carbomethoxy-3[β]-(4-fluorophenyl) tropane (CFT) labeled with (^{18}F) and (^{11}C) *(27–29)*. DAT binding is estimated by an uptake ratio or distribution volume ratio between the striatum and a reference tissue such as cerebellum.

A number of studies have demonstrated DAT binding ligands as effective markers of nigrostriatal dopaminergic degeneration in aging and parkinsonism *(30,31)*. Striatal DAT binding indices decline by 6% per decade in normal controls, reflecting cell loss associated with aging process. DAT imaging reliably differentiates PD subjects from normal volunteers and other PD-like syndromes, and the degree of striatal binding correlates inversely with clinical measures of PD severity *(27,28,31)*. DAT binding in the orbitofrontal cortex also is significantly lower in nondemented patients with early PD and correlates negatively with scores for mentation and depression *(29)*. The reduction in mesocortical or mesolimbic function may contribute to the mental and behavioral impairment observed in PD.

DAT imaging reveals the same pattern of neuron degeneration in PD as FDOPA. However, a number of dual tracer experiments in the same sets of subjects have shown that reductions in DAT binding are larger than those in FDOPA *(7,8,16,23)*. This confirms the absence of upregulation in DAT imaging as compared with FDOPA imaging. It also has been observed that PET imaging with DAT can provide images with better resolution and higher signal-to-noise ratio over FDOPA *(9,11)*. We have improved the ability of DAT imaging to track the time course of disease onset and evolution by introducing a brain mapping strategy *(32)*. DAT may be a more sensitive marker for assessing nigrostriatal cell loss in parkinsonism and its restoration by dopaminergic therapies.

2.2. POSTSYNAPTIC DA FUNCTION

Altered postsynaptic DA function in PD has also been investigated with a number of PET radioligands. It is known that dopaminergic transmission is facilitated mainly by D1 and D2 receptors in the striatum. Striatal D1 receptor binding is usually measured with [^{11}C]SCH23390 and remains normal in early PD patients who are not on drug therapy *(33)*. However, D1 binding seems to be reduced by 10% if a patient receives treatment with levodopa for several years *(34)*.

Striatal D2 receptor binding is most often estimated using [^{11}C]raclopride (RAC) and is mildly elevated in early phases of PD, untreated with antiparkinsonian medications *(34–36)*. The elevation is particularly pronounced in the putamen. This upregulation is reversed with DA replacement therapy at more advanced stage *(37)*. Striatal D2 binding remains unchanged in the putamen but reduced by 16% in the caudate after continued medical treatment *(34)*.

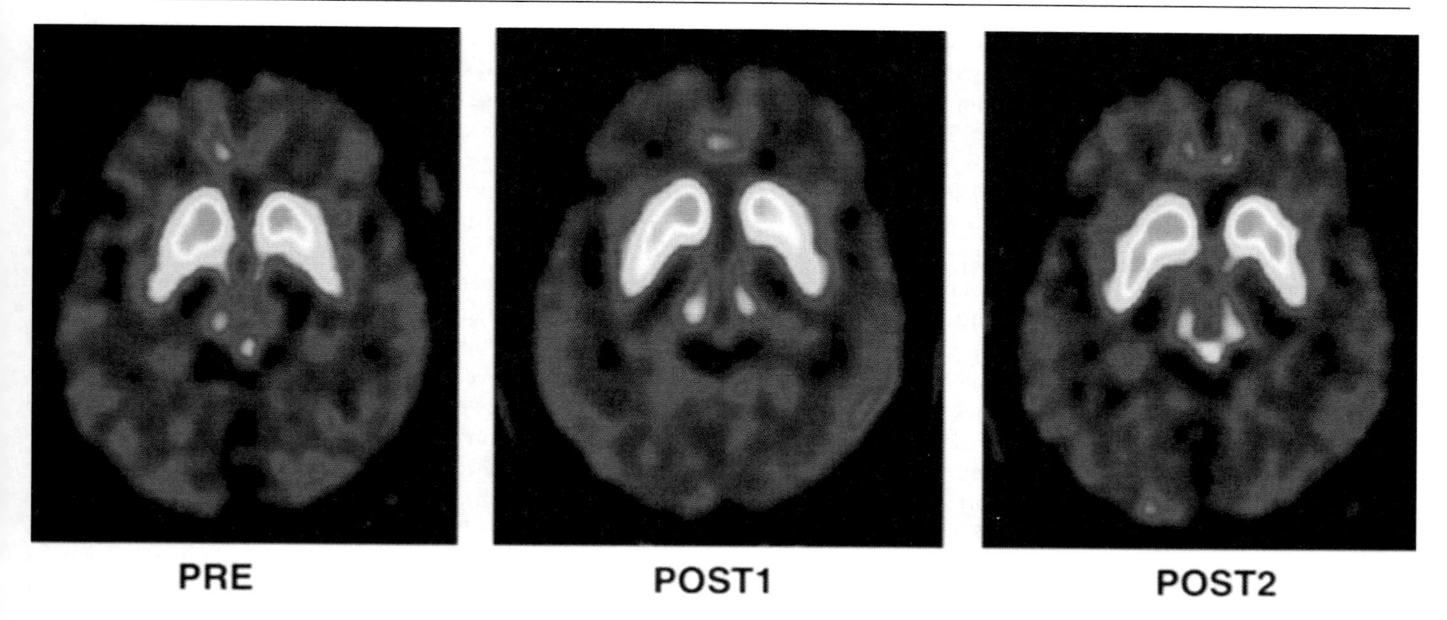

Fig. 1. FDOPA images of one patient with advanced PD scanned at baseline (PRE) and at 1 (POST1) and 2 (POST2) yr after embryonic DA cell implantation in the bilateral putamen. Both postoperative scans show gradual increases in FDOPA uptake, particularly in the putamen, where DA loss was the largest preoperatively. *See* color version on Companion CD.

Because RAC has a low affinity to D2 receptors, PET imaging with this tracer has been widely used to measure DA release under interventions that modulate dopaminergic systems. The amount of release is estimated by the percentage reduction in RAC binding as a result of the competition between external stimulation and endogenous DA. It has been reported that striatal RAC binding relative to baseline is reduced in PD after acute levodopa administration, most notably in the posterior putamen (18%), followed by the anterior putamen (12%), and the caudate nucleus (6%) *(38)*. The magnitude of reduction is correlated with the drug-free disability of motor function. This gradient of DA release is consistent with the topographic pattern of DA lesions portrayed by presynaptic PET imaging markers. Another study has recently demonstrated that pharmacological challenge with methamphetamine produces significantly reduced DA release in striatum, but normal levels of prefrontal DA release in advanced PD compared with the normal controls *(39)*. Putamen DA release in PD is correlated with residual DA storage capacity measured by FDOPA uptake.

Release of endogenous DA can also be induced with behavior modulation tasks. Compared with a resting baseline, RAC binding is reduced significantly throughout the striatum in healthy volunteers during the execution of a sequential finger movement *(40)*. PD patients show smaller reductions in RAC binding with the same motor paradigm in striatal areas less affected by the disease process. This is consistent with the notion that deficiency in synaptic DA is contributing to the impaired performance of sequential movement in PD.

In addition, RAC is also useful in investigating the mechanism of motor fluctuation in PD patients treated for several years with levodopa *(41)*. It is reported that 1 h after a dose of orally administered levodopa, the estimated increase in the synaptic level of DA is three times higher in fluctuators than in stable responders receiving the same drug regimen. Increased level of synaptic DA is maintained only in stable responders after 4 h. The rapid increase in synaptic DA observed in fluctuators suggests that increased DA turnover might play a role in levodopa-related motor complications.

2.3. CEREBRAL BLOOD FLOW (CBF) AND METABOLIC IMAGING

Lost dopaminergic projections in PD cause profound changes in resting and activated brain hemodynamics and metabolism by disturbing the normal function of striato–pallido–thalamocortico–striatal pathways. PET imaging with $[^{15}O]H_2O$ and $[^{18}F]$fluorodeoxyglucose (FDG) have been used to measure regional CBF (rCBF) and glucose metabolism in PD. However, general results show that rCBF and glucose metabolism at resting states are either normal or increased in parts of the basal ganglia and decreased in the selected cortical areas contralateral to the more affected limbs *(42–44)*. Bilaterally affected PD patients show more diffuse abnormality in the cortex.

$[^{15}O]H_2O$/PET imaging has been mainly used in brain activation studies to investigate impaired motor function in PD under different stimuli as compared with normal volunteers. Brain mapping algorithms such as statistical parametric mapping (SPM) are used to localize significant changes in rCBF relative to baseline. For example, reduced activation is observed in PD patients in the contralateral putamen and the anterior cingulate, supplementary motor area (SMA), and dorsolateral prefrontal cortex (DLPFC) when performing simple motor tasks *(45–47)*. These cortical areas receive their inputs mainly from the striatum and the impaired activations in SMA and DLPFC may explain the difficulty the PD patients have in initiating and executing this kind of movement. In contrast, increased activation is observed in the precuneus and premotor and parietal cortices when PD patients perform long sequential finger movements *(48)*. More complex movements lead to greater activations in the premotor and parietal cortices with additional increases in the anterior cingulate and SMA. This means that cortical areas receiving less input from the striatum are recruited in PD patients to activate SMA and prefrontal cortex. Increased cortical activity from the other areas is a good example of brain plasticity in neurological disorders.

Brain network analysis is a more powerful strategy to map abnormal topographic organizations of cerebral function associated with PD. This is based on the use of principal component analysis to identify functional connectivity among a set of brain areas *(49,50)*. These areas form a distinct brain network in which local functional indices vary simultaneously as a single entity. Using FDG/PET data, we have identified a PD-related metabolic covariance pattern (PDRP) marked by hypermetabolism in the putamen and thalamus, cerebellum, and pons and hypometabolism in the cortical motor cortex, such as premotor, pre-SMA, and posterior parietal areas *(51–53)*. A simple index can be computed prospectively to measure the expression of PDRP network activity in each individual brain. Network activity is positively correlated with disease severity ratings in PD and is highly sensitive in discriminating PD from other atypical parkinsonism.

Network analysis has also been used in [^{15}O]H_2O/PET data to generate unique topographic patterns of rCBF activations associated with behavior components of explicit motor sequence learning in the early stages of PD *(54,55)*. The covariance pattern related to acquisition performance exhibits increased activation in the left DLPF, ventral prefrontal, and rostral premotor areas but not in the striatum (activated in normal subjects). The covariance pattern related to retrieval performance shows increased activation in the right DLPF and bilaterally in the precuneus, premotor, and posterior parietal cortices. These results show that in early stages of PD, networks for sequence learning incite additional cortical activations to compensate for striatal dysfunction.

In summary, PET imaging studies with a variety of radiotracers have revealed specific and characteristic functional abnormalities underlying PD. Presynaptic dopaminergic function and cerebral topographic organizations are altered not only in the striatum but also in certain areas of the cortex. Striatal indices of impaired presynaptic DA function and activity of an abnormal metabolic brain network are correlated with clinical ratings of motor dysfunction in PD. PET imaging affords an objective way to evaluate disease progression and assess the efficacy of experimental therapeutics.

3. MEDICAL AND SURGICAL TREATMENTS OF PD

PD is one of the primary neurodegenerative disorders associated with a progressive loss of nigrostriatal DA neurons and a reduction in striatal DA. This leads to impaired function on a set of relay stations as part of a cortical-basal ganglia motor loop. Pharmacological treatment is always the first course of action at the early phases of the illness. Medication therapy provides adequate control of symptoms over several years by restoring presynaptic DA production and release in the brain. However, long-term treatment is limited by increasing disability and the development of motor fluctuations and abnormal involuntary movement known as dyskinesia *(56)*. It has been suspected that the dyskinesias may result from erratic increase in storage and release of DA and its subsequent interaction with postsynaptic DA receptors.

A number of surgical interventions are viable options to give further symptomatic relief and minimize any drug-induced complications. This is intended to protect or restore dopaminergic transmission by repairing dysfunctional basal ganglia circuits. Different neurosurgical approaches have been extensively discussed elsewhere by other investigators *(57,58)*. Basically there are three general types. The traditional method is to make an ablative lesion to disrupt part of the circuit. However, this method is irreversible postoperatively. The second method is deep brain stimulation (DBS) where an electrode is implanted in the central part of the brain to electronically modulate the circuit. This method has received more attention because it is reversible and adjustable postoperatively *(59)*. Ablative lesion and DBS have been performed at different parts of the thalamus, pallidum, and subthalamus to compensate for the biochemical effect of DA deficiency. The third method is brain restorative therapies such as nerve growth factor infusion or DA cell transplantation along the nigrostriatal pathway. These techniques all use the principle of stereotactic functional neurosurgery to determine the optimal location of the target using image guidance and microelectrode recording.

PET imaging markers described above provide important functional basis for introducing effective interventions and directly assessing their therapeutic efficacy. Imaging evaluation is absolutely necessary because clinical observation is most likely to be insensitive to incremental improvement in the brain function. This can be performed by measuring longitudinal changes in functional indices in PD patients before and after the treatment. The patients should be off dopaminertic medications for at least 12 h before PET scanning to minimize any confounding effects.

In recent years, many imaging studies have demonstrated an association between relative regional changes in brain function induced by treatment and corresponding clinical performance. It has been reported that impaired rCBF activations in regions involved in simple motor tasks can be restored by levodopa infusion and DBS of the internal globus pallidus (GPi) *(60,61)*. UPDRS motor ratings are improved by more than 34% and rCBF increases correlate with independent measures of enhanced motor behavior. Both therapies also establish specific but similar relationships in motor sequence learning *(55,62)*. Other brain-activation experiments have shown that stimulation of subthalamic nucleus significantly alleviates principal motor symptoms in PD *(63–65)*. However, this intervention seems to cause some degree of cognitive impairment in the patients.

In addition, FDG/PET imaging has also been used to examine the effect of these treatments on metabolic substrates. We have shown that subthalamic lesion and GPi DBS have respectively reduced metabolism in several relay stations of the basal ganglia motor circuit that are overactive in PD and increased metabolism in the premotor area and the cerebellum *(66–68)*. Treatment-induced metabolic changes are associated with clinical improvement given by UPDRS motor ratings.

An alternative way for evaluating the surgical outcome of cerebral brain function is the use of neural network analysis. It is based on an algorithm to calculate the global expression of a PD-related topographic pattern associated with abnormal metabolism. For instance, we have consistently shown that the network activity of PDRP is decreased in PD patients after effective treatments: subthalamotomy, GPi-DBS, and levodopa infusion *(66–69)*. The interval changes in network activity mediated by these interventions are all correlated significantly with the improvement in clinical outcome.

However, the restorative therapies try to correct the biochemical defect in PD by promoting the survival of host neurons in the nigrostriatal circuit or replacing the lost dopaminergic neurons. Glial cell line-derived neurotrophic factor (GDNF) is a potent neurotrophic factor with restorative effects in a wide variety of rodent and primate models of PD. In a recent study, GDNF has been delivered directly into the putamen of 5 PD patients in a safety trial *(70)*. After 1 yr, there is a 40% improvement in the off-medication motor UPDRS and a 60% improvement in the activities of daily living score. Medication-induced dyskinesias are reduced by 64% and not observed in PD patients off medication during chronic GDNF delivery. PET scans show a significant 28% increase in putamen FDOPA uptake after 18 mo. This suggests a direct effect of GDNF on DA function and supports it as a viable treatment for PD.

4. DA NEURON TRANSPLANTATION:

Transplantation of embryonic DA neurons into the nigrostriatal pathway is a fundamental way to treat patients with advanced PD that is complicated by motor fluctuations and dyskinesias. It is assumed that healthy DA cells can reinnervate the host neurons to functionally compensate the biochemical abnormalities of PD but also to arrest its progression. In animal models of PD, fetal nigral transplants have been shown to survive grafting into the striatum, provide extensive striatal reinnervation, and improve motor function.

The principle and surgical strategy of fetal cell implantation in humans have been described in detail *(71,72)*. Briefly, fetal grafts derived from human embryonic mesencephalic tissue are stereotactically implanted into the striatum of patients with PD. The tissue is harvested 6 to 9 wk after conception, and the transplantation typically requires three to four donors per side. Cyclosporine may be administered for a period of time to provide immune suppression. Clinical outcome is evaluated by standardized ratings during off states at baseline and at regular postoperative intervals following transplantation. Functional imaging provides a valuable adjunct to clinical evaluation when assessing the efficacy of this treatment.

4.1. IMAGING WITH FDOPA

Animal models of parkinsonism induced with a neurotoxin 1-methyl-4-phenyl-1,2,3,6-tetrahydropyridine (MPTP) have played an important role in the development of cell transplantation surgery *(73,74)*. This has been performed mainly in primates and mice to provide controlled striatal lesions to test various aspects of transplantation strategy. The functional restoration of the DA innervation of striatum has been investigated in MPTP-lesioned Gottingen minipigs after grafting of fetal pig mesencephalic neurons *(75)*. Pigs received bilateral grafts to the striatum of tissue blocks harvested from fetal pig mesencephalon with and without immunosuppressive treatment after grafting. Neurons marked by tyrosine hydroxylase (TH) were counted in the grafts by stereological methods.

The MPTP persistently reduced the relative FDOPA activity in striatum by 60%. Grafting restored the rate of FDOPA decarboxylation and normalized the performance in motor function at 3 and 6 mo after surgery. The biochemical and functional recovery was associated with survival of about 100,000 TH-positive graft neurons in each hemisphere at the end of 6 mo. Immunosuppression did not induce a greater recovery of FD-OPA uptake or increase the number of TH-positive graft neurons or the volumes of the grafts. Pig ventral mesencephalic allografts can restore functional DA innervation in adult MPTP-treated minipigs, in agreement with other animal experiments and an early study in two PD patients induced by MPTP *(76)*.

Since early 1990s, many studies in PD patients have been performed to evaluate the safety and efficacy of fetal nigral transplantation into the striatum. This has evolved from unilateral transplantation *(77–80)* to bilateral transplantation *(81–83)* for a more complete functional recovery. Patients received transplants in caudate and putamen and exhibited significant clinical benefits along with increased FDOPA uptake in the grafted areas. Doses of medications often were reduced, resulting in the resolution of drug-related complications. One serial study followed six patients for 24 mo after bilateral fetal nigral transplantation in the putamen *(84,85)*. Activities of daily living, motor, and total UPDRS scores during the off state improved significantly compared with baseline. Mean total UPDRS off score improved 32%. Mean putamenal FDOPA uptake on PET image increased significantly at 6 and 12 mo relative to baseline. This increase correlated with clinical improvement as demonstrated in an earlier study with unilateral grafts *(78)*. Two patients died 18 mo after transplantation from causes unrelated to the surgery. Histopathological analysis of their brains confirmed survival of TH immunoreactive cells and abundant reinnervation of the putamen *(86)*.

Five parkinsonian patients were transplanted bilaterally into the putamen and caudate *(87)*. To increase graft survival, the lipid peroxidation inhibitor tirilazad mesylate was administered to the tissue before implantation and intravenously to the patients for 3 d thereafter. During the second postoperative year, the mean daily levodopa dose was reduced by 54% and the mean UPDRS motor score was reduced by 40%. At 10–23 mo after grafting, a mean 60% increase of FDOPA uptake in the putamen, and 24% increase in the caudate were observed compared with preoperative values. The pattern of motor recovery did not differ from other previous studies with putamen grafts alone. The amount of mesencephalic tissue implanted in each putamen and caudate was 42% and 50% lower, respectively, compared with previously transplanted patients. Despite this reduction in grafted tissue, the magnitudes of symptomatic relief and graft survival were very similar. These results suggest that tirilazad mesylate may improve survival of grafted DA neurons in patients.

Most previous clinical trials on DA cell transplantation have used a small number of patients with an open-label design that may suffer from a potential placebo effect or investigator bias. The first double-blind, placebo-controlled surgical trial of human embryonic DA cell implantation in PD has been described *(88)*. Forty patients with severe PD (mean age 57 yr; mean duration 14 yr) were randomly assigned to either transplantation or placebo surgery. In the transplant recipients, cultured mesencephalic tissue from four embryos was implanted into the putamen bilaterally without immunosuppression. The sham group had a mock surgery but without fetal cell implantation. Among younger patients (age ≤60 yr), standardized tests revealed significant improvement in total motor UPDRS (30%, $p < 0.01$) in the transplantation group as compared with the sham-surgery group. There was no significant improvement in older patients in the transplantation group, nor was any change in the sham group regardless of age. Fiber outgrowth from the transplanted neurons was detected at postmortem examination in one transplant recipient who died from unrelated causes during the first postoperative year.

One of the concerns is whether this trial has any effect on the cognitive status of transplant recipients. Analysis of detailed neuropsychological data in our transplantation cohort before and 1 y after surgery showed that postsurgical change in cognitive function was not significantly different between real and sham surgery groups *(89)*. Most neuropsychological measures in both groups remained unchanged at the 1-yr follow-up. Hence, embryonic DA neurons can be implanted safely into the putamen bilaterally without impairing cognition in patients with PD.

Using PET imaging with FDOPA, we first evaluated the recovery of nigrostriatal dopaminergic function in 39 surviving patients scanned before and 1 yr after surgery *(90)*. Investigators who were blinded to treatment status and clinical outcome analyzed the images. We also determined the effects of age on the interval changes in FDOPA. After unblinding, we detected a significant increase in FDOPA uptake in the bilateral putamen of the group receiving implants compared with the placebo surgery patients (40%, $p < 0.01$). Increases in putamen FDOPA uptake were similar in both younger and older (age >60 yr) transplant recipients. Significant reductions in putamen uptake were evident in younger placebo-operated patients (–6.5%, $p < 0.05$) but not in their older counterparts. This reduction reflected an ongoing disease process over time. PET changes were significantly correlated with clinical outcome only in the younger patient subgroup ($r = 0.58, p < 0.01$). Our results suggest that patient age does not affect graft viability or development in the first postoperative year. However, host age may alter the time course of the downstream functional changes that are necessary for appreciable clinical benefit.

The survival and function of the graft were also evaluated at the second postoperative year in our transplant group (Table 1). UPDRS scores at the first and second years improved by 20% and 30%, respectively, relative to baseline (Fig. 2). Concurrently, putamen FDOPA uptake increased significantly by 40% at the first year and 44% at the second year (Fig. 3). Both measures seemed to show slight improvements at the second year compared to the first year although the differences did not reach significance (Table 1).

To complement the ROI results described above, we also mapped the interval changes in FDOPA uptake over time using SPM (Fig. 4; Table 2). Comparing the first postoperative year to preoperative baseline, we localized the relative increase to the posterior putamen (left: $x = -28$, $y = -2$, $z = 2$ mm, $Z_{max} = 4.7$; right $x = 30$, $y = -8$, $z = 2$ mm, $Z_{max} = 4.6$; $p < 0.001$). The same patterns of FDOPA increase were seen comparing the second year to the baseline. Relative to the first year, SPM was able to reveal further increase at the second year in the most posterior portion of the ventral putamen, with a lower statistical significance ($p < 0.01$). This additional engraftment might be responsible for the continued clinical improvement at two years post implantation. No significant differences were seen in the nonimplanted caudate in all of these comparisons.

In summary, the therapeutical efficacy of fetal cell transplantation in PD has been proven consistently despite the absence of immunosuppressive treatment and ongoing disease progression. Current clinical trials with FDOPA PET imaging suggest that the apparent symptomatic relief is associated with the establishment of synaptic connections between the grafted neurons and the host brain. The double-blinded trial shows absence of a measurable placebo effect based on UPDRS changes at 1 yr after the surgery. Although young patients seem to show functional recovery in the first postoperative year, it remains to be seen whether the older transplant recipients will improve clinically during later follow-up. In accordance with previous reports by other investigators, some of patients in our transplant cohort have also demonstrated sustained graft survival and function beyond 4 to 6 yr after transplantation.

4.2. IMAGING OF DA TRANSPORTER

PET imaging with DAT radioligands may be used as molecular markers to assess striatal presynaptic function in PD patients after DA cell transplantation. This has been demonstrated in a unilateral rat model of neurotransplantation with $[^{11}C]$CFT and microPET scanner *(91)*. Parkinsonian lesions were created by injecting 6-hydroxydopamine. DAT binding in the lesioned striatum was reduced to 15% to 35% of the unoperated side. After grafting with non-DA cells from dorsal mesencephalon, the binding remained to levels observed before transplantation and rats had no behavioral recovery. In contrast, after DA neuronal transplantation, behavioral recovery occurred only after the specific DAT binding had increased to 75% to 85% of the intact side.

DAT radioligand has also been used to investigate the role of implanting DA neurons in the clinical manifestation of levodopa-induced dyskinesia in a rat model of PD *(92)*. Dyskinesia was induced gradually during the course of 1 mo with a low dose of levodopa in 6-hydroxydopamine-treated rats who subsequently received embryonic ventral mesencephalic tissue in the striatum. Dyskinesia improved significantly in the grafted rats after another 3 mo of continued levodopa treatment. The severity of residual dyskinesia correlated negatively with the density of DA nerve terminals in the striatum and this complication became resolved when DAT binding was more than 10–20% of normal.

At present, there have been no systematic human trials of DA cell transplantation using DAT imaging. A human study has compared the striatal uptake of DAT with FDOPA in six

Table 1
Mean Values and Interval Changes in Clinical Outcome and Putamen FDOPA Uptake

Mean Values			*Interval Changes (%)*		
PRE	*POST1*	*POST2*	*POST1/PRE*	*POST2/PRE*	*POST2/POST1*
Putamen FDOPA Uptake			Δ Putamen FDOPA Uptake		
0.53 ± 0.02	0.72 ± 0.04	0.73 ± 0.05	40.3 ± 10.0*b*	43.6 ± 11.5*b*	2.6 ± 4.0
Total Motor UPDRS			Δ Total Motor UPDRS		
60.9 ± 5.3	49.2 ± 5.9	42.2 ± 4.5	−19.6 ± 6.8*a*	−28.8 ± 5.3*c*	−2.6 ± 12.4

a$p > 0.01$
b$p < 0.001$
c$p < 0.0001$

Imaging and clinical data (mean ± standard error) are measured at baseline (PRE) and 1 (POST1) and 2 (POST2) years following embryonic dopamine cell implantation in 19 transplant recipients. Δ refers to the percent change between two conditions (A − B) × 100/B with the statistical significance calculated from paired t-test.

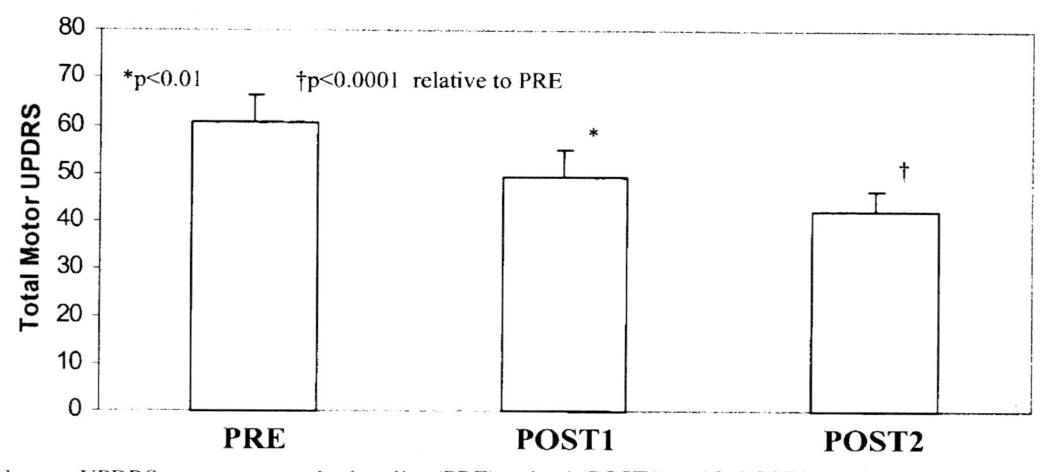

Fig. 2. Total motor UPDRS scores measured at baseline (PRE) and at 1 (POST1) and 2 (POST2) yr after embryonic DA cell implantation. Data are mean from 19 transplant recipients taken 12 h after antiparkinsonian medications. The error bars represent the mean standard error.

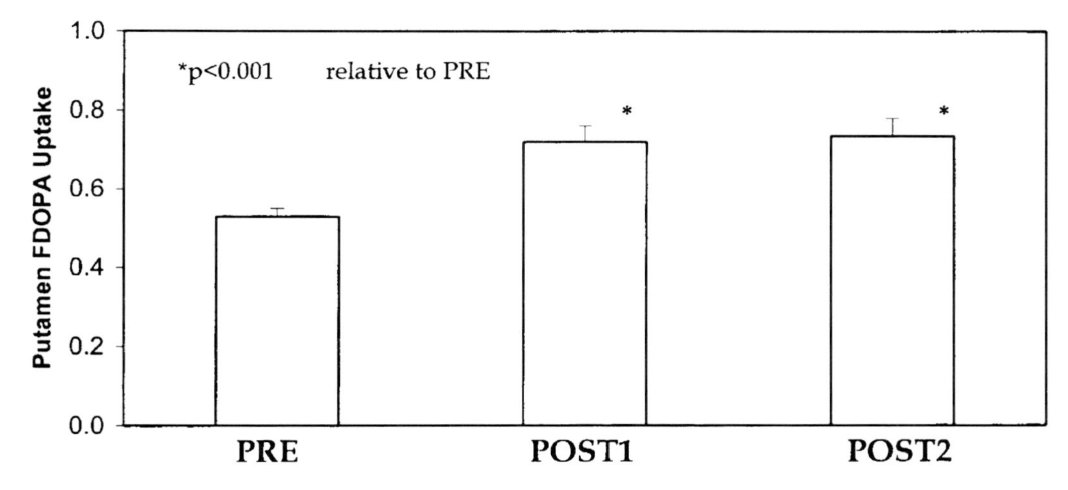

Fig. 3. Putamen FDOPA update measured at baseline (PRE) and at 1 (POST1) and 2 (POST2) yr after embryonic DA cell implantation. Data are mean specific binding from 19 transplant recipients calculated at 95 min after tracer injection. The error bars represent the mean standard error.

patients with PD grafted with fetal mesencephalic cells *(93)*. There was no change in DAT binding in the grafted putamen, despite a significant increase of FDOPA uptake. Clinical and FDOPA uptake changes after the grafts were correlated with the amount of ventral mesencephalic tissue used for implantation. The clinical benefit induced by the graft seems to be more related to increased dopaminergic activity than improved dopaminergic innervation in the host striatum. It is important to

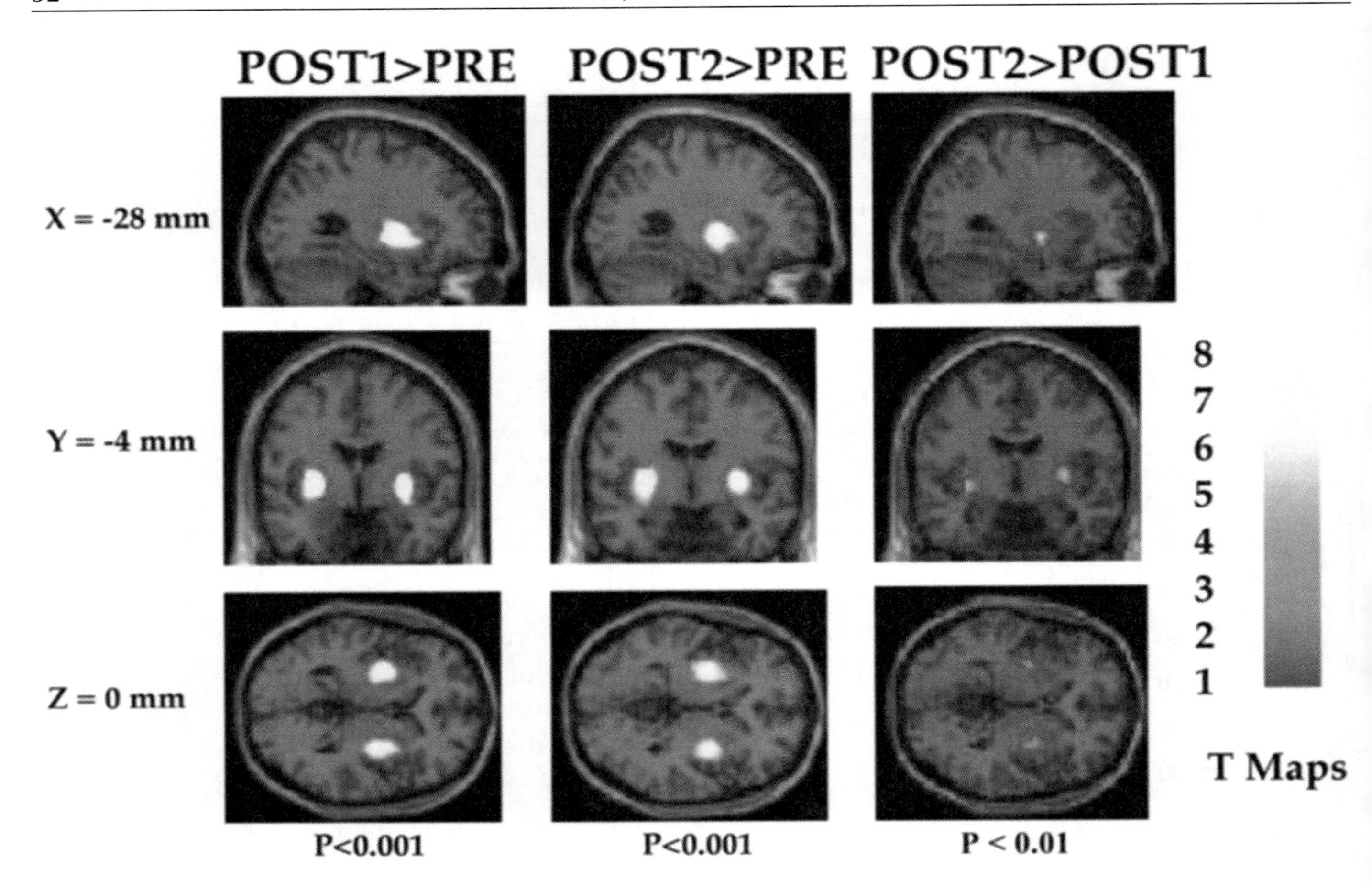

Fig. 4. Mapping of FDOPA uptake at baseline (PRE) and at 1 (POST1) and 2 (POST2) yr after embryonic DA cell implantation. Data from 19 transplant recipients are compared by paired tests using SPM. Transplantation results in significantly increased FDOPA uptake in the bilateral posterior putamen at both postoperative time points. Comparison of POST2 and POST1 shows significant bilateral increase in the most posterior part of ventral putamen, reflecting continued innervation 2 yr after implantation. *See* color version on Companion CD.

Table 2
Statistical Characteristics of Brain Regions With Significant Increases in FDOPA Uptake

	POST1 > PRE				*POST2 > PRE*				*POST2 > POST1*			
Regions	Z_{max}	X	Y	Z	Z_{max}	X	Y	Z	Z_{max}	X	Y	Z
Left Putmen	4.7	–28	–2	2	4.9	–28	–4	0	3.3	–30	–6	–4
Right Putamen	4.6	30	–8	2	5.2	32	–2	2	3.4	32	–6	4

Data represent Z scores and Talairach coordinates (mm) of peaks within the significant clusters detected by statistical parametric mapping of FDOPA binding from PET images (Fig. 4).

note that this is only a preliminary study and has not been replicated. More work is necessary to determine whether FDOPA is the optimal tracer to evaluate grafted PD patients.

4.3. IMAGING OF POSTSYNAPTIC DA RECEPTORS

D1 and D2 receptor imaging may also provide a dopaminergic molecular basis of functional recovery in the nigrostriatal system after neural transplantation. An increase in the number of functional D1 and D2 DA receptors is likely to indicate a more extensive integration of the graft with the host brain. PET with [^{11}C]SCH 23390 and RAC has been used to examine the effect of donor stage on the survival and function of embryonic striatal grafts in the adult rat brain *(94)*. The grafts from the youngest donors showed a significant increase in D1 and D2 receptor binding when compared with the lesion-alone group. This increase was correlated with the improved motor function in the rats. Striatal grafts from younger donors produced greater behavioral recovery than grafts prepared from older embryos due to the increased proportion of viable tissue within the grafts.

PET with RAC is capable of measuring synaptic DA release from embryonic nigral transplants. This was performed in a PD patient demonstrating the sustained and marked clinical improvement, as well as gradual increase of FDOPA uptake to normal during a period of 10 yr after grafting in the unilateral putamen *(95)*. RAC binding to DA D2 receptors was measured with saline or methamphetamine infusion. Binding at baseline was normal in the grafted putamen but upregulated in the nongrafted putamen and caudate. Drug-induced binding reduction was also restored to the normal level in the grafted putamen. These results suggest that graft has successfully normalized DA storage and D2 receptor occupancy in the denervated striatum.

4.4. IMAGING WITH FUNCTIONAL BRAIN ACTIVATION

$H_2{}^{15}O$ with PET can be a useful tool to study the recovery of striatocortical functional systems in patients treated with fetal cell implantation. It is reported that recovery of movement-related cortical function has been delayed in PD after striatal dopaminergic grafting *(96)*. The cortical activation was still impaired at 6.5 mo after transplantation, although motor symptoms and mean striatal DA storage capacity had significantly improved. At 18 mo after surgery, there was further significant clinical improvement without any more increase in striatal FDOPA uptake. The surgery significantly improved rostral supplementary motor and dorsal prefrontal cortical activation during performance of joystick movements, in agreement with the results of subthalamic nucleus DBS *(63,65)*. These data suggest that it is not enough for the graft to simply deliver DA and that functional integration of the grafted neurons within the host brain is necessary to produce substantial clinical recovery in PD.

Presently, this type of study can also be performed using functional magnetic resonance imaging to measure alternation in regional cerebral blood volume (rCBV). rCBV is an indicator of neuronal metabolism as it is linked to the cerebral oxygen consumption varying in response to motor tasks. Two PD patients with clinically excellent outcome after transplantation were examined this way with a repetitive motor task *(97)*. Activation was recorded consistently in the part of the putamen receiving the graft, suggesting that the formation of neural connections between host and transplanted tissues as the same paradigm also activates the putamen in normal volunteers.

It is known that DA release in response to amphetamine challenge induces a significant increase in rCBV in specific parts of the corticostriatal circuitry. One double-blind study in a unilateral rat model of PD reported complete absence of rCBV activation to amphetamine in striatum and cortex ipsilateral to the lesion *(98)*. Cell graft in the lesioned striatum resulted in appreciable amphetamine-induced activation in the striatum and sensorimotor cortex, similar in magnitudes to those in the contralateral intact hemisphere. No response was evident in control rats with sham surgery. This study demonstrates that behavioral restoration in animal models parallels observation from functional MRI data.

4.5. IMAGING WITH METABOLIC MARKERS

FDG/PET imaging can be very useful in mapping the metabolic brain response to neuronal cell implantation in the human trials of PD patients. This has been proven repeatedly with other neurosurgical therapies summarized earlier. In the first open-label human trial for stroke *(99)*, 12 patients with chronic basal ganglia infarction and motor impairment were investigated before and 6 and 12 mo after stereotactic implantation of cultured human neuronal cells. Alterations in glucose metabolism in the stroke area at 6 and 12 mo after implantation correlated positively with motor performance measures. The results suggest improved local cellular function or engraftment of implanted cells in some of these patients.

5. POTENTIAL COMPLICATIONS AFTER TRANSPLANTATION

Complications related to the intrastriatal transplantation of human embryonic mesencephalic tissue are usually mild and transient as reported by most of clinical trials with PD. All patients have bilateral dyskinesias before grafting that are greatly decreased a few months after the surgery because of concurrent reduction in dopaminergic drugs. However, as the number of graft recipients increases, dyskinesias in the absence of or with only minimal amounts of dopaminergic medication have been reported after the surgery. One study examined five patients during the course of 1–3 yr after unilateral implantation in the caudate and putamen *(100)*. There was a moderate increase in FDOPA uptake in the grafted putamen along with a different degree of bilateral improvement in motor skills. Delayed asymmetrical dyskinesias were observed in three patients on the side contralateral to the graft. It was speculated that this might reflect increased presynaptic storage and release of DA induced by the graft in the host striatum.

One recent study reported that mild to moderate dyskinesias increased during postoperative off-phases in 14 patients who were followed long-term after grafting *(101)*. Varied implantation targets had been employed in this group of PD patients. Dyskinesia severity was not related to the magnitude of graft-derived dopaminergic innervation measured by FDOPA uptake. This retrospective study seems to suggest that dyskinesias are not associated with the excessive growth of grafted dopaminergic neurons. However, the experimental design in that report is somewhat limited as a result of the large variability in grafting strategy and dyskinesia symptoms, as well as the absence of controls from the same transplant cohort who were free of this complication.

Persistent dyskinesias have been observed during off states in 5 of 33 patients who received fetal transplants as part of the double-blind trial *(88)*. This phenomenon took place during the second postoperative year after marked improvement of clinical signs in the first year. Using FDOPA PET, we sought to determine whether this complication resulted from specific alterations in DA function after DA cell implantation *(102)*. Caudate and putamen FDOPA uptake in five dyskinetic patients were compared with those from 12 age- and disease duration-matched transplant recipients who were free of dyskinesia. We found that FDOPA uptake did not differ at baseline between both groups. However, putamen FDOPA uptake was significantly increased ($p < 0.005$) in dyskinetic transplant recipients at 12 and 24 mo after transplantation. SPM analysis revealed that hyperactive areas were predominantly localized to two zones within the left putamen. In addition to the posterodorsal zone in which a prominent reduction in FDOPA uptake was present at baseline, the dyskinetic group also a relative increase ventrally, in which preoperative dopaminergic input was relatively preserved. Postoperative FDOPA uptake did not reach supranormal values over the 24-mo period. These results suggest that unbalanced increases in dopaminergic function can complicate the outcome of neuronal transplantation for parkinsonism.

Although there appears to be a link between suboptimal recovery of presynaptic function and graft-induced dyskinesia, its exact mechanism remains unknown. The onset of dyskinesias is also likely to come from interactions with postsynaptic receptors. Although striatal D1 and D2 receptor binding did not change in drug-induced dyskinesia *(103)*, PET imaging with [^{11}C]diprenorphine indicated that binding to opioid receptors in the striatum and thalamus was reduced compared with nondyskinetic patients.

$H_2{}^{15}O$ PET studies of patients with focal limb dyskinesias showed that rCBF after oral levodopa were increased during dyskinesias in putamen, and motor, premotor and dorsal prefrontal cortices *(103)*. Dyskinesias might be associated with derangement of basal ganglia opioid transmission, resulting in overactivity of basal ganglia-frontal projections. Further PET imaging studies may help find better ways to treat graft-induced dyskinesia or improve the transplanting strategy to prevent the occurrence of this potential side effect.

6. NEW DEVELOPMENT AND PROSPECTIVE

DA cell transplantation is one of the most fundamental neuroprotective and restorative therapies to treat PD. However, a number of challenges still remain that limit its widespread clinical applications *(104)*, including the development of sustainable sources of DA tissue, introduction of neurotrophins, improved transplantation strategy, as well as comprehensive imaging methods to better evaluate the postoperative recovery in the brain function.

6.1. ALTERNATIVE SOURCES OF DA TISSUE

The major obstacle is that this cell replacement strategy needs large amounts of human fetal mesencephalic tissue to achieve appreciable and sustainable therapeutic effects. This may not be widely available because of ethic and practical considerations. There is a growing interest to find other sources of viable DA cells for use in humans. A postmortem study has presented histological evidence of fetal pig neural cell survival and growth over several months after transplantation into the striatum of a PD patient *(105)*. Pig neurons extended axons from the graft sites into the host brain. It has also been demonstrated that transplantation of embryonic porcine ventral mesencephalic tissue into PD patients is well tolerated and free of serious adverse events *(106)*. The tissue was deposited unilaterally in the caudate and putamen of 12 patients with cyclosporine immunosuppression. Despite a slight improvement of 20% in clinical UPDRS ratings, there was no change in FDOPA uptake at 1 yr after transplantation as compared to that at baseline. This is similar to early trials of human embryonic allografts that transplanted small amounts of tissue. The absence of increase in FDOPA uptake could also reflect low survival rates of DA neurons as a result of immunoreactions. Nevertheless, the surgery is safe with no evidence of transmission of porcine endogenous retrovirus measured from peripheral blood mononuclear cells.

Stem cells derived from human embryos may hold promise as a virtually unlimited source of self-renewing progenitors for transplantation. One study has described that transplanting small amounts of undifferentiated mouse embryonic stem cells unilaterally into the rat striatum leads to their proliferation into fully differentiated DA neurons *(98)*. These DA neurons caused gradual and sustained behavioral recovery of DA-mediated motor asymmetry. Parallel increase in DAT binding was seen in the grafted striatum (equal to 75–90% of the intact side) and correlated with the number of neurons counted at postmortem in the graft. In contrast, there was much less DAT activity in sham controls. These findings demonstrate that transplanted embryonic stem cells can develop spontaneously into DA neurons to restore cerebral function and behavior in an animal model of PD. Other types of substitute DA cells and DA neurons cultivated from human embryonic cells are currently under investigation and may become available in the near future *(74,107)*. A new therapy closely related to DA transplantation is the use of genetic manipulation to normalize DA production and delivery in the brain. This pioneering development has been actively tested in animal models of PD *(108,109)*. However, clinical trials of these cell and gene therapies in patients can be contemplated only after vigorous evaluations in experimental parkinsonism.

6.2. USE OF NERVE GROWTH FACTORS

The second obstacle is that DA cells have low survival rates in culture and after grafting in the brain. Neurotrophic growth factors such as GDNF have neuroprotective effects on dopaminergic neurons both in vitro and in vivo *(73)*. An animal study has been performed in which 6-hydroxydopamine lesions were created in the left medial forebrain bundle of rats *(110)*. Mesencephalic tissue was suspended in solutions with GDNF before transplantation into the left striatum. This treatment enhanced graft-induced compensation of amphetamine-stimulated rotations and recovery of striatal DAT binding of [^{11}C]RTI-121 measured by PET. It significantly prolonged graft survival as shown in post mortem analysis.

Exposure of human fetal nigral tissue to GDNF may enhance survival of stored dopaminergic cells and promote graft survival. A human study described that cells stored with GDNF had a 30% increase in survival time compared with those without GDNF *(111)*. Two PD patients received bilateral putaminal implants of fetal dopaminergic cells exposed to GDNF. FDOPA uptake in the putamen of these patients was more than doubled relative to baseline 12 mo after surgery. This substantial increase is much larger than those reported in previous human trials with DA neurons alone. Therefore, GDNF is not only a viable neuroprotective agent, but also effective in improving the survival and functional activity of mesencephalic grafts.

6.3. IMPROVEMENT IN TRANSPLANTATION STRATEGY

Current cell transplantation surgery in PD has been mainly on reinnervating the striatum. Evidence from preclinical data suggests that simultaneous intrastriatal and intranigral grafts may produce a more complete functional recovery. A recent study evaluated three patients who received implants of fetal mesencephalic tissue in putamen and substantia nigra bilaterally *(112)*. Clinical improvement was noted in total UPDRS along with an increase in the mean FDOPA uptake in the putamen and nigra 12 mo after surgery. It remains to be seen whether this double transplant strategy is better than others in clinical trials with more patients.

The overall goal of DA cell transplantation is to repair the nigrostriatal pathway and fully restore dopaminergic function in the striatocortical circuits. It has been suggested that optimal outcome of this intervention is age-dependent and may vary according to preoperative clinical status and biochemical deficit of the patient. This calls for a surgical strategy that is tailored to the time course of the disease in individual patients. PET imaging with presynaptic markers offers an objective criteria to select patients and optimize the design of transplantation strategy. This process might be facilitated by the use of a three-dimensional gradient map quantifying the absolute loss of striatal binding in PD relative to normal *(32)*. The map is superimposed on an MRI brain template so that DA cells could be delivered locally to achieve a more homogeneous reinnervation after grafting.

6.4. EVALUATION WITH MULTIPLE IMAGING MARKERS

To establish whether the transplantation fully restores the cerebral brain function in PD patients, it is necessary to evaluate its efficacy by multitracer PET imaging. Previous studies have documented both striatal and frontal reductions in presynaptic DA function and their relationships with motor and cognitive impairment *(15,29)*. The capacity to release endogenous DA in these areas has also been proven by measuring reduction in postsynaptic D2 receptor binding induced by pharmacological challenges *(39)*. The successful DA transplantation should maximally increase and sustain the levels of these parameters of dopaminergic function postoperatively.

PET FDG or $H_2{}^{15}O$ activation studies are ideal to investigate the graft-mediated recovery of metabolic or blood flow function and its role in improving motor and cognitive impairment in patients with PD. The best way is likely to involve acquiring presynaptic DA imaging, FDG, and rCBF activation data in the same patients before and after transplantation. Increasingly, imaging of brain activation can be performed using functional MRI with appropriate stimuli. Together, this hybrid approach would provide an unique opportunity to study the correlation among multiple indices of brain function and their respective relationships with clinical and behavior improvement.

7. CONCLUSION

Fetal nigral tissue can be transplanted into the striatum bilaterally in patients with advanced PD safely and with little morbidity. Human trials in conjunction with PET imaging have demonstrated consistent long-term clinical benefit and increased FDOPA uptake. Neurotrophins such as glial cell line-derived neurotrophic factor can promote survival and growth of implanted dopaminergic cells. Clinical improvement appears to be related to the survival and function of transplanted fetal tissue. In addition, DA transporter seems to be a more valuable imaging marker for quantifying dopaminergic nerve terminals. The recovery of cortical function may be better investigated by mapping DA-regulated metabolic and hemodynamic changes in the striatum and associated brain circuitry. Functional imaging with PET will continue to play a key role in evaluating and optimizing this surgical therapy for successful use in the treatment of PD and other brain disorders resulting from neurodegeneration.

ACKNOWLEDGMENTS

This project has been funded by National Institute of Health grants RO1 NS 32368, RO1 NS 35069, and P50 NS 38370. We are grateful to Drs. Thomas Chaly and Abdel Belakhlef and Mr. Claude Margouleff for their excellent technical support in radiochemistry and PET brain imaging. Von Pillai and John Okulski have assisted in data analysis and presentation described in this contribution.

REFERENCES

1. Patlak C, Blasberg R. Graphical evaluation of blood-to-brain transfer constants from multiple-time uptake data. Generalizations. J Cereb Blood Flow Metab 1985;5:584–590.
2. Logan, J. Graphical analysis of PET data applied to reversible and irreversible tracers. Nucl Med Biol 2000;27:661–670 (Record as supplied by publisher).
3. Friston KJ, et al. Statistical parametric maps in functional imaging: A general linear approach. Human Brain Mapping 1995;2:189–210.
4. Worsley KJ, et al. A unified statistical approach for determining significant signals in images of cerebral activation. Human Brain Mapping 1996;4:58–73.
5. Kish SJ, Shannak K, Hornykiewicz O. Uneven pattern of dopamine loss in the striatum of patients with idiopathic Parkinson's disease. Pathophysiologic and clinical implications. N Engl J Med, 1988;318:876–880.
6. Fearnley J, Lees A. Ageing and Parkinson's disease: substantia nigra regional selectivity. Brain 1991;114:2283–2301.
7. Tedroff J, et al. Regulation of dopaminergic activity in early Parkinson's disease. Ann Neurol 1999;46:359–365.
8. Rinne JO, et al. [18F]FDOPA and [18F]CFT are both sensitive PET markers to detect presynaptic dopaminergic hypofunction in early Parkinson's disease. Synapse 2001;40:193–200.
9. Ishikawa T, et al. Comparative nigrostriatal dopaminergic imaging with iodine-123-beta CIT- FP/SPECT and fluorine-18-FDOPA/PET. J Nucl Med 1996;37:1760–1765.
10. Dhawan V, et al. Comparative analysis of striatal FDOPA uptake in Parkinson's disease: ratio method vs. graphical approach. J Nucl Med 2002;43:1324–1330.
11. Ishikawa T, et al. Clinical significance of striatal DOPA decarboxylase activity in Parkinson's disease. J Nucl Med, 1996;37: 216–222.
12. Vingerhoets FJ, et al. Reproducibility and discriminating ability of fluorine-18-6-fluoro-L-Dopa PET in Parkinson's disease. J Nucl Med 1996;37:421–426.
13. Morrish PK, Sawle GV, Brooks DJ. Regional changes in [18F]dopa metabolism in the striatum in Parkinson's disease. Brain 1996;119:2097–2103.
14. Vingerhoets FJ, et al. Which clinical sign of Parkinson's disease best reflects the nigrostriatal lesion? Ann Neurol 1997;41:58–64.
15. Rinne JO, et al. Cognitive impairment and the brain dopaminergic system in Parkinson disease: (18F)fluorodopa positron emission tomographic study. Arch Neurol 2000;57:470–475.
16. Ribeiro MJ, et al. Dopaminergic function and dopamine transporter binding assessed with positron emission tomography in Parkinson disease. Arch Neurol 2002;59:580–586.
17. Vingerhoets, F., et al. Longitudinal fluorodopa positron emission tomographic studies of the evolution of idiopathic parkinsonism. Ann Neurol 1994;36:759–764.
18. Morrish PK, et al. Measuring the rate of progression and estimating the preclinical period of Parkinson's disease with [18F]dopa PET. J Neurol Neurosurg Psychiatry 1998;64:314–319.
19. Ito K, et al. Statistical parametric mapping with 18FDOPA PET shows bilaterally reduced striatal and nigral dopaminergic function in early Parkinson's disease. J Neurol Neurosurg Psychiatry 1999;66:754–758.
20. Rakshi JS, et al. Frontal, midbrain and striatal dopaminergic function in early and advanced Parkinson's disease A 3D [18F]dopa-PET study. Brain 1999;122:1637–1650.
21. Ishikawa, T, et al. Fluorodopa positron emission tomography with an inhibitor of catechol-O- methyltransferase: effect of the plasma 3-O-methyldopa fraction on data analysis. J Cereb Blood Flow Metab 1996;16:854–863.

22. Kish SJ, et al. Striatal 3,4-dihydroxyphenylalanine decarboxylase in aging: disparity between postmortem and positron emission tomography studies? Ann Neurol 1995;38:260–264.
23. Lee CS, et al. In vivo positron emission tomographic evidence for compensatory changes in presynaptic dopaminergic nerve terminals in Parkinson's disease. Ann Neurol 2000;47:493–503.
24. Wilson JM, et al. Differential changes in neurochemical markers of striatal dopamine nerve terminals in idiopathic Parkinson's disease. Neurology 1996;47:718–726.
25. Chaly T, et al. Radiosynthesis of [18F] N-3-fluoropropyl-2-beta-carbomethoxy-3-beta-(4-iodophenyl) nortropane and the first human study with positron emission tomography. Nucl Med Biol 1996;23: 999–1004.
26. Marek K, et al. Dopamine transporter brain imaging to assess the effects of pramipexole vs levodopa on Parkinson disease progression. JAMA 2002;287:1653–1661.
27. Rinne JO, et al. Usefulness of a dopamine transporter PET ligand (18F)beta-CFT in assessing disability in Parkinson's disease. J Neurol Neurosurg Psychiatry 1999;67:737–741.
28. Ilgin N, et al., PET imaging of the dopamine transporter in progressive supranuclear palsy and Parkinson's disease. Neurology 1999; 52:1221–1226.
29. Ouchi Y, et al. Alterations in binding site density of dopamine transporter in the striatum, orbitofrontal cortex, and amygdala in early Parkinson's disease: compartment analysis for beta-CFT binding with positron emission tomography. Ann Neurol 1999;45:601–610.
30. Volkow ND, et al. Dopamine transporters decrease with age. J Nucl Med 1996;37:554–559.
31. Kazumata K, et al. Dopamine transporter imaging with fluorine-18-FPCIT and PET. J Nucl Med 1998;39:1521–1530.
32. Ma,Y, et al. Parametric mapping of (18F)FPCIT binding in early stage Parkinson's disease: a PET study. Synapse 2002;45:125–133.
33. Ouchi Y, et al. Presynaptic and postsynaptic dopaminergic binding densities in the nigrostriatal and mesocortical systems in early Parkinson's disease: a double-tracer positron emission tomography study. Ann Neurol 1999;46:723–731.
34. Turjanski N, Lees AJ, Brooks DJ. In vivo studies on striatal dopamine D1 and D2 site binding in L-dopa-treated Parkinson's disease patients with and without dyskinesias. Neurology 1997;49:717–723.
35. Rinne JO, et al. PET study on striatal dopamine D2 receptor changes during the progression of early Parkinson's disease. Mov Disord 1993;8:134–138.
36. Antonini A, et al. [11C]raclopride and positron emission tomography in previously untreated patients with Parkinson's disease: Influence of L-dopa and lisuride therapy on striatal dopamine D2-receptors. Neurology 1994;44:1325–1329.
37. Antonini A, et al., Long-term changes of striatal dopamine D2 receptors in patients with Parkinson's disease: a study with positron emission tomography and [11C]raclopride. Mov Disord 1997; 12:33–38.
38. Tedroff J, et al. Levodopa-induced changes in synaptic dopamine in patients with Parkinson's disease as measured by (11C)raclopride displacement and PET. Neurology 1996;46:1430–1436.
39. Piccini P, Pavese N, Brooks DJ. Endogenous dopamine release after pharmacological challenges in Parkinson's disease. Ann Neurol 2003;53:647–653.
40. Goerendt IK, et al. Dopamine release during sequential finger movements in health and Parkinson's disease: a PET study. Brain 2003; 126:312–325.
41. de la Fuente-Fernandez R, et al. Biochemical variations in the synaptic level of dopamine precede motor fluctuations in Parkinson's disease: PET evidence of increased dopamine turnover. Ann Neurol 2001;49:298–303.
42. Otsuka M, et al. Striatal blood flow, glucose metabolism and 18FDOPA uptake: difference in Parkinson's disease and atypical parkinsonism. J Neurol Neurosurg Psychiatry 1991;54:898–904.
43. Eidelberg D, et al. Regional metabolic correlates of surgical outcome following unilateral pallidotomy for Parkinson's disease. Ann Neurol 1996;39:450–459.
44. Hu MT, et al. Cortical dysfunction in non-demented Parkinson's disease patients: a combined (31)P-MRS and (18)FDG-PET study. Brain 2000;123:340–352.
45. Playford ED, et al., Impaired mesial frontal and putamen activation in Parkinson's disease: a positron emission tomography study. Ann Neurol 1992;32:151–161.
46. Jahanshahi M, et al. Self-initiated versus externally triggered movements. I. An investigation using measurement of regional cerebral blood flow with PET and movement-related potentials in normal and Parkinson's disease subjects. Brain 1995;118:913–933.
47. Samuel M, et al. Motor imagery in normal subjects and Parkinson's disease patients: an $H_2{}^{15}O$ PET study. Neuroreport 2001;12:821–828.
48. Catalan MJ, et al. A PET study of sequential finger movements of varying length in patients with Parkinson's disease. Brain 1999;122: 483–495.
49. Alexander G, Moeller J. Application of the scaled subprofile model to functional imaging in neuropsychiatric disorders: a prinicipal component approach to modeling brain function in disease. Human Brain Mapping 1994;2:1–16.
50. Zuendorf G, et al. Efficient principal component analysis for multivariate 3D voxel-based mapping of brain functional imaging data sets as applied to FDG-PET and normal aging. Hum Brain Mapp 2003;18:13–21.
51. Eidelberg D, et al. Assessment of disease severity in parkinsonism with fluorine-18-fluorodeoxyglucose and PET. J Nucl Med 1995;36: 378–383.
52. Eidelberg D, et al., Early differential diagnosis of Parkinson's disease with 18F-fluorodeoxyglucose and positron emission tomography. Neurology 1995;45:1995–2004.
53. Moeller JR, et al. Reproducibility of regional metabolic covariance patterns: comparison of four populations. J Nucl Med 1999;40:1264–1269.
54. Nakamura T, et al. Functional networks in motor sequence learning: abnormal topographies in Parkinson's disease. Hum Brain Mapp 2001;12:42–60.
55. Carbon M, et al. Learning networks in health and Parkinson's disease: reproducibility and treatment effects. Hum Brain Mapp 2003;19:197–211.
56. Fahn S. The spectrum of levodopa-induced dyskinesias. Ann Neurol, 2000;47(4 Suppl 1):S2–S9; discussion S9–S11.
57. Obeso JA, et al. Surgical treatment of Parkinson's disease. Baillieres Clin Neurol 1997;6:125–145.
58. Starr PA, Vitek JL, Bakay RA. Ablative surgery and deep brain stimulation for Parkinson's disease. Neurosurgery 1998;43:989–1013; discussion 1013–1015.
59. Haslinger B, et al. Differential modulation of subcortical target and cortex during deep brain stimulation. Neuroimage 2003;18:517–524.
60. Feigin A, et al. Effects of levodopa infusion on motor activation responses in Parkinson's disease. Neurology 2002;59:220–226.
61. Fukuda M, et al. Functional correlates of pallidal stimulation for Parkinson's disease. Ann Neurol 2001;49:155–164.
62. Fukuda M, et al. Pallidal stimulation for parkinsonism: improved brain activation during sequence learning. Ann Neurol 2002;52:144–152.
63. Limousin P, et al. Changes in cerebral activity pattern due to subthalamic nucleus or internal pallidum stimulation in Parkinson's disease. Ann Neurol 1997;42:283–291.
64. Siebner HR, et al. Changes in handwriting resulting from bilateral high-frequency stimulation of the subthalamic nucleus in Parkinson's disease. Mov Disord 1999;14:964–971.
65. Ceballos-Baumann AO, et al. A positron emission tomographic study of subthalamic nucleus stimulation in Parkinson disease: enhanced movement-related activity of motor-association cortex and decreased motor cortex resting activity. Arch Neurol 1999;56:997–1003.
66. Su PC, et al. Metabolic changes following subthalamotomy for advanced Parkinson's disease. Ann Neurol 2001;50:514–520.
67. Trost M, et al. Evolving metabolic changes during the first postoperative year after subthalamotomy. J Neurosurg 2003;99:872–878.

68. Fukuda M, et al. Networks mediating the clinical effects of pallidal brain stimulation for Parkinson's disease: a PET study of resting-state glucose metabolism. Brain 2001;124:1601–1609.
69. Feigin A, et al. Metabolic correlates of levodopa response in Parkinson's disease. Neurology 2001;57:2083–2088.
70. Gill SS, et al. Direct brain infusion of glial cell line-derived neurotrophic factor in Parkinson disease. Nat Med 2003;9:589–595.
71. Tabbal S, Fahn S, Frucht S. Fetal tissue transplantation [correction of transplanation] in Parkinson's disease. Curr Opin Neurol 1998;11: 341–349.
72. Lindvall O. Rationales and strategies of fetal neural transplantation in PD. In: Krauss JK, Jankovic J, Grossman RG, editors. Surgery for Parkinson's Disease and Movement Disorders. Philadelphia: Lippincott Williams & Wilkins; 2001:194–209.
73. Date I, et al. Efficacy of pretransection of peripheral nerve for promoting the survival of cografted chromaffin cells and recovery of host dopaminergic fibers in animal models of Parkinson's disease. Neurosci Res 1994;20:213–221.
74. Subramanian T, et al. Polymer-encapsulated PC-12 cells demonstrate high-affinity uptake of dopamine in vitro and 18F-Dopa uptake and metabolism after intracerebral implantation in nonhuman primates. Cell Transplant 1997;6:469–477.
75. Dall AM, et al. Quantitative [18F]fluorodopa/PET and histology of fetal mesencephalic dopaminergic grafts to the striatum of MPTP-poisoned minipigs. Cell Transplant 2002;11:733–746.
76. Widner H, et al. Bilateral fetal mesencephalic grafting in two patients with parkinsonism induced by 1-methyl-4-phenyl-1,2,3,6-tetrahydropyridine (MPTP). N Engl J Med 1992;327:1556–1563.
77. Lindvall O, et al. Grafts of fetal dopamine neurons survive and improve motor function in Parkinson's disease. Science 1990;247:574–577.
78. Remy P, et al. Clinical correlates of [18F]fluorodopa uptake in five grafted parkinsonian patients. Ann Neurol 1995;38:580–588.
79. Wenning GK, et al. Short- and long-term survival and function of unilateral intrastriatal dopaminergic grafts in Parkinson's disease. Ann Neurol 1997;42:95–107.
80. Levivier M, et al. Intracerebral transplantation of fetal ventral mesencephalon for patients with advanced Parkinson's disease. Methodology and 6-month to 1-year follow-up in 3 patients. Stereotact Funct Neurosurg 1997;69:99–111.
81. Freed CR, et al. Survival of implanted fetal dopamine cells and neurologic improvement 12 to 46 months after transplantation for Parkinson's disease. N Engl J Med 1992;327:1549–1555.
82. Hagell P, et al. Sequential bilateral transplantation in Parkinson's disease: effects of the second graft. Brain 1999;122:1121–1132.
83. Mendez I, et al. Neural transplantation cannula and microinjector system: experimental and clinical experience. Technical note. J Neurosurg 2000;92:493–499.
84. Freeman TB, et al. Bilateral fetal nigral transplantation into the postcommissural putamen in Parkinson's disease. Ann Neurol 1995; 38:379–388.
85. Hauser RA, et al. Long-term evaluation of bilateral fetal nigral transplantation in Parkinson disease. Arch Neurol 1999;56:179–187.
86. Kordower JH, et al. Fetal nigral grafts survive and mediate clinical benefit in a patient with Parkinson's disease. Mov Disord 1998;13: 383–393.
87. Brundin P, et al. Bilateral caudate and putamen grafts of embryonic mesencephalic tissue treated with lazaroids in Parkinson's disease. Brain 2000;123 (Pt 7):1380–90.
88. Freed CR, et al. Transplantation of embryonic dopamine neurons for severe Parkinson's disease. N Engl J Med 2001;344:710–719.
89. Trott CT, et al. Cognition following bilateral implants of embryonic dopamine neurons in PD: A double blind study. Neurology 2003; 60:1938–1943.
90. Nakamura T, et al. Blinded positron emission tomography study of dopamine cell implantation for Parkinson's disease. Ann Neurol 2001;50:181–187.
91. Brownell AL, et al. In vivo PET imaging in rat of dopamine terminals reveals functional neural transplants. Ann Neurol 1998;43:387–390.
92. Lee CS, et al. Embryonic ventral mesencephalic grafts improve levodopa-induced dyskinesia in a rat model of Parkinson's disease. Brain 2000;123:1365–1379.
93. Cochen V, et al. Transplantation in Parkinson's disease: PET changes correlate with the amount of grafted tissue. Mov Disord 2003; 18:928–932.
94. Fricker RA, et al., The effects of donor stage on the survival and function of embryonic striatal grafts in the adult rat brain. II. Correlation between positron emission tomography and reaching behaviour. Neuroscience 1997;79:711–721.
95. Piccini P, et al. Dopamine release from nigral transplants visualized in vivo in a Parkinson's patient. Nat Neurosci 1999;2:1137–1140.
96. Piccini P, et al. Delayed recovery of movement-related cortical function in Parkinson's disease after striatal dopaminergic grafts. Ann Neurol 2000;48:689–695.
97. Bluml S, et al. Activation of neurotransplants in humans. Exp Neurol 1999;158:121–125.
98. Bjorklund LM, et al. Embryonic stem cells develop into functional dopaminergic neurons after transplantation in a Parkinson rat model. Proc Natl Acad Sci USA 2002;99:2344–2349.
99. Meltzer CC, et al. Serial [18F] fluorodeoxyglucose positron emission tomography after human neuronal implantation for stroke. Neurosurgery 2001;49:586–591; discussion 591–592.
100. Defer GL, et al. Long-term outcome of unilaterally transplanted parkinsonian patients. I. Clinical approach. Brain 1996;119:41–50.
101. Hagell P, et al. Dyskinesias following neural transplantation in Parkinson's disease. Nat Neurosci 2002;5:627–628.
102. Ma Y, et al. Dyskinesia after fetal cell transplantation for parkinsonism: a PET study. Ann Neurol 2002;52:628–634.
103. Brooks DJ, et al. Neuroimaging of dyskinesia. Ann Neurol 2000;47(4 Suppl 1):S154–S158; discussion S158–S159.
104. Bjorklund A, et al. Neural transplantation for the treatment of Parkinson's disease. Lancet Neurol 2003;2:437–445.
105. Deacon T, et al. Histological evidence of fetal pig neural cell survival after transplantation into a patient with Parkinson's disease. Nat Med 1997;3:350–353.
106. Schumacher JM et al. Transplantation of embryonic porcine mesencephalic tissue in patients with PD. Neurology 2000;54:1042–1050.
107. Prasad KN, et al. Efficacy of grafted immortalized dopamine neurons in an animal model of parkinsonism: a review. Mol Genet Metab 1998;65:1–9.
108. During MJ, et al. Subthalamic GAD gene transfer in Parkinson disease patients who are candidates for deep brain stimulation. Hum Gene Ther 2001;12:1589–1591.
109. Kang UJ, Lee WY, Chang JW. Gene therapy for Parkinson's disease: determining the genes necessary for optimal dopamine replacement in rat models. Hum Cell 2001;14:39–48.
110. Sullivan AM, Pohl J, Blunt SB. Growth/differentiation factor 5 and glial cell line-derived neurotrophic factor enhance survival and function of dopaminergic grafts in a rat model of Parkinson's disease. Eur J Neurosci 1998;10:3681–3688.
111. Mendez I, et al. Enhancement of survival of stored dopaminergic cells and promotion of graft survival by exposure of human fetal nigral tissue to glial cell line—derived neurotrophic factor in patients with Parkinson's disease. Report of two cases and technical considerations. J Neurosurg 2000;92:863–869.
112. Mendez I, et al. Simultaneous intrastriatal and intranigral fetal dopaminergic grafts in patients with Parkinson disease: a pilot study. Report of three cases. J Neurosurg 2002;96:589–596.

Alzheimer's Disease

II

6 Neurotoxicity of the Alzheimer's β-Amyloid Peptide

Spectroscopic and Microscopic Studies

DAVID R. HOWLETT, PhD

SUMMARY

Alzheimer's disease (AD) is characterized pathologically by the presence in the brain of extracellular deposits of the β-amyloid or Abeta (Aβ) protein and intracellular neurofibrillary tangles composed of hyperphosphorylated tau. Molecular genetic data, largely based on familial AD patients, led to the "The Amyloid Hypothesis" being proposed, where the aggregation and deposition of Aβ is a pathogenic feature of the disease. The Aβ protein generally is described as a 39- to 43-amino acid peptide, although that barely hints at the complexities encountered in trying to understand the process by which it folds, aggregates, is deposited and, at some stage, is believed to bring about neuronal cell degeneration. This chapter, therefore, explores how many microscopic and spectroscopic techniques have been used in attempts to understand the Aβ fibrillization process and to provide a means of generating agents capable of halting Aβ deposition and hence disease progression in AD.

Key Words: Aβ; beta-amyloid; β-amyloid; Alzheimer's disease; neurotoxicity; β-sheet formation; fibrillization.

1. INTRODUCTION

A century ago, Alois Alzheimer described the histopathological analysis of the brain of an unusual case of dementia, reporting the presence of a "peculiar substance" (Alzheimer 1907 Über eine eigenartige Erkrankung der Hirnrinde. Centralblatt für Nervenheilkunde und Psychiatrie 30, pp. 177–179). That peculiar substance has now been subjected to a vast armamentarium of protein biophysical technologies in attempts to unravel the fundamental mysteries of the etiology of Alzheimer's disease (AD). The lack of any means of halting disease progression and the inevitably of its outcome has been an important driving force in providing the support for several decades of study of the biochemical manifestations of AD.

From: *Bioimaging in Neurodegeneration*
Edited by P. A. Broderick, D. N. Rahni, and E. H. Kolodny

The "peculiar substance" is the β-amyloid -protein (Aβ, also known as beta-A4, amyloid β, ABeta, βAP), which is derived from the proteolytic cleavage of the amyloid precursor protein (APP) *(1)*, a much larger membrane-spanning protein of uncertain biological function. However, because Aβ is the major component of senile plaques in AD, it has been postulated that it is responsible for the neuronal cell loss observed in the vicinity of the plaques. The role of Aβ in the pathogenesis of AD has been widely reviewed and will not be repeated in this present chapter, although it is necessary to touch on the basic tenet of the Amyloid Hypothesis *(2)* in that aggregated Aβ protein is deposited in senile plaques and that somewhere between proteolytic cleavage from APP and deposition in plaques the Aβ protein is the cause of the neurodegeneration.

In discussing the makeup of senile plaque, the term "Aβ proteins" is used because the amyloid plaque is composed of not just the 39 to 43 amino acid peptides reported in early studies *(1,3,4)* but of a whole series of N- and C-terminally truncated species *(5)*. Cerebrospinal fluid (CSF) also contains many differing-length Aβ forms *(6)*. In general, the longer forms of Aβ fibrillize faster and although Aβ 1-40 is the major species in the CSF and blood vessels *(7)*, Aβ 1-42 is the major form in plaques *(8)*.

Biochemical studies have investigated the behavior of the full-length Aβ 1-40 and Aβ 1-42 amino acid peptides plus many shorter fragments used because of their increased solubility. Of the two longer forms, a number of factors have led to Aβ 1-40 being widely studied: (1) it exhibits amyloidogenicity and neurotoxicity (*see* Sections 2.1. and 3.1.); (2) it possesses the ability to be initially disaggregated so that the kinetics of the fibrillization process can be studied in a sensible time window, which is a major plus for studying the fibrillization process; (3) it is relatively easy to synthesize (compared with Aβ1-42); and (4) it was more readily available commercially in the first half of the 1990s.

The earliest studies of Aβ in vitro provided a warning of the confusion that has persisted to this present day in terms of the actions of the peptide at the cellular level in that both trophic and toxic responses were observed on central nervous system neurons in culture. When initially exposed to synthetic Aβ, hippocampal neurons showed enhanced survival and increased neurite outgrowth *(9–11)* although after several days in culture, neuronal degeneration was observed *(11)*. In vivo, however, Aβ was reported to be directly neurotoxic *(12,13)*. Some clarity regarding the in vitro actions of Aβ has been provided by various studies showing that the cytotoxic properties of the peptide are dependent on its aggregation state *(14–17)*.

The aggregation state of the Aβ peptide and its relationship to neuronal cell death in AD has been the subject of hundreds of reports in the scientific literature. A major requirement in such studies is the ability to be able to determine the biophysical state of the peptide. This forms the basic thread of this chapter—how imaging techniques, from spectroscopy to microscopy, have aided our understanding of the biophysical and biological properties of Aβ.

2. THE Aβ AGGREGATION PROCESS

As a title, "The Aβ Aggregation Process" may be somewhat misleading because it implies that we understand the process. There is no doubt that synthetic Aβ 1-40 peptide, when dissolved in aqueous medium and incubated at 37°C, undergoes conformational changes resulting in the appearance of long straight fibrils. Figure 1 shows electron micrographs of two examples of fibrils formed from Aβ 1-40 with differing fibril diameters and periodicities (Fig 1A) and possible splitting or unraveling (Fig 1B). In both micrographs, but particularly at the 44-h time point (Fig 1A), there are numerous examples of what appear to be shorter, often globular structures that may represent early protofibrillar forms. Both electron microscopy and atomic force microscopy (AFM) have been widely used to image visually the changes that occur when Aβ is incubated in vitro although this tells us little about the biophysical process involved or what is happening at the molecular level.

With both full-length and truncated Aβ peptides, various solvent systems have been used to improve solubilization of the peptide. Additionally, to provide the starting material for studying the aggregation process, the consensus of opinion is that peptide devoid of any β-sheet structure is required. Several solvents (e.g., hexafluoroisopropanol, trifluorethanol, dimethylsulphoxide) have been used, although it is known that organic solvents do affect Aβ aggregation *(18)*. Fezoui et al *(19)* have claimed that initial dissolution of Aβ 1-40 in sodium hydroxide solution (2 m*M*) followed by sonication and lyophilization resulted in peptide solutions with enhanced solubility and fibrillization reproducibility. Certainly, circular dichroism (CD) spectroscopy and AFM suggested that peptide prepared in this way had superior properties for fibrillization studies *(19)*. Because the behavior of Aβ peptides is so dependent on initial solubilization method and even the particular batch of Aβ, it is obviously difficult to reconcile these data with the behavior of endogenous Aβ proteins in the CNS. Only time will tell which, if any, resembles the fate of Aβ cleaved from APP in the AD brain.

The development of the ability to be cytotoxic to neuronal-type cultures has been repeatedly shown to be associated with the formation of multimeric forms of Aβ. However, the precise conformational changes and molecular reorganization of the peptide that accompany the acquisition of that property is still somewhat unclear. In recent years, much attention has focussed on small oligomers, protofibrils, and amyloid-derived diffusable ligands (ADDLs) as being the neurotoxic forms of Aβ, and various analytical techniques have been used to monitor the formation of these species and their relationship to neuronal cell death.

2.1. THE AD AMYLOID FIBRIL ASSEMBLY PROCESS

Unfortunately, many Aβ fibrillization reports only describe the biophysical changes in the peptide and so tell us nothing about biological properties, that is, neurotoxicity. However, much of our basic understanding of the fibrillization process stems from this literature and hence it is vital that it is discussed.

Aβ fibril formation is proposed as being a nucleation-dependent mechanism where the nucleus grows by the incorporation of additional protein molecules. The rate-limiting step in this process is the formation of the Aβ nucleus *(20)*. The precise nature of the Aβ nucleus is unknown, although data from quasi-elastic light-scattering studies have suggested the formation of micelles in solution on route to fibril formation *(21)*. Light-scattering data from these authors *(22,23)* have suggested that Aβ forms dimers and tetramers. There also is evidence that during the nucleation process, fibrillization can occur by the addition of Aβ monomer to nonamyloidogenic seed *(20)*. Furthermore, fibril growth can be accelerated by the addition of preformed fibrils, that is, by seeding *(24)*, suggesting that the Aβ nucleus may simply be preformed mature fibrils.

The kinetics of fibril extension have been studied by using surface plasmon resonance (SPR) where fibrillar Aβ is immobilized onto the sensor chip *(25)*. The application of monomeric Aβ to the SPR flow cell resulted in increases in the SPR signal. Parallel AFM studies confirmed fibril extension. The use of SPR allowed the authors to determine that the extension reaction and linear dissociation phase were consistent with a first-order kinetic model and, by using this model, the critical monomer concentration and equilibrium association constants were calculated. However, care should be taken with the interpretation of data employing immobilized Aβ because the strong interaction between amyloid proteins and surfaces such as mica, used in AFM studies, can result in the formation of fibrils morphologically different from the intact amyloid fibrils *(26)*.

Understanding the molecular structure of the Aβ peptide has been thwarted by its noncrystalline, insoluble nature that renders it unsuitable for high resolution techniques such as X-ray crystallography and solution phase nuclear magnetic resonance (NMR). Because the fibrils are noncrystalline, diffraction patterns are difficult to resolve, although early studies did suggest an anti-parallel β-sheet arrangement *(27,28)*. Fourier-transform infrared (FTIR) spectroscopy data confirms the β-sheet structure of the amyloid fibrils and also is supportive of an anti-parallel alignment of the β-strands *(29,30)*. More recently, X-ray diffraction studies on orientated fibrillar bundles have confirmed the cross-β diffraction pattern characteristic of amyloid fibrils *(31,32)*. The data from these studies suggests the presence of hydrogen bonded parallel chains in β-strand conformation running parallel to the long axis of the fibril thus forming β-sheet ribbon-like structures. Reviewing X-ray dif-

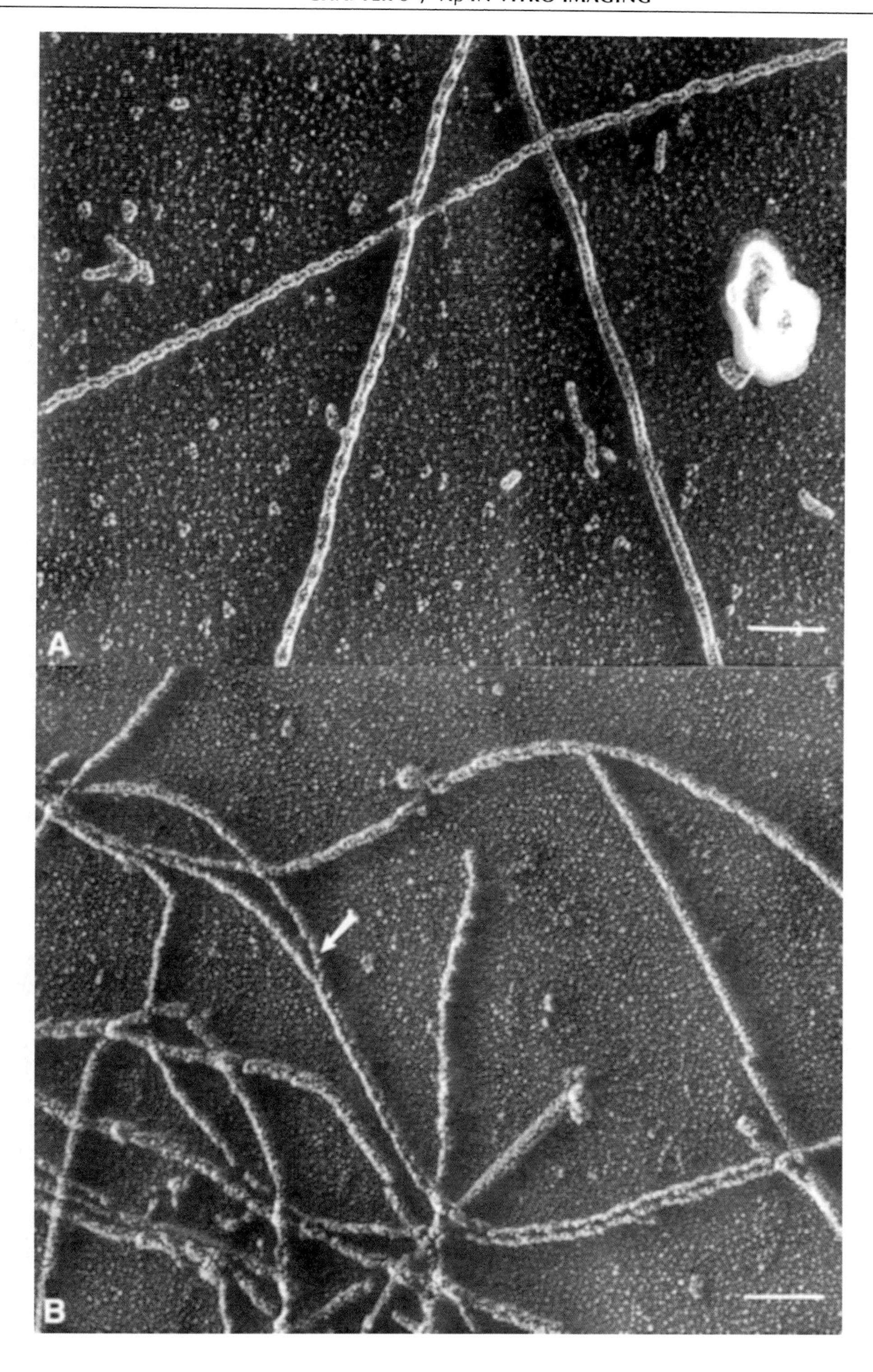

Fig. 1. Freeze-dried, rotary low-angle replica of Aβ 1–40. Aβ 1–40 (batch ZK051 Bachem UK Ltd) was dissolved in 0.1% acetic acid at 0.46 m*M*. Further dilutions to 11.6 μ*M* were made in phosphate-buffered saline. Peptide solutions were incubated for **(A)** 44 h and **(B)** 9 d at 37°C in aqueous solution pH 7.4. Possible split fibre arrowed. Magnification (A,B) ×168,000. Scale bar, 100 nm

fraction and electron microscope (EM) data, Perutz et al. *(33)* have suggested that amyloid fibrils may exist as water-filled nanotubes. Negatively stained electron micrographs typically show unbranched straight fibrils with periodic twists, often of length greater than 1um (e.g., ref. *17*). The diameter of these fibrils (10–12 nm) is similar to those found in amyloid plaque *(34,35)*. The periodicity of the fibril twists has been noted by both EM *(17)* and by AFM *(36,37)* and may reflect either the composition being a series of composite units or a helical twist in the fibril.

The data described above show that being able to monitor the fibrillization process by multiple analytical means, therefore, is almost a prerequisite for gaining a thorough understanding of the process and, subsequently, how inhibitors might act. In studying the effects of melatonin, for instance, Pappolla and colleagues *(38)* have used EM to demonstrate the inhibition of fibril production during the course of 2–6 h (Aβ 1-42) or 48 h (Aβ 1-40) at a melatonin to Aβ ratio of 1:2.5. Under the stated condition, CD spectroscopy showed that freshly dissolved peptides contained some β-sheet structure and that the addition of melatonin to either peptide resulted in an immediate increase in random coil conformation and a decrease in β-sheet formation with time. Such early events describing interactions between melatonin and Aβ were not observed in EM studies. Furthermore, NMR spectroscopy revealed that the effects of melatonin were dependent on structural interactions between melatonin and Aβ rather than being associated with the antioxidant effects of the former. Similar data have been reported by Skribanek et al *(39)* who, by use of EM, CD spectroscopy, and electrospray mass spectrometry, were able to show hydrophobic interactions between melatonin and Aβ 1-40. Attempts to pin down further the site of interaction by employing proteolytic cleavage in line with mass-spectrometry suggested that the interaction might be between the hormone and the 29-40 segment of Aβ 1-40 *(39)*. These two studies *(38,39)*, therefore, demonstrate the benefits of studying Aβ fibrillization by an array of techniques.

2.2. FACTORS REGULATING Aβ AGGREGATION

As commented earlier, many Aβ aggregation studies have investigated the behavior of Aβ peptides in aqueous buffer/salt solutions, attempting to draw correlations with plaque formation in the brain, almost ignoring the fact that in vivo conditions will be very different to simple inorganic buffers or salt solutions. This is maybe surprising knowing how easy it is to affect Aβ fibrillization by factors such as pH, ion content, temperature, and metals. However, the fundamental question remains as to what affects and regulates Aβ fibrillization and plaque formation in the brains of patients with AD.

It is perhaps not surprising that pH value plays an important role in the determination of Aβ peptide conformation. Some years ago it was shown that the pH-dependent polymerization of Aβ 1-40 results in the formation of species of peptide with different physical and biological properties *(40)*, although to complicate matters further, the nature of that species varied from batch to batch of peptide (Fig. 2; Howlett, unpublished). Although all five batches of Aβ 1-40 shown in Fig. 2 produced some degree of oligomer formation, albeit with slightly different pH optima, as illustrated by the immunoassay data, only batch MW0241 was cytotoxic in subsequent cell-based testing in IMR 32 human neuroblastoma cells (data not shown). This is obviously a very major problem in the overall interpretation of studies of the physical behavior of Aβ peptides when comparing data between laboratories using differing starting materials.

Other groups have also noted effects of pH on Aβ fibrilllization. Antzutkin et al. *(41)* compared Aβ 10-35 and 1-42 incubated at room temperature at pH 3.7 and 7.4 with the peptide solutions being constantly agitated. Using EM and scanning transmission EM, they found that Aβ 10-35 formed single protofilaments at pH 3.7 but paired or higher order bundles at pH 7.4. Solid-state NMR indicated parallel β-sheet organization at both pH values. With Aβ 1-42, parallel β-sheet and unpaired protofilaments were also observed with Aβ 1-42 at both pH 3.7 and 7.4. Solid-state NMR data with Aβ 1-40 fibrils and site-directed spin-labeling experiments with both Aβ 1-40 and Aβ 1-42 fibrils are also consistent with in-register parallel β-sheets *(42,43)*. Interestingly, the solid-state NMR data suggest that the structure of Aβ 1-42 is not significantly different than that of Aβ 1-40 and does not offer any evidence to explain the greater amyloidogenicity of the longer peptide *(44)*.

A quasielastic light-scattering study by Lomakin and colleagues *(21)* showed that at low pH, Aβ 1-40 does form fibrils that resemble those found in senile plaques. Analysis of the process suggested that above a certain concentration peptide micelles formed and that fibrils may nucleate within these micelles. This technique, therefore, aids the kinetic analysis of Aβ fibrillization.

Metals appear to play an important role in the fibrillization of Aβ in the brain and abnormal interactions between metals and the amyloid peptide may provide a basis for understanding the development of AD (for review, *see* ref. *45*). Zinc, copper, and iron all are found at high levels in brain regions normally associated with AD pathology, particularly the neocortex. These metal ions induce Aβ aggregation and the interactions between Aβ and metal ions have thus become the potential target for therapeutic intervention. Raman spectroscopy has been used to explore these interactions and indicates that the amino acid residues involved in both Cu(II) and Zn(II) binding to Aβ are the histidine residues at positions 6, 13, and 14 *(46)*. Iron, however, binds primarily to the phenolic oxygen atom of the tyrosine residue at position 10 but also to the carboxylate groups of the glutamate and aspartate side chains. The binding of Cu(II) to the three histidine residues has been confirmed by an NMR and electron spin resonance (ESR) study, also suggesting the involvement of the tyrosine residue at position 10 *(47)*.

A number of studies have tried to link findings of senile plaque-associated proteins with Aβ fibrillization, largely on the basis that these proteins may facilitate heterogeneous nucleation. These include proteins such as apolipoprotein E, alpha-1-antichymotrypsin, heparan sulphate proteoglycan, laminin, and acetylcholinesterase.

The enzyme acetylcholinesterase (AChE) is found in close proximity to both senile and diffuse plaques and with cerebrovascular amyloid *(48,49)*. In senile plaques it is found mainly associated with the plaque core, suggesting that it may play a role in mature fibril formation. Indeed, EM studies have shown that AChE accelerates the formation of Aβ 1-40 fibrils in vitro *(50)*. Using short peptide fragments (Aβ 12-28 and Aβ 25-35), these workers demonstrated that AChE-amyloid complexes form during the fibrillization process, data consistent with the close association of AChE and Aβ in the senile plaque *(51)*. The effects of AChE on β-sheet content of Aβ 1-40 prior to fibril formation has been demonstrated by parallel monitoring with CD spectroscopy and electron microscopy *(52)*. Thus, it would appear that AChE is capable of inducing a conformational change in the Aβ peptide that facilitates subsequent fibrillization.

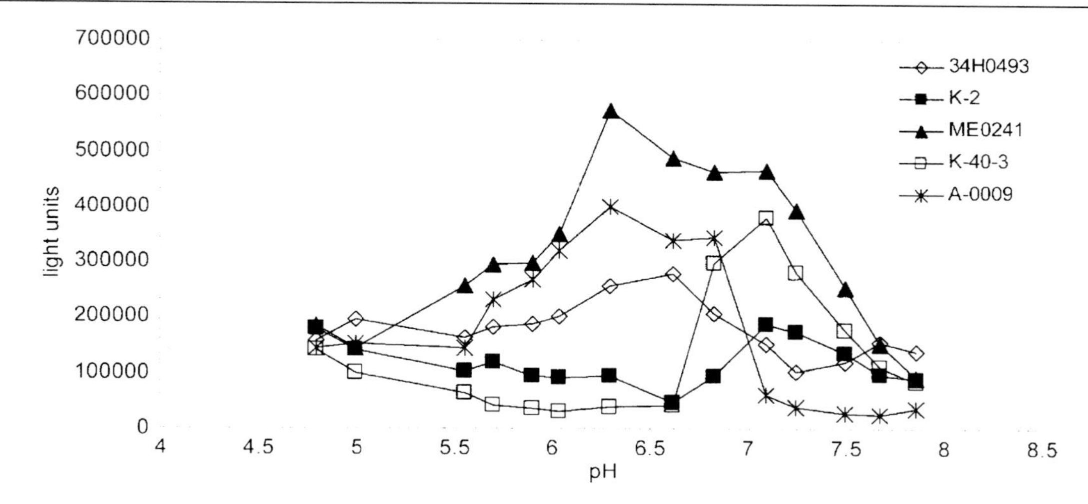

Fig. 2. pH-Dependent fibrillization of differing batches of Aβ 1–40. Aβ 1–40 peptides were dissolved in 0.1% acetic acid at 0.46 m*M*. Further dilutions to 11.6 μ*M* were made in phosphate-buffered saline. Peptide solutions were incubated overnight at 37°C before being assessed in oligomer-specific immunoassay *(107)*. Batches of Aβ1–40: 34H0493 (Sigma Aldrich); K-2 (K-Biologicals); ME0241 (California Peptide Research); K-40–3 (U.S.Peptide Inc); and A-0009 (Wherl GmbH). Data are from a single representative experiment that was repeated three times. Standard errors of the mean varied by less than 15%.

Laminin is an extracellular matrix component that accumulates in senile plaques *(53)* and has been shown to regulate Aβ fibrillization in vitro. At a molar ratio of approximately 1:200 inhibitor to peptide, laminin has been reported to prevent mature fibril formation by Aβ1-40 although nonthioflavin positive amorphous aggregates were noted by electron microscopy *(54)*, suggesting an interaction at a polymeric stage of the aggregation process. An inhibition of fibril production by laminin, as demonstrated by transmission electron microscopy, is mirrored by prevention of Aβ neurotoxicity in rat hippocampal neurons *(55)*. Laminin also inhibits the fibrillization of the E22Q Dutch amyloid mutant but does not prevent AChE accelerated fibril formation *(56)*. Hence, the presence of AChE may erode any anti-amyloidogenic properties of laminin when in close proximity to senile plaque.

All types of amyloid are closely associated with different types of proteoglycan suggesting that after binding to the amyloidogenic protein, the glycosoaminoglycan (GAG) promotes a shift toward β-sheet structure. In AD and in Down syndrome, both diffuse and senile plaques show colocalization with GAGs *(57–59)*, leading to the proposition that GAGs facilitate growth and stabilization of Aβ fibrils. The addition of GAG to Aβ results in a rapid transition to β-sheet structure, suggesting that the GAG acts as a scaffold for the assembly of Aβ fibrils *(60)*. The binding interactions between Aβ and GAG chain have also been examined by using a series of chondroitin sulfate-derived mono- and disaccharides *(61)*. EM and fluorescence spectroscopy showed that the monosaccharide was the smallest unit for Aβ binding but that a disaccharide is necessary to promote the transition of protofibril into fibril. Studies such as this are fundamental in understanding fibril growth and the stage(s) at which interruptions of the fibrillization process may prove to be therapeutically beneficial.

α1-Antichymotrypsin (ACT) is an acute phase protein of the serine protease family. It has been shown to colocalize with senile plaques and sites of vascular angiopathy *(62,63)*. ACT is produced in astrocytes, possibly as a response to the deposition of Aβ, but it may also influence the deposition. When ACT was added to Aβ 1-40 or to a number of analogs, including the E22Q Dutch amyloidoses mutant form, mature fibrils, as viewed by electron microscopy, appeared to disassemble *(64)*, suggesting that ACT may cause structural rearrangements within the aggregated Aβ resulting in enhanced proteolysis.

2.3. THE USE OF FLUORESCENCE TECHNIQUES

In studying the fibrillization process, fluorescence methods provide both sensitivity and specificity. The association of fluorescently tagged Aβ peptide with a nontagged protein counterpart by fluorescence anisotrophy has been used to monitor the early stages of the Aβ aggregation process *(65)*. This technique uses the relationship between the rates of tumbling and orientation of a fluorescent species in solution, where the orientation is a function of particle size. Using time-resolved fluorescence anisotropy, Allsop and colleagues *(65)* were able to demonstrate changes in fluorescence signal after as little as 30 min of incubation whereas conventional fluorescence assays using thioflavin T showed no changes until after at least 7 h of incubation. Interestingly, in this report, the appearance of small "protofibril-like" structures were noted by electron microscopy after 30 min of incubation. Unfortunately, no parallel data on cell viability was presented in this report.

The early stages of the pathway from monomer/dimer to fibril formation also have been explored by fluorescence resonance energy transfer using native and fluorescently tagged Aβ derivatives *(66)*. Despite the modification of the Aβ molecule as a result of the fluorescence tag, the aggregation profiles did not differ from those of the native Aβ peptides. Fluorescence resonance energy transfer (FRET) as a technique lends itself very usefully to the study of Aβ fibril formation because the distance between fluorophores means that it can operate in the 10–100 Å window close to the diameter of typical amyloid fibrils. FRET data support the belief that a dimer is the smallest Aβ structure occurring under physiological conditions *(66)*.

A number of fluorescent probes have been used for following the Aβ fibrillization process. Ethyldiaminonaphalene-1-sulfonic acid (EDANS) is extremely susceptible to quenching by water and can therefore be used as a highly sensitive agent for monitoring changes in the peptide conformation of EDANS-Aβ associated with water accessibility to the fluorophore *(67)*. Thus, in a nonaggregated state, water is accessible to the EDANS and thus the fluorescence is quenched. When Aβ assembly occurs, water access is restricted and thus the fluorescence increases. These workers, by using EDANS-Aβ as energy acceptor and tryptophan-labeled Aβ as fluorescence energy donor have shown that it is possible to study the kinetics of Aβ fibrillization by FRET. This report also highlights the advantages of using parallel complementary analytical techniques in that the fibrillization process was monitored by CD spectroscopy, dynamic light scattering, and AFM, allowing the authors to describe three separate Aβ association reactions. These reactions occurred at endosomal pH and within time frames similar to that observed for endosomal transit in vivo leading the authors to conclude that the small spherical aggregate forms detected in vitro might correspond to the soluble neurotoxic Aβ species secreted by neurons *(68)*.

Many studies quoted above report the formation of small oligomeric or globular species of Aβ on route to long mature fibrils. The differences between laboratories and techniques are exemplified by a fluorescence correlation spectroscopy study presented by Tjernberg and colleagues (69). Fluorescence correlation spectroscopy should allow the simultaneous detection of Aβ species with different molecular weights at low nanomolar concentrations. However, using rhodamine labeled Aβ 1-40 they reported that the polymerization of the labeled peptide proceeded very rapidly from monomer/dimer to large aggregate without any small oligomeric or even micelle intermediate. Although the concentration dependence of rhodamine-Aβ 1-40 was similar to that reported for native Aβ 1-40 *(70)*, the rapidness of the fibrillization process (40 min to maximum) is obviously in contrast to AFM data, where small globular forms have been reported and even the more amyloidogenic Aβ 1-42 takes days to form fibrils *(71,72)*.

As noted above, fluorescently tagged Aβ has been used in many studies of peptide fibrillization. Although it is claimed, as with 6-(*N*-(7-nitrobenz-2-oxa-1,3-diazol-4-yl)amino) hexanoic acid-Aβ (NBD-Aβ; ref. *73*), that the presence of the tag does not influence the folding and polymerization process, when numerous data demonstrate how readily the fibrillization process is affected, even by the particular synthetic batch of peptide, it becomes obvious that it is preferable to work with the native protein. The dramatic increase in the fluorescence of thioflavin T that accompanies binding to amyloid has been used for many years for imaging amyloid fibril formation and. has recently been used in combination with total internal reflection fluorescence microscopy to visualize amyloid formation of β2-microglobulin *(74)*. Although yet to be applied to Aβ studies, total internal reflection fluorescence microscopy has the advantage over other fluorescence techniques in that it does not require the protein concerned to be tagged.

A major objective of any in vitro study employing synthetic Aβ peptides should be to work with peptide conformations and structures that mirror those observed in the AD brain. As noted earlier, the size of the straight mature fibrils of synthetic Aβ 1-40 reported in vitro is similar to that found in senile plaques in vivo *(17)*. In the AD brain, diffuse deposits of Aβ are also found, often being referred to as "preamyloid" on the belief that they are the precursors of the senile plaque. The amyloid in diffuse plaques is amorphous in nature, is non-Congophilic, and contains little fibrillar Aβ *(75)*. The lack of ordered β-sheet secondary structure for the protein in these deposits obviously precludes the use of a number of analytical techniques that have been used for studying Aβ fibrils. Like EDANS, NBD is a fluorophore that is readily quenched by water and, in combination with unlabeled Aβ, NBD-Aβ 1-40 has been used in fluorescence emission and depolarization spectroscopy, in parallel with CD spectroscopy and EM to study unstructured aggregates in vitro and to provide evidence that the amorphous Aβ deposits are indeed precursors of the fibrillar protein found in senile plaques *(73)*.

3. BIOLOGICAL STUDIES

3.1. NEUROTOXICITY OF Aβ

Despite the large numbers of reports describing the Aβ fibrillization process from a biophysical standpoint, the vast majority are simply studies of the aggregation of a synthetic peptide and provide no link to any biological feature associated with the polymerization. When attempts have been made to correlate molecular changes with a biological response, the latter usually takes the form of cytotoxicity in cultured cells, probably because (1) the presence of Aβ in senile plaques suggested that the neurodegeneration observed in AD arose from a neurotoxic effect of the peptide and (2) the early reports described effects in cultured cells *(11,76)*. Although criticisms have been directed towards the use of MTT assays as relevant means of assessing the physiological effects of Aβ *(77)*, this feature is the biological correlate for many studies of Aβ fibrillization.

Characteristically, Aβ neurotoxicity (MTT assay) has been shown to be associated with the presence of fibrils and protofibrils on visualization by EM *(17,78)*. In parallel to the development of neurotoxicity, the presence of β-sheet structure, as reflected in the major amide I component of infrared spectra, was shown to increase with incubation time of Aβ 1-40 *(17,78)*. Somewhat contrary to this, it has been demonstrated that Aβ 1-40 polymerized at acidic pH is characterized under EM by amorphous sheet-like aggregates that, in MTT assays, were not neurotoxic *(40)*.

Several different hypotheses that have been put forward to explain the mechanism behind the neurotoxic properties of Aβ. These include: (1) the formation of calcium channels in cell membranes *(79)*; (2) an interaction between Aβ and binding targets such as the receptor for advanced glycation end-products (*80*), the serpin enzyme complex (*81*), or endoplasmic reticulum Aβ binding protein (*82*); (3) membrane disruption associated with partial assembly of Aβ within the cell membrane *(83)* and (4) direct or indirect free radical production *(84,85)*.

An alternative theory to explain the aggregation and deposition of Aβ in AD centers around the observation that oxidative damage to phospholipids precedes and may actually stimulate the fibrillization of Aβ. Phospholipids and gangliosides have been shown to enhance β-sheet formation *(86–88)*. Interactions of Aβ 1-40 with ganglioside-containing membranes determined by polarized attenuated total internal reflection Fourier transform infrared spectroscopy suggested the formation of anti-parallel β sheet with the binding being dependent upon the nature of the lipid matrix (89). Koppaka et al (90) also used polarized attenuated total internal reflection Fourier transform infrared spectroscopy to monitor the rate of accumulation, the conformation, and the orientation of Aβ peptide interacting with an oxidatively damaged phospholipid bilayer. The accumulation of β-sheet conformation Aβ 1-42 was accelerated by oxidatively damaged phospholipids compared with nonoxidized or saturated phospholipids supporting a role of lipid peroxidation in AD pathogenesis.

Although many in vitro studies have been conducted with the full length 40- and 42-amino acid Aβ peptides, in CSF and particularly in brains of patients with AD, much of the amyloid protein is N-terminally truncated *(5)*. However, studies employing short Aβ peptides have generally opted for nonnaturally occurring fragments with enhanced solubility rather than selecting the physiologically relevant forms. Some reports on N-terminally truncated fragments do exist. Pike and co-workers *(91)* reported that N-terminal deletion resulted in enhanced peptide aggregation with fibrillar morphology, CD spectra showing β-sheet structure and neurotoxicity in cultures of rat hippocampal neurons. Of particular interest is the observation that truncation of the entire β- to α- secretase fragment, as in Aβ 17-40 and Aβ 17-42, produced an amyloidogenic, β-sheet forming neurotoxic form of Aβ *(91)*. Thus, β-secretase cleavage of APP within the Aβ segment should not necessarily preclude the formation of a neurotoxic species and, indeed, Aβ 17-42 has been reported in AD diffuse plaque *(92)*. In contrast, by monitoring fibril formation using EM, CD spectroscopy, and cell toxicity in PC 12 cells, Seilheimer et al. *(93)* were able to demonstrate that C-terminally truncated Aβ peptides show a progressive decline in amyloidogenicity and neurotoxicity. Thus, Aβ 1-42 was more fibrillogenic than shorter forms although considerable lot-to-lot variability was noted in neurotoxic responses, data supported by CD spectroscopy where poorly toxic forms were largely random coil in structure. A considerable proportion of the Aβ in the brains of patients with AD has been found to begin with a pyroglutamate at position 3 *(94)*, a modification rendering the peptide resistant to most aminopeptidase activity *(95)*. The truncated peptides Aβ 3(pyroGlu)-40 and Aβ 3(pyroGlu)-42 exhibit virtually identical CD spectroscopy profiles to the native peptides and have comparable toxic effects in cortical cell culture systems *(96)*. Hence, the reduced clearance and build-up of Aβ 3(pyroGlu)-42 in particular in senile plaques may play a significant role in cell death in AD.

Assuming that the Amyloid Hypothesis is correct and that the aggregated Aβ is a major pathogenic feature in AD, the mechanism by which Aβ elicits the cell death characteristic of the neurodegeneration in the disease state is unclear. Although apoptosis has been proposed as a means through which neuronal death may occur with Aβ *(97,98)*, there is little evidence of apoptosis in AD. A number of studies have investigated the interactions between Aβ and lipid membranes, proposing some sort of alteration in the membrane activity. By binding Aβ 29-40 and Aβ 29-42 to lipid, Demeester et al. *(99)* were able to show by CD spectroscopy and FTIR spectroscopy that the induction of apoptosis in mouse Neuro 2A neuroblastoma and rat PC12 cells accompanied an α-helix to β-sheet change in peptide conformation. The data concerning Aβ 29-40/42 peptides raises the possibility that in normal APP processing, after intramembranal gamma-secretase cleavage at the C-terminus of Aβ, fibrillization of the peptide may begin within the membrane.

The relationship between secondary structure and biological properties has also been demonstrated for Aβ 25-35. In aqueous solution, Aβ 25-35 shows a correlation between the development of neurotoxicity and β-sheet formation as assessed by FTIR and CD spectroscopy *(100)*. In contrast, Aβ 35-25 shows neither β-sheet formation in solution nor any biological effect while amidating the C-terminus of Aβ 25-35, thus locking the peptide fragment in a random coil formation, also prevents any neurotoxicity developing. Although receptor interactions have been proposed for Aβ *(80)*, the observation that all-D and all-L stereoisomers of Aβ 25-35 and Aβ 1-42 have virtually identical properties, in terms of CD spectra, EM appearance, and neurotoxic effects, effectively rules out a stereospecific ligand-receptor interaction and points more to the fibrillar features of the Aβ peptide being the vital determinant of biological effect *(101)*.

Although Aβ 25-35 is neurotoxic, obviously this fragment does not have the histidine residues at positions 6, 13, and 14, or the tyrosine residue at position 10, which are considered to be important metal binding sites. Nevertheless, Aβ 25-35 does result in the generation of hydrogen peroxide, which might suggest that there is at least one other suitable metal binding site between residues 25 and 35. The methionine residue at position 35 is an obvious candidate and substitution of Met35 in Aβ (25-35) by leucine, norleucine, lysine, or tyrosine residues results in peptides that neither aggregate nor are neurotoxic. However, when Met35 is replaced by aspartate, serine or cysteine residues aggregation and neurotoxicity are retained *(102)*.

Nevertheless, care needs to be taken in viewing any studies of Aβ 25-35 because it has been shown that there are differences in the neurotoxic properties of this short fragment compared to those shown by full-length Aβ 1-42 *(103)*. Not only is Aβ 25-35 more rapidly toxic that Aβ 1-42, the short fragment also lacks the metal binding properties of the full length peptide *(104)*. Varadarajan and colleagues *(103)* used a combination of electron paramagnetic resonance, metal binding assays, and EM to explore the relationship between structure and cell toxicity. Spin trapping showed that both Aβ 25-35 and Aβ 1-42 are capable of inducing cellular toxicity by membrane protein oxidation. However, the lack of the strong Cu(II) binding sites (histidines at positions 6, 13, and 14) precludes the role of transition metal ions in the neurotoxic process elicited by Aβ 25-35. A number of papers have acknowledged the role of the methionine at position 35 in determining the neurotoxic effects of the peptides. The secondary structure of Aβ 1-42 around Met35 probably contributes to the oxidative stress and neurotoxic properties of the peptide. Replacement of the Ile31 residue by Pro breaks the proposed helical interaction between these two amino acids, as shown by CD spectroscopy, resulting in a loss of aggregatory and consequent neurotoxic properties *(105)*.

3.2. INHIBITORS OF Aβ AGGREGATION

Belief in the Amyloid Hypothesis as a major pathogenic feature in the development of AD has led to many attempts to discover small molecules capable of inhibiting the fibrillization process. One of the first compounds demonstrated as being capable of preventing Aβ 1-40 fibrillization was the water-soluble oligosaccharide β-cyclodextrin (106). When Aβ 1-40 was preincubated with β-cyclodextrin and then added to PC12 rat phaeochromocytoma cells, the decrease in cell viability normally associated with aggregated Aβ was attenuated. Electrospray mass spectrometry suggested that the aromatic moieties of the tyrosine and phenylalanine residues within the Aβ molecule had complexed with the hydrophobic cavity of the cyclic oligosaccharide, thus preventing fibrillization *(106)*. It has not always been possible to demonstrate such clear interactions between peptide and inhibitors. For instance, a series of benzofuran molecules were shown by electron microscopy to prevent the formation of the large straight fibrils characteristic of Aβ 1-40 aggregation *(107)*. These compounds also prevented the production of a neurotoxic species of Aβ, but attempts to identify a binding site for the compounds with the peptide by liquid scintillation and NMR spectroscopy were prevented by the self-aggregation of the inhibitors and the apparent formation of Aβ-inhibitor aggregate complexes. Similar difficulties in understanding Aβ-inhibitor interactions were also noted with other compounds (108).

Although somewhat cumbersome, cell-based neurotoxicity assays have been used in attempts to identify compounds that inhibit the formation of toxic forms of Aβ. A relatively miscellaneous series of inhibitors, discovered by screening in a PC12 cell viability assay, have been reported *(109)*. The lead compound is this report was shown by EM to prevent mature fibril formation by Aβ 1-42, but nonfibrillar sheet like polymeric structures were still observed. The behavior of the peptide may, however, have been influenced by the derivatized Aβ used in the assays. SPR studies showed that the compound bound to mature fibrils but not monomeric Aβ *(109)*, suggesting an interaction at a polymer stage. A further study *(110)* reported on a series of imidazopyridoindoles that inhibited the fibrillization of Aβ 25-35, again in a PC12-based neurotoxicity assay. These researchers also describe the use of CD spectroscopy to follow the time-dependent changes in peptide conformation prepared under varying conditions. Their data demonstrate very ably how peptide concentration, pH, temperature, and salt concentration can influence the Aβ fibrillization process and suggests that a lack of a basic understanding of the factors controlling this process are at the root of the differing data describing Aβ aggregation products.

SPR data have demonstrated the complex relationship between inhibition of fibrillization and binding of inhibitor to the short Aβ 10-35 fragment *(111)*. These researchers found an excellent correlation between SPR-determined affinity for binding to Aβ and prevention of cell toxicity for a series of short peptides modelled around the KLVFF hydrophobic core of Aβ. However, other inhibitors of fibrillization and fibril-dependent toxicity, such as melatonin and rifampicin, did not show any apparent binding. NMR spectroscopy data *(38)* indicate that melatonin binds to Aβ 1-40 and Aβ 1-42 peptides. The SPR data, therefore, may point to differences in the conformation of the Aβ 10-35 when bound to the SPR chip playing an important role in determining the binding of inhibitors.

One of the few compounds to show inhibition of Aβ accumulation in transgenic mice is the copper/zinc chelator clioquinol. When administered orally for 9 wk to Tg2576 APP transgenic mice, clioquinol decreased brain Aβ content by almost 50% *(112)*. At the in vitro level, clioquinol inhibits and reverses copper- and zinc-induced Aβ 1-40 and Aβ 1-42 aggregation. NMR spectroscopy showed that Cu^{2+} bound to the histidines of Aβ, as witnessed by a broadening of the NMR peaks. Subsequent addition of clioquinol reversed this effect consistent with removal of Cu^{2+} from the metal binding sites.

The inhibitory activity of some fibrillization inhibitors has been claimed to be associated with a free-radical scavenging property of the compounds. ESR spectrometric studies using N-tert-butyl-β-phenylnitrone (PBN) as a spin trapping agent indicate that rifampicin-type compounds inhibit the production of Aβ 1-40 induced hydroxyl radicals *(113)*. Other groups, however, have proposed that Aβ does not produce free radicals and that the ESR spectroscopy in such studies is actually detecting the hydrolysis and decomposition of PBN *(114)*.

Further confusion over the mechanism of action of so-called fibrillization inhibitors is further exemplified by studies such as that described by Kim et al. *(115)*. The subject of this work, Congo red, binds to all types of amyloid fibrils and its shift in absorbance and birefringence under polarized light has long been a characteristic used for determining the presence of amyloid. Despite Congo red being a pathological determinant of senile plaques in AD, in vitro, Congo red can either enhance or inhibit Aβ fibrillization (see ref. *115* and references therein). The behavior of Aβ in the presence of Congo red has, however, been partially explained by Kim and colleagues *(115)* with a dimeric amyloidogenic light chain variable domain. Using isothermal titration calorimetry to monitor the thermodynamics of Congo red binding to amyloidogenic protein, they demonstrated that at low ratios of Congo red to protein, being an enthalpy-driven process, Congo red favors a structurally perturbed, aggregation-prone species. Thus, at low Congo red to protein ratios, enhancement of fibrillization occurs. Conversely, at high Congo red to protein ratios, protein unfolding can occur. This study does highlight the potential complexities of Aβ-fibrillization inhibitor interactions and the obvious need for thorough evaluation of potential Aβ fibrillization inhibitors.

Any derivatization may affect the behavior of Aβ peptides, as may the deposition of Aβ onto the support surfaces used in electron microscopy and AFM studies. Recent data using AFM has visualized protofibril formation by Aβ on a mica and oxidized graphite surface *(116,117)*. Koppaka et al. *(90)* have pointed out the similarities between their FTIR data demonstrating the effects of oxidatively damaged lipid membranes on Aβ fibrillization and AFM data, raising the possibility that the nature of the AFM surface may be influencing peptide aggregation behavior. The behavior of Aβ 25-35 in solution, where the formation of protofibrils has been reported, can be monitored by small-angle neutron scattering using ^2H_2O to enhance scattering contrast and highlight the protofibrils in solution *(118)*. In this study, Aβ 25-35 was shown to fibrillize and be toxic to PC12 cells. Using EM, we observed that a series of *N*-methylated derivatives of Aβ 25-35 interfered with the ability of Aβ 25-35 to form neurotoxic species. In particular, NMeGly33 was shown by small-angle neutron scattering to be effective in preventing protofibril formation when added at the initiation of the polymerization process at a ratio of 1:1 with Aβ 25-35. Although data were not presented to support this, the equimolar effectiveness of NMeGly33 suggests an interaction with Aβ 25-35 at the earliest stage in the fibrillization process.

N-methylated fragments of Aβ 1-40 have also been reported to prevent fibrillization of the full-length peptide in a molar ratio of inhibitor to peptide of 2:10 *(119)*. Using NMR and CD spectroscopy, it has been demonstrated that *N*-methylated Aβ 16-20 adopts an extended Aβ strand conformation and that in this state, possessing solubility in both aqueous and organic solvents, the peptide has enhanced membrane permeability properties.

3.3. SHORT FRAGMENTS

To overcome the solubility difficulties encountered in working with full length Aβ peptides (1-40 and 1-42), many groups have turned to shorter fragments of the amyloid peptide. The hydrophobic core region in particular has been studied, largely because it is believed that residues 17-21 are essential for the polymerization of the full-length peptide *(27,120)*. The importance of this region in the fibrillization process is also supported by data showing that short peptides based on this hydrophobic region can inhibit the aggregation of the full-length peptide, presumably by complexing with that region of the full length peptide *(121,122)*. Although short Aβ fragments do form fibrils and lend themselves to NMR characterization, there are, as reported with Aβ 16-22, inadequacies such as not being able to determine supramolecular structure on such short fragments *(123)*.

Numerous groups have worked with the Aβ 25-35 peptide based on the knowledge that many of the neurotoxic effects of full-length Aβ 1-40 or 1-42 are retained within this short fragment. The Aβ 31-35 region in particular is interesting in that 3D.NMR has suggested α-helical conformation *(124,125)* whereas X-ray diffraction points to a reverse turn at Gly33 *(126)*. Characteristically, a reverse turn can initiate β-sheet formation *(127)* and hence ultimately lead to fibril formation. X-ray diffraction coupled with EM has suggested that the reverse turn proposed for the Aβ 31-35 domain is exposed in oligomeric forms of the peptide and that this may facilitate binding to the tachykinin receptor *(126)*. The issue of whether Aβ is a tachykinin receptor ligand arose from the recognition of structural similarity between Aβ and tachykinin and the fact that substance P attenuated Aβ neurotoxicity *(11)*. Other data have not particularly supported this concept *(128,129)*. The Aβ-tachykinin issue has been explored more recently using EM, CD, and NMR spectroscopy by comparing Aβ 25-35, Aβ 25-35-amide, and their Nle35 and Phe31 analogs *(130)*. The use of these fragments was based on the assumption that a true tachykinin would have an amide C-terminus and a Phe residue at position 31 (i.e. an aromatic residue five amino acids from the C-terminus). The combination of the analytical techniques confirmed the tendency for the native Aβ 25-35 to form β-sheet structure (and fibrils), features that were absent in the more tachykinin-like peptides.

The importance of the Met35 residue and oxidized Met35 has been explored in a CD and NMR spectroscopy study where it was shown that although Aβ 1-40 and Aβ 1-40 Met35(O) both adopt a random coil conformation when freshly dissolved in water, the oxidized methionine peptide does not aggregate, whereas during the course of several days, the nonoxidized peptide takes on characteristic β-sheet structure *(131)*. NMR data suggested that the presence of the oxidized Met35 prevented the formation of a helical region between residues 28 and 36 thus preventing random coil to β-sheet transition. Support for this suggestion has come from Fourier transform ion cyclotron resonance electrospray mass-spectrometry, a technique that allows kinetics to be studied at nano- and micromolar concentrations *(132)*. Data on the behavior of Aβ 1-42 Met35(O) is somewhat controversial, with studies showing either no effect of methionine oxidation *(93,103)* or inhibition of fibrillization *(133)*. This is in contrast to data with Aβ1-40 where oxidation of Met35 does not inhibit the propensity of the peptide to form fibrils. Although this difference in behavior of the two oxidized peptide is apparent in vitro, whether there is any pathologic relevance to these findings is not known. However, it has been reported in a Raman spectroscopy study of senile plaque cores that there is extensive methionine oxidation and co-ordination of Zn(II) and Cu(II) with histidine residues, confirming data in vitro *(134)*. Interestingly, treatment of the plaques with a metal chelator resulted in a greater heterogeneity of the β-sheet structure supporting the role of metal ions in plaque formation.

Aβ 10-35 is an attractive proposition for study because of its enhanced solubility (compared with Aβ 1-40 and 1-42), ability to form fibrils and to add to plaques *(135)*. There have been a number of solid-state NMR studies of Aβ 10-35 peptide designed to analyze the end product of the fibrillization process. For instance, ^{13}C-labelled Aβ 10-35 data supports the existence of parallel β-sheets with H bonds along the fibril axis *(136)*.

3.4. PROTOFIBRILS

One of the major issues in the Aβ aggregation/fibrillization field in the last decade has been the identification of the toxic species. Tissue from AD brain, almost exclusively studied at autopsy, shows amyloid plaques surrounded by regions of degenerating neurons. In vitro data with synthetic Aβ peptides repeatedly demonstrates that when incubated under particular conditions, various multimeric species of peptide are formed. Furthermore, these species, formed under conditions determined by their exponents, have invariably been shown to be neurotoxic. Both in vivo and in vitro, therefore, it is far from clear which is the "killer" form of Aβ or indeed whether multiple species of peptide share the guilt. A review of the Aβ assembly literature is beyond the scope of this present chapter —the interested reader is referred to recent reviews (e.g., refs. *137,138*). However, the basic picture is that the earliest Aβ toxicity studies pointed to fibrillar Aβ as the toxic form *(139,140)*, which was supported and extended by numerous other studies demonstrating that "ageing" of the Aβ peptide was necessary for neurotoxicity to be observed (e.g., refs. *14,15,17*). Questions were obviously asked about the assembly process from the smallest Aβ species up to mature fibrils and the intermediate protofibrillar form of Aβ peptide was reported by *(23,141)*. The small protofibrillar forms of Aβ were subsequently shown to be neurotoxic *(78,142,143)*. Perhaps not surprisingly, the process of protofibril growth is not simple/uniform. Using multi-angle light scattering and AFM, Nichols et al. *(144)* have shown how salt concentration affects protofibril extension. In the absence of salt, protofibril growth occurs purely by the deposition of monomer on the ends of the protofibrils with no increase in protofibril diameter. In contrast, in the presence of salt, lateral association of protofibrils occurs. Unfortunately this report does not relate the two types of protofibril growth to their biological properties.

The existence of protofibrillar Aβ structures has been shown by atomic force microscopy where the absorption of Aβ monolayers onto a gold surface has been used to explore early fibril structure *(145)*. The small globular species of Aβ visualized by AFM in this study correspond to the small oligomeric forms isolated both from brain and from synthetic Aβ preparations *(146,147)*. The presence of multiple globular species formed during the incubation of Aβ 1-42 and detected by AFM has also been reported by Parbhu et al. *(71)*, although these workers reported that Aβ fibrils were only detected after prolonged incubation (>24 h) at high concentration (3 mg/mL) in phosphate-buffered saline, pH 7.4. It is possible that the lack of fibrils in this latter study is associated with the removal of large Aβ aggregates (nuclei?) by the methodology employed in that the samples were centrifuged immediately after dissolving the peptide and before the incubation.

In many studies, globular structures are seen as one of the earliest structures observed when Aβ peptide solutions are incubated in vitro. The study by Goldsbury et al. *(72)* provides a good example of the use of EM and scanning TEM to monitor the development of Aβ protofibrils and fibrils. They describe a process of change from globule to protofibril to long straight fibril, occurring over the course of several days, with the increase in fibril formation being paralleled by increases in thioflavin T fluorescence. Scanning transmission electron microscopy, which facilitates quantification of protein mass and thus allows mass per length determination showed that under the conditions employed, after 2 d incubation, a broad distribution of globular particle size was apparent with individual particles containing 50 to 200 Aβ molecules. At the same time point, however, protofibrils were of a single size, calculated to be 19 kDa per nm. Further incubation resulting in mature fibril formation did not result in additive changes in the mass per length of the structure reinforcing the belief that the protofibrils are the precursors of mature fibrils.

In parallel to the protofibril we have witnessed the emergence of the amyloid-derived-diffusable-ligand or ADDL *(68,148)* as a multimeric neurotoxic form of Aβ. AFM studies of small oligomers and ADDLs have invariably described them as nonfibrillar globular Aβ forms *(68,148,149)*. EM studies have also reported other globular forms of aggregated Aβ peptide *(150,151)*.

To add credence to the belief that any of these synthetic aggregates is relevant to AD, their existence in AD brain needs to be determined. Some evidence for this has emerged from immunohistochemical studies where antibodies specific for small soluble Aβ oligomers *(152)* and ADDLs *(153)* have been shown to label immunoreactive deposits in AD brain.

4. CONCLUSIONS

The application of the many techniques described in this chapter have no doubt greatly extended our knowledge of the biophysical behavior and properties of synthetic Aβ peptide. The attainment of amyloid status confers, on this molecule properties, such as insolubility and protease resistance, that naturally limit analysis of endogenous Aβ in its native environment. Consequently, any techniques that operate with the peptide in a solvent or aqueous mobile phase are almost certainly guilty of inducing conformational changes in the peptide. This obviously applies to endogenous as well as synthetic Aβ.

The last decade has seen acceptance by many workers that the fibrillization of synthetic Aβ produces conformational changes that result in a species of peptide that induces some form of cell death when added to cultures of human, rat or mouse neurons, although we have yet to see this property linked conclusively to the neurodegeneration observed in AD.

ACKNOWLEDGMENTS

Dr. Kevin Jennings, GlaxoSmithKline, is gratefully thanked for the electron micrographs of Aβ 1-40.

REFERENCES

1. Glenner GG, Wong CW. Alzheimer's disease: initial report of the purification and characterization of a novel cerebrovascular amyloid protein. Biochem Biophys Res Commun 1984;120:885–890.
2. Hardy J, Allsop D. Amyloid deposition as the central event in the aetiology of Alzheimer's disease. Trends Pharmacol Sci 1991;12:383–388.
3. Masters CL, Simms G, Weinman NA, Multhaup G, McDonald BL, Beyreuther K. Amyloid plaque core protein in Alzheimer's disease and Down syndrome. Proc Natl Acad Sci USA 1985;82:4245–4249.
4. Kang J, Lemaire HG, Unterbeck A, Salbaum JM, Masters CL, Grzeschik KH, et al. The precursor of Alzheimer's disease amyloid A4 protein resembles a cell-surface receptor. Nature 1987;325:733–736.
5. Roher AE, Lowenson JD, Clarke S, Wolkow C, Wang R, Cotter RJ, et al. Structural alterations in the peptide backbone of beta-amyloid core protein may account for its deposition and stability in Alzheimer's disease. J Biol Chem 1993;268:3072–3083.
6. Vigo-Pelfrey C, Lee D, Keim P, Lieberburg I, Schenk DB. Characterization of beta-amyloid peptide from human cerebrospinal fluid. J Neurochem 1993;61:1965–1968.
7. Seubert P, Vigo-Pelfrey C, Esch F, Lee M, Dovey H, Davis D, et al. Isolation and quantification of soluble Alzheimer's beta-peptide from biological fluids. Nature 1992;359:325–327.
8. Iwatsubo T, Odaka A, Suzuki N, Mizusawa H, Nukina N, Ihara Y. Visualisation of A beta 42(43) and A beta 40 in senile plaques with end-specific A beta monoclonals: Evidence that an initially deposited species is A beta 42(43). Neuron 1994;13:45–53.
9. Whitson JS, Glabe CG, Shintani E, Abcar A, Cotman CW. Beta-amyloid protein promotes neuritic branching in hippocampal cultures. Neurosci Lett 1990;110:319–324.
10. Whitson JS, Selkoe DJ, Cotman CW. Amyloid beta protein enhances the survival of hippocampal neurons in vitro. Science 1989;243:1488–1490.

11. Yankner BA, Duffy LK, Kirschner DA. Neurotrophic and neurotoxic effects of amyloid beta protein: reversal by tachykinin neuropeptides. Science 1990;250:279–282.
12. Frautschy SA, Baird A, Cole GM. Effects of injected Alzheimer beta-amyloid cores in rat brain. Proc Natl Acad Sci USA 1991;88:8362–8366.
13. Kowall NW, Beal MF, Busciglio J, Duffy LK, Yankner BA. An in vivo model for the neurodegenerative effects of b-amyloid and protection by substance P. Proc Natl Acad Sci U S A 1991;88:7247–7251.
14. Pike CJ, Burdick D, Walencewicz AJ, Glabe CG, Cotman CW. Neurodegeneration induced by beta-amyloid peptides in vitro: the role of peptide assembly state. J Neurosci 1993;13:1676–1687.
15. Pike CJ, Walencewicz AJ, Glabe CG, Cotman CW. In vitro aging of beta-amyloid protein causes peptide aggregation and neurotoxicity. Brain Res 1991;563:311–314.
16. May PC, Gitter BD, Waters DC, Simmons LK, Becker GW, Small JS et al. beta-Amyloid peptide in vitro toxicity: lot-to-lot variability. Neurobiol Aging 1992;13:605–607.
17. Howlett DR, Jennings KH, Lee DC, Clark MSG, Brown F, Wetzel R, et al. Aggregation state and neurotoxic properties of Alzheimer beta- amyloid peptide. Neurodegeneration 1995;4:23–32.
18. Shen CL, Murphy RM. Solvent effects on self-assembly of beta-amyloid peptide. Biophys J 1995;69:640–651.
19. Fezoui Y, Hartley DM, Harper JD, Khurana R, Walsh DM, Condron MM, et al. An improved method of preparing the amyloid beta-protein for fibrillogenesis and neurotoxicity experiments. Amyloid 2000;7: 166–178.
20. Jarrett JT, Berger EP, Lansbury PT, Jr. The carboxy terminus of the beta amyloid protein is critical for the seeding of amyloid formation: implications for the pathogenesis of Alzheimer's disease. Biochemistry 1993;32:4693–4697.
21. Lomakin A, Chung DS, Benedek GB, Kirschner DA, Teplow DB. On the nucleation and growth of amyloid beta-protein fibrils - detection of nuclei and quantitation of rate constants. Proc Natl Acad Sci USA 1996;93:1125–1129.
22. Lomakin A, Benedek GB, Teplow DB, Lomakin A, Benedek GB, Teplow DB et al. Monitoring protein assembly using quasielastic light scattering spectroscopy. Methods Enzymol 1999;309:429–459.
23. Walsh DM, Lomakin A, Benedek GB, Condron MM, Teplow DB. Amyloid beta-protein fibrillogenesis - detection of a protofibrillar intermediate. J Biol Chem 1997;272:22364–22372.
24. Harper JD, Lansbury PT. Models of amyloid seeding in Alzheimer's-disease and scrapie - mechanistic truths and physiological consequences of the time- dependent solubility of amyloid proteins. Ann Rev Biochem 1997;66:385–407.
25. Hasegawa K, Ono K, Yamada M, Naiki H. Kinetic modeling and determination of reaction constants of Alzheimer's beta-amyloid fibril extension and dissociation using surface plasmon resonance. Biochemistry 2002;41:13489–13498.
26. Goldsbury C, Kistler J, Aebi U, Arvinte T, Cooper GJS. Watching amyloid fibrils grow by time-lapse atomic force microscopy. J Mol Biol 1999;285:33–39.
27. Kirschner DA, Inouye H, Duffy LK, Sinclair A, Lind M, Selkoe DJ. Synthetic peptide homologous to beta protein from Alzheimer disease forms amyloid-like fibrils in vitro. Proc Natl Acad Sci USA 1987;84:6953–6957.
28. Fraser PE, Nguyen JT, Surewicz WK, Kirschner DA. pH-dependent structural transitions of Alzheimer amyloid peptides. Biophys J 1991;60:1190–1201.
29. Halverson KJ, Sucholeiki I, Ashburn TT, Lansbury PT. Location of 8-sheet-forming sequences in amyloid proteins by FTIR. J Am Chem Soc 1991;113:6701–6703.
30. Hilbich C, Kisters-Woike B, Reed J, Masters CL, Beyreuther K. Aggregation and secondary structure of synthetic amyloid beta A4 peptides of Alzheimer's disease. J Mol Biol 1991;218:149–163.
31. Sunde M, Serpell LC, Bartlam M, Fraser PE, Pepys MB, Blake CCF. Common core structure of amyloid fibrils by synchrotron x-ray-diffraction. J Mol Biol 1997;273:729–739.
32. Sunde M, Blake C. The structure of amyloid fibrils by electron-microscopy and x- ray-diffraction. Adv Protein Chem 1997;50:123–159.
33. Perutz MF, Finch JT, Berriman J, Lesk A. Amyloid fibers are water-filled nanotubes. Proc Natl Acad Sci USA 2002;99:5591–5595.
34. Merz PA, Wisniewski HM, Somerville RA, Bobin SA, Masters CL, Iqbal K. Ultrastructural morphology of amyloid fibrils from neuritic and amyloid plaques. Acta Neuropathol 1983;60:113–124.
35. Narang HK. High-resolution electron microscopic analysis of the amyloid fibril in Alzheimer's disease. J Neuropath Exp Neurol 1980;39:621–631.
36. Stine WB, Snyder SW, Ladror US, Wade WS, Miller MF, Perun TJ, et al. The nanometer-scale structure of amyloid-beta visualized by atomic force microscopy. J Protein Chem 1996;15:193–203.
37. Harper JD, Lieber CM, Lansbury PT. Atomic-force microscopic imaging of seeded fibril formation and fibril branching by the Alzheimer's-disease amyloid-beta protein. Chem Biol 1997;4:951–959.
38. Pappolla M, Bozner P, Soto C, Shao H, Robakis NK, Zagorski M et al. Inhibition of Alzheimer beta-fibrillogenesis by melatonin. J Biol Chem 1998;273:7185–7188.
39. Skribanek Z, Balaspiri L, Mak M. Interaction between synthetic amyloid-beta-peptide (1–40) and its aggregation inhibitors studied by electrospray ionization mass spectrometry. J Mass Spectrom 2001;36:1226–1229.
40. Howlett DR, Ward RV, Bresciani L, Jennings KH, Christie G, Allsop D, et al. Identification of a high molecular weight toxic species of beta-amyloid. Alzheimer's Reports 1999;2:171–177.
41. Antzutkin ON, Leapman RD, Balbach JJ, Tycko R. Supramolecular structural constraints on Alzheimer's beta-amyloid fibrils from electron microscopy and solid-state nuclear magnetic resonance. Biochemistry 2002;41:15436–15450.
42. Petkova AT, Ishii Y, Balbach JJ, Antzutkin ON, Leapman RD, Delaglio F, et al. A structural model for Alzheimer's beta-amyloid fibrils based on experimental constraints from solid state NMR. Proc Natl Acad Sci USA 2002;99:16742–16747.
43. Torok M, Milton S, Kayed R, Wu P, McIntire T, Glabe CG, et al. Structural and dynamic features of Alzheimer's A beta peptide in amyloid fibrils studied by site-directed spin labeling. J Biol Chem 2002;277:40810–40815.
44. Jarrett JT, Berger EP, Lansbury PT, Jr. The C-terminus of the beta protein is critical in amyloidogenesis. Ann N Y Acad Sci 1993;695:144–148.
45. Bush AI. The metallobiology of Alzheimer's disease. Trends Neurosci 2003;26:207–214.
46. Miura T, Suzuki K, Kohata N, Takeuchi H. Metal binding modes of Alzheimer's amyloid beta-peptide in insoluble aggregates and soluble complexes. Biochemistry 2000;39:7024–7031.
47. Curtain CC, Ali F, Volitakis I, Cherny RA, Norton RS, Beyreuther K, et al. Alzheimer's disease amyloid-beta binds copper and zinc to generate an allosterically ordered membrane-penetrating structure containing superoxide dismutase-like subunits. J Biol Chem 2001; 276:20466–20473.
48. Ulrich J, Meier-Ruge W, Probst A, Meier E, Ipsen S. Senile plaques: staining for acetylcholinesterase and A4 protein: a comparative study in the hippocampus and entorhinal cortex. Acta Neuropathol (Berl) 1990;80:624–628.
49. Geula C, Greenberg BD, Mesulam MM. Cholinesterase activity in the plaques, tangles and angiopathy of Alzheimer's disease does not emanate from amyloid. Brain Research 1994;644:327–330.
50. Inestrosa NC, Alvarez A, Perez CA, Moreno RD, Vicente M, Linker C et al. Acetylcholinesterase accelerates assembly of amyloid-beta-peptides into alzheimers fibrils—possible role of the peripheral site of the enzyme. Neuron 1996;16:881–891.
51. Alvarez A, Opazo C, Alarcon R, Garrido J, Inestrosa NC. Acetylcholinesterase promotes the aggregation of amyloid-beta-peptide fragments by forming a complex with the growing fibrils. J Mol Biol 1997;272:348–361.
52. Bartolini M, Bertucci C, Cavrini V, Andrisano V. Beta-amyloid aggregation induced by human acetylcholinesterase: inhibition studies. Biochem Pharmacol 2003;65:407–416.
53. Murtomaki S, Risteli J, Risteli L, Koivisto UM, Johansson S, Liesi P. Laminin and its neurite outgrowth-promoting domain in the brain

in Alzheimer's disease and Down's syndrome patients. J Neurosci Res 1992;32:261–273.
54. Bronfman FC, Garrido J, Alvarez A, Morgan C, Inestrosa NC. Laminin inhibits amyloid-beta-peptide fibrillation. Neurosci Lett 1996;218:201–203.
55. Morgan C, Bugueno MP, Garrido J, Inestrosa NC. Laminin affects polymerization, depolymerization and neurotoxicity of Abeta peptide. Peptides 2002;23:1229–1240.
56. Bronfman FC, Alvarez A, Morgan C, Inestrosa NC. Laminin blocks the assembly of wild-type A beta and the Dutch variant peptide into Alzheimer's fibrils. Amyloid 1998;5:16–23.
57. Snow AD, Mar H, Nochlin D, Kimata K, Kato M, Suzuki S, et al. The presence of heparan sulfate proteoglycans in the neuritic plaques and congophilic angiopathy in Alzheimer's disease. Am J Pathol 1988;133:456–463.
58. Snow AD, Castillo GM. Specific proteoglycans as potential causative agents and relevant targets for therapeutic intervention in alzheimers- disease and other amyloidoses. Amyloid-Int J Exp Clin Invest 1997;4:135–141.
59. Snow AD, Mar H, Nochlin D, Sekiguchi RT, Kimata K, Koike Y et al. Early accumulation of heparan sulfate in neurons and in the beta-amyloid protein-containing lesions of Alzheimer's disease and Down's syndrome. Am J Pathol 1990;137:1253–1270.
60. McLaurin J, Franklin T, Kuhns WJ, Fraser PE. A sulfated proteoglycan aggregation factor mediates amyloid-beta peptide fibril formation and neurotoxicity. Amyloid-Int J Exp Clin Invest 1999;6:233–243.
61. Fraser PE, Darabie AA, McLaurin JA. Amyloid-beta interactions with chondroitin sulfate-derived monosaccharides and disaccharides. implications for drug development. J Biol Chem 2001;276: 6412–6419.
62. Abraham CR, Selkoe DJ, Potter H. Immunochemical identification of the serine protease inhibitor, a1-antichymotrypsin in the brain amyloid deposits of Alzheimer's disease. Cell 1988;52:487–501.
63. Rozemuller JM, Eikelenboom P, Stam FC, Beyreuther K, Masters CL. A4 protein in Alzheimer's disease: primary and secondary cellular events in extracellular amyloid deposition. J Neuropathol Exp Neurol 1989;48:674–691.
64. Fraser PE, Nguyen JT, McLachlan DR, Abraham CR, Kirschner DA. Alpha 1-antichymotrypsin binding to Alzheimer A beta peptides is sequence specific and induces fibril disaggregation in vitro. J Neurochem 1993;61:298–305.
65. Allsop D, Swanson L, Moore S, et al. Fluorescence anisotropy: a method for early detection of Alzheimer beta-peptide (Abeta) aggregation. Biochem Biophy Res Commun 2001;285:58–63.
66. Garzon Rodriguez W, Sepulveda Becerra M, Milton S, Glabe CG. Soluble amyloid Abeta-(1–40) exists as a stable dimer at low concentrations. J Biol Chem 1997;272:21037–21044.
67. Gorman PM, Yip CM, Fraser PE, Chakrabartty A. Alternate aggregation pathways of the Alzheimer beta-amyloid peptide: a beta association kinetics at endosomal pH. J Mol Biol 2003;325:743–757.
68. Lambert MP, Barlow AK, Chromy BA, et al. Diffusible, nonfibrillar ligands derived from a-beta(1–42) are potent central-nervous-system neurotoxins. Proc Natl Acad Sci USA 1998;95:6448–6453.
69. Tjernberg LO, Pramanik A, Bjorling S, et al. Amyloid beta-peptide polymerization studied using fluorescence correlation spectroscopy. Chem Biol 1999;6:53–62.
70. Lomakin A, Chung DS, Benedek GB, Kirschner DA, Teplow DB. On the nucleation and growth of amyloid beta-protein fibrils—detection of nuclei and quantitation of rate constants. Proc Natl Acad Sci USA 1996;93:1125–1129.
71. Parbhu A, Lin H, Thimm J, Lal R. Imaging real-time aggregation of amyloid beta protein (1–42) by atomic force microscopy. Peptides 2002;23:1265–1270.
72. Goldsbury CS, Wirtz S, Muller SA, et al. Studies on the in vitro assembly of A beta 1–40: Implications for the search for A beta fibril formation inhibitors. J Struct Biol 2000;130:217–231.
73. Huang TH, Fraser PE, Chakrabartty A. Fibrillogenesis of Alzheimer Abeta peptides studied by fluorescence energy transfer. J Mol Biol 1997;269:214–224.
74. Ban T, Hamada D, Hasegawa K, Naiki H, Goto Y. Direct observation of amyloid fibril growth monitored by thioflavin T fluorescence. J Biol Chem 2003;278:16462–16465.
75. Yamaguchi H, Hirai S, Shoji M, Harigaya Y, Okamoto Y, Nakazato Y. Alzheimer type dementia: diffuse type of senile plaques demonstrated by beta protein immunostaining. Prog Clin Biol Res 1989; 317:467–474.
76. Yankner BA, Dawes LR, Fisher S, Villa-Komaroff L, Oster-Granite ML, Neve RL. Neurotoxicity of a fragment of the amyloid precursor associated with Alzheimer's disease. Science 1989;245:417–420.
77. Abe K, Saito H. Amyloid beta protein inhibits cellular MTT reduction not by suppression of mitochondrial succinate dehydrogenase but by acceleration of MTT formazan exocytosis in cultured rat astrocytes. Neurosc Res 1998;31:295–305.
78. Ward RV, Jennings KH, Jepras R, et al. Fractionation and characterization of oligomeric, protofibrillar and fibrillar forms of beta-amyloid peptide. Biochem J 2000;348:137–144.
79. Arispe N, Pollard HB, Rojas E. beta-Amyloid Ca(2+)-channel hypothesis for neuronal death in Alzheimer disease. Mol Cell Biochem 1994;140:119–125.
80. Yan SD, Chen X, Fu J, et al. Rage and amyloid-beta peptide neurotoxicity in Alzheimer's-disease. Nature 1996;382:685–691.
81. Joslin G, Krause JE, Hershey AD, Adams SP, Fallon RJ, Perlmutter DH. Amyloid-beta peptide, substance P, and bombesin bind to the serpin-enzyme complex receptor. J Biol Chem 1991;266:21897–21902.
82. Yan SD, Fu J, Soto C, et al. An intracellular protein that binds amyloid-beta peptide and mediates neurotoxicity in alzheimers-disease. Nature 1997;389:689–695.
83. Yip CM, McLaurin J. Amyloid-beta peptide assembly: A critical step in fibrillogenesis and membrane disruption. Biophys J 2001;80:1359–1371.
84. Hensley K, Carney JM, Mattson MP, et al. A model for beta-amyloid aggregation and neurotoxicity based on free radical generation by the peptide: relevance to Alzheimer disease. Proc Natl Acad Sci USA 1994;91:3270–3274.
85. Huang X, Cuajungco MP, Atwood CS, et al. Cu(II) potentiation of Alzheimer abeta neurotoxicity. Correlation with cell-free hydrogen peroxide production and metal reduction. J Biol Chem 1999; 274:37111–37116.
86. Choo Smith LP, Garzon Rodriguez W, Glabe CG, Surewicz WK. Acceleration of amyloid fibril formation by specific binding of Abeta-(1–40) peptide to ganglioside-containing membrane vesicles. J Biol Chem 1997;272:22987–22990.
87. McLaurin J, Chakrabartty A. Characterization of the interactions of Alzheimer beta-amyloid peptides with phospholipid membranes. Eur J Biochem 1997;245:355–363.
88. McLaurin J, Fraser PE, Franklin T, Chakrabartty A. Phosphatidylnositol and inositol, possible nucleation seeds for amyloid fibril growth. J Neurochem 1997;69:S 46.
89. Matsuzaki K, Horikiri C. Interactions of amyloid beta-peptide (1–40) with ganglioside-containing membranes. Biochemistry 1999;38: 4137–4142.
90. Koppaka V, Axelsen PH. Accelerated accumulation of amyloid beta proteins on oxidatively damaged lipid membranes. Biochemistry 2000;39:10011–10016.
91. Pike CJ, Overman MJ, Cotman CW. Amino-terminal deletions enhance aggregation of beta-amyloid peptides in vitro. J Biol Chem 1995;270:23895–23898.
92. Gowing E, Roher AE, Woods AS, et al. Chemical characterization of A beta 17–42 peptide, a component of diffuse amyloid deposits of Alzheimer disease. J Biol Chem 1994;269:10987–10990.
93. Seilheimer B, Bohrmann B, Bondolfi L, Muller F, Stuber D, Dobeli H. The toxicity of the Alzheimer's beta-amyloid peptide correlates with a distinct fiber morphology. J Struct Biol 1997;119:59–71.
94. Kuo YM, Emmerling MR, Woods AS, Cotter RJ, Roher AE. Isolation, chemical characterization, and quantitation of A beta 3-pyroglutamyl peptide from neuritic plaques and vascular amyloid deposits. Biochem Biophys Res Commun 1997;237:188–191.
95. Mcdonald JK, Barrett AM. Mammalian Proteases. London: Academic Press; 1986.

96. Tekirian TL, Yang AY, Glabe C, Geddes JW. Toxicity of pyroglutaminated amyloid beta-peptides 3(pE)-40 and-42 is similar to that of A beta 1–40 and-42. J Neurochem 1999;73:1584–1589.
97. Forloni G, Chiesa R, Smiroldo S, et al. Apoptosis mediated neurotoxicity induced by chronic application of beta amyloid fragment 25–35. Neuroreport 1993;4:523–526.
98. Cotman CW, Whittemore ER, Watt JA, Anderson AJ, Loo DT. Possible role of apoptosis in alzheimers-disease. Ann N Y Acad Sci 1994;747:36–49.
99. Demeester N, Baier G, Enzinger C, et al. Apoptosis induced in neuronal cells by C-terminal amyloid beta-fragments is correlated with their aggregation properties in phospholipid membranes. Mol Membr Biol 2000;17:219–228.
100. Buchet R, Tavitian E, Ristig D, et al. Conformations of synthetic beta peptides in solid state and in aqueous solution: relation to toxicity in PC12 cells. Biochim Biophys Acta 1996;1315:40–46.
101. Cribbs DH, Pike CJ, Weinstein SL, Velazquez P, Cotman CW. All-d-enantiomers of beta-amyloid exhibit similar biological properties to all-l-beta-amyloids. J Biol Chem 1997;272:7431–7436.
102. Pike CJ, Walencewiczwasserman AJ, Kosmoski J, Cribbs DH, Glabe CG, Cotman CW. Structure-activity analyses of beta-amyloid peptides - contributions of the beta-25–35 region to aggregation and neurotoxicity. J Neurochem 1995;64:253–265.
103. Varadarajan S, Kanski J, Aksenova M, Lauderback C, Butterfield DA. Different mechanisms of oxidative stress and neurotoxicity for Alzheimer's A beta(1–42) and A beta(25–35). J Am Chem Soc 2001;123:5625–5631.
104. Atwood CS, Moir RD, Huang XD, et al. Dramatic aggregation of Alzheimer a-beta by Cu(II) is induced by conditions representing physiological acidosis. J Biol Chem 1998;273(21):12817–12826.
105. Kanski J, Aksenova M, Schoneich C, Butterfield DA. Substitution of isoleucine-31 by helical-breaking proline abolishes oxidative stress and neurotoxic properties of Alzheimer's amyloid beta-peptide (1–42). Free Rad Biol Med 2002;32:1205–1211.
106. Camilleri P, Haskins NJ, Howlett DR. beta-Cyclodextrin interacts with the Alzheimer amyloid beta-A4 peptide. FEBS Lett 1994;341: 256–258.
107. Howlett DR, Perry AE, Godfrey F, et al. Inhibition of beta-amyloid peptide fibrillization by a novel series of benzofurans. Biochem J 1999;340:283–289.
108. Howlett DR. Ab oligomerization: a therapeutic target for Alzheimer's disease. Curr Med Chem 2001;1:25–38.
109. Bohrmann B, Adrian M, Dubochet J, et al. Self-assembly of beta-amyloid 42 is retarded by small molecular ligands at the stage of structural intermediates. J Struct Biol 2000;130:232–246.
110. Reixach N, Crooks E, Ostresh JM, Houghten RA, Blondelle SE. Inhibition of beta-amyloid-induced neurotoxicity by imidazopyridoindoles derived from a synthetic combinatorial library. J Struct Biol 2000;130:247–258.
111. Cairo CW, Strzelec A, Murphy RM, Kiessling LL. Affinity-based inhibition of beta-amyloid toxicity. Biochemistry 2002;41:8620–8629.
112. Cherny RA, Atwood CS, Xilinas ME, et al. Treatment with a copper-zinc chelator markedly and rapidly inhibits beta-amyloid accumulation in Alzheimer's disease transgenic mice. Neuron 2001;30: 665–676.
113. Tomiyama T, Shoji A, Kataoka K, et al. Inhibition of amyloid beta protein aggregation and neurotoxicity by rifampicin. Its possible function as a hydroxyl radical scavenger. J Biol Chem 1996;271:6839–6844.
114. Turnbull S, Tabner BJ, El Agnaf OMA, Twyman LJ, Allsop D. New evidence that the Alzheimer beta-amyloid peptide does not spontaneously form free radicals: an ESR study using a series of spin-traps. Free Rad Biol Med 2001;30:1154–1162.
115. Kim YS, Randolph TW, Manning MC, Stevens FJ, Carpenter JF. Congo red populates partially unfolded states of an amyloidogenic protein to enhance aggregation and amyloid fibril formation. J Biol Chem 2003;278:10842–10850.
116. Harper JD, Wong SS, Lieber CM, Lansbury PT, Jr. Assembly of A beta amyloid protofibrils: an in vitro model for a possible early event in Alzheimer's disease. Biochemistry 1999;38:8972–8980.
117. Kowalewski T, Holtzman DM. In situ atomic force microscopy study of Alzheimer's beta-amyloid peptide on different substrates: new insights into mechanism of beta-sheet formation. Proc Natl Acad Sci USA 1999;96:3688–3693.
118 Doig AJ, Hughes E, Burke RM, Su TJ, Heenan RK, Lu J. Inhibition of toxicity and protofibril formation in the amyloid-beta peptide beta(25–35) using N-methylated derivatives. Biochem Soc Trans 2002;30:537–542.
119. Gordon DJ, Tappe R, Meredith SC. Design and characterization of a membrane permeable N-methyl amino acid-containing peptide that inhibits A beta(1–40) fibrillogenesis. J Peptide Res 2002;60: 37–55.
120. Hilbich C, Kisters-Woike B, Reed J, Masters CL, Beyreuther K. Substitutions of hydrophobic amino acids reduce the amyloidogenicity of Alzheimer's disease beta A4 peptides. J Mol Biol 1992;228:460–473.
121. Tjernberg LO, Naslund J, Lindqvist F, et al. Arrest of beta-amyloid fibril formation by a pentapeptide ligand. J Biol Chem 1996;271:8545–8548.
122. Soto C, Kindy MS, Baumann M, Frangione B. Inhibition of alzheimers amyloidosis by peptides that prevent beta-sheet conformation. Biochem Biophys Res Comm 1996;226:672–680.
123. Balbach JJ, Ishii Y, Antzutkin ON, et al. Amyloid fibril formation by A beta(16–22), a seven-residue fragment of the Alzheimer's beta-amyloid peptide, and structural characterization by solid state NMR. Biochemistry 2000;39:13748–13759.
124. Sticht H, Bayer P, Willbold D, et al. Structure of amyloid A4-(1–40)-peptide of Alzheimer's disease. Eur J Biochem 1995;233:293–298.
125. Coles M, Bicknell W, Watson AA, Fairlie DP, Craik DJ. Solution structure of amyloid beta-peptide(1–40) in a water-micelle environment - is the membrane-spanning domain where we think it is. Biochemistry 1998;37:11064–11077.
126. Bond JP, Deverin SP, Inouye H, El Agnaf OMA, Teeter MM, Kirschner DA. Assemblies of Alzheimer's peptides A beta 25–35 and A beta 31–35: reverse-turn conformation and side-chain interactions revealed by X-ray diffraction. J Struct Biol 2003;141:156–170.
127. Munoz V, Thompson PA, Hofrichter J, Eaton WA. Folding dynamics and mechanism of beta-hairpin formation. Nature 1997;390:196–199.
128. Rovero P, Patacchini R, Renzetti AR, et al. Interaction of amyloid beta protein (25–35) with tachykinin receptors. Neuropeptides 1992;22:99–101.
129. Lee JM, Weinstein DA, Kowall NW, Beal MF. Inability of β-amyloid (25–35) to bind to central nervous system neurokinin 1 receptors. Drug Dev Res 1992;27: 441–444.
130. El-Agnaf OA, Irvine GB, Fitzpatrick G, Glass WK, Guthrie DS. Comparative studies on peptides representing the so-called tachykinin-like region of the Alzheimer A beta peptide [A beta(25–35)]. Biochem J 1998;336:419–427.
131. Watson AA, Fairlie DP, Craik DJ. Solution structure of methionine-oxidized amyloid b-peptide (1–40). Does oxidation affect conformational switching. Biochemistry 1998;37:12700–12706.
132. Palmblad M, Westlind-Danielsson A, Bergquist J. Oxidation of methionine 35 attenuates formation of amyloid beta-peptide 1–40 oligomers. J Biol Chem 2002;277:19506–19510.
133. Hou LM, Kang I, Marchant RE, Zagorski MG. Methionine 35 oxidation reduces fibril assembly of the amyloid A beta-(1–42) peptide of Alzheimer's disease. J Biol Chem 2002;277:40173–40176.
134. Dong J, Atwood CS, Anderson VE, et al. Metal binding and oxidation of amyloid-beta within isolated senile plaque cores: Raman microscopic evidence. Biochemistry 2003;42:2768–2773.
135. Lee JP, Stimson ER, Ghilardi JR, et al. 1H NMR of A beta amyloid peptide congeners in water solution. Conformational changes correlate with plaque competence. Biochemistry 1995;34:5191–5200.
136. Saito H. Conformation-dependent 13C chemical-shifts—a new means of conformational characterization as obtained by high-resolution solid-state 13C NMR. Magn Reson Chem 1986;24:835–852.
137. Kirkitadze MD, Bitan G, Teplow DB. Paradigm shifts in Alzheimer's disease and other neurodegenerative disorders: the emerging role of oligomeric assemblies. J Neurosci Res 2002;69:567–577.

138. Hardy J, Selkoe DJ. The amyloid hypothesis of Alzheimer's disease: progress and problems on the road to therapeutics. Science 2002;297:353–356.
139. Pike CJ, Walencewicz AJ, Glabe CG, Cotman CW. Aggregation-related toxicity of synthetic beta-amyloid protein in hippocampal cultures. Eur J Pharmacol 1991;207:367–368.
140. Roher AE, Ball MJ, Bhave SV, Wakade AR. Beta-amyloid from Alzheimer disease brains inhibits sprouting and survival of sympathetic neurons. Biochem Biophys Res Commun 1991;174:572–579.
141. Harper JD, Wong SS, Lieber CM, Lansbury PT. Observation of metastable a-beta amyloid protofibrils by atomic force microscopy. Chem Biol 1997;4:119–125.
142. Hartley DM, Walsh DM, Ye CPP, et al. Protofibrillar intermediates of amyloid beta-protein induce acute electrophysiological changes and progressive neurotoxicity in cortical neurons. J Neurosci 1999;19:8876–8884.
143. Walsh DM, Hartley DM, Kusumoto Y, et al. Amyloid beta-protein fibrillogenesis - Structure and biological activity of protofibrillar intermediates. J Biol Chem 1999;274:25945–25952.
144. Nichols MR, Moss MA, Reed DK, et al. Growth of beta-amyloid(1–40) protofibrils by monomer elongation and lateral association. Characterization of distinct products by light scattering and atomic force microscopy. Biochemistry 2002;41:6115–6127.
145. Blackley HK, Patel N, Davies MC, et al. Morphological development of beta(1–40) amyloid fibrils. Exp Neurol 1999;158:437–443.
146. Dyrks T, Dyrks E, Monning U, Urmoneit B, Turner J, Beyreuther K. Generation of beta A4 from the amyloid protein precursor and fragments thereof. FEBS Lett 1993;335:89–93.
147. Roher AE, Chaney MO, Kuo YM, et al. Morphology and toxicity of a-beta-(1–42) dimer derived from neuritic and vascular amyloid deposits of Alzheimer's-disease. J Biol Chem 1996;271:20631–20635.
148. Oda T, Wals P, Osterburg HH, et al. Clusterin (apoJ) alters the aggregation of amyloid beta-peptide (a-beta(1–42)) and forms slowly sedimenting a-beta complexes that cause oxidative stress. Exp Neurol 1995;136:22–31.
149. Stine WB, Dahlgren KN, Krafft GA, Ladu MJ. In vitro characterization of conditions for amyloid-beta peptide oligomerization and fibrillogenesis. J Biol Chem 2003;278:11612–11622.
150. Westlind-Danielsson A, Arnerup G. Spontaneous in vitro formation of supramolecular beta-amyloid structures, "beta amy balls", by beta-amyloid 1–40 peptide. Biochemistry 2001;40:14736–14743.
151. Hoshi M, Sato M, Matsumoto S, et al. Spherical aggregates of beta-amyloid (amylospheroid) show high neurotoxicity and activate tau protein kinase I/glycogen synthase kinase-3 beta. Proc Natl Acad Sci USA 2003;100:6370–6375.
152. Kayed R, Head E, Thompson JL, et al. Common structure of soluble amyloid oligomers implies common mechanism of pathogenesis. Science 2003;300:486–489.
153. Gong Y, Chang L, Viola KL, et al. Alzheimer's disease-affected brain: presence of oligomeric A beta ligands (ADDLs) suggests a molecular basis for reversible memory loss. Proc Natl Acad Sci USA 2003;100:10417–10422.

7 Functional Imaging and Psychopathological Consequences of Inflammation in Alzheimer's Dementia

Jan Versijpt, MD, PhD, Rudi A. Dierckx, MD, PhD, and Jakob Korf, PhD

SUMMARY

In the past decades, our understanding of the central nervous system has evolved from one of an immune-privileged site, to one where inflammation is pathognomonic for some of the most prevalent neurodegenerative diseases, including Alzheimer's dementia. Inflammation, whether in the brain or periphery, is almost always a secondary response to a primary pathogen. In Alzheimer's dementia, inflammation is considered as a secondary response after impaired processing and precipitation of amyloid, which will ensue and likely cause additional neurone loss. In this chapter, visualization tools for these neuro-inflammatory processes, both structural and mainly functional, are critically reviewed and discussed. From the basic neuro-inflammatory mechanisms and its biochemical characteristics, the role of its potential mediators playing a key role in neurodegenerative disorders has been explored, with Alzheimer's disease as a prototype. Whereas structural imaging shows merely late anatomical consequences of an inflammatory response, functional imaging is a strong potential candidate to bridge this gap between in vitro and in vivo knowledge. A number of radioligands have been recently explored that allow the early in vivo visualization of inflammatory responses and, as such, open a promising window on both understanding as well as possible clinical management of inflammatory neurodegenerative disorders. Magnetic resonance imaging and ^{99m}Tc-ECD (ethylcysteinate dimer) single-photon emission tomography scans yielded conclusive results as to the exclusion of other pathologies and confirmation of diagnosis. ^{57}Co single-photon emission tomography scanning did not reveal any regional raised uptake, ongoing tissue decay, or inflammation, irrespective of the type of dementia, the depth or extent of perfusion defects, the presence of atrophy on magnetic resonance imaging, or the results of neuropsychological tests. The use of PK11195 to detect cerebral inflammation was more positive. For instance, the mean (^{123}I)iodo-PK11195 uptake was higher in Alzheimer's dementia as compared with controls in many neocortical regions, particularly in the frontal and right mesotemporal regions. A significant correlation was found between regional increased (^{123}I)iodo-PK11195 uptake and cognitive deficits. (^{123}I)iodo-PK11195 and (^{11}C)-PK-11195 are cellular disease activity markers that allow one to make an in vivo assessment of microglial inflammation in Alzheimer's dementia inflammation.

Key Words: Alzheimer; cobalt; PET; SPECT; inflammation; microglia; tryptophan; depression.

1. CLINICAL ASPECTS OF ALZHEIMER'S DEMENTIA

Available evidence indicates that in the diagnosis of Alzheimer's dementia, mild dementia is rarely diagnosed and even moderately severe dementia is underrecognized in clinical practice *(1)*. The specificity of the Mini Mental State Examination is good (96%) but the sensitivity is poor (63%), indicating that by itself the test (using standard cut off score of 24) will fail to detect a substantial proportion of early dementia cases. The typical syndrome of Alzheimer's dementia includes three aspects, the first of which includes deficits in cognitive functioning that cause an amnestic type of memory defect marked by difficulties in learning and recalling new information, a progressive language disorder beginning with anomia and progressing to fluent aphasia, disturbances of visual–spatial skills manifested by environmental disorientation and difficulties in copying figures, the inability to do motor tasks despite intact motor function (apraxia), and the inability to recognize persons, places, or objects, despite intact sensory functions (agnosia). There are usually deficits in executive function (planning, judgment, and insight) and the patient typically is unaware of memory or cognitive compromise. All cognitive deficits progressively worsen. The second aspect includes neuropsychiatric and behavioral disturbances, such as personality changes, delusions, hallucinations, and misidentifications. Apathy is present early in the clinical course, with diminished interest and reduced concern. Agitation becomes

From: *Bioimaging in Neurodegeneration*
Edited by P. A. Broderick, D. N. Rahni, and E. H. Kolodny

increasingly common as the illness advances and is a frequent precipitant of nursing home placement. Depressive symptoms are present in more than 50% of patients, and approx 25% exhibit delusions. The third aspect includes difficulties with activities of daily living, which manifest early in the disease and affect functions such as handling money, using the telephone, and driving (instrumental), and later, difficulties with dressing, feeding, and toileting (basic). Motor system abnormalities are (still) absent in Alzheimer's dementia until the final few years of the disease; focal abnormalities, gait changes, or seizures occurring early in the clinical course of dementia make the diagnosis of Alzheimer's dementia unlikely. Patients with Alzheimer's dementia usually survive 7–10 yr after onset of symptoms, with a range of 2–20 yr, and typically die from bronchitis or pneumonia. *(2,3)*. In clinical practice, neuroimaging should be obtained to identify vascular contributions to the dementia syndrome and to identify other intracranial pathology.

Functional imaging with positron emission tomography (PET) or single-photon emission tomography (SPECT), which measure regional cerebral blood flow or energy metabolism, are helpful, particularly when clinical features of Alzheimer's dementia are ambiguous (4).

It is often stated that the correct diagnosis of Alzheimer's dementia relies on postmortem examination. The current criteria for the pathologic diagnosis of Alzheimer's dementia require the presence of both senile neuritic plaques and neurofibrillary tangles in excess of the abundance anticipated for age-matched healthy controls *(5)*. Neuritic plaques consist of a ventral core of amyloid protein surrounded by astrocytes, microglia, and dystrophic neurites often containing paired helical filaments. Neurofibrillary tangles are the second major histopathological feature of Alzheimer's dementia. They contain paired helical filaments of abnormally phosphorylated tau protein that occupy the cell body and extend it into the dendrites. However, there are more than amyloid plaques and neurofibrillary tangles. At least a third of Alzheimer's dementia cases exhibit significant cerebrovascular pathology. Central amyloid angiopathy, microvascular degeneration affecting the cerebral endothelium and smooth muscle cells, basal lamina alterations, hyalinosis, and fibrosis are often seen in Alzheimer's dementia. These changes may be accompanied by perivascular denervation that is causal in the cognitive decline in Alzheimer's dementia. Peripheral vascular diseases, such as long-standing hypertension, atrial fibrillation, coronary artery disease, and diabetes, could further modify the cerebral circulation in such a way that a sustained hypoperfusion or oligemia has an impact upon the aging brain. In addition to the classic histopathological features, Alzheimer's dementia also is characterized by reductions of synaptic density, loss of neurones, and granulovascular degeneration in hippocampal neurones. Neuronal loss or atrophy in the nucleus basalis, locus coeruleus, and raphe nucleus of the brain stem leads to deficits in cholinergic, noradrenergic, and serotonergic transmitters, respectively, with the cholinergic deficit as the most consistent neurochemical abnormality. Central in the pathogenesis of Alzheimer's dementia is the amyloid protein, which is derived from the amyloid precursor protein, and its abnormal cleavage (to the amyloid β-peptide, Aβ) finally leads to the deposition of neuritic plaques *(6)*.

2. INFLAMMATION IN ALZHEIMER DEMENTIA

The previous concept of the brain as an immunologically privileged organ appears to be no longer tenable. This concept stems from the observations that the brain lacks a lymphatic drainage system (now considered to be a lymph-like system *[7]*) that has an unusual tolerance to transplanted tissue, and the idea that lymphocytes were excluded from the central nervous system (CNS) by blood–brain and blood–cerebrospinal fluid barriers. As such, the immune and nervous system were considered different compartments acting autonomously in their contribution to physiological homeostasis *(8)*. Contemporary research learned that the blood–brain barrier (BBB) is, under certain conditions, less restrictive to the migration of monocytes, lymphocytes, or natural killer cells, irrespective of antigen specificity *(9)*. Nevertheless, the CNS inflammation threshold is still considered higher as compared with that of the periphery, leading to a delay between peripheral and CNS inflammation during a general inflammatory status. For example, the rapid recruitment of neutrophils in the CNS is virtually absent, and monocytes are only recruited after a delay of several days. The reason for this higher threshold is at least threefold. First, because only activated T lymphocytes traverse the BBB, it is the small pool of peripherally activated T cells that enter the CNS for immune surveillance *(10)*. Yet, without peripheral T cell activation, antigens escape detection; thus, brain transplants survive despite an antigen mismatch *(11)*. Second, there is an active suppression of antigen expression leading to T lymphocytes not recognizing their target nor activating inflammatory mechanisms *(12)*. Third, the adhesion molecule expression, essential in cell–cell contacts during inflammatory cell migration, is low on cerebral endothelial cells *(13)*. CNS immune responses usually take milder courses, and it is not clear yet whether this relative deficit is explicable solely by the lack of immunological structures, or is compounded by counterregulatory mechanisms. Recent evidence indicates that CNS immune responses are indeed downregulated, with a key role for electrically active neurones *(14)*. Both in vitro and in vivo studies have clearly established that astrocytes and microglia (brain resident macrophages), in addition to peripherally originated immune cells, can initiate an inflammatory cascade within the CNS *(15)*. Also, all components of the complement system are found in the brain and are produced by astrocytes, microglia, and, surprisingly, neurones.

There is now substantial epidemiological evidence of the involvement of inflammation in Alzheimer's dementia. There are now about 20 reports on the incidence of Alzheimer's dementia in populations with a long antiinflammatory drug consumption history. Nearly all of these studies showed a lower AD incidence with a decrease of 50% or a delay in onset of 5–7 yr, and, in one prospective study, the relative risk fell with increasing duration of drug use *(16)*. Clinical trials with indomethacin or propentofylline, another agent with antiinflammatory properties, showed both a significant cognitive improvement *(17,18)*, whereas one study on diclofenac and one recent study on hydroxychloroquine did not demonstrate a positive effect on the progression of the disorder *(19,20)*. Alzheimer's dementia shows an apolipoprotein E (ApoE) genotype susceptibility with ApoE4 as a risk factor. Interestingly, ApoE4 seems essential for Amyloid precursor protein-induced microglial activation and the expression of several inflammatory indicators *(21)*. Also, several genetic cytokine risk factors have already been identified with a decreased cytokine activity associated with a reduced risk and delayed onset *(22)*.

Aβ protein precipitation and the ensuing neurodegeneration are the most likely sources for inflammation in AD. From this point, a nearly bewildering number of inflammatory mediators come into play, each characterized by an abundance of amplifying and dampening loops, as well as multiple interactions with other subsystems. Like a web, all these inflammatory pathways make it likely for one set of mediators to induce most of the others. For this reason, the selection of any particular starting point for explaining the Alzheimer's dementia inflammatory mechanism must be taken as a matter of convenience *(23)*.

The cellular source of Aβ is still a matter of debate. Microglia are able to synthesize Aβ in response to nerve injury or even Aβ itself *(24)*. As for toxicity, although it has been hypothesized that the presence of amyloid is the direct cause of AD pathology, the in vivo confirmation of Aβ toxicity has not yet been conclusively demonstrated, whereas there is even evidence that Aβ may be neuroprotective *(25)*. Moreover, in one study, a fulminant hippocampal neurone loss by neuritic plaques was only observed in the presence of microglia *(26)*. As for Aβ phagocytosis, the degradation rate seems limited and results in the release of potentially neurodestructive compounds *(27)*. However, that microglia can remove amyloid was strongly suggested by the demonstration of Aβ colocalized with microglia in Aβ-immunized transgenic mice, where amyloid was apparently cleared and this removal also protected against a further cognitive decline *(28)*. Interestingly, C1q, a complement protein that binds Aβ, apparently inhibits the microglial Aβ uptake *(29)*. The fact that activated microglia are the sole and consistent accompaniment of neuritic plaques and not readily found in association with nonneuritic plaques suggests their pivotal role in the promotion of neuritic plaque formation, a role similar to that ascribed to peripheral macrophages in systemic amyloidosis *(30)*. Aβ may also act in a feedforward mechanism to maintain microglial activation, directly and indirectly by stimulating cytokine production, thereby rendering neurones subject to deleterious effects of activated microglia *(31,32)*.

It should be emphasized that the theory of inflammation as a primary disease-aggravating hallmark, opposed to a secondary or even a disease-ameliorating factor, remains a hypothesis. One should be aware that our current knowledge of microglia is still incomplete, speculative, and mainly based upon in vitro observations rather than in vivo studies *(33)*. Indeed, B or T cells and immunoglobulins (Igs) are not readily detectable in the Alzheimer's dementia brain and are found only in very small amounts in relation to amyloid plaques (without IgM/IgA) *(34)*. Likewise, although the presence of leukocytes has been demonstrated, their role in Alzheimer's dementia has not been established *(35)*. As such, the evidence for an antigen-driven acquired immune response in Alzheimer's dementia, with T cells eliminating amyloid and B cells producing Aβ-specific antibodies, is not as overt as in well-established neuro-inlfammatory diseases (e.g., multiple sclerosis; *36*).

3. NEURO-INFLAMMATORY IMAGING

Visualizing neuro-inflammation in Alzheimer's dementia is of interest, first for clarifying the pathophysiology, second for selecting patient subgroups that are more eligible for antiinflammatory treatment, and finally for monitoring patients during trials with these antiinflammatory agents. Here we review and discuss current neuro-inflammatory imaging modalities, both structural and functional. Structural imaging aims to describe in detail the spatial relationship of neurodegenerative and inflammatory consequences like mass effects, edema, vascular congestion, thrombosis, petechial hemorrhages, secondary demyelinization, gliosis, and finally neuronal destruction, necrosis, or atrophy, as well as visualizing other (nonspecific) structural changes.

3.1. COMPUTED TOMOGRAPHY (CT) IMAGING AND MAGNETIC RESONANCE IMAGING (MRI)

CT and, to a greater extent, MRI (gadolinium-enhanced) with its excellent soft-tissue contrast resolution (used mainly for the evaluation of white matter and posterior fossa) are able to detect CNS changes caused by mostly localized inflammatory and degenerative processes *(37)*. The degenerative processes and inflammation must already be at an advanced stage before they can be depicted by one of these imaging modalities, whose sensitivity is poor at the early stages of AD (when anatomical changes are not detectable yet). But, in chronic processes, these modalities may also detect structural changes that do not reflect the actual state of disease activity. For instance, Ketonen and Tuite have noted already that early AIDS dementia encephalitis was characterized at postmortem by scattered microglial nodules and HIV-infected multinucleated giant cells located primarily in the white matter and correlating with the severity of dementia. Both CT and MRI are too insensitive to detect these microglial nodules, and, for this reason, the neuroimaging appearance early in the disease is usually normal *(38)*. Kim et al. *(39)* noted that CT and MRI are able to detect general or basal ganglia atrophy and white matter lesions that appear to increase in severity with the progression of the HIV infection. However, these authors also concluded that CT and MRI are unable to relate neurological signs or positive findings on neuropsychological tests to clinical dementia in HIV. In addition, these imaging modalities show poor correlation with histopathological findings. Although MRI is useful in the workup of patients with dementia because it shows the presence of space-occupying lesions, ventricular dilatation, cerebral atrophy, widening of sulci, or infarcts, this technique is not of particular value in the direct diagnosis of Alzheimer's dementia, although promising results have been made with volumetric measurements of the (para)hippocampal and amygdala region *(40)*.

Cecil et al. *(41)* reviewed the newer structural or metabolic imaging tools in brain inflammation and concluded that proton MR spectroscopy is a sensitive and specific imaging tool in Creutzfeldt-Jakob disease, herpes simplex encephalitis, and AIDS, indicating its usefulness in longitudinal studies for predicting and monitoring the response to therapy *(41)*. Likewise, Bitsch et al. *(42)* found that the measured increases of choline and myo-inositol corresponded to the histopathologically verified glial proliferation and the infiltration of subcortical grey matter structures with foamy macrophages. Recently, Rovaris et al. *(43)* reported on the value of magnetization transfer imaging in measuring brain involvement in systemic immune-mediated diseases. It was found that magnetization transfer imaging provides information about brain damage with increased pathological specificity and detects subtle microscopic abnormalities in the normal brain tissue, which go undetected with conventional scanning. However, in some immune-mediated diseases microscopic brain tissue damage seemed to be absent despite macroscopic MRI lesions or clinical evidence of CNS involvement *(43)*.

3.2. FUNCTIONAL IMAGING USING RADIOPHARMACEUTICALS

Nuclear medicine provides several techniques for the detection of inflammation. Studies demonstrating inflammatory lesions were reported as early as in 1959, when Athens et al. *(44)* labeled leukocytes by intravenous injection of diisopropylfluoro-phospate labeled with ^{32}P and demonstrated skin blisters in volunteers . Classically, scintigraphic imaging of inflammation has been done with ^{67}Gallium-citrate, radiolabeled leukocytes, nanocolloids, nonspecific human immunoglobulins (HIGs), and ^{18}F-deoxyglucose (FDG). Uptake mechanisms included direct binding to relevant inflammatory cells or proteins (radiolabeled leukocytes, ^{67}Gallium-citrate, HIG) over hyperemia, and binding to lactoferrin excreted *in loco* by leukocytes or to siderophores produced by microorganisms (^{67}Gallium-citrate). In addition, nonspecific local increases in blood supply, extravasation through vessels with increased permeability may give rise to expansion of the local interstitial fluid space (^{67}Gallium-citrate, nanocolloid, HIG). Finally, high glucose uptake is often seen in inflammatory cells (FDG-PET) *(45)*, but inflammatory processes in CNS tissue cannot easily be distinguished because of the high rate of energy metabolism in otherwise unaffected tissue (even in AD). Radiolabeled leukocytes used in cerebral ischemia to detect inflammation accumulated well in massive infarcts with severe neurological impairments and little improvement *(46)* but are of little use in Alzheimer's dementia. This is because of the minor hemodynamic and permeability changes (little or no vasodilatation), the slow cellular turnover, and the predominant mononuclear cell infiltrate of chronic processes.

Often, semiquantitative analyses are based on a regional normalization of radioactivity and the cerebellum as reference region and thus normalization factor. A regional rather than a global normalization (with whole brain as normalization factor) may be preferred because a region-specific normalization is known to be more sensitive for diseases in which various regions are pathophysiologically involved, as in Alzheimer's dementia *(47)*. Although some reports described the pathological involvement of the cerebellum in Alzheimer's dementia *(48)*, this region was chosen as the normalization region because it has both low pathologic susceptibility and absence or at least minimal presence of upregulated inflammatory mediators in the cerebellum *(49)*. A previous study concluded already that the cerebellum is the more appropriate choice of reference region in the quantification of perfusion SPECT in primary degenerative dementia *(50)*. With regard to perfusion SPECT imaging, the cerebellum was shown to be scintigraphically uninvolved *(51)*.

3.3. FUNCTIONAL CO-IMAGING

Attempts have been made to visualize inflammation by means of cobalt radioisotopes. Both in vivo and in vitro experiments have shown that Ca^2 accumulates in the damaged nerve cell body and degenerating axons by two mechanisms: (1) a passive influx caused by a shortage of ATP following ischemia, resulting in the disappearance of the membrane potential, and (2) neuronal and glial uptake by divalent cation-permeable kainate-activated non-*N*-methyl-D-aspartate glutamate receptor-operated channels in the membrane *(52–57)*. ^{57}Co (SPECT) and ^{55}Co (PET), both as Ca^{2+}-analogs, can reflect Ca^{2+}-influx in ischemically or neurotoxically damaged cerebral tissue. In this way, both ^{57}Co SPECT and ^{55}Co PET have been shown capable of visualizing focal neurodegenerative changes, reactive gliosis, endangered brain tissue, and/or ongoing neuronal tissue decay, including inflammatory lesions in various brain diseases, for example, multiple sclerosis, trauma, tumors, and stroke *(58–64)*. Moreover, the time sequence of the ^{55}Co-load seems to correlate well with cell death and glial proliferation in the ipsilateral thalamus after supratentorial ischemic stroke in an experimental rat stroke model *(65)*. The visualization of these inflammatory processes in AD can be expected to occur by means of the final common pathway of the Ca^{2+}-homeostasis-disturbance in both neuronal degeneration and inflammation *(66–70)*. The limitations of ^{57}Co SPECT and ^{55}Co PET should also be mentioned here. Because of the long physical half-life of 270 d of ^{57}Co, only a limited dose can be injected which is responsible for the low count rate and the resulting low statistics. Alternatively, the PET-radionuclide ^{55}Co has been used (physical half-life 17.5 h). Moreover, whether divalent radioactive Co visualizes specific aspects of neuronal damage or BBB integrity is still uncertain. To what extent ^{57}Co and ^{55}Co really visualize calcium-mediated processes (in vivo) and therefore reflect identical molecular uptake mechanisms has yet to be determined, although the cerebral uptake of intravenously administered radioactive ^{45}Ca and ^{60}Co in neuronal damage is highly similar *(53)*. Finally, the exact cellular site of accumulation of radioactivity is, as yet, not known. As for inflammatory imaging, however, it is interesting to note that calcium may also accumulate in activated leukocytes and that for both ^{55}Co and ^{57}Co only 12% of the total fraction is in its free form while the remainder is bound to leukocytes or plasma proteins *(71–73)*.

A pilot study by Oosterink et al. *(74)* with PET and ^{55}Co suggested that this technique could generate additional specific information, which cannot be obtained with conventional neuroimaging techniques like perfusion and ^{18}FDG PET. A recent study, however, concluded that ^{57}Co SPECT was not able to show any regional raised uptake in AD patients, irrespective of the depth or extent of the associated perfusion defects, the presence of atrophy on MRI, or the neuropsychological test results *(75)*. Moreover, the long physical half-life with its resulting low count rate and statistics and the incomplete knowledge about the specific cellular uptake mechanisms all limit the application of Cobalt radioisotopes.

3.4. RADIOLABELED RECEPTOR-SPECIFIC PROTEINS AND PEPTIDES

During the last decade, there has been a shift in scintigraphic inflammatory imaging from large proteins with nonspecific uptake mechanisms, including receptors, to receptor-specific small proteins and peptides (i.e., mediators of the inflammatory response), allowing the noninvasive detection of specific cells and tissues *(76)*. Monoclonal antibodies (against leukocyte antigens or endothelial adhesion molecules) were the first example of this new class of radiopharmaceuticals, and in the past few years several new radiolabeled receptor ligands have been developed *(77)*. As such, radiolabeled monoclonal antibodies against granulocyte and lymphocyte antigens, adhesion molecules, cytokines, chemokines, chemotactic peptides, and macrophages were developed *(78)*. For example, Paul et al. *(79)* recently reported the detection and quantitation of neuro-inflammation through BBB permeability changes during experimental allergic encephalomyelitis using a radiolabeled tuftsin analog. Tuftsin is a tetrapeptide derived from the Fc portion of IgG that promotes chemotaxis and phagocytosis of neutrophils, monocytes, and macrophages by binding to receptors on these cells. In addition, this radiopharmaceutical enabled the authors to successfully monitor glucocorticoid suppression of inflammation, recording a typical dose-response to increasing steroid concentrations.

In nuclear medicine, radiopharmaceuticals have emerged as a promising class of agents with attractive characteristics for scintigraphic detection of inflammation. Theoretically, the high binding affinity for their receptors—expressed in inflammatory tissue—facilitates the retention of these agents in inflammation, whereas the small size permits rapid clearance from blood and other nontarget tissues. Whether this small radiolabeled receptor-binding agent specifically localizes in an inflammatory focus depends first on the receptor expression in the particular inflammatory response. Second, the receptors should be accessible for the ligand, and third, the interaction of the ligand and its receptor should be characterized by high affinity and specificity (shown by autoradiography, receptor blockade, or studies with control agents). For neuroimaging studies in pathologies without BBB breakdown, the ligand should also easily penetrate the BBB.

3.5. IMAGING OF MICROGLIA IN ALZHEIMER DEMENTIA

PK11195 (1-[2-chlorophenyl]-*N*-[1-methyl-propyl]-3-isoquinoline carboxamide) is a specific and selective high-affinity ligand for the peripheral benzodiazepine receptor and, in this way, can be used as a marker for neuro-inflammatory lesions. The peripheral benzodiazepine receptor is structurally and pharmacologically distinct from the central benzodiazepine receptor (associated with γ-aminobutyric acid-regulated chloride-channels) and earned its name based on its localization outside the CNS and its high affinity for several 1,4-benzodiazepines. It has neither anxiolytic nor spasmolytic activity or interactions with other receptors and has been classified as an antagonist or partial agonist *(80)*. As such, Banati et al. *(81)* showed an increased PK11195 binding to activated microglia after facial nerve axotomy, a lesion causing a retrograde neuronal reaction without nerve cell death with a rapid proliferation and activation of microglia while keeping the BBB intact. The peak of PK11195 binding was observed 4 d after the peripheral nerve lesion, which is consistent with the well-known time course of microglial activation. Moreover, photoemulsion microautoradiography confirmed the restriction of PK11195 binding to activated (i.e., peripheral benzodiazepine receptor-expressing) microglia, where the full transformation of microglia into parenchymal phagocytes is not necessary to reach maximal levels of PK11195 binding. It was concluded that PK11195 is a well-suited marker of microglial activation in areas of subtle brain pathology, without BBB disturbance, or the presence of macrophages *(81)*. The peripheral benzodiazepine receptor is found in highest concentrations in kidneys, colon membranes, heart, steroid hormone-producing cells of the adrenal cortex, ovaries, and testes, and several cell types of the immune system, such as mast cells and macrophages. It is also present in low concentrations throughout the brain, primarily associated with the choroid plexus, ependymal linings, and glial cells. Although the specific function of the peripheral benzodiazepine receptor remains unknown, it is generally accepted to be involved in lipid metabolism and/or transport, heme biosynthesis, cell proliferation, or ion channel functions *(82)*. Its immunomodulatory role includes the ability to induce monocyte chemotaxis, modulate cytokine expression and superoxide generation, and stimulate antibody-producing cell formation *(83)*. Interestingly, the peripheral benzodiazepine receptor has the ability to reflect neuronal injury, neurotoxicity, and inflammatory lesions without BBB damage, by a rise in the number of binding sites in the case of activated microglia *(84,85)*, as previously indicated autoradiographically for AD *(86,87)*.

In vivo visualization of the human peripheral benzodiazepine receptor has been performed with ^{11}C-radiolabeled PK11195 for PET in various diseases like glial neoplasms, ischemic stroke, multiple sclerosis, Rasmussen's encephalitis, Alzheimer's dementia, and Parkinson's disease, producing a signal of activated microglia unrelated to the influx of blood-borne macrophages *(88–91)*. The potential of this approach was shown in multiple sclerosis, where significant ^{11}C-PK11195 binding was detected in areas where MRI did not show any abnormalities. For instance, PK11195-related signals were localized in deafferentiated grey matter regions such as the lateral geniculate body (to which the optic nerve projects) and visual cortex of patients with previous optic neuritis. ^{11}C-PK11195 PET has also been applied in early and mild dementia patients revealing an increased regional binding in the entorhinal, temporoparietal, and cingulate cortex. Moreover, serial volumetric MRI scans revealed that areas with high ^{11}C-PK11195 binding subsequently showed the highest rate of atrophy up to 12–24 mo later, indicating that the presence of a local immune response in cortical areas did indeed reflect an active disease process associated with tissue loss. Comparison with FDG-PET revealed that areas with high ^{11}C-PK11195 binding were also characterized by decreased regional glucose use. In one patient with isolated memory impairment without dementia, the pattern of atrophy as seen by volumetric MRI imaging was predicted by the initial distribution of increased ^{11}C-PK11195 binding *(92)*.

Recently, PK11195 radiolabeled with iodine for SPECT has become available. [^{123}I]-labeled iodo-PK11195 is a suitable agent for the visualizations of the peripheral benzodiazepine receptor and indirectly for the imaging of neuro-inflammatory lesions *(93)*. In a recent pilot study, [^{123}I]iodo-PK11195 was also applied in Alzheimer's dementia, which showed a distinct difference in ligand uptake between Alzheimer's dementia patients and controls, indicating the pathophysiological involvement of microglia in frontal, temporal, and parietal cortical regions that were pathognomonically compromised in patients with Alzheimer's dementia *(93)*. Moreover, inverse correlations were found between regional [^{123}I]iodo-PK11195 uptake values and cognitive test results. Mean uptake values were increased in various neocortical regions pathognomonically compromised in Alzheimer's dementia, and significance was particularly reached in frontal neocortical regions. Although somewhat unexpected, this is in concordance with a very recent study where an intense immunoreactivity for the immune and inflammatory mediator CD40L, expressed on microglia and involved in microglia-dependent neurone death, was found throughout the frontal cortex of AD patients *(94)*. Also, this frontal increase in [^{123}I]iodo-PK11195 uptake could possibly indicate the progression together with the spreading of active inflammation towards more frontal regions in patients already at an advanced stage of the disease, although the mean mini mental state examination score in that study was at a moderate level of 19. This advanced neuropathological stage is in concordance with the frontal perfusion deficits observed in the present study, deficits that typically are observed later in the course of the disease *(95)*. Concerning this progression towards more frontal regions, recent biopsy results also showed that the progressive neurological impairment in Alzheimer's dementia patients is accompanied by a significant increase in senile plaques, neurofibrillary tangles, and microglial cell activation in the frontal cortex *(96)*. However, group analyses should be carefully interpreted because there is a marked heterogeneity in Alzheimer's dementia patients concerning stage of the disease, progression pattern, predominant topographical lesion, and cognitive subtype, with a substantial overlap between Alzheimer's dementia and other neurodegenerative conditions *(95–99)*. Such a heterogeneity may contribute to the rather large range of neuropsychological scores of the Alzheimer's dementia patients and may also be reflected in the higher variability of [^{123}I]iodo-PK11195 uptake in Alzheimer's dementia patients as compared with controls. Concerning this heterogeneity, behavioral as well as cognitive variability has been correlated with PET and SPECT findings *(99)*. Two subgroups with distinct progression rates were already segregated by neuropsychological and cerebral metabolic profiles, in which one rapidly deteriorating group had a significantly greater impairment in executive functions attributed to the frontal lobe and a concomitant greater frontal hypometabolism revealed by PET scanning *(100)*. Where age difference between AD patients and controls may explain at least some of the perfusion SPECT findings, it cannot explain the increased [^{123}I]iodo-PK11195 uptake in Alzheimer's dementia patients because age-related increases in ^{11}C PK11195 uptake have been described only in the thalamus, and no age-related effect at all was found in the present study *(92,101)*. Moreover, the age discrepancy between Alzheimer's dementia patients and controls probably leads to an underestimation of the actual [^{123}I]iodo-PK11195 uptake as a result of the fact that atrophy was not taken into account; atrophy is more prominent in the older AD group, particularly in the left mesotemporal region, because this area, encompassing the hippocampus, is known for its substantial atrophy in Alzheimer's dementia patients *(102)*.

The literature reviewed here and other reports indicate that the radioligand PK11195, developed both for SPECT and PET, can be considered as a highly sensitive cellular marker for the functional monitoring of microglia *in vivo*, useful for the visualization of chronic neurodegeneration without BBB breakdown nor other imaging findings.

4. INFLAMMATION AND BRAIN SEROTONIN

As indicated in Section 1, psychopathological symptoms include apathy, agitation, delusions, and depressive symptoms. Depressive symptoms are present in half of patients, and another 25% may suffer delusions. Particularly, agitation and depression are seen in somatic illnesses with inflammatory characteristics (e.g., interferon treatment in hepatitis C, myocardial infarction; refs. *103–105*). In these psychosomatic disorders, brain serotonin may play a crucial role as for example, selective serotonin uptake inhibitors alleviate concomitant psychopathology *(106)*. The cause of these symptoms may in part be explained as the result of awareness of the deteriorating state of patient or by suboptimal functions of, for example, the frontal cortex or parts of the limbic system. The limbic system has classically been associated with mood disorders. In addition, emerging psychopathology may be considered as the direct result of inflammation on brain function, presumably mediated by cytokines *(107,108)*. Here, we describe an additional consequence of cerebral inflammation on neural function via the induction of the enzyme indoleamine 2,3-dioxygenase in microglia. Similar to peripheral benzodiazepine receptor, indoleamine 2,3-dioxygenase is also in an apparent dormant state in the noninflamed brain but becomes induced after microglia recruitment *(109)*. As a consequence, tryptophan catabolism via the kynurenine pathway becomes highly activated. The synthesis of cerebral serotonin is highly dependent on the access of tryptophan from the blood circulation. This is caused in part by the relatively low abundance of this essential amino acid in food and in part because the rate-limiting synthesizing enzyme of serotonin has a high Km in the same order of its (free) concentration in the brain. Increased catabolism of tryptophan often may lead therefore to a depletion of serotonin. So regional inflammation in, for example, limbic brain regions (including the frontal cortex) may evoke some of the symptoms seen in Alzheimer's dementia patients. A schematic representation of this idea is shown in Fig. 1. A detailed discussion on the possible involvement of inflammation in affective illness can be read in Korf et al. *(105)*.

ACKNOWLEDGMENTS

Our studies were supported by the Netherlands Heart Foundation (grant 96.015), the Dutch Technology Foundation

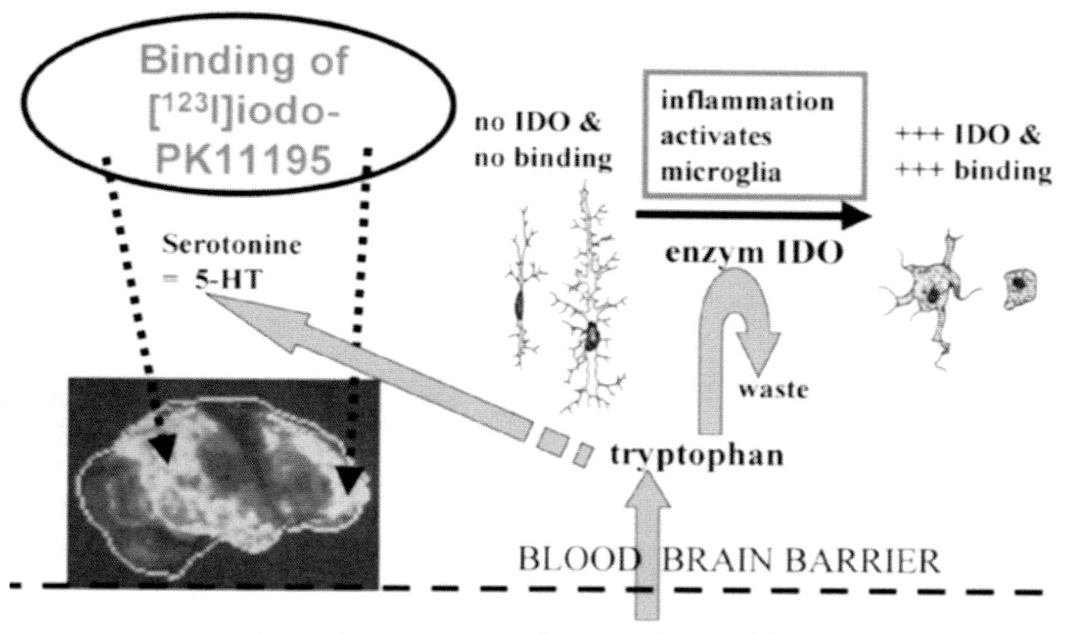

Fig. 1. Schematic representation of the consequences of activation of brain microglia on the increased expression of the peripheral benzodiazepine receptor and the metabolism of tryptophan and the neurotransmitter serotonin (5-hydroxytryptamine). Brain scan indicates areas of glucose hypometabolism and increased PK11195 binding (in frontal and temporal cerebral cortex). *See* color version on Companion CD.

(STW, grant GGN22.2741), and De Internationale Stichting Alzheimer Onderzoek (International Foundation Alzheimer Research, project cerebral inflammation and neuroreceptors in Alzheimer's disease, 1996-1998).

REFERENCES

1. Callahan CM, Hendrie HC, Tierney WM. Documentation and evaluation of cognitive impairment in elderly primary care patients. Ann Intern Med 1995;122:422–423.
2. Cummings JL, Cole G. Alzheimer disease. JAMA 2002;287:2335–2338.
3. Wolfson C, Wolfson B, Asgharian M, et al. A reevaluation of the duration of survival after the onset of dementia. N Engl J Med 2001;344:1111–1116.
4. Knopman DS, Dekosky ST, Cummings JL, et al. Practice parameter: diagnosis of dementia (an evidence-based review). Report of the Quality Standards Subcommittee of the American Academy of Neurology. Neurology 2001;56:1143–1153.
5. The National Institute on Aging and Reagan Institute Working Group on Diagnostic Criteria for the Neuropathological Assessment of Alzheimer's Disease, Consensus Recommendations for the Postmortem Diagnosis of Alzheimer's Disease. Neurobiol Aging 1997;18: S1–S2.
6. Versijpt J. Functional neuroinflammatory and serotonergic imaging in Alzheimer's disease: validation and preliminary clinical findings. Thesis, University of Groningen, 2003:5–177.
7. Cserr HF, Knopf PM. Cervical lymphatics, the blood-brain barrier and the immunoreactivity of the brain: a new view. Immunol Today 1992;13:507–512.
8. Barker CF, Billingham RE. Immunologically privileged sites. Adv Immunol 1977;25:1–54.
9. Mucke L, Eddleston M. Astrocytes in infectious and immune-mediated diseases of the central nervous system. FASEB J 1993;7:1226–1232.
10. Hickey WF, Hsu BL, Kimura H. T-lymphocyte entry into the central nervous system. J Neurosci Res 1991;28:254–260.
11. Head JR, Griffin WS. Functional capacity of solid tissue transplants in the brain: evidence for immunological privilege. Proc R Soc Lond B Biol Sci 1985;224:375–387.
12. Hart MN, Fabry Z. CNS antigen presentation. Trends Neurosci 1995;18:475–481.
13. Lassmann H, Rossler K, Zimprich F, Vass K. Expression of adhesion molecules and histocompatibility antigens at the blood-brain barrier. Brain Pathol 1991;1:115–123.
14. Neumann H, Wekerle H. Neuronal control of the immune response in the central nervous system: linking brain immunity to neurodegeneration. J Neuropathol Exp Neurol 1998;57:1–9.
15. Owens T, Renno T, Taupin V, Krakowski M. Inflammatory cytokines in the brain: does the CNS shape immune responses? Immunol Today 1994;15:566–570.
16. In 't Veld BA, Ruitenberg A, Hofman A, et al. Nonsteroidal antiinflammatory drugs and the risk of Alzheimer's disease. N Engl J Med 2001;345:1515–1521.
17. Rogers J, Kirby LC, Hempelman SR, et al. Clinical trial of indomethacin in Alzheimer's disease. Neurology 1993;43:1609–1611.
18. Rother M, Erkinjuntti T, Roessner M, Kittner B, Marcusson J, Karlsson I. Propentofylline in the treatment of Alzheimer's disease and vascular dementia: a review of phase III trials. Dement Geriatr Cogn Disord 1998;9S1:36–43.
19. Scharf S, Mander A, Ugoni A, Vajda F, Christophidis N. A double-blind, placebo-controlled trial of diclofenac/misoprostol in Alzheimer's disease. Neurology 1999;53:197–201.
20. van Gool WA, Weinstein HC, Scheltens PK, Walstra GJ. Effect of hydroxychloroquine on progression of dementia in early Alzheimer's disease: an 18-month randomised, double-blind, placebo-controlled study. Lancet 2001;358:455–460.
21. Egensperger R, Kosel S, von Eitzen U, Graeber MB. Microglial activation in Alzheimer disease: association with APOE genotype. Brain Pathol 1998;8:439–447.

22. McGeer PL, McGeer EG. Polymorphisms in inflammatory genes and the risk of Alzheimer disease. Arch Neurol 2001;58:1790–1792.
23. Akiyama H, Barger S, Barnum S, et al. Inflammation and Alzheimer's disease. Neurobiol Aging 2000;21:383–421.
24. Bitting L, Naidu A, Cordell B, Murphy GM Jr. β-amyloid peptide secretion by a microglial cell line is induced by β-amyloid-(25–35) and lipopolysaccharide. J Biol Chem 1996;271:16084–16089.
25. Mattson MP, Cheng B, Culwell AR, Esch FS, Lieberburg I, Rydel RE. Evidence for excitoprotective and intraneuronal calcium-regulating roles for secreted forms of the b-amyloid precursor protein. Neuron 1993;10:243–254.
26. Giulian D, Haverkamp LJ, Yu JH, et al. Specific domains of b-amyloid from Alzheimer plaque elicit neuron killing in human microglia. J Neurosci 1996;16:6021–6037.
27. Paresce DM, Chung H, Maxfield FR. Slow degradation of aggregates of the Alzheimer's disease amyloid b-protein by microglial cells. J Biol Chem 1997;272:29390–29397.
28. Ingram DK. Vaccine development for Alzheimer's disease: a shot of good news. Trends Neurosci 2001;24:305–307.
29. Webster SD, Yang AJ, Margol L, Garzon-Rodriguez W, Glabe CG, Tenner AJ. Complement component C1q modulates the phagocytosis of Ab by microglia. Exp Neurol 2000;161:127–138.
30. Mackenzie IRA, Hao CH, Munoz DG. Role of microglia in senile plaque formation. Neurobiol Aging 1995;16:797–804.
31. Weldon DT, Rogers SD, Ghilardi JR, et al. Fibrillar b-amyloid induces microglial phagocytosis, expression of inducible nitric oxide synthase, and loss of a select population of neurons in the rat CNS in vivo. J Neurosci 1998;18:2161–2173.
32. Meda L, Cassatella MA, Szendrei GI, et al. Activation of microglial cells by b-amyloid protein and interferon-g. Nature 1995;374:647–650.
33. Rozemuller JM, Van Muiswinkel FL. Microglia and neurodegeneration (see comment). Eur J Clin Invest 2000;30:469–470.
34. Eikelenboom P, Stam FC. Immunoglobulins and complement factors in senile plaques. Acta Neuropathol 1982;57:242.
35. Myllykangas-Luosujarvi R, Isomaki H. Alzheimer's disease and rheumatoid arthritis. Br J Rheumatol 1994;33:501–502.
36. Marx F, Blasko I, Pavelka M, Grubeck-Loebenstein B. The possible role of the immune system in Alzheimer's disease. Exp Gerontol 1998;33:871–881.
37. Sze G, Zimmerman RD. The magnetic resonance imaging of infections and inflammatory diseases. Radiol Clin North Am 1988;26:839–859.
38. Ketonen L, Tuite MJ. Brain imaging in human immunodeficiency virus infection. Semin Neurol 1992;12:57–69.
39. Kim DM, Tien R, Byrum C, Krishnan KR. Imaging in acquired immune deficiency syndrome dementia complex (AIDS dementia complex): a review. Prog Neuropsychopharmacol Biol 1996;20: 349–370.
40. Scheltens P. Early diagnosis of dementia: neuroimaging. J Neurol 1999;246:16–20.
41. Cecil KM, Lenkinski RE. Proton MR spectroscopy in inflammatory and infectious brain disorders. Neuroimaging Clin N Am 1998;8:863–880.
42. Bitsch A, Bruhn H, Vougioukas V, J, et al. Inflammatory CNS demyelination: histopathologic correlation with in vivo quantitative proton MR spectroscopy. AJNR Am J Neuroradiol 1999;20:1619–1627.
43. Rovaris M, Viti B, Ciboddo G, et al. Brain involvement in systemic immune mediated diseases: magnetic resonance and magnetisation transfer imaging study. J Neurol Neurosurg Psychiatry 2000;68: 170–177.
44. Athens JW, Mauer AM, Ashenbrucker H, Cartwright GE, Wintrobe MM. Leukokinetic studies. I. A method for labeling leukocytes with diisopropylfluorophosphate (DFP^{32}). Blood 1959;14:303–333.
45. Corstens FH, van der Meer JW. Nuclear medicine's role in infection and inflammation. Lancet 1999;354:765–770.
46. Stevens H, Van de Wiele C, Santens P, et al. Cobalt-57 and Technetium-99m-HMPAO-labeled leukocytes for visualization of ischemic infarcts. J Nucl Med 1998;39:495–498.
47. Syed GM, Eagger S, Toone BK, Levy R, Barrett JJ. Quantification of regional cerebral blood flow (rCBF) using $^{99}Tc^{m}$-HMPAO and SPECT: choice of the reference region. Nucl Med Commun 1992; 13:811–816.
48. Joachim CL, Morris JH, Selkoe DJ. Diffuse senile plaques occur commonly in the cerebellum in Alzheimer's disease. Am J Pathol 1989;135:309–319.
49. Rozemuller JM, Stam FC, Eikelenboom P. Acute phase proteins are present in amorphous plaques in the cerebral but not in the cerebellar cortex of patients with Alzheimer's disease. Neurosci Lett 1990; 119:75–78.
50. Talbot PR, Lloyd JJ, Snowden JS, Neary D, Testa HJ. Choice of reference region in the quantification of single-photon emission tomography in primary degenerative dementia. Eur J Nucl Med 1994;21:503–508.
51. Pickut BA, Dierckx RA, Dobbeleir A, et al. Validation of the cerebellum as a reference region for SPECT quantification in patients suffering from dementia of the Alzheimer type. Psychiatry Res 1999;90:103–112.
52. Linde R, Laursen H, Hansen AJ. Is calcium accumulation post-injury an indicator of cell damage? Acta Neurochir 1995;66:15–20.
53. Gramsbergen JBP, Veenma-van der Duin L, Loopuit L, Paans AMJ, Vaalburg W, Korf J. Imaging of the degeneration of neurons and their processes in rat or cat brain by $^{45}CaCl_2$ autoradiography or $^{55}CoCl_2$ positron emission tomography. J Neurochem 1988;50:1798–1807.
54. Dubinsky JM. Examination of the role of calcium in neuronal death. Ann N Y Acad Sci 1993;679:34–40.
55. Gibbons SJ, Brorson JR, Bleakman D, Chard PS, Miller RJ. Calcium influx and neurodegeneration. Ann N Y Acad Sci 1993;679:22–33.
56. Hartley DM, Kurth MC, Bjerkness L, Weiss JH, Choi DW. Glutamate receptor-induced $^{45}Ca^{2+}$ accumulation in cortical cell correlates with subsequent neuronal degeneration. J Neurosci 1993; 13:1993–2000.
57. Müller T, Möller T, Berger T, Schnitzer J, Kettenmann H. Calcium entry through kainate receptors and resulting potassium-channel blockade in Bergmann glial cells. Science 1992;256:1563–1566.
58. Pruss RM, Akeson RL, Racke MM, Wilburn JL. Agonist-activated cobalt uptake identifies divalent cation-permeable kainate receptors on neurons and glial cells. Neuron 1991;7:509–518.
59. Williams LR, Pregenzer JF, Oostveen JA. Induction of cobalt accumulation by excitatory amino acids within neurons of the hippocampal slice. Brain Res 1992;581: 181–189.
60. De Reuck J, Stevens H, Jansen H, et al. Cobalt-55 positron emission tomography of ipsilateral thalamic and crossed cerebellar hypometabolism after supratentorial ischaemic stroke. Cerebrovasc Dis 1999;9:40–44.
61. Jansen HM, Willemsen AT, Sinnige LG, et al. Cobalt-55 positron emission tomography in relapsing-progressive multiple sclerosis. J Neurol Sci 1995;132:139–145.
62. Jansen HM, van der Naalt J, van Zomeren AH, et al. Cobalt-55 positron emission tomography in traumatic brain injury: a pilot study (published erratum appears in J Neurol Neurosurg Psychiatry 1996 Jul;61(1):12[illegible]). J Neurol Neurosurg Psychiatry 1996;60:221–224.
63. Jansen HM, Dierckx RA, Hew JM, Paans AM, Minderhoud JM, Korf J. Positron emission tomography in primary brain tumours using Cobalt-55. Nucl Med Commun 1997;18:734–740.
64. Stevens H, Van de Wiele C, Santens P, et al. Cobalt-57 and Technetium-99m-HMPAO-labeled leukocytes for visualization of ischemic infarcts. J Nucl Med 1998;39:495–498.
65. Iizuka H, Sakatani K, Young W. Neural damage in the rat thalamus after cortical infarcts. Stroke 1990;21:1485–1488.
66. Thibault O, Porter NM, Chen K-C, et al. Calcium dysregulation in neuronal aging and Alzheimer's disease: history and new directions. Cell Calcium 1998;24:417–433.
67. Pascale A, Etcheberrigaray R. Calcium alterations in Alzheimer's disease: pathophysiology, models and therapeutic opportunities. Pharmacol Res 1999;39:81–88.
68. Holscher C. Possible causes of Alzheimer's disease: amyloid fragments, free radicals, and calcium homeostasis. Neurobiol Dis 1998; 5:129–141.
69. Eckert A, Forstl H, Zerfass R, Hartmann H, Muller W. Lymphocytes and neutrophils as peripheral models to study the effect of beta-amyloid on cellular calcium signaling in Alzheimer's disease. Life Sci 1996;59:499–510.

70. Mattson M, Rydel R, Lieberburg I, Smith-Swintosky V. Altered calcium signaling and neuronal injury: stroke and Alzheimer's disease as examples. Ann N Y Acad Sci 1993;679:1–21.
71. Haverstick DM, Gray LS. Increased intracellular Ca^{2+} induced Ca^{2+} influx in human T lymphocytes. Mol Biol Cell 1993;4:173–184.
72. Clementi E, Martino G, Grimaldi LM, Brambilla E, Meldolesi J. Intracellular Ca^{2+} stores of T lymphocytes: changes induced by in vitro and in vivo activation. Eur J Immunol 1994;24:1365–1371.
73. Jansen HM, Knollema S, van-der-Duin LV, Willemsen AT, Wiersma A, Franssen EJ, et al. Pharmacokinetics and dosimetry of cobalt-55 and cobalt-57. J Nucl Med 1996;37:2082–2086.
74. Oosterink BJ, Jansen HML, Paans AMJ, et al. Cobalt-55 positron emission tomography in senile dementia: a pilot study. J Cereb Blood Flow Metab 1995;15:S794.
75. Versijpt J, Decoo D, Van Laere KJ, et al. ^{57}Co SPECT, ^{99m}Tc-ECD SPECT, MRI and neuropsychological testing in senile dementia of the Alzheimer type. Nucl Med Commun 2001;22: 713–719.
76. Van der Laken CJ, Boerman OC, Oyen WJG, van de Ven MTP, van der Meer JWM, Corstens FHM. Scintigraphic detection of infection and inflammation: new developments with special emphasis on receptor interaction. Eur J Nucl Med 1998;25:535–546.
77. Rennen HJ, Boerman OC, Oyen WJ, Corstens FH. Imaging infection/inflammation in the new millennium. Eur J Nucl Med 2001; 28:241–252.
78. Chianelli M, Mather SJ, Martin-Comin J, Signore A. Radiopharmaceuticals for study of inflammatory processes: a review. Nucl Med Commun 1997;18:437–455.
79. Paul C, Peers SH, Woodhouse LE, Thornback JR, Goodbody AE, Bolton C. detection and quantitation of inflammation in the central nervous system during experimental allergic encephalomyelitis using the radiopharmaceutical ^{99m}Tc-RP128. J Neurosci Methods 2000;98: 83–90.
80. Parola AL, Yamamura HI, Laird HE. Peripheral-type benzodiazepine receptors. Life Sci 1993; 52:1329–1342.
81. Banati RB, Myers R, Kreutzberg GW. PK ('peripheral benzodiazepine—binding sites in the CNS indicate early and discrete brain lesions: microautoradiographic detection of [^{3}H]PK11195 binding to activated microglia. J Neurocytol 1997;26:77–82.
82. Zisterer DM, Williams DC. Peripheral-type benzodiazepine receptors. Gen Pharmacol 1997; 29:305–314.
83. Zavala F, Taupin V, Descamps-Latscha B. In vivo treatment with benzodiazepines inhibits murine phagocytic oxidative metabolism and production of interleukin 1, tumor necrosis factor and interleukin-6. J Pharmacol Exp Ther 1990;255:442–450.
84. Guilarte TR, Kuhlmann AC, O'Callaghan JP, Miceli RC. Enhanced expression of peripheral benzodiazepine receptors in trimethyltin-exposed rat brain: a biomarker of neurotoxicity. Neurotoxicology 1995;16:441–450.
85. Banati RB, Newcombe J, Gunn RN, et al. The peripheral benzodiazepine binding site in the brain in multiple sclerosis: quantitative in vivo imaging of microglia as a measure of disease activity. Brain 2000;123:2321–2337.
86. Kuhlmann AC, Guilarte TR. Cellular and subcellular localization of peripheral benzodiazepine receptors after trimethyltin neurotoxicity. J Neurochem 2000;74:1694–1704.
87. Diorio D, Welner SA, Butterworth RF, Meaney MJ, Suranyi CB. Peripheral benzodiazepine binding sites in Alzheimer's disease frontal and temporal cortex. Neurobiol Aging 1991;12:255–258.
88. Groom GN, Junck L, Foster NL, Frey KA, Kuhl DE. PET of peripheral benzodiazepine binding sites in the microgliosis of Alzheimer's disease. J Nucl Med 1995;36:2207–2210.
89. Junck L, Olson JM, Ciliax BJ, et al. PET imaging of human gliomas with ligands for the peripheral benzodiazepine binding site. Ann Neurol 1989; 26:752–758.
90. Ramsay SC, Weiller C, Myers R, et al. Monitoring by PET of macrophage accumulation in brain after ischaemic stroke (letter). Lancet 1992;339:1054–1055.
91. Banati RB, Cagnin A, Myers R, et al. In vivo detection of activated microglia by [^{11}C]PK11195–PET indicates involvement of the globus pallidum in idiopathic Parkinson's disease. Movement Disord 1999;5:S56.
92. Cagnin A, Brooks DJ, Kennedy AM, et al. In-vivo measurement of activated microglia in dementia. Lancet 2001;358:461–467.
93. Versijpt J, Dumont F, Thierens H, et al. Biodistribution and dosimetry of [^{123}I]iodo-PK11195: a potential agent for SPET imaging of the peripheral benzodiazepine receptor. Eur J Nucl Med 2000;27: 1326–1333.
94. Calingasan NY, Erdely HA, Anthony AC. Identification of CD40 ligand in Alzheimer's disease and in animal models of Alzheimer's disease and brain injury. Neurobiol Aging 2002;23:31–39.
95. Report of the Therapeutics and Technology Assessment Subcommittee of the American Academy of Neurology: Assessment of brain SPECT. Neurology 1996;46:278–285.
96. Di Patre PL, Read SL, Cummings JL, et al. Progression of clinical deterioration and pathological changes in patients with Alzheimer disease evaluated at biopsy and autopsy. Arch Neurol 1999;56:1254–1261.
97. Cummings JL. Cognitive and behavioral heterogeneity in Alzheimer's disease: seeking the neurobiological basis. Neurobiol Aging 2000;21:845–861.
98. Perl DP, Olanow CW, Calne D. Alzheimer's disease and Parkinson's disease: distinct entities or extremes of a spectrum of neurodegeneration? Ann Neurol 1998;44:S19–S31.
99. Waldemar G. Functional brain imaging with SPECT in normal aging and dementia. Methodological, pathophysiological, and diagnostic aspects. Cerebrovasc Brain Metab Rev 1995;7:89–130.
100. Mann UM, Mohr E, Gearing M, Chase TN. Heterogeneity in Alzheimer's disease: progression rate segregated by distinct neuropsychological and cerebral metabolic profiles. J Neurol Neurosurg Psychiatry 1992;55:956–959.
101. Cagnin A, Gerhardt A, Banati RB. In vivo imaging of neuroinflammation. Eur Neuropsychopharm 2002;12:581–586.
102. Eagger S, Syed GM, Burns A, Barret JJ, Levy R. Morphologic (CT) and functional (rCBF-SPECT) correlates in Alzheimer's disease. Nuclear Med Commun 1992;13:644–647.
103. Dieperink E, Willenbring M, Ho SB. Neuropsychiatric symptoms associated with hepatitis C and interferon alpha: a review. Am J Psychiatry 2000;157:867–876.
104. Murr C, Widner B, Sperner-Unterweger B, Ledochowski M, Schuber C, Fuchs B. Immune reaction links disease progression in cancer patients with depression. Medical Hypotheses 2000;55:137–140.
105. Korf J, Klein H, Versijpt J, Den Boer JA, Ter Horst GJ. Considering depression as a consequence of activation of the inflammatory response system. Acta Neuropsychiatrica 2002;14:1–10.
106. Musselman DL, Lawson DH, Gumnick JF, et al. Paroxetine for the prevention of depression induced by high-dose interferon alfa. N Engl J Med 2000;344:961–960.
107. Leonard BE. The immune system, depression and the action of antidepressants. Prog Neuro-Psychopharmacol Biol Psychiatry 2001; 25:767–780.
108. Yirmiya R. Behavioral and psychological effects of immune activation: implications for 'depression due to a general medical condition'. Current Trends Psychiatry 1997;10:470–476.
109. Russo S, Kema IP, Fokkema MR, et al. Tryptophan as a link between psychopathology and somatic states. Psychosom Med 2003;65: 665–671.

8 Neurotoxic Oxidative Metabolite of Serotonin

Possible Role in Alzheimer's Disease

LADISLAV VOLICER, MD, PhD, MONIKA Z. WRONA, PhD,
WAYNE MATSON, PhD, AND GLENN DRYHURST, PhD

SUMMARY

Excessive generation of reactive oxygen species and reactive nitrogen species in the Alzheimer's disease (AD) brain and the presence of serotonin in the brain regions that are damaged in this disorder may lead to oxidation of serotonin to tryptamine-4,5-dione, which may possess neurotoxic properties. Efforts to detect this dione in brain tissue or cerebrospinal fluid of AD patients have been so far unsuccessful. However, it may be of relevance that the expression of NAD(P)H:quinone oxidoreductase 1 (NQO1), an enzyme that acts to protect against oxidative stress caused by xenobiotic quinones, is localized not only to neurofibrillary tangles in the AD brain but also to the cytoplasm of hippocampal neurons. In contrast, very little NQO1 is present in the same neuronal populations in age-matched controls. The expression of NQO1 in the AD brain not only provides additional support for excessive production of oxygen free radicals but also, possibly, for a role of a quinone such as tryptamine-4,5-dione.

Key Words: Serotonin; tryptamine-4,5-dione; neurotoxicity; free radicals; Alzheimer's disease.

1. INTRODUCTION

Alzheimer's disease (AD) is a neurodegenerative brain disorder that is the most common form of senile dementia. The AD brain exhibits region-specific patterns of amyloid plaque deposition, neurofibrillary tangles accumulation, and neuronal death. The limbic system and association areas of the cortex exhibit the most profound neuronal damage in AD. In contrast, other cortical areas, such as the somatosensory cortex and cerebellum, are largely spared *(1–3)*.

The increasing incidence of AD and other dementias is a worldwide problem. It has been estimated that in the year 2000, the worldwide number of people with dementia was approx 25 million and that this number will increase to 63 million by 2030 and 114 million by 2050 *(4)*. Currently, almost 50% of the people afflicted with dementia live in Asia, 30% in Europe, and only 12% in North America. There is presently no effective treatment for AD and related dementias.

AD is normally diagnosed by symptoms of clinical dementia. However, positive confirmation of AD can presently be obtained only at autopsy by the presence of senile plaques and neurofibrillary tangles *(5)*. At least two proteins appear to be involved in the pathogenesis of AD, β-amyloid (Aβ), a 42 amino acid peptide that forms the core of senile plaques, and hyperphosphorylated tau, which forms neurofibrillary tangles, although it is possible that α-synuclein also plays a role *(6)*. The relative importance of Aβ and tau is not completely resolved, but increasing evidence suggests that Aβ, either directly or indirectly, is primarily responsible for mediating the nerve cell dysfunction and degeneration observed in AD. For example, Aβ can mediate neurotoxicity. Furthermore, genetic mutations that cause AD are accompanied by increased production of the amyloidogenic form of Aβ *(7)*. In addition, it has been found that immunization of AD patients against Aβ, which decreases amyloid plaques in transgenic mice *(8)*, slows the progression of dementia in those patients who develop antibodies compared with patients without such antibodies *(9)*.

2. OXIDATIVE AND NITRATIVE STRESS IN AD

Damage resulting from excessive production of reactive oxygen species (ROS), a condition known as oxidative stress *(10)*, is evident in the AD brain in the form of oxidized proteins and lipids, and nuclear, and mitochondrial deoxyribonucleic acid *(11–21)*. Such oxidative damage has a regional distribution that parallels the density of senile plaques, being greatest in the parietal lobe and hippocampus and low in the cerebellum *(15)*. In view of the corresponding locations of senile plaques and oxidative damage in the AD brain, it is interesting that Aβ is not only neurotoxic *(22–24)*, but the underlying mechanism involves increased ROS generation and oxidative stress *(25)*. Furthermore, the neurotoxicity of Aβ in vitro is blocked by several antioxidants *(26)*, including α-tocopherol (vitamin E; ref. *27*), curcumin *(28)*, and melatonin *(29)*. It has been sug-

From: *Bioimaging in Neurodegeneration*
Edited by P. A. Broderick, D. N. Rahni, and E. H. Kolodny

gested that one way that Aβ may mediate increased ROS production is by impairing the glutathione (GSH) antioxidant system *(30)*. However, it has long been recognized that reactive microglia are closely associated with neuritic and core plaques in the AD brain *(31,32)* and that Aβ stimulates microglia to produce ROS *(33,34)*. Abnormalities of antioxidant mechanisms in AD brain tissue have been reported *(20,35)*, and melatonin levels in postmortem cerebrospinal fluid exhibit a negative correlation with Braak stages of AD and are decreased even in individuals with only early neuropathalogical changes in the temporal cortex *(36)*.

Increased oxidative stress may affect brain function in AD by several mechanisms. These include mitochondrial dysfunction, disturbed energy homeostasis, excitotoxicity, apoptosis, and advanced glycation *(20)*. There is also evidence that Aβ causes cytoskeletal perturbations that result in apoptosis *(37)* and that this is mediated by lipid peroxidation *(38)*. Aβ-mediated oxidative deoxyribonucleic damage and cell death may be mediated by activation of c-Jun N terminal kinase *(39)*. Oxidative stress also promotes protein fibrillization, which is crucial for the neurotoxicity of Aβ *(6)*.

When stimulated by Aβ, microglia not only generate ROS such as superoxide (O_2^{-}; ref. *40*) but also express inducible nitric oxide synthase (iNOS) and hence generate nitric oxide (NO; refs. *41* and *42*). NO reacts very rapidly with O_2^{-} to form peroxynitrite ($ONOO^{-}$ refs. *43* and *44*). Peroxynitrite is a highly reactive species that nitrates tyrosine residues of proteins *(45)*, and such modified proteins are markedly increased in the AD brain in regions that undergo severe neurodegeneration, such as the hippocampus and some neocortical areas *(46,47)*. Excessive production of NO and peroxynitrite, reactive nitrogen species (RNS), is often described as nitrative stress.

3. REGIONAL SPECIFICITY OF NEURODEGENERATION IN AD

Neurons that are lost in AD involve multiple neurotransmitter systems located either entirely within the association areas of the cortex, the hippocampus, and amygdala or project from subcortical cell bodies and connect to these structures. The latter long neurons include cholinergic *(48,49)*, noradrenergic *(50,51)*, and serotonergic *(52,53)* neurons that project from the nucleus basalis of Meynert, locus ceruleus, and raphe nucleus, respectively, to the association areas of the cortex and hippocampus/amygdala. Such a selective pattern of neurodegeneration in AD argues against a random appearance of lesions and points to a progression of the disease from defined starting points along selected connecting neuronal pathways, that is, a system degeneration *(54)*. Indeed, it has been suggested *(55)* that AD is initiated by chronic attack on axon terminals of long serotonergic and noradrenergic neurons that project from the brain stem at the points where they innervate blood capillaries in the association areas of the cortex and hippocampus. Hardy et al. *(55)* have speculated that such a chronic attack on serotonergic and noradrenergic terminals might involve an environmental toxicant or virus, delivered by the circulatory system, or an autoimmune response. In essence, such a system degeneration concept implies that a toxin enters serotonergic and/or noradrenergic terminals, where they impinge on blood capillaries and then evoke a retrograde degeneration of these and connected neurons by transneuronal transfer of the toxin. Such a concept, in principal, could provide a rationale for the anatomic selectivity of the neurodegeneration that occurs in AD.

4. THE SEROTONERGIC SYSTEM IN AD

Many lines of evidence indicate that the serotonergic system is affected in AD. Thus, serotonergic cells in the raphe nuclei of AD patients contain large numbers of neurofibrillary tangles *(56)* and the number of neurons synthesizing 5-hydroxytryptamine (5-HT; serotonin) is lower than in controls *(57)*. Levels of 5-HT and its major metabolite, 5-hydroxyindole-3-acetic acid (5-HIAA), are decreased in brains of individuals with AD at autopsy *(58)*, and uptake of 5-HT into cortical tissue, obtained during neurosurgery, is lower in patients with AD than in controls *(48)*. The importance of the serotonergic system for cognitive function is indicated by increased cognitive impairment in patients with AD induced by depletion of L-tryptophan *(59)*, the amino acid precursor of 5-HT. AD also affects several 5-HT receptors. In the frontal cortex, for example, 5-HT levels correlate negatively, whereas $5\text{-HT}^{1}{}_{A}$ receptor density correlates positively, with the progression of dementia *(60)*. 5-HT_{2A} receptor density is also decreased in several cortical areas of the AD brain compared with age-matched controls *(61)*. In contrast, 5-HT_4 receptors are unchanged in the frontal and temporal cortex of AD patients *(62)*. 5-HT receptor changes or of the 5-HT plasmalemmal transporter may also contribute to some behavioral symptoms of AD because, for example, 5-HT_{1A} receptor density in the temporal cortex is inversely correlated with aggressive behavior *(63)* and prolactin response to a fenfluramine challenge is increased in patients exhibiting aggressive behavior *(64)*. In addition, a polymorphism in the promoter region of the 5-HT transporter is correlated with the presence of psychotic symptoms and aggressive behavior in AD patients *(65)*.

The serotonergic system may also be involved in the pathogenesis of AD because stimulation of 5-HT receptors may lead to decreased production of amyloidogenic peptides. Thus, stimulation of 5-HT_{2c} receptors by dexnorfenfluramine increases secretion of amyloid precursor protein (APP) metabolite APP(s) in vitro and increases levels of these metabolites, but not of Aβ, in the cerebrospinal fluid (CSF) of guinea pigs *(66)*. However, chronic treatment of guinea pigs with dexnorfenfluramine increases CSF levels of APP(s) but decreases levels of Aβ *(66)*. Similarly, a 5-HT_4 agonist, prucalopride, increases the secretion of APP(s) from Chinese hamster ovary cells *(67)*. Thus, decreased serotonergic activity, which is a normal consequence of AD, may potentiate formation of neurotoxic amyloidogenic peptides.

5. COULD AN ABNORMAL METABOLITE OF 5-HT PLAY A ROLE IN THE PATHOGENESIS OF AD?

Senile plaques, extraneuronal markers for degenerating nerve terminals, have Aβ at their cores and are surrounded by activated microglia that produce O_2^{-} and NO and thence

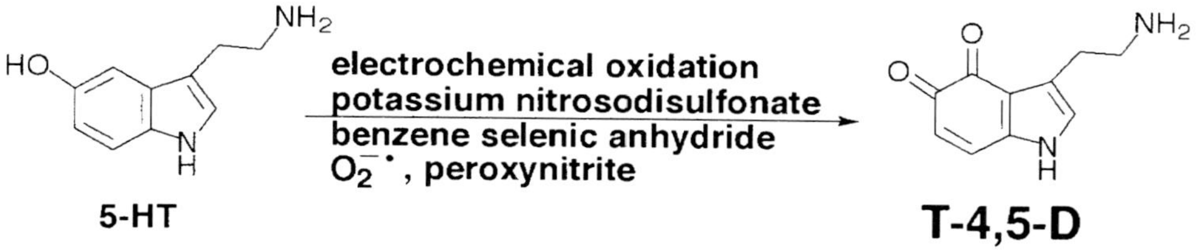

5-HT → T-4,5-D

electrochemical oxidation
potassium nitrosodisulfonate
benzene selenic anhydride
$O_2^{-\cdot}$, peroxynitrite

Fig. 1.

peroxynitrite. Evidence for neuronal damage caused by ROS and RNS is clearly evident in those areas of the brain that degenerate in AD, areas that are innervated with serotonergic neurons. This raises the possibility that the neurotransmitter 5-HT might be oxidized by ROS and/or RNS and, therefore, that one or more products of such oxidation might possess neurotoxic properties. Furthermore, at sites of central nervous system damage, during ischemia, and in the course of inflammatory reactions, 5-HT is released from both blood platelets and serotonergic terminals *(68–70)* and appears to play a role in the selective neurodegeneration resulting from ischemia *(70)*. In this context, mixed vascular and AD pathologies are quite common *(71)*, and there has long been a suspicion that a breakdown or impairment of the blood–brain barrier may play a causative role in the neurodegeneration that occurs in this disorder *(72)*. Furthermore, there is evidence that platelet levels of 5-HT in AD patients are greatly reduced compared with age-matched controls *(73,74)*.

The preceding lines of evidence indicate that ROS and RNS are generated at the sites of neuronal damage in AD and that these species contribute to such damage. However, the presence of serotonergic neurons and fibers, the possible loss of blood–brain barrier integrity and the availability of 5-HT, both from neuronal and blood platelet sources, raise the possibility that oxidation of 5-HT may lead to endogenous toxins that contribute to the neurodegeneration in AD.

An oxidation product of 5-HT that appears to possess neurotoxic properties is tryptamine-4,5-dione (T-4,5-D, Fig. 1). This compound was first reported in 1987 as a product of the electrochemical oxidation of ≤30 μ*M* 5-HT in aqueous 0.01 *M* HCl solutions *(75)*. The electrochemical oxidation of higher concentrations of 5-HT (millimolar) both in acidic solution and at neutral pH also forms T-4,5-D but as a relatively minor constituent of a very complex mixture of products *(76–78)*. Nevertheless, the controlled potential electro-oxidation of ≤30 μ*M* 5-HT in 0.01 *M* HCl can, under appropriate conditions, result in the rapid and quantitative formation of T-4,5-D *(75)*. Indeed, most of the studies aimed at exploring the biological properties of T-4,5-D have used electrochemical methods to synthesize very dilute solutions of the dione in acidic solution that are subsequently adjusted to give the desired pH and/or concentration. Only relatively recently has a procedure been developed to isolate T-4,5-D as a pure solid. This method is based upon oxidation of 5-HT with potassium nitrosodisulfonate and purification by preparative scale reversed-phase high-performance liquid chromatography (HPLC; ref. *79*). Other investigators have used the same oxidant to prepare solutions of T-4,5-D from 5-HT *(80)*. Solutions of T-4,5-D can also be produced by the oxidation of 5-HT with benzeneselenic anhydride *(81)*. Oxidation of 5-HT with hypoxanthine/xanthine oxidase *(82)* or xanthine/xanthine oxidase *(83)*, systems that generate $O_2^{-\cdot}$, also produce T-4,5-D in high yield. Furthermore, peroxynitrite oxidizes 5-HT producing T-4,5-D as a major reaction product *(83)*.

6. STABILITY OF T-4,5-D

Most of the studies aimed at exploring the biological properties of T-4,5-D, both in vitro and in vivo, have used an electrochemical method to prepare a dilute solution of the dione by quantitative oxidation of 5-HT in, for example, 0.01 *M* HCl, which is then adjusted to the desired pH. However, even very low concentrations of T-4,5-D (≤30 μ*M*) in 0.01 *M* HCl are only moderately stable *(75,81,84)* and both increasing concentration and pH enhance the instability of the dione. The stability of T-4,5-D in aqueous solution has recently been studied in some detail *(85)*. At relatively low concentrations (≤200 μ*M*) T-4,5-D is most stable in artificial CSF (aCSF; pH 6–6.5), in which it decomposes ≤10% during the course of 24 h, forming primarily 3-(2-aminoethyl)-6-[3'-(2-aminoethyl)-indol-4',5'-dione-7'-yl]-5-hydroxyindole-4,7-dione (**10**, Fig. 2). In phosphate buffer or 0.5 *M* NH_4Cl solutions at pH 7.4 and in acidic solution (e.g., 0.01 or 0.5 *M* HCl), such concentrations of T-4,5-D also decompose to **10,** although more rapidly than in aCSF. As the concentration of T-4,5-D is increased in all of these media, its decomposition becomes more rapid and shifts towards formation of 7,7'-bi-(5-hydroxytryptamine-4-one) (**6**) and its autoxidation product 7,7'-bitryptamine-4,5-dione (**7**, Fig. 2). At 20 m*M* concentrations in aCSF or at pH 7.4, T-4,5-D very rapidly decomposes to a dark, uncharacterized polymeric precipitate. However, in 0.01 *M* HCl solution, ≥20 m*M* T-4,5-D rapidly and almost quantitatively dimerizes to **6**, Fig. 2). The initial reaction of T-4,5-D, which leads to the ultimate formation of **6**, **7**, and **10**, is the nucleophilic addition of water to the C(7)-position of the dione to form 4,5,7-trihydroxytryptamine (**1**, Fig. 2A). Oxidation of **1** by T-4,5-D and/or by molecular oxygen forms the radicals **2** and **3** (Fig. 2B). Subsequent reactions of these radicals leads to the formation of **6**, **7**, and **10** by the pathways summarized in Fig. 2–f.

7. NEUROTOXICITY OF T-4,5-D

In view of the well-known serotonergic neurotoxicity of 5,6- and 5,7-dihydroxytryptamine, which undergo intraneuronal autoxidation to toxic quinones *(86.87)*, it seems possible that the oxidized form of 4,5-dihydroxytryptamine, i.e., T-4,5-D, might also possess neurotoxic properties. This was

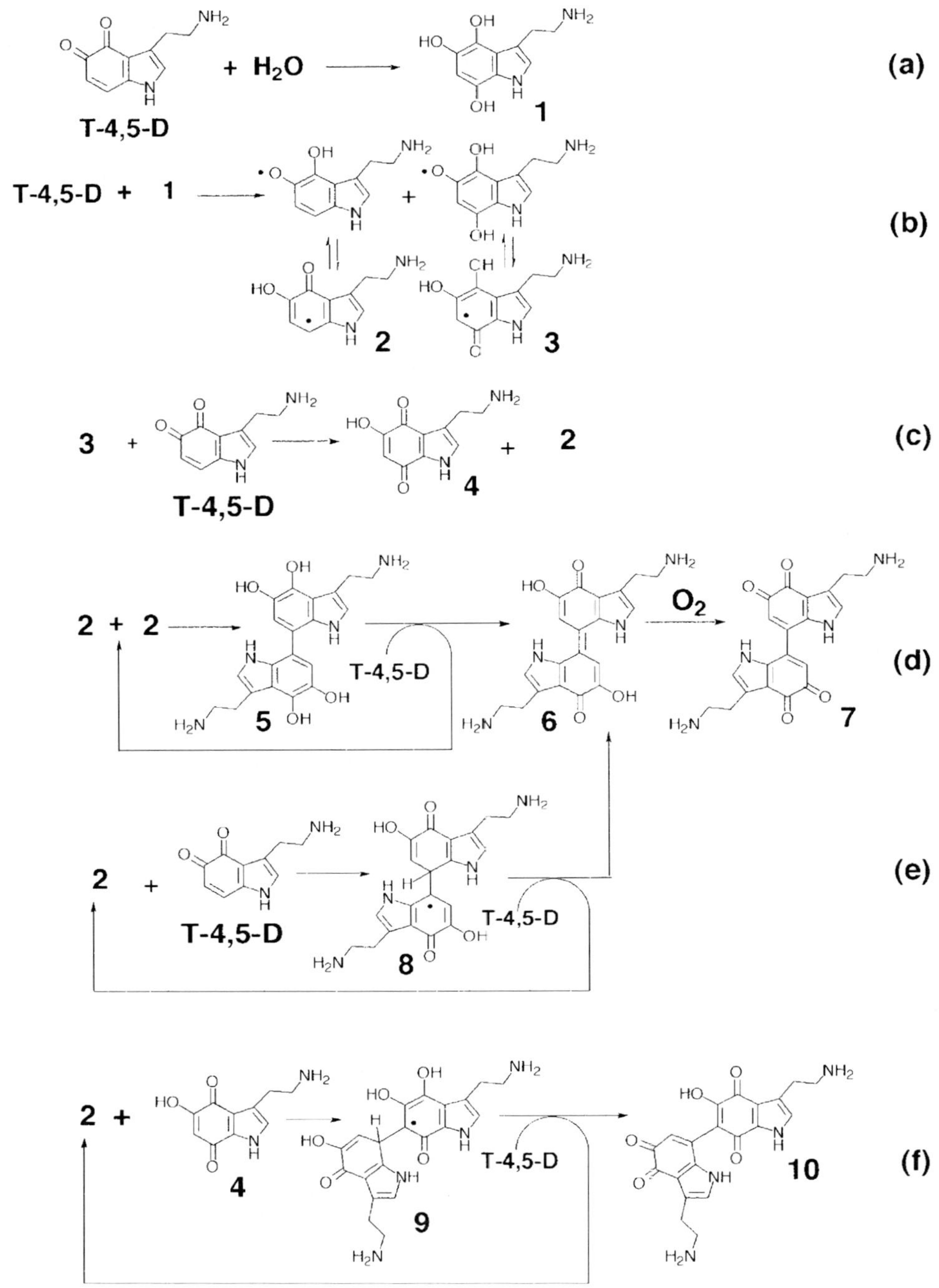

Fig. 2.

first explored in vivo by microinjecting relatively high concentrations of T-4,5-D into the cerebral ventricles or brain tissue of rats *(88)*. In this study, a very dilute solution of 5-HT in 0.01 *M* HCl was electrochemically oxidized to produce T-4,5-D, which was then concentrated to as high as 21 m*M* by means of an ion-exchange procedure. Fink-Heimer staining showed that intracerebroventricular injection of T-4,5-D resulted in terminal degeneration that was most profound in layers I and III of the insular cortex, layer I of the cingulate cortex, and the molecular layer of the dentate gyrus. Argyrophilic and probably degenerating neurons were most frequently subjacent to the granule cell layer of the dentate gyrus, layers II, III, and IV of the entorhinal cortex and throughout the insula. Injection of T-4,5-D directly into the hippocampus indicated that neurotoxicity was dose dependent and produced axon terminal degeneration and neuronal argyrophilia in sectors CA1 and CA3 and in the dentate gyrus. Argyrophilic neurons also were observed in layers II, III, and IV of ipsi- and contralateral entorhinal cortices. Injections of T-4,5-D into the anterior and posterior cingulate cortices produced neurodegeneration in the caudate and anterior thalamic nuclei and the contralateral cortex. These results indicate that T-4,5-D targets the limbic structures that are affected in AD.

Intraventricular administration of T-4,5-D to rats also results in changes of neurotransmitter concentrations *(89)*. Thus, 5-HT and 5-HIAA levels were decreased in the hippocampus, striatum, and prefrontal cortex 7 and 14 d after administration of T-4,5-D but not 3 d after administration of the dione. T-4,5-D had no effect on levels of L-tryptophan and L-tyrosine in these three brain structures. Fourteen days after T-4,5-D administration, the activity of tryptophan hydroxylase was reduced in the cortex. However, administration of T-4,5-D did not alter the binding of paroxetine, a specific inhibitor of the 5-HT plasmalemmal transporter, to nerve terminals. These results indicate that T-4,5-D produces depletion of 5-HT without eliminating serotonergic nerve terminals.

Similar results were obtained using in vitro superfusion experiments *(81)*. Furthermore, T-4,5-D significantly increased the basal efflux of 5-HT from both rat hippocampal and striatal fragments without altering the basal release of dopamine and its metabolite 3,4-dihydroxyphenylacetic acid from striatal fragments. Continuous perfusion of T-4,5-D did not modify the effect of KCl on either 5-HT or DA release from these brain areas. In in vitro incubation experiments, T-4,5-D evoked 5-HT efflux from rat hippocampus in a dose-dependent fashion. The 5-HT uptake inhibitor fluoxetine partially blocked the T-4,5-D-induced 5-HT release, whereas the monoamine oxidase inhibitor pargyline significantly inhibited 5-HIAA efflux but did not modify T-4,5-D-stimulated release of 5-HT. These results suggest that T-4,5-D does not modulate the membrane depolarizing effect of KCl but potentiates release of 5-HT, possibly by mediating the release of vesicular 5-HT into the cytoplasm.

Incubation of T-4,5-D with cultured embryonic chick brain neurons also evokes dose-dependent neurotoxicity *(90)*. Interestingly, such neurotoxic properties of electrochemically synthesized T-4,5-D completely disappear after 24 h, which is indicative of the instability of the dione. The addition of GSH to the freshly prepared solution of T-4,5-D significantly increased the in vitro neurotoxicity of the dione *(90)*. Because T-4,5-D reacts avidly with GSH and other sulfur nucleophiles producing, initially, 7-*S*-thioethers, the latter observation implies that, in vitro, 7-*S*-glutathionyl-T-4,5-D is a more potent neurotoxin than T-4,5-D. Interestingly, GSH blocks the neurotoxic effects of 5,6- and 5,7-dihydroxytryptamine. Furthermore, while the 5-HT uptake inhibitor fluoxetine blocks the neurotoxicity of 5,6- and 5,7-dihydroxytryptamine, it has no effect on the neurotoxicity of T-4,5-D or its 7-*S*-glutathionyl conjugate *(90)*.

Although the evidence presented above suggests that T-4,5-D is neurotoxic both in vivo and in vitro, it is important to note that these studies were conducted before the stability of the dione had been elucidated. As reviewed in Fig. 2, T-4,5-D decomposes to **6/7** and **10**, the actual decomposition product being dependent on the experimental conditions. Thus, some of the neurotoxic effects attributed to T-4,5-D, particularly when high concentrations of the dione were administered into rat brain, may have actually been evoked by its decomposition products.

Despite this cautionary note, a key question concerns possible mechanisms by which T-4,5-D might evoke neurotoxic effects. In vitro studies indicate that when incubated with intact rat brain mitochondria, T-4,5-D uncouples respiration and inhibits state 3 *(91)*. Experiments with rat brain mitochondrial membranes confirm that T-4,5-D irreversibly inhibits NADH-coenzyme Q_1 reductase (complex I) and cytochrome c oxidase (complex IV; ref. *91*). T-4,5-D also strongly inhibits the mitochondrial pyruvate dehydrogenase and α-ketoglutarate dehydrogenase complexes *(80)*. In vitro, T-4,5-D irreversibly inactivates tryptophan hydroxylase *(92)*. The inhibitory effects of T-4,5-D on mitochondrial respiratory enzymes could contribute, in part at least, to the neurotoxic properties of the dione. The inhibition/inactivation of mitochondrial complex I and complex IV respiration, the pyruvate dehydrogenase and α-ketoglutarate dehydrogenase complexes, and tryptophan hydroxylase all appear to be the result of the covalent attachment of T-4,5-D to active site cysteinyl residues of these proteins.

Covalent attachment of T-4,5-D to protein sulfhydryl groups could also influence 5-HT binding protein *(93)* or inhibit the autoreceptor mechanism by blocking the tonic inhibition of endogenous 5-HT. Thus, the negative feedback pathway for 5-HT release from nerve endings involves binding of 5-HT to the 5-HT_{1b} presynaptic receptor and subsequent activation of the guanine nucleotide-binding regulatory protein (G protein), Gi *(94)*. G proteins are abundant proteins that play key roles in the regulation of signal transduction pathways, ion channels and neurotransmitter release *(95)* and contain cysteine residues that should be susceptible to modification by T-4,5-D. Indeed, exposure of striatal synaptosomes or the major guanine nucleotide binding proteins Gi and Go to [3H]T-4,5-D results in radiolabeling of proteins with apparent molecular weights equivalent to the α-subunits of Gi and Go *(96)*. The binding of [35S]guanosine-5'-*O*-(3-thiotriphosphate) to Gi and Go and pertussis toxin-catalyzed [32P]ADP-ribosylation of the G protein α-sub-units are both inhibited in a dose-dependent manner by T-4,5-D *(96)*. Such findings suggest that T-4,5-D may also exert neurotoxic effects through its interactions with G proteins.

8. ABNORMAL FORMS OF 5-HT AND 5-HYDROXYTRYPTOPHAN IN THE CSF OF PATIENTS WITH AD

The development of HPLC with multichannel coulometric (electrochemical) detectors *(97)* provides a powerful method to study the biogenic amine neurotransmitters and metabolites in homogenates from brain tissue and in cerebrospinal fluid. These systems use an array-cell concept that uses coulometric detector electrodes set at different electrochemical potentials. Several electrodes are set across a potential region so that analytes that might coelute from the HPLC column are detected only at particular electrodes because of differences in their oxidation or reduction potentials. Coulometric flow-through detector electrodes allow a compound to display a maximum response (oxidation or reduction current) at a particular detector and then, because of exhaustive electrochemical oxidation or reduction at this detector, no response at the next detector. The band-spreading, dead volume, and pressure characteristics of coulometric array detectors prevent degradation of the chromatographic separation at the last sensor in the array and a significant difference in the registration time between the first and last electrode. The detector response (current) obtained with flow-through coulometric electrodes is often sharper than that obtained with amperometric detectors for many compounds, including the biogenic amine neurotransmitters and their metabolites. HPLC-coulometric array detector systems allow separation of compounds across both a voltage axis and a time axis. In practice, it is possible to resolve co-eluting peaks with current-potential curves differing by 30–40 mV. Thus, over a potential range of 0–600 mV, the number of compounds that can be detected separately in a chromatogram can be increased by as much a 20-fold compared with a single detector system.

Since the development of the early six-sensor embodiment of coulometric electrode array concepts used to study the 5-HT system in CSF from AD subjects *(97,98)*the technology has evolved. Current commercial 16-sensor gradient systems and associated software (Coularray TM, ESA Inc. Chelmsford, MA) are capable of resolving *ca.* 2000 compounds from biological matrices at picogram levels. Such multivariable measurement capabilities have been used in a search mode to determine the kynurenic acid deficit in Huntington's disease *(99,100)*. Use of all data from an assay (whether known or unidentified compounds) has been employed to define categorical metabonomic differences between dietary restricted vs control rats *(101–104)*, between mitochondria from diabetic and control rats *(105,106)*, within subjects prior to during and after extended bed rest *(107)*, and between subjects with AD and controls *(108)*.

In the initial studies of the 5-HT system in AD analysis of CSF from patients with AD using HPLC with a six-channel coulometric array detector revealed that the concentrations of 5-HT and 5-HIAA were lower than those measured in the CSF of control subjects *(98)*. However, the coulometric array detector revealed that the CSF from AD patients, but not from controls, contained additional components that eluted closely to or co-eluted with 5-HT and 5-HIAA but had slightly different electrochemical signatures. To evaluate these compounds the array was configured as a "gate" system *(97)* in which sensors were set alternately at reducing (–370 mV) and oxidizing (380 mV) potentials on the product reduction and oxidation plateaus of 5-HT. In this mode a signal was observed in CSF from AD subjects on the first reduction sensor for compounds closely eluting with 5-HT and 5-hydroxytryptophan (5-HTP). Interestingly, electrochemical oxidation of 5-HT in 0.01 *M* HCl yielded a compound with a similar chromatographic retention time and electrochemical behavior at the coulometric array detector as one of the co-eluting compounds present in the CSF of AD patients. Similarly, electrochemical oxidation of 5-HTP in 0.01 *M* HCl yielded a compound that showed similar behavior upon HPLC-coulometric array detector analysis as another co-eluting component present in the CSF of AD patients.

In a separate study of CSF from a subject with AD, potentiometric measurements were made on site across samples from the rostrocaudal gradient (7th to 23rd ml). Potentials measured in a cell containing a carbon microelectrode against Ag/AgCl (+197mV vs NHE) were from –146 mV to –120 mV in the 7th through 18th ml and up to +204 mV in the 20th to 23rd ml. Voltammetric scans of 10 μ*M* 5-HT in CSF in the same cell indicated measurable oxidation current for 5-HT at ca 5–10% of maximum at +204 mV. Thus, it was concluded that it was thermodynamically feasible for oxidation of 5-HT to occur in vivo.

These observations suggest that T-4,5-D, a product of the electrochemical oxidation of 5-HT *(75)*, might be an aberrant metabolite of the latter neurotransmitter in the CSF of patients with AD. Similarly, the electrochemical oxidation of 5-HTP forms tryptophan-4,5-dione *(109)* (TP-4,5-D), suggesting that this compound might be present in the CSF of AD patients. However, T-4,5-D and TP-4,5-D are highly electrophilic compounds that react avidly with both sulfur and nitrogen nucleophiles *(81,110)* and, hence, it is extremely unlikely that these diones could be detected in the free state in brain tissue or CSF. Indeed, when CSF from AD patients and controls was reanalyzed by HPLC with a 16-channel coulometric array detector, a signal for T-4,5-D could not be detected *(84)*. However, five peaks exhibited significant differences between AD patients and controls. One of these peaks corresponded to 4-hydroxyphenylacetic acid, although the compounds responsible for the other four remain to be identified.

In conclusion, despite strong evidence implicating excessive generation of ROS and RNS in the brain with AD, the presence of 5-HT in the brain regions that are damaged in this disorder and the ease of oxidation of 5-HT to T-4,5-D, which apparently possesses neurotoxic properties, efforts to detect this dione in brain tissue or CSF of AD patients have thus far been unsuccessful. However, it may be of relevance that the expression of NAD(P)H:quinone oxidoreductase 1 (NQO1), an enzyme that acts to protect against oxidative stress caused by xenobiotic quinones, is localized not only to neurofibrillary tangles in the AD brain but also to the cytoplasm of hippocampal neurons *(111)*. In contrast, very little NQO1 is present in the same neuronal populations in age-matched controls. The expression of NQO1 in the AD brain not only provides additional support for the excessive production of oxygen free radicals but also, possibly, for a role for a quinone such as T-4,5-D.

REFERENCES

1. Pearson RCA, Esiri MM, Hiorns RW, Wilcock GK, Powell TPS. Anatomical correlates of the distribution of the pathological changes in the neocortex in Alzheimer disease. Proc Nat Acad Sci USA 1985;82:4531–4534.
2. Henderson VR, Finch CE. The neurobiology of Alzheimer's disease. J Neurosurg 1989;70:325–353.
3. Braak H, Braak E. Pathology of Alzheimer's disease. In: Calne DB, editor. Neurodegenerative diseases. Philadelphia: Saunders; 1994: 585–613.
4. Wimo A, Winblad B, Aguero-Torres H, Von Strauss E. The magnitude of dementia occurrence in the world. Alzheimer Dis Assoc Disord 2003;17:63–67.
5. McKhann G, Drachman D, Folstein M, Katzman R, Price D, Stadlan EM. Clinical diagnosis of Alzheimer's disease: report of the NINCDS-ADRDA Work Group under the auspices of Department of Health and Human Services Task Force on Alzheimer's Disease. Neurology 1984;34:939–944.
6. Trojanowski JQ, Mattson MP. Overview of protein aggregation in single, double and triple neurodegenerative brain amyloidoses. NeuroMol Med 2003;4:1–6.
7. Citron M, Westaway D, Xia W, et al. Mutant presenilins of Alzheimer's disease increase production of 42-residue amyloid-β protein in both transfected cells and transgenic mice. Nat Med 1997;3:67–72.

8. Schenk D, Barbour R, Dunn W, et al. Immunization with amyloid-β attenuates Alzheimer-disease-like pathology in the PDAPP mouse. Nature 1999;400:173–177.
9. Hock C, Konietzko U, Streffer JR, et al. Antibodies against β-amyloid slow cognitive decline in Alzheimer's disease. Neuron 2003;38:547–554.
10. Halliwell B. Reactive oxygen species and the central nervous system. J Neurochem 1992;59:1609–1623.
11. Smith CD, Carney JM, Starke-Reed PE, Oliver CN, Stadtman ER, Floyd RA. Excess brain protein oxidation and enzyme dysfunction in normal aging and in Alzheimer disease. Proc Natl Acad Sci USA 1991;88:10540–10543.
12. Balazs L, Leon M. Evidence of an oxidative challenge in the Alzheimer brain. Neurochem Res 1994;19:1131–1137.
13. Mattson MP. Cellular actions of β-amyloid precursor protein and its soluble and fibrillogenic derivatives. Physiol Rev 1997;77:1081–1132.
14. Smith MA, Rottkamp CA, Nunomura A, Raina AK, Perry G. Oxidative stress in Alzheimer's disease. Biochim Biophys Acta 2000; 1502:139–144.
15. Hensley K, Hall N, Subramanian R, et al. Brain regional correspondence between Alzheimer's disease histopathology and biomarkers of protein oxidation. J Neurochem 1995;65:2146–2156.
16. Lovell MA, Ehmann WD, Butler SM, Markesbery WR. Elevated thiobarbituric acid-reactive substances and antioxidant enzyme activity in the brain in Alzheimer's disease. Neurology 1995;45:1594–1601.
17. Smith MA, Perry G, Richey PL, et al. Oxidative damage in Alzheimer's. Nature 1996;382:120–121.
18. Lyras L, Cairns NJ, Jenner A, Jenner P, Halliwell B. An assessment of oxidative damage to proteins, lipids, and DNA in brain from patients with Alzheimer's disease. J Neurochem 1997;68:2061–2069.
19. Sayre LM, Zelasko DA, Harris PL, Perry G, Salomon RG, Smith MA. 4-Hydroxynonenal-derived advanced lipid peroxidation end products are increased in Alzheimer's disease. J Neurochem 1997;68:2092–2097.
20. Retz W, Gsell W, Münch G, Rösler M, Riederer P. Free radicals in Alzheimer's disease. J Neural Transm 1998;54:221–236.
21. Beal MF. Oxidative damage in neurodegenerative diseases. Neuroscientist 1997;3:21–27.
22. LaFerla FM, Tinkle BT, Bieberich CJ, Haudenschild CC, Jay G. The Alzheimer's A beta peptide induces neurodegeneration and apoptotic cell death in transgenic mice. Nat Genet 1995;9:21–30.
23. Loo DT, Copani A, Pike CJ, Whittemore ER, Walencewicz AJ, Cotman CW. Apoptosis is induced by β-amyloid in cultured central nervous system neurons. Proc Natl Acad Sci USA 1993;90:7951–7955.
24. Mattson MP. Cellular actions of β-amyloid precursor protein and its soluble and fibrillogenic derivatives. Physiol Rev 1997;77:1081–1132.
25. Hensley K, Carney JM, Mattson MP, et al. A model for β-amyloid aggregation and neurotoxicity based on free radical generation by the peptide: Relevance to Alzheimer disease. Proc Natl Acad Sci USA 1994;91:3270–3274.
26. Butterfield DA. β-amyloid-associated free radical oxidative stress and neurotoxicity: Implications for Alzheimer's disease. Chem Res Toxicol 1997;10:495–506.
27. Yatin SM, Aksenov M, Butterfield DA. The antioxidant vitamin E modulates amyloid β-peptide-induced creatine kinase activity inhibition and increased protein oxidation: Implications for the free radical hypothesis of Alzheimer's disease. Neurochem Res 1999;24: 427–435.
28. Lim GP, Chu T, Yang FS, Beech W, Frautschy SA, Cole GM. The curry spice curcumin reduces oxidative damage and amyloid pathology in an Alzheimer transgenic mouse. J Neurosci 2001;21:8370–8377.
29. Matsubara E, Bryant-Thomas T, Quinto JP, et al. Melatonin increases survival and inhibits oxidative and amyloid pathology in a transgenic model of Alzheimer's disease. J Neurochem 2003;85: 1101–1108.
30. Cardoso SM, Oliveira CR. Glutathione cycle impairment mediates A beta-induced cell toxicity. Free Radic Res 2003;37:241–250.
31. Frackowiak J, Wisniewski HM, Wegeil J, Merz GS, Iqbal K, Wang KC. Ultrastructure of the microglia that phagocytose amyloid and the microglia that produce β-amyloid fibrils. Acta Neuropathol (Berl) 1992;84:225–233.
32. McGeer PL, Itagaki S, Boyes BE, McGeer EG. Reactive microglia are positive for HLA-DR in the substantia nigra of Parkinson's and Alzheimer's disease brains. Neurology 1988;38:1285–1291.
33. Qin L, Liu Y, Cooper C, Liu B, Wilson B, Hong JS. Microglia enhance β-amyloid-peptide-induced toxicity in cortical and mesencephalic neurons by producing reactive oxygen species. J Neurochem 2002;83:973–983.
34. Banati RB, Gehrmann J, Schubert P, Kreutzberg GW. Cytotoxicity of microglia. Glia 1993;7:111–118.
35. Volicer L, Wells JM, McKee AC, Kowall NW. Enhanced inhibition of free radical-induced deoxyribose breakdown by Alzheimer brain homogenates. Neurosci Lett 1999;270:169–172.
36. Zhou JN, Liu RY, Kamphorst W, Hofman MA, Swaab DF. Early neuropathological Alzheimer's changes in aged individuals are accompanied by decreased cerebrospinal fluid melatonin levels. J Pineal Res 2003;35:125–130.
37. Sponne I, Fifre A, Drouet B, , et al. Apoptotic neuronal cell death induced by the non-fibrillar amyloid-beta peptide proceeds through an early reactive oxygen species-dependent cytoskeleton perturbation. J Biol Chem 2003;278:3437–3445.
38. Butterfield DA, Lauderback CM. Lipid peroxidation and protein oxidation in Alzheimer's disease brain: potential causes and consequences involving amyloid β-peptide-associated free radical oxidative stress. Free Radic Biol Med 2002;32:1050–1060.
39. Jang JH, Surh YJ. beta-Amyloid induces oxidative DNA damage and cell death through activation of c-Jun N terminal kinase. Ann NY Acad Sci 2002;973:228–236.
40. Colton CA, Snell J, Chernyshev O, Gilbert DL. Induction of superoxide anion and nitric oxide in cultured microglia. Ann NY Acad Sci 1994;738:54–63.
41. Li M, Sunamoto M, Ohnishi K, Ichimori Y. β-Amyloid protein-dependent nitric oxide production from microglial cells and neurotoxicity. Brain Res 1996;720:93–100.
42. Goodwin JL, Uemura E, Cunnick JE. Microglial release of nitric oxide by the synergistic action of β-amyloid and IFN-γ. Brain Res 1995;692:207–214.
43. Beckman JS, Chen J, Ischiropoulos H, Crow P. Oxidative chemistry of peroxynitrite. Meth. Enzymol 1994;233:299–240.
44. Koppenol WH, Moreno JJ, Pryor WA, Ischiropoulos H, Beckman JS. Peroxynitrite, a cloaked oxidant formed by nitric oxide and superoxide. Chem Res Toxicol 1992;5:834–842.
45. Yi D, Smythe GA, Blount BC, Duncan MW. Peroxynitrite-mediated nitration of peptides: characterization of the products by electrospray and combined gas chromatography-mass spectrometry. Arch Biochem Biophys 1997;344:253–259.
46. Smith MA, Harris PLR, Sayre LM, Beckman JS, Perry G. Widespread peroxynitrite-mediated damage in Alzheimer's disease. J Neurosci. 1997;17:2653–2657.
47. Hensley K, Maidt ML, Yu Z, Sang H, Markesbery WR, Floyd RA. Electrochemical analysis of protein nitrotyrosine and dityrosine in the Alzheimer brain indicates region-specific accumulation. J Neurosci 1998;18:8126–8132.
48. Bowen DM, Allen SJ, Benton JS, et al. Biochemical assessment of serotonergic and cholinergic dysfunction in cerebral atrophy in Alzheimer's disease. J Neurochem 1983;41:266–272.
49. Davies P, Maloney AFJ. Selective loss of cerebral cholinergic neurones in Alzheimer's disease. Lancet 1976;2:1403.
50. Mann DMA, Yates PO, Hawkes J. The noradrenergic system in Alzheimer and multi-infarct dementias. J Neurol Neurosurg Psychiatry 1982;45:113–119.
51. Tomlinson BE, Irving D, Blessed G. Cell loss in locus ceruleus in senile dementia of the Alzheimer type. J Neurol Sci 1981;49:418–421.
52. Yamamoto T, Hirano A. Nucleus raphe dorsalis in Alzheimer's disease: Neurofibrillary tangles and loss of large neurones. Ann Neurol 1985;17:573–577.
53. Palmer AM, Francis PT, Benton JS, et al. Presynaptic serotonergic dysfunction in patients with Alzheimer's disease. Brain Res 1987;48:8–15.

54. Saper CB, Wainer BH, German DC. Axonal and transneuronal transport in the transmission of neurobiological disease: potential role in system degenerations, including Alzheimer's disease. Neuroscience 1987;23:389–398.
55. Hardy J, Adolfssun R, Alafuzoff L, et al. Transmitter deficits in Alzheimer's disease. Neurochem Int 1985;7:545–563.
56. Curcio CA, Kemper T. Nucleus raphe dorsalis in dementia of the Alzheimer type: Neurofibrillary changes and neuronal packing density. J Neuropathol Exp Neurol 1984;43:359–368.
57. Kovacs GG, Kloppel S, Fischer I, et al. Nucleus-specific alteration of raphe neurons in human neurodegenerative disorders. Neuroreport 2003;14:73–76.
58. Arai H, Kosaka K, Iizuka R. Changes of biogenic amines and their metabolites in postmortem brains from patients with Alzheimer type dementia. J Neurochem 1984;43:388–393.
59. Porter RJ, Lunn BS, Walker LL, Gray JM, Ballard CG, O'Brien JT. Cognitive deficit induced by acute tryptophan depletion in patients with Alzheimer's disease. Am J Psychiatry 2000;157:638–640.
60. Lai MKP, Tsang SWY, Francis PT, et al. Postmortem serotoninergic correlates of cognitive decline in Alzheimer's disease. Neuroreport 2002;13:1175–1178.
61. Versijpt J, Van Laere KJ, Dumont F, et al. Imaging of the 5-HT2A system: age-, gender-, and Alzheimer's disease-related findings. Neurobiol Aging 2003;24:553–561.
62. Lai MK, Tsang SW, Francis PT, et al. [3H]GR113808 binding to serotonin 5-HT4 receptors in the postmortem neocortex of Alzheimer disease: a clinicopathological study. J Neural Transm 2003;110:779–788.
63. Lai MKP, Tsang SWY, Francis PT, et al. Reduced serotonin 5-HT1A receptor binding in the temporal cortex correlates with aggressive behavior in Alzheimer disease. Brain Res 2003;974:82–87.
64. Lanctôt KL, Herrmann N, Eryavec G, van Reekum R, Reed K, Naranjo CA. Central serotonergic activity is related to the aggressive behaviors of Alzheimer's disease. Neuropsychopharmacology 2002;27:646–654.
65. Sukonick DL, Pollock BG, Sweet RA, et al. The *5-HTTPR * S/ * L* polymorphism and aggressive behavior in Alzheimer disease. Arch Neurol 2001;58:1425–1428.
66. Arjona AA, Pooler AM, Lee RK, Wurtman RJ. Effect of a 5-HT(2C) serotonin agonist, dexnorfenfluramine, on amyloid precursor protein metabolism in guinea pigs. Brain Res 2002;951:135–140.
67. Lezoualc'h F, Robert SJ. The serotonin 5-HT4 receptors and the amyloid precursor protein processing. Exp Gerontol 2003;38:159–166.
68. Pletscher A. The 5-hydroxytryptamine system of blood platelets: physiology and pathophysiology. Int J Cardiol 1987;14:177–188.
69. Saruhashi Y, Hukuda S, Maeda T. Evidence for a neural source of acute accumulation of serotonin in platelets in the injured spinal cord of rats. J. Neurotrauma 1991;8:121–128.
70. Globus MYT, Wester P, Busto R, Dietrich WD. Ischemia-induced extracellular release of serotonin plays a role in CA1 neuronal death in rats. Stroke 1992;23:1595–1601.
71. Goulding JMR, Signorini DF, Chattergee S, et al. Inverse relationship between Braak stage and cerebrovascular pathology in Alzheimer-predominant dementia. J Neurol Neurosurg Psychiatry 1999;67:654–657.
72. Wardlaw JM, Sandercock PAG, Dennis MS, Starr J. Is breakdown of the blood-brain barrier responsible for lacunar stroke, leukoaraiosis, and dementia? Stroke 2003;34:806–812.
73. Kumar AM, Sevush S, Kumar M, Ruiz J, Eisdorfer C. Peripheral serotonin in Alzheimer's disease. Neuropyschobiology 1995;32:9–12.
74. Inestrosa NC, Alarcon R, Arriagada J, Donoso A, Alvarez J. Platelet of Alzheimer patients: increased counts and subnormal uptake and accumulation of [^{14}C]5-hydroxytryptamine. Neurosci Lett 1993;163: 8–10.
75. Wrona MZ, Dryhurst G. Oxidation chemistry of 5-hydroxytryptamine. I. Mechanism and products formed at micromolar concentrations. J Org Chem 1987;52:2817–2825.
76. Wrona MZ, Dryhurst G. Electrochemical oxidation of 5-hydroxytryptamine in acidic aqueous solution. J Org Chem 1989;54:2718–2721.
77. Wrona MZ, Dryhurst G. Oxidation chemistry of 5-hydroxytryptamine. II. Mechanisms and products formed at millimolar concentrations in acidic aqueous solution. J Electroanal Chem 1990;278:249–267.
78. Wrona MZ, Dryhurst G. Electrochemical oxidation of 5-hydroxytryptamine in aqueous solution at physiological pH. Bioorg Chem 1990;18:291–317.
79. Jiang X-R, Dryhurst G. Inhibition of the α-ketoglutarate dehydrogenase and pyruvate dehydrogenase complexes by a putative aberrant metabolite of serotonin, tryptamine-4,5-dione. Chem Res Toxicol 2002;15:1242–1247.
80. Chen J-C, Crino PB, Schnepper PW, To ACS, Volicer L. Increased serotonin efflux by a partially oxidized serotonin: tryptamine-4,5-dione. J Pharm Exp Ther 1989;250:141–148.
81. Cai P, Snyder JK, Chen J-C, Fine R, Volicer L. Neurotoxicity of free-radical-mediated serotonin neurotoxin in embryonic chick brain neurons. Eur J Pharmacol 1996;303:109–114.
82. Chen J-C, Volicer L. Tryptamine-4,5-dione, a serotonin derivative produced by hypoxanthine and xanthine oxidase. Soc Neurosci Abstr 1990;16:529.
83. Wrona MZ, Dryhurst G. Oxidation of serotonin by superoxide radical: implications to neurodegenerative brain disorders. Chem Res Toxicol 1998;11:639–650.
84. Volicer L, Chen J-C, Crino PB, et al. Neurotoxic properties of a serotonin oxidation product: possible role in Alzheimer's disease. Prog Clin Biol Res 1989;317:453–465.
85. Wrona MZ, Jiang X-R, Kotake Y, Dryhurst G. Stability of the putative neurotoxin tryptamine-4,5-dione. Chem Res Toxicol 2003;16:493–501.
86. Baumgarten HG, Jenner S, Klemm HP. Serotonin neurotoxins: recent advances in the mode of administration and molecular mechanism of action. J Physiol (Paris) 1981;77:309–314.
87. Tabatabaie T, Dryhurst G. Molecular mechanisms of action of 5,6- and 5,7-dihydroxytryptamine. In: Kostrzewa RM, editor. Highly Selective Neurotoxins. Basic and Clinical Applications. Totowa, NJ: Humana Press; 1998:269–291.
88. Crino PB, Vogt BA, Chen J-C, Volicer L. Neurotoxic effects of partially oxidized serotonin: tryptamine-4,5-dione. Brain Res 1989; 504:247–257.
89. Chen J-C, Schnepper PW, To A, Volicer L. Neurochemical changes in rat brain after intraventricular administration of tryptamine-4,5-dione. Neuropharmacology 1992;31:215–219.
90. Chen JC, Fine RE, Squicciarini J, Volicer L. Neurotoxicity of free-radical-mediated serotonin neurotoxin in cultured embryonic chick brain neurons. Eur J Pharmacol 1996;303:109–114.
91. Jiang X-R, Wrona MZ, Dryhurst G. Tryptamine-4,5-dione, a putative endotoxic metabolite of the superoxide-mediated oxidation of serotonin, is a mitochondrial toxin: possible implications in neurodegenerative brain disorders. Chem Res Toxicol 1999;12:429–436.
92. Wrona MZ, Dryhurst G. A putative metabolite of serotonin, tryptamine-4,5-dione, is an irreversible inhibitor of tryptophan hydroxylase: possible relevance to the serotonergic neurotoxicity of methamphetamine. Chem Res Toxicol 2001;14:1184–1192.
93. Tamir H, Liu K-P. On the nature of the interaction between serotonin and serotonin binding protein: effects of nucleotides, ions, and sulfhydryl reactions. J Neurochem 1982;38:135–141.
94. Okada F, Tokumitsu Y, Nomura Y. Pertussis toxin attenuates 5-hydroxytryptamine1A receptor-mediated inhibition of forskolin-stimulated adenylate cyclase activity in rat hippocampal membranes. J Neurochem 1989;52:1566–1569.
95. Gilman AG. G proteins: Transducers of receptor-generated signals. Ann Rev Biochem 1987;56:615–649.
96. Fishman JB, Rubins JB, Chen J-C, Dickey BF, Volicer L. Modification of brain guanine nucleotide-binding regulatory proteins by tryptamine-4,5-dione, a neurotoxic derivative of serotonin. J Neurochem 1991;56:1851–1854.
97. Matson WR, Langlais P, Volicer L, Gamache PH, Bird E, Mark KA. n-Electrode three dimensional liquid chromatography with electrochemical detection for determination of neurotransmitters. Clin Chem 1984;30:1477–1488.
98. Volicer L, Langlais PJ, Matson WR, Mark KA, Gamache PH. Serotoninergic system in dementia of the Alzheimer type: abnormal

forms of 5-hydroxytryptophan and serotonin in cerebrospinal fluid. Arch Neurol 1985;42:1158–1161.

99. Beal MF, Matson WR, Swartz KJ, Gamache PH, Bird ED. Kynurenine pathway measurements in Huntington's disease striatum: evidence for reduced formation of kynurenic acid. J Neurochem 1990;55:1327–1339.
100. Beal MF, Matson WR, Storey E, Milbury P, Ryan EA, Ogawa T, Bird ED. Kynurenic acid concentrations are reduced in Huntington's disease cerebral cortex. J Neurol Sci 1992;108:80–87.
101. Shi H, Vigneau-Callahan KE, Shestopalov AI, Milbury PE, Matson WR, Kristal BS. Characterization of diet-dependent metabolic serotypes: primary validation of male and female serotypes in independent cohorts of rats. J Nutr 2002;132:1039–1046.
102. Shi H, Vigneau-Callahan KE, Shestopalov AI, Milbury PE, Matson WR, Kristal BS. Characterization of diet-dependent metabolic serotypes: proof of principle in female and male rats. J Nutr 2002;132:1031–1038.
103. Shi H, Vigneau-Callahan KE, Matson WR, Kristal BS. Attention to relative response across sequential electrodes improves quantitation of coulometric array. Anal Biochem 2002;302:239–245.
104. Vigneau-Callahan KE, Shestopalov AI, Milbury PE, Matson WR, Kristal BS. Characterization of diet-dependent metabolic serotypes: analytical and biological variability issues in rats. J Nutr 2001;131:924S-932S.
105. Kristal BS, Vigneau-Callahan KE, Matson WR. Simultaneous analysis of the majority of low-molecular-weight, redox-active compounds from mitochondria. Anal Biochem 1998;263:18–25.
106. Kristal BS, Vigneau-Callahan KE, Moskowitz AJ, Matson WR. Purine catabolism: links to mitochondrial respiration and antioxidant defenses? Arch Biochem Biophys 1999;370:22–33.
107. Milbury PE. CEAS generation of large multi-parameter metabolic data bases for determining process involvement of biological molecules. In: Acworth IN, Naoi M, Parvez H, Parvez S, editors. Progess in HPLC-HPCE, Vol 6. .The Netherlands: VS Press; 1997:127–144.
108. Matson WR. Method of diagnosing or categorizing or categorizing disorders from biochemical profiles. US Patent 6,210,970 (April 3, 2001).
109. Wu Z, Dryhurst G. 7-*S*-Glutathionyl-tryptophan-4,5-dione: formation from 5-hydroxytryptophan and reactions with glutathione. Bioorg Chem 1996;24:127–149.
110. Jiang XR, Wrona MZ, Alguindigue SS, Dryhurst G. Reactions of the putative neurotoxin tryptamine-4,5-dione with L-cysteine and other thiols. Chem Res Toxicol 2004;17:35–369. .
111. Raina AK, Templeton DJ, Deak JC, Perry G, Smith MA. Quinone reductase (NQO1), a sensitive redox indicator, is increased in Alzheimer's disease. Redox Report 1999;4:23–27.

9 Predicting Progression of Alzheimer's Disease With Magnetic Resonance

Kejal Kantarci, MD and Clifford R. Jack, Jr., MD

SUMMARY

Advances in the field of molecular biology concerning Alzheimer's disease (AD) generate the possibility of useful therapeutic interventions in the near future. The major beneficiaries of disease-modifying treatments that are currently under development will be those patients who have early pathological involvement. Improved methods for early diagnosis and noninvasive surrogates of disease severity in AD are crucial for early therapeutic interventions and for measuring their effectiveness. Various quantitative magnetic resonance (MR) techniques that measure the anatomic, biochemical, microstructural, functional, and bloodflow changes are being evaluated as possible surrogate measures of disease progression. Cross-sectional and longitudinal studies indicate that MR-based volume measurements are potential surrogates of disease progression in AD, starting from the preclinical stages. The recent development of amyloid imaging tracers for positron emission tomography has been a major breakthrough in the field of imaging markers for AD. Efforts to image plaques also are underway in MRI. As with indirect MR measures, the approaches that directly image the pathological substrate will need to undergo a validation process by longitudinal studies to prove their usefulness as predictors of AD.

Key Words: Alzheimer's disease; mild cognitive impairment; magnetic resonance imaging; MRI; magnetic resonance spectroscopy; MRS, diffusion weighted imaging; DWI; arterial spin labeling MRI; functional MRI.

1. INTRODUCTION

The incidence and prevalence of Alzheimer's disease (AD) increases significantly with aging. With improvements in healthcare and increasing life expectancy, AD is becoming a significant public health problem. Early diagnosis and treatment of this condition are crucial to sustaining the quality of life of many elderly individuals and their families. Although there are no proven treatments that can reverse the pathology of AD, therapies that slow down disease progression on clinical grounds generate the prospect for preventive interventions. Surrogate biomarkers that are sensitive to disease progression starting from the earliest stages are critically needed for disease-specific preventive therapies. Quantitative magnetic resonance (MR) techniques that measure the anatomic, biochemical, microstructural, functional, and bloodflow changes are being evaluated as possible surrogates of disease progression with cross-sectional and longitudinal studies. This chapter will focus on quantitative MR techniques as potential surrogates of disease progression in AD. We will first discuss cross-sectional studies in patients with AD and people who are at higher risk of developing AD. These studies lay the groundwork for longitudinal studies evaluating quantitative MR techniques as predictors of disease progression, which will be discussed in the latter half of this chapter.

2. CROSS-SECTIONAL STUDIES OF QUANTITATIVE MR TECHNIQUES IN PEOPLE WITH THE CLINICAL DIAGNOSIS OF AD

Progression of the neurofibrillary pathology of AD follows a hierarchical topographical course in the brain. It involves the medial temporal limbic cortical structures: the entorhinal cortex and hippocampus during the earliest stages and later progressing to paralimbic cortical areas, involving the neocortex at the later stages of the disease *(1)*. This orderly anatomical progression of the neurofibrillary pathology is important in evaluating potential surrogate MR markers for early diagnosis and disease progression, specifically for methods using measurements from predetermined anatomic regions of interest. Because each MR marker measures a certain feature of AD pathology, the strategic choice of regions to study must be based on the progression of the target pathology and the stage of the disease being studied. It is expected that this approach would identify different MR measurements from different brain

From: *Bioimaging in Neurodegeneration*
Edited by P. A. Broderick, D. N. Rahni, and E. H. Kolodny

regions of interest that would be sensitive to pathological progression at different stages of disease severity.

Memory impairment is the earliest symptom of AD. In keeping with that, medial temporal lobe limbic cortical regions, which are essential for episodic memory function, are involved with the pathology of AD early in the course of disease *(1)*. Neuron and tissue loss correlate closely with the neurofibrillary pathology of AD, and atrophy is the resultant macroscopic change *(2)*. For this reason, the medial temporal lobe is an attractive target for MR-based measurements. Volume measurements from different medial temporal lobe structures have been extensively studied to differentiate patients with AD from cognitively normal elderly *(3–12)*. Of these, the entorhinal cortex and hippocampus volumes are generally considered to be the most accurate in differentiating patients clinically diagnosed as AD from normal. The abilities of these medial temporal lobe volume measurements to discriminate patients with AD from normal do not differ significantly *(13,14)*. However, greater difficulty with MRI boundary definition of the entorhinal cortex compared to hippocampus is noted, which results in better test–retest reproducibility of hippocampal measurements (Fig. 1).

Proton MR spectroscopy (^{1}H MRS) is a diagnostic imaging technique that is sensitive to the changes in the brain at the cellular level. With ^{1}H MRS, several of the major proton-containing metabolites in the brain are measured during a common data acquisition period. The metabolite *N*-acetyl aspartate (NAA) is a marker for neuronal integrity. NAA decreases in a variety of neurological disorders, including AD *(15–19)*. The decrease of NAA or the NAA/creatine (Cr) ratio shows a regional variation in AD *(20–22)*. In patients with mild-to-moderate AD, NAA/Cr levels are lower than normal in the posterior cingulate gyrus, whereas they are normal in the medial occipital lobe including the visual cortex (Fig. 2; ref. *23*). This regional pattern is in agreement with the distribution of the neurofibrillary pathology, and the associated neuron loss in people with mild-to-moderate AD, indicating that regional NAA/Cr levels are potential surrogates for disease progression. Another metabolite that is consistently found to be abnormal in people with AD is myo-inositol (mI) or mI/Cr ratios *(21,22,24)*. The mI peak consists of glial metabolites, which are responsible for osmoregulation *(25,26)*. Elevated mI levels correlate with glial proliferation in inflammatory central nervous system demyelination *(27)*. It is thought that the elevation of the mI peak also is related to glial proliferation and astrocytic activation in AD *(21,23,24,28)*. Another metabolite peak of interest in the ^{1}H MRS of the brain in AD is the choline (Cho) peak. Although some studies identified elevated Cho and Cho/Cr ratios, some reported normal levels in people with AD compared with normal *(29)*. The largest amount of Cho in the brain is in the Cho-bound membrane phospholipids that are precursors of Cho and acetylcholine synthesis. It has been postulated that the elevation of Cho peak is the consequence of membrane phosphatidylcholine catabolism to provide free choline for the chronically deficient acetylcholine production in AD *(29,30)*.

Diffusion-weighted MR imaging (DWI) is sensitive to the microscopic structural changes in the brain through measuring the diffusivity of water molecules. The apparent diffusion coefficient measurements of DWI indicate that the diffusivity of water is higher in the hippocampus and white matter regions in patients with AD than cognitively normal elderly (Fig. 3; refs. *31–34*). Elevation of the apparent diffusion coefficients in the brains of people with AD is attributed to the expansion of the extracellular space owing to the loss of neuron cell bodies and dendrites in the gray matter and Wallerian degeneration in the white matter. Another MR technique that is sensitive to the mobility of water molecules is magnetization transfer (MT) MRI. The MT ratio of immobile protons to free protons in the hippocampus and in the whole brain is lower in patients with AD than normal *(35–37)*. DWI and MT MRI are both sensitive to the ultrastructural changes in the brains of people with AD. Longitudinal studies are needed for determining their usefulness in early diagnosis and disease progression.

Cerebral blood volume MR measurements using contrast agents indicate a reduction in temporoparietal blood volume in patients with AD *(38–40)*. Another technique sensitive to cerebral blood flow but does not require injection of contrast agents is arterial spin labeling (ASL). Significant blood flow reductions were identified in the temporal, parietal, frontal, and posterior cingulate cortices of patients with AD relative to controls *(41)*. ASL is an appealing technique for blood flow measurements because it does not require contrast injection or ionizing radiation. Studies comparing the accuracy of ASL to nuclear medicine imaging modalities as surrogate markers for blood flow changes in AD are needed.

Changes in cognitive function accompany and may even precede the MR-detectable microscopic and macroscopic structural changes related to AD pathology in the brain. For this reason, functional imaging methods are of interest for early diagnosis. Measurements of brain activation with functional MRI (fMRI) show that activation patterns are different in people with AD compared with cognitively normal elderly using activation paradigms such as visual saccades, visual and motor responses, semantic processing, angle discrimination, and memory *(42–50)*.

All MR measurements discussed in this section are sensitive to a certain feature of AD pathology in people who are clinically diagnosed as AD. Autopsy studies, however, indicate that the pathology of AD precedes the clinical diagnosis of dementia, perhaps by decades. One way of evaluating MR markers for early AD pathology is through studying risk groups, which will be discussed in the next section.

3. CROSS-SECTIONAL STUDIES OF QUANTITATIVE MR TECHNIQUES IN PEOPLE WHO ARE AT AN ELEVATED RISK OF PROGRESSING TO AD

Although tangles and plaques are the pathological signatures of AD, they also are commonly encountered in individuals who are not clinically demented *(51,52)*. Typically this pathology manifests as clinical AD only after a certain quantitative threshold is reached. By the time the individual is diagnosed with AD, a significant synapse and neuron loss had already occurred *(53)*. The most favorable stage for disease-modifying therapies in AD is before the irreversible damage takes place. People who possess early AD pathology but not yet

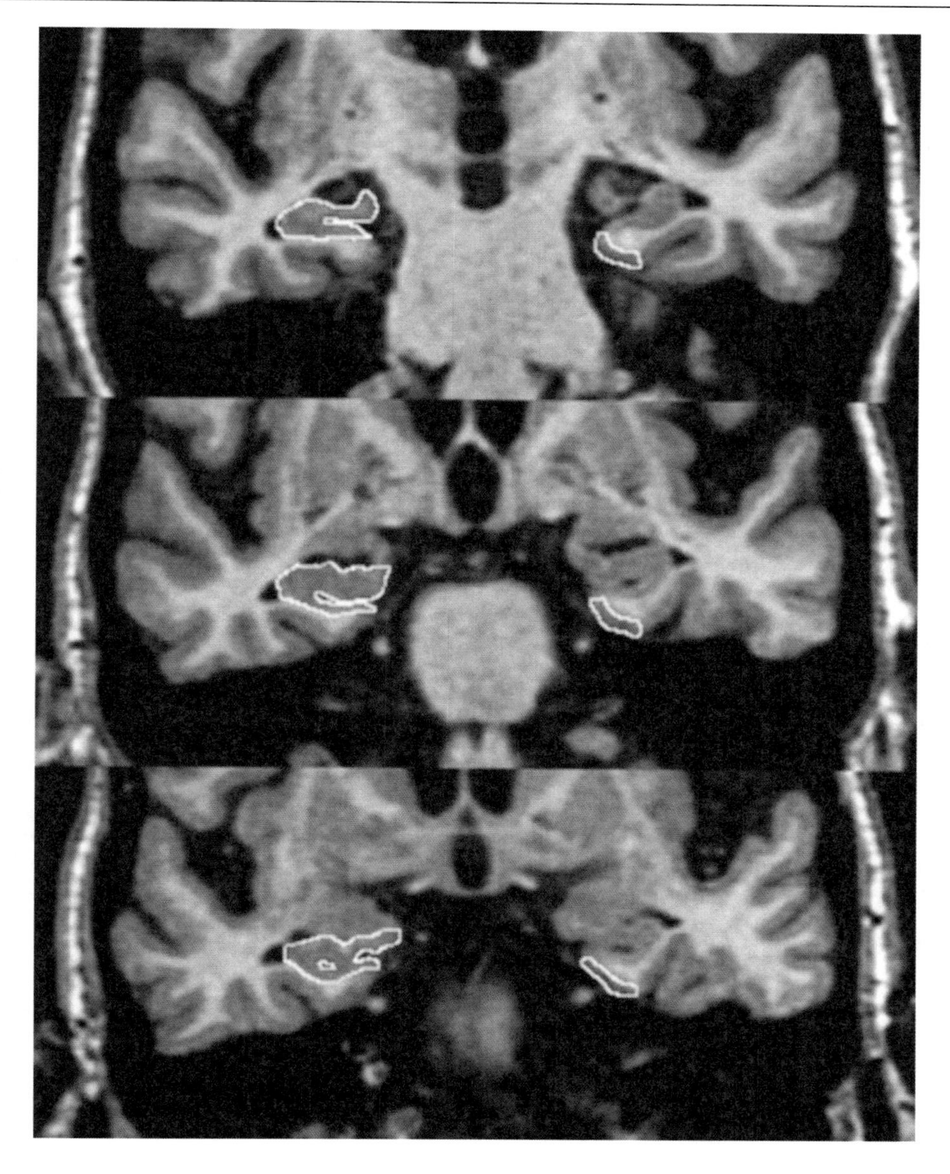

Fig. 1. Hippocampal and entorhinal cortex boundary definitions. The figure is composed of three oblique coronal images subvolumed to include only the temporal lobe regions. From top to bottom in the collage, the images progress from posterior to anterior. Tracings of the right hippocampus indicate hippocampal volume boundaries. The tracings on the left indicate entorhinal cortex volume. (Reprinted with permission from ref. *14*.)

demented are of particular interest for preventive therapies and for determining surrogate markers of early pathology. One way of identifying people who possess early AD pathology is through studying risk groups. Aging is the strongest risk factor for AD. Higher risk groups in the aging population are composed of individuals who have a greater probability of developing AD than their peers. The higher risk groups are identified either through clinical examination or family history and genetic testing.

Memory impairment is the earliest symptom of AD. Many elderly individuals with memory impairment, however, do not meet the clinical criteria for dementia. The syndrome of mild cognitive impairment (MCI) was defined on clinical grounds to identify these people with memory impairment who are not clinically demented *(54)*. Recently, these individuals have been subclassified as amnestic MCI, single-nonmemory domain, and multiple-domain MCI *(55)*. Only the outcome of people with amnestic MCI has been validated with longitudinal studies, revealing that people with amnestic MCI have a higher risk of developing AD than their cognitively normal peers *(56)* and that most of the people with amnestic MCI will progress to AD during their lifetime *(57)*. In accordance with this, autopsy studies show that most people with amnestic MCI have early AD pathology in the limbic cortical structures responsible for memory function *(53,57)*. People with amnestic MCI therefore are an important clinical group for preventive trials and for evaluating surrogate MR markers for early diagnosis and monitoring disease progression at the early stages of the disease.

In line with the autopsy findings, MR-based volume measurements indicate that the hippocampus and entorhinal cortex

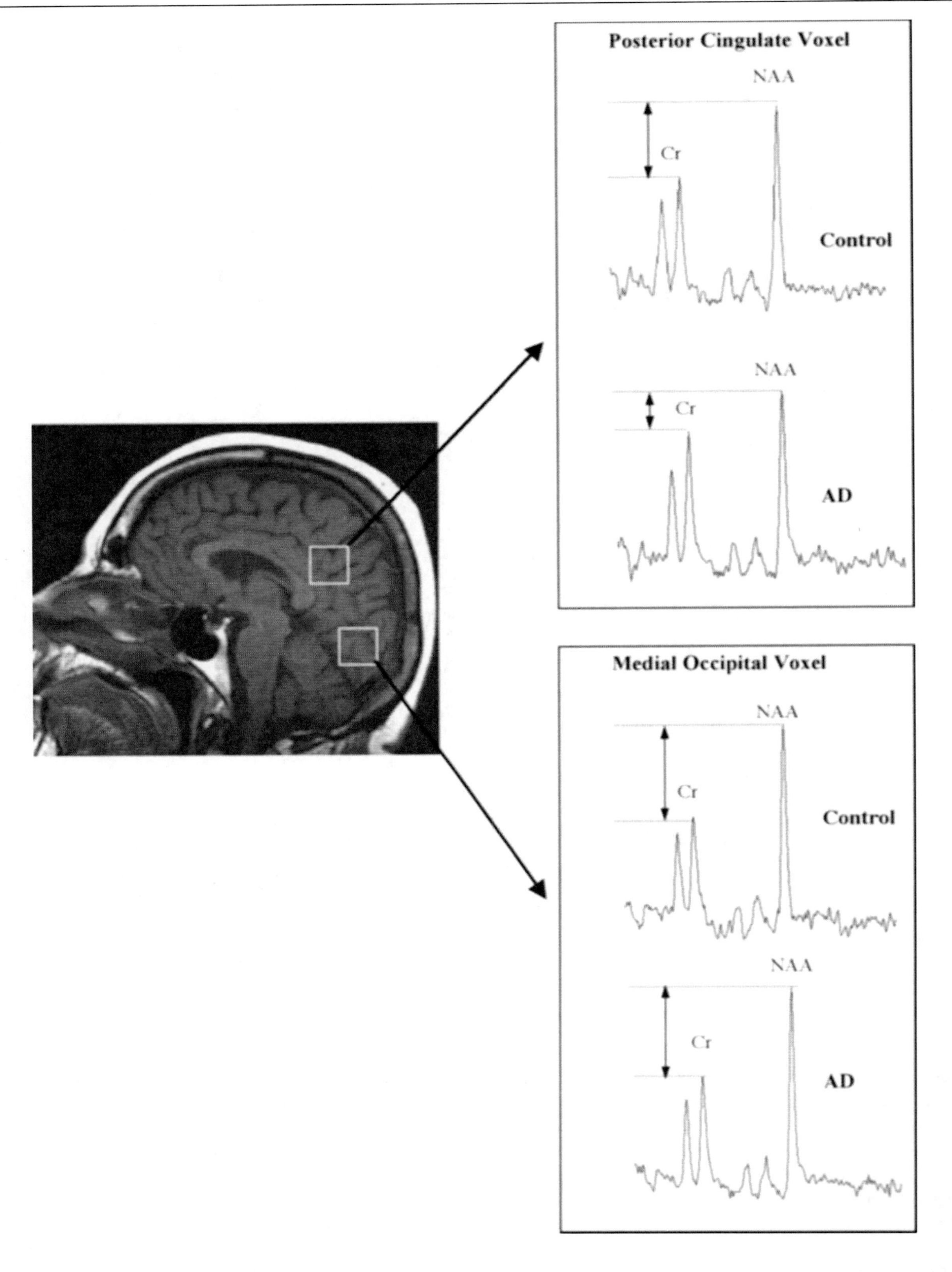

Fig. 2. Regional NAA/Cr decrease in AD. Examples of proton spectra obtained from the posterior cingulate and medial occipital voxel with an echo time of 135 ms in a control subject and an AD patient. In patients with mild-to-moderate AD, NAA/Cr levels are lower than normal in the posterior cingulate gyri and inferior precunei, a region that is involved with the pathology of AD early in the disease course, while they are normal in the medial occipital lobe including the visual cortex.

volumes of people with mild impairment syndromes are smaller than normal *(58–61)*. Furthermore, similar to people with AD, ^{1}H MRS measurements of the posterior cingulate gyri mI/Cr ratios *(21,62)* and DWI measurements of the hippocampal ap-

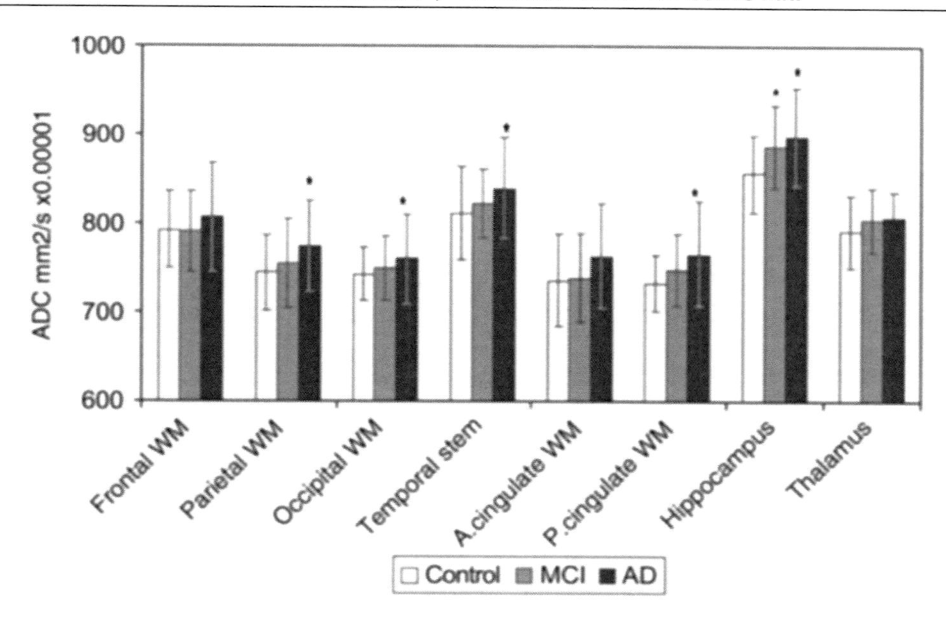

Fig. 3. Apparent diffusion coefficients (ADCs) from different regions in the brains of controls, patients with mild cognitive impairment (MCI) and Alzheimer's disease (AD). The bar graph shows the means and the error bars show the standard deviations of ADC ($mm^2/s \times 10^{-6}$) in the control, MCI, and AD subjects from the eight different regions of interest (ROI) in the brain. *, The parietal, posterior (P) cingulate, temporal stem, and occipital WM and the hippocampal ADC are higher in AD patients than in controls, and hippocampal ADC are higher in MCI patients than in controls ($p < 0.05$). (Reprinted with permission from ref. *34*.)

parent diffusion coefficients *(63)* are higher, whereas MT MRI measurements of temporal lobe MT ratios are lower than normal *(64,65)* in people with MCI. These MR measurements are in agreement with the clinically transitional nature of MCI, residing between normality and dementia. An important consideration in evaluating imaging markers for early AD is that some members of any cognitively normal control group will likely have preclinical AD pathology. The expectation from imaging markers therefore should not be complete separation of cognitively impaired individuals from normal individuals. Some degree of overlap between the clinically identified groups is expected given that preclinical AD pathology is not uncommon in the elderly.

A recent fMRI study showed that people with MCI and AD have a similarly low medial temporal lobe activation on a memory task when compared with cognitively normal elderly individuals *(66)*. Memory dysfunction is the earliest symptom of AD, and it is common to both MCI and AD. Longitudinal studies are needed to test whether fMRI is sensitive to the functional changes before memory impairment becomes clinically apparent, such as in cognitively normal people who are destined to develop MCI or AD in the future.

The apo E ε4 allele increases the risk of developing AD in a dose-dependent fashion and also lowers the average age of disease onset *(67)*. Quantitative MR studies, which have investigated the association between apo E genotype and MR measurements, report conflicting findings. In them, some researchers found smaller whole brain and medial temporal lobe volumes in patients with AD possessing the apo E ε4 allele *(68,69)*, whereas others did not find any difference in the hippocampal volumes of apo E ε4-positive and -negative individuals *(70–72)*. One ^{1}H MRS study showed that metabolic alterations in postmortem brains of patients with AD are exaggerated by apo E ε4 *(73)*; another, however did not identify such an effect *(74)*. The differences in age, disease duration, and the number of subjects in these studies may be responsible for the discrepancies.

4. MR PATHOLOGIC CORRELATION STUDIES REFLECTING PATHOLOGIC STAGE ACROSS THE ENTIRE SEVERITY SPECTRUM

Histopathological findings are considered to be the "gold standard" in evaluating surrogate markers for diagnosis and disease progression in AD. Few studies have correlated quantitative MR measurements with the histopathologic diagnosis and staging so far. The correlation between antemortem MR measurements of the hippocampal volumes and postmortem Braak and Braak staging (Fig. 4; ref. *1*) indicate that hippocampal atrophy, although not specific for AD, is a fairly sensitive marker of pathologic stage *(75)* and hippocampal neurofibrillary tangle burden *(76)*. MR-based hippocampal volume measurements on postmortem samples further show a strong correlation between hippocampal volumes and neuron numbers, validating technique's sensitivity to hippocampal neurodegeneration *(77)*.

Neurofibrillary pathology and associated neuron loss are present in some elderly individuals with normal cognition. This pathology in normal elderly is usually confined to the medial temporal lobe corresponding to the Braak stages I and II. MR-based hippocampal volumetry on postmortem scans of the Nun study participants suggest that hippocampal volumes may be useful in identifying early pathology of AD in nondemented individuals *(78)*. Both antemortem and postmortem MR studies indicate that MR-based hippocampal volumetry is a valid marker for the pathologic stages of AD, regardless of the clinical diagnosis.

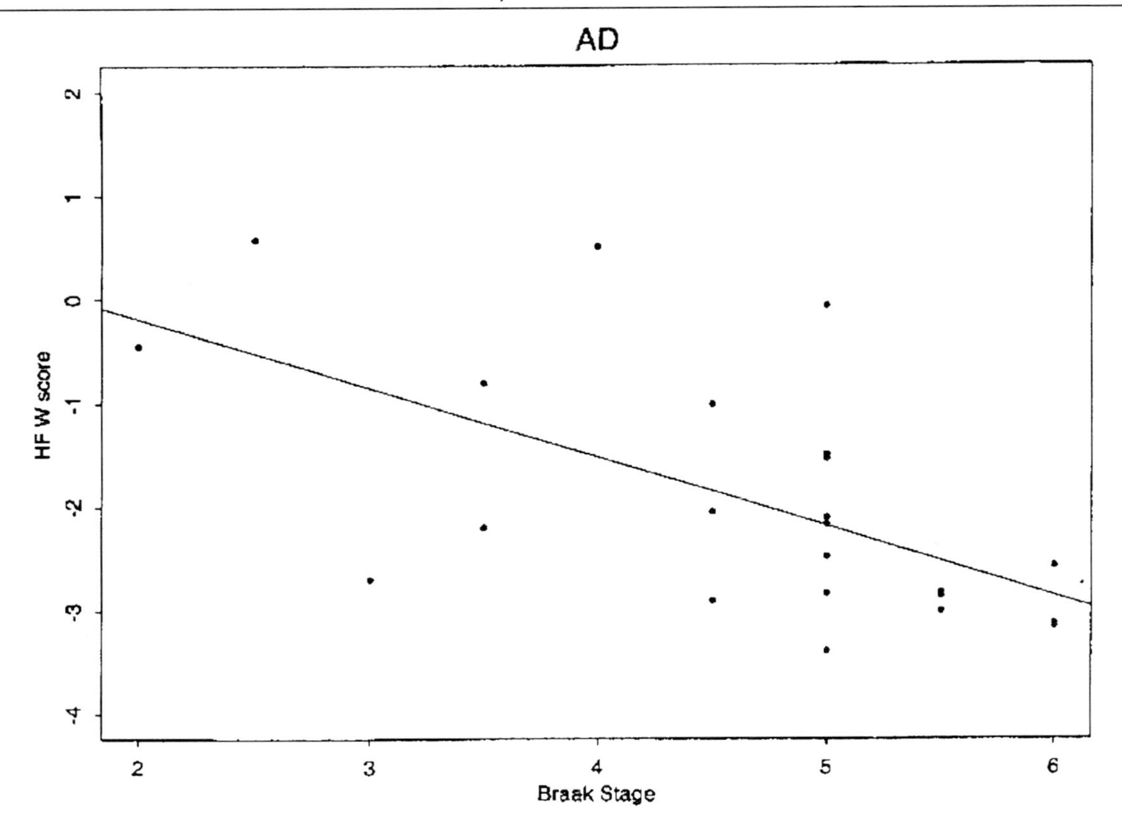

Fig. 4. Antemortem MR-based hippocampal volume measurements correlate with postmortem Braak and Braak staging in subjects with isolated AD pathology. (Reprinted with permission from ref. *75*.)

5. LONGITUDINAL STUDIES PREDICTING FUTURE PROGRESSION TO AD IN COGNITIVELY NORMAL ELDERLY AND IN RISK GROUPS USING QUANTITATIVE MR TECHNIQUES

The value of quantitative MR techniques for predicting future progression to AD both in cognitively normal elderly and in risk groups is assessed through longitudinal studies that test whether baseline MR measurements can predict clinical outcome in these individuals after several years of follow-up. People with mild impairment syndromes are an attractive group to study for identifying quantitative MR techniques for predicting clinical outcome because most of them eventually progress to AD. MR-based medial temporal lobe, hippocampal, and entorhinal cortex volumetry is predictive of subsequent progression to AD in people with mild impairment syndromes *(79–81)*. However, because patients with MCI progress to AD at different rates, MR-based volumetry was also tested for predicting the rate of progression to AD in people with MCI. Kaplan-Meier analysis performed on 80 patients with MCI who were followed at an average of 32.6 mo indicate that patients with a smaller hippocampal volume at baseline progress to AD faster than the ones with larger volumes (Fig. 5; ref. *79*). Furthermore, longitudinal studies on cognitively normal elderly people indicate that temporal lobe volume loss may mark the beginning of the disease process as much as six years prior to dementia onset *(82)*.

6. LONGITUDINAL AND SERIAL MR MEASUREMENTS CORRELATING WITH CLINICAL DISEASE PROGRESSION ACROSS THE ENTIRE SEVERITY SPECTRUM

The neurodegenerative pathology of AD causes progressive atrophy and deformation of the brain overtime. Serial MR measurements that are sensitive to this change can potentially be useful for tracking the pathological progression. The earliest and most severe atrophy during the progression of AD takes place in the medial temporal lobe. Serial MR-based volume measurements of the hippocampus in healthy young volunteers in whom no change is expected show that hippocampal volumetry is a reproducible technique with a quite low coefficient of variation (0.28%; ref. *83*). MR-based hippocampal measurements on serial scans of 24 patients with AD and 24 age- and gender-matched controls showed approx 2.5 times higher rate of hippocampal volume loss in patients with AD compared with normal elderly *(83)*. Another study, which followed 27 patients with AD and 8 controls for 3 yr, showed a statistically nonsignificant trend towards accelerated volume loss in the AD group compared with controls *(84)*.

As discussed in the previous section, hippocampal volume loss on MRI correlates with the pathologic progression, which begins in the medial temporal lobe years before the clinical diagnosis of AD. One study followed 30 nondemented elderly individuals annually during a mean period of 42 mo *(82)*.

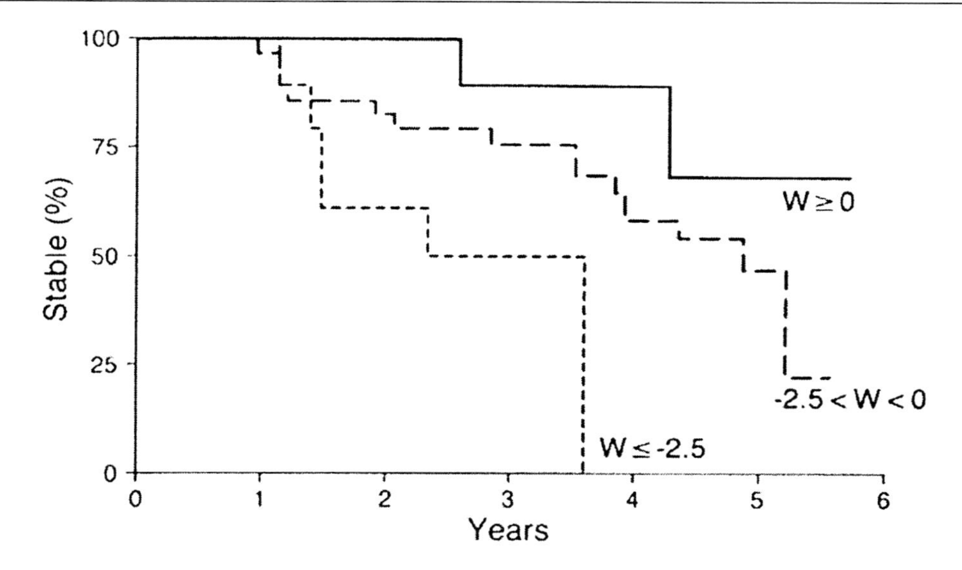

Fig. 5. Hippocampal W score and crossover from MCI to AD. Kaplan-Meier curves of patients whose hippocampal W score at baseline is ≥0 (n = 13), 0 > W > –2.5 (n = 54), and ≤ –2.5 (n = 13). Patients with a lower hippocampal W score at baseline progress to AD faster than the ones with higher W scores. (Reprinted with permission from ref. *79*.)

Twelve of these subjects cognitively declined and eventually were diagnosed as AD during the follow-up. Although these subjects who declined had significantly lower baseline hippocampal volumes than the ones who did not decline, the rate of hippocampal volume loss was not different between the two groups. The rate of temporal lobe volume loss, on the other hand, was higher in the decliners than in the stables. This result was interpreted to indicate that an individual starts showing signs of cognitive impairment only after atrophy extends beyond the hippocampus to the temporal lobe neocortex.

However, another study that followed a group of normal elderly individuals identified higher rates of hippocampal atrophy in normal elderly who cognitively declined and progressed to MCI or AD than the ones who were clinically stable *(85)*. The normal elderly subjects in this study were younger than the subjects of the former study (78 vs. 86 yr; ref. *82*). The risk for AD increases with age. An older cohort of cognitively normal elderly subjects is more likely to have a higher proportion of incipient AD cases that have not yet declared themselves. In line with this argument, the hippocampal atrophy rates of the stable elderly people were higher in the older than the younger subjects (–2.09% vs –1.7% per year), suggesting that undetected preclinical AD pathology in older subjects may confound estimates of atrophy rates in normally aging elderly cohorts.

Rates of hippocampal atrophy in people with MCI are higher than those in normal elderly patients *(85)*. Furthermore, rates of hippocampal atrophy of MCI subjects who progress to AD in the future are higher than MCI subjects who do not progress. People with MCI progress to AD at different rates, and although most people with MCI progress to AD during their lifetime, some do not. Baseline hippocampal volumetry in MCI provides predictive information on the risk rate of future progression to AD. It is possible that measurements of the rate of hippocampal atrophy would increase the diagnostic accuracy of baseline measurements for predicting the risk and rate of progressing to AD in individual patients with MCI.

Another region that was studied with serial measurements is the corpus callosum volumes. Serial measurements of the corpus callosum volumes in 21 patients with AD and 10 elderly controls indicate that the rate of total corpus callosum, splenium, and rostrum atrophy in patients with AD is higher than normal. It was thought that corpus callosum atrophy reflects loss of intracortical projecting neurons in the neocortex. If so, corpus callosum volumetry may be a potential marker for tracking neocortical pathology. Relation of the atrophic change to the pathologic involvement however needs to be demonstrated with postmortem studies *(86)*.

Automated or semiautomated MR volumetry techniques that are less labor-intensive than tracing specific regions of interest are being used to track the structural changes that take place within the brain during the progression of AD. One of the most promising techniques used for this purpose is the brain boundary shift integral (BBSI) developed by Fox and Freeborough *(87)*. The BBSI measures the change in brain boundaries first by spatially coregistering 3D scans acquired at different points in time, then measuring the intensity difference over the 3D surface of the brain in the combined data. Thus, the longitudinal shrinkage of the whole brain volume and the expansion of the ventricular volume can be measured. Using BBSI, brain atrophy rates on serial scans of 18 patients with AD were 2.37% ±1.11% per year compared with 0.41% ±0.47% per year in an age- and gender-matched control group *(88)*. Furthermore, the rate of atrophy correlated with the cognitive decline in AD based on the Mini-Mental State Examination scores implying the relevance of this marker to clinical progression *(89)*. Another study that calculated the annual rate of change in the whole brain, temporal lobes, and the ventricle volumes using automated techniques identified higher rates of temporal lobe and whole brain atrophy and ventricular enlargement in 14 people with AD than 14 age- and gender-matched controls. Combining temporal lobe atrophy rate with ventricular enlargement rate, the discriminant function completely separated normal elderly and people with AD *(90)*.

One drawback of global cerebral atrophy measurements is that this phenomenon is not specific to AD. A variety of neurological disorders can cause global or regional atrophy in the brain. For example, cerebrovascular disease, a common disorder in the elderly, may confound the rates of cerebral atrophy in a patient with AD. Measurements of regional atrophy may be more specific to the pathological process of AD. The topology of AD pathology in the brain varies with disease progression. The rate of atrophy at specific strategic regions such as the hippocampus may be most significant during the early stages. However, as the pathology spreads to the neocortex, rate of atrophy in other regions of the brain may be more significant than the rate of hippocampal atrophy. Because of this regional variability in the rate of atrophy, deformation of the brain is nonlinear. A fluid registration model developed by Freeborough and Fox *(91)* uses a viscous fluid model to compute a deformation field throughout the baseline image at the voxel level, giving an estimate of volume change occurring at each voxel with serial scans over time. Statistical parametric mapping was used to compare the deformation computed by fluid registration in 10 people with mild and 12 people with moderate AD and 4 presymptomatic cases with a family history of autosomal dominant early onset AD who progressed to AD during the follow-up *(92)*. There was a regional variation in the rate of atrophy with increasing disease severity, which correlated with the pathologic progression of the disease. Increased rates of hippocampal and medial parietal lobe atrophy were identified in presymptomatic and mildly affected patients. Increased atrophy rates spread to the temporal and parietal lobes in people with mild to moderate AD. Atrophy rates of the frontal lobe were increased only in people with moderate AD.

Regional structural markers for disease progression may vary with the pathologic stages of AD. Measuring global atrophy rates and statistical mapping of the difference from normal at different stages of disease severity may reveal the best regional measurement for each stage of disease progression. Longitudinally followed cognitively normal elderly and people with MCI would be of interest in such studies for understanding the earliest structural changes that may be useful as surrogate markers for therapies.

The only other MR technique that has been used for serial measurements in AD is ^{1}H MRS. Three such longitudinal studies have been published so far. Although one study identified a longitudinal decrease in NAA levels in 12 people with AD *(93)*, another did not reveal a change in medial temporal lobe NAA/Cr levels in 13 people with AD *(94)*. Both studies identified a correlation between cognitive decline and NAA levels. One recent study identified a significant decrease in left but not right hippocampal NAA levels in 8 people with AD *(95)*. Longitudinal ^{1}H MRS studies in larger subject groups are necessary to clarify these discrepancies.

7. MONITORING THERAPEUTIC EFFICACY WITH SERIAL MR MEASUREMENTS

Clinical and neuropsychological measures are the standards for assessing the progression of AD in a living person. Measuring the effect of a disease-modifying agent on a specific pathological process, however, requires valid markers of the pathology. MR-based volumetry is a sensitive marker for the pathologic progression of AD *(80)*. Two studies showed that MR-based volumetry techniques in AD may have enough power to measure the rate of structural change in the brain in a clinical trial setting if the magnitude of treatment effect is >10% *(88,96)*.

The feasibility of MR-based volumetry as a treatment outcome measure in AD was tested in a multisite therapeutic trial of milameline, a centrally active muscarinic agonist *(97)*. Using a centrally coordinated quality control program for MRI, the hippocampal volume measurements were found to be consistent across sites, validating the feasibility of multisite acquisition MR-based volumetry in AD. This study, however, did not prove that MR-based volumetry is a valid biomarker of therapeutic efficacy because therapeutic efficacy was not demonstrated (the trial was not completed due to a projected lack of efficacy). The validity of MR-based volumetry as a surrogate marker for therapeutic efficacy in AD remains to be tested in a positive disease-modifying drug trial.

In another centrally acting muscarinic agonist (Xanomeline) trial, brain Cho/Cr ratios were measured with ^{1}H MRS two times within the trial period of 6 mo *(98)*. Cho/Cr levels declined in 10 AD patients who were taking xanomeline and were stable in two AD patients who were taking placebo. The decline in Cho/Cr ratio was attributed to decreased free Cho levels based on the hypothesis that muscarinic agonists reduce neuronal membrane breakdown by reducing the cellular requirement for free Cho for acetylcholine synthesis. Although this study suggests that ^{1}H MRS may be useful in detecting therapeutic response to cholinergic agonists, long-term trials in larger groups are required to validate ^{1}H MRS as a therapeutic outcome measure in AD.

8. FUTURE DIRECTIONS

A significant advancement in the field of imaging in AD has been the development of amyloid imaging tracers for PET *(99,100)*. Efforts to image plaques also are underway in MRI. β-amyloid plaques contain significantly elevated levels of metal ions, which accelerate the T2* relaxation rate of the plaques. This property enables visualization of β-amyloid plaques as foci of decreased intensity on T2*-weighted MR microscopy of postmortem brain specimens in patients with AD *(101)*. However, neither the accelerated T2* relaxation rate nor the presence of metal ions are specific to plaques. Focal areas of accelerated T2* relaxation may be present in normal individuals as well as in a variety of pathologic conditions. Furthermore, endogenous iron is contained in the hemoglobin in blood vessels and in microhemorrhages. This necessitates development of MR contrast agents that specifically bind to the β-amyloid plaques and selectively enhance these plaques on MRI *(102,103)*. A novel approach in this regard is molecular probes that carry the MR contrast agent gadolinium-diethylenetriaminepentaacetic acid across the blood–brain barrier and bind specifically to β-amyloid plaques *(102)*. With high-resolution MRI of gadolinium-diethylenetriamine-pentaacetic acid-labeled β-amyloid plaques, it would be possible to noninvasively monitor the pathological progression of AD in vivo by directly imaging the pathology itself.

Advances in imaging the amyloid pathology of AD creates new research questions. Although amyloid plaques are one of the key mechanisms in AD, the social and economic impact of the disease on both the individual and the society depends on the clinical dementia severity. An important research question is how well the amyloid plaque density correlates with the clinical disease severity. Autopsy data show that synaptic loss is a better correlate of disease severity than the amyloid plaque burden. Amyloid imaging will require studies correlating clinical severity with amyloid plaque density. Another area that needs to be explored is amyloid imaging for therapy decisions. Amyloid plaques are common in cognitively normal elderly people *(104)*. The stepwise approach to validate indirect MR measures for predicting the pathological progression of AD, will have to be applied to direct measures of amyloid burden as well.

ACKNOWLEDGMENTS

This work was supported by grants NIH-NIA AG11378, AG16574, and AG06786.

REFERENCES

1. Braak H, Braak E. Neuropathological staging of Alzheimer's disease. Acta Neuropathol (Berl) 1991;82:239–259.
2. Seab JB, Jagust WJ, Wong STS, et al. Quantitive NMR measurements of hippocampal atrophy in Alzheimer's disease. Magn Reson Med 1988;8:200–208.
3. Kesslak JP, Nalcioglu O, Cotman CW. Quantification of magnetic resonance scans for hippocampal and parahippocampal atrophy in Alzheimer's disease. Neurology 1991;41:51–54.
4. Jack CR Jr., Petersen RC, O'Brien PC, Tangalos EG. MR based hippocampal volumetry in the diagnosis of Alzheimer's disease. Neurology 1992;42:183–188.
5. Convit A, de Leon MJ, Golomb J, et al. Hippocampal atrophy in early Alzheimer's disease: anatomic specificity and validation. Psychiatr Q 1993;64:371–387.
6. Frisoni GB, Bianchetti A, Geroldi C, Trabucchi M. Measures of medial temporal lobe atrophy in Alzheimer's disease. J Neurol Neurosurg Psychiatry 1994;57:1438–1439.
7. Lehericy S, Baulac M, Chiras J, et al. Amygdalohippocampal MR volume volume measurements in the early stages of Alzheimer's disease. AJNR Am J Neuroradiol 1994;15:927–937.
8. de Leon MJ, George AE, Golomb J, et al. Frequency of hippocampal formation atrophy in normal aging and Alzheimer's disease. Neurobiol Aging 1997;18:1–11.
9. Jack CR, Petersen RC, Xu Y, et al. Medial temporal atrophy on MRI in normal aging and very mild Alzheimer's disease. Neurology 1997;49:786–794.
10. Juottonen K, Laasko MP, Insausti R, et al. Volumes of the entorhinal and perirhinal cortices in Alzheimer's disease. Neurobiol Aging 1998;19:15–22.
11. Golebiowski M, Barcikowska M, Pfeffer A. Magnetic resonance imaging-based hippocampal volumetry in patients with dementia of the Alzheimer type. Dementia Geriatric Cognitive Disord 1999;10:284–288.
12. Bobinski M, deLeon MJ, Convit A, et al. MRI of entorhinal cortex in Alzheimer's Disease. Lancet 1999;353:38–40.
13. Juottonen K, Laakso MP, Partanen K, Soininen H. Comparative MR analysis of the entorhinal cortex and hippocampus in diagnosing Alzheimer disease. AJNR, Am J Neuroradiol 1999;20:139–144.
14. Xu Y, Jack CR Jr., O'Brien PC et al. Usefulness of MRI measures of entorhinal cortex versus hippocampus in AD. Neurology 2000;54:1760–1767.
15. Klunk WE, Panchalingam K, Moosy J, McClure RJ, Pettegrew JW. N-acetyl-L-aspartate and other amino acid metabolites in Alzheimer's disease brain: a preliminary proton nuclear magnetic resonance study. Neurology 1992;42:1578–1585.
16. Shonk TK, Moats RA, Gifford PG, et al. Probable Alzheimer's disease: diagnosis with proton MR spectroscopy. Radiology 1995;195:65–72.
17. Meyerhoff DJ, MacKay S, Norman D, Van Dyke C, Fein G, Weiner MW. Axonal injury and membrane alterations in Alzheimer's disease suggested by in vivo proton magnetic resonance spectroscopic imaging. Ann Neurol 1994;36:40–47.
18. Kwo-On-Yuen PF, Newmark RD, Budinger TF, Kaye JA, Ball MJ, Jagust WJ. Brain N-acetyl-l-aspartic acid in Alzheimer's disease: a proton magnetic resonance spectroscopy study. Brain Res 1994;667:167–174.
19. Tsai G, Coyle JT. N-acetylaspartate in neuropsychiatric disorders. Prog Neurobiol 1995;46:531–540.
20. Jessen F, Block W, Träber F, et al. Proton MR spectroscopy detects a relative decrease of N-acetyl aspartate in the medial temporal lobe of patients with AD. Neurology 2000;55:684–688.
21. Kantarci K, Jack CR, Xu YC, et al. Regional metabolic patterns in mild cognitive impairment and Alzheimer's disease, a ^{1}H MRS study. Neurology 2000;55:210–217.
22. Schuff N, Capizzano AA, Du AT, et al. Selective reduction of N-acetylaspartate in medial temporal and parietal lobes in AD. Neurology 2002;58:928–935.
23. Miller BL, Moats RA, Shonk T, Earnst T, Wooley S, Ross BD. Alzheimer disease: depiction of increased cerebral myo-inositol with proton MR spectroscopy. Radiology 1993;187:433–437.
24. Huang W, Alexander GE, Chang L, et al. Brain metabolite concentration and dementia severity in Alzheimer's disease. A ^{1}H MRS study. Neurology 2001;57:626–632.
25. Brand A, Richter-Landsberg C, Leibfritz D. Multinuclear NMR studies on the energy metabolism of glial and neuronal cells. Devel Neurosci 1993;15:289–298.
26. Urenjak J, Williams SR, Gadian DG, et al. Proton nuclear magnetic resonance spectroscopy unambiguously identifies different neural cell types. J Neurosci 1993;13:981–989.
27. Bitsch A, Bruhn H, Vougioukas V, et al. Inflammatory CNS demyelination: histopathologic correlation with in vivo quantitative proton MR spectroscopy. Am J Neuroradiol 1999;20:1619–1627.
28. Ernst T, Chang L, Melchor R, Mehringer M. Frontotemporal dementia and early Alzheimer disease: differentiation with frontal lobe H-1 MR spectroscopy. Radiology 1997;203:829–836.
29. MacKay S, Ezekiel F, Di Sclafani V, et al. Alzheimer's disease and subcortical ischemic vascular dementia: evaluation by combining MR imaging segmentation and H-1 MR spectroscopic imaging. Radiology 1996;198:537–545.
30. Wurtman RJ, Blusztajn JK, Marie JC. "Autocannibalism" of choline-containing membrane phospholipids in the pathogenesis of Alzheimer's disease. Neurochem Int 1985;7:369–372.
31. Hanyu H, Sakurai H, Takasaki M, Shindo H, Abe K. Diffusion – weighted MR imaging of the hippocampus and temporal white matter in Alzheimer's disease. J Neurol Sci 1998;156:195–200.
32. Sandson TA, Felician O, Edelman RR, Warach S. Diffusion-weighted magnetic resonance imaging in Alzheimer's Disease. Dement Geriatr Cogn Disord 1999;10:166–171.

33. Kantarci K, Jack CR, Xu YC, et al. Regional diffusivity of water in mild cognitive impairment and Alzheimer's disease. Radiology 2001;219:101–107.
34. Bozzali M, Falini A, Franceschi M, et al. White matter damage in Alzheimer's disease assessed in vivo using diffusion tensor magnetic resonance imaging. J Neurol Neurosurg Psychiatry 2002; 72:742–746.
35. Hanyu H, Asano T, Iwamoto T, Takasaki M, Shindo H, Abe K. Magnetization transfer measurements of the hippocampus in patients with Alzheimer's disease, vascular dementia, and other types of dementia. AJNR Am J Neuroradiol 2000;21:1235–1242.
36. Bozzali M, Franceschi M, Falini A et al. Quantification of tissue damage in AD using diffusion tensor and magnetization transfer MRI. Neurology 2001;57:1135–1137.
37. van der Flier WM, van den Heuvel DMJ, Weverling-Rijnsburger AWE, et al. Cognitive decline in AD and mild cognitive impairment is associated with global brain damage. Neurology 2002;59:874–879.
38. Maas LC, Harris GJ, Satlin A, English CD, Lewis RF, Renshaw PF. Regional cerebral blood volume measured by dynamic susceptibility contrast MR imaging in Alzheimer's disease: a principal components analysis. J Magn Reson Imaging. 1997;7:215–219.
39. Harris GJ, Lewis RF, Satlin A, et al. Dynamic susceptibility contrast MR imaging of regional cerebral blood volume in Alzheimer disease: a promising alternative to nuclear medicine. AJNR, Am J Neuroradiol 1998;19:1727–1732.
40. Bozzao A, Floris R, Baviera ME, Apruzzese A, Simonetti G. Diffusion and perfusion MR imaging in cases of Alzheimer's disease: Correlations with cortical atrophy and lesion load. AJNR Am J Neuroradiol 2001;22:1030–1036.
41. Alsop DC, Detre JA, Grossman M. Assessment of cerebral blood flow in Alzheimer's disease by spin-labeled magnetic resonance imaging. Ann Neurol 2000;47:93–100.
42. Thulborn K, Martin C, Voyvodic J. Functional MR imaging using a visually guided saccade paradigm for comparing activation patterns in patients with probable Alzheimer's disease and in cognitively able elderly volunteers. AmJ Neuroradiol 2000;21:524–531.
43. Buckner R, Snyder A, Sanders A, Raichle M, Morris J. Functional brain imaging of young, nondemented, and demented older adults. J Cogn Neurosci 2000;12:24–34.
44. Johnson S, Saykin A, Baxter L, et al. The relationship between fMRI activation and cerebral atrophy: comparison of normal aging and Alzheimer disease. NeuroImage 2000;11:179–187.
45. Saykin A, Flashman L, Frutiger S, et al. Neuroanatomic substrates of semantic memory impairment in Alzheimer's disease: patterns of functional MRI activation. J Int Neuropsychol Soc 1999;5:377–392.
46. Prvulovic D, Hubl D, Sack A, et al. Functional imaging of visuospatial processing in Alzheimer's disease. NeuroImage 2002;17:1403–1414.
47. Kato T, Knopman D, Liu H. Dissociation of regional activation in mild AD during visual encoding. Neurology 2001;57:812–816.
48. Rombouts S, Barkhof F, Veltman D, et al. Functional MR imaging in Alzheimer's disease during memory encoding. Am J Neuroradiol 2000;21:1869–1875.
49. Small S, Perera G, Delapaz R, Mayeux R, Stern Y. Differential regional dysfunction of the hippocampal formation among elderly with memory decline and Alzheimer's disease. Ann Neurol 1999;45:466–472.
50. Sperling R, Bates J, Chua E, et al. fMRI studies of associative encoding in young and elderly controls and mild Alzheimer's disease. J Neurol Neurosurg Psychiatry 2003;74:44–50.
51. Price JL, MorrisJC. Tangles and plaques in nondemented aging and preclinical Alzheimer's disease. Ann Neurol 1999;45:358–368.
52. Schmitt FA, Davis DG, Wekstein DR, Smith CD, Ashford JW, Markesbery WR. "Preclinical" AD revisited. Neuropathology of cognitively normal older adults. Neurology 2000;55:370–376.
53. Kordower JH, Chu Y, Stebbins GT, et al. Loss and atrophy of layer II entorhinal cortex neurons in elderly people with mild cognitive impairment. Ann Neurol 2001;49:202–213.
54. Petersen RC, Smith GE, Waring SC, Ivnik RJ, Tangalos EG, Kokmen E. Mild cognitive impairment clinical characterization and outcome Arch Neurol 1999;56: 303–308.
55. Petersen RC, Doody R, Kurz A, Mohs RC, Morris JC, Rabins PV, et al. Current concepts in mild cognitive impairment. Arch Neurol 2001;58:1985–1992.
56. Petersen RC, Stevens JC, Ganguli M. Tangalos EG, Cummings JL, DeKosky ST. Practice parameter: early detection of dementia: Mild cognitive impairment (an evidence based review). Report of the Quality Standards Subcommittee of the American Academy of Neurology. Neurology 2001;56:1133–1142.
57. Morris JC, Storandt M, Miller JP, McKeel DW, Price JL, Rubin EH, Berg L. Mild cognitive impairment represents early-stage Alzheimer disease. Arch Neurol 2001;58:397–405.
58. Du AT, Schuff N, Amend D, et al. Magnetic resonance imaging of the entorhinal cortex and hippocampus in mild cognitive impairment and Alzheimer's disease. J Neurol Neurosurg Psychiatry. 2001;71:431–432.
59. Krasuski JS, Alexander GE, Horwitz B, et al. Volumes of medial temporal lobe structures in patients with Alzheimer's disease and mild cognitive impairment. Biol Psychiatry 1998;43:60–68.
60. Dickerson BC, Goncharova I, Sullivan MP, et al. MRI-derived entorhinal and hippocampal atrophy in incipient and very mild Alzheimer's disease. Neurobiol Aging. 2001;22:747–754.
61. De Santi S, de Leon MJ, Rusinek H, et al. Hippocampal formation glucose metabolism and volume losses in MCI and AD. Neurobiol Aging 2001;22:529–539.
62. Catani M, Cherubini A, Howard R. [1]H MR spectroscopy differentiates mild cognitive impairment from normal brain aging. Neuroreport 2001;12:2315–2317.
63. Kantarci K, Jack CR, Xu YC, et al. Regional diffusivity of water in mild cognitive impairment and Alzheimer's disease. Radiology 2001;219:101–107.
64. Kabani NJ, Sled JG, Shuper A, Chertkow H. Regional magnetization transfer ratio changes in mild cognitive impairment. Magn Reson Med 2002;47:143–148.
65. van der Flier WM, van den Heuvel DMJ, Weverling-Rijnsburger AWE, et al. Cognitive decline in AD and mild cognitive impairment is associated with global brain damage. Neurology 2002;59:874–879.
66. Machulda MM, Ward HA, Borowski B, et al. Comparison of memory fMRI response among Normal, MCI, and Alzheimer's patients. Neurology 2003;61:500–506.
67. Yasuda M, Mori E, Kitagaki H, et al. Apolipoprotein E epsilon 4 allele and whole brain atrophy in late-onset Alzheimer's disease. Am J Psychiatry 1998;155:779–784.
68. Geroldi C, Pihlajamaki M, Laasko MP, et al. APOE-ε4 is associated with less frontal and more medial temporal lobe atrophy in AD. Neurology 1999;53:1825–1832.
69. Hashimoto M, Yasuda M, Tanimukai S, et al. Apolipoprotein E _4 and the pattern of regional brain atrophy in Alzheimer's disease. Neurology 2001;57:1461–1466.
70. Jack CR, Petersen RC, Xu Y, et al. Hippocampal atrophy and apolipoprotein E genotype are independently associated with Alzheimer's disease. Ann Neurol 1998;43:303–310.
71. Reiman EM, Uecker A, Caselli RJ, et al. Hippocampal volumes in cognitively normal persons at genetic risk for Alzheimer's disease. Ann Neurol 1998;44:288–291.
72. Barber R, Gholkar A, Scheltens P, et al. Apolipoprotein E epsilon4 allele, temporal lobe atrophy, and white matter lesions in late-life dementias. Arch Neurol 1999;56:961–965.
73. Klunk WE, Panchalingam K, McClure RJ, Stanley A, Pettegrew JW. Metabolic alterations in postmortem Alzheimer's disease brain are exaggerated by Apo-E4. Neurobiol Aging 1998;19:511–515.
74. Kantarci K, Smith GE, Ivnik RJ, et al. ^{1}H MRS, cognitive function, and apolipoprotein E genotype in normal aging, mild cognitive impairment and Alzheimer's disease. J Int Neuropsychol Soc 2002;8:934–942.
75. Jack CR, Jr., Dickson DW, Parisi JE, et al. Antemortem MRI findings correlate with hippocampal neuropathology in normal aging and dementia. Neurology. 2002;58:750–757.
76. Silbert LC, Quinn JF, Moore MM, et al. Changes in premorbid brain volume predict Alzheimer's disease pathology. Neurology 2003;61:487–492.

77. Bobinski M, de Leon MJ, Wegiel J, et al. The histological validation of post mortem magnetic resonance imaging-determined hippocampal volume in Alzheimer's disease. Neuroscience 2000;95(3):721–725.
78. Goesche KM, Mortimer JA, Smith CD, Markesbery WR, Snowdon DA. Hippocampal volume as an index of Alzheimer neuropathology. Findings from the Nun Study. Neurology 2002;58:1476–1482.
79. Jack CR, Petersen RC, Xu Y, et al. Prediction of AD with MRI-based hippocampal volume in mild cognitive impairment. Neurology 1999;52:1397–1403.
80. Visser PJ, Scheltens P, Verhey FR, et al. Medial temporal lobe atrophy and memory dysfunction as predictors for dementia in subjects with mild cognitive impairment. J Neurol 1999;246:477–485.
81. Killiany RJ , Gomez-Isla T, Moss M, et al. Use of structural Magnetic Resonance Imaging to predict who will get Alzheimer's disease. Ann Neurol 2000;47: 430–439.
82. Kaye JA, Swihart T, Howieson D, et al. Volume loss of the hippocampus and temporal lobe in healthy elderly persons destined to develop dementia. Neurology 1997;48:1297–1304.
83. Jack CR, Petersen RC, Xu Y, et al. Rate of medial temporal lobe atrophy in typical aging and Alzheimer's disease. Neurology 1998;51:993–999.
84. Laakso MP, Lehtovirta M, Partanen K, Riekkinen PJ, Soininen H. Hippocampus in Alzheimer's disease: a 3-yr follow-up MRI study. Biol Psychiatry 2000;47:557–561.
85. Jack CR, Petersen RC, Xu Y, et al. Rates of hippocampal atrophy correlate with change in clinical status in aging and AD. Neurology 2000;55: 484–489.
86. Teipel SJ, Bayer W, Alexander GE, et al. Progression of corpus callosum atrophy in Alzheimer's disease. Arch Neurol 2002;59:243–248.
87. Fox NC, Freeborough PA. Brain atrophy progression measured from registered serial MRI: validation and application to Alzheimer's disease. J Magn Reson Imaging 1997;7:1069–1075.
88. Fox NC, Cousens S, Scahill R, Harvey RJ, Rossor MN. Using serial registered brain magnetic resonance imaging to measure disease progression in Alzheimer disease: power calculations and estimates of sample size to detect treatment effects. Arch Neurol 2000;57:339–344.
89. Fox NC, Scahill RI, Crum WR, Rossor MN. Correlation between rates of brain atrophy and cognitive decline in AD. Neurology 1999;52:1687–1689.
90. Wang D, Chalk JB, Rose SE, et al. MR image-based measurement rates of change in volumes of brain structures. Part II: Application to a study of Alzheimer's disease and normal aging. Magn Reson Imaging. 2002;20:41–48.
91. Freebrough PA, Fox NC. Modeling brain deformations in Alzheimer disease by fluid registration of serial 3D MR images. J Computer Assisted Tomogr 1998;22:838–843.
92. Scahill RI, Schott JM, Stevens JM, Rossor MN, Fox NC. Mapping the evolution of regional atrophy in Alzheimer's disease: unbiased analysis of fluid-registered serial MRI. Proc Natl Acad Sci USA 2002;99:4703–4707.
93. Adalsteinsson E, Sullivan EV, Kleinhans N, Spielman DM, Pfefferbaum A. Longitudinal decline of the neuronal marker N-acetyl aspartate in Alzheimer's disease. Lancet 2000;355:1696–1697.
94. Jessen F, Block W, Träber F, et al. Decrease of N-acetylaspartate correlates with cognitive decline of AD patients. Neurology 2001;57:930–932.
95. Dixon RM, Bradley KM, Budge MM, Styles P, Smith AD. Longitudinal quantitative proton magnetic resonance spectroscopy of the hippocampus in Alzheimer's disease. Brain 2002;125:2332–2341.
96. Bradley KM, Bydder GM, Budge MM, et al. Serial brain MRI at 3–6 month intervals as surrogate marker for Alzheimer's disease. Br J Radiol 2002;75:506–513.
97. Jack CR, Jr., Slomkowski M, Gracon S, et al. MRI as a biomarker of disease progression in a therapeutic trial of milameline for AD. Neurology 2003;60:253–260.
98. Satlin A, Bodick N, Offen WW, Renshaw PF. Brain proton magnetic resonance spectroscopy (^{1}H-MRS) in Alzheimer's disease: changes after treatment with Xanomeline, an M_1 selective cholinergic agonist. Am J Psychiatry 1997;154:1459–1461.
99. Shoghi-Jadid K, Small GW, Agdeppa ED, et al. Localization of neurofibrillary tangles and beta-amyloid plaques in the brains of living patients with Alzheimer's disease. Am J Geriatric Psychiatry. 2002;10:24–34.
100. Bacskai BJ, Hickey GA, Skoch J, et al. Four-dimensional multiphoton imaging of brain entry, amyloid binding, and clearance of an amyloid-beta ligand in transgenic mice. Proc Natl Acad Sci USA 2003;100:12462–12467.
101. Benveniste H, Einstein G, Kim KR, Hulette C, Johnson A. Detection of neuritic plaques in Alzheimer's disease by magnetic resonance microscopy. Proc Natl Acad Sci USA 1999;96:14079–14084.
102. Poduslo JF, Wengenack TM, Curran GL, et al. Molecular targeting of Alzheimer's amyloid plaques for contrast-enhanced magnetic resonance imaging. Neurobiol Dis 2002;11:315–329.
103. Wadghiri YZ, Sigurdsson EM, Sadowski M, et al. Detection of Alzheimer's amyloid in transgenic mice using magnetic resonance microimaging. Magn Reson Med 2003;50: 293–302.
104. Guillozet AL, Weintraub S, Mash DC, Mesulam MM. Neurofibrillary tangles, amyloid, and memory in aging and mild cognitive impairment. Arch Neurol 2003;60:729–736.

10 Stages of Brain Functional Failure in Alzheimer's Disease

In Vivo Positron Emission Tomography and Postmortem Studies Suggest Potential Initial Reversibility and Later Irreversibility

STANLEY I. RAPOPORT, MD

SUMMARY

In vivo imaging of brain glucose metabolism and blood flow by means of positron emission tomography (PET) or functional magnetic resonance imaging (fMRI) suggests that brain functional failure in the course of Alzheimer's disease occurs in two general stages. In the first stage, evident in patients who are mildly demented as well as in presymptomatic genetically at-risk but affected subjects, the brain can be almost normally activated despite reduced resting state reductions in metabolism or flow. The second stage, found in moderately-to-severely demented patients, is largely irreversible, as the brain's ability to respond to activation is severely reduced. Biopsies and postmortem studies of the brains of patients with Alzheimer's disease suggest that the staged reductions of brain metabolism and flow in life, accompanying dementia progression, correspond to staged changes in synaptic structure and in the mitochondrial oxidative phosphorylation that is coupled to synaptic activity. In the first stage, where downregulation of oxidative phosphorylation is potentially reversible, pharmacotherapy should be directed to maintain or ameliorate synaptic functional integrity.

Key Words: Alzheimer's disease; positron emission tomography; brain, energy; metabolism; oxidative phosphorylation; mitochondria; cytochrome oxidase; activation; glucose; blood flow; synapse; staging; neurofibrillary tangles.

1. INTRODUCTION

The mammalian brain is distinguished by a high rate of glucose consumption and by high activities of enzymes involved in mitochondrial oxidative phosphorylation (OXPHOS). OXPHOS produces ATP, which during synaptic activation is largely consumed by the Na/K–ATPase pump and by receptor-mediated signaling processes *(1,2)*. Because OXPHOS supports synaptic activity, in vivo brain imaging of regional cerebral metabolic rates for glucose ($rCMR_{glc}$) and of regional cerebral blood flow (rCBF), which is coupled to $rCMR_{glc}$, can be used to estimate synaptic functional activity at rest and in response to activation. In Alzheimer's disease (AD), cerebral metabolic and flow decrements have been localized and quantified by means of positron emission tomography (PET) and have been correlated with characteristic cognitive and behavioral deficits.

In this chapter, I discuss studies largely undertaken with in vivo PET that suggest that functional failure in the course of AD occurs in two general stages. The first, evident in patients who have been diagnosed as mildly demented as well as in presymptomatic but affected subjects, is potentially reversible but certainly treatable for the maintenance or amelioration of cognitive integrity. The second stage is largely irreversible. Biopsies and postmortem studies of brains from patients with AD suggest that the staged reductions of metabolism and flow accompanied by dementia progression correspond to staged changes in synaptic markers and in the mitochondrial OXPHOS that is coupled to synaptic activity.

2. IN VIVO BRAIN IMAGING

2.1. RESTING-STATE BRAIN METABOLISM IN DIAGNOSED AD PATIENTS: CORRELATIONS WITH COGNITION AND BEHAVIOR

Cross-sectional PET studies of subjects diagnosed as having AD demonstrate resting-state (eyes covered, ears plugged with cotton) reductions in $rCMR_{glc}$, measured with ^{18}F-fluoro-2-deoxy-D-glucose. The reductions occur throughout the neocortex in proportion to dementia severity *(3,4)*. They represent intrinsic reductions in metabolism per gram of brain tissue, since they remain statistically significant albeit diminished after correction for brain atrophy (Fig. 1; refs. *5,6*). In contrast, statistically significant reductions in atrophy-uncorrected $rCMR_{glc}$ that are found with healthy aging lose significance after atrophy correction *(7)*.

Metabolic reductions noted in the AD brain are not uniformly distributed. Generally, neocortical association areas are

From: *Bioimaging in Neurodegeneration*
Edited by P. A. Broderick, D. N. Rahni, and E. H. Kolodny

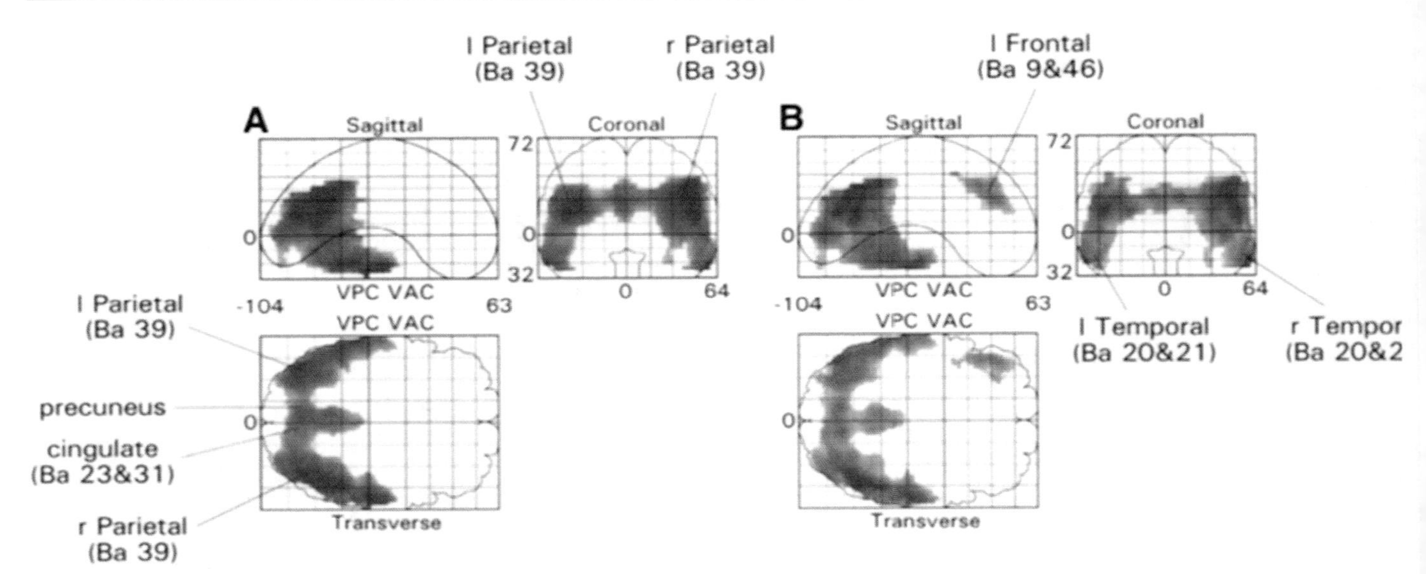

Fig. 1. Statistical parametric views of brain regions where $rCMR_{glc}$ is significantly less in patients with AD than in controls, before (left, **A**) and after (right, **B**) atrophy correction. Projections, presented with a Z threshold of 2.3 (top) and 3.6 (bottom), show involvement of association cortical regions in AD brain. No area with a significant reduction in $rCMR_{glc}$ before correction became normal after correction, whereas some areas (e.g., left frontal cortex) showed a significant reduction only after correction. BA, Brodmann area; l, left; r, right; PVE, partial volume effect; VAC, anterior commissure; VPC, posterior commissure. From ref. *5*.

affected earlier and more severely than are primary visual, auditory, motor, and somatosensory areas, whereas thalamic and basal ganglia nuclei are relatively spared *(3,4)*. In the AD association neocortex, postmortem studies indicate loss of large pyramidal neurons and their long intracortical and intercortical axons. Functional interactions between them, measured in life as interregional correlation $rCMR_{glc}$ coefficients, are thereby disrupted *(8)*. Thus, AD has been suggested to represent a "cortical disconnection" syndrome *(9)*.

The brain regions that are metabolically affected earliest and most severely in AD contain high densities of intraneuronal neurofibrillary tangles (NFTs; refs. *10,11*) and constitute a forebrain "association system." The regions include association neocortex, posterior hippocampus, corticobasal nucleus of the amygdaloid complex, transentorhinal cortex, entorhinal cortex and nucleus basalis of Meynert, with the transentorhinal and entorhinal cortices likely being affected earliest *(12–14)*. Comparative neuroanatomical studies and the fossil record suggest that the forebrain "association system" rapidly expanded during primate evolution, and thus that AD is a specifically human, "phylogenic" disease *(15)*. Supporting this is the fact that no natural animal model exists for AD.

Patterns of resting-state metabolic reductions in PET scans from individual AD patients are quite variable but correspond to patterns of cognitive stimulation and behavioral abnormalities in each patient, as predicted from earlier lesion studies *(16,17)*. In one study of AD, four independent $rCMR_{glc}$ patterns were identified with a "principal components" analysis, after converting the patient's $rCMR_{glc}$ values to *Z* scores based on the patient's global mean metabolism and its standard deviation (Fig. 2; ref. *18*). The profile of the largest subgroup, group 1 (45% of the patients), resembled metabolic deficits usually reported as typical of AD *(19–21)*, consisting of hypometabolism in parietotemporal regions with relative sparing of frontal, primary sensorimotor, and subcortical areas. Group 2 (25% of patients) had bilateral hypometabolism in paralimbic structures—orbitofrontal, anterior cingulate, and anterior insular cortex. In group 3 (15% of patients) there were reductions in the left hemisphere, and group 4 (15% of patients) resembled group 1 in their metabolic changes but with more severe frontal lobe reductions. Major behavioral changes in group 1 patients included depression; in group 2 patients, agitation, inappropriate behavior, and personality change; in group 3 patients, depression; and in group 4 patients, psychosis and inappropriate behavior. Patients in group 1 differed from those in groups 2 and 3 by having predominant impairment on parietally mediated compared with frontal lobe cognitive tests and on right (visuospatial) compared with left (verbal ability and fluency) tests, with the opposite pattern seen in groups 2 and 3.

In mildly-to-moderately demented AD patients, abnormal left–right hemispheric metabolic asymmetries or abnormal frontal–parietal metabolic gradients (Fig. 2) are correlated with specific cognitive "discrepancies" (*see* **Note 1**; refs. *22–26*). Patients with a lower right-sided than left-sided $rCMR_{glc}$ on average have worse scores on visuospatial tests (range drawing) than on language tests (syntax comprehension), whereas the reverse is true in patients with a relatively lower left-sided $rCMR_{glc}$. Furthermore, a metabolic asymmetry in a mildly demented AD patient having only a memory deficit usually predicts the visuospatial-language "discrepancy" (difference in rank-ordered cognitive test scores) that appears 1–3 yr later. Subjects with a lower right- than left-sided metabolic rate later demonstrate significantly worse visuospatial than language abilities and vice versa. Similarly, early-appearing abnormal frontal–parietal metabolic ratios predict patterns of later-appearing deficits in cognitive functions mediated by the frontal and parietal lobes *(27)*.

This predictability implies that (1) resting-state metabolic measurements can be more sensitive markers of neocortical

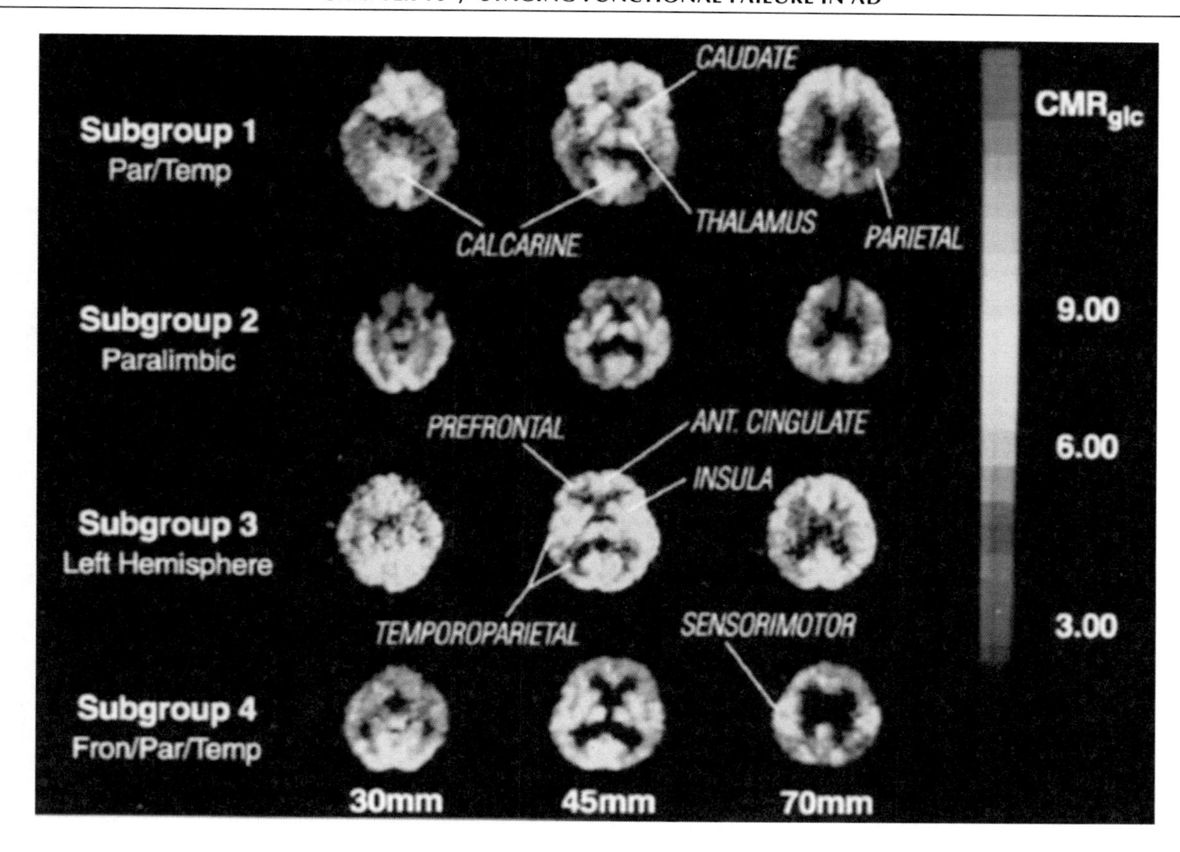

Fig. 2. PET scans from individual patients with AD that best characterize four independent subgroups (*see* text). Three planes are shown for each subject: left, at level of orbitofrontal cortex; middle, at level of basal ganglia; right, at level of centrum semiovale. Values for $rCMR_{glc}$ are color-scaled in units of mg/100 g brain/min. Reprinted with permission from ref. *18*. Copyright 1990 American Psychiatric Press. *See* color version on Companion CD.

dysfunction in AD than available psychometric tests; and (2) once a metabolic asymmetry is established, its direction usually will be maintained throughout much of the subsequent course of disease. The latter point was confirmed in 11 patients with AD who were followed for as many as 4 yr *(23,24)*. A maintained direction of asymmetry implies, in turn, that metabolic rates, once they start to fall in a given patient, continue to decline at equivalent velocities in both hemispheres. Constant velocities may arise because AD neurodegeneration, once begun, appears to be a "first-order" process involving membrane instability and altered membrane phospholipid composition *(28,29)*.

Regional PET metabolic rates in mildly demented AD patients frequently overlap metabolic rates in healthy controls. Such overlap limits our ability to use resting PET data to reliably identify a minimally affected subject as having AD. Discriminant analysis combined with multiple regression may overcome this limitation, by providing probabilities regarding the likelihood of an individual PET scan being similar to scans from a defined disease or age-matched control group *(30,31)*.

2.2. ACTIVATION PET IN PATIENTS DIAGNOSED WITH AD

The resting-state brain metabolic and flow reductions in early AD are caused by intrinsic brain changes resulting in reduced energy demand rather than by reduced delivery of glucose or oxygen delivery to the brain by blood flow. This was confirmed by measuring the blood oxygen extraction fraction at rest, as well as rCBF during activation with PET. In the former instance, AD patients with white matter hyperintensities on MRI, indicative of leukoencephalopathy and local ischemia, had an increased oxygen extraction fraction, whereas AD patients without hyperintensities had a normal oxygen extraction fraction (**Note 2**; refs. *32–34*).

The use of $H_2^{15}O$ was combined with PET to quantify rCBF in occipitotemporal visual association regions that subserve object recognition in patients with AD and healthy controls while they performed a control or a face-matching task *(35,36)*. rCBF during the control task (a button was pressed alternately with right and left thumbs in response to a neutral visual stimulus) was subtracted from rCBF during the face-matching task (the button was pressed with the appropriate thumb after the subject decided whether the right or left face was to be matched) to produce a mean "difference" image of ΔrCBF values (Fig. 3). Patients with AD could perform the face-matching task as accurately as control subjects (85 ± 8% [SD] vs 92 ± 5% correct choices, respectively), although their reaction times during the task were more variable than in the controls, 3.34 ± 1.46 s compared with 2.07 ± 0.54 s. As illustrated in the lateral views of the brain of Fig. 3, rCBF increments during face matching occurred in equivalent occipital-temporal and occipital-parietal visual brain areas of the AD patients and the controls.

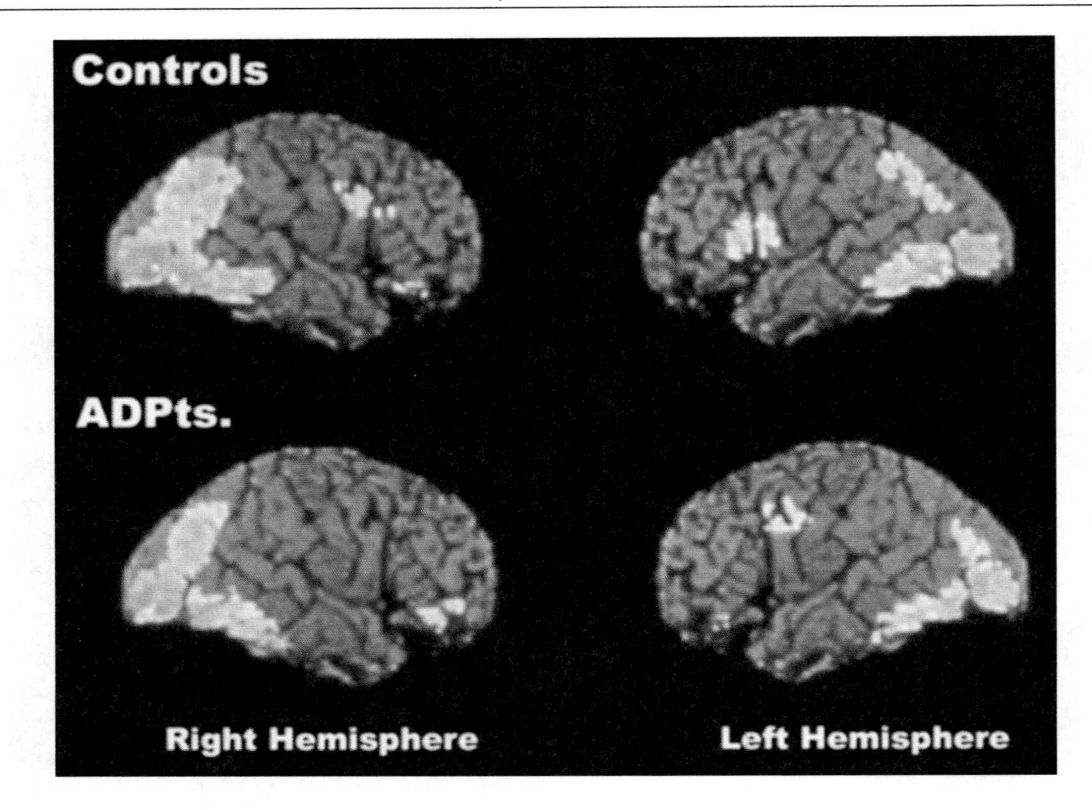

Fig. 3. Brain blood flow increments in control subjects and subjects with AD during face-matching task compared with resting condition. Colored pixels identify rCBF elevations >30% above baseline, with intensity representing significance above $p < 0.01$. Occipital temporal and occipital parietal visual pathways are activated, as well as frontal lobe regions. There is no difference in the geographic extent of activation between groups. From ref. *35*. *See* color version on Companion CD.

During the "control task," baseline rCBF was lower in occipitotemporal regions (Brodmann areas 17, 19, and 37) of patients with AD than of control subjects (Table 1). Nevertheless, during the face-matching task, the mean flow increment ΔrCBF (mL/100 g/min) did not differ between patients and controls. Thus, the affected AD brain areas, which in the resting state showed reduced rCBF, could be almost "normally" activated during the face recognition task *(36)*. Because coupling between ΔrCBF and $\Delta rCMR_{glc}$ is unchanged in normotensive AD patients without leukoencephalopathy *(37)*, glucose and oxygen delivery to brain by blood is not rate limiting. This conclusion is consistent with PET evidence that the oxygen extraction fraction is normal in AD patients who are otherwise healthy *(19,32)*.

Like rCBF responses, $rCMR_{glc}$ responses to stimulation are within normal limits in mildly demented AD patients *(38,39)*. rCMRglc was measured in the resting-state (eyes covered/ears blocked) in AD and control subjects and in the same PET session during audiovisual stimulation by a movie not requiring subject compliance. In the controls, audiovisual stimulation significantly increased rCMRglc in visual and auditory cortical areas. In mildly demented patients, rCMRglc increments in these regions were within two standard deviations (95 percentile limits) of the mean control response. The increments fell below these limits in moderately demented patients and were further reduced in severely demented patients.

Brain responsiveness to the anticholinesterase drug, donepezil, also appears to decline in stages in relation to dementia progression. In one clinical trial, donepezil produced significant "cognitive enhancement" on the cognitive portion of the Alzheimer Disease Assessment Scale in mildly but not in moderately demented AD patients *(40)*.

Graded activation paradigms, in which a parameter such as stimulus intensity or task difficulty is varied discretely while rCBF is measured, can be used to construct stimulus-rCBF response curves that can be interpreted in terms of synaptic "efficacy" *(36,41,42)*. Using such a "parametric" approach, patterned alternating flashes of red light grids were presented to control and AD subjects at frequencies between 0 and 15 Hz by means of goggles, and rCBF was measured at each frequency with $H_2{}^{15}O$ and PET (Fig. 4; refs. *43–45*). In the controls, striate cortex ΔrCBF increased linearly between stimulus frequencies of 0 and 8 Hz, then fell at the higher frequencies, whereas lateral association cortex ΔrCBF continued to increase between 0 and 15 Hz. In mildly demented AD patients, ΔrCBF at frequencies above 4 Hz was significantly less than in the controls in both regions; differences from control were even greater in moderately-to-severely demented patients.

The biphasic frequency-rCBF response curve in the striate cortex to flash stimulation has been ascribed to dropout of synaptic responses of the parvocellular visual system at frequencies above 8 Hz, whereas high frequency-responding synapses

Table 1
rCBF in Occipitotemporal Visual Association Cortex (Brodmann Areas 19 and 37) in Patients With AD and Normal Volunteers Performing a Control or a Face-Matching Task

Task	*Control subjects* (n = *13*)	*AD patients* (n = *11*)
	rCBF (mL/100 g/min)	
Baseline control task	48.4 ± 0.9	40.1 ± 1.1*
Face-matching task	53.1 ± 0.9	44.1 ± 1.2*
Difference Δ rCBF, face matching - control task	4.7 ± 0.2	4.0 ± 0.4

*Mean ± SE differs significantly from control mean, $p < 0.001$
From Rapoport and Grady *(36)*.

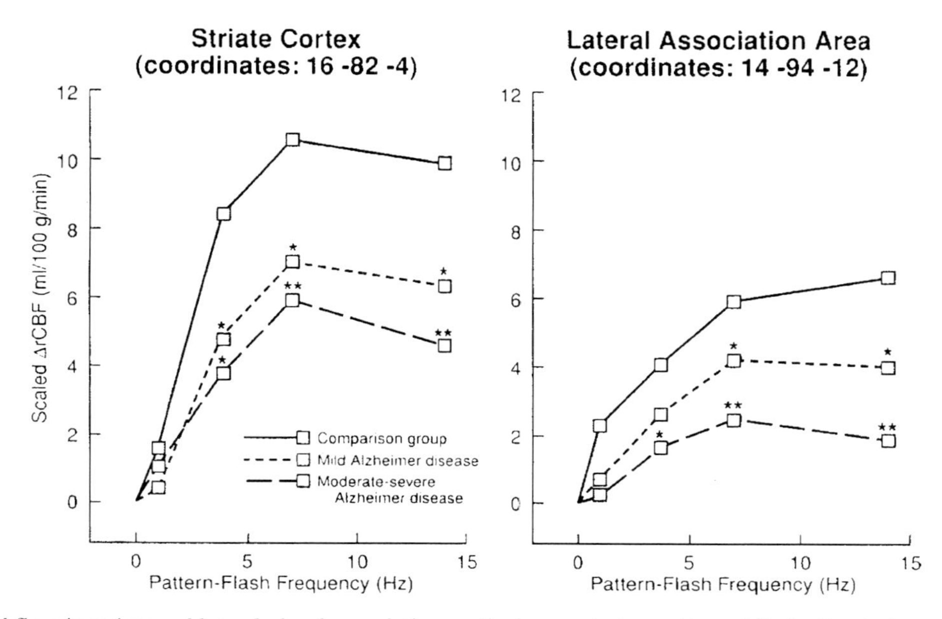

Fig. 4. Scaled blood flow in striate and lateral visual association cortical areas during patterned flash stimulation at different frequencies, in control subjects and AD patients with mild or moderate-to-severe dementia. Scaling corrects for global flow differences between groups. Asterisks indicate significant difference from control mean ($p < 0.05$). Coordinates in 3D Talairach space *(122)*. Adapted from ref. *43* with permission.

in the magnocellular visual system continue to be activated at frequencies greater than 8 Hz *(46,47)*. Both visual systems coexist within the striate cortex, but only magnocellular cells are present in the middle temporal region, V5/MT. This latter region could be activated at 1 Hz (when movement was apparent) in the controls but not in the AD patients *(43–45)*. Thus, high-frequency responding and presumably more metabolically dependent synapses *(48)* are more vulnerable in AD than are low-frequency responding synapses, supplying further evidence for early synaptic dysfunction (*see* Section 4).

2.3. RESTING AND ACTIVATION PET IN SUBJECTS AT RISK FOR AD

Advances in molecular genetics and epidemiology have identified a number of genes whose expression or mutations are risk factors for AD. Mutations of the amyloid precursor protein (APP) gene on chromosome 21, or of presenilin 1 and 2 genes on chromosome 14 and chromosome 1, respectively, or the presence of the Apo E4 allele whose gene is on chromosome 19, are genetic risk factors for AD *(49)*. Patients with these mutations, <2% of the AD population, tend to develop AD earlier and more frequently than do marker-free individuals, whereas individuals with Apo E4 have a higher risk for later onset AD.

In 1962, Kral *(50)* identified certain elderly who complained of or had subtle memory changes as having "benign senescent forgetfulfulness." Currently, such changes are seen in a more sinister light. They are called "mild cognitive impairment" (MCI) and are considered a presymptomatic phase of AD *(51)*. The conversion rate of subjects with MCI to AD varies from 24% to 80%, depending on inclusion criteria for diagnosis *(52)*, but PET can help to identify affected patients. In one study, temporoparietal $rCMR_{glc}$ deficits on a first scan predicted who in a group of MCI subjects subsequently developed AD and who did not, with an accuracy of 75% *(53)*.

2.3.1. Resting PET and Other Imaging Studies

As noted above, discriminant analysis combined with multiple regression can be used to provide probabilistic statements regarding the likelihood of an individual PET scan being similar to scans from a disease or age-matched control group *(30,31)*. This was demonstrated in a study in which a discriminant function, derived from $rCMR_{glc}$ data from identified AD patients and controls, was used to classify subjects in either category with 87% accuracy. When later applied to brain metabolic data from a genetically at-risk subject with recent memory complaints, the function indicated that the subject had a significant AD brain metabolic pattern. A diagnosis of AD was confirmed one year later, when parietal lobe $rCMR_{glc}$ was found reduced in a second PET scan and the patient had demonstrated cognitive decline *(31)*.

Subtle but statistically significant regional PET abnormalities have been reported in asymptomatic subjects genetically at risk for AD, further suggesting that PET measurements are more sensitive to early neuropathology than are cognitive tests. In one study, subjects with only a memory impairment were divided according to their Apo E4-allele status: Apo E4 heterozygotes and noncarriers of this allele *(54)*. Compared with the noncarriers, the heterozygotes had reduced mean $rCMR_{glc}$ in the parietal cortex; both groups had a mean higher parietal $rCMR_{glc}$ than patients affected with AD. In another study, middle-aged cognitively normal Apo E4 homozygotes had an AD-like group PET pattern on cortical projection maps of $rCMR_{glc}$ *(55)*. A third study noted that unaffected subjects at risk because of a chromosome-14 linkage, an APP mutation, or other genetic factors, had reduced mean global and parietotemporal $rCMR_{glc}$ compared with age-matched controls *(56)*. In each of the above studies, there was considerable overlap of brain metabolic values between the at-risk group and controls.

A resting-state "AD-like" $rCMR_{glc}$ PET pattern recently was reported in Apo E4 carriers compared with noncarriers between 20 and 39 years of age, decades before the age of maximum risk for symptomatic disease *(57)*. These results, and evidence of a low hippocampal volume in Apo E4 carriers in their 40s *(58)*, may mean that a so-called AD-like metabolic pattern in Apo E4 carriers is a trait marker of the carrier status but not a state marker of ongoing AD pathology.

2.3.2. Activation Imaging

Activation PET and fMRI have demonstrated "recruitment" of additional regions during task performance in subjects genetically at risk for AD compared with those not at risk and in older compared with younger healthy subjects. Older compared with young subjects performing a memory task were shown with PET to recruit additional brain regions, suggesting greater effort on their part *(59,60)*. A synaptic basis for this phenomenon is supported by evidence that recruitment could be reduced by pretreatment with the cholinesterase inhibitor, physostigmine *(61,62)*, and that brain tissue from healthy elderly subjects nevertheless shows loss of the postsynaptic dendritic spine marker, drebrin, from pyramidal neurons *(63)*. Increased recruitment also has been observed in asymptomatic subjects at-risk for AD because they carried the Apo E4 allele *(64,65)*.

3. POSTMORTEM AND BIOPSY STUDIES OF THE BRAINS OF PATIENTS WITH AD

3.1. PROGRESSIVE LOSSES OF SYNAPTIC MARKERS AND OF OXPHOS

3.1.1. Synapse Loss in AD

It generally is agreed that dementia symptoms in patients with AD are most closely related to brain synaptic changes. In the brains of patients with AD, synaptic structures are lost disproportionately to neurons, and the extent of loss correlates with dementia severity prior to death. The loss is accompanied by reduced markers for cholinergic, dopaminergic, glutamatergic, and serotonergic transmission, as well as a reduction in postsynaptic spine marker drebrin *(36,63,66–75)*. Despite evidence for a primary synaptic loss in AD, the basis for this loss remains conjectural. One suggestion is that diffusible oligomeric assemblies of β-amyloid peptide accumulate at the synapse and are toxic *(76–78)*. This suggestion, however, is not supported by observations that many transgenic mouse models of AD show no change in presynaptic bouton number in relation to β-amyloid accumulation *(79)* or to synaptophysin loss *(80)*. Other suggested causes are disruption of calcium homeostasis, glutamate accumulation and excitotoxicity, and oxidative damage *(81,82)*.

3.1.2. Synapse–OXPHOS Relations

Synaptic activity requires ATP, which is synthesized by mitochondrial OXPHOS. Thus, measuring markers of OXPHOS in brain tissue can indirectly elucidate the synaptic dysfunction that underlies declining brain metabolism and blood flow in living patients. Mitochondrial OXPHOS is mediated by enzymes in the electron transport chain *(83)*. One enzyme, cytochrome oxidase (COX, complex V), consists of eight subunits, three of which (COX I-III) are encoded by mitochondrial DNA (mtDNA) and five (COX IV-VIII) by nuclear DNA (nDNA). Transcription of the eight-subunit COX complex takes place when mitochondria are in the vicinity of the cell nucleus *(84)*, as illustrated in Fig. 1 for a brain pyramidal neuron *(85–90)*. The mitochondria then are carried by fast axonal transport to synaptic sites, particularly dendritic spines, where they synthesize ATP. After mitochondrial ribonucleic acid (mRNA) and protein components of OXPHOS have decayed *(91)*, the mitochondria are returned to the vicinity of the cell nucleus for renewed OXPHOS transcription or translation.

3.1.3. Selective OXPHOS Downregulation in AD

The first data column in Table 2 illustrates that activities of COX and NADH dehydrogenase, enzymes regulating OXPHOS within mitochondria, as well as mRNA levels for subunits of these and other OXPHOS enzymes *(83)*, are reduced in the midtemporal cortex of the postmortem AD brain. In contrast, statistically significant reductions are absent in the AD motor cortex, which is comparatively spared of AD pathology *(10,11)*. Levels of mRNA for the mitochondrial (mt)-encoded ND1 and ND4 subunits of NADH dehydrogenase and for the COX I and COX III subunits also are significantly decreased. Levels of the nuclear (n) DNA-encoded mRNA for COX IV and for the β-subunit of ATP synthase (ATPsyn.β) are decreased as well.

Thus, the mRNA reductions noted in the midtemporal AD cortex are specific to OXPHOS and occur whether the mRNA is derived from mitochondria or from the nucleus (**Note 3**; refs. *92–95;* Chandrasekaran K, Rapoport SI, unpublished observations). The reductions are unrelated to mitochondrial dropout or reduced transcription of non-OXPHOS related mtDNA, as net levels of mtDNA-encoded 12S rRNA and total mtDNA are normal. Nor are they related to a general reduction in nuclear transcription, as nDNA-encoded mRNA for β-actin, lactic acid dehydrogenase-B, and 28S rRNA are unchanged *(73,74,96–98)*.

A comparable pattern of reduced COX activity and reduced levels of mRNAs for COX subunits coded for by nDNA or

Table 2
Mitochondrial and Nuclear DNA Markers in the Midtemporal Cortex of AD Brain and in Monkey Lateral Geniculate Nucleus (LGN)[a]

Marker	*% decrease in AD vs control temporal cortex (refs.)*	*% decrease in LGN TTX treated vs control monkey[1]*
OXPHOS markers		
COX enzyme activity	20–25 *(73)*	23 ± 1*b*
NADH dehydrogenase enzyme activity	40 *(121)*	n.d.
COX protein	n.d.	23 ± 2
COX I mRNA (mtDNA)*c*	58 ± 3 *(98)*	49 ± 3
COX III mRNA (mtDNA)	54 ± 5 *(98)*	n.d.
ND1 mRNA (mtDNA)	50–60 *(98)*	n.d.
ND4 mRNA (mtDNA)	60 ± 8 *(96)*	n.d.
COX IV mRNA (nDNA)	40 ± 8*(98,120)*	18 ± 3
COX VIII mRNA (nDNA)	n.d.	29 ± 3
ATPsyn.β mRNA (nDNA)	50–60 *(98)*	n.d.
Non-OXPHOS markers		
mtDNA	n.s.*d*	26 ± 4
12S rRNA (mtDNA)	n.s. *(98)*	n.d.
28S rRNA (nDNA)	n.s. *(98)*	n.d.
β-actin mRNA (nDNA)	n.s. *(98)*	n.d.
LDH-B mRNA (nDNA)	n.s. *(98)*	n.d.

*a*7 d after injecting tetrodotoxin into vitreous humor of contralateral eye.
*b*mean ± SEM; n.s., not significant; n.d., not determined
*c*Parenthesis identifies whether mRNA is encoded by mitochondrial DNA (mtDNA) or nuclear DNA (nDNA).
*d*Tamataini M, Chandrasekaran K, Filburn CR, unpublished observations, 1987; *4 98*;
AD, Alzheimer's disease; ATPsyn.β, ATP synthase subunit β; COX, cytochrome oxidase; COX I, III, IV, VIII, cytochrome oxidase subunits I-VIII; LDH-B, lactate dehydrogenase subunit B; LGN, lateral geniculate nucleus; NADH, reduced nicotinamide adenine dinucleotide; ND1, ND4, subunits of NADH dehydrogenase; OXPHOS, oxidative phosphorylation; TTX, tetradotoxin.

mtDNA was found in the monkey lateral geniculate nucleus 3–7 d after tetrodotoxin was injected into the contralateral eye to reduce retinal input to the nucleus (2nd data column, Table 2; ref. *99*). The COX reductions could be reversed within days after tetrodotoxin was discontinued, allowing synaptic activity in the nucleus to recover *(85)*. These changes suggest that OXPHOS downregulation in the temporal cortex of the AD brain (column 1) is potentially reversible as well. Downregulation in either case may follow changes in the precursor pool for mitochondrial peptides or proteins *(100)* or in transcriptional or post-transcriptional factors produced by the nuclear genome *(97,101)*.

3.1.4. Staging of OXPHOS Downregulation in AD

OXPHOS declines in a graded manner in individual pyramidal neurons of the AD midtemporal association cortex, in relation to the accumulation of NFTs but not senile (neuritic) plaques. This was shown in a study using *in situ* hybridization to measure intracellular levels of mtDNA-encoded COX III mRNA and 12S rRNA and of poly(A)+ mRNA coded by mtDNA plus nDNA (Fig. 5; ref. *73*). Intracellular NFT density, evidenced by tau protein abnormally phosphorylated at its 396 and 404 serine sites, was quantified with an appropriate antibody.

Pyramidal neurons without NFTs showed reduced COX III mRNA levels but normal mRNA levels for mtDNA-encoded 12S rRNA, compared with NFT-free neurons in control brain or motor cortex of AD brain (Figs. 6 and 7). When NFTs were present but filled less than 50% of the pyramidal cell cytoplasm, COX III mRNA was further reduced whereas poly(A)+ mRNA and 12S rRNA levels remained unchanged. When more than 50% of the neuronal cytoplasm was filled with NFTs, poly(A)+ mRNA and 12S rRNA levels were reduced as well, indicative of nontranscribing, dying neurons (**Note 4**).

In contrast to the significant correlation between NFT density and COX III expression in individual pyramidal neurons of the AD brain, COX expression is not correlated with how close the neurons are to senile (neuritic) plaques *(102)*. This, as well as evidence that regional cortical NFT density but not plaque density correlates with regional reductions in $rCMR_{glc}$ in life *(11)*, argue that senile plaques, which contain condensed insoluble polymeric β-amyloid peptide, do not by themselves decrease brain energy metabolism in AD. In agreement is evidence that COX I and COX IV protein levels in the postmortem hippocampus of healthy elderly and AD patients are unrelated to the intensity of local plaque accumulation *(103)* and that brain COX III mRNA levels are normal in the transgenic V717F PDAPP mouse, despite marked neuritic plaque accumulation *(104)*.

4. DISCUSSION

In vivo brain imaging of patients with AD demonstrates staged reductions in resting-state brain glucose metabolism and blood flow in relation to dementia severity, more so in association than primary cortical regions. During cognitive or psychophysical stimulation, blood flow and glucose metabolism in affected brain regions can increase to the same extent in mildly

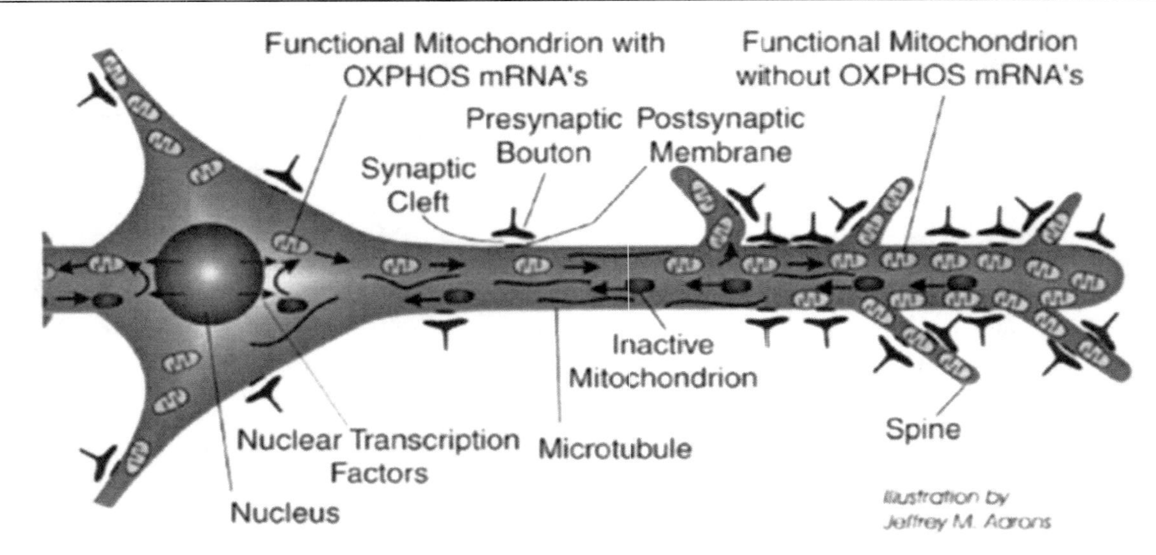

Fig. 5. Representation of mitochondrial recycling in principal dendrite of brain pyramidal neuron. Transcription of subunits for enzymes of OXPHOS, requiring mtDNA and nDNA and nucleus-derived transcription factors, occurs when mitochondria are near cell nucleus. After transcription and enzyme constitution, the mitochondria are carried by fast axonal transport to synaptic sites (largely dendritic spines) to synthesize ATP for synaptic activity. After some OXPHOS mRNAs and protein subunits decay, the mitochondria are returned to perinuclear region for replenishment. From ref. *90* with permission from F. P. Graham Publishing Co.

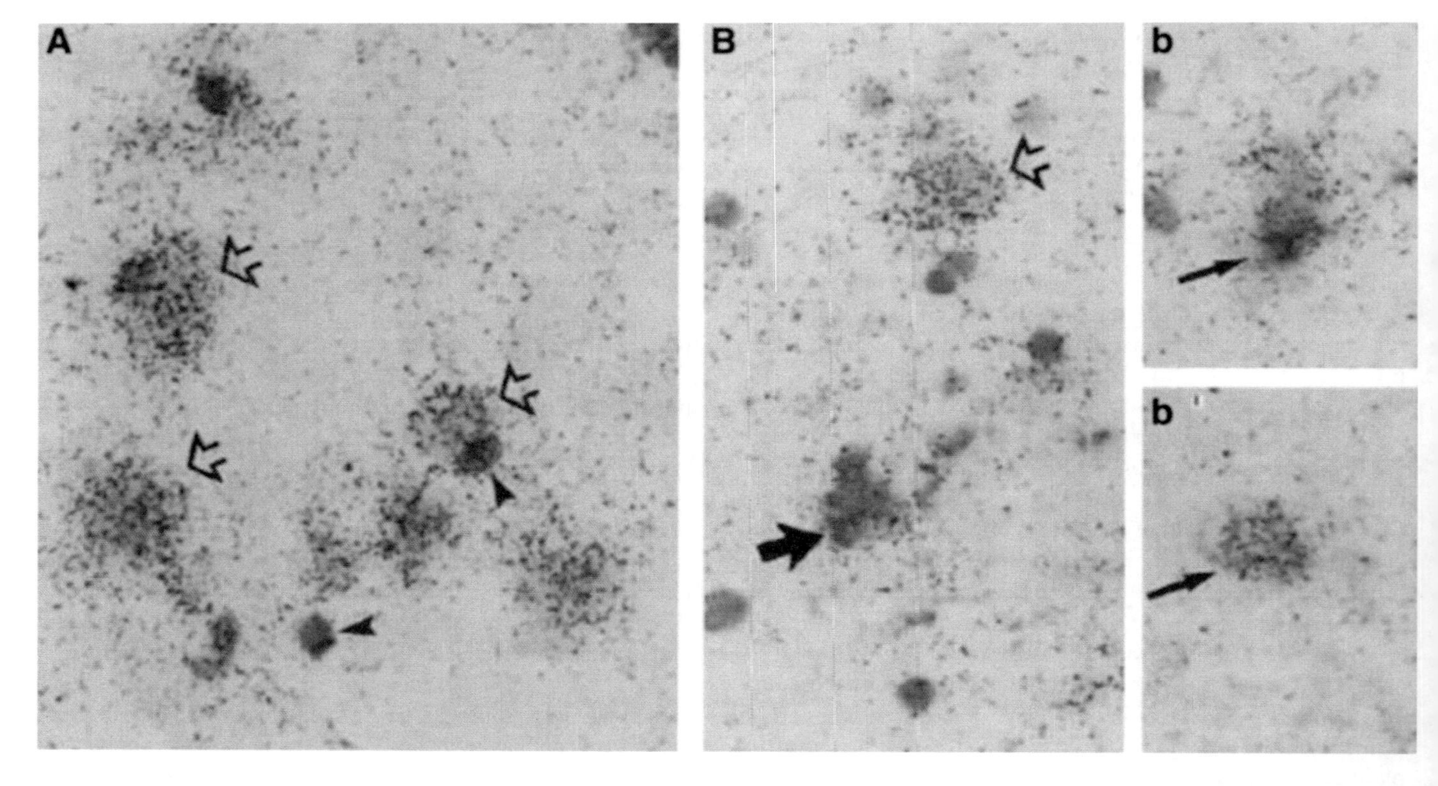

Fig. 6. *In situ* hybridization of cytochrome oxidase subunit III (COX III) mRNA (signaled by grain counts) and immunostaining (brown color) in relation to two stages of neurofibrillary tangle accumulation in pyramidal neurons from midtemporal cortex. **(A)** Control brain. NFT-free neurons (open arrows) are stained with cresyl violet, as are glial nuclei (arrowheads). **(B)** Brain with AD. A NFT-free neuron (open arrow) shows more COX III mRNA grains than a neuron (small filled arrow) in which NFTs fill less than half of cell body (early stage) and many more grains than a neuron (big filled arrow) where NFTs fill >50% of cell body (late stage). **(b)** Early stage NFT-bearing neuron has more grains than late-stage NFT-bearing neuron in B. **(b')** NFT-bearing neuron from same AD section From ref. *73*. Copyright © 1996. Reprinted by permission of John Wiley & Sons, Inc. *See* color version on Companion CD.

demented AD patients as in age-matched controls, suggesting, together with evidence for a normal oxygen extraction fraction, that energy availability as the result of reduced blood flow is not rate-limiting at this early stage of disease. Responses to stimulation decline, however, with dementia severity, and are markedly reduced in severely demented patients. To the extent

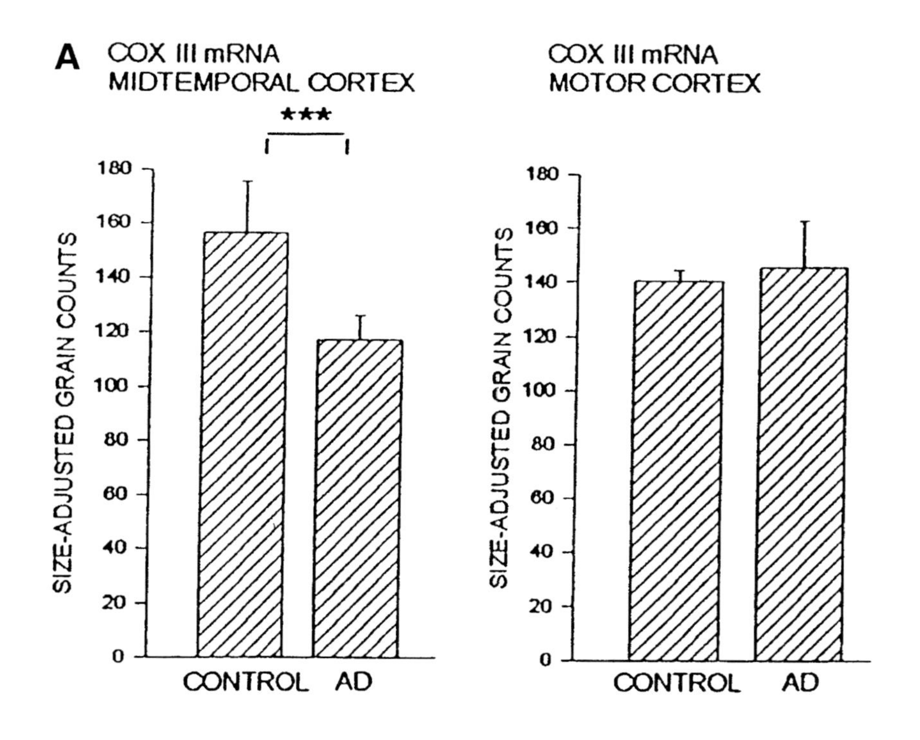

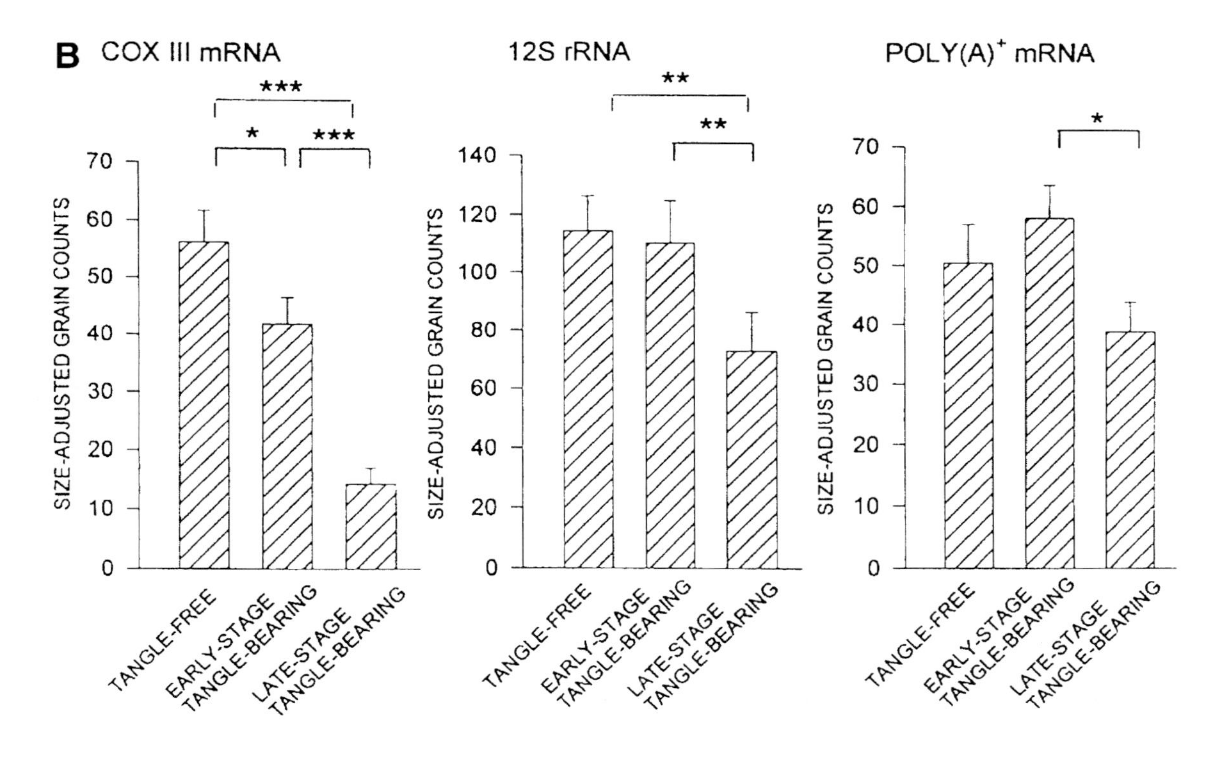

Fig. 7. Relation between neurofibrillary tangle accumulation and cytochrome oxidase subunit III (COX III) mRNA level in pyramidal neurons of AD brain. In (**A**) compared with control brain, COX III mRNA is reduced significantly in NFT-free neurons of midtemporal but not of motor cortex (n = 3–4) from AD brain. In (**B**) COX III mRNA in AD midtemporal cortex declines progressively as NFTs fill cell cytoplasm. With >50% filling, the decline is accompanied by statistically significant declines in non-OXPHOS mitochondrial 12S RNA and total poly(A)$^+$ mRNA. Grain counts give mRNA levels. Statistical significance: ***$p < 0.001$, **$p < 0.01$, *$p < 0.05$. From ref. *73*. Copyright © 1996. Reprinted by permission of John Wiley & Sons, Inc.

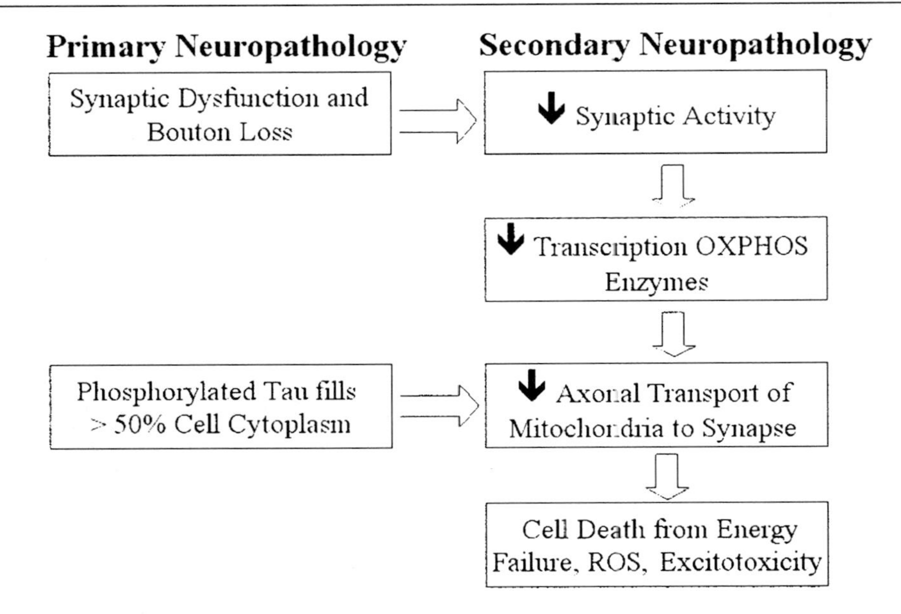

Fig. 8. Model for correlated progression of synaptic and OXPHOS changes in the course of AD. *See* text for discussion

that brain metabolism and blood flow are markers of synaptic function *(1,2,48)*, the in vivo stages of brain responsiveness to stimulation likely reflect stages in synaptic loss and dysfunction in the brain itself. These stages are schematized in Fig. 8.

In the brain, OXPHOS enzyme expression normally is coordinated with the expression of nuclear genes involved in ATP production in a number of mitochondrial disorders *(105,106)* and with the activity of Na,K-ATPase, suggesting common transcriptional regulatory mechanisms *(107)*. Ours and other's data suggest that a similar coordinated downregulation of OXPHOS, glucose delivery, and energy consumption applies to the early stage of AD. In AD, downregulation of OXPHOS is accompanied by downregulation of brain protein levels of the GLUT1 glucose transporter in capillaries and of the GLUT3 glucose transporter in neurons, and of the neuron-specific α3 subunit of brain Na,K-ATPase *(108–111)*.

A normal activation response in mildly demented AD patients, despite reduced resting state brain metabolism or flow, is consistent with a physiologically and potentially reversible downregulation of OXPHOS, the likely cause of which is reduced energy demand by dysfunctional synapses *(112)*. Activation studies showing recruitment of additional brain regions in pre-symptomatic subjects genetically at risk for AD, and reduced high frequency responses to visual stimuli in mildly demented subjects, further support some synaptic dysfunction in early stages of AD.

In summary (Fig. 8), in the early stage of AD, brain glucose metabolism, OXPHOS and energy consumption appear to be coordinately downregulated, which we propose is a normal "physiological" response to reduced energy demand by dysfunctional synapses. To date, however, the causes of this dysfunction are not agreed on. Because non-OXPHOS transcription appears intact, downregulation at this early point may be reversible and the brain may remain amenable to appropriate pharmacological intervention *(40,62)*. As disease progresses to loss of functional responsiveness, NFTs containing phosphorylated tau accumulate in neurons, thereby removing the nonphosphorylated tau required for axonal transport of mitochondria between synapses and cell cytoplasm. When NFTs fill a critical fraction of the cell cytoplasm *(113)*, axonal transport of mitochondria is so disrupted that insufficient ATP is produced to maintain synaptic structure and function *(114)*. The result is energy deprivation, accompanied by increased neuronal vulnerability to excitotoxicity and by accumulation of reactive oxygen species, leading to cell death *(107,115–119)*.

5. NOTES

1. Dementia severity can be stratified according to the Mini-Mental State Examination in this paper, a score of 30–21 defines mild dementia, of 20–11 moderate dementia, and of 10–0 severe dementia.
2. In hypertensive subjects, the oxygen extraction fraction may be increased, suggesting rate-limited delivery of oxygen to the brain caused by insufficient blood flow. If they develop dementia, brain demand can be reduced due to cell death, and the oxygen extraction fraction is no longer elevated.
3. The disproportionate decrease in mtDNA-encoded COX subunit mRNA and other OXPHOS mRNAs in the absence of changes in mtDNA-encoded 12S rRNA in the AD brain may have been due to the comparatively short half-lives of the OXPHOS markers.
4. The discrepancy between selective downregulation of OXPHOS in postmortem AD cortex (Table 2) and loss of general transcriptional ability in cortical pyramidal neurons whose cytoplasm is more than 50% filled with NFTs can be explained if NFT-filled pyramidal cells in the postmortem cortical sample were incapable of significant transcription. Transcription by the remaining "viable cells" then would account for the observed selective down-regulation of OXPHOS in the samples.

REFERENCES

1. Sokoloff L. Relationship between functional activity and energy metabolism in the nervous system: whether, where and why? In: Lassen NA, Ingvar DH, Raichle ME, Friberg L, eds. Brain Work and Mental Activity. Quantitative Studies with Radioactive Tracers. Alfred Benzon Symposium. Vol. 31. Copenhagen: Munksgaard; 1991:52–67.
2. Purdon AD, Rapoport SI. Energy requirements for two aspects of phospholipid metabolism in mammalian brain. Biochem J 1998; 335:313–318.
3. Kumar A, Schapiro MB, Grady C, et al. High-resolution PET studies in Alzheimer's disease. Neuropsychopharmacology 1991;4:35–46.
4. Rapoport SI. Positron emission tomography in Alzheimer's disease in relation to disease pathogenesis: a critical review. Cerebrovasc. Brain Metab Rev 1991;3:297–335.
5. Ibanez V, Pietrini P, Alexander GE, et al. Regional glucose metabolic abnormalities are not the result of atrophy in Alzheimer's disease. Neurology 1998;50:1585–1593.
6. Bokde AL, Pietrini P, Ibanez V, et al. The effect of brain atrophy on cerebral hypometabolism in the visual variant of Alzheimer disease. Arch Neurol 2001;58:480–486.
7. Ibanez V, Pietrini P, Furey ML, et al. Resting state brain glucose metabolism is not reduced in normotensive healthy men during aging, after correction for brain atrophy. Brain Res Bull 2004;63:147–154.
8. Horwitz B, Grady CL, Schlageter NL, Duara R, Rapoport SI. Intercorrelations of regional cerebral glucose metabolic rates in Alzheimer's disease. Brain Res 1987;407:294–306.
9. Morrison JH, Hof PR, Campbell MJ, et al. Cellular pathology in Alzheimer's disease: implications for corticocortical disconnection and differential vulnerability. In: Rapoport SI, Petit H, Leys D, Christen Y, eds. Imaging, Cerebral topography and Alzheimer's disease, Research and Perspectives in Alzheimer's disease. Berlin: Fondation Ipsen, Springer-Verlag; 1990:19–40.
10. Lewis DA, Campbell MJ, Terry RD, Morrison JH. Laminar and regional distributions of neurofibrillary tangles and neuritic plaques in Alzheimer's disease: A quantitative study of visual and auditory cortices. J Neurosci 1987;7:1799–1808.
11. DeCarli C, Atack JR, Ball MJ, et al. Post-mortem regional neurofibrillary tangle densities but not senile plaque densities are related to regional cerebral metabolic rates for glucose during life in Alzheimer's disease patients. Neurodegeneration 1992;1:113–121.
12. Rapoport SI. Brain evolution and Alzheimer's disease. Rev Neurol (Paris) 1988;144:79–90.
13. Rapoport SI. Integrated phylogeny of the primate brain, with special reference to humans and their diseases. Brain Res Rev 1990; 15:267–294.
14. Braak H, Braak E, Bohl J. Staging of Alzheimer-related cortical destruction. Eur Neurol 1993;33:403–408.
15. Rapoport SI. Hypothesis: Alzheimer's disease is a phylogenic disease. Med Hypoth 1989;29:147–150.
16. Wieser HG. Depth recorded limbic seizures and psychopathology. Neurosci Biobehav Rev. 1983;7:427–440.
17. Benton A. Visuoperceptual, visuospatial, and visuoconstructive disorders. In: Heilman KM, Valenstein E, eds. Clinical Neuropsychology. New York: Oxford University Press; 1985:151–185.
18. Grady CL, Haxby JV, Schapiro MB, et al. Subgroups in dementia of the Alzheimer type identified using positron emission tomography. J Neuropsychiatry Clin Neurosci 1990;2:373–384.
19. Frackowiak RSJ, Pozzilli C, Legg NJ, et al. Regional cerebral oxygen supply and utilization in dementia. A clinical and physiological study with oxygen-15 and positron tomography. Brain 1981;104:753–778.
20. Friedland RP, Budinger TF, Ganz E, et al. Regional cerebral metabolic alterations in dementia of the Alzheimer type: Positron emission tomography with [18F]fluorodeoxyglucose. J Comp Assist Tomogr 1983;7:590–598.
21. McGeer PL, Kamo H, Harrop R, et al. Positron emission tomography in patients with clinically diagnosed Alzheimer's disease. Can Med Assoc J 1986;134:597–607.
22. Haxby JV, Grady CL, Duara R, Schlageter N, Berg G, Rapoport SI. Neocortical metabolic abnormalities precede nonmemory cognitive defects in early Alzheimer's-type dementia. Arch Neurol 1986;43: 882–885.
23. Grady CL, Haxby JV, Horwitz B, et al. Longitudinal study of the early neuropsychological and cerebral metabolic changes in dementia of the Alzheimer type. J Clin Exp Neuropsychol 1988;10:576–596.
24. Grady CL, Sonies B, Haxby J, Luxenberg J, Friedland R, Rapoport S. Cerebral metabolic asymmetries predict decline in language performance in dementia of the Alzheimer type (DAT). J. Clin Exp Neuropsychol 1988;10:39.
25. Haxby JV, Grady CL, Koss E, et al. Longitudinal study of cerebral metabolic asymmetries and associated neuropsychological patterns in early dementia of the Alzheimer type. Arch Neurol 1990;47:753–760.
26. Folstein MF, Folstein SE, McHugh PR. "Mini Mental State." A practical method for grading the cognitive state of patients for the clinician. J Psychiatr Res 1975;12:189–198.
27. Haxby JV, Grady CL, Koss E, et al. Heterogeneous anterior-posterior metabolic patterns in dementia of the Alzheimer type. Neurology 1988;38:1853–1863.
28. Ginsberg L, Atack JR, Rapoport SI, Gershfeld NL. Regional specificity of membrane instability in Alzheimer's disease brain. Brain Res 1993;615:355–357.
29. Ginsberg L, Rafique S, Xuereb JH, Rapoport SI, Gershfeld NL. Disease and anatomic specificity of ethanolamine plasmalogen deficiency in Alzheimer's disease brain. Brain Res 1995;698:223–226.
30. Clark CM, Ammann W, Martin WR, Ty P, Hayden MR. The FDG/PET methodology for early detection of disease onset: a statistical model. J Cereb Blood Flow Metab 1991;11:A96–A102.
31. Azari NP, Pettigrew KD, Schapiro MB, et al. Early detection of Alzheimer's disease: A statistical approach using positron emission tomographic data. J Cereb Blood Flow Metab 1993;13:438–447.
32. Yamaji S, Ishii K, Sasaki M, et al. Changes in cerebral blood flow and oxygen metabolism related to magnetic resonance imaging white matter hyperintensities in Alzheimer's disease. J Nucl Med 1997;38: 1471–1474.
33. Yao H, Sadoshima S, Kuwabara Y, Ichiya Y, Fujishima M. Cerebral blood flow and oxygen metabolism in patients with vascular dementia of the Binswanger type. Stroke 1990;21:1694–1699.
34. Fujii K, Sadoshima S, Okada Y, et al. Cerebral blood flow and metabolism in normotensive and hypertensive patients with transient neurologic deficits. Stroke 1990;21:283–290.
35. Grady CL, Haxby JV, Horwitz B, et al. Activation of cerebral blood flow during a visuoperceptual task in patients with Alzheimer-type dementia. Neurobiol Aging 1993;14:35–44.
36. Rapoport SI, Grady CL. Parametric in vivo brain imaging during activation to examine pathological mechanisms of functional failure in Alzheimer disease. Int J Neurosci 1993;70:39–56.
37. Ricciardi E, Furey ML, Giovacchini G, et al. Correlation between regional cerebral blood flow (rCBF) and glucose metabolism (rCMRglc) during rest and audiovisual stimulation is maintained in Alzheimer's disease. Soc Neurosci Abstr 2001;27:426.9.
38. Pietrini P, Furey ML, Alexander GE, et al. Association between brain functional failure and dementia severity in Alzheimer's disease: resting versus stimulation PET study. Am J Psychiatry 1999;156:470–473.
39. Pietrini P, Alexander GE, Furey ML, et al. Cerebral metabolic response to passive audiovisual stimulation in patients with Alzheimer's disease and healthy volunteers assessed by PET. J Nucl Med 2000;41:575–583.
40. Rogers SL, Farlow MR, Doody RS, Mohs R, Friedhoff LT. A 24-week, double-blind, placebo-controlled trial of Donezepil in patients with Alzheimer's disease. Donepezil Study Group. Neurology 1998;50:136–145.
41. Pellmar T. Electrophysiological correlates of peroxide damage in guinea pig hippocampus in vitro. Brain Res 1986;364:377–381.
42. VanMeter JW, Maisog JM, Zeffiro TA, Hallett M, Herscovitch P, Rapoport SI. Parametric analysis of functional neuroimages: application to a variable-rate motor task. Neuroimage 1995;2:373–383.

43. Mentis MJ, Horwitz B, Grady CL, et al. Visual cortical dysfunction in Alzheimer's disease evaluated with a temporally graded "stress test" during PET. Am. J. Psychiatry 1996;153:32–40.
44. Mentis MJ, Alexander GE, Grady CL, et al. Frequency variation of a pattern-flash visual stimulus during PET differentially activates brain from striate through frontal cortex. Neuroimage 1997;5:116–128.
45. Mentis MJ, Alexander GE, Krasuski J, et al. Increasing required neural response to expose abnormal brain function in mild versus moderate or severe Alzheimer's disease: PET study using parametric visual stimulation. Am J Psychiatry 1998;155:785–794.
46. Fox PT, Raichle ME. Stimulus rate determines regional brain blood flow in striate cortex. Ann Neurol 1985;17:303–305.
47. Livingstone M, Hubel D. Segregation of form, color, movement, and depth: anatomy, physiology, and perception. Science 1988;240:740–749.
48. Sokoloff L. Energetics of functional activation in neural tissues. Neurochem Res 1999;24:321–329.
49. Saunders AM. Gene identification in Alzheimer's disease. Pharmacogenomics 2001;2:239–249.
50. Kral VA. Senescent forgetfulness: benign and malignant. Can Med Assoc J 1962;86:257–260.
51. Petersen RC. Mild Cognitive Impairment: Aging to Alzheimer's Disease. New York: Oxford; 2003:269.
52. Rapoport SI. Mild cognitive impairment: aging to Alzheimer's Disease (book review). N Engl J Med 2003;349:1393.
53. Arnaiz E, Jelic V, Almkvist O, et al. Impaired cerebral glucose metabolism and cognitive functioning predict deterioration in mild cognitive impairment. Neuroreport 2001;12:851–855.
54. Small GW, Mazziotta JC, Collins MT, et al. Apolipoprotein E type 4 allele and cerebral glucose metabolism in relatives at risk for familial Alzheimer disease. JAMA 1995;273:942–947.
55. Reiman EM, Caselli RJ, Yun LS, et al. Preclinical evidence of Alzheimer's disease in persons homozygous for the epsilon 4 allele for apolipoprotein E. N Engl J Med 1996;334:752–758.
56. Kennedy AM, Frackowiak RSJ, Newman SK, et al. Deficits in cerebral glucose metabolism demonstrated by positron emission tomography in individuals at risk of familial Alzheimer's disease. Neurosci Lett 1995;186:17–20.
57. Reiman EM, Chen K, Alexander GE, et al. Functional brain abnormalities in young adults at genetic risk for late-onset Alzheimer's dementia. Proc Natl Acad Sci USA 2004;101:284–289.
58. Tohgi H, Takahashi S, Kato E, et al. Reduced size of right hippocampus in 39- to 80-year-old normal subjects carrying the apolipoprotein E epsilon4 allele. Neurosci Lett 1997;236:21–24.
59. Grady CL. Age-related changes in cortical blood flow activation during perception and memory. Ann NY Acad Sci 1996;777:14–21.
60. Grady CL. Age-related differences in face processing: a meta-analysis of three functional neuroimaging experiments. Can J Exp Psychol 2002;56:208–220.
61. Furey ML, Pietrini P, Haxby JV, et al. Cholinergic stimulation alters performance and task-specific regional cerebral blood flow during working memory. Proc Natl Acad Sci USA 1997;94:6512–6516.
62. Ricciardi E, Pietrini P, Rapoport SI, Schapiro MB, Furey ML. Effect of cholinergic modulation on brain response to visual working memory (VWM) as task difficulty increases in young and older subjects. Soc Neurosci Abstr 2003;29:189.16.
63. Hatanpää K, Isaacs KR, Shirao T, Brady DR, Rapoport SI. Loss of proteins regulating synaptic plasticity in normal aging of the human brain and in Alzheimer disease. J Neuropathol Exp Neurol 1999;58:637–643.
64. Bookheimer SY, Strojwas MH, Cohen MS, et al. Patterns of brain activation in people at risk for Alzheimer's disease. N Engl J Med 2000;343:450–456.
65. Burggren AC, Small GW, Sabb FW, Bookheimer SY. Specificity of brain activation patterns in people at genetic risk for Alzheimer disease. Am J Geriatr Psychiatry 2002;10:44–51.
66. Davies CA, Mann DMA, Sumpter PQ, Yates PO. A quantitative morphometric analysis of the neuronal and synaptic content of the frontal and temporal cortex in patients with Alzheimer's disease. J Neurol Sci 1987;78:151–164.
67. Mann DMA, Marcyniuk B, Yates PO, Neary D, Snowden JS. The progression of the pathological changes of Alzheimer's disease in frontal and temporal neocortex examined both at biopsy and at autopsy. Neuropathol Appl Neurobiol 1988;14:177–195.
68. Greenamyre JT, Maragos WF, Albin RL, Penney JB, Young AB. Glutamate transmission and toxicity in Alzheimer's disease. Prog Neuropsychopharmacol Biol Psychiatry 1988;12:421–430.
69. Scheff SW, DeKosky ST, Price DA. Quantitative assessment of cortical synaptic density in Alzheimer's disease. Neurobiol Aging 1990;11:29–37.
70. DeKosky ST, Scheff SW. Synapse loss in frontal cortex biopsies in Alzheimer's disease: Correlation with cognitive severity. Ann Neurol 1990;27:457–464.
71. Terry RD, Masliah E, Salmon DP, et al. Physical basis of cognitive alterations in Alzheimer's disease: Synapse loss is the major correlate of cognitive impairment. Ann Neurol 1991;30:572–580.
72. Harigaya Y, Shoji M, Shirao T, Hirai S. Disappearance of actin-binding protein, drebrin, from hippocampal synapses in Alzheimer's disease. J Neurosci Res 1996;43:87–92.
73. Hatanpää K, Brady DR, Stoll J, Rapoport SI, Chandrasekaran K. Neuronal activity and early neurofibrillary tangles in Alzheimer's disease. Ann Neurol 1996;40:411–420.
74. Rapoport SI, Hatanpää K, Brady DR, Chandrasekaran K. Brain energy metabolism, cognitive function and down-regulated oxidative phosphorylation in Alzheimer disease. Neurodegeneration 1996;5:473–476.
75. Yao PJ, Zhu M, Pyun EI, et al. Defects in expression of genes related to synaptic vesicle trafficking in frontal cortex of Alzheimer's disease. Neurobiol Dis 2003;12:97–109.
76. Selkoe DJ. Alzheimer's disease is a synaptic failure. Science 2002;298:789–791.
77. Wang SS, Becerra-Arteaga A, Good TA. Development of a novel diffusion-based method to estimate the size of the aggregated Abeta species responsible for neurotoxicity. Biotechnol Bioeng 2002;80:50–59.
78. Gong Y, Chang L, Viola KL, et al. Alzheimer's disease-affected brain: Presence of oligomeric Aβ ligands (ADDLs) suggests a molecular basis for reversible memory loss. Proc Natl Acad Sci USA 2003;100:10417–10422.
79. Rutten BP, Wirths O, Van de Berg WD, et al. No alterations of hippocampal neuronal number and synaptic bouton number in a transgenic mouse model expressing the beta-cleaved C-terminal APP fragment. Neurobiol Dis 2003;12:110–120.
80. Sato M, Kawarabayashi T, Shoji M, et al. Neurodegeneration and gliosis in transgenic mice overexpressing a carboxy-terminal fragment of Alzheimer amyloid-beta protein precursor. Dement Geriatr Cogn Disord 1997;8:296–307.
81. Barger SW, Basile AS. Activation of microglia by secreted amyloid precursor protein evokes release of glutamate by cystine exchange and attenuates synaptic function. J Neurochem 2001;76:846–854.
82. Mattson MP, Gary DS, Chan SL, Duan W. Perturbed endoplasmic reticulum function, synaptic apoptosis and the pathogenesis of Alzheimer's disease. Biochem Soc Symp 2001;67:151–162.
83. Lehninger AL, Nelson DL, Cox MM. Principles of Biochemistry. New York: Worth Press, 1993:446–448 and 669–683.
84. Attardi G, Schatz G. Biogenesis of mitochondria. Annu Rev Cell Biol 1988;4:289–333.
85. Wong-Riley MTT. Cytochrome oxidase: an endogenous metabolic marker for neuronal activity. Trends Neurosci 1989;12:94–101.
86. Hevner RF, Wong-Riley MTT. Neuronal expression of nuclear and mitochondrial genes for cytochrome oxidase (CO) subunits analyzed by in situ hybridization: Comparison with CO activity and protein. J Neurosci 1991;11:1942–1958.
87. Chandrasekaran K, Stoll J, Brady DR, Rapoport SI. Localization of cytochrome oxidase (COX) activity and COX mRNA in the hippocampus and entorhinal cortex of the monkey brain: correlation with specific neuronal pathways. Brain Res 1992;579:333–336.
88. Chandrasekaran K, Stoll J, Rapoport SI, Brady DR. Localization of cytochrome oxidase (COX) activity and COX mRNA in the perirhinal and superior temporal sulci of the monkey brain. Brain Res 1993;606:213–219.

89. Rapoport SI. Functional brain imaging in the resting state and during activation in Alzheimer's disease. Implications for disease mechanisms involving oxidative phosphorylation. Ann N Y Acad Sci 1999;893:138–153.
90. Rapoport SI. Coupled reductions in brain oxidative phosphorylation and synaptic function can be quantified and staged in the course of Alzheimer disease. Neurotox Res 2003;5:385–398.
91. Gelfand R, Attardi G. Synthesis and turnover of mitochondrial ribonucleic acid in HeLa cells: the mature ribosomal and messenger ribonucleic acid species are metabolically unstable. Mol Cell Biol 1981;1:497–511.
92. Liu LI, Rapoport SI, Chandrasekaran K. Regulation of mitochondrial gene expression in differentiated PC12 cells. Ann N Y Acad Sci 1999;893:341–344.
93. Chrzanowska-Lightowlers ZM, Preiss T, Lightowlers RN. Inhibition of mitochondrial protein synthesis promotes increased stability of nuclear-encoded respiratory gene transcripts. J Biol Chem 1994;269:27322–27328.
94. Gaines G, Rossi C, Attardi G. Markedly different ATP requirements for rRNA synthesis and mtDNA light strand transcription versus mRNA synthesis in isolated human mitochondria. J Biol Chem 1987;262:1907–1915.
95. Micol V, Fernandez-Silva P, Attardi G. Functional analysis of in vivo and in organello footprinting of HeLa cell mitochondrial DNA in relationship to ATP and ethidium bromide effects on transcription. J Biol Chem 1997;272:18896–18904.
96. Fukuyama R, Hatanpää K, Rapoport SI, Chandrasekaran K. Gene expression of ND4, a subunit of complex I of oxidative phosphorylation in mitochondria, is decreased in temporal cortex of brains of Alzheimer's disease patients. Brain Res 1996; 713:290–293.
97. Chandrasekaran K, Hatanpää K, Brady DR, Rapoport SI. Evidence for physiological down-regulation of brain oxidative phosphorylation in Alzheimer's disease. Exp Neurol. 1996;142:80–88.
98. Chandrasekaran K, Hatanpää K, Rapoport SI, Brady DR. Decreased expression of nuclear and mitochondrial DNA-encoded genes of oxidative phosphorylation in association neocortex of Alzheimer disease. Brain Res Mol Brain Res 1997;44:99–104.
99. Hevner RF, Wong-Riley MTT. Mitochondrial and nuclear gene expression for cytochrome oxidase subunits are disproportionately regulated by functional activity in neurons. J Neurosci 1993;13: 1805–1819.
100. Liu S, Wong-Riley M. Nuclear-encoded mitochondrial precursor protein: intramitochondrial delivery to dendrites and axon terminals of neurons and regulation by neuronal activity. J Neurosci 1994;14: 5338–5351.
101. Scarpulla RC. Nuclear respiratory factors and the pathways of nuclear-mitochondrial interaction. Trends Cardiovasc Med 1996; 6:39–45.
102. Hatanpää K, Chandrasekaran K, Brady DR, Rapoport SI. No association between Alzheimer plaques and decreased levels of cytochrome oxidase subunit mRNA, a marker of neuronal energy metabolism. Brain Res Mol Brain Res 1998;59:13–21.
103. Nagy Z, Esiri MM, LeGris M, Matthews PM. Mitochondrial enzyme expression in the hippocampus in relation to Alzheimer-type pathology. Acta Neuropathol (Berl) 1999;97:346–354.
104. Irizarry MC, Soriano F, McNamara M, et al. Abeta deposition is associated with neuropil changes, but not with overt neuronal loss in the human amyloid precursor protein V717F (PDAPP) transgenic mouse. J Neurosci 1997;17:7053–7059.
105. Heddi A, Lestienne P, Wallace DC, Stepien G. Mitochondrial DNA expression in mitochondrial myopathies and coordinated expression of nuclear genes involved in ATP production. J Biol Chem 1993;268:12156–12163.
106. Wallace DC. Mitochondrial diseases in man and mouse. Science 1999;283:1482–1488.
107. Hevner RF, Duff RS, Wong-Riley MT. Coordination of ATP production and consumption in brain: Parallel regulation of cytochrome oxidase and Na+,K+-ATPase. Neurosci Lett 1992;138:188–192.
108. Harik SI, Mitchell MJ, Kalaria RN. Ouabain binding in the human brain. Effects of Alzheimer's disease and aging. Arch Neurol 1989;46:951–954.
109. Kalaria RN, Harik SI. Reduced glucose transporter at the blood-brain barrier and in cerebral cortex in Alzheimer disease. J Neurochem 1989;53:1083–1088.
110. Simpson IA, Chundu KR, Davies-Hill T, Honer WG, Davies P. Decreased concentrations of GLUT1 and GLUT3 glucose transporters in the brains of patients with Alzheimer's disease. Ann Neurol 1994;35:546–551.
111. Chauhan NB, Lee JM, Siegel GJ. Na,K-ATPase mRNA levels and plaque load in Alzheimer's disease. J Mol Neurosci 1997;9:151–166.
112. Rapoport SI. Deux stades, réversible et irréversible, de l'insuffisance fonctionnelle dans le cerveau Alzheimerien. In: Christien Y, ed. De la Neurophysiologie à la Maladie D'Alzheimer. Symposium en Hommage à Yvon Lamour. Marseille: Solal; 1997:165–172.
113. Sheetz MP, Steuer ER, Schroer TA. The mechanism and regulation of fast axonal transport. Trends Neurosci 1989;12:474–478.
114. Yaffe MP. The machinery of mitochondrial inheritance and behavior. Science 1999;283:1493–1497.
115. Mecocci P, MacGarvey U, Beal MF. Oxidative damage to mitochondria DNA is increased in Alzheimer's disease. Ann Neurol 1994;36:747–751.
116. Greene JG, Greenamyre JT. Bioenergetics and glutamate excitotoxicity. Prog Neurobiol 1996;48:613–634.
117. Beal MF. Mitochondrial dysfunction in neurodegenerative diseases. Biochim Biophys Acta 1998;1366:211–223.
118. Calingasan NY, Uchida K, Gibson GE. Protein-bound acrolein: a novel marker of oxidative stress in Alzheimer's disease. J Neurochem 1999;72:751–756.
119. Sorensen L, Ekstrand M, Silva JP, et al. Late-onset corticohippocampal neurodepletion attributable to catastrophic failure of oxidative phosphorylation in MILON mice. J Neurosci 2001;21: 8082–8090.
120. Chandrasekaran K, Giordano T, Brady DR, Stoll J, Martin LJ, Rapoport SI. Impairment in mitochondrial cytochrome oxidase gene expression in Alzheimer disease. Brain Res Mol Brain Res 1994; 24:336–340.
121. Parker Jr WD, Parks J, Filley CM, Kleinschmidt-DeMasters BK. Electron transport chain defects in Alzheimer's disease brain. Neurology 1994;44:1090–1096.
122. Talairach J, Tournoux P. Co-planar stereotaxic atlas of the human brain. New York: Thieme Medical Publishers, Inc; 1988:122.

EPILEPSY III

11 Neocortical Epilepsy

α-Methyl-L-Tryptophan and Positron Emission Tomography Studies

Jun Natsume, MD, PhD, Andrea Bernasconi, MD, and Mirko Diksic, PhD

SUMMARY

A review of the use of a radioactively labeled tracer, α-methyl-L-tryptophan (α-MTrp), proposed for the study of the brain serotonin synthesis in normal and diseased brain, is presented. Serotonin is one of many brain neurotransmitters shown to be involved in many brain processes, and an alteration of its biochemistry has been proposed to be present in many brain diseases and disorders. Labeled α-MTrp in normal brain and probably in affective disorders can be used for the measurement of the regional serotonin synthesis, but in some diseases like epilepsy the uptake of the tracer is probably related more to the altered tryptophan metabolism, mainly via the kynurenine pathway. A summary of the published data indicates that positron emission tomography after injection of α-[^{11}C]MTrp could be very valuable in determining the epileptic focus; however, it is not possible to get identification in all cases evaluated.

Key Words: Cortical malformations; epileptic foci identification; α-methyl-L-tryptophan; positron emission tomography; tracer imaging; tuberous sclerosis complex.

1. SEROTONIN AND EPILEPSY: FROM ANIMAL MODELS TO HUMAN EPILEPSY

The involvement of serotonergic systems in epilepsy has been reported in animal models *(1–9)* and in human brains *(10–12)*. The synthesis of serotonin (5-HT) from the essential amino acid L-tryptophan (L-Trp) is a two-step enzymatic process. Tryptophan is hydroxylated through the action of tryptophan hydroxylase (EC 1.14.16.4) and molecular oxygen, with tetrahydrobiopterin as a coenzyme. Tryptophan hydroxylase is considered to be the rate-limiting enzyme in 5-HT synthesis *(13)*; it is found exclusively in 5-HT neurons *(14)*. The product of the Trp hydroxylation reaction, 5-hydroxy-L-tryptophan (5-HTP), undergoes decarboxylation through the action of aromatic amino acid decarboxylase (EC 1.4.3.4), which also is found in catecholaminergic neurons and which also catalyzes the decarboxylation of 3,4-dihydroxy-L-phenylalanine. 5-HT is metabolized to the 5-hydroxyindoleacetic acid (5-HIAA) by monoamine oxidase and is promptly removed from the central nervous system into the cerebrospinal fluid (CSF) and then into the blood *(15)* by a probenecid-sensitive mechanism. Despite a large number of investigations performed on the relation between concentrations of lumbar CSF 5-HIAA and different brain disorders, it is not generally accepted that the concentration of the 5-HT metabolite in the lumbar CSF has a linear relationship to the brain 5-HT synthesis; instead, it most likely represents a combination of processes, including peripheral 5-HT metabolism and clearance of 5-HIAA from the brain and CSF *(15,16)*.

The genetically epilepsy-prone rat (GEPR) is a widely used model for generalized epilepsy. Dailey et al. *(5)* measured serotonin levels in various regions of the GEPR and found a significant decrease in serotonin content in the hippocampus and five other discrete brain areas. The seizure susceptibility in GEPR is reported to be caused by abnormalities in several neurotransmitter systems including the serotonergic system *(8)*. By contrast to the GEPR, increased levels of serotonin is reported in experimental models of partial epilepsy *(7,17)*. In animal models of temporal lobe epilepsy (TLE), serotonin in hippocampus is increased in the acute period of repetitive limbic seizures and status epilepticus *(7)*.

Increased levels of serotonin metabolites have also been reported in human partial epilepsy *(10–12)*. Louw et al. *(10)* reported increased levels of 5-HIAA in temporal cortex and hippocampus obtained from TLE patients with various etiologies, including hippocampal sclerosis, cortical dysplasia, and brain tumor. Pintor et al. *(11)* also revealed increase of serotonin and 5-HIAA in the temporal neocortex surgically removed from patients with TLE. Broderick et al. and Pacia et al. performed microvoltammetry for specimens from patients with neocortical and mesial TLE *(18,19)*. They showed that the concentration of serotonin in the hippocampus was higher in patients with neocortical TLE than that in patients with mesial

From: *Bioimaging in Neurodegeneration*
Edited by P. A. Broderick, D. N. Rahni, and E. H. Kolodny

TLE and that the serotonin concentration in the temporal neocortex was similar in patients with neocortical and mesial TLE patients. Trottier et al. *(12)* examined the morphology and the laminar distribution of the serotonin innervation of the resected cerebral cortex of four patients with extratemporal lobe epilepsy. They revealed serotonergic hyperinnervation in the dysplastic epileptogenic tissues of two patients with cortical dysplasia, whereas the serotonin innervation was normal in two patients with cryptogenic epilepsy.

Two radically opposed hypotheses are proposed to interpret the role of serotonin in epilepsy. Serotonin may act as an inhibitory or excitatory mediator in seizure activity. The implied increased rate of serotonin synthesis might represent a compensatory mechanism to decrease the level of cortical excitability *(11)*. Many studies with animal models indicate the anticonvulsant effect of serotonin *(3,6,9)*. Lesions of the midbrain raphe delete forebrain serotonin and decrease the seizure threshold *(1)*. Conversely, electrical stimulation of the raphe, producing release of serotonin, inhibits kindled amygdaloid seizures *(2,3)*. Furthermore, it has been shown that exogenous serotonin can inhibit epileptiform activity in the CA1 region of the hippocampus of GEPR by acting on the 5-HT_{1A} receptor subtype *(9)*. On the other hand, serotonergic system might be contributing to the increased excitability of the focus *(11)*. Some investigators have reported that serotonin has effects of the inhibition of the release of γ-aminobutyric acid (GABA) and the increase in potency of the action of NMDA *(20–22)*.

2. α-MTRP METHOD

The α-methyl-L-tryptophan (α-MTrp) method was originally developed for the measurements of the brain regional 5-HT synthesis *(23)*. The method is based upon the trapping of the tracer and/or its metabolite, labeled α-methylserotonin (α-M5-HT). The tracer, labeled α-MTrp, is a substrate for tryptophan hydroxylase, and the product this reaction is acted upon by aromatic amino acid decarboxylase to give α-M5-HT *(23,24)*. In addition, it has been reported that α-MTrp is also a substrate for the tryptophan pyrrolase *(25)*, which more or less is nonexistent in the normal brain *(26)* but gets activated in some pathologies, such as inflammation and stroke *(27)*. Because of this the interpretation of the brain tissue trapping of α-MTrp must be done with caution in some pathologies, for example, epilepsy *(28)*. α-MTrp is not incorporated into proteins *(23,29)*. Moreover, it does not undergo any peripheral or central metabolism, making the determination of the plasma input function rather simple. The above-mentioned characteristics have given us an opportunity to develop a simple three-compartmental biological model, which can be used to quantify autoradiographic data. The calculation of the 5-HT synthesis rate requires a knowledge of the lumped constant (LC), which has been measured in the rat brain *(23,30)*.

Tryptophan is transported into brain by a system that is saturable, stereospecific, and active towards all large neutral amino acids *(31,32)*. Thus, in addition to tryptophan, the carrier transports phenylalanine, tyrosine, leucine, isoleucine, valine, methionine, cysteine, histidine, threonine, glutamine, asparagine, and serine. In addition to being transported by a specific carrier, tryptophan can also enter the brain by a passive diffusion *(33)*. The α-MTrp as tracer seems to be transported by the same charier as tryptophan. In the plasma, tryptophan is to a large extend bound to the plasma proteins but, as shown by many investigations, the plasma-free tryptophan relates the best to the tryptophan concentration in the brain and to the brain serotonin synthesis *(33–35)*.

Mathematically the α-MTrp biological model *(23,30)* can be represented by a set of differential equations with constant coefficients. This set of differential equations can be solved by several methods. The solution of this system by the Laplace transformation method gives the total radioactivity in the tissue compartments as a function of the plasma input function and the brain trapping constant K* [mL/g/min]. Knowing the constant K^*, the LC, and the plasma free-tryptophan concentration (C_p; nmol g^{-1}) is all that is required to estimate the rate of 5-HT synthesis (R; nmol g^{-1} min^{-1}) $[R = K^* \cdot C_p/LC]$. We determined the LC for the α-MTrp model in a large number of brain structures. It was shown that the regional LC estimates, measured in vivo in the rat brain, can be taken as being constant all over the rat brain and an average value is 0.42 ± 0.07 *(30)*.

In the measurements in the patients with epilepsy, only K* is calculated because of a possibility that the trapping, at least in part, relates to the activity of tryptophan pyrrolase *(28)*. However, when a statistical comparison such as statistical parametric mapping with a proportional scaling is used the comparisons obtained by using K* functional maps are the same as those obtained with other derived parameters *(34)*.

3. α-MTRP POSITRON EMISSION TOMOGRAPHY (PET) STUDIES IN HUMAN NEOCORTICAL EPILEPSY

α-[^{11}C]methyl-L-tryptophan (α-[^{11}C]MTrp) has been developed as a tracer for PET measurement of serotonin synthesis in vivo *(23)*. Reports suggest that α-[^{11}C]MTrp PET is a useful tool in the localization of the epileptogenic area in partial epilepsy *(28,35–38)*. However, the possibility also exists that in some pathological conditions, this tracer may accumulate locally via the kynurenine pathway, in which case its uptake and trapping would not be related to serotonin synthesis *(39,40)*.

In the following paragraphs we will briefly review and discuss the studies that have been performed concerning the application of α-[^{11}C]MTrp in human neocortical epilepsy.

3.1. TUBEROUS SCLEROSIS COMPLEX (TSC) AND CORTICAL MALFORMATIONS

TSC is an autosomal-dominant genetic disorder characterized by lesions of the skin and multiple organs, including central nervous system hamartoma *(41)*, and severe seizures that are frequently resistant to antiepileptic drug therapy. Surgical resection of seizure focus is performed, and good outcome is obtained at least in some patients with TSC *(42–44)*. The suboptimal results are attributable to the fact that these patients often have multiple cortical abnormalities and multifocal interictal EEG abnormalities. Therefore, differentiation of epileptogenic and nonepileptogenic lesions is sometimes difficult, and the identification of the most active tubers increases the likelihood for successful surgical treatment. Chugani et al. performed α-[^{11}C]MTrp PET scans in patients with TSC and revealed that α-[^{11}C]MTrp uptake was increased only in epileptogenic tubers, whereas the other nonepileptogenic tubers showed decreased uptake *(28)*. Ictal EEG data available in eight patients showed seizure onset corresponding to foci of increased α-[^{11}C]MTrp uptake in four patients. Asano et al. also calculated uptake ratios between tubers and normal cortex and determined the optimal uptake ratio for detecting epileptogenic tubers *(45)*. They concluded that cortical tubers with α-[^{11}C]MTrp uptake greater than or equal to normal cortex (uptake ratio 0.98) are related to epileptiform activity. It has been shown that ictal single photon imaging computed tomography is a useful technique for localization of the epileptogenic focus in several forms of epilepsy *(46)*, and the usefulness of the ictal single photon imaging computed tomography also is reported in patients with TSC *(47,48)*. However, the question if hyperperfused regions may represent areas of secondary propagation rather than the primary focus *(48)* remains unanswered. Also, in case of very short-lasting seizures, the tracer might fail to accumulate to the seizure focus *(49,50)*. In such cases, α-[^{11}C]MTrp PET may be a powerful tool for localization of the epileptogenic area before the invasive intracranial EEG monitoring.

Patients with TSC often present with various degree of mental retardation and autism *(51–54)*. Asano et al. *(55)* focused on autism in patients with TSC and revealed increase of the α-[^{11}C]MTrp uptake in the caudate nuclei in patients with TSC and autism. They hypothesized that increased quinolinic acid, a kynurenine pathway metabolite of tryptophan discussed below, in the caudate nuclei or adjacent subependymal nodules of autistic children with TSC might be the mechanism of the increased uptake of α-[^{11}C]MTrp.

The usefulness of α-[^{11}C]MTrp PET is also reported in patients with other malformations of cortical development *(35,38)*. Fedi et al. performed α-[^{11}C]MTrp PET studies in patients with focal cortical dysplasia (FCD) or cryptogenic partial epilepsy. They found increased α-[^{11}C]MTrp uptake at the seizure focus in four of seven patients with FCD (Fig. 1). Juhasz et al. studied 27 children with neocortical epilepsy, including 8 patients with developmental abnormalities on magnetic resonance imaging (MRI; ref. *38*). α-[^{11}C]MTrp PET showed abnormalities in six of eight patients with developmental abnormalities. The sensitivity of α-[^{11}C]MTrp PET in patients with FCD seems to be equivalent to that in patients with TSC. FCD was firstly reported by Taylor et al. *(56)*. This malformation, which is related to abnormal neuronal proliferation, is a frequent cause of medically refractory partial epilepsy, and some patients may achieve good seizure outcome after its surgical resection *(57–59)*. However, the epileptogenic abnormality extends beyond the area of maximal structural abnormality visible on MRI *(60)*. In FCD patients, by showing increased uptake area corresponding to the seizure focus, α-[^{11}C]MTrp PET may help identifying the extent of the epileptogenic area and consequently improve the surgical planning *(37)*.

3.2. CRYPTOGENIC EPILEPSY

The investigation and treatment of patients with epilepsy has been revolutionized by the advent of MRI, which has been demonstrated to be a precise and reliable indicator of pathologic findings underlying epilepsy in many patients. MRI has had a major impact in epilepsy surgery by helping defining cerebral structural damage and consequently in delineating the extent of the epileptogenic zone, that is, the site of seizure onset. However, conventional MRI is normal in a considerable number of patients with neocortical epilepsy *(61)*. Successful surgery is possible in these patients, but invasive EEG monitoring is frequently needed in the presurgical evaluation to identify the seizure focus. Functional neuroimaging techniques, including ^{18}F-deoxyglucose (FDG)-PET, are useful to localize nonobvious lesions and seizure foci in patients with normal MRI *(62,63)*. In a recent study, Fedi et al. showed that α-MTrp PET can reveal focal or regional increased uptake in patients with cryptogenic partial epilepsy *(35)*. Juhasz et al. performed α-[^{11}C]MTrp PET in 19 patients with normal MRI and showed focal abnormalities in nine patients *(38)*. Although sensitivity of α-[^{11}C]MTrp in the patients with cryptogenic epilepsy is lower than that in patients with cortical dysplasia (CD) or TSC, increased α-MTrp uptake on PET is highly specific for the seizure onset. Furthermore, α-[^{11}C]MTrp PET can detect epileptogenic cortical areas when both MRI and FDG-PET fail to provide adequate localizing information *(35,38)*. Therefore, increased focal uptake of α-[^{11}C]MTrp in such patients is a valuable addition to the current methods of investigation.

3.3. COMPARISON OF α-[^{11}C]MTRP WITH FDG-PET AND OTHER PET TRACERS

FDG-PET can detect focal areas of decreased glucose metabolism that are consistent with epileptogenic cortex in patients with neocortical partial epilepsy *(62,64)*. However, compared with classic mesial TLE, nonlesional neocortical epilepsies are much more likely to have normal interictal FDG-PET studies *(64,65)*. Moreover, glucose hypometabolism in patients with neocortical epilepsies is often widespread and extends beyond epileptogenic regions as that in mesial temporal lobe epilepsy *(62)*. Because not all cortical hypometabolic areas need to be resected to achieve seizure freedom, there has been an effort to develop other PET tracers that are specific to epileptogenic activities, including [^{11}C]flumazenil (FMZ) and α-[^{11}C]MTrp. Few studies, however, were dedicated to direct comparison of FDG-PET, and α-MTrp PET findings in same patients are performed *(35,38)*. Fedi et al. reported four patients (one patient with CD and three patients with cryptogenic epilepsy) with normal FDG-PET findings and increased α-[^{11}C]MTrp uptake *(35)*. Juhasz et al. performed the comparison in 27 patients and showed that the sensitivity of α-[^{11}C]MTrp PET for seizure onset is lower, but its specificity (100%) was higher ($p < 0.0001$) than that of hypometabolism on FDG-PET (sensitivity 73.2%, specificity 62.7%; ref. *38*). α-[^{11}C]MTrp PET abnormalities are smaller than corresponding FDG-PET hypometabolic regions *(24)*. Therefore, the location of increased α-[^{11}C]MTrp uptake can be a useful guide for the localization of seizure focus, especially in patients in whom FDG-PET show extensive areas of glucose hypometabolism.

FMZ is a specific antagonist at the benzodiazepine-binding site of $GABA_A$ receptors *(66)*, and FMZ PET provides a useful in vivo marker of $GABA_A$ receptor binding *(67)*. Focal changes in central benzodiazepine receptor distributions can be detected in partial epilepsies with FMZ PET *(63,68)*. Savic et al. described focal decreased FMZ binding in patients with frontal lobe epilepsy with or without lesions on MRI *(69)*. It appears

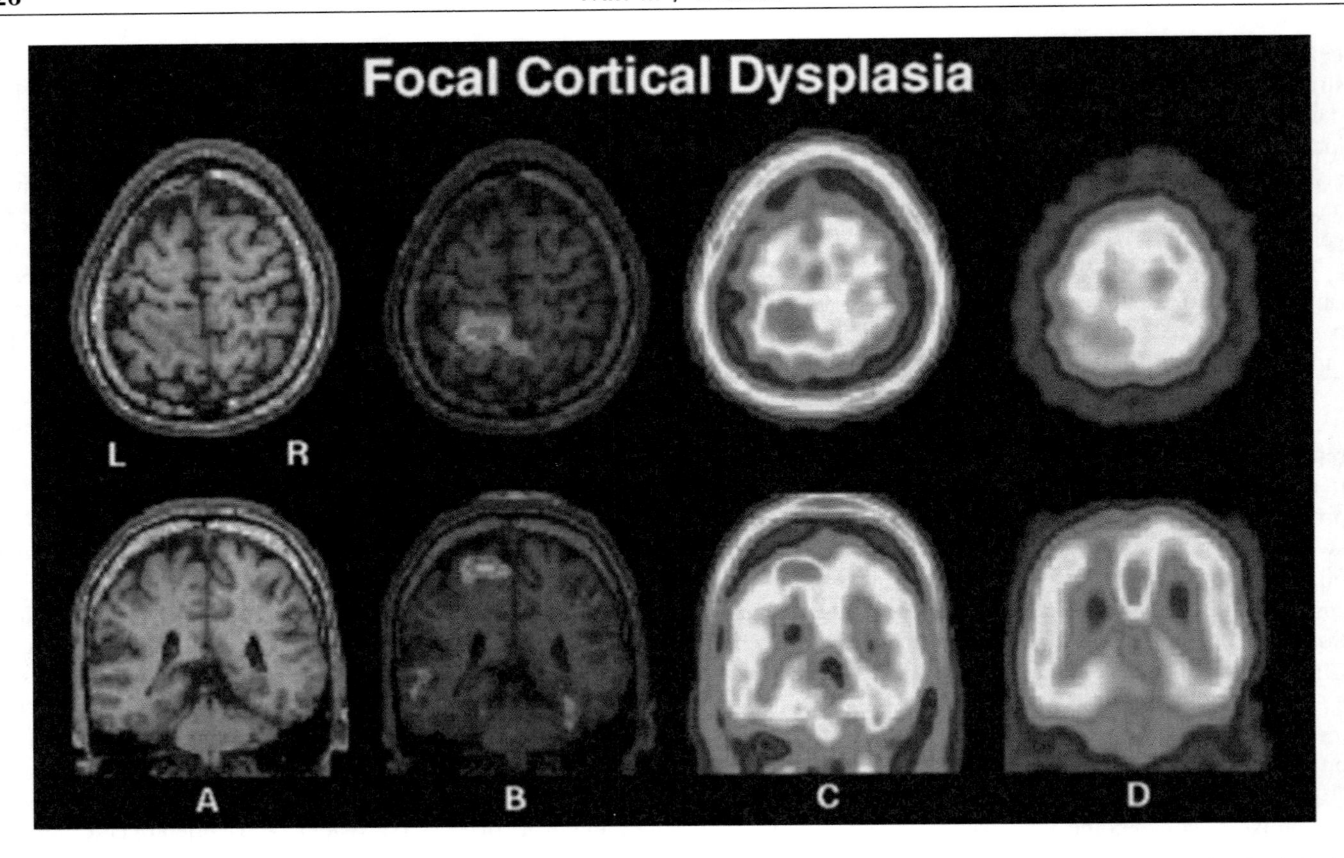

Fig. 1. MRI (**A**) and [^{11}C]-α-methyl-L-tryptophan (α-MTrp) PET (**C** and **D**) in a 24-yr-old-man with an extensive left parietal dysplastic lesion and medically intractable seizures. Intericial EEG recording demonstrated slow waves and sharp activity over the left centroparietal region. (A) The axial and coronal plane T1-weighted images show abnormal thickness of the paracentral lobule with lack of normal interdigitation of the white matter and pachygyrta. (B and C) α-MTrp-PET shows increased uptake within the dysplastic lesion. (**D**) Intericial FDG-PET shows decreased uptake corresponding to structural abnormality. (Reproduced by permission from Fedi et al., Neurology 2001;57:1629–1636.) *See* color version on Companion CD.

that the area with reduced benzodiazapine (BZD) receptor density is smaller than the corresponding hypometabolic region on FDG-PET. However, some studies have found localized increases of FMZ binding. Richardson et al. described reduction of FMZ binding in patients with lesions, and increase, decrease or both in FMZ binding in patients with normal MRI *(70,71)*. It is reported that decreased FMZ binding is correlated with the seizure onset *(72,73)*. However, in patients with multiple lesions on the MRI, FMZ PET may not be able to differentiate between epileptogenic and nonepileptogenic lesions because FMZ binding reduction may also reflect the neuronal loss at the structural abnormalities *(74)*. This indicates that presently α-[^{11}C]MTrp PET has a unique advantage in detecting epileptogenic lesions compared to other PET modalities.

Increased [^{11}C]methionine uptake has been reported at the ictal onset zones in patients with CD*(75)*. We performed preliminary study to compare the findings of methionine and α-[^{11}C]MTrp PET in neocortical epilepsy patients *(76)*. Our results showed that the regional uptake of α-[^{11}C]MTrp and methionine does not correlate and that the increased uptake of α-[^{11}C]MTrp and methionine are located in different sites in the same lesions. The lack of correlation between α-[^{11}C]MTrp and methionine suggest that these tracers are related to independent tissue process even though both tracers share the same tissue transporter. Therefore, α-[^{11}C]MTrp and methionine PET study may provide complementary information about different aspects of the epileptogenic lesions.

Recently PET tracer for 5-HT_{1A} receptor imaging has been developed and applied for studying epilepsy *(77)*. In temporal lobe epilepsy patients, low [^{18}F]FCWAY volume of distribution was found in the epileptogenic temporal lobe, midbrain raphe, and ipsilateral thalamus *(77)*. This finding might reflect reduced receptor concentration, lower affinity, or increased occupancy of the receptors by serotonin. This technique offers an interesting possibility to study neocortical partial epilepsy with the tracer and compare the finding with that of α-[^{11}C]MTrp PET that could elucidate the role of serotonin in the pathophysiology of neocortical epilepsy.

3.4. α-[^{11}C]MTRP PET AND EEG FINDINGS

Juhasz determined the spatial comparison of α-[^{11}C]MTrp PET abnormalities with intracranial EEG recordings *(38)*. The cortex adjacent to the region showing increased α-[^{11}C]MTrp uptake is most often the site of seizure onset on the intracranial EEG recordings, although parts of the region with increased α-[^{11}C]MTrp uptake itself are also commonly epileptogenic. Consequently, the location of increased α-[^{11}C]MTrp uptake can be a useful guide for the placement of intracranial electrodes.

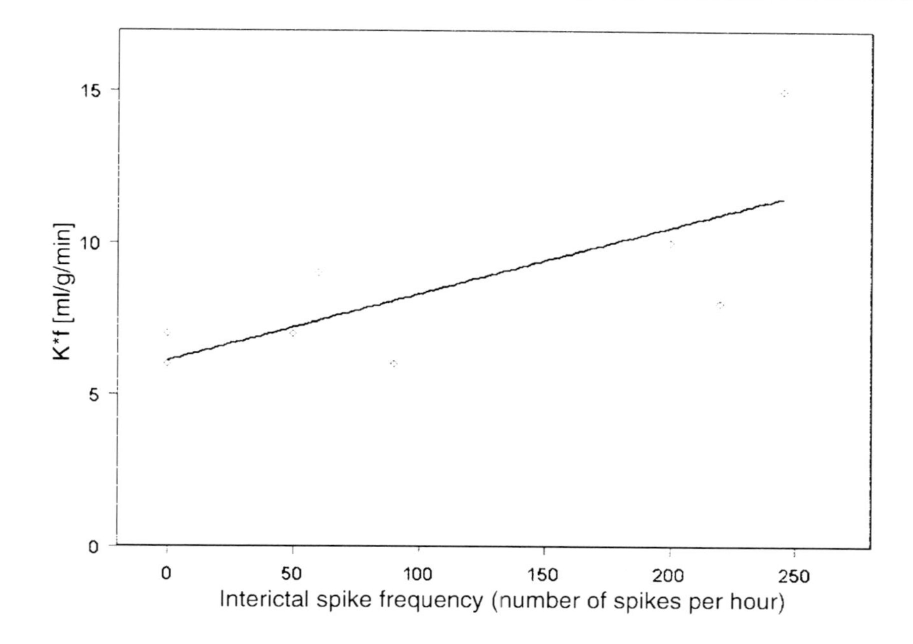

Fig. 2. Correlation between the number of intericial spikes per hour in the presumed epileptogenic focus and the K* values in the epileptogenic area in patients with tuberous sclerosis ($n = 8$). Spearman's correlation coefficient = 0.6; $p < 0.05$. (Reproduced by permission from Fedi et al. Epilepsy 2003;53:203–213.)

By comparing scalp EEG monitoring during the PET scans with the α-[^{11}C]MTrp uptake in patients with TSC *(37)*, Fedi et al. showed that increased uptake was mostly seen in patients with frequent interictal spikes and found a significant correlation between α-[^{11}C]MTrp uptake and the frequency of interictal spikes (Fig. 2). Although frequent epileptiform discharges may increase perfusion and glucose metabolism as a result of increase local synaptic activity *(78–81)*, decreased glucose hypometabolism at the site of increased α-[^{11}C]MTrp uptake, this increase is possibly not caused by change of general synaptic activity, but more specific epileptogenic activity. Juhasz et al. examined the correlation between α-[^{11}C]MTrp uptake and spike frequency in patients with cortical dysplasia or cryptogenic epilepsy, but they could not confirm the correlation in their patients *(38)*. The reason for the inconsistency of the finding is not clear. It might be because of the difference of underlying pathology or the location of the lesions might explain the inconsistent findings.

4. ALTERNATIVE MECHANISM OF INCREASED α-[^{11}C]MTRP UPTAKE VIA THE KYNURENINE PATHWAY

In addition to metabolism by tryptophan hydroxylase, tryptophan is metabolized through the kynurenine pathway. Although metabolites concentrations of this pathway are low under normal circumstance, under pathologic conditions, as in areas of ischemic injury and in the tubers for instance, the activity of this pathway may be markedly upregulated *(27,39,40)*. Quinolinic acid, a tryptophan metabolite of the kynurenine pathway, is a known convulsant through its action as an agonist at the NMDA receptor *(82,83)*. Experimental data have shown that application of quinolinic acid induces epileptiform discharges *(84,85)*. The increased transport of α-[^{11}C]MTrp, in this context, is intriguing. The increased uptake of α-[^{11}C]MTrp at the epileptogenic focus may represent increased activity of the kynurenine pathway, with formation of proconvulsant products. Chugani et al. analyzed resected tissues from patients with TSC for tryptophan hydroxylase activity, 5-HIAA, 5-HT, and quinolinic acid and found that quinolinic acid was fivefold higher in the epileptogenic tuber than in the nonepileptogenic tuber *(86)*. These results suggest that, at least in patients with TSC, increased α-[^{11}C]MTrp uptake may represent increased metabolism by means of the kynurenine pathway rather than serotonin synthesis.

5. CONCLUSION

α-[^{11}C]MTrp, a new technique allowing examination of serotonergic system, has been helpful in detecting the epileptogenic focus in patients with epilepsy. However, there is no yet clear understanding of the biological meaning of trapping this tracer in an epileptic tissue. This method has been particularly effective in patients with diffuse or multifocal structural abnormalities related to malformations of cortical development and when anatomical imaging failed to identify a structural lesion. This is mainly due to its high specificity and sensitivity, compared with the more widely used FDG-PET, to detect abnormalities related to microscopic dysplastic tissue. Therefore, radioligands representing serotonergic activity show great promise for presurgical evaluation of patients with intractable medically seizures. However, further research is warranted to better elucidate how PET findings and neurochemical changes of resected tissue relate to surgical outcome. Ultimately, new radioligands have the potential to contribute to the understanding the relation between the serotonergic abnormalities and dysfunctional circuits in human epilepsy.

REFERENCES

1. Kuhar MJ, Aghajanian GK, Roth RH. Tryptophan hydroxylase activity and synaptosomal uptake of serotonin in discrete brain regions after midbrain raphe lesions: correlations with serotonin levels and histochemical fluorescence. Brain Res 1972;44:165–176.
2. Samanin R, Valzelli L, Gumulka W. Inhibitory effect of midbrain raphe stimulation on cortical evoked potentials in rats. Psychopharmacologia 1972;24:373–379.
3. Kovacs DA, Zoll JG. Seizure inhibition by median raphe nucleus stimulation in rat. Brain Res 1974;70:165–169.
4. Dailey JW, Reigel CE, Mishra PK, Jobe PC. Neurobiology of seizure predisposition in the genetically epilepsy-prone rat. Epilepsy Res 1989;3:3–17.
5. Dailey JW, Mishra PK, Ko KH, Penny JE, Jobe PC. Serotonergic abnormalities in the central nervous system of seizure-naive genetically epilepsy-prone rats. Life Sci 1992;50:319–326.
6. Yan QS, Jobe PC, Dailey JW. Evidence that a serotonergic mechanism is involved in the anticonvulsant effect of fluoxetine in genetically epilepsy-prone rats. Eur J Pharmacol 1994;252:105–112.
7. Cavalheiro EA, Fernandes MJ, Turski L, Naffah-Mazzacoratti MG. Spontaneous recurrent seizures in rats: amino acid and monoamine determination in the hippocampus. Epilepsia 1994;35:1–11.
8. Statnick MA, Dailey JW, Jobe PC, Browning RA. Abnormalities in brain serotonin concentration, high-affinity uptake, and tryptophan hydroxylase activity in severe-seizure genetically epilepsy-prone rats. Epilepsia 1996;37:311–321.
9. Salgado-Commissariat D, Alkadhi KA. Effects of serotonin on induced epileptiform activity in CA1 pyramidal neurons of genetically epilepsy prone rats. Brain Res 1996;743:212–216.
10. Louw D, Sutherland GR, Glavin GB, Girvin J. A study of monoamine metabolism in human epilepsy. Can J Neurol Sci 1989;16:394–397.
11. Pintor M, Mefford IN, Hutter I, Pocotte SL, Wyler AR, Nadi NS. Levels of biogenic amines, their metabolites, and tyrosine hydroxylase activity in the human epileptic temporal cortex. Synapse 1990;5:152–156.
12. Trottier S, Evrard B, Vignal JP, Scarabin JM, Chauvel P. The serotonergic innervation of the cerebral cortex in man and its changes in focal cortical dysplasia. Epilepsy Res 1996;25:79–106.
13. Eccleston DJ, Ashcroft GW, Crawford TBB. 5-Hydroxyindole metabolism in rat brain. A study of intermediate metabolism using the technique of tryptophan loading. II Applications and drug studies. J Neurochem 1965;12:493–503.
14. Kuhar MJ, Aghajanian GH, Roth RH. Tryptophan hydroxylase activity and synaptosomal uptake of serotonin in discrete brain regions after midbrain raphe lesions: correlation with serotonin levels and histochemical fluorescence. Brain Res 1972;44:165–176.
15. Burns D, London J, Brunswick DJ, Pring M, Garfinkel D, Rabinowitz JL, Mendels J. A kinetic analysis of 5-hydroxyindoleacetic acid excretion from rat brain and csf. Biol Psychiat 1976;11:125–157
16. Wolf WA, Youdim MBH, Kuhn DM. Does brain 5-HIAA indicate serotonin release or monoamine oxidase activity? Eur J Pharmacol 1985; 109:381–387.
17. King JT Jr, LaMotte CC. El mouse as a model of focal epilepsy: a review. Epilepsia 1989;30:257–265.
18. Broderick PA, Pacia SV, Doyle WK, Devinsky O. Monoamine neurotransmitters in resected hippocampal subparcellations from neocortical and mesial temporal lobe epilepsy patients: in situ microvoltammetric studies. Brain Res 2000;878:48–63.
19. Pacia SV, Doyle WK, Broderick PA. Biogenic amines in the human neocortex in patients with neocortical and mesial temporal lobe epilepsy: identification with in situ microvoltammetry. Brain Res 2001; 899:106–111.
20. Reynolds JN, Baskys A, Carlen PL. The effects of serotonin on *N*-methyl-D-aspartate and synaptically evoked depolarizations in rat neocortical neurons. Brain Res 1988;456:286–292.
21. Nedergaard S, Engberg I, Flatman JA. Serotonin facilitates NMDA responses of cat neocortical neurones. Acta Physiol Scand 1986; 128:323–325.
22. Flint RS, Murphy JM, Calkins PM, McBride WJ. Monoamine, amino acid and cholinergic interactions in slices of rat cerebral cortex. Brain Res Bull 1985;15:197–202.
23. Diksic M, Nagahiro S, Sourkes TL, Yamamoto YL. A new method to measure brain serotonin synthesis in vivo. I. Theory and basic data for a biological model. J Cereb Blood Flow Metab 1990;10:1–12.
24. Gharib A, Balende C, Sarda N, Weissmann D, Plenevaux A, Luxen A, et al. Biochemical and autoradiographic measurements of brain serotonin synthesis rate in the freely moving rat: a reexamination of the α-methyl-L-tryptophan method. J Neurochem 1999;72:2593–2600.
25. Sourkes TL. A-methyltryptophan and its actions on tryptophan metabolism. Fed Proc 1971;30:897–903.
26. Saito K, Nowak TS Jr, Suyama K, et al. Kynurenine pathway enzymes in brain: responses to ischemic brain injury versus systemic immune activation. J Neurochem 1993;61:2061–2070.
27. Heyes MP, Saito K, Crowley JS, et al. Quinolinic acid and kynurenine pathway metabolism in inflammatory and non-inflammatory neurological disease. Brain 1992;115:1249–1273.
28. Chugani DC, Chugani HT, Muzik O, et al. Imaging epileptogenic tubers in children with tuberous sclerosis complex using α-[11C]methyl-L-tryptophan positron emission tomography. Ann Neurol 1998;44:858–866.
29. Roberge AG, Missala K, Sourkes TL. α-Methyltryptophan: effects on synthesis and degradation of serotonin in the brain. Neuropharmacology 1972;11:197–209.
30. Vanier M, Tsuiki K, Grdisa M, Worsley K, Diksic M. Determination of the lumped constant for the α-methyltryptophan method of estimating the rate of serotonin synthesis. J Neurochem 1995;64:624–635.
31. Oldendorf WH, Szabo J. Amino acid assignment to one of three blood-brain barrier amino acid carriers. Am J Physiol 1976;230:94–98.
32. Pardridge WM, Oldendorf WH. Transport of metabolic substrates through the blood-brain barrier. J Neurochem 1977;28:5–12.
33. Takada A, Grdisa M, Diksic M, Gjedde A, Yamamoto YL. Rapid steady-state analysis of the blood-brain transfer of L-Trp in rat, with special reference to the plasma protein bindig. Neurochem Int 1993; 23:351–359.
34. Friston KJ, Frith CD, Liddle PF, Dolan RJ, Lammertsma AA, Frackowiak RSJ. The relationship between global and local changes in PET studies. J Cereb Blood Flow Metab 1990;10:458–466.
35. Fedi M, Reutens D, Okazawa H, et al. Localizing value of α-methyl-L-tryptophan PET in intractable epilepsy of neocortical origin. Neurology 2001;57:1629–1636.
36. Natsume J, Kumakura Y, Bernasconi N, et al. A-[11C] methyl-L-tryptophan and glucose metabolism in patients with temporal lobe epilepsy. Neurology 2003;60:756–761.
37. Fedi M, Reutens DC, Andermann F, et al. α-[11C]-Methyl-L-tryptophan PET identifies the epileptogenic tuber and correlates with interictal spike frequency. Epilepsy Res 2003;52:203–213.
38. Juhasz C, Chugani DC, Muzik O, et al. α-methyl-l-tryptophan PET detects epileptogenic cortex in children with intractable epilepsy. Neurology 2003;60:960–968.
39. Saito K, Nowak TS, Jr., Suyama K, et al. Kynurenine pathway enzymes in brain: responses to ischemic brain injury versus systemic immune activation. J Neurochem 1993;61:2061–2070.
40. Chugani DC, Muzik O. A[C-11]methyl-L-tryptophan PET maps brain serotonin synthesis and kynurenine pathway metabolism. J Cereb Blood Flow Metab 2000;20:2–9.
41. Gomez MR. Tuberous Sclerosis. 2nd ed. New York: Raven Press; 1988.
42. Avellino AM, Berger MS, Rostomily RC, Shaw CM, Ojemann GA. Surgical management and seizure outcome in patients with tuberous sclerosis. J Neurosurg 1997;87:391–396.
43. Bebin EM, Kelly PJ, Gomez MR. Surgical treatment for epilepsy in cerebral tuberous sclerosis. Epilepsia 1993;34:651–657.
44. Guerreiro MM, Andermann F, Andermann E, et al. Surgical treatment of epilepsy in tuberous sclerosis: strategies and results in 18 patients. Neurology 1998;51:1263–1269.

45. Asano E, Chugani DC, Muzik O, et al. Multimodality imaging for improved detection of epileptogenic foci in tuberous sclerosis complex. Neurology 2000;54:1976–1984.
46. O'Brien TJ, So EL, Mullan BP, et al. Subtraction ictal SPECT co-registered to MRI improves clinical usefulness of SPECT in localizing the surgical seizure focus. Neurology 1998;50:445–454.
47. Koh S, Jayakar P, Resnick T, Alvarez L, Liit RE, Duchowny M. The localizing value of ictal SPECT in children with tuberous sclerosis complex and refractory partial epilepsy. Epileptic Disord 1999;1:41–46.
48. Koh S, Jayakar P, Dunoyer C, et al. Epilepsy surgery in children with tuberous sclerosis complex: presurgical evaluation and outcome. Epilepsia 2000;41:1206–1213.
49. Ebner A, Buschsieweke U, Tuxhorn I, Witte OW, Seitz RJ. Supplementary sensorimotor area seizure and ictal single-photon emission tomography. Adv Neurol 1996;70:363–368.
50. Runge U, Kirsch G, Petersen B, et al. Ictal and interictal ECD-SPECT for focus localization in epilepsy. Acta Neurol Scand 1997;96:271–276.
51. Jambaque I, Cusmai R, Curatolo P, Cortesi F, Perrot C, Dulac O. Neuropsychological aspects of tuberous sclerosis in relation to epilepsy and MRI findings. Dev Med Child Neurol 1991;33:698–705.
52. Hunt A, Shepherd C. A prevalence study of autism in tuberous sclerosis. J Autism Dev Disord 1993;23:323–339.
53. Smalley SL. Autism and tuberous sclerosis. J Autism Dev Disord 1998;28:407–414.
54. Gillberg IC, Gillberg C, Ahlsen G. Autistic behaviour and attention deficits in tuberous sclerosis: a population-based study. Dev Med Child Neurol 1994;36:50–56.
55. Asano E, Chugani DC, Muzik O, et al. Autism in tuberous sclerosis complex is related to both cortical and subcortical dysfunction. Neurology 2001;57:1269–1277.
56. Taylor DC, Falconer MA, Bruton CJ, Corsellis JA. Focal dysplasia of the cerebral cortex in epilepsy. J Neurol Neurosurg Psychiatry 1971;34:369–387.
57. Palmini A, Andermann F, Olivier A, Tampieri D, Robitaille Y. Focal neuronal migration disorders and intractable partial epilepsy: results of surgical treatment. Ann Neurol 1991;30:750–757.
58. Kloss S, Pieper T, Pannek H, Holthausen H, Tuxhorn I. Epilepsy surgery in children with focal cortical dysplasia (FCD): results of long-term seizure outcome. Neuropediatrics 2002;33:21–26.
59. Kral T, Clusmann H, Blumcke I, et al. Outcome of epilepsy surgery in focal cortical dysplasia. J Neurol Neurosurg Psychiatry 2003;74: 183–188.
60. Hirabayashi S, Binnie CD, Janota I, Polkey CE. Surgical treatment of epilepsy due to cortical dysplasia: clinical and EEG findings. J Neurol Neurosurg Psychiatry 1993;56:765–770.
61. Siegel AM, Jobst BC, Thadani VM, et al. Medically intractable, localization-related epilepsy with normal MRI: presurgical evaluation and surgical outcome in 43 patients. Epilepsia 2001;42:883–888.
62. da Silva EA, Chugani DC, Muzik O, Chugani HT. Identification of frontal lobe epileptic foci in children using positron emission tomography. Epilepsia 1997;38:1198–1208.
63. Ryvlin P, Bouvard S, Le Bars D, et al. Clinical utility of flumazenil-PET versus [^{18}F]fluorodeoxyglucose-PET and MRI in refractory partial epilepsy. A prospective study in 100 patients. Brain 1998; 121:2067–2081.
64. Henry TR, Sutherling WW, Engel J Jr., et al. Interictal cerebral metabolism in partial epilepsies of neocortical origin. Epilepsy Res 1991;10:174–182.
65. Henry TR, Mazziotta JC, Engel J Jr. The functional anatomy of frontal lobe epilepsy studied with PET. Adv Neurol 1992;57:449–463.
66. Olsen RW, McCabe RT, Wamsley JK. GABAA receptor subtypes: autoradiographic comparison of GABA, benzodiazepine, and convulsant binding sites in the rat central nervous system. J Chem Neuroanat 1990;3:59–76.
67. Maziere M, Hantraye P, Prenant C, Sastre J, Comar D. Synthesis of ethyl 8-fluoro-5,6-dihydro-5-[11C]methyl-6-oxo-4H-imidazo [1,5-a] [1,4]benzodiazepine-3-carboxylate (RO 15.1788–11C): a specific radioligand for the in vivo study of central benzodiazepine receptors by positron emission tomography. Int J Appl Radiat Isot 1984;35: 973–976.
68. Savic I, Persson A, Roland P, Pauli S, Sedvall G, Widen L. In-vivo demonstration of reduced benzodiazepine receptor binding in human epileptic foci. Lancet 1988;2:863–866.
69. Savic I, Thorell JO, Roland P. [11C]flumazenil positron emission tomography visualizes frontal epileptogenic regions. Epilepsia 1995;36:1225–1232.
70. Richardson MP, Koepp MJ, Brooks DJ, Duncan JS. 11C-flumazenil PET in neocortical epilepsy. Neurology 1998;51:485–492. 71. Hammers A, Koepp MJ, Richardson MP, Hurlemann R, Brooks DJ, Duncan JS. Grey and white matter flumazenil binding in neocortical epilepsy with normal MRI. A PET study of 44 patients. Brain 2003; 126:1300–1318.
72. Muzik O, da Silva EA, Juhasz C, et al. Intracranial EEG versus flumazenil and glucose PET in children with extratemporal lobe epilepsy. Neurology 2000;54:171–179.
73. Arnold S, Berthele A, Drzezga A, et al. Reduction of benzodiazepine receptor binding is related to the seizure onset zone in extratemporal focal cortical dysplasia. Epilepsia 2000;41:818–824.
74. Richardson MP, Koepp MJ, Brooks DJ, Fish DR, Duncan JS. Benzodiazepine receptors in focal epilepsy with cortical dysgenesis: an 11C-flumazenil PET study. Ann Neurol 1996;40:188–198.
75. Sasaki M, Kuwabara Y, Yoshida T, et al. Carbon-11-methionine PET in focal cortical dysplasia: a comparison with fluorine-18-FDG PET and technetium-99m-ECD SPECT. J Nucl Med 1998; 39:974–977.
76. Rosa P, Andermann F, Natsume J, et al. PET studies using amino acids tracers: multimodal analysis. Epilepsia 2000;41(Suppl 7):58–59.
77. Toczek MT, Carson RE, Lang L, et al. PET imaging of 5-HT1A receptor binding in patients with temporal lobe epilepsy. Neurology 2003;60:749–756.
78. Hajek M, Antonini A, Leenders KL, Wieser HG. Epilepsia partialis continua studied by PET. Epilepsy Res 1991;9:44–48.
79. Sztriha L, Pavics L, Ambrus E. Epilepsia partialis continua: follow-up with 99mTc-HMPAO-SPECT. Neuropediatrics 1994;25:250–254.
80. Volkmann J, Seitz RJ, Muller-Gartner HW, Witte OW. Extrarolandic origin of spike and myoclonus activity in epilepsia partialis continua: a magnetoencephalographic and positron emission tomography study. J Neuroimaging 1998;8:103–106.
81. Sheth RD, Riggs JE. Persistent occipital electrographic status epilepticus. J Child Neurol 1999;14:334–336.
82. Lapin IP, Prakhie IB, Kiseleva IP. Excitatory effects of kynurenine and its metabolites, amino acids and convulsants administered into brain ventricles: differences between rats and mice. J Neural Transm 1982;54:229–238.
83. Perkins MN, Stone TW. An iontophoretic investigation of the actions of convulsant kynurenines and their interaction with the endogenous excitant quinolinic acid. Brain Res 1982;247:184–187.
84. Gusel WA, Mikhailov IB. Effect of tryptophan metabolites on activity of the epileptogenic focus in the frog hippocampus. J Neural Transm 1980;47:41–52.
85. Vezzani A, Stasi MA, Wu HQ, Castiglioni M, Weckermann B, Samanin R. Studies on the potential neurotoxic and convulsant effects of increased blood levels of quinolinic acid in rats with altered blood-brain barrier permeability. Exp Neurol 1989;106:90–98.
86. Chugani DC, Heyes MP, Kuhn DM, Chugani HT. Evidence that α-[C-11]methyl-L-tryptophan PET traces tryptophan metabolism via the kynurenine pathway in tuberous sclerosis complex. Soc Neurosci Abstr 1998;24:1757.

12 Pediatric Cortical Dysplasia

Positron Emission Tomography Studies

Bharathi Dasan Jagadeesan, MD, Csaba Juhász, MD, PhD, Diane C. Chugani, PhD, and Harry T. Chugani, MD

SUMMARY

Malformations of cortical development (MCDs) constitute an important cause of seizures in children and often form the underlying basis for many of the epilepsy syndromes found in childhood. Functional neuroimaging with positron emission tomography (PET) plays a critical role in the neuroimaging of children with epilepsy, including imaging of children with suspected or known MCDs such as cortical dysplasia. In this chapter, we begin with a focused discussion on the continuing relevance of well-established PET techniques such as 2-deoxy-2[^{18}F]fluoro-D-glucose (FDG) PET and [^{11}C]flumazenil (FMZ) PET in the light of newer advances in structural imaging techniques. We discuss the role of these techniques in the detection, classification, and prognostication of cortical dysplasias causing pediatric epilepsy and the electrophysiologic and clinical correlates of specific imaging findings with these techniques. The limitations of each of these techniques also are discussed in some detail, emphasizing the need for further refinements in PET imaging of cortical dysplasias. Thereafter, we discuss the exciting new developments in PET imaging of cortical dysplasias, especially the advent of α[^{11}C]-methyl-L-tryptophan (AMT) PET imaging in multifocal cortical dysplasias, tuberous sclerosis, and in imaging of patients with failed epilepsy surgery. Finally, we conclude the chapter with a brief discussion on other promising new tracers and their likely roles in PET imaging of cortical dysplasia in pediatric epilepsy.

Key Words: Cortical dysplasia; epilepsy; FDG PET; FMZ PET; dual pathology; infantile spasms.

1. INTRODUCTION

Cortical dysplasias account for the majority of brain developmental malformations associated with pediatric epilepsy. Cortical dysplasias encompass the entire range of neuronal migration disorders and constitute a condition known as malformations of cortical development (MCDs). On neuropathology, MCDs were found to have a spectrum of morphologic features associated with multiple putative etiologic features, including genetic and environmental influences. A newer classification of cortical dysplasias based on the primary step in neocortical development that is affected for each disorder is given in Table 1 *(1)*.

MCDs encompass lesions ranging from those that are very easily evident on structural neuroimaging, such as schizencephaly and bilateral diffuse polymicrogyria, to those that may be detected only on histological sections such as microscopic focal cortical dysplasias and excessive single white matter neurons. The role of functional neuroimaging in children with cortical dysplasias is threefold: (1) To enable the detection of focal or multifocal neuropathology in children with epilepsy caused by cortical dysplastic lesions and normal structural imaging. This will aid in disease detection and further management including surgical intervention in these children. (2) To enable further subclassification and phenotyping of epileptic syndromes with multifactorial etiology such as infantile spasms and Lennox–Gastaut syndrome, with the purpose of tailoring treatment according to focal or multifocal etiologies in individual patients. (3) To study the functional correlates of lesions irrespective of whether or not they are visible on structural neuroimaging. This involves the documentation of functional abnormalities extending beyond structural lesions including perilesional and remote functional cortical and subcortical abnormalities and dual pathology. These remote abnormalities may denote emerging or established secondary epileptogenic foci. Thus, functional imaging can contribute to a deeper understanding of the mechanisms of epileptogenesis, seizure spread, secondary neuronal injury, and also cognitive abnormalities found in patients with different cortical malformations.

Positron emission tomography (PET) is a functional neuroimaging technique that has revolutionized the management of some disorders associated with cortical dysplasias, such as infantile spasms, and significantly enhanced our ability

From: *Bioimaging in Neurodegeneration*
Edited by P. A. Broderick, D. N. Rahni, and E. H. Kolodny

Table 1
Classification System for MCDs

I. Malformations due to abnormal neuronal and glial proliferation.
  A. Generalized
    1. Decreased proliferation (microlissencephaly): Microcephaly with simplified gyral pattern or microlissencephaly with thin cortex, microlissencephaly with thick cortex.
    2. Increased proliferation/abnormal proliferation: (none known)
  B. Focal or multifocal.
    1. Decreased proliferation (none known),
    2. Increased and abnormal proliferation (megalencephaly and hemimegalencephaly),
    3. Abnormal proliferation:
      a. Nonneoplastic: focal cortical dysplasia,
      b. Neoplastic (but associated with disordered cortex).
II. Malformations caused by abnormal neuronal migration
  A. Generalized (e.g., Classical lissencephaly [type 1], subcortical band heterotopia [agyria ± pachygyria-band spectrum], Cobblestone dysplasia [type 2 lissencep]), Lissencephaly: other types, Heterotopia.)
  B. Focal or multifocal malformations of neuronal migration (e.g., Focal or multifocal heterotopia, Focal or multifocal heterotopia with organizational abnormality of the cortex, Excessive single ectopic white matter neurons.)
III. Malformations caused by abnormal cortical organization.
  A. Generalized: bilateral diffuse polymicrogyria.
  B. Focal or multifocal (e.g., Bilateral partial polymicrogyria, schizencephaly, focal or multifocal cortical dysplasia [no balloon cells], microdysgenesis.)
IV. Malformations of cortical development, not otherwise classified.

to detect cortical dysplasias in patients with epilepsy while deepening our understanding of the pathophysiology of this group of developmental malformations of the brain. PET allows the noninvasive, in vivo assessment and quantification of diverse biochemical processes in the brain (both globally and regionally) such as glucose consumption, oxygen consumption, neurotransmitter release, postsynaptic receptor binding, and neurotransmitter transport. Furthermore, PET allows the assessment of such processes under various physiologic states, including at rest and after neuronal activation states. More importantly, it also enables the detection of quantitative alterations in biochemical processes in various disease. These unique features of PET have made it an invaluable tool in the study of the diverse conditions using different tracers that target specific biochemical abnormalities. In the present review, we address the role of PET in cortical dysplasia and epilepsy starting with the most widely used tracer, 2-deoxy-2[^{18}F]fluoro-D-glucose (FDG).

2. FDG PET IN CORTICAL DYSPLASIA

In general, the abnormalities found on interictal FDG PET studies of patients with cortical dysplasia consist of focal or multifocal areas of decreased FDG uptake (Fig. 1; refs. *2–5*). Decreased FDG uptake can reflect diminished neuronal activity, resulting in decreased glucose use in regions of cortical dysplasia, which may not always be macroscopically evident. However, ectopic areas of dysplastic cortical tissue, which are found in the subcortical white matter, typically appear as areas of increased FDG uptake located beneath the overlying cortical

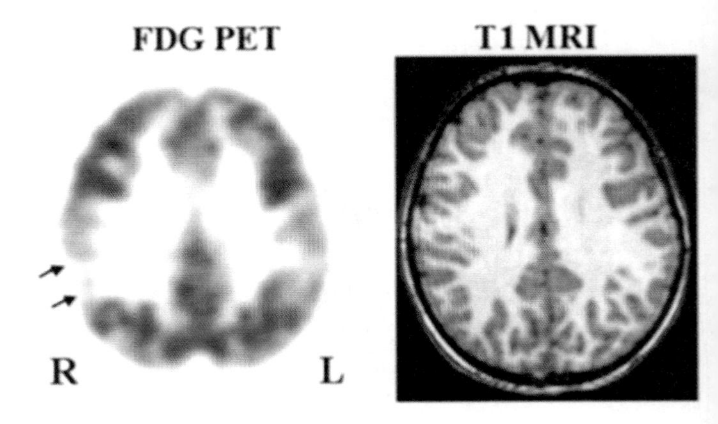

Fig. 1. FDG PET image showing focal hypometabolism in a patient with normal structural MRI. (Reproduced with permission from: Juhasz C, Chugani HT. Imaging the epileptic brain with PET. Neuroimaging Clin North Am 2003;13:705–716.)

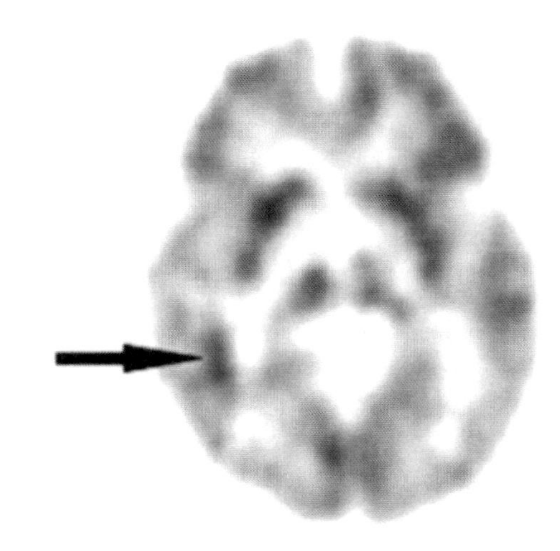

Fig. 2. FDG PET showing focally increased tracer uptake (black arrow) over heterotopic cortical malformation with hypometabolism in the overlying cortex. This malformation was not detected initially on the MRI.

activity (Fig. 2; refs. *6,7*). Focal increases in FDG uptake may also be present in otherwise hypometabolic cortical areas if clinical or subclinical electroencephalographic seizures occur during the period of FDG uptake into the brain after tracer injection and may also be found when there is frequent interictal spiking *(8,9)*. Thus, all patients with epilepsy should be subjected to scalp EEG recording during the period of FDG uptake, which lasts up to 60 min after injection and FDG PET findings must always be interpreted in correlation with the EEG data thus obtained.

2.1. FDG PET IN "NONLESIONAL" NEOCORTICAL EPILEPSY CAUSED BY MICROSCOPIC CORTICAL DYSPLASIAS

Structural neuroimaging with magnetic resonance imaging (MRI) continues to be paradoxically normal in a significant number of patients with cortical developmental malformations despite the rapid advances in MRI image acquisition and analysis over the past decade *(10–12)*. Patients with such normal

MRI studies who present with partial seizures with or without secondary generalization are considered to have nonlesional or cryptogenic partial epilepsy. However, it has been shown that successful seizure outcome is possible in many of these patients when epileptogenic cortex is resected following invasive monitoring with subdural grid EEG or depth electrodes *(13)*. FDG PET has been extensively studied in this group of patients and has proved to be of immense value by: (1) serving as an alternative to invasive EEG monitoring when it reveals highly circumscribed focal cortical abnormalities that are in agreement with ictal scalp EEG findings and seizure semiology, especially in very young children where subdural EEG grid placement should be avoided if possible; (2) guiding preoperative subdural grid placement for EEG monitoring in the localization of epileptogenic cortex when PET abnormalities are clearly lateralized to a single hemisphere and localized to a relatively confined region of the cortical surface (typically encompassing one or two lobes); and (3) helping to avoid the need for invasive monitoring when ictal scalp EEG monitoring does not reveal highly focal onset of seizures and PET suggests multifocal abnormalities involving both hemispheres or multiple lobes of the brain. These applications of FDG PET are strongly suggested by the studies of Hong et al. *(14)*, who reported on 41 patients with nonlesional epilepsy using FDG PET, scalp ictal EEG, and ictal single-photon emission computed tomography. All patients underwent surgical intervention, and FDG PET was found to have a sensitivity of 42.9% (31.3% for extratemporal epilepsy and 70% for temporal neocortical epilepsy) for localizing the epileptic focus when compared with findings from subdural grid placement. Good surgical outcome was found in 33 patients, including cessation of seizures in 16. The authors suggested that FDG PET is a useful diagnostic tool in the localization of epileptic foci in patients with nonlesional neocortical epilepsy and that FDG PET findings and ictal EEG findings provide complementary and independent information in the localization of epileptogenic brain regions. Similarly, Won et al. *(10)* compared MRI with FDG PET and ictal single-photon emission computed tomography in 118 patients with intractable partial epilepsy, including 26 with extratemporal lobe epilepsy. They found that in patients with extratemporal lobe epilepsy, MRI findings agreed with those of pathologic findings in 50%, whereas FDG PET was concordant in 71%. In their study, of the 16 patients in whom MRI was completely normal but FDG PET was abnormal, four were found to have cortical dysplasia, nine were found to have microdysgenesis, and one was found to have focal neuronal loss.

Neocortical dysplasias also constitute an important cause of childhood temporal lobe epilepsy. Porter et al. *(15)* reviewed the neuropathologic features of temporal lobectomy specimens from 33 pediatric patients with refractory temporal lobe epilepsy and found evidence for cortical dysplasia in 64% of the patients. It has been shown that FDG PET can reliably localize neocortical dysplastic areas in patients with nonlesional temporal lobe epilepsy *(16)*. However, temporal lobe neocortical dysplasias may not only occur alone but may coexist with hippocampal sclerosis in affected children *(17,18)*; MRI may reveal only the coexisting hippocampal atrophy in such cases. Diehl et al. *(19)* studied FDG uptake patterns in 23 patients with temporal lobe epilepsy and hippocampal sclerosis, including 13 with coexisting cortical dysplasia, which was not evident on preoperative MRI *(19)*. They found that patients with cortical dysplasia and hippocampal sclerosis had a different pattern of glucose hypometabolism than patients with hippocampal sclerosis alone; the former have their greatest decreases in FDG uptake in the lateral temporal lobe, whereas the latter show maximum decreases in the anterior temporal lobe. Thus, FDG PET may aid in the localization of epileptogenic dysplastic cortical regions in patients with temporal and extratemporal refractory partial epilepsies and normal MRI studies.

2.2. FDG PET IN CHILDHOOD EPILEPSY SYNDROMES

Epileptic syndromes are defined on the basis of the clustering together of particular clinical findings with characteristic EEG changes in patients afflicted with these conditions. The clinical characteristics could include the type of seizures, cognitive abnormalities, and predominance of seizures in particular age groups. However, a direct correlation between clinical epilepsy syndromes and underlying neuropathology may not always be present. Syndromes such as infantile spasms and Lennox–Gastaut syndrome have been noted to have a broad spectrum of underlying neuropathology associated with them *(20–22)*. In these syndromes, the amenability of individual patients to surgical intervention and the need for better prognostication dictate the need for accurate localization of underlying neuropathology by imaging. Structural imaging studies seldom meet these requirements completely and functional neuroimaging with FDG PET has greatly filled this gap in our ability to understand and treat these patients.

2.2.1. West Syndrome

This syndrome is characterized by the triad of infantile tonic/myoclonic spasms, hypsarrhythmia on EEG, and developmental arrest; this syndrome manifests in early infancy and peaks between 4 and 6 mo of age *(23)*. PET studies have dramatically altered the management of infantile spasms, which were considered to be generalized seizures resulting from complex corticosubcortical interactions. Using FDG PET, it has been found that most infants diagnosed with "cryptogenic" spasms have, in fact, focal or multifocal cortical regions of decreased (or, occasionally increased) glucose use that are often consistent with areas of cortical dysplasia missed by MRI (Fig. 3; refs. *24,25*). Most importantly, it has been shown that when a single region of abnormal glucose use is apparent on PET, corresponding to the EEG focus, surgical removal of the PET focus results not only in control of otherwise intractable seizures but also in complete or partial reversal of the associated developmental delay. When bilateral multifocal areas of hypometabolism are found, if all seizures arise from a single area, resective surgery may ameliorate the epilepsy, but there may be no significant improvement in cognitive development. However, most patients with bilateral multifocal lesions are not surgical candidates because no single focus of seizure origin can be found by EEG monitoring. When a generalized and symmetric hypometabolic pattern is found on FDG PET, then neurometabolic or neurogenetic disorders are likely to constitute the underlying pathology and lesional etiology is less likely (Fig. 4). Hence the pattern of glucose hypometabolism on FDG PET not only enables evaluation for surgery but also enables a

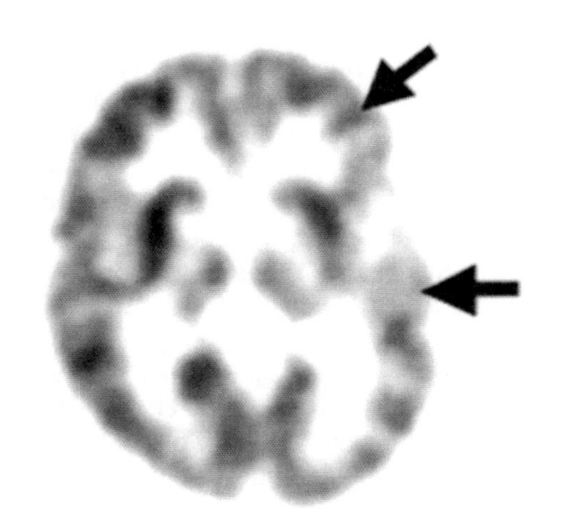

Fig. 3. FDG PET study in a child with infantile spasms showing focal cortical hypometabolism (black arrows) in the left frontal-temporal cortex corresponding to an area of cortical dysplasia.

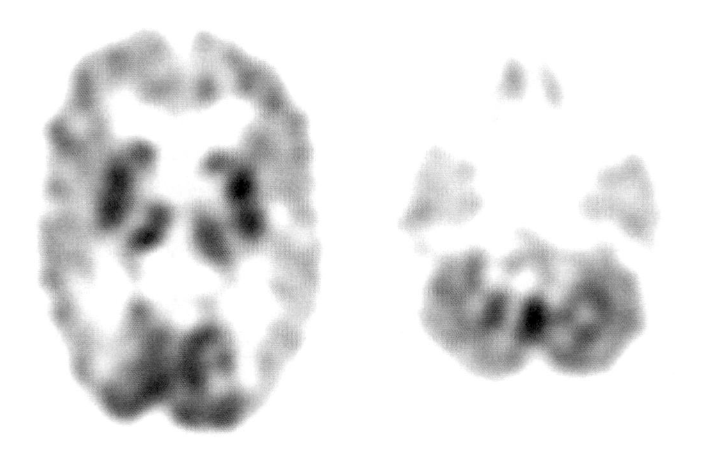

Fig. 4. FDG PET study showing diffuse cerebral cortical and cerebellar hypometabolism in a child with infantile spasms. The calcarine cortex appears to be relatively spared. This child was later diagnosed with a metabolic disorder (carbohydrate-deficient glycoprotein syndrome).

better understanding of the underlying pathophysiology in individual children with West syndrome.

In a recent study of 10 children with symptomatic and 7 children with cryptogenic infantile spasms, most children with favorable outcome had transient focal cortical hypometabolism at the onset of the spasms, which reduced in extent or disappeared when the FDG PET scan was repeated after a year *(26)*. These transient hypometabolic lesions were found to be equally common in patients with cryptogenic and symptomatic etiology. Furthermore, it was found that disappearance of occipital hypometabolic foci was especially common. It is, therefore, likely that in these children regional hypometabolism represents malfunctioning cortex rather than structural lesions and hence the surgical approach to children with infantile spasms and extensive cortical hypometabolism may need to be carefully addressed using intracranial electrodes. FDG PET studies can also show hypermetabolism in the lenticular nuclei and in the brainstem area, suggesting that these structures may be associated with the basic mechanisms of infantile spasms *(27)*.

2.2.2. Lennox–Gastaut Syndrome

Children with Lennox–Gastaut syndrome have a triad consisting of multiple seizure types, developmental delay, and 1- to 2.5-Hz generalized "slow" spike and wave EEG pattern. Using FDG PET, four metabolic subtypes have been identified in this disorder: unilateral focal, unilateral diffuse and bilateral diffuse hypometabolism, and normal patterns *(28,29)*. Focal cortical hypometabolism may be related to cortical dysplasia *(30)*, and patients with the unilateral focal and unilateral diffuse patterns may be occasionally considered for cortical resection when ictal EEG findings are concordant with the area of hypometabolism.

2.3. FUNCTIONAL CORRELATES OF FDG PET ABNORMALITIES IN PATIENTS WITH CORTICAL DYSPLASIA

Comparison of MRI and FDG PET findings in patients with cortical dysplasia in whom a clear structural abnormality is depicted on MRI reveals that the abnormalities in FDG PET often extend well beyond the spatial limits of the structural lesion. The extent of perilesional glucose metabolism found in these patients with intractable lesional epilepsy has been shown to have a correlation with the lifetime number of seizures: the greater the lifetime number of seizures, the more extensive the area of FDG hypometabolism *(31)*. On the other hand, focal FDG PET abnormalities are uncommon in children with new onset partial epilepsy (after the third unprovoked seizure; ref. *32*). Further, repeated seizures result in cortical glucose hypometabolism remote from the EEG focus, a finding supported by studies showing recovery of metabolism in remote brain regions after successful elimination of the primary focus *(33,34)*. These findings altogether suggest that interictal cortical hypometabolism is a dynamic functional change, and a significant part of it is likely to be the consequence rather than cause of recurring seizures. An understanding of the exact mechanisms underlying such changes in FDG uptake may enhance our knowledge on the mechanisms of secondary epileptogenesis and cognitive decline in patients with epilepsy. Alternatively, remote abnormalities may represent dual pathology and guide the placement of subdural electrodes when seizures are accompanied by multiple sites of origin in ictal scalp EEG.

Another interesting finding to emerge out of the comparison of MRI and FDG PET findings is the phenomenon of varying patterns of metabolic abnormalities found in anatomically similar cortical lesions. For example, in patients with perisylvian dysgenesis, the dysplastic cortex was found to have increased, decreased, or a heterogenous pattern of interictal FDG uptake *(35)*. It has been suggested that such variations may reflect differences in the timing of brain injury in these patients and, also variations in synaptic density and cytoarchitecture in the presence of similar macroscopic appearance. The implication of these variations for patient management and prognostication are yet to be addressed fully, but are likely to be significant in at least a subset of patients with lesional epilepsy.

The extensive and nonspecific nature of FDG PET abnormalities in many patients with extratemporal cortical dysplasia has led to the search for other, more specific tracers, which can clearly differentiate those abnormal brain areas that are the primary cause for epilepsy from those areas that show changes in function as a consequence of epilepsy.

3. GABA RECEPTOR IMAGING WITH FLUMAZENIL (FMZ) PET IN CHILDREN WITH CORTICAL DYSPLASIA AND EPILEPSY

γ-Aminobutyric acid (GABA) is the major inhibitory neurotransmitter in the human brain and GABAergic mechanisms play a key role in regulating central nervous system excitability and susceptibility to seizures *(36)*. The action of GABA is mediated in part by the $GABA_A$ receptor complex. $[^{11}C]$flumazenil (FMZ) is a benzodiazepine antagonist that binds to the α subunit of the $GABA_A$ receptor. Hence, PET imaging of $GABA_A$ receptor binding can be performed using the tracer FMZ. Patients undergoing FMZ PET should not take drugs (such as benzodiazepines) that directly interact with the $GABA_A$ receptors. Among antiepileptic medications, chronic vigabatrin treatment has been shown to be associated with decreased regional $GABA_A$ receptor binding in young children with partial seizures or infantile spasms *(37)*; however, similar effects have not been reported in adults whose $GABA_A$ binding capacity is generally lower than that of children under normal circumstances *(38)*. Indeed, age-related changes in FMZ binding have been reported in humans and, in normal adult volunteers, baseline FMZ binding values are 25–50% lower (depending on the brain region) than those in children approx 2 yr of age (measured in the nonepileptic hemisphere of children with epilepsy).

The utility of FMZ PET in cortical dysplasias can again be studied in terms of its utility in (1) detection of primary (and secondary) epileptic foci in patients with partial epilepsies and normal structural imaging, (2) study of epilepsy syndromes with a view to improve subclassification and prognostication; and (3) study of the clinical and electrophysiologic correlations of FMZ PET abnormalities regardless of the visualization or nonvisualization of structural brain malformations on MRI.

3.1. FMZ PET IN THE DETECTION OF EPILEPTIC FOCI

Initial observations suggested that FMZ PET may be more sensitive and specific than FDG PET in the detection of epileptic foci in temporal lobe epilepsy; these observations were followed by similar findings on the utility of FMZ PET in frontal lobe epilepsy *(39)*. This initiated the use of FMZ PET in patients with cortical dysplasias and intractable epilepsy. Ryvlin et al. *(40)* compared the utility of structural MRI, FDG PET, and FMZ PET in 100 patients with intractable partial seizures, including 34 with normal MRI, and found that FMZ PET outperformed FDG PET in the detection of cortical epileptogenic foci in extratemporal lobe neocortical epilepsies with normal structural imaging, including neocortical epilepsies due to focal cortical dysplasias. In a subsequent study, Muzik et al. *(41)* compared the performance of FMZ and FDG PET with each other in the detection of neocortical epileptogenic foci in children with extratemporal lobe epilepsy using the gold standard of intracranial EEG monitoring. They found that FMZ PET is significantly more sensitive than FDG PET for the detection of cortical regions of seizure onset and frequent spiking. Therefore, the application of PET, in particular FMZ PET, in guiding subdural electrode placement in refractory extratemporal lobe epilepsy can enhance coverage of the epileptogenic zone. The exact nature of the change in FMZ uptake likely depends on the nature of the underlying cortical malformation since either decreased or increased regional FMZ binding has been shown in epileptogenic dysplastic cortical regions *(42)*. The objective technique of statistical parametric mapping has been used in the analysis of FMZ PET images for localization of epileptogenic foci, but further studies comparing SPM analysis results with intracranial EEG and surgical outcome are required *(43)*. Interestingly, a recent study suggests that in patients with cortical dysplasias undergoing FMZ PET studies, abnormal periventricular FMZ binding may be found deep in the white matter *(44)*. These findings may reflect disordered neuronal migration, which accounts for cortical dysplasia. However, further studies with detailed histological comparisons are required to substantiate this hypothesis.

3.2. FMZ PET AND DUAL PATHOLOGY

Dual pathology refers to the coexistence of hippocampal sclerosis and an epileptogenic neocortical lesion *(45)*, and this condition is present in 15–35% of patients with intractable epilepsy, including those with a history of infantile spasms *(46–48)*. Unsuspected dual pathology is considered one of the major factors responsible for poor surgical outcome in such patients. Our group has previously reported that visual as well as quantitative analysis of FMZ PET sensitively detected decreased benzodiazepine receptor binding in the hippocampus as well as the neocortex (often consistent with subtle cortical dysplasia) in patients with dual pathology *(49)*. Thus, in addition to its role in detecting epileptogenci foci in children with normal MRI and cortical dysplasia, FMZ PET can also play a useful role in the detection of additional neocortical lesions in patients with temporal lobe epilepsy due to hippocampal sclerosis (Fig. 5).

3.3. FMZ PET IN CHILDHOOD EPILEPSY SYNDROMES

In children with infantile spasms associated with cortical dysplasia, FMZ PET can indicate regions of decreased (or increased) FMZ binding consistent with subtle cortical abnormalities not detected by MRI (Fig. 6; ref. *50*). A potential advantage of using FMZ PET in small children is the high regional FMZ binding that allows better visualization of some brain regions such as the medial temporal lobe *(51)*. The exact role of FMZ PET as compared to FDG PET in children with West syndrome, however, requires further study to determine whether FMZ PET offers additional localizing information. Data on subcortical abnormalities in $GABA_A$ binding in children with West syndrome are as yet unavailable but may be potentially interesting in view of the documented abnormalities in lenticular nuclei of children with West syndrome using FDG PET and the occurrence of thalamic and other subcortical abnormalities in adults with temporal lobe epilepsy who undergo FMZ PET imaging *(52,53)*.

3.4. CLINICAL AND ELECTROPHYSIOLOGICAL CORRELATES OF FMZ PET ABNORMALITIES

In patients with extratemporal epilepsy, FMZ PET abnormalities are found to be less extensive than FDG PET abnormalities and also more specific in the delineation of cortical

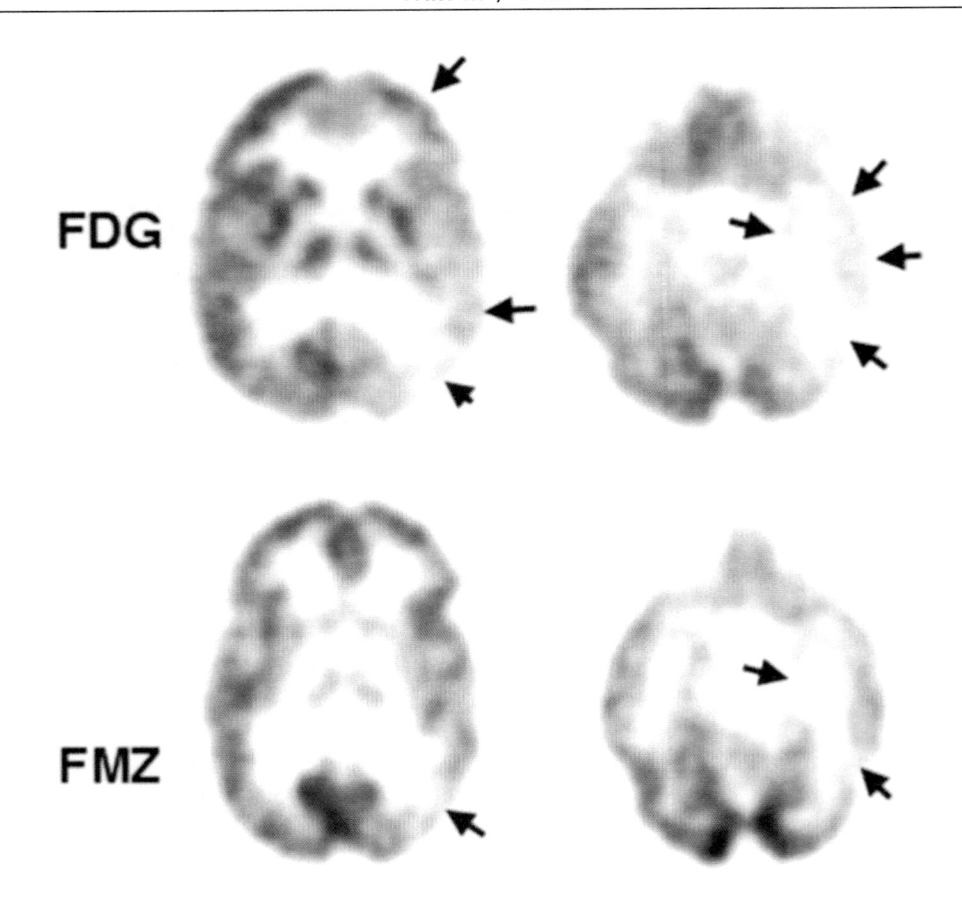

Fig. 5. FDG PET (top row) and FMZ PET (bottom row) images from a child with intractable epilepsy having dual pathology (i.e., hippocampal in addition to neocortical involvement). Note that the abnormalities noted on FDG PET are more extensive than those noted in the FMZ PET study. (Reproduced with permission from: Juhasz C, Chugani HT. Imaging the epileptic brain with PET. Neuroimaging Clin North Am 2003;13:705–716.)

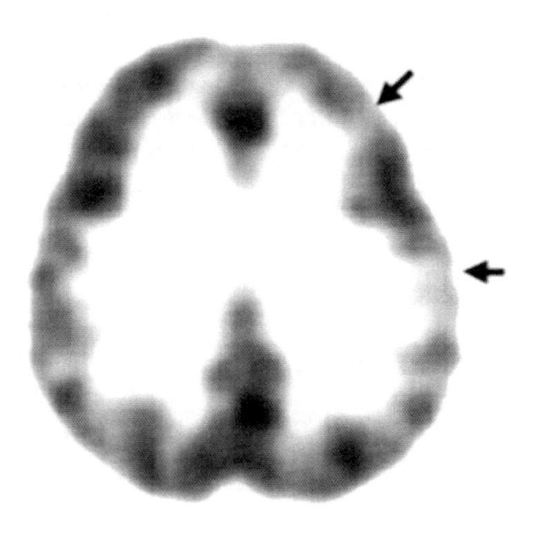

Fig. 6. FMZ PET study of a child with infantile spasms showing focal regions of decreased $GABA_A$ receptor binding in the left frontal and parietal cortex.

areas of seizure onset and frequent interictal spiking as defined by intracranial EEG (Fig. 5; *54,55*). A recent study has demonstrated that the extent of nonresected areas with decreased FMZ binding in the lobe of seizure onset is a good predictor of seizure outcome after cortical resection, that is, poor surgical outcome is associated with nonresection of cortex showing decreased flumazenil binding in the lobe of seizure onset as shown on EEG *(56)*. Thus, FMZ PET is an excellent complementary imaging method in patients with normal MRI and extensive glucose hypometabolism on FDG PET.

In children with epilepsy and a readily apparent lesion on structural imaging, that is, lesional neocortical epilepsy, it has been demonstrated that decreased FMZ binding present in the lesion frequently extended into the adjacent cortex *(57)*. Furthermore, these perilesional FMZ PET defects often showed an eccentric and not concentric distribution *(57)*. Subdural EEG recordings after grid placement showed that the ictal onset almost always involved the perilesional FMZ abnormality (perilesional epileptogenicity) and that electrodes on the MRI defined structural lesion were often electrographically silent *(57,58)*. Resection of the perilesional FMZ abnormalities with spiking resulted in excellent seizure-free outcome. The nature of these perilesional abnormalities also differs from those found with FDG PET in that their extent appears to be independent of seizure number prior to the scan.

In addition to the occurrence of perilesional abnormalities, remote FMZ PET abnormalities can occur outside the lobe of seizure onset *(56–60)*. These remote FMZ PET abnormalities are associated with an early age at seizure onset and often appear in ipsilateral, synaptically connected regions from the lesion area. These remote cortical regions are commonly involved in early seizure propagation *(57,60)*. Based on these findings, we hypothesize that some remote FMZ PET abnormalities may represent secondary epileptic foci, although it is yet to be determined whether dependent secondary epileptic foci can be distinguished from independent secondary epileptic foci by neuroimaging. However, the finding of remote cortical abnormalities, if present, leads to a decrease in the specificity of FMZ PET abnormalities for the localization of the primary epileptogenic region since these remote areas are not always epileptogenic and, therefore, should not be resected without EEG evidence of ictal involvement. Limited specificity of FDG, and, to a lesser degree, FMZ PET has fueled the search for newer tracers that will be even more specific in their ability to identify epileptic cortical regions in children with epilepsy and eliminate the need for invasive intracranial EEG monitoring. Newer tracers are also needed to expand our knowledge of the mechanisms of seizure spread, secondary generalization and secondary epileptogenesis in patients with epilepsy.

4. NEWER PET TRACERS IN THE STUDY OF CORTICAL DYSPLASIA

4.1. SEROTONIN IMAGING IN CORTICAL DYSPLASIA

Imaging serotonin metabolism in epileptic patients has developed along two different directions, (1) the imaging of tryptophan metabolism along the serotonin pathway and (2) the imaging of serotonin receptor binding.

4.1.1. Imaging Tryptophan Metabolism Along the Serotonin Pathway

Recently, the tracer α[^{11}C]-methyl-L-tryptophan (AMT) has emerged as an important tool in the study of abnormalities in serotonergic neurotransmission in epileptic patients. In particular, AMT PET has emerged as the best currently available method for the localization of epileptogenic tuber(s) in patients with tuberous sclerosis and intractable seizures *(61,62)*. Unlike most other tracers, which show predominantly decreased uptake in epileptogenic cortical areas interictally, AMT PET scanning reveals locally increased uptake of AMT in and/or around epileptogenic tubers while showing normal or decreased uptake in nonepileptogenic tubers. Studies of resected epileptic tissues have suggested that the likely mechanism for this increased AMT uptake is the accumulation of tryptophan metabolites of the kynurenine pathway *(63)*. Local increases in tryptophan metabolism through this pathway can occur due to immune activation of the rate-limiting enzyme indoleamine-2,3-dioxygenase, and results in the production of the excitotoxic and convulsant *N*-methyl-D-aspartate (NMDA) agonist quinolinic acid. Hence, the increased AMT uptake found in epileptogenic cortical lesions may reflect increased quinolinic acid production rather than increased serotonin synthesis.

Fedi et al. *(64)* used interictal AMT PET to study 18 patients with neocortical epilepsy, including 7 with cortical dysplasia identified on structural MRI and 11 with partial epilepsy who had normal MRI and FDG-PET studies *(64)*. They found increased AMT uptake corresponding to the epileptogenic area in seven patients (four with MRI evident cortical dysplasia and three with cryptogenic partial epilepsy). Furthermore, they found that the AMT uptake in the epileptic focus correlated with the frequency of interictal spikes on scalp EEG. Subsequently, we studied 27 children with extratemporal lobe epilepsy, including 7 with cortical dysplasia using AMT PET, FDG PET, structural MRI and prolonged ictal and interictal scalp EEG monitoring in all patients, including subdural grid monitoring and postoperative histopathological examination of resected tissue *(65)*. Focal cortical increase of AMT uptake was objectively identified in 15 patients (55%). Occurrence of increased AMT uptake did not correlate with interictal spike frequency in this latter study. Of the three patients with normal MRI and histologically proven cortical dysplasia, AMT PET identified the lesion in two, and FDG PET in two. In those patients with structurally evident cortical dysplasia, FDG abnormalities were found to be widespread and nonspecific in all the cases but one, whereas AMT PET localized specifically the epileptogenic dysplastic region correctly. Overall, we found that although AMT PET may be less sensitive than FDG PET in children with extratemporal neocortical epilepsies, it is considerably more specific for the epileptogenic region. Additionally, the finding of focally increased AMT uptake in cortical regions, in the absence of abnormalities in structural MRI or FDG PET, likely indicates underlying cortical malformations (Fig. 7). Given its highly specific nature, this focal uptake may guide preoperative grid placement for EEG monitoring or perhaps eliminate the need for such monitoring if it is highly concordant with ictal scalp EEG and MRI findings. In addition, we found AMT PET to be a useful imaging modality to localize nonresected epileptic cortex in patients with a previously failed cortical resection, including those with focal cortical dysplasia (Fig. 8). This is particularly important since other tracers such as FDG or FMZ are rarely useful in such a clinical setting *(66)*. Therefore, further evaluation of this promising new tracer as an alternative to invasive EEG monitoring in children with known or suspected cortical dysplasia is highly warranted. In the study of the role of serotonergic pathways in epilepsy, alternatives to AMT PET are beginning to emerge in the form of tracers that can be used to image serotonin receptor binding in the brain.

4.1.2. Serotonin Receptor Imaging in Cortical Dysplasia

PET imaging studies using the tracer [^{18}F]trans-4-fluoro-*N*-2-[4-(2-methoxyphenyl)piperazin-1-yl]ethyl]-*N*-(2-pyridyl) cyclo-hexane-carboxamide [^{18}F]FCWAY, a selective 5-HT$_{1A}$ receptor antagonist, have found cortical and subcortical (including raphe nuclei) changes of 5-HT$_{1A}$ receptors in patients with developmental malformations of the neocortex. The cortical abnormalities in tracer uptake have been found in areas corresponding to the anatomically malformed area as well as other anatomically normal areas *(67)*. The abnormalities in tracer binding in subcortical regions are especially interesting since they are consistent with a role of the serotonergic system in the epileptic network to include subcortical structures. The ^{11}C analog of [^{18}F]FCWAY, [carbonyl-^{11}C]WAY-100635 (^{11}C-WAY), is also used for studying serotonin receptor binding in the brain *(68)*. A tracer of particular interest with possible applications in epilepsy is the putative metabolite of ^{11}C-WAY and selective radioligand [carbonyl-^{11}C]desmethyl-WAY-100635 (^{11}C-DWAY), which has been shown to have higher brain uptake and radioactive signal than ^{11}C-WAY(PPP).

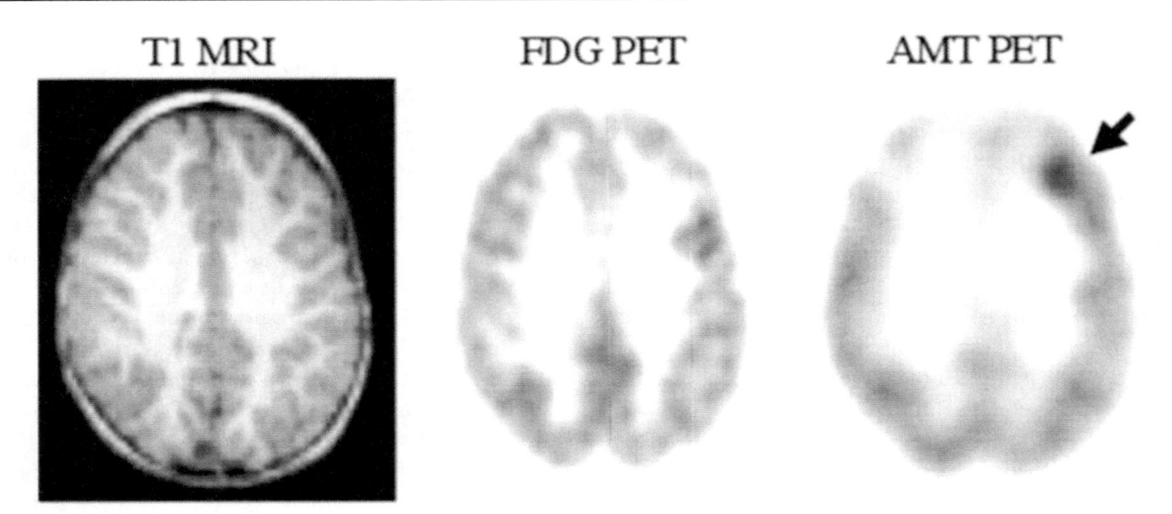

Fig. 7. AMT PET study showing a well-defined focus of increased AMT uptake corresponding to a region of focal cortical dysplasia, which could not be detected on structural MR imaging (until after the PET scans) and which showed up as an area of mildly decreased glucose uptake on the FDG PET study.

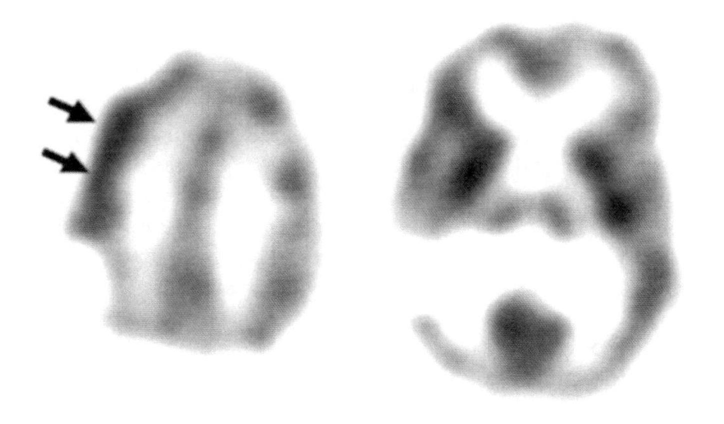

Fig. 8. AMT PET study in a child with recurrent seizures following a right parietal-temporal resection. The AMT PET images indicate nonresected epileptogenic cortex, which can be seen as an area of increased tracer uptake, in the right frontal cortex. (Reproduced with permission from: Juhasz C, et al. Evaluation with α-[11C]methyl-L-tryptophan positron emission tomography for reoperation after failed epilepsy surgery. Epilepsia 2004;45:124–130.)

4.2. OTHER TRACERS IN CORTICAL DYSPLASIA AND EPILEPSY

Increased uptake of [11C]methionine has been shown in a few cases of histologically verified epileptogenic cortical dysplasias *(69,70)* where FDG PET showed decreased glucose metabolism. The exact mechanism for increased uptake of methionine is as yet undetermined, although it may be to the result of ongoing epileptic activity, disruption of the blood–brain barrier in regions showing increased uptake, or a subacute gliotic reaction *(71)*. Increased uptake may also be the result of an immature pattern of amino acid transport *(72)* in developmentally abnormal dysplastic tissue.

Other tracers that have been applied and may be potentially useful in detecting epileptogenic dysplastic cortex include [11C]carfentanil for mu-opiate receptors *(73)* and [11C] doxepin *(74)* for histamine H_1 receptors, but these have to be tested on a larger number of patients to determine their utility against that of well established imaging techniques such as MRI and FDG PET.

REFERENCES

1. Barkovich AJ, Kuzniecky RI, Dobyns WB, Jackson GD, Becker LE, Evrard P. A classification scheme for malformations of cortical development. Neuropediatrics 1996;27:59–63.
2. Williamson PD, Van Ness PC, Wieser HG, Quesney LF. Surgically remediable extratemporal syndromes. In: Engel J. Jr. ed. Surgical Treatment of the Epilepsies. New York: Raven Press; 1993:65–76.
3. Williamson PD. Frontal lobe epilepsy: some clinical characteristics. In: Jasper HH, Riggio S, Goldman-Rakic, PS, eds. Epilepsy and the Functional Anatomy of the Frontal Lobe. New York: Raven Press; 1995:127–152.
4. Swartz BE; Khonsari A; Brown C, Mandelkern M, Simpkins F, Krisdakumtorn T. Improved sensitivity of 18FDG-positron emission tomography scans in frontal and 'frontal plus' epilepsy. Epilepsia 1995;36:388–395.
5. Casse R, Rowe CC, Newton M, Berlangieri SU, Scott AM. Positron emission tomography and epilepsy. Mol Imaging Biol 2002;4:338–351.
6. Rintahaka PJ, Chugani HT, Mess C, et al. Hemimegalencephaly: Evaluation with positron emission tomography. Pediatr Neurol 1993;9:1–28.
7. Volder AG, Gadisseux JF, Michel CJ, Maloteaux JM, Bol AC, Grandin CB, et al. Brain glucose utilisation in band heterotopia:synaptic activity of "double cortex." Pediatr Neurol 1994; 11:290–294.
8. Engel J Jr, Kuhl DE, Phelps ME. Patterns of human local cerebral glucose metabolism during epileptic seizures. Science 1982;218:64–66.
9. Engel J Jr, Kuhl DE, Phelps ME, Mazziotta JC. Interictal cerebral glucose metabolism in partial epilepsy and its relation to EEG changes. Ann Neurol 1982;12:510–517.
10. Won JW, Chang KH, Cheon JE, et al. Comparison of MR imaging with PET and ictal SPECT in 118 patients with intractable epilepsy. Am J Neuroradiol 1999;20:593–599.
11. Russo GL, Tassi L, Cossu M, et al. Focal cortical resection in malformations of cortical development. Epileptic Disord 2003;5(suppl 2):S115–23.
12. Oertzen VJ, Urbach H, Junbluth S, et al. Standard magnetic resonance imaging is inadequate for patients with refractory focal epilepsy. J Neurol Neurosurg Psychiatry 2002;73:643–647.

13. Siegel Am, Jobst BC, Thadani VM, Rhodes CH, Lewis PJ, Roberts DW, Williamson PD. Medically intractable, localization-related epilepsy with normal MRI: presurgical evaluation and surgical outcome in 43 patients. Epilepsia 2001;42:833–838.
14. Hong KS, Lee SK, Kim JY, Lee DS, Chung CK. Pre-surgical evaluation and surgical outcome of 41 patients with non-lesional neocortical epilepsy. Seizure 2002;11:184–192.
15. Porter BE, Judkins AR, Clancy RR, Duhaime A, Dlugos DJ, Golden JA. Dysplasia: a common finding in intractable pediatric temporal lobe epilepsy. Neurology 2003;61:365–368.
16. Kim YK, Lee DS, Lee SK, et al. Differential features of metabolic abnormalities between medial and lateral temporal lobe epilepsy: quantitative analysis of (18)F-FDG PET using SPM. J Nucl Med 2003;44:1006–1012.
17. Bocti C, Robitaille Y, Diadori P, Lortie A, Mercier C, Bouthillier A, Carmant L. The pathological basis of temporal lobe epilepsy in childhood. Neurology 2003;60:191–195.
18. Krishnan B, Armstrong DL, Grossman RG, Zhu ZQ, Rutecki PA, Mizrahi EM. Glial cell nuclear hypertrophy in complex partial seizures. J Neuropathol Exp Neurol 1994;53:502–507.
19. Diehl B, Lapresto E, Najm I, et al. Neocortical temporal FDG-PET hypometabolism correlates with temporal lobe atrophy in hippocampal sclerosis associated with microscopic cortical dysplasia. Epilepsia 2003;44:559–564.
20. Jellinger K. Neuropathological aspects of infantile spasms. Brain Dev 1987;9:349–357.
21. Meencke HJ. Morphological aspects of etiology and the course of infantile spasms (West syndrome). Neuropediatrics 1985;16:59–66.
22. Meencke HJ, Janz D. Neuropathological findings in primary generalized epilepsy: a study of eight cases. Epilepsia 1984;25:8–21.
23. Gibbs EL, Fleming MM, Gibbs FA. Diagnosis and prognosis of hypsarrhythmia and infantile spasms. Pediatrics 1954;13:66–73.
24. Chugani HT, Shields WD, Shewmon DA, Olson DM, Phelps M, Peacock WJ. Infantile spasms: I. PET identifies focal cortical dysgenesis in cryptogenic cases for surgical treatment. Ann Neurol 1990;27:406–441.
25. Chugani HT, Shewmon DA, Shields WD, Sankar R, Comair Y, Vinters HV, et al. Surgery for intractable infantile spasms: neuroimaging perspectives. Epilepsia 1993;34:764–771.
26. Metsahonkala L, Gaily E, et al. Focal and global cortical hypometabolism in patients with newly diagnosed infantile spasms. Neurology 2002;58:1646–1651.
27. Chugani HT, shewmon DA, Sankar R, Chen BC, Phelps ME. Infantile spasms: II. Lenticular nuclei and brain stem activation on positron emission tomography. Ann Neurol 1992;31:212–219.
28. Chugani HT, Mazziotta JC, Engel J Jr, Phelps ME. The Lennox–Gastaut syndrome: metabolic subtypes determined by 2-deoxy-2[^{18}F]fluoro-d-glucose positron emission tomography. Ann Neurol 1987;21:4–13.
29. Theodore WH, Rose D, Patronas N, Sato S, Holmes M, Bairamian D, et al. Cerebral glucose metabolism in the Lennox–Gastaut syndrome. Ann Neurol 1987;21:14–21.
30. Kobayashi K, Ohtsuka Y, Ohno S, et al. Clinical spectrum of epileptic spasms associated with cortical malformation. Neuropediatrics 2001;32:236–244.
31. Juhász C, Chugani DC, Muzik O, Watson C, Shah J, Shah A, et al. Electroclinical correlates of flumazenil and fluorodeoxyglucose PET abnormalities in lesional epilepsy. Neurology 2000;55:825–834.
32. Gaillard WD, Kopylev L, Weinstein S, Conry J, Pearl PL, Spanaki MV, et al. Low incidence of abnormal (18)FDG-PET in children with new-onset partial epilepsy: a prospective study. Neurology 2002;58:717–722.
33. Hajek M, Wieser HG, Khan N, Antonini A, Schrott PR, Maguire P, et al. Preoperative and postoperative glucose consumption in mesiobasal and lateral temporal lobe epilepsy. Neurology 1994; 44:2125–2132
34. Spanaki MV, Kopylev L, DeCarli C, Gaillard WD, Liow K, Fazilat S, et al. Postoperative changes in cerebral metabolism in temporal lobe epilepsy. Arch Neurol 2000;57:1447–1452.
35. Bogaert PV, David P, Gillian CA et al. Perisylvian dysgenesis Clinical. EEG. MRI and glucose metabolism features in 10 patients. Brain 1998;121:2229–2238.
36. Sivilotti L, Nistri A. GABA receptor mechanisms in the central nervous system. Prog Neurobiol 1991;36;35–92.
37. Juhász C, Muzik O, Chugani DC, Shen C, Janisse J, Chigani HT. Prolonged vigabatrin treatment modifies developmental changes of GABA(A)-receptor binding in young children with epilepsy. Epilepsia 2001;42:1320–1326.
38. Verhoeff NP, Petroff OA, Hyder F, et al. Effects of vigabatrin on the GABAergic system as determined by [123I]iomazenil SPECT and GABA MRS. Epilepsia 1999;40:1443–1448.
39. Savic I, Thorell JO, Roland P. [^{11}C]flumazenil positron emission tomography visualizes frontal epileptogenic regions. Epilepsia 1995;36:1225–1232.
40. Ryvlin P, Bouvard S, Le Bars D, et al. Clinical utility of flumazenil-PET versus [^{18}F]fluorodeoxyglucose-PET and MRI in refractory partial epilepsy. A prospective study in 100 patients. Brain 1998;121: 2067–2081.
41. Muzik O, da Silva EA, Juhász C, et al. Intracranial EEG versus flumazenil and glucose PET in children with extratemporal lobe epilepsy. Neurology 2000;54;171–179.
42. Richardson MP, Koepp MJ, Brooks DJ, Duncan JS. ^{11}C-flumazenil PET in neocortical epilepsy. Neurology 1998;51;845–892.
43. Richardson MP, Koepp MJ, Brooks DJ, Duncan JS. 11C-flumazenil PET in neocortical epilepsy. Neurology 1998:485–492.
44. Hammers A, Koepp MJ, Richardson MP, Hurlemann R, Brooks DJ, Duncan JS. Grey and white matter flumazenil binding in neocortical epilepsy with normal MRI. A PET study of 44 patients. Brain 2003; 126:1300–1318.
45. Levesque MF, Nakasato N, Vinters HV, Babb TL. Surgical treatment of limbic epilepsy associated with extrahippocampal lesions: the problem of dual pathology. J Neurosurg 1991;75:364–370.
46. Raymond AA, Fish DR, Stevens JM, Cook MJ, Sisodiya SM, Shorvon SD. Association of hippocampal sclerosis with cortical dysgenesis in patients with epilepsy. Neurology 1994;44:1841–1845.
47. Lawn N, Londono A, Sawrie S, Morawetz R, Martin R, Gilliam F, et al. Occipitoparietal epilepsy, hippocampal atrophy, and congenital developmental abnormalities. Epilepsia 2000;41:1546–1553.
48. Vernet O, Farmer JP, Montes JL, Villemure JG, Meagher-Villemure K. Dysgenetic mesial temporal sclerosis: an unrecognized entity. Childs Nerv Syst 2000;16:719–723.
49. Juhász C, Nagy F, Muzik O, Watson C, Shah J, Chugani HT. [^{11}C]Flumazenil PET in patients with epilepsy with dual pathology. Epilepsia 1999;40:566–574.
50. Juhász C, chugani HT, Muzik O, Chugani DC. Neuroradiological assessment of brain structure and function and its implication in the pathogenesis of West syndrome. Brain Dev 2001;23:488–495.
51. Chugani DC, Muzik O, Juhász C, Janisse JJ, Ager J, Chugani HT. Postnatal maturation of human $GABA_A$ receptors measured with positron emission tomography. Ann Neurol 2001;49:618–626.
52. Juhász C, Nagy F, Watson C, da Silva EA, Muzik O, Chugani DC, et al. Glucose and [^{11}C]flumazenil positron emission tomography abnormalities of thalamic nuclei in temporal lobe epilepsy. Neurology 1999;53:2037–2045.
53. Newberg AB, Alavi A, Berlin J, Mozley PD, O'Connor M, Sperling M. Ipsilateral and contralateral thalamic hypometabolism as a predictor of outcome after temporal lobectomy for seizures. J Nucl Med 2000;41:1964–1968.
54. Arnold S, Berthele A, Drzezga A et al. Reduction of benzodiazepine receptor binding is related to the seizure onset zone in extratemporal focal cortical dysplasia. Epilepsia 2000;41:818–824.
55. Muzik O, da Silva E, Juhász C, Chugani DC, Shah J, Nagy F, et al. Intracranial EEG vs. flumazenil and glucose PET in children with extratemporal lobe epilepsy. Neurology 2000;54:171–179.
56. Juhász C, Chugani DC, Muzik O, Shah A, Shah J, Watson C, et al. Relationship of flumazenil and glucose PET abnormalities to neocortical epilepsy surgery outcome. Neurology 2001;56:1650–1658.
57. Juhász C, Chugani DC, Muzik O, Watson C, Shah J, Shah A, et al. Electroclinical correlates of flumazenil and fluorodeoxyglucose PET abnormalities in lesional epilepsy. Neurology 2000;55:825–835.
58. Asano E, Chugani DC, Juhász C, Muzik O, Philip S, Shah J, et al. Epileptogenic zones in tuberous sclerosis complex: subdural EEG versus MRI and FDG PET. Epilepsia 2000;41(Suppl 7):128.

59. Savic I, Blomqvist G, Halldin C, Litton JE, Gulyas B. Regional increases in [^{11}C]flumazenil binding after epilepsy surgery. Acta Neurol Scand 1998;97:279–286.
60. Juhász C, Chugani DC, Muzik O, Watson C, Shah J, Shah A, et al. Relationship between EEG and positron emission tomography abnormalities in clinical epilepsy. J Clin Neurophysiol 2000;17:29–42.
61. Asano E, Chugani DC, Muzik O, et al. Multimodality imaging for improved detection of epileptogenic foci in tuberous sclerosis complex. Neurology 2000;54:1976–1984.
62. Fedi M, Reutens DC, Andermann F, et al. α-[^{11}C]-Methyl-L-tryptophan PET identifies the epileptogenic tuber and correlates with inter-ictal spike frequency. Epilepsy Res 2003;52:203–213.
63. Chugani DC, Muzik O. α[C-11]methyl-l-tryptophan PET maps brain serotonin synthesis and kynurenine pathway metabolism. J Cereb Blood Flow Metab 2000;20:2–9.
64. Fedi M, Reutens D, Okazawa H, et al. Localizing value of α-methyl-L-tryptophan PET in intractable epilepsy of neocortical origin. Neurology 2001;57:1629–1636.
65. Juhász C, Chugani Dc, Muzik, O et al. α-methyl-L-tryptophan PET detects epileptogenic cortex in children with intractable epilepsy. Neurology 2003;60:960–968.
66. Juhász C, Chugani DC, Padhye, UN et al. Evaluation with α-[^{11}C]Methyl-l-tryptophan positron emission tomography for reoperation after failed epilepsy surgery. Epilepsia 2004;45:124–130.
67. Dufournel D, Merlet I, Costes N, LeBars D, Faillenot I, Lavenne F, et al. In vivo PET study of 5-HT_{1A} receptors in malformations of cortical development. Epilepsia 2002;43:354.
68. Martinez D, Hwang D, Mawlawi O, et al. Differential occupancy of somatodendritic and postsynaptic 5HT(1A) receptors by pindolol: a dose-occupancy study with [^{11}C]WAY 100635 and positron emission tomography in humans. 2001;24:209–229.
69. Andree B, Halldin C, Pike VW, Gunn RN, Olsson H, Farde L. The Pet radioligand [carbonyl-(11)C]desmethyl-WAY-100635 binds to 5HT(1A) receptors and provides a higher radioactive signal than [carbonyl-(11)C]WAY-100635 in the human brain. J Nucl Med 2002;43:292–303.
70. Sasaki M, Kuwabara Y, Yoshida T, Fukumura T, Morioka T, Nishio S, et al. Carbon-11-methionine PET in focal cortical dysplasia: a comparison with fluorine-18-FDG PET and technetium-99m-ECD SPECT. J Nucl Med 1998;39:974–977.
71. Madakasira PV, Simkins R, Narayanan T, Dunigan K, Poelstra RJ, Mantil J. Cortical dysplasia localized by [^{11}C]methionine positron emission tomography: case report. AJNR Am J Neuroradiol 2002; 23:844–846.
72. O'Tuama LA, Phillips PC, Smith QR, Uno Y, Dannals RF, Wilson AA, et al. L-methionine uptake by human cerebral cortex: maturation from infancy to old age. J Nucl Med 1991;32:16–22.
73. Frost JJ, Mayberg HS, Fisher RS, et al. Mu-opiate receptors measured by positron emission tomography are increased in temporal lobe epilepsy. Ann Neurol 1988;23:231–237.
74. Iinuma K, Yokoyama H, Otsuki T, Yanai K, Watanabe T, Ido T, et al. Histamine H1 receptors in complex partial seizures. Lancet 1993;341:238.

13 Bioimaging L-Tryptophan in Human Hippocampus and Neocortex

Subtyping Temporal Lobe Epilepsy

Steven V. Pacia, MD and Patricia A. Broderick, PhD

SUMMARY

We have previously demonstrated serotonin (5-HT) alterations in the hippocampi (HPC) and neocortices of patients with mesial temporal lobe epilepsy (MTLE) as compared with those presenting with neocortical temporal lobe epilepsy (NTLE). Now, we extend these findings by analysis of L-tryptophan (L-Trp) concentrations, recorded within the same patient population. We used an electrochemical technology, that is, neuromolecular imaging (NMI) with a novel miniature sensor, the carbon-based BRODERICK PROBE®. These sensors detected L-Trp in resected HPC subparcellations as well as in temporal lobe neocortex of patients diagnosed with either MTLE or NTLE. Five patients were classified as NTLE and nine as MTLE based on magnetic resonance imaging (MRI) and intracranial EEG evaluations. HPC subparcellations studied in 12 of 14 patients were (1) granular cells of the dentate gyrus (DG); (2) polymorphic layer of the DG; and (3) HPC pyramidal layer. Layer IV of temporal neocortex was studied in all 14 patients. A specific oxidation potential voltametrically provided the signature for L-Trp. The results showed that in granular cells of DG and, in pyramidal layer, L-Trp concentrations were significantly higher in MTLE patients in contrast to NTLE patients ($p < 0.05$, Mann–Whitney rank sum). Taken together with our recently published data, L-Trp concentrations were inversely proportional to 5-HT concentrations in these neuroanatomic substrates. In polymorphic layer of the DG, there was a trend toward lower 5-HT and higher L-Trp concentrations in MTLE patients. In neocortical layer IV, NTLE patients had significantly lower L-Trp concentrations than MTLE patients; however, 5-HT concentrations were similar in both groups ($p < 0.01$, Mann–Whitney rank sum). Our results indicate markedly different alterations in L-Trp and 5-HT synthesis and metabolism in the epileptogenic temporal lobes of patients with MTLE as compared with NTLE. Considering that various L-Trp metabolites may have proconvulsant or anticonvulsant properties, these alterations have important implications in the pathogenesis of both types of epilepsy.

Key Words: Epilepsy; neurotransmitters; serotonin; L-tryptophan;, monoamine; precursors; human hippocampus; human neocortex; in vivo microvoltametry (in vivo electrochemistry); miniature carbon sensors; neuromolecular imaging (NMI).

1. INTRODUCTION

L-tryptophan (L-Trp) plays a significant role in human epilepsy, but the mechanism of its activity is still under investigation. L-Trp is the precursor of several important neuroactive substances. It is not only involved in the synthesis of serotonin (5-HT) but is also a precursor to both melatonin and kynurenine. Although the effects of 5-HT on neuronal excitability may depend on which 5-HT receptors are activated *(1)*, in most studies, 5-HT decreases the excitability and sensitivity of postsynaptic neurons to epileptogenic stimuli. Waterhouse et al. demonstrated that an increase in 5-HT concentration retards the generation of seizures *(2)*. In fact, the destruction of serotonergic terminals facilitates kindled seizures *(3)*. Long-term animal model studies in both the genetically epilepsy-prone rat and the audiogenic seizure mouse show that low serotonin levels are directly related to seizure susceptibility *(4,5)*. A study by Neuman also supports the anticonvulsant effectiveness of serotonin by suggesting that it may be involved in suppressing certain proconvulsant stimuli *(6)*.

Lu and Gean conducted a study on endogenous serotonin and epileptiform activity *(7)*. The study found that the antiepileptic action of fluoxetine, a serotonin reuptake blocker, is enhanced by the mediation of 5-HT1A receptors and that seizure activity induced by a γ–aminobutyric acid ($GABA_A$) antagonist is suppressed by 5-HT. Activation of 5-HT1A receptors leads to a membrane hyperpolarization and inhibition of seizure activity due to an increase in the conduction of K^+, which inhibits excessive cell firing and reduces population spikes *(8)*.

Another important byproduct of L-Trp metabolism is melatonin, a hormone released by the pineal gland that inhibits brain

From: *Bioimaging in Neurodegeneration*
Edited by P. A. Broderick, D. N. Rahni, and E. H. Kolodny

glutamate receptors and nitric oxide production, suggesting that it may exert a neuroprotective and antiexcitotoxic effect. A scavenger of hydroxyl and peroxyl radicals, melatonin has also been shown to increase the activity of brain glutathione peroxidase, an antioxidative enzyme that metabolizes H_2O_2 to H_2O, in rodents *(9–12)*. A study performed by Kabuto et al. showed that injection of melatonin prevents the occurrence of epileptic discharges induced by iron ion. This suppressive effect of melatonin against the development of epileptic discharges is believed to be mediated through free radical scavenging and antioxidative enzyme activation *(13)*. Melatonin also increases brain levels of GABA, a potent inhibitory neurotransmitter of seizure activity *(10,14)*. A study by Munoz-Hoyos et al. reported that low levels of melatonin associated with decreased levels of monoamines have been shown to increase sensitivity to audiogenic and photogenic seizures due to decreased levels of GABA inhibition in the brain *(14)*. Just as the metabolism of L-Trp is complex, its effects on epileptogenic cortex may also be complex. When L-Trp is converted via the kynurenine pathway, a portion forms quinolinic acid, an intermediate that causes seizure-related damage *(15)*. Quinolinic acid is a known activator for *N*-methyl-D-aspartate (NMDA)-type glutamate receptors *(12)*. Glutamate, an excitatory amino acid neurotransmitter, is known to induce seizure activity *(16)*. NMDA receptors can be selectively activated by quinolinic acid, thereby causing epileptic activity *(17)*. Conversely, another kynurenine pathway intermediary is kynurenic acid, an NMDA receptor antagonist that may have anticonvulsant properties. The complex relationship between NMDA compounds and epilepsy has been reviewed in detail *(3)*.

Despite the potential importance of L-Trp in epilepsy, few studies have examined concentrations of L-Trp in epileptogenic tissue. The potential importance of L-Trp in human epilepsy has been highlighted by one clinical trial. Bibileishvili studied the effects of L-Trp treatment in 83 patients with epilepsy, 54 who had temporal lobe electrophysiologic abnormalities, and found significant seizure reductions *(18)*. This effect was more apparent for nocturnal seizures, generalized tonic-clonic seizures, and complex partial seizures. The antiepileptic effect was thought to be mediated through a serotonergic mechanism exerted during nonslow wave sleep.

By examining the concentration of L-Trp and its product, 5-HT, specific pathways by which L-Trp inhibits or provokes seizure activity can be understood. The present study investigates L-Trp and its relationship to 5-HT concentrations in specific hippocampal (HPC) subparcellations and temporal neocortex in patients with seizures that originate in either of these neuroanatomic substrates. This study is the first to distinguish significant differences in the L-Trp neurochemistry of human temporal lobe epilepsy.

2. METHODS

2.1. PATIENT CLASSIFICATION

Patients who had temporal lobectomies for intractable seizures were studied. Patients were classified as having mesial temporal lobe epilepsy (MTLE) if pathologic examination of the resected temporal lobe revealed severe HPC neuronal loss and gliosis and if examination of the neocortex revealed no other etiology for the patient's epilepsy. Nine patients were classified as MTLE based on these features. Five patients were classified as having neocortical temporal lobe epilepsy (NTLE) based on the lack of HPC atrophy on magnetic resonance imaging and demonstration of seizure onset in temporal neocortex during chronic intracranial EEG study with lateral temporal subdural grid electrodes and multiple baso-mesial temporal subdural strip electrodes.

We used HPC from MTLE patients and neocortex from NTLE patients as controls for our study. Although this tissue cannot be proven to be completely normal brain tissue, these cortical regions demonstrated no epileptogenicity during long-term intracranial EEG study and no neuropathologic abnormalities. Thus, our controls are as close to normal human brain tissue controls as possibly available. Moreover, we did not consider peritumoral tissue or tissue from neurosurgical patients that is adjacent to other pathologies, a good choice for normal controls. It is important to remember that intracranial EEG studies to rule out epileptogenicity are not routinely performed in these other neurosurgical patients.

2.2. TISSUE PREPARATION

Figure 1 shows a longitudinal section of a HPC specimen resected *en bloc*, and Fig. 2 shows a section of the anterolateral temporal neocortex, also resected en bloc. Each specimen was placed in Ringer's Lactate (McGaw Inc. Irvine, CA) and studied with *in situ* microvoltametry. Specimen size for HPC was 2 by 15 by 20 mm^3; specimen size for neocortex was approximately twice as large. Visualization of the cytoarchitectonic features in HPC was performed with the OptiVISOR® (Optical Glass Binocular Magnifier, Donegan Optical Company, Lenexa, KS). The subfields of the pyramidal layer were recognizable by using density differences characteristic between CA1 as compared with CA2 as previously described *(19,20)*. Moreover, because the CA1 subfield of the pyramidal layer is immediately adjacent to the prosubiculum, this subparcellation was used as an initial marker for CA1 subfield recognition.

2.3. STUDY PROTOCOL

Resected temporal lobe tissue from 14 epilepsy patients was examined in a medium of Ringer's lactate buffer solution. Electrochemical detection studies of L-Trp in MTLE and NTLE patients were performed by placing the BRODERICK PROBE® stearate sensor in separate hippocampal subparcellations as follows: (1) within granular cells of the dentate gyrus (DG), (2) polymorphic layer of the DG, and (3) HPC pyramidal layer. Only eight of the nine MTLE patients and four of five NTLE patients had HPC tissue available for analysis. In each study, a Ag/Ag/Cl reference and an auxiliary electrode were placed in contact with the surface of the specimen. In further separate studies, the same protocol was used for the detection of L-Trp in gray matter of neocortex. Specifically, Layer IV of neocortex, the band of Baillarger, was studied. Resected tissue did not exhibit signs of hypoxia and/or anoxia. These signs are easily recognizable because catecholamines and indoleamines are significantly increased as a result of these conditions *(24)*.

After the triple microelectrode assembly (microindicator, reference, and auxiliary) was put in place for each HPC subparcellation, and for neocortex, Layer IV, scanning poten-

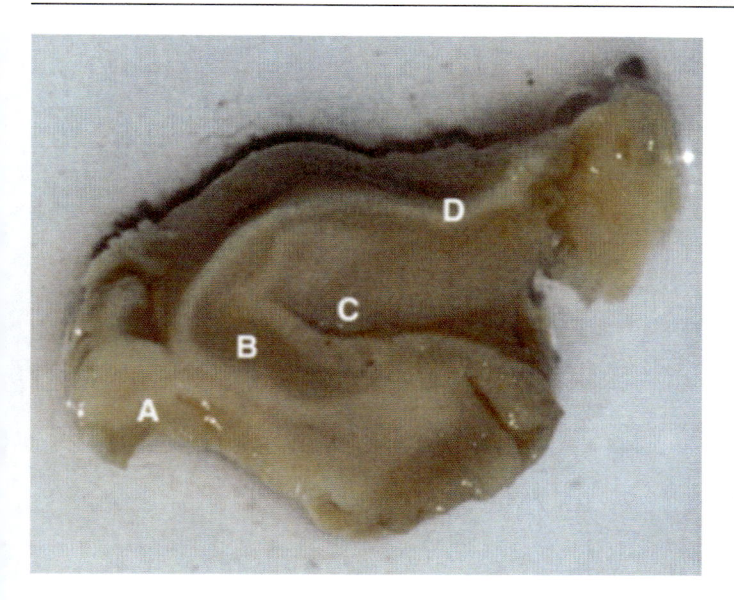

Fig. 1. A longitudinal section of the hippocampus resected en bloc just prior to NMI. A, subiculum; B, pyramidal cell layer; C, granular cell layer of the DG; D, alveus.

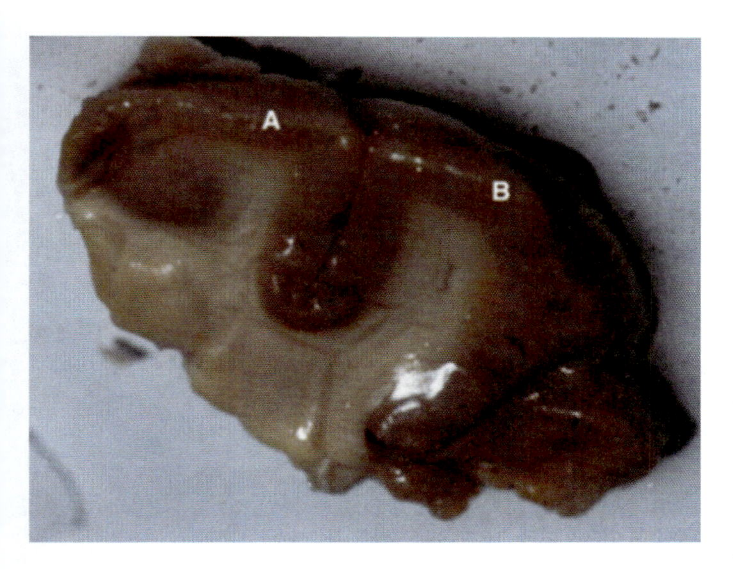

Fig. 2. A section of the anterolateral temporal neocortex through the inferior and middle temporal gyrus. A, Inferior temporal gyrus; B, middle temporal gyrus.

tials (Eapp) in mv were applied to the microindicator via a CV37 potentiostat (BAS, Lafayette, IN) at a scan rate of 10 mV/s, with an initial applied potential of 0.2 V and an ending applied potential of 0.99 V. Sensitivity was set at 5 nA/V; time constant was set at one second. The CV37 detector was electrically connected to a Minigard surge suppressor (Jefferson Electric Magnetek, New York, NY), which was then connected to an electrical ground in isolation. What is unique about this technology is that it enables the detection of 5-HT as well as L-Trp in separate signals in the same recording within less than 1 min. In addition, each individual electrochemical signal is detected within seconds and signals are widely separated by over 300 mV for facile identification and interpretation. Detailed methods are published *(21–29)*. Importantly, especially for human studies, are controlled data, gleaned from Dr. Tierno's laboratory in New York University Medical School, which revealed that each component of the three-microelectrode assembly does not promote bacterial growth whether or not microelectrodes were previously underwent gamma irradiation by Sterigenics (Charlotte, NC).

2.4. IDENTIFICATION OF ELECTROCHEMICAL SIGNALS

Figure 3 shows a recording, drawn from original raw data and recorded in resected HPC tissue from an epilepsy patient. On the x-axis, the oxidation potential in millivolts is delineated; current generated is depicted on the y-axis. Note that the oxidation potential at which 5-HT generates its maximum current, at a peak oxidation potential of +0.290 V (experimentally derived standard error is ±0.015 V). Note also that the oxidation potential at which L-Trp generates its maximum current, at a peak oxidation potential of +0.675 V (experimentally derived is ±0.025 V).

2.5. VALIDATION FOR IDENTIFICATION OF ELECTROCHEMICAL SIGNALS, IN VITRO, *IN SITU*, AND IN VIVO

It is critical to have insight into the accurate identification and validation of electrochemical signatures. Most relevant is the fact that the biogenic amine, 5-HT, and precursor, L-Trp, have amine groups that are protonated at neutral pH and, therefore, exist as cations, whereas metabolites of the monoamines are deprotonated at neutral pH and exist as anions *(30)*. Thus, 5-HT is detected without interference at the same oxidation potential from its metabolite, 5-hydroxyindoleacetic acid, catecholamines, or their metabolites or uric acid, which is a constituent of brain with similar electroactive properties to those of 5-HT. Although, L-Trp is not a metabolite of but a precursor to 5-HT it too exists as a cation, therefore precluding metabolites of other chemicals to be oxidized either at the same oxidation potential as L-Trp or to be oxidized at all. Neuropeptides such as somatostatin and dynorphin A exhibit different oxidative profiles than does L-Trp.

2.6. STATISTICAL ANALYSIS

To investigate differences in continuous variables between the two groups, MTLE and NTLE, the Mann–Whitney rank sum test was used for nonnormally distributed variables. p values <0.05 were considered significant.

3. RESULTS

Table 1 lists L-Trp concentrations for the eight MTLE and four NTLE patients in HPC subparcellations. In the present analysis, for these same 12 patients, L-Trp was significantly increased in the granular cell layer and pyramidal layer of the hippocampi in MTLE patients compared to the same region in NTLE patients ($p < 0.05$, Mann–Whitney rank sum test). There was a trend toward higher levels of L-Trp in the polymorphic layers of the hippocampi in MTLE patients but this failed to reach statistical significance. Table 2 lists the L-Trp concentrations in anterolateral temporal neocortex layer IV for all 14 patients studied (9 MTLE and 5 NTLE). L-Trp concentrations were found to be significantly higher in the neocortex of MTLE patients when compared with neocortex of NTLE patients ($p < 0.01$, Mann–Whitney rank

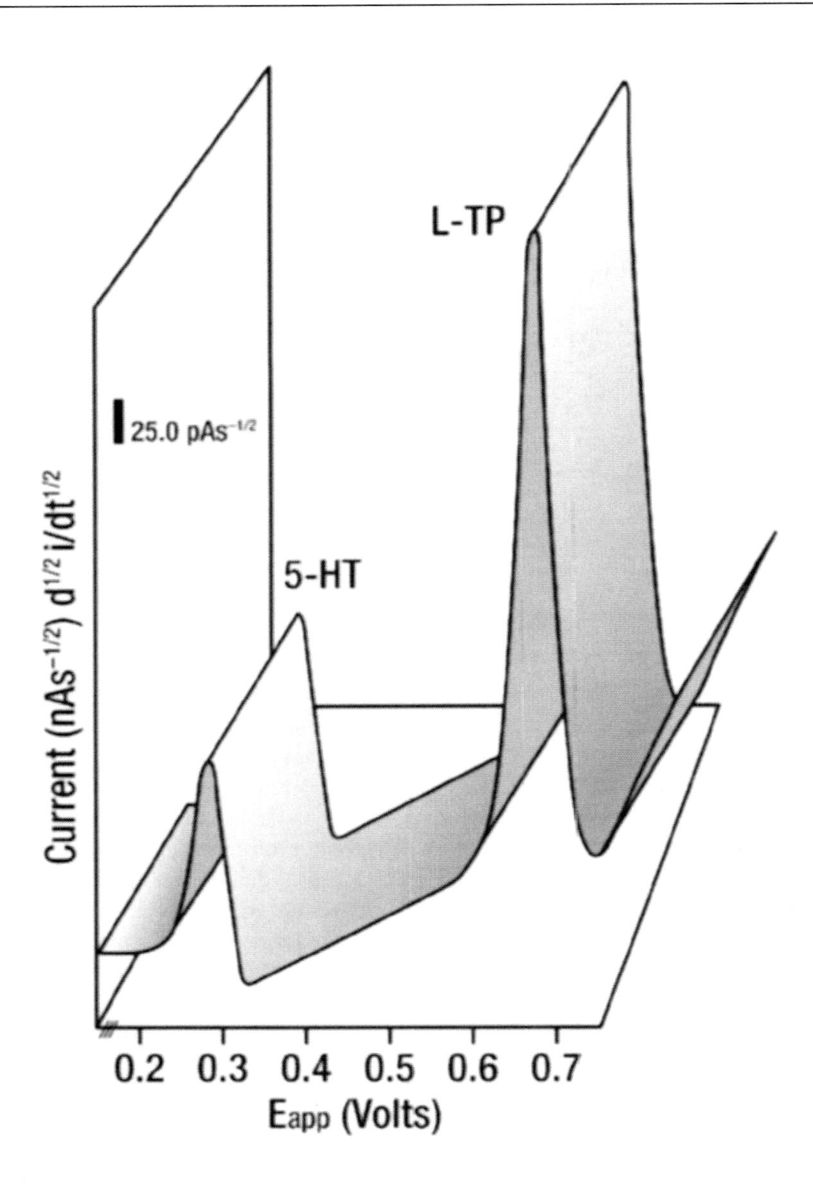

Fig. 3. A recording, drawn from original raw data, recorded in resected HPC tissue from an epilepsy patient, *in situ*. On the *x*-axis, the oxidation potential in millivolts; on the *y*-axis, current. Note that the oxidation potential at which 5-HT generates its maximum current is at a peak oxidation potential of +0.290 V (experimentally derived standard error is ±0.015 V). Note also that the oxidation potential at which L-Trp generates its maximum current is at a peak oxidation potential of +0.675 V (experimentally derived standard error is ±0.025 V).

sum test). Also, consistent with prior studies *(31)*, L-Trp values were in the micromolar range.

Interestingly, L-Trp concentrations were consistently lower in all HPC subparcellations and neocortex in NTLE vs MTLE patients. This deficiency in L-Trp in NTLE patients compared with MTLE patients appears to be more dramatic in the neocortices of MTLE vs NTLE. With regard to HPC subparcellations, the deficiency in L-Trp is more marked in the granular cells of the dentate gyrus vs pyramidal cell layer vs the polymorphic layer.

The mean L-Trp concentrations in HPC for the eight MTLE patients were 5.95 ± 1.13 for the granular cell layer of the dentate gyrus, 4.64 ±2.65 for the polymorphic layer, and 6.40 ± 1.71 for the pyramidal cell layer. For the four NTLE patients the mean L-Trp concentrations were 3.99 ± 1.22 for the granular cell layer of the dentate gyrus, 3.74 ± 0.80 for the polymorphic layer, and 4.92 ± 1.00 for the pyramidal cell layer. In layer IV of lateral temporal neocortex the mean L-Trp concentration was 4.65 ± 0.81 for the MTLE group and 2.22 ± 0.87 for the NTLE group.Analysis of the 5-HT concentrations for these 12 patients with one other were previously reported and revealed significantly lower concentrations in the granular cell and pyramidal cell layer (CA1) of the MTLE patients compared with the same regions analyzed in the NTLE group *(21)*. Conversely, 5-HT concentrations were similar in layer IV of anterolateral temporal neocortex for 14 patients studied (nine MTLE, five NTLE patients; ref. *15*). Therefore, taken together with previous data, an inverse relationship between L-Trp and 5HT in all hippocampal subparcellations is clear whereas this inverse relationship between L-Trp and 5-HT was not demonstrated in temporal neocortex.

Table 1
5-HT and L-Trp Concentrations in HPC Subparcellations

	Granular cells of DG		*Polymorphic layer*		*Pyramidal cell layer*	
	5-HT (n*M*)[a]	L-Trp (μ*M*)[b]	5-HT (n*M*)	L-Trp (μ*M*)	5-HT (n*M*)[c]	L-Trp (μ*M*)[d]
MTLE (patient #)						
1	1.4	2.70	1.5	2.75	1.5	6.40
2	2.5	2.92	NA	NA	1.3	2.98
3	1.3	8.75	1.4	6.25	1.4	17.8
4	1.3	5.28	1.4	1.00	1.2	2.65
5	2.3	4.50	3.0	2.68	1.5	7.00
6	1.6	7.68	2.9	4.30	2.1	4.60
7	2.2	3.98	2.2	2.98	1.8	5.48
8	2.1	11.8	3.2	12.5	1.3	4.35
NTLE (patient #)						
10	1.5	2.67	1.9	2.66	1.3	7.64
11	3.6	2.90	5.1	3.60	2.0	3.80
12	3.6	7.65	4.7	6.05	4.0	5.13
13	3.2	2.77	2.5	2.65	1.5	3.10

Mann–Whitney rank sum test for 5-HT and L-Trp concentration MTLE vs NTLE.

[a] Concentrations of 5-HT were significantly greater for NTLE vs MTLE in granular cells of the DG ($p = 0.055$, $p < 0.05$).

[b] Concentrations of 5-HT were significantly greater for NTLE vs MTLE in pyramidal cell layer (CA1 Cell Layer [CA1 subfield] ($p = 0.095$, $p < 0.05$).

[c] Concentrations of L-Trp were significantly greater for MTLE vs NTLE in granular cells of the DG ($p = 0.024$, $p < 0.01$).

[d] Concentrations of L-Trp were significantly greater for MTLE vs NTLE in pyramidal cell layer [CA1–4 subfields] ($p = 0.055$, $p < 0.05$).

NA, not available.

Table 2
5-HT and L-Trp Concentrations in Temporal Lobe Neocortex (Layer IV)

	*5-HT concentration (n*M*)*	*L-Trp concentration (μM)*[a]
MTLE (patient #)		
1	1.5	2.85
2	1.7	2.58
3	1.3	9.30
4	1.3	5.15
5	1.9	3.85
6	2.0	2.74
7	3.5	7.60
8	3.2	5.16
9	3.6	2.66
NTLE (patient #)		
10	2.0	1.20
11	2.1	0.48
12	4.9	2.76
13	1.7	1.30
14	2.2	5.35

Mann–Whitney rank sum test for 5-HT and L-Trp concentration MTLE vs NTLE.

5-HT concentrations in MTLE vs NTLE patients were not significantly different from each other ($p = 0.689$, $p > 0.05$).

[a] L-Trp concentrations were significantly higher in temporal neocortex (layer IV) of MTLE vs NTLE patients ($p = 0.06$, $p < 0.05$).

4. DISCUSSION

Our results indicate markedly different alterations in L-Trp and 5-HT synthesis and metabolism in the epileptogenic temporal lobes of patients with MTLE compared with those with NTLE. The results distinguish MTLE from NTLE in HPC by showing an inversely proportional relationship between L-Trp and 5-HT in all subparcellations. In contrast, the results distinguish MTLE from NTLE in neocortex by significant differences in L-Trp only.

The metabolism of tryptophan and its precise role in other neurologic disorders such as AIDS- dementia complex, Huntington's disease, and stroke is still being investigated *(32)*.

Of the L-Trp available for nonprotein synthesis, a portion is converted to 5-HT in the serotonergic nerve terminals whereas a significant amount may be metabolized via the kynurenine pathway *(32)*. In a recent article, a linear relationship was demonstrated between the bioavailability of L-Trp and 5-HT concentrations in rat hippocampus *(33)*. The increased L-Trp in the granular cell neurons of the dentate gyrus and pyramidal layer of our MTLE patients may be related to the reduced 5-HT concentrations possibly through dysfunction of the L-Trp hydroxylase-dependent conversion of L-Trp to 5-HT or due to decreased use by potentially injured or hypoactive serotonergic neurons in the mesial temporal lobe. Altered tryptophan hydroxylase activity has been demonstrated in the genetically epilepsy-prone rat *(34)*. Excess L-Trp is preferentially metabolized via the kynurenine pathway, but whether the metabolism prefers quinolinic acid or kynurenic acid is not known.

L-Trp was also relatively deficient in layer IV of the temporal neocortex of our NTLE patients compared with our MTLE patients. Interestingly, 5-HT concentrations were similar in these two groups, raising the possibility of either a compensatory increase in L-Trp hydroxylase activity and L-Trp use or increased use of L-Trp by the kynurenine pathway. During the synthesis of NAD or NADP from L-Trp via the kynurenine pathway, several intermediates are generated in the process known as kyurenines. These compounds have varied and important effects on neurons. Considering that approx 80% of nonprotein L-Trp metabolism is through the kynurenine pathway, the potential exists for excitotoxicity to the neocortex of these patients, especially if there is increased conversion of L-Trp to quinolinic acid *(12,17)*. The potential importance of kynurenine pathway manipulation is highlighted in a recent study by Scharfman and Ofer, in which epileptiform activity in rat entorhinal/hippocampal slices was suppressed by pretreatment with L-kynurenine *(35)*. This effect resulted from increased kynurenic acid formation. One clinical study did analyze kynurenine acid levels in patients with seizures but only in serum and cerebrospinal fluid *(31)*. In these patients, kynurenine levels in cerebrospinal fluid were unremarkable and serum levels were actually reduced. Additionally, two other products generated by the kynurenine pathway, 3-hydroxykynurenine and 3-hydroxyanthanilic acid, are implicated in the generation of free radicals, which may contribute to neuronal injury or death *(36)*.

The difference in L-Trp/5-HT profiles between our MTLE and NTLE patients may also be accounted for by 5-HT receptor subtype expression differences in the two regions. Recent studies indicate that 5-HT has a predominately anticonvulsant effect in epilepsy when mediated via its 5-HT1A receptor *(30,37,38)*. Bobula et al., studying rat CA1 pyramidal neurons, showed that serotonin can inhibit epileptiform activity in a variety of accepted epilepsy models and that inhibition of epileptiform bursts by serotonin may be mediated by activation of both the 5-HT1A and 5HT1B receptor subtypes *(1)*. However, this same study revealed activation of epileptiform activity in frontal cortex with 5-HT2 receptor activation, although antagonism of 5-HT2 receptor sites in one study of rat hippocampus failed to lower seizure thresholds *(33)*. Regardless of which receptor subtypes are being activated in our patients, the availability of extracellular L-Trp could be driven by 5-HT use, which is a function of receptor density. Whether these receptor subtype differences have a clinically important effect in human epilepsy is unknown.

One prior study showed increased 5-HT production in actively spiking cortex using an amygdala kindling model *(39)*. However, this study is not comparable to ours as tissue was taken from our patients under general anesthesia, where there would not be actively spiking or seizing cortical regions prior to resection. A recent PET study using α-methyl-L-tryptophan in neocortical foci found increased tracer uptake in epileptogenic cortex that correlated with increased interictal spike frequencies *(40)*. From these studies, it cannot be determined whether this increased L-Trp uptake was caused by increased 5-HT or increased kynurenine metabolite production. Regardless, increased L-Trp uptake may correlate with our findings of L-Trp reductions and possible increased L-Trp use in our NTLE patients. Another PET study in TLE patients revealed increased α-methyl-L-tryptophan uptake in the hippocampi of NTLE patients when compared with patients with MTLE and controls *(6)*. Our finding of a relative LTP deficiency in the hippocampi of our NTLE patients may explain this compensatory increase in LTP uptake.

Considering the growing evidence for the importance of L-Trp in experimental models of temporal lobe epilepsy, our findings of high L-Trp concentrations in the hippocampi of MTLE patients and low L-Trp concentrations in the temporal neocortices of NTLE patients are clinically relevant. The present studies using simultaneous sampling of L-Trp and 5-HT allow a unique advantage to decipher mechanistically the relationship between this monoamine neurotransmitter and its amino acid precursor.

ACKNOWLEDGMENTS

This work was supported by the Parents Against Childhood Epilepsy (PACE) Foundation; NIH/NIGMS SO6 GM08168; and PSC/CUNY Awards # RF 64282-00-33. We thank Karen Schulz of Humana Press for artistic assistance.

REFERENCES

1. Bobula B, Zahorodna A, Bijak M. Different receptor subtypes are involved in the serotonin-induced modulation of epileptiform activity in rat frontal cortex in vitro. J Physiol Pharmacol 2001;52:265–274.
2. Waterhouse BD. Electrophysiological assessment of monamine synaptic function in neuronal circuits of seizure susceptible brains. Life Sci 1986;39:807–818.
3. Lerner Natoli M. Serotonin and kindling development. Int J Neurosci 1987;36:139–151.
4. Jobe PC, Dailey JW, Reigel CE. Noradrenergic and serotonergic determinants of seizure susceptibility and severity in genetically epilepsy prone rats. Life Sci 1986;39:775–782.
5. Laird HE, Dailey JW, Jobe PC. Neurotransmitter abnormalities in genetically epileptic rodents. Fed Proc 1984;43:2505–2509.
6. Neuman BS. Suppression of penicillin-induced epileptiform activity by noxious stimulation: mediation by 5-hydroxytryptamine. Electroencephal Clin Neurophysiol 1986;64:546–555.
7. Lu KT, Gean PW. Endogenous serotonin inhibits epileptiform activity in rat hippocampal CA1 neurons via 5-hydroxytryptamine$_{1A}$ receptor activation. Neuroscience 1998;86:729–737.
8. Andrade R, Nicoll RA. Pharmacologically distinct actions of serotonin on single pyramidal neurons of the rat hippocampus recorded in vitro. J Physiol 1987;394:99–124.

9. Barlow-Walden LR, Reigter RJ, Abe M, et al. Melatonin stimulates brain glutathione peroxidase activity. Neurochem Int 1995;26:497–502.
10. Luow D, Sutherland GR, Glavin GB, Girvin J. A study of monamine metabolism in human epilepsy. Can J Neurol Sci 1989;16:394–397.
11. Reiter RJ, Tan DX, Poeggeler B, Menendez PA. Melatonin as a free radical scavenger: implications for aging and age-related diseases. Ann NY Acad Sci 1994;719:1–12.
12. Stone TW. Endogenous neurotoxins from tryptophan. Toxicon 2001; 39:61–73.
13. Kabuto H, Yokoi I, Norio O. Melatonin inhibits iron-induced epileptic discharges in rats by suppressing peroxidation. Epilepsia 1998; 39:237–243.
14. Munoz-Hoyos A, Sanchez-Forte M, Molina-Carballo A, et al. Melatonin's role as an anticonvulsant and neuronal protector: experimental and clinical evidence. J Child Neurol. 1998;13:501–509.
15. Noguchi KK. NMDA antagonist-induced neurotoxicity and psychosis. In: Massaro EJ, Schardein JL, Broderick PA, Schlaepfer TE, Mattson JL, eds. Handbook of Neruotoxicology, Vol 2. (Totowa, NJ: Humana Press Inc.; 2001.
16. Hayashi T. A physiological study of epileptic seizures following cortical stimulation in animals and its application to human clinics. Jpn J Physiol 1952;3:46–64.
17. Stone TW. Kynurenic acid antagonists and kynurenine pathway inhibitors. Expert Opin Invest. Drugs 2001;10:633–645.
18. Bibileishvili S. Amino acid tryptophan, as an antiepileptic agent. Klin Med (Mosk) 1980;58:91–106.
19. Afifi AK, Bergman RA. Brain Neuroscience. Baltimore: Urban and Schwarzenberg Inc.; 1980.
20. Mathern GW, Babb TL, Armstrong DL. Hippocampal sclerosis. In: Engel J Jr., Pedley TA, editors. Epilepsy: A Comprehensive Textbook. Philadelphia: Lippincott-Raven Pub.; 1997:1807–1817.
21. Broderick PA, Pacia SV, Doyle WK, Devinsky O. Monoamine neurotransmitters in resected hippocampal subparcellations from neocortical and mesial temporal lobe epilepsy patients: in situ microvoltammetric studies., Brain Res 2000;878:48–63.
22. Broderick PA. Microsensors detect neuroadaptation by cocaine: serotonin released in motor basal ganglia is not rhythmic with movement. In Massaro EJ, Broderick PA, Mattson JL, Schardein JL, Schlaepfer TE, editors. Handbook of Neurotoxicology, Vol. 2. Totowa, NJ: Humana Press; 2001, pp. 323–367.
23. Broderick PA. Rat striatal dopamine release mechanisms of cocaine. NIDA Res. Mono. Ser. 1986;75:367–370.
24. Broderick PA, Gibson GE. Dopamine and serotonin in rat striatum during *in vivo* hypoxic-hypoxia . Metab Br Dis 1989;4: 143–153.
25. Broderick PA. State-of-the-art microelectrodes for *in vivo* voltammetry. Electroanalysis 1990;2:241–251.
26. Broderick PA. Microelectrodes and their use in cathodic electrochemical current arrangement with telemetric application. European Patent 90914306.7, 1996.
27. Broderick PA. Microelectrodes and their use in an electrochemical arrangement with telemetric application. US Patent 5,938,903, 1999.
28. Broderick PA. In vivo electrochemical studies of gradient effects of (sc) cocaine on dopamine and serotonin release in dorsal striatum of conscious rats, Pharmacol Biochem Behav 1993;46:973–984.
29. Broderick PA, Pacia SV. Identification, diagnosis, and treatment of neuropathologies, neurotoxicities, tumors, and brain and spinal cord injuries using microelectrodes with microvoltammetry. U.S. 10/118,571 and Foreign Patent pending # PCTUS021/11244, 2002.
30. Coury LA Jr, Huber EW, Heineman WR. Applications of modified electrodes in the voltammetric determination of catecholamine neurotransmitters . Biotechnology 1989;11: 1–37.
31. Heyes MP, Saito K, Devinsky O, Nadi SN. Kynurenine pathway metabolites in cerebrospinal fluid and serum in complex partial seizures. Epilepsia 1994;35:251–257.
32. Stone TW. Kynurenines in the CNS: from endogenous obscurity to therapeutic importance. Prog Neurobiol 2001;64:185–218.
33. Van der Stelt HM, Olivier B, Westenburg HGM. Effects of availability of tryptophan on serotonin levels in the dorsal hippocampus of rats. Society for Neuroscience 32nd annual meeting, Orlando, FL, 2002.
34. Statnick MA, Dailey JW, Jobe PC, Browning RA. Abnormalities in brain serotonin concentration, high affinity uptake, and tryptophan hydroxylase activity in severe-seizure genetically epilepsy-prone rats. Epilepsia 1996;37:311–321.
35. Scharfman HE, Ofer A. Pretreatment with L-kynurenine, the precursor to the excitatory amino acid antagonist kynurenic acid, suppresses epileptiform activity in combined hippocampal/entorhinal slices. Neurosci Lett 1997;224:115–118.
36. Moroni F. Tryptophan metabolism and brain function: focus on kynurenine and other indole metabolites. Eur J Pharmacol. 1999;375:87–100.
37. Gentsch K, Heinemann U, Schmitz B, Behr J. Fenfluramine blocks low Mg2+-induced epileptiform activity in rat entorhinal corx. Epilepsia 2001;35:251–257.
38. Salgado-Commissariat D, Alkadhi KA. Serotonin inhibits epileptiform discharge by activation of 5-HT_{IA} receptors in CA1 pyramidal neurons, Neuropharmacology 1997;36:1705–1712.
39. Shouse MN, Staba RJ, Ko PY, Saquib SF, Farber PR. Monoamines and seizures: microdialysis findings in locus ceruleus and amygdala before and during amygdala kindling. Brain Res 2001;892:176–192.
40. Fedi M, Reutens D, Okazawa H, Andermann F, Boling W, Dubeau F, et al. Localizing value of -methyl-L-tryptophan PET in intractable epilepsy of neocortical origin. Neurology 2001;57:1629–1636.
41. Natsume J, Kumakura Y, Bernasconi N, et al. α-(^{11}C) Methyl-l-tryptophan and glucose metabolism in patients with temporal lobe epilepsy. Neurology 2003;60:756–761.
42. Pacia SV, Doyle WK, Broderick PA. Biogenic amines in the human neocortex in patients with neocortical and mesial temporal lobe epilepsy: identification with in situ microvoltammetry. Brain Res 2001;899–:106–111.
43. Watanabee K, Ashby CR Jr, Katsumoriand H, . Minabe Y. The effect of acute administration of various selective 5-HT receptor antagonists on focal hippocampal seizures in freely-moving rats. Eur J Pharmacol 2000;398:239–246.

14 In Vivo Intrinsic Optical Signal Imaging of Neocortical Epilepsy

Sonya Bahar, PhD, Minah Suh, PhD, Ashesh Mehta, MD, PhD, and Theodore H. Schwartz, MD

SUMMARY

Intrinsic optical signal (IOS) imaging is a technique for measuring changes in blood flow, metabolism, and cellular swelling associated with neuronal activity. The combined spatial and temporal resolution, in addition to the ability to sample large areas of cortex simultaneously, make it a powerful technique for brain mapping. IOS has only recently been applied systematically to the study of epilepsy. This chapter will explore the utility and feasibility of mapping interictal spikes, ictal onsets, offsets, and horizontal propagation using IOS imaging in acute and chronic animal models of epilepsy. The implementation of IOS imaging in the operating room during neurosurgical procedures will be discussed as well as technical challenges that currently restrict this translational work to a very few centers.

Key Words: Epilepsy; optical imaging; intrinsic signal; rat; ferret; seizure; ictal; interictal; iron; tetanus toxin; human; surgery.

1. BACKGROUND

1.1. Historical Background and Physiological Origin of the Intrinsic Optical Signal

Intrinsic optical changes in neural tissue were first observed by Hill and Keynes (*1*; *see* also refs. *2* and *3*) in response to an applied train of stimuli and later in response to single electrical stimuli (*4*; *see* ref. *5* for review). Only recently, however, has the intrinsic optical signal (IOS) been systematically applied to the study of the nervous system. A decrease in light reflectance has been shown to correlate spatially with increased electrophysiological activity *(6–8)*.

The IOS has several distinct advantages: first, it offers high spatial resolution, to a level of approx 50 μm *(7, 9)*. In contrast, the spatial resolution of functional magnetic resonance imaging (fMRI) is 1 mm *(10)*. Second, because it does not involve the application of potentially phototoxic dyes, it is a minimally invasive technique and thus well suited to applications involving the intraoperative imaging of human patients or long-term in vivo animal studies. In contrast to voltage-sensitive dyes and calcium-sensitive dyes, however, the time-course of the change in light reflectance is considerably slower than the time-course of single action potentials or field potentials *(11,12)*. This suggests that the intrinsic signal demonstrates physiological changes caused by the underlying electrophysiology rather than reflecting the electrophysiology itself *(11,12)*.

The IOS derives from changes in the light absorbance properties of electrophysiologically active neural tissue, caused by focal alterations in blood flow, oxygenation of hemoglobin, and scattering of light *(13,14)*. The amplitude and time-course of the IOS are dependent on the wavelength of incident light *(13,14)*. Multiwavelength imaging in striate cortex during visual activation has demonstrated that the IOS recorded at different wavelengths represents separate cortical processes, each associated indirectly with neuronal activity. Around the isosbestic wavelengths of hemoglobin, roughly 480–590 nm, the signal is most sensitive to blood volume *(13)*. At higher wavelengths, from 600–650 nm, the signal is dominated by the oxygenation state of hemoglobin. Finally, at wavelengths greater than 660 nm, the light-scattering component created by fluid shifts becomes dominant *(14)*. Although the signal is never "pure," and at each wavelength there are contributions from each component, a high spatial correlation with neuronal activity has been well demonstrated in multiple animal preparations in various regions of cortex.

The exact physiological components of the IOS recorded at each wavelength are not yet fully understood. There is general agreement that as neurons become activated, they increase their metabolic demand causing an increase in the concentration of deoxyhemoglobin *(14–16)*. This "initial dip" in local oxygenation is spatially highly colocalized with neuronal activity and begins within 100 ms of neuronal activation *(13)*. Simultaneously, the opening of sodium, potassium, and calcium channels causes a change in the volume of cells and the extracellular

From: *Bioimaging in Neurodegeneration*
Edited by P. A. Broderick, D. N. Rahni, and E. H. Kolodny

fluid as well as swelling of glia, which buffer extracellular potassium. These fluid shifts are also spatially well colocalized with neuronal activity. Approximately 300–500 ms later, the arterial microvasculature begins to dilate and cause an increase in blood volume, which is not as well localized with neuronal activity *(14)*. Finally, between 0.5 and 1.5 s later, there is an activity-dependent increase in blood flow that delivers oxygenated blood to the area. This increase in blood flow causes an increase in oxygenation because the incoming oxygenated hemoglobin overwhelms the metabolic needs of the neurons *(17)*. This later component causes an inverted optical signal in the larger blood vessels and is the basis of the blood oxygen level-dependent (BOLD) signal imaged with low-tesla fMRI (i.e., 1.5 T; refs. *15,16,18*). Colocalization of this process with neuronal activity is less precise and often seen in the draining veins *(16)*.

1.2. USE OF THE IOS IN IMAGING FUNCTIONAL ARCHITECTURE

Because of its high spatial resolution and slow time course, the IOS was initially applied to the mapping of functional architecture in the cortex. In this arena, the IOS proved spectacularly successful. In fact, it was using this technique that it was first demonstrated that iso-orientation domains in the cat visual cortex are arranged in a pinwheel-like structure *(7)*. Numerous other studies have since been conducted in the visual cortex using the IOS in the mouse *(19)*, cat *(6,20–23)*, ferret *(24–27)*, tree shrew *(28,29)* and nonhuman primate *(30,31)*. Because the signal is extremely small (~0.1%), these studies rely on signal averaging and trial repetition to improve the signal-to-noise ratio. In general, trials in which the cortex is activated by a known stimulus are divided by, or subtracted from, trials without activation *(32)*. Novel techniques for image processing such as Fourier transform (FT) and principal component analysis (PCA)-based methods are also being investigated to eliminate the need for image division *(33,34)*.

Figure 1 shows the orientation columns and ocular dominance columns imaged from a ferret. In such a typical functional architecture study, images are taken from the exposed cortex of the ferret while it is viewing moving gratings at different angles on a video screen through each eye while the other eye is covered. To map the areas that respond to a given orientation, the IOS obtained during stimulus with a given condition (grating at a certain angle) is divided by a signal averaged from the responses of the animal to a variety of stimuli (gratings at all different angles, the so-called "cocktail blank"). Ocular dominance columns are obtained by dividing all images acquired when one eye is open by those acquired when the other is open.

Other regions of functional cortex, ranging from the rodent barrel cortex *(13,35–37)* to somatosensory cortex in the nonhuman primate *(38–40)* have been mapped as well. Technical advances in the development of an "artificial dura" have made chronic imaging of awake, behaving animals a reality *(41)*.

1.3. USE OF IOS FOR IMAGING DYNAMIC EVENTS: FROM SPREADING DEPRESSION TO EPILEPSY

In addition to its use in mapping static functional architecture, however, the IOS also can be used for the study of dynamic events. The critical difference between imaging static architecture and dynamic processes is that multiple trials cannot be averaged together. Events such as spreading depression and seizures are not only dynamic, in that they move across the cortex over time, but sequential events are not necessarily stereotypical and can spread along different pathways and at different rates. Because events cannot be averaged, the signal would be very difficult to see using image division techniques if it were the same size as physiologic sensory activation. Luckily, the IOS associated with spreading depression, interictal spikes, and ictal events is anywhere from 5 to 500 times as large as found with physiologic sensory activation. Even so, averaging is occasionally possible, even with dynamic events, as in the case of certain models of interictal spiking, as will be discussed below.

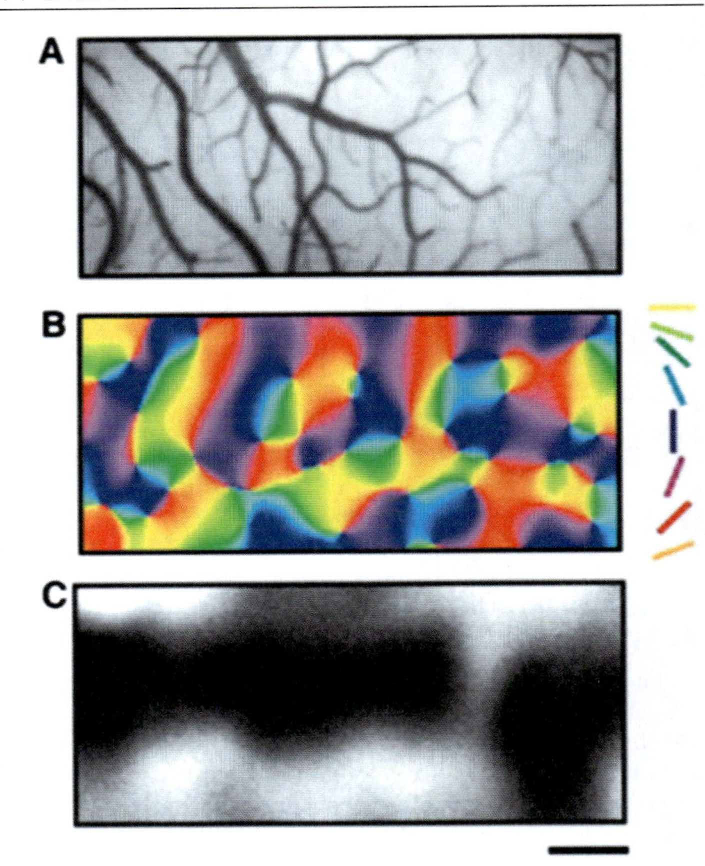

Fig. 1. Optical imaging of intrinsic signals reveals the functional architecture in ferret visual cortex. **(A)** Blood vessel pattern of the surface of the visual cortex imaged at 546 nm. **(B)** The angle map is generated by color-coded vectorial summation of each single condition (0, 45, 90, 135, 180, 225, 270, 315, and 360°) map on a pixel-by-pixel basis. **(C)** Ocular dominance columns are obtained by dividing images acquired with one eye covered by those acquired with the other eye covered. Imaging performed at 707 nm. Scale bar, 1 mm.

1.3.1. Spreading Depression

Cortical spreading depression (CSD), first studied by Leão *(42,43)*, consists of a sharp drop in the DC potential ("DC shift", typically 10-20 mV) that spreads across the cortex at a rate of up to 10 mm/min *(44)* and also can be accompanied by a depression of the EEG and suppression of evoked potentials *(42,43)*. CSD can occur in response to noxious stimuli such as direct mechanical injury *(42,43,45,46)*, application of glutamate *(47)*, elevated potassium *(48)*, electrical stimulation *(43)*, or hypoxia *(49)*.

CSD has been studied with IOS imaging in the visual system *(50)*, rat hippocampal slices *(51,52)*, and the rat neocortex in vivo *(46,53)*. The initial CSD-related IOS response consists of a decrease in light reflectance in vivo (often observed as an increase in light transmittance in slice experiments). This is followed by a sharp rise in reflectance (decrease in transmittance) that propagates in a wave-like fashion through the tissue; the leading edge of this "wave" coincides spatially with the DC shift *(49,51)*.

1.3.2. Epilepsy

Epilepsy is a chronic neurological disease that affects as much as 2% of the population with recurrent seizures *(54)*. Seizures, or "ictal events," consist of the paroxysmal, synchronous, rhythmic firing of a population of pathologically interconnected neurons capable of demonstrating high-frequency oscillatory activity called "fast ripples" (250–500 Hz; refs. *55–57*), synchronized by axo-axonal gap junctions, field effects or interneurons *(58–60)*. These events are caused by an imbalance in excitatory and inhibitory mechanisms leading to both hypersynchrony and hyperexcitability *(61)*. The role of inhibition in the etiology of epileptiform events, however, is controversial and quite relevant to optical imaging studies. Because epileptiform activity is characterized by a hyperexcitable, hypersynchronous state, classic hypotheses often supposed a decrease in inhibitory tone in the region of the epileptic focus. In agreement with this theory, administration of GABA-A antagonists such as penicillin or bicuculline induces acute epileptiform events in vitro and in vivo *(62–70)*. Likewise, losses of several subtypes of interneurons are reported in both experimental and human temporal lobe epilepsy *(71)*. However, there is a great deal of evidence that inhibitory mechanisms are functionally unaltered or even increased in several non-lesional chronic in vivo and in vitro models *(72–76)*, as well as human epileptic tissue, studied both in vivo and in vitro *(73,77,78)*.

The role of inhibitory processes in the prevention of epileptic events also is unclear. In certain models of acute epilepsy, electrophysiologic recordings from brain surrounding the epileptic focus demonstrate upregulated inhibition, the so-called "inhibitory surround." First described by Prince in the acute disinhibition model , the inhibitory surround develops several minutes after the initiation of bursting in the focus and was thought to be generated by long-range horizontal inhibitory connections, presumably mediated by basket cells, which are recruited by bursting cells in the focus *(79–83)*. Alternatively, long-range horizontal excitatory connections recently have been shown to powerfully recruit local inhibitory circuitry in response to focal repetitive stimulation in tangential slices *(84,85)*. Surround inhibition has been hypothesized to play a role in preventing interictal to ictal transition, secondary generalization of focal ictal events, and may explain the profound interictal focal hypometabolism/hypoperfusion found in most functional imaging studies performed in chronic human epilepsy *(72,82,86)*. Although not well-described in human epilepsy or experimental chronic models of epileptogenesis, more recent experimental evidence demonstrates an inhibitory surround restricting the size of cortex capable of generating fast ripples in the hippocampus of kainate treated rats *(56)*.

When seizures are not occurring, surface recordings from patients with chronic epilepsy, or intracortical field potential (f.p.) recordings adjacent to experimentally induced epilepsy in laboratory animals, show abnormal paroxysmal events in a large population of neurons called interictal spikes (IIS). The IIS generally consists of a high amplitude surface negativity (1–5 mV) lasting 50–200 ms followed by a slow wave with no behavioral correlate *(87)*. The intracellular event underlying the IIS, explored in a variety of animal models, consists of a large paroxysmal depolarizing shift (PDS) with a superimposed burst of action potentials that occurs in a variable percentage of the adjacent neurons depending on the etiology of the epilepsy, followed by an afterhyperpolarization *(62,88–92)*. The transition to an ictal event, or seizure, occurs when the afterhyperpolarization gradually disappears and is replaced by further depolarization *(88,89)*.

1.3.3. IOS Imaging of Neocortical Epilepsy

Because epileptiform events involve synchronous activity in a large population of neurons with associated changes in blood flow, metabolism, and shifts in extracellular ions and fluids, the associated IOS should be enormous. In fact, Wilder Penfield, a pioneer epilepsy neurosurgeon, commented in the 1930s that he could grossly perceive focal dilation in local vasculature associated with neocortical seizures in the operating room with the naked eye *(93)*. IOS imaging has been applied to the study of epileptogenesis in the slice preparation *(94–97)*, isolated guinea pig whole brain *(98)*, in vivo rat *(99)*, ferret *(100)*, nonhuman primate *(101)*, and human *(102)*. The first report of using IOS imaging to map epilepsy in vivo was performed in humans by Haglund et al. in 1992 *(102)*. They triggered afterdischarges with a bipolar stimulating electrode and optically recorded the IOS at 610 nm. In addition to noting the large amplitude of the signal, they described an "inverted" optical signal recorded from the adjacent brain, which was hypothesized to represent either surround inhibition or shunting of blood flow from adjacent cortex. The first report of in vivo imaging of pharmacologically induced spontaneous (not electrically triggered) epileptiform events with simultaneous electrophysiological monitoring was by Schwartz and Bonhoeffer *(100)*. In their study in ferret visual cortex at 707 nm, both interictal and ictal events were induced pharmacologically. With simultaneous field potential recordings, the precise time and morphology of the epileptiform event was correlated with the optical signal. They also confirmed the inverted optical signal from surrounding cortex and demonstrated a decrease in neuronal activity with single-unit recordings. In this work, they also described the first imaging of a mirror focus in contralateral homotopic cortex. Chen et al. *(99)* reported IOS imaging of penicillin-induced seizures in the rat and raised the possibility that the IOS may be able to predict seizure onset as early as 1 min before an electrographic event, which has not been confirmed by other groups. Finally, Schwartz *(27)* reported simultaneous optical imaging of epileptiform events and functional architecture in ferret visual cortex.

1.4. THEORETICAL BASIS FOR CLINICAL UTILITY OF IOS IMAGING IN THE TREATMENT OF EPILEPSY

The epilepsies are categorized based on the location of the region of brain involved at the onset of the seizure. Focal or "partial" seizures begin in localized, abnormal areas of the brain. These epileptic events can secondarily generalize by propagating through more normal areas and sequentially recruiting additional neuronal populations. Currently, the only known cure for epilepsy is surgical removal of the epileptic focus. It is estimated that there are more than 100,000 currently untreated surgical candidates with partial epilepsy with 5000 new candidates each year *(103)*. Partial epilepsies can start in either the medial temporal lobe structure or the neocortex. Surgery is only curative in partial epilepsy, and the key to successful outcome is the ability to localize the epileptic focus. IOS imaging is most useful in imaging the neocortex since it is easily exposed and hence only neocortical epilepsy will be discussed in this chapter.

The current gold standard in mapping epilepsy uses electrophysiologic recordings from the surface of the brain. Surface electrographic techniques such as electroencephalography (EEG) and electrocorticography (ECoG) record field waves generated by membrane currents of coactive neurons passing though the extracellular space. These field potentials reflect the linear sum of fields generated by current sources (current from the intracellular space to the extracellular space) and sinks (current from the extracellular space to the intracellular space; ref. *104*). The predominant sources of these fields are slow synaptic and nonsynaptic events. Because the principles of field recording are based on volume conduction, and the cortex can be modeled as a convoluted dipole layer *(105)*, the relationship between the size of the potential at a given point and the distance of the point from the generator is not straightforward. This severely undermines the localizing value of field recordings *(105)*. In addition, differentiating volume conduction from neurophysiological propagation with surface recordings is quite difficult *(106)*. Action potentials contribute little to field recordings unless the electrode is in very close proximity to a large population of synchronously bursting neurons. However, simultaneous extracellular and surface recordings from models of chronic epileptogenesis have shown that surface recording techniques do not localize the activity of bursting "epileptic" neurons. Hence, surface spikes may be a propagated phenomenon only loosely co-localized with the epileptogenic region *(90)*.

This raises the intriguing possibility that IOS imaging may actually be a better, or at least complementary, technique for localizing the epileptic focus and thus may potentially help guide surgical resections. Preliminary evidence for this conclusion was reported by Schwartz and Bonhoeffer *(100)*, who found that the surface ECoG was insensitive to interictal spikes that could be recorded with both a local field potential and IOS imaging. Likewise, the report by Chen et al. *(99)* that IOS may be a useful predictor of epileptiform events increases the potential clinical utility of the technique. Optically recorded changes in cytosolic free calcium in vivo were also noted approx 20 s before electrophysiological events, further supporting the notion that optical imaging techniques may be more sensitive to pre-epileptiform events and useful in predicting seizure onset *(107)*.

In the remainder of this chapter, we will present recent advances in the imaging of interictal and ictal events in vivo in the ferret and rat (Subheading 2) and discuss the current state of the use of IOS imaging for intraoperative human imaging (Subheading 3). Our focus will be the work done in our laboratory.

2. OPTICAL IMAGING OF ANIMAL MODELS OF EPILEPSY

2.1. BACKGROUND

Because the bulk of the experiments in this study were performed on animal models of epileptogenesis, it will be helpful to review our current understanding of the pathophysiology of those models relevant to this study. Because few animals develop spontaneous seizures, none of these models are fully trustworthy as an imitation of clinical epilepsy *(108,109)*. However, each provides a well-controlled environment for exploring different aspects of epileptogenesis. Acute models, in particular, have been extensively studied but lack several of the key components that characterize more chronic models and human epileptogenesis . We use acute models to address questions regarding the relationship between the optical signal and the underlying electrophysiology because they have been extensively studied and their pathophysiology is well understood. Chronic models are useful to investigate specific questions relevant to human epileptogenesis that cannot be studied in patients for ethical reasons.

2.1.1. Acute in Vivo Rodent Models

Acute models of neocortical interictal events in the rodent usually involve either the focal intracortical application of GABA-A antagonists, such as penicillin or bicuculline methiodide (BMI), or cryogenic injury *(109,110)*. Neocortical ictal events, however, can be precipitated either by application of 4-aminopyridine (4-AP) or direct cortical stimulation producing afterdischarges *(111–113)*.

2.1.1.1. The Disinhibition Model

Intracortical injection of GABA-A antagonists induces stereotypical IIS at a frequency of 0.2–0.7 Hz in many animals, including the nonhuman primate, cat, ferret, rabbit, rat, and mouse *(62,64,100,109,110,114–116)*. Intracellular recordings from within the focus reveal that a majority of neurons sampled exhibit a paroxysmal depolarization shift (PDS) simultaneous with the IIS recorded from the local f.p. *(62–64,83)*. Intracellular recordings from neurons at variable distances from the center of the focus reveal varying degrees of activity ranging from a sustained PDS in the center, to a truncated PDS at the margin, to subthreshold EPSPs, to brief EPSPs, followed by an IPSP and prolonged IPSPs in the surrounding cortex, representing the so-called "inhibitory surround" *(81–83,117)*. Whether surround inhibition is an essential component of the epileptic focus or only found in the acute pharmacologically disinhibited focus is controversial. Although the acute disinhibited model of epileptogenesis has been historically the most widely used model, it has several disadvantages. Although there is some evidence that GABAergic circuitry and pharmacology may be reduced in human epilepsy, it is not necessary for epileptogenesis and is even upregulated in many models of chronic epileptogenesis (*see* Subheading 1.3.2.; ref. *72*). The intensity of the focus as manifested in the large percentage of neuronal participation in the IIS is not seen in neurons in human foci studied with extracellular recordings during neurosurgical operations *(118–120)*. Likewise, the inhibitory surround has not been described in either human or chronic models of epileptogenesis, although few studies have directly examined this question *(91,120)*.

2.1.1.2. The 4-AP Model

4-AP is a potent convulsant when applied to the neocortex *(121)*. It has many mechanisms of action, most of which act to increase synaptic transmission. At doses less than 10 μM, it acts by blocking slowly inactivating potassium currents *(122)* and enhances the release of synaptic neurotransmitters *(123,124)*. At higher doses, it enhances calcium currents at synaptic terminals *(125)*. In contrast to its effect in the slice preparation, in which interictal events are produced, application to the neocortex in vivo generates tonic-clonic ictal electrographic seizures *(111,112,126)*. Events last anywhere from 60–300 s with interictal periods of 5–20 min *(111,112,126)*. Ictal events begin focally at the point of application and spread horizontally in a symmetric, reproducible fashion *(100,127,128)*. In contrast to interictal models, intracellular recordings reveal little correlation between depolarizing potentials and surface paroxysmal discharges *(126)*. These findings are reminiscent of more chronic models of epilepsy as well the human situation. The 4-AP model is the best acute model of ictal events and is particularly useful for examining the relative localization of ictal onset versus offset as well as the relationship between the optical recording and surface electrophysiological recording of horizontally propagating seizures.

2.2. IOS IMAGING OF INTERICTAL SPIKES

We have performed IOS imaging of IIS using the acute disinhibition model in both ferrets and rats. It was originally unclear if such a slow signal like the IOS had sufficient temporal resolution to map the IIS, which spreads across the cortex at a rate of approx 80 mm/s *(67,129)*. In fact, the electrophysiologic spread of the IIS is far too rapid to be resolved with the IOS. However, as will be demonstrated, if the time between each IIS is sufficient for the intrinsic signal to rise and fall back to near baseline, and the lateral spread of each IIS is smaller than the region of exposed cortex, the spatial extent of each IIS can be easily resolved with the IOS.

2.2.1. Animal Surgery

For the ferrets, anesthesia was induced with a mixture of ketamine (15–30 mg/kg i.m.) and xylazine (1.5–2.0 mg/kg i.m.) supplemented with atropine (0.15 mg/kg i.m). After tracheotomy, animals were ventilated with 60–70% N_2O, 30–40% O_2, and 1.4–1.6% halothane (halothane was reduced to 0.8–0.9% for the imaging) and placed in a stereotactic frame. End-tidal CO_2 was maintained at 3.2–3.8% and animals were hydrated with 2 mL/kg/h dextrose/Ringer's solution and paralyzed with intravenous gallamine triethiodide (30 mg/kg/h). For the rat studies, adult male Sprague–Dawley rats (250–375 g) were initially anesthetized with an intraperitoneal (i.p.) injection of a cocktail of 90 mg/kg ketamine and 4.0 mg/kg xylazine. After induction, the trachea was cannulated and anesthesia maintained with i.p. injection of 1.3 g/kg urethane. Animals were placed in a stereotactic frame and hydrated with i.p. 2 mL/kg/h dextrose/Ringer's solution. Supplementary anesthesia was administered on occasion as needed, depending on the animal's reflex response to a toe-pinch. Oxygen saturation and end-tidal CO_2 were monitored and kept constant at 100% and 3.5%, respectively. In both preparations, dexamethasone (0.1 mg/kg; Steris Laboratories, Phoenix, AZ) and atropine (0.5 g/kg; Atroject; Burns, Rockville, NY) were administered subcutaneously. ECG and rectal temperature were continually monitored and temperature was maintained at 37°C by means of a homeothermic blanket system (Harvard Apparatus, Holliston, MA). For the ferrets, a craniotomy was performed over visual cortex with a high-speed dental drill and the dura was opened. For the rats, the skull was thinned and a small hole made in the skull and dura over the hindpaw somatosensory area (Fig. 2).

2.2.2. Electrophysiology and Epileptogenesis

Epidural ECoG was performed with two electrodes on either side of the craniotomy, approx 5 mm from the epileptic focus. In both preparations, two glass micropipets were advanced into layers II-III with micromanipulators. One micropipet was filled with 1% NaCl for field potential recording. A second micropipette with a tip resistance of 4–6 MΩ, filled with a solution of bicuculline methiodide (5 mM in 165 mM NaCl, pH 3.0), was positioned less than 1 mm from the field potential electrode. ECoG and f.p. signals were amplified, bandpass filtered (1–100 Hz), and digitized at 200 Hz (ferret) or 2000 Hz (rat) and recorded onto a PC.

Interictal foci were induced by iontophoresis of BMI using a current of –15 to –20 nA for retention and +50 to +500 nA for release depending on the resistance of the micropipette tip. Positive currents were maintained until stereotypical IIS were recorded (after ~5 min).

2.2.3. IOS Imaging

2% agar and a glass coverslip were placed over the cortex for stabilization. For the ferret experiments, the brain was illuminated by a halogen lamp filtered to 707 ± 10 nm through two fiberoptic light guides. For the rat experiments, illumination was at 546 ± 10 nm, 605 ± 10 nm, 630 ±10 nm, and 700 ± 10 nm. In the ferret, the optical reflectance signal was recorded at 2 Hz with a cooled CCD camera (ORA 2001, Optical Imaging Inc., Germantown, NY) equipped with a tandem lens *(130)*, focused approx 500 µm beneath the cortical surface. For the rat, the optical signal was recorded with a 10-bit video camera (Imager 3001, Optical Imaging Inc., Germantown, NY) at 10 Hz.

Image processing was done with custom software written in either IDL (Research Systems, Inc) or MATLAB (The MathWorks, Inc.) was used to generate the epilepsy maps. Blank-divided (BD) maps were produced by dividing each frame acquired during epileptic conditions from control images obtained with a negative holding current. Spike-triggered (ST) epilepsy maps were obtained by dividing single frames following the epileptic event by the single frame preceding the event. These maps could be averaged over multiple events since the IIS is so stereotypical. To determine the spatial extent of each epileptiform event, we used the method of normalized threshold analysis developed by Chen-Bee et al. *(131)* and others *(35,132)*. For a given set of images, the dynamic range of pixel intensities (minimum to maximum) was determined. Each pixel was then evaluated to determine whether it was a given percentile of the dynamic range below the median pixel intensity. The median was used instead of the mean because it is less

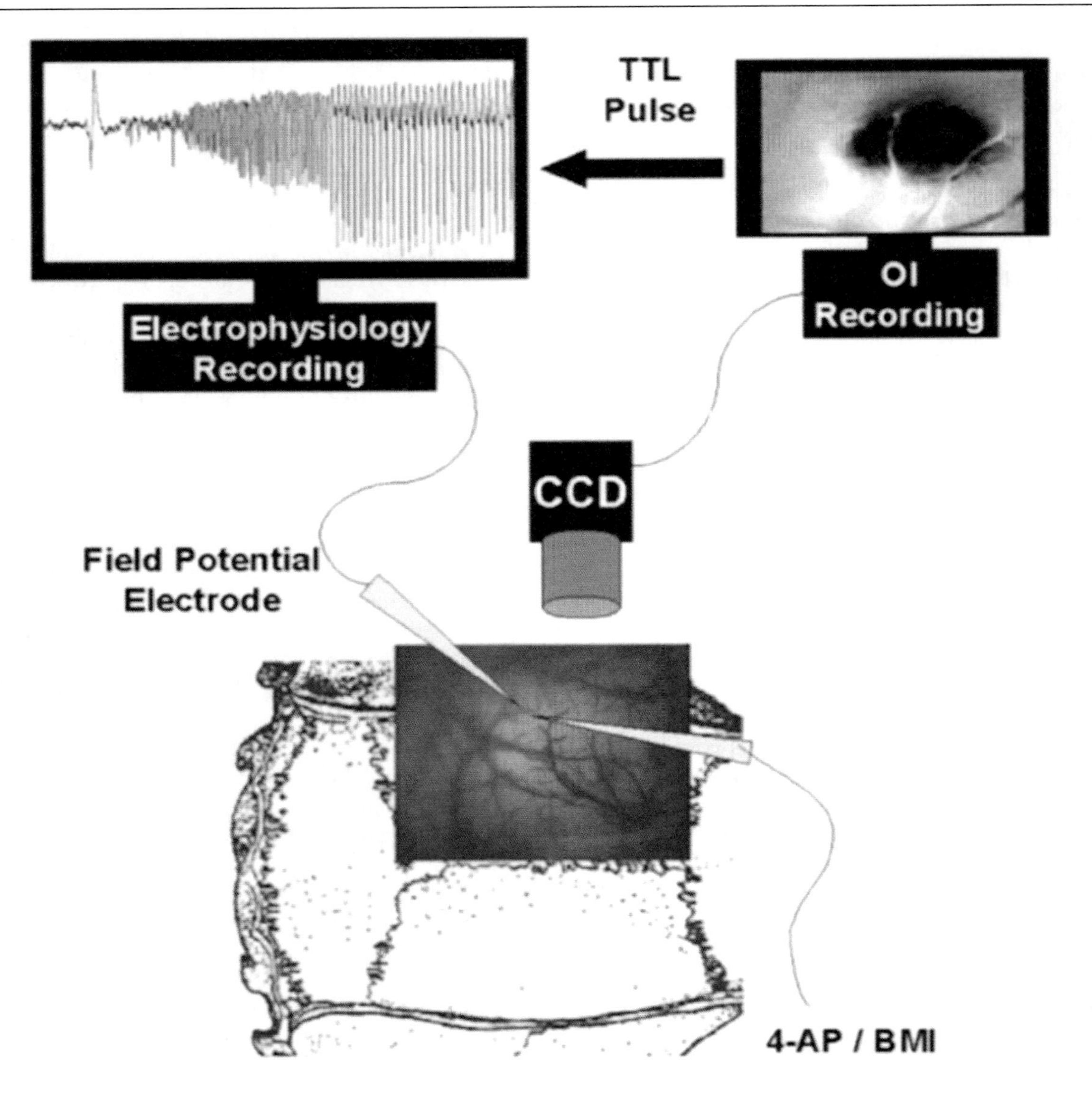

Fig. 2. Schematic diagram of the experimental setup for in vivo optical imaging in the rat neocortex. At bottom, the rat brain is exposed and one side of the skull over the neocortex is thinned. A small hole is made in the thinned skull and the dura below, and a field potential electrode is inserted, along with a second electrode, through which a pharmacological agent may be injected to cause ictal or interictal events. The field potential signal is continuously recorded. Images are collected by the CCD camera placed and digitized onto a PC. A TTL pulse from the imaging computer is fed into the computer that records the electrophysiology mark each frame acquisition, so that the imaging and the interictal or ictal events can be temporally correlated during off-line analysis.

sensitive to outliers. Note that this procedure also can be used to determine the size of the region of increased light reflectance (i.e., inverted optical signal in the putative inhibitory surround), by evaluating which pixels are at a given percentile of the dynamic range *above* the median. The determination of spatial extent as just defined can be performed using any given percentile of the dynamic range ("threshold"). With higher thresholds, the criterion for determining whether a given pixel is part of the seizure is more stringent, and thus as the threshold is increased, the seizure will appear "smaller." Hence, the calculated area is somewhat arbitrary.

2.2.4. Single Wavelength Imaging in the Ferret

In the first report of simultaneous electrophysiology and IOS imaging of seizures in vivo, Schwartz and Bonhoeffer *(100)* observed a clear change in reflectance of light associated with each IIS (Fig. 3). As the IIS developed, and its amplitude increased in size, the area, and magnitude of the change in reflectance also increased. Figure 3 shows blank-divided maps of an IIS focus developing in ferret visual cortex. A clear change in reflectance of light can be seen after each IIS. Note that the IIS is seen in the local f.p. but not in the adjacent ECoG. Although the large amplitude of the signal makes imaging a single IIS possible, by averaging over multiple spikes, with spike-triggered image division, the signal-to-noise ratio is improved and time-course of the optical signal is apparent (Fig. 4). The optical signal begins within 500 ms after the IIS and peaks at 1 s (Fig. 4). These studies provoked the following questions, which we explored in the rat. First, if the signal is

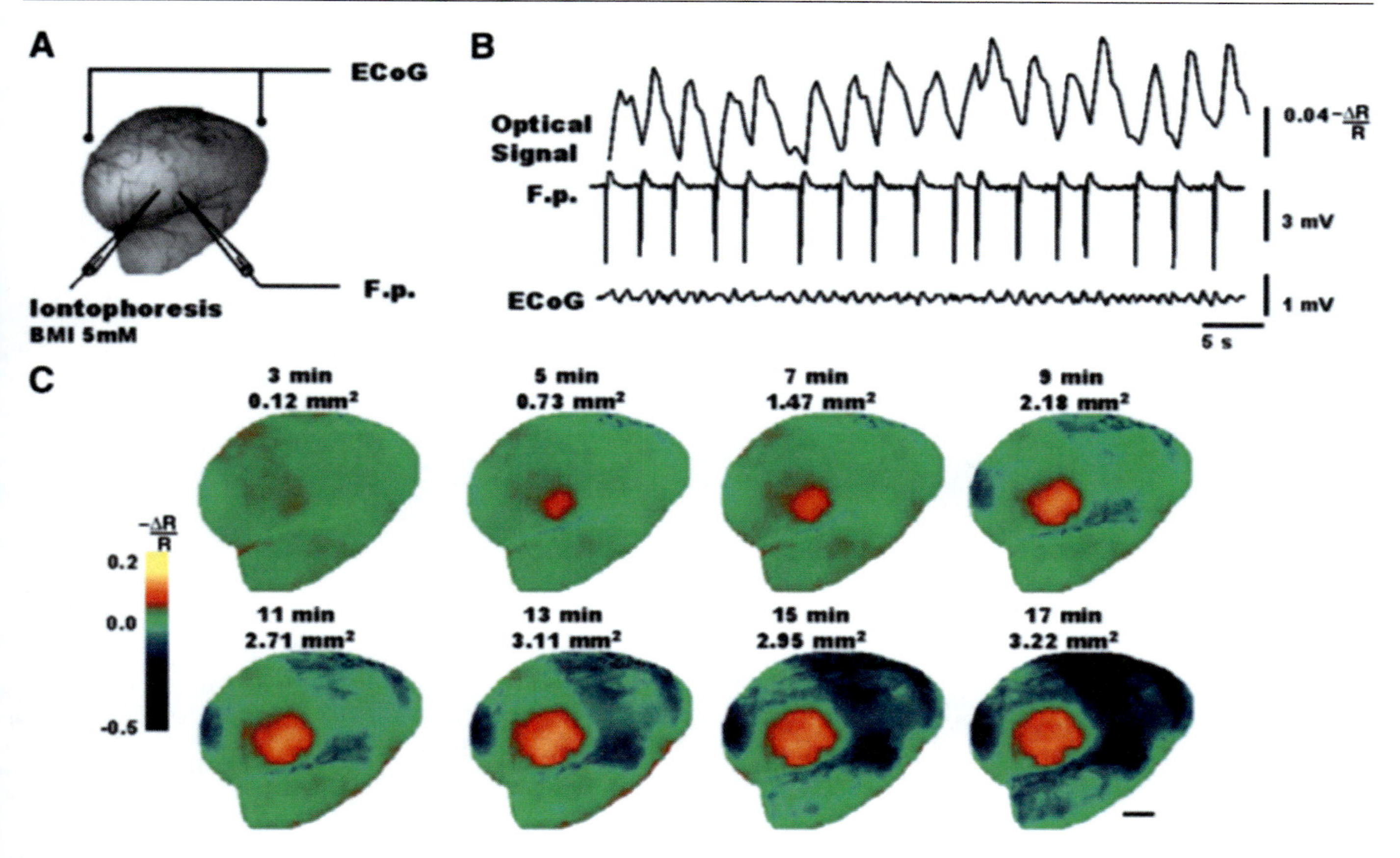

Fig. 3. (**A**) Simultaneous optical signal, f.p., and ECoG recording show that (**B**) each IIS has a spatially discrete optical correlate. Note that (A) the ECoG, located beyond the limits of the optical signal, (**B**) does not record the interictal events supporting the conclusion that the optical signal indicates the spatial limits of the electrophysiological event. (**C**) BD maps, each averaged over 1 min of recording (~ 21 spikes). During the earliest, small amplitude IIS, the mean area of the focus for all experiments was 0.12 ± 0.02 mm^2, with a minimum of 0.08 mm^2. The area of the focus then increased in size during the next several minutes and finally stabilized at a mean size of 2.84 ± 1.59 mm^2, corresponding with an increase in the amplitude of the f.p. spike. The area of the IIS was derived from the BD maps by thresholding to a pixel value one standard deviation above the pixel values from the area of the focus during control conditions. Notice that the inverted optical signal in the surrounding cortex also increases in intensity and area as the focus develops. Scale bar, 1 mm. (Reprinted with the permission of *Nature Medicine*.)

apparent as early as 500 ms after the IIS, with faster imaging it might be visible even sooner. This would be unusual because the IOS is thought to be a slow signal, as demonstrated in the functional architecture studies. Perhaps in epilepsy, the signal is more rapid? Second, these ferret studies were only performed at one wavelength and the relationship between the wavelength of incident light and epileptiform events had not been explored.

2.2.5. Multiwavelength Imaging in the Rat

Figure 5 demonstrates each individual frame recorded at 10 Hz (100 ms/frame) using spike-triggered image division in rat neocortex at four different wavelengths in the same animal. What is immediately apparent is that the change in reflectance occurs as early as 100 ms after the IIS, regardless of the wavelength recorded. At 546 nm, the tissue darkening in the focus is not seen in the blood vessels, and it takes longer to evolve and appears to spread more widely. An inverted signal is appreciated in the surrounding cortex. At higher wavelengths, the darkening is more focal and there is an inverted signal in the draining veins as well as the surrounding cortex. Also, at higher wavelengths, the signal in the focus develops and dissipates more rapidly. Figure 6 is a graphical representation of the amplitude of the change in light reflectance in the focus and surround as well as the extent of spread of the optical signal recorded at each wavelength. As demonstrated in Fig. 6A, the amplitude of reflectance change in the focus is greatest at 546 nm (0.4%; $n = 7$ rats) and least at 700 nm (0.05%; $n = 9$ rats). When the data are normalized to amplitude, it is clear that the signal increases earlier and more quickly at higher wavelengths (Fig. 6B). A statistically significant change in light reflectance (paired t-test, $p < 0.05$) is observed as early as 100 ms for four rats at 546 nm, eight rats at 605 nm, seven rats at 630 nm, and five rats at 700 nm (Fig. 6C). Likewise, the peak of the reflectance change occurred 1–2 s after the IIS for 605 nm, 630 nm, and 700 nm and 2–3 s after the IIS for 546 nm.

The amplitude and latency of the inverted optical signal in the surround, on the other hand, was similar at different wavelengths (Fig. 6D), which can be seen even when normalized to the maximal amplitude of the inverted signal (Fig. 6E). This inverted signal peaked 1–2 s after the IIS for all wavelengths but also was recorded as early as 100 ms after the IIS. A statistically

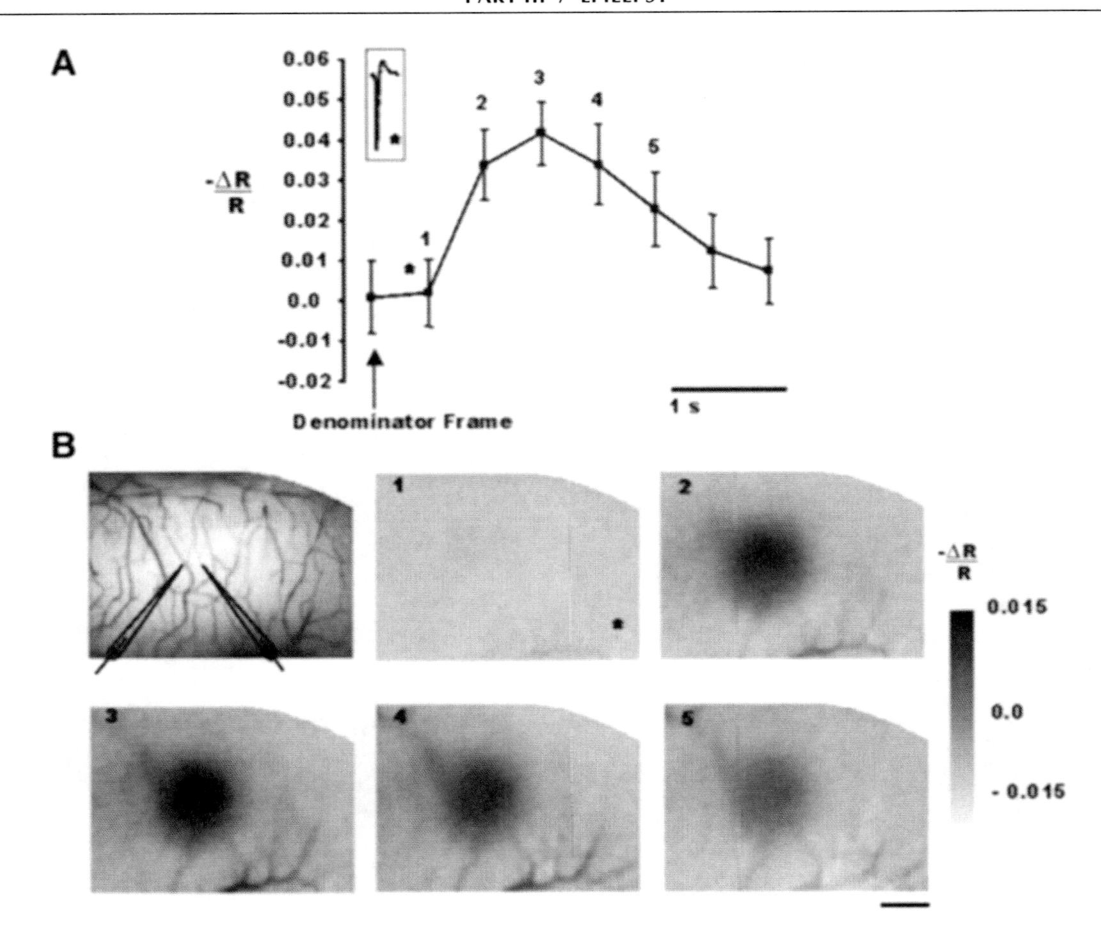

Fig. 4. (**A**) The change in reflectance (-ΔR/R) as a function of time for the 0.5 s before and 3 s after each IIS is determined with spike triggered image division. The error bars show standard deviation over 116 different spikes. The asterisk indicates the time of occurrence of the IIS. The numbers in (A) correspond to the numbers in the panels of (**B**). The top left panel in (B) shows the location of the electrodes and the blood vessel configuration. Subsequent images show spike-triggered epilepsy maps obtained by dividing each camera frame after the spike by the frame prior to the spike (the "denominator frame"). Divided images are averaged more than 116 spikes. Averaging increases signal-to-noise ratio, though maps can often be seen from a single interictal spike. Scale bar in (B) indicates 1 mm. (Reprinted with the permission of *Nature Medicine*.)

significant change in reflectance in the surround (paired *t*-test, $p < 0.05$) was observed at 100 ms for three rats at 546 nm, six rats at 605 nm, five rats at 630 nm, and two rats at 700 nm (Fig. 6F).

Although in Fig. 5 the change in reflectance of light appears more diffuse at 546 nm and more focal with increasing wavelength, this was not always the case. Figure 6G demonstrates the average area of spread of the change in reflectance in the focus in all animals as a function of time. Although the IOS signal spreads more rapidly at higher wavelengths, as seen in the steeper slope of the curves at 605, 630, and 700 nm, the maximal area (peak) is not significantly different between 546, 605. and 630 nm when averaged among animals (analysis of variance [ANOVA]). The optical signal recorded at 700 nm, however, was clearly more focal than the other wavelengths.

2.2.6. IOS of a Mirror Focus

In several animal experimental models of epilepsy, mirror foci have been described in which an independent epileptic focus develops in a homotopic location in the contralateral hemisphere. Although not described in humans, the etiology likely depends on cross-callosal kindling. In acute models of epilepsy, rapid cross-callosal spread of epileptiform events has been described *(133,134)*. Schwartz and Bonhoeffer *(100)* investigated this phenomenon optically. Using acute focal disinhibition with BMI iontophoresis in ferret somatosensory cortex, the contralateral homotopic area of cortex was imaged and spike-triggered image division was performed based on the timing of the IIS that occurred contralateral to the imaging. Figure 7 shows the results of their experiments. A small epileptic focus could be seen optically in the cortex contralateral to the iontophoresis. Contralateral hemisphere maps, triggered to ipsilateral spikes, showed a clear increase in the optical signal during the first 0.5–1 s, with an average amplitude of 0.4 ± 0.18% (Fig. 7). As expected from single-unit recordings in homotopic foci *(133,134)*, the optical signal was smaller and slightly more delayed than the signal recorded ipsilateral to a focus. Nevertheless a clear, well-circumscribed focus could be seen, demonstrating that it is possible to visualize epileptic events and detect a focus even if it is comparatively weak and not directly caused by epileptogenic agents.

2.3. IOS OF ICTAL EVENTS

As described in Subheading 2.1.1.2, the 4-AP model is a particularly useful acute model of ictal events for examining the initiation, spread and offset of seizures as well as the rela-

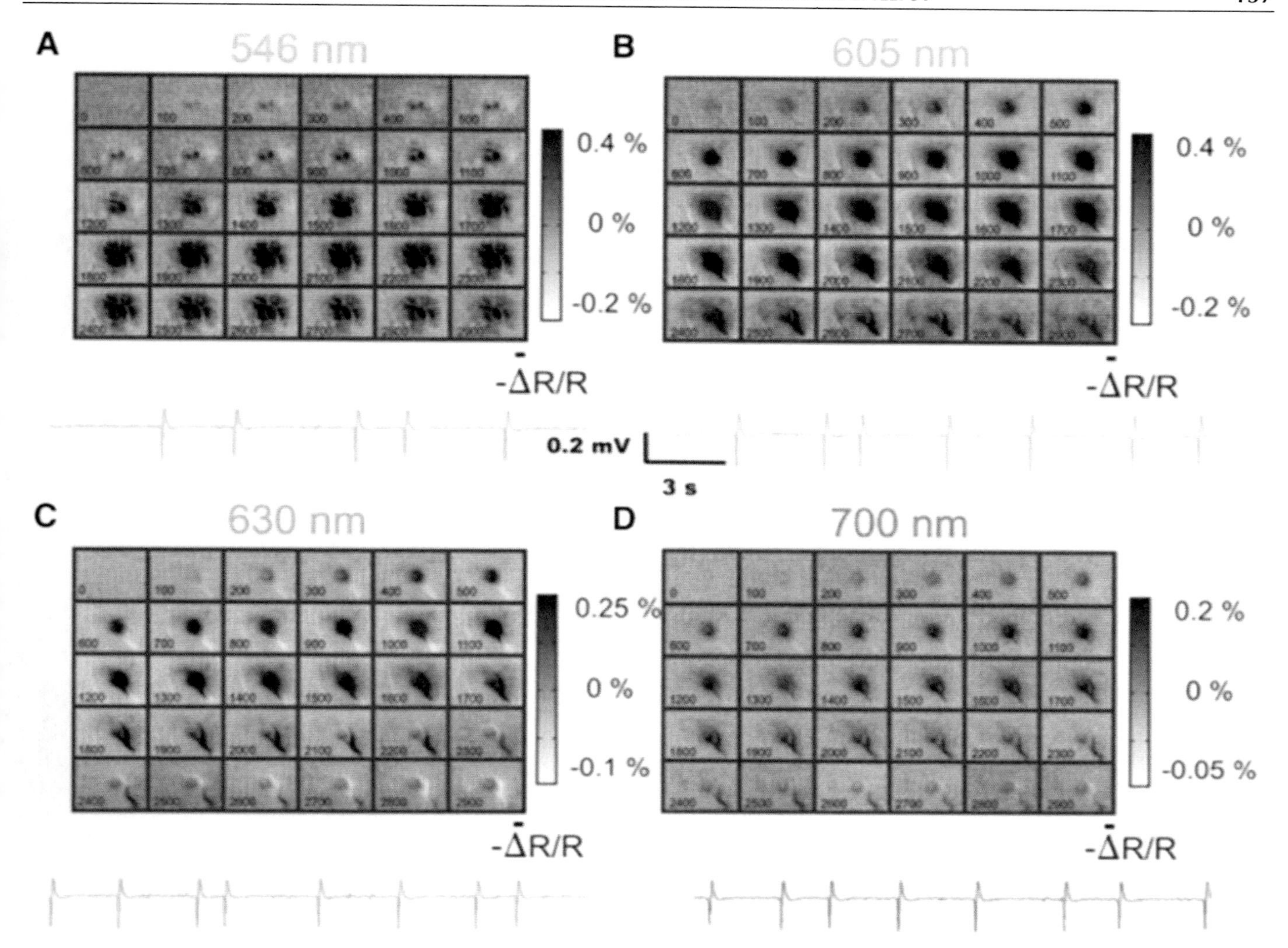

Fig. 5. Multiwavelength fast imaging of IIS. Averaging multiple spikes (100–200) with spike triggered image division at (**A**) 546 ± 10 nm, (**B**) 605 ± 10 nm, (**C**) 630 ± 10 nm, and (**D**) 700 ± 10 nm demonstrate that the intrinsic signal change is evident within 100 ms of the IIS. Note that although there is more blood vessel artifact at higher wavelengths, the intrinsic signal change also appears to be more focal. Draining veins have an early inverted signal higher wavelengths. Examples of IIS field potential recordings are shown at the bottom of each panel. The number at the bottom left of each image indicates the time after the spike occurrence, in units of milliseconds. Scale bar, 1 mm.

tionship between the optical recording at different wavelengths and surface electrophysiological recording of horizontally propagating seizures. Once again, it was initially not clear if the IOS had sufficient temporal resolution to image the horizontal spread of a seizure and whether the ictal onset zone (site of initiation) could be determined prior to lateral propagation.

2.3.1. Methods

The animal preparation and surgical technique are identical to the induction of IIS, except that a glass micropipette filled with a solution of 25 m*M* 4-AP and attached to a Nanoject II oocyte injector (Drummond Scientific, Broomall PA) was positioned less than 1 mm from the field potential electrode. Electrographic seizures were induced by the injection of 0.5 µL of 4-AP (25 m*M* in 1% NaCl) into cortical layers II-III in increments of 50 nL. Once seizures were initiated, they were often stereotypical in form (though exhibiting several different onset morphologies, as discussed below), but varied in duration between animals, from 60 to 300 s. The seizures occurred periodically at intervals of 5–20 min for up to 3 h.

2.3.2. IOS Imaging in Ferret Neocortex

Schwartz and Bonhoeffer *(100)* optically recorded the spread of a 4-AP seizure through ferret visual cortex. They reported that with a frame rate of 2 Hz, at 707 nm, the site of initiation could be localized to a region as small as 1 mm^2 (Fig. 8; ref. *100*). This finding is significant because successful surgery for epilepsy often requires the identification of the ictal onset zone, which must be removed to eliminate seizures *(135)*. Because the rate of horizontal propagation was relatively slow, the IOS appears to be an excellent technique to map seizure propagation. The change in reflectance is almost ten times larger for seizures (~50%) than for the IIS, so no signal averaging is required. In fact, given the variability in the morphology of each individual ictal event, signal averaging would eliminate any inter-seizure variability. Technically, one can divide all

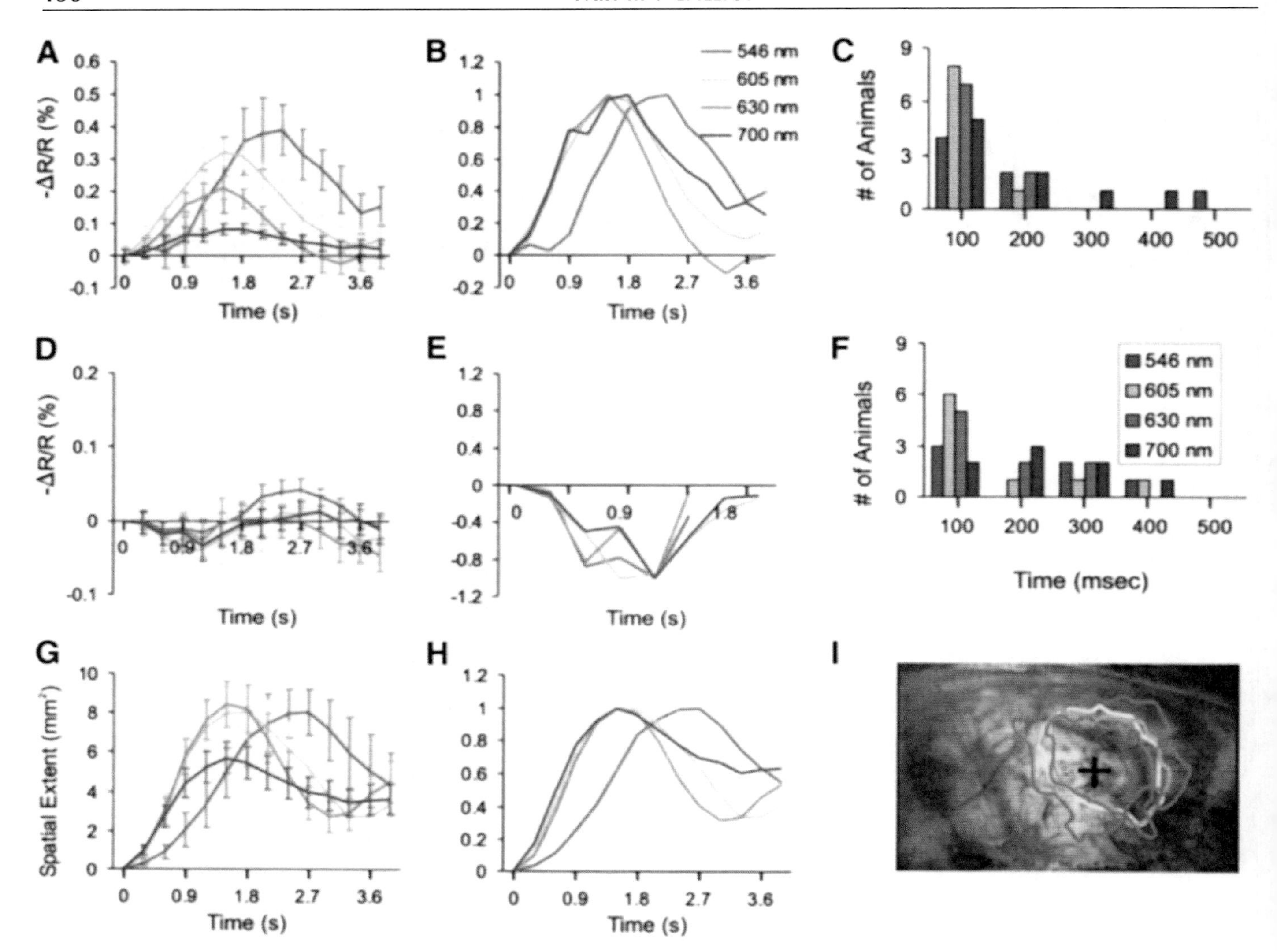

Fig. 6. **(A)** Percent change in reflection of light in the focus as a function of time and wavelength after each IIS. The amplitude of the IOS is clearly greatest when recorded at 546 nm, followed by 605 nm, 630 nm. and 700 nm. **(B)** Data are normalized to the maximum change in reflectance to show that the rise in the signal is fastest at higher wavelengths. **(C)** The earliest statistically significant change in reflectance of light was seen at 100 ms regardless of the wavelength. **(D)** Percent change in reflection of light from the surround as a function of time and wavelength following each IIS. An inhibitory signal is seen in all wavelengths and the amplitude of this inverted IOS is equivalent, regardless of wavelength. **(E)** Data are normalized to the maximum negative change in reflectance to show that temporally there is no wavelength dependence in this signal. **(F)** The earliest statistically significant inverted change in reflectance of light in the surround was seen at 100 ms, regardless of the wavelength. However, in some experiments this negative signal occurred later than the positive signal in the focus. **(G)** The extent of spread (area) of the change in reflectance in the focus over time is less wavelength-dependent. Although the maximal area was achieved earlier at 605 and 630 nm and the rate of lateral spread was faster, the maximal area of spread was similar to 546 nm. At 700 nm, the area of spread was clearly smaller. **(H)** Normalization to the maximal area shows the slower propagation at 546 nm. These data were averaged over multiple animals. **(I)** Thresholding the maximal area of spread in one animal, however, reveals that in this animal, the signal at 546 clearly propagates further than at other wavelengths. *See* color version on Companion CD.

images acquired during the seizure by a "blank" image taken before the seizure onset, generating a "movie" of the seizure (Fig. 8). Schwartz and Bonhoeffer (2001) also noted a region of inverted optical signal surrounding the evolving seizure that disappeared as the seizure spread horizontally. Whether this inverted signal truly represents neuronal inhibition or merely shunting of oxygenated blood to the more active focus is still not clear, although it will be examined in more detail below.

2.3.3. Multiwavelength IOS Imaging of Acute Seizures in Rat Neocortex

To investigate the relationship between the electrophysiology and the IOS signal in the acute seizure model, we imaged 67 seizures in 6 rats at several wavelengths. As with the IIS, the largest amplitude of IOS from the seizure was recorded at lower wavelengths. The average (S.E.) maximal values of $-\Delta R/R$ (%) were 13.70 ± 2.06% at 546 nm (21 seizures in 6 rats), 4.97 ±

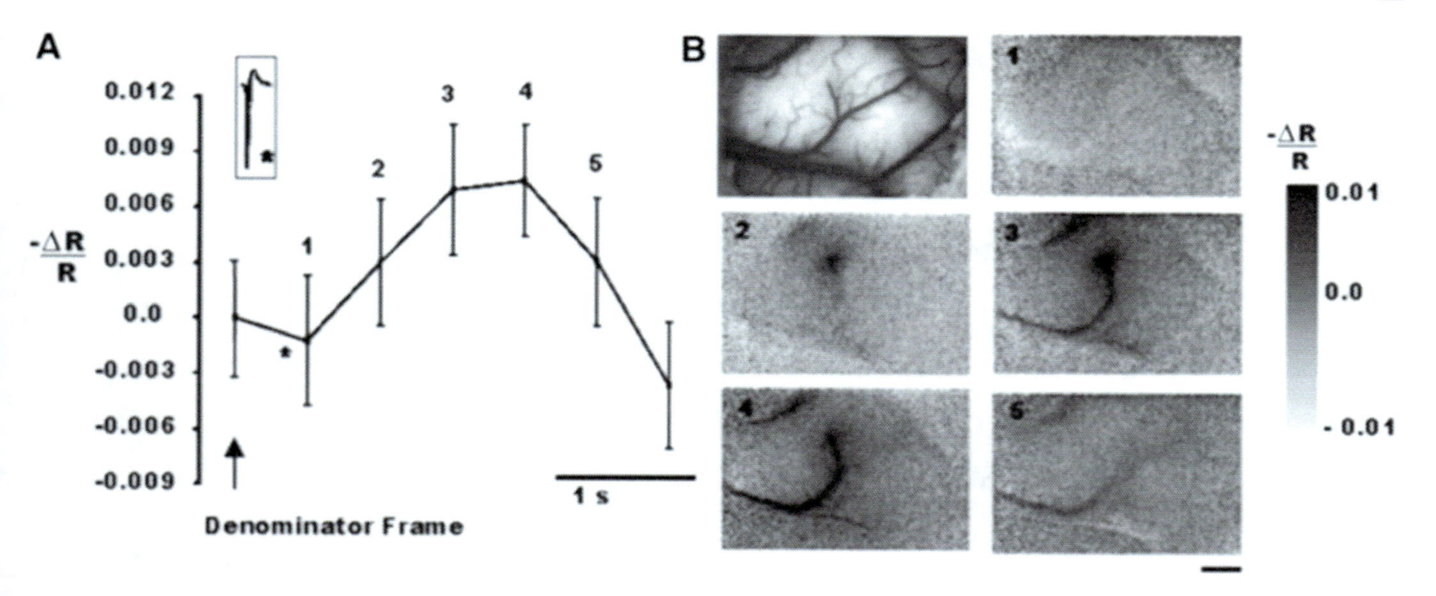

Fig. 7. (**A**) Secondary, homotopic focus epilepsy maps from the contralateral hemisphere were obtained by triggering the image division to the timing of the spikes in the ipsilateral hemisphere and averaging over multiple spikes. *Y*-axis shows-(R/R as a function of time for the 0.5 s before and 3.5 s after each spike. Error bars (± SD) were calculated from 142 consecutive spikes. (*) refers to the time of the spike and the numbers refer to the images in (B). (**B**) ST epilepsy maps are obtained by dividing each frame after the spike by the frame prior to the spike (denominator frame) and averaging over 142 spikes. The focus is well localized and appears prior to the signal from the blood vessels. Scale bar, 1 mm. (Reprinted with permission from *Nature Medicine*.)

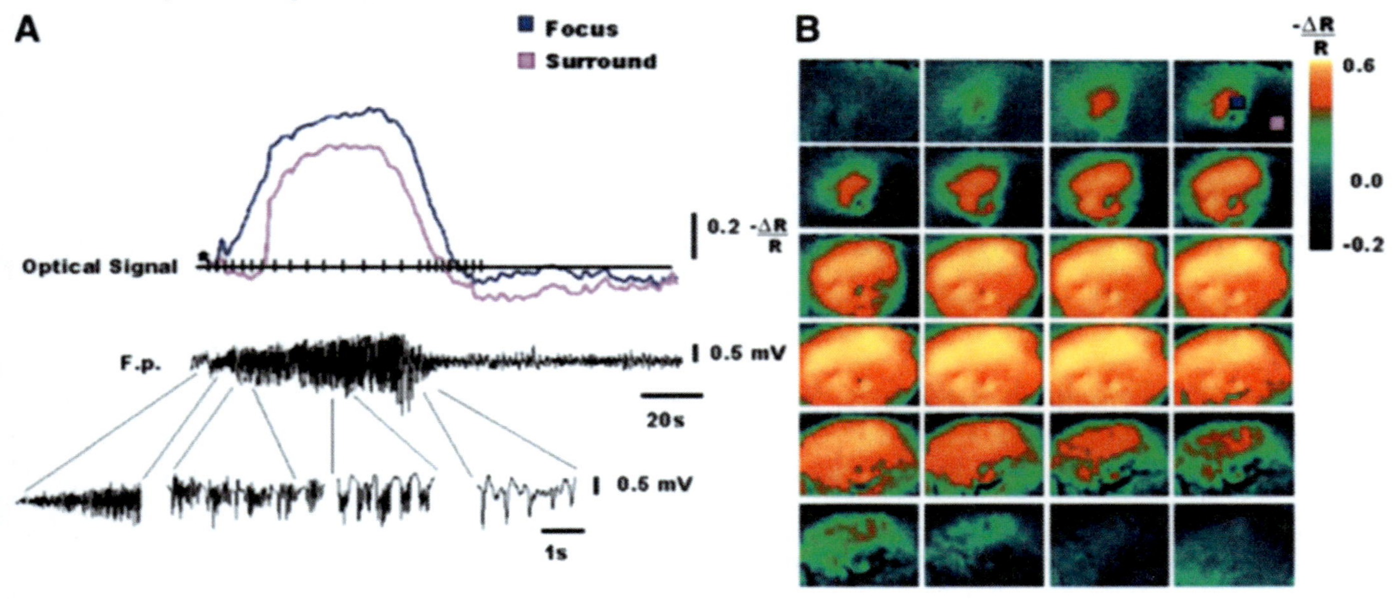

Fig. 8. (**A**) Optical signal from the focus (blue) and surround (pink) during an ictal event induced by focal application of 4-AP in ferret cortex. Ictal events last 40–80 s with interictal periods of 2–4 min. Each image is generated by dividing a single frame by a denominator frame that occurs prior to the event. (*) refers to the time of the denominator frame for the images in (B). The signal in the surround shows transient inhibition until the event propagates past. (**B**) The timing of each image corresponds with the tickmarks in (A). Images were chosen to show the onset and offset of the ictal event. The dimples in the signal represent the locations of the two micropipettes. Scale bars, 1 mm. (Reprinted with permission from *Nature Medicine*.)

0.79% at 605 nm (15 seizures in 6 rats), 3.80 ± 0.71% at 630 nm (15 seizures in 6 rats), and 1.90 ± 0.38 % at 700 nm (16 seizures in 5 rats; ANOVA, $p < 0.0001$; post-hoc Student-Neumann-Keuls (SNK) test significant for 546 nm vs 700 nm, 630 nm and 605). As discussed earlier, the signal at 546 nm is near the isosbestic wavelength of hemoglobin and is more sensitive to blood volume. Wavelengths from 600 to 650 nm image oxy/deoxyhemoglobin and >660 nm include signal from light scattering. It is not clear why the amplitude of the ictal signal is less in the rat than in the ferret. Perhaps imaging though the intact bone and dura in the rat may have attenuated the signal, although the percent change should not be affected.

Figure 9 demonstrates ictal events recorded at three different wavelengths. One can clearly see that each seizure is slightly

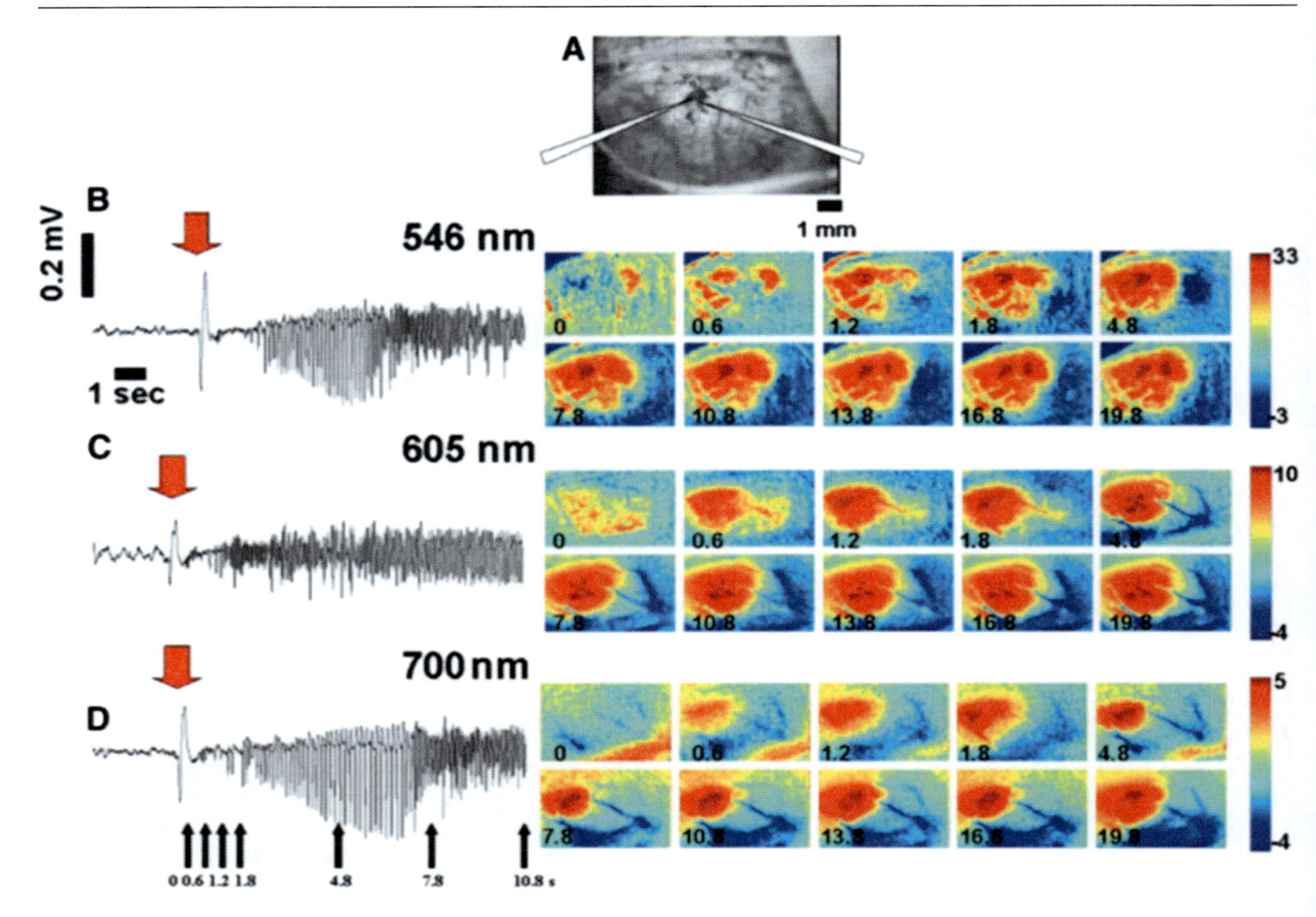

Fig. 9. (**A**) Thin skull 546-nm image or blood vessel pattern with position of f.p. electrode and 4-AP pipette. (**B–D**) The f.p. recording of the seizures is on the left and the corresponding imaging on the right. Each ictal event differs slightly from the others. The timing of the images noted on the image and with black arrows under (B). The red arrow indicates the onset of the seizure and the "denominator frame." Recordings were made at 546 nm (B), 605 nm (C), and 700 nm (D). Scale bar units show –ΔR/R (%).

different electrographically. In the example shown in Fig. 9, the optical signal at 546 nm appears to be less focal than at higher wavelengths. However, using the thresholding technique described in Subheading 2.2.3., on average, there was no statistical difference in the maximal area of spread comparing seizures imaged at 546 nm (20 seizures in 6 rats), 605 nm (15 seizures in 6 rats), 630 nm (15 seizures in 6 rats), and 700 nm (15 seizures in 6 rats). The inverted optical signal also was apparent at all four wavelengths. The average maximal (S.E.) inverted optical signal in the surround (here described as –ΔR/R, so that an increase in reflectance is a negative number) was –4.56 ± 0.85% at 546 nm (21 seizures in 6 rats), –9.62 ± 1.51% at 605 nm (15 seizures in 6 rats), –7.45 ± 1.39 % at 630 nm (15 seizures in 6 rats), and –3.98 ± 0.56 % at 700 nm (16 seizures in 5 rats) (ANOVA, p value of 0.0019). A post-hoc SNK test showed that the increase in reflectance at 605 was significantly greater than that at both 700 and 546 nm ($p < 0.01$).

2.3.4. Electrographic Variability and IOS Variability

In human epilepsy, two common electrographic patterns of ictal onset have been defined, originally in medial temporal lobe recordings from humans, but also found in the neocortex: (1) periodic spiking and (2) low-voltage fast activity (LVFA; refs. *136–140*). Periodic spiking involves high amplitude repetitive spike-and-wave events that can occur in low (1–2 Hz) or at higher frequencies (10–20 Hz) in the few seconds before or immediately at seizure onset, which may then evolve into LVFA or continue as periodic spiking *(139,140)*. Periodic spiking is characterized by single-unit burst and suppression patterns. Corresponding in time to the "spike" and the "wave," respectively, that are less likely to propagate or secondarily generalize, and resemble recurrent IIS *(136–138)*. This type of ictal onset has been correlated with an increase in inhibitory tone based on single unit recordings *in situ* in human hippocampus *(77,136)*, although contrary data have been demonstrated in resected hippocampal specimens *(141)*. LVFA, also called the "recruiting rhythm," on the other hand, is believed to represent periods of disinhibition, based on in situ human hippocampal single-unit recordings, in which the field oscillations occur at much higher frequencies (>30 Hz) compared with during periodic spiking *(136–139,141)*. This type of activity is though to propagate more readily to adjacent brain and become symptomatic *(137,138)*.

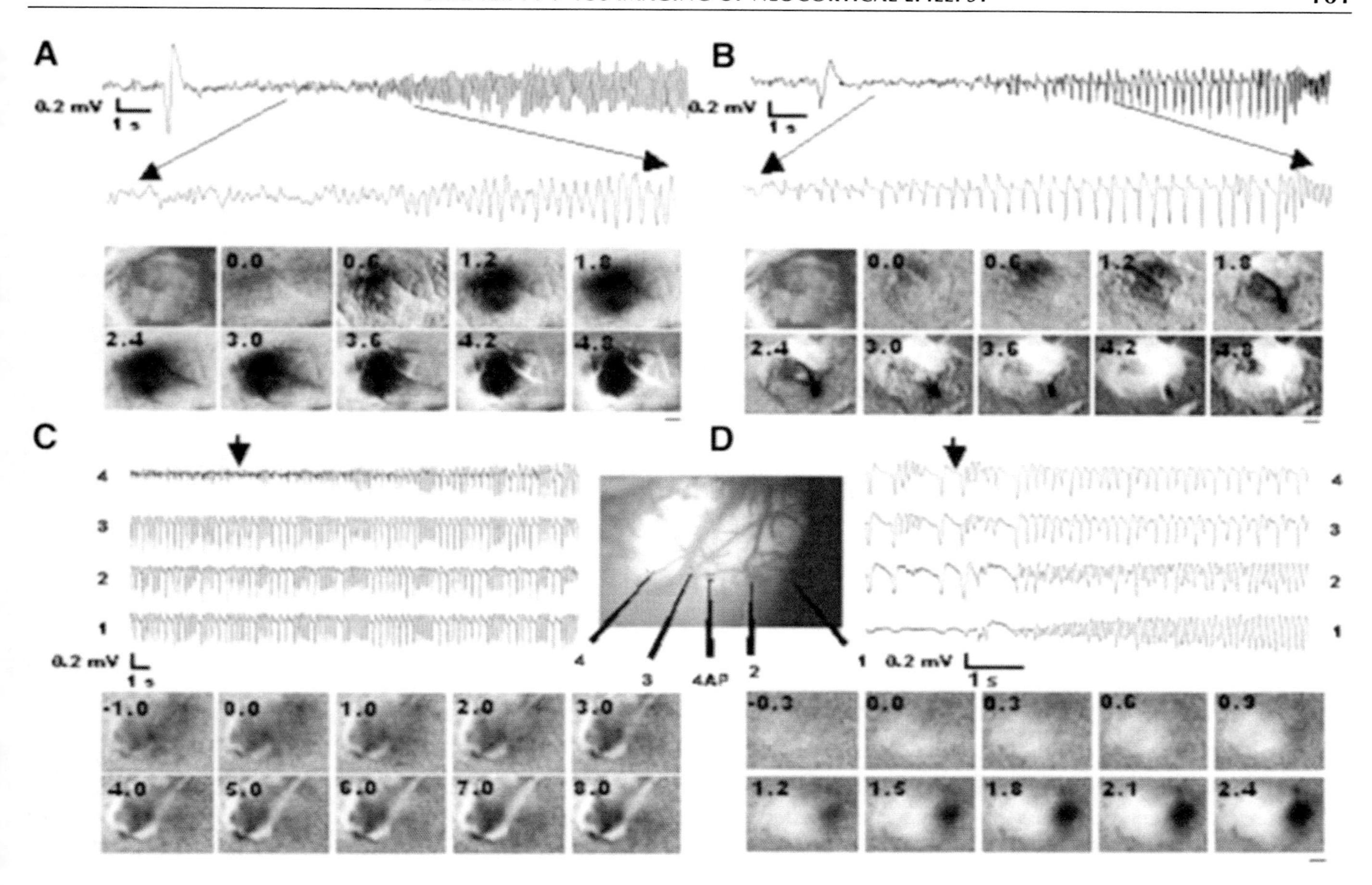

Fig. 10. Most acute 4-AP-induced seizures start with a single large spike followed by attenuation. Then, either (**A**) LVFA or (**B**) periodic spiking ensues. Optical imaging demonstrates that (A) LVFA causes a more rapid spread with less surrounding inhibition compared with (B) periodic spiking. Using a multicontact linear array of electrodes, we can image seizures originating sequentially at several different sites as inhibitory activity shifts accordingly. In (**C**) the onset is at electrode 4 and in (**D**) the onset is at electrode 1. This complex interaction is not apparent from the electrophysiology. The number at the top left of each image indicates the time after the electrographically defined seizure onset, in units of seconds. Scale bar, 1 mm. Seizures shown here are imaged with incident light of wavelengths 605 nm.

With the 4-AP model, seizures generally begin with a large spike followed either by LVFA (Fig. 10A) or periodic spiking (Fig. 10B). Each seizure is slightly different and sometimes the initial spike is absent and the length of time spent in either LVFA or periodic spiking is quite variable. Accordingly, the optical signal is variable. Even when imaging at a single wavelength, the dynamic relationship between excitation and inhibition, the focus and the surround, is quite evident. In Fig. 10A, the seizure begins with LVFA and the optical signal spreads rapidly with almost no negative signal in the surround (Fig. 10C). In contrast, the seizure in Fig. 10B begins with periodic spiking and the optical signal spreads less rapidly and there is a large inverted signal (Fig. 10D). This raises the tantalizing possibility that the optical signal is quite sensitive to excitatory and inhibitory activity and may show us the dynamic topography of this relationship, providing a map of the variability of the electrographic event.

Further support for this finding comes from simultaneous multicontact f.p. recordings and IOS imaging of ictal events. In Fig. 10C, we see a seizure developing in one area of cortex, while another seizure is occurring in a separate area of cortex. In Fig. 10D, a seizure begins within yet another area of the cortex. The corresponding optical images accurately demonstrate the spatial specificity of each of these events and the shifting foci. The topography of each set of optical maps is clearly different as is the electrographic recording. Each method provides complementary yet different information about the same event.

2.3.5. Ictal Onset Versus Offset

In human epilepsy, the area of ictal termination is not always identical to the area of initiation. The significance of this phenomenon is not well understood, but it may have prognostic significance for successful surgical treatment *(142)*. Several theories exist to explain why and how seizures terminate. One possibility is a return of inhibitory function, which eventually terminates the ictal discharge *(137,138)*. As seizures terminate, neuronal synchronization returns *(143)*. Other theories of seizure termination involve hyperpolarization via the Na-K pump *(144)*, rundown of transmitter release *(145)*, pacemaker failure *(129)*, or a large depolarizing shift, not involving neuronal inhibition, similar to spreading depression *(146)*.

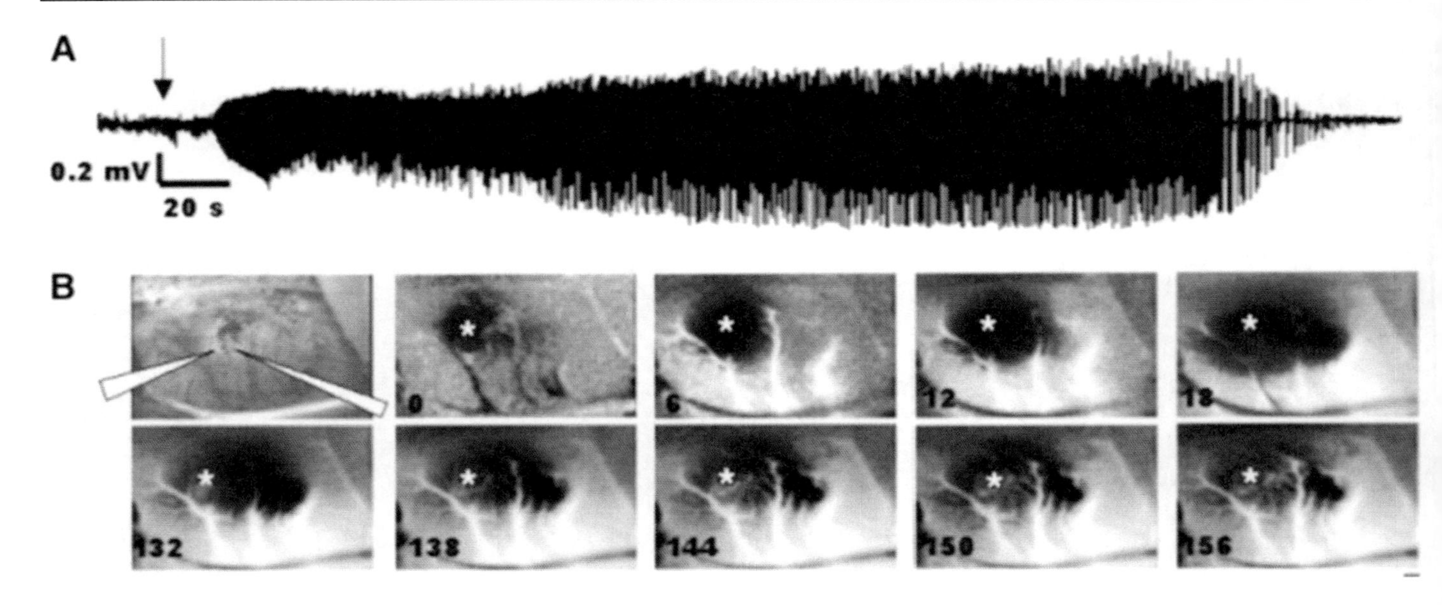

Fig. 11. **(A)** Field potential recording from 4-AP induced seizure show that ictal onsets appear electrographically different than offsets but the spatial extent of this difference is not known. **(B)** The intrinsic signals show that the area of onset (asterisk) is spatially not co-localized with the area of offset (*see* last image). The numbers in images correspond to the time after the onset (arrow). Scale bar, 1 mm. *See* color version on Companion CD.

In acute 4-AP seizures, we have observed that the location of seizure onset (determined as the centroid of the darkened region at the time of its first statistically significant rise above baseline, with the areal extent of the region determined by the thresholding method described above) often differs from the location of offset. This is illustrated in ferret cortex in Fig. 8 and in rat cortex in Fig. 11. We clearly do not see any evidence of spreading depression, which causes a large IOS change, and thus our findings contradict that theory. We also do not see an inverted signal that terminates the seizure, so a rise in inhibition may also not be critical. Regardless of the precise mechanism of seizure termination, IOS imaging is an ideal technique for investigating the relationship between ictal onset and offset zones.

2.4. IOS IMAGING OF THE RELATIONSHIP BETWEEN EPILEPTIFORM EVENTS AND FUNCTIONAL ARCHITECTURE

One of the best clinical examples of the interaction between functional brain architecture and epilepsy is "pattern-sensitive epilepsy", in which epileptic events are triggered by patterned visual stimuli *(147,148)*. Another example is the periodicity of horizontal spread of epileptic events, thought to reflect the columnar architecture of the cortex and the anisotropy of long-range horizontal connections *(66,67)*. To investigate this relationship, Schwartz *(27)* performed simultaneous IOS recording of an interictal focus as well as orientation and spatial frequency columns in ferret visual cortex. Depending on which column was injected with the epileptogenic agent, patterned visual stimuli of a particular orientation and spatial frequency could trigger each IIS *(27)*. Within the focus, the orientation and spatial frequency columns were severely distorted. In the adjacent surrounding cortex, however, the columnar architecture was not only preserved but could be easily mapped with IOS. This implies that in spite of the altered cerebral hemodynamics caused by the epileptic focus, the IOS was still accurate at mapping functional architecture within a millimeter of the focus (Fig. 12). This may prove very important for human mapping of functional architecture in patients with epilepsy adjacent to eloquent cortex.

2.5. ANIMAL MODELS OF CHRONIC EPILEPSY

Chronic neocortical foci are thought to represent a model of epileptogenesis that more closely approximates the situation in the human. Seizures develop during the course of days to months rather than seconds to minutes and last from weeks to years. Although the precise mechanism of neocortical epileptogenesis in chronic models is unknown, the etiology is thought to be multifactorial with evidence for alterations in local disinhibition *(149)*, intrinsic membrane properties favoring hyperexcitability in pyramidal neurons *(150)*, axonal sprouting in layer 5 leading to enhanced recurrent excitation *(151)*, glial regulation of extracellular potassium *(152)*, and neurotransmitter-specific neurons or receptors *(153)*. Chronic models are generally produced with the subpial injection of metal compounds such as aluminum hydroxide, cobalt, tungsten, or iron into somatosensory cortex *(109,154)*. Chronic foci tend to be more diffuse than acute foci and may produce multiple areas of epileptogenesis *(91,155)*. Surface recordings show that population spikes rarely reach 1 mV compared with spikes of 3–5 mV in acute foci *(62,63,91,155)*. The lower amplitude of interictal spikes reflects the smaller percentage of cells that participate in each interictal event *(90,91,155,156)*. Electrophysiologic mapping of chronic foci with single cell recordings has demonstrated that anywhere from 10% to 50% of the sampled populations exhibit spontaneous bursting activity simultaneous with surface potentials *(90,91,118,155,157)*. There is also a variable and inconsistent correlation between

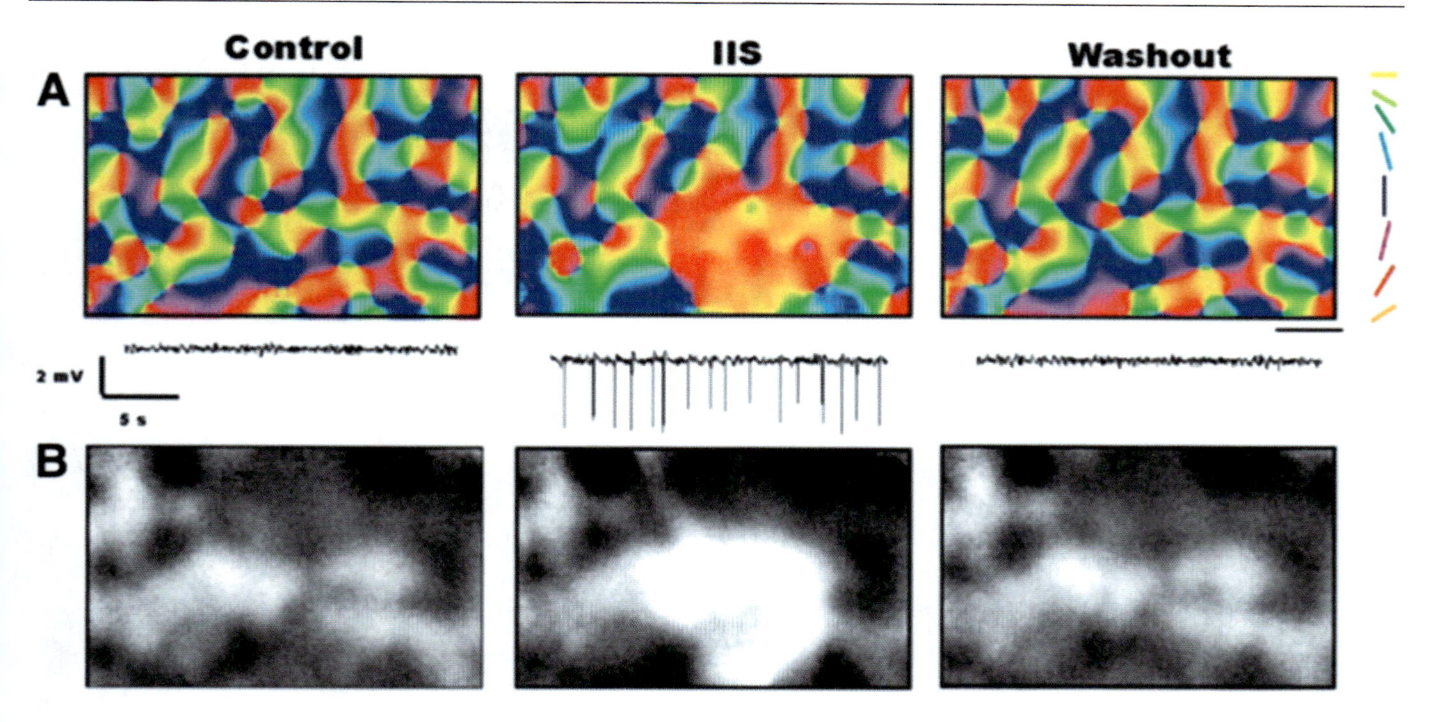

Fig. 12. (A) Angle map generated before, during and after iontophoresis of BMI into ferret cerebral cortex elicits IISs. The intrinsic signal within the focus is distorted by the occurrence of the IIS. The intrinsic signal from the surrounding cortex, however, is unaltered. The dominant yellow color in the focus indicates that 0° stimuli were more likely to trigger spikes than other orientations. Sample of the f.p. recording simultaneous with the imaging is shown under (B). Scale bar, 1 mm. **(B)** Spatial frequency maps from the same animal demonstrate that low-spatial frequency stimuli are more likely to trigger IIS than high-spatial frequency stimuli and that these maps are also distorted within the IIS focus but preserved in the surrounding brain. (Reprinted with permission from *Cerebral Cortex*.) *See* color version on Compannion CD.

surface potentials and single neuron action potentials indicating that the size of the epileptic aggregate varies from burst to burst and separate discharges begin at different sites *(90, 91,119,155)*. Electrophysiologic studies in chronic foci have also failed to report a significant amount of cellular inhibition from the neurons in the surrounding cortex, although few of these studies have specifically examined this issue *(91, 120,155)*. It has been hypothesized that the diffuse, less intense chronic focus does not strongly recruit recurrent inhibitory circuitry *(91,155)*. Overall, chronic models tend to approximate findings in chronic human epilepsy and provide a better model for investigations into epileptogenesis. We have concentrated on two models of chronic epileptogenesis in our IOS studies in the rat.

2.5.1. The $FeCl_2$ Model

Subpial microliter injections of iron salts are a reliable model of chronic interictal and ictal epileptogenesis which are particularly effective in rodents *(158–161)*. This model is particularly relevant to post-traumatic or post-hemorrhagic human epilepsy that emerges after the extravasation of blood and deposition of iron into the neuropil. The mechanism of action is not clear but may involve the free radical intermediates of oxygen and peroxidative stress on cell components as well as inhibition of Na^+/K^+ adenosine triphosphatase *(162)*. Isolated spikes appear in 90% of animals after 72 h and stabilize at 30 d with a frequency of three spikes/min *(158–161)*. Spike-and-wave complexes accompanied by behavioral manifestation appear in 75–100% of animals starting at 30 d and stabilize at 90 d *(159,161,163)*. Both types of events remain stable for up to 12 mo. Epileptiform activity contralateral to the injection has also been reported *(163)*. Histologic analysis of the resulting lesion reveals hemosiderin-laden macrophages, fibroblasts, gliosis, neuronal cell loss, and a decrease in the number of dendritic spines *(159,164)*. These findings are reminiscent of those found both in aluminum hydroxide-induced animal models and human epilepsy *(165,166)*. In vivo microdialysis reveals similar changes in extracellular amino acids as in chronic human epilepsy *(163)*.

2.5.2. The Tetanus Toxin Model

Injection of tetanus toxin to induce chronic epilepsy in the rat has been studied both in the hippocampus and in the neocortex *(167–173)*. Tetanus toxin acts by blocking exocytosis preferentially in inhibitory interneurons, thereby decreasing GABAergic inhibition *(174–177)*. Rapid but patchy spreading to adjacent and contralateral cortex occurs though axonal and transsynaptic mechanisms *(178,179)*. Injection of nanogram quantities into the cortex of the rat leads to spontaneous interictal events within 3–5 d in almost 100% of the animals at a rate of 0.2–2 spikes/s *(170)*. In the hippocampus, epileptiform events are multifocal, often arising at a distance from the injection site *(171)*. Behavioral seizures have been reported in as many as 92% of injected rats *(180)*. Spontaneous independent events arise in contralateral homotopic cortex occurs several days later. These events last for longer than 7 mo, long after the toxin has been cleared from the tissue given a half-life of only a few days, indicating that long-lasting plastic changes cause chronic epileptogenicity *(181)*. In

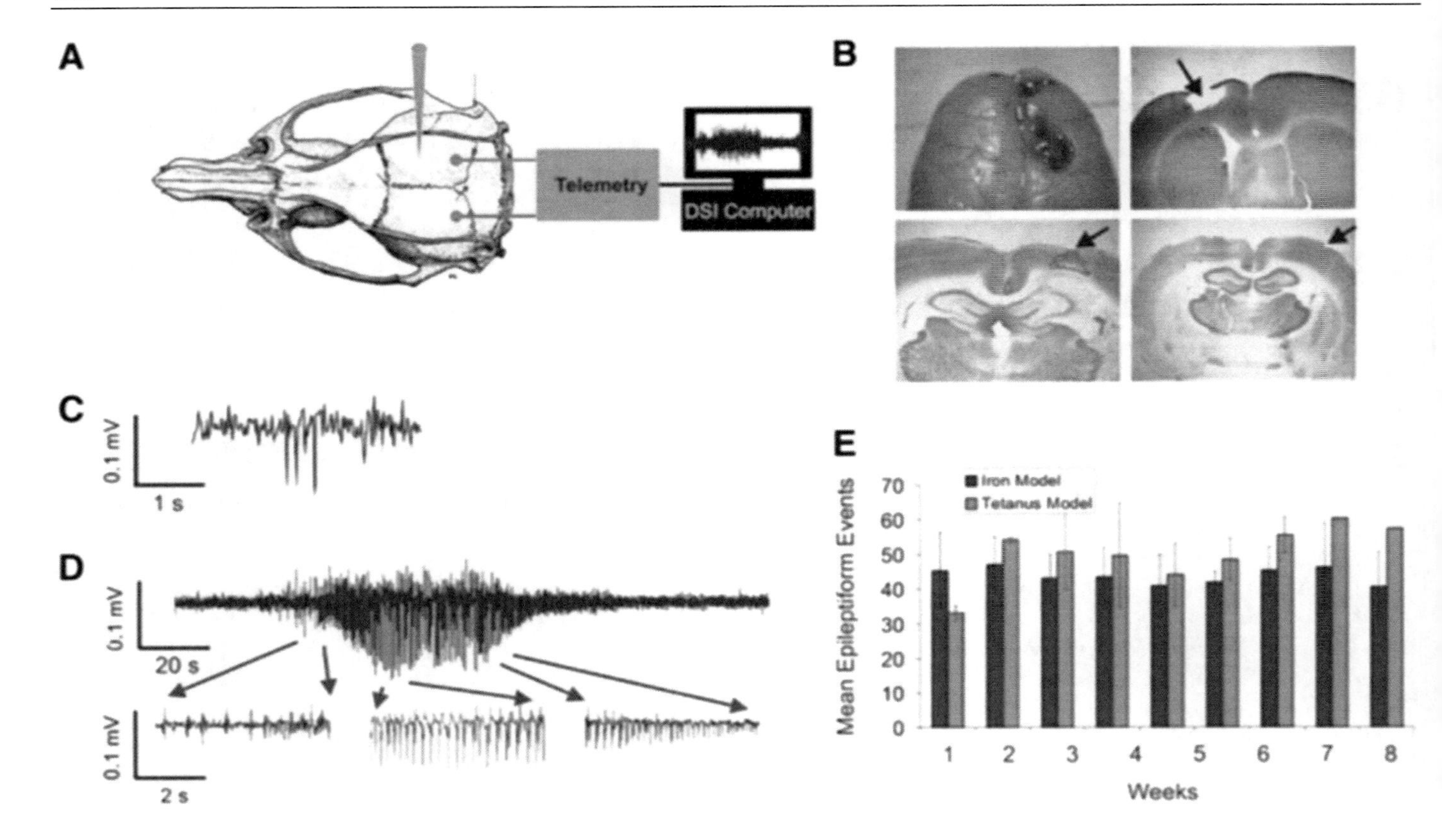

Fig. 13. Using video-EEG monitoring and telemetry **(A),** we can record chronic epileptiform events in rats, which are stable for several months **(E)**. **(B)** Whereas the injection of iron creates a cavitary lesion seen grossly (top left) and with Nissl (top right) and parvalbumin (lower left) staining, tetanus toxin is nonlesional (lower right). Arrows indicate site of injections. Both interictal **(C)** and ictal **(D)** events are recorded during chronic video-EEG monitoring and quantified using custom-written seizure detection algorithms (*see* Section 3.3.). Scale bar, 1 mm. *See* color version on Companion CD.

vitro intracellular recordings reveal a characteristic PDS during field recordings of interictal events *(170)*. Histologically, there is no cell loss or gliosis *(180,182)*.

2.5.3. Methods

Under ketamine/xylazine anesthesia, rats were placed in a stereotactic frame and a small parasagittal skin incision and a trephine hole (0.5 mm in diameter) were made under sterile conditions, 2.5 mm lateral and 1.5 mm rostral to the bregma over the hindpaw sensorimotor cortex. Then, 2.5 μL of 400 m*M* $FeCl_2$ or 50 ng/0.5 μL tetanus toxin was injected 1–2 mm below the surface through the hole using an oocyte injector. Two stainless-steel epidural screw electrodes (0.5 mm in diameter) were implanted into opposite sides of cranial bone, 3 mm lateral and 3 mm rostral to the bregma and 4 mm lateral and 4 mm caudal to the bregma for chronic ECoG recording (Fig. 13A). These electrodes were soldered to the distal wires of a two-lead telemetry system (DSI) and a transmitter was placed under the skin in a pocket between the shoulder blades. The surgical field was irrigated with antibiotic solution and closed. The animals were awakened and placed in a facility for chronic behavioral and video-EEG monitoring for epileptic activity. Signals were digitized onto a PC. Quantification of interictal and ictal events was performed with the use of a custom-made seizure detection algorithms written in MATLAB (The MathWorks, Inc.) The signals were bandpass filtered (3 to 70 Hz) and the signal energy (power of EEG) associated with the EEG is computed. Each peak that crosses a fixed threshold is counted as one seizure. IISs are also counted using a custom made C^{++} program. Animals with consistent interictal and ictal spikes with or without behavioral events (paroxysmal twitching in the musculature contralateral to the epileptogenic agent with or without secondary generalization) were used for further study. These animals were prepared for imaging as in the acute epileptogenesis models.

Rats with chronic epilepsy from iron or tetanus toxin injection occasionally do not have spontaneous seizures while under anesthesia. Events can be triggered by one of two methods. Because the epileptogenic agent is injected into the area of somatosensory cortex corresponding with hindpaw representation, focal peripheral stimulation of the hindpaw with an S48 stimulator (Grass Telefactor, W. Warwick, RI) delivered through an SIU-7 to apply a 1 mA, 5-Hz stimulus of 2-s duration has successfully triggered epileptiform events. Alternatively, one can administer BMI, 1 mg/kg i.p. every 10 min.

Rats were injected with a lethal dose of pentobarbital sodium (120 mg/kg, i.p.) and perfused transcardially with saline followed by 4% paraformaldehyde and 0.1% glutaraldehyde. After removal of the brains, they were postfixed and stored overnight in a 100 m*M* phosphate buffer (pH 7.4) containing 30% sucrose. Serial coronal sections were cut through the injection site of the iron or tetanus

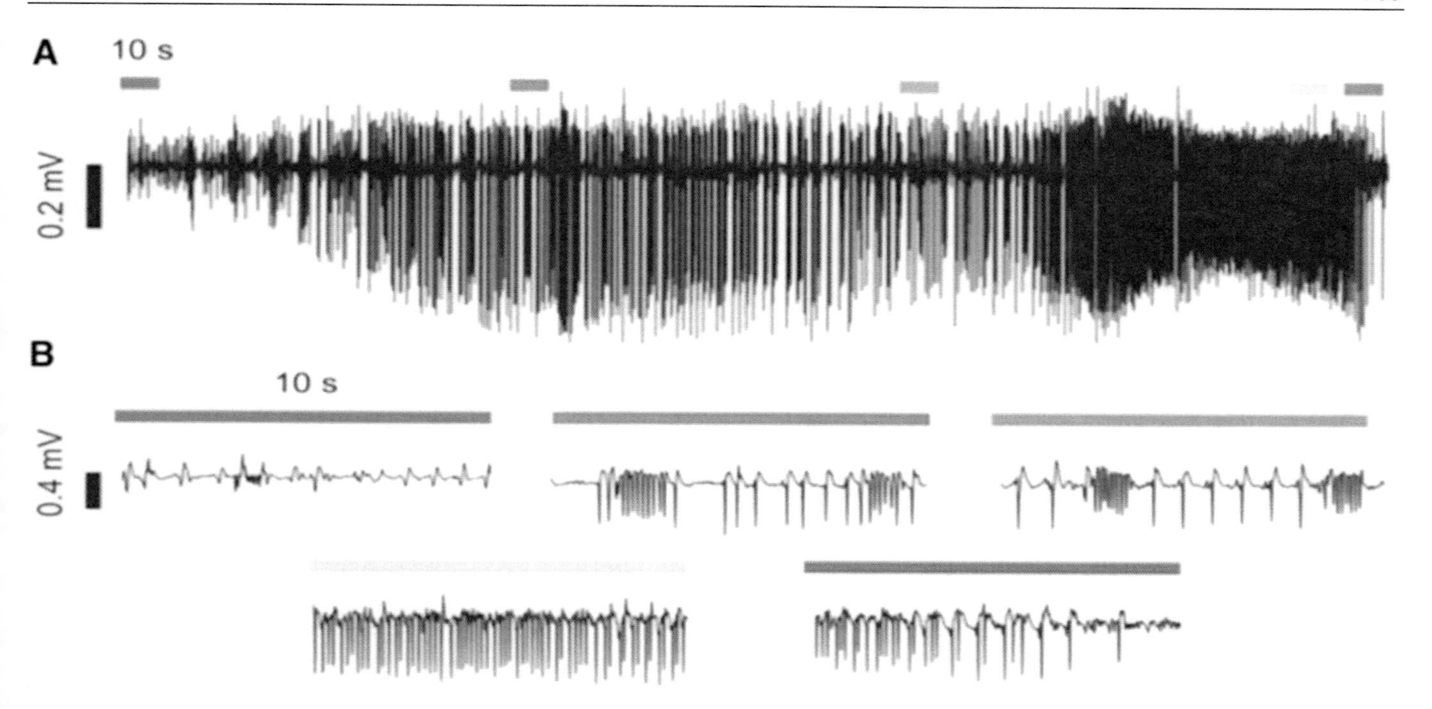

Fig. 14. Interictal spikes and ictal events occur under general anesthesia in chronically epileptic rats. (**A**) Field potential recording from a rat with tetanus-toxin induced epilepsy demonstrates a stereotypical ictal event occurring while under anesthesia. (**B**) Expanded regions from (A) reveal the build-up and diminution of recurrent spike-and-wave activity. *See* color version on Companion CD.

toxin using a freezing microtome (Leica) at 40 μm. Sections were serially stained for Nissl, parvalbumin, glial fibrillary acidic protein, somatostatin, and γ–aminobutyric acid (Fig. 13B).

2.5.4. IOS Imaging of Chronic Neocortical Epilepsy

The iron and tetanus toxin models are not only useful as models of chronic neocortical epilepsy, but also as models of lesional versus nonlesional epilepsy. Whereas iron injection creates a cavitary lesion in the brain, the tetanus toxin injection does not disrupt the cortical architecture (Fig. 13B). Despite these histologic differences, both models generate a stable epileptic focus exhibiting both spontaneous IIS (Fig. 13C) and ictal events (Fig 13D) that are consistent in frequency over time (Fig. 13E). Even under general anesthesia, these animals exhibit epileptiform events suitable for imaging (Fig. 14).

IOS imaging of IIS and ictal events was performed in both models to map the areas of epileptic activity with respect to the injection site of the epileptogenic agent. Our hypothesis was that IIS and ictal onsets would arise from different areas within a single animal, demonstrating a shifting focus, and that the dynamic interaction between excitation and inhibition responsible for this variability would be apparent in the IOS maps. Figure 15A demonstrates the relative locations of the ictal onsets, ictal offsets, and IISs in a single animal relative to the architecture of the pial blood vessels seen though the thinned skull. The IOS clearly demonstrates that each epileptiform event arises from a different area of cortex adjacent to the lesion, with an inter-event distance as far as 8 mm. Figure 15B shows the electrophysiology and IOS imaging at 546 nm during single ictal event. As in the 4-AP acute seizure, chronic seizures also manifest periodic spiking and fast activity. In this example, the seizure begins with periodic spike-and-wave activity and the IOS demonstrates a darkened focus contained by a ring of inverted optical signal in the surrounding cortex. As the ictal event progresses to faster activity with no intervening "wave," the inverted activity disappears and the darkening in the focus spreads to adjacent cortex. These experiments provide further evidence that the IOS can reveal complementary data about excitation and inhibition that form the basis of the electrographic data recorded with field electrodes. Future experiments with multiple single-unit and field recordings and voltage-sensitive dyes will help us understand the complex relationship between imaging and electrophysiology.

3. HUMAN IMAGING IN THE OPERATING ROOM

3.1. BACKGROUND

The past few years have seen a rapid growth in brain imaging techniques used in a clinical setting. Chief among these have been techniques such as fMRI and positron emission tomography. The fast time course and high spatial resolution observed in laboratory studies using the IOS quickly sparked interest in applying this technique to intraoperative imaging. Although high-resolution IOS requires an exposed cortex, which makes it more invasive than other imaging techniques, its promise of extremely high spatial resolution offers an attractive alternative for intraoperative cortical mapping in humans, and also, as we shall see below, the possibility of localizing interictal and ictal epileptiform events during epilepsy surgery.

The IOS was first used intraoperatively by MacVicar and colleagues in 1990 *(183)* for the imaging of stimulation-evoked cortical activation. This study was soon followed by that of Haglund et al. *(102)*, who imaged both stimulation-evoked

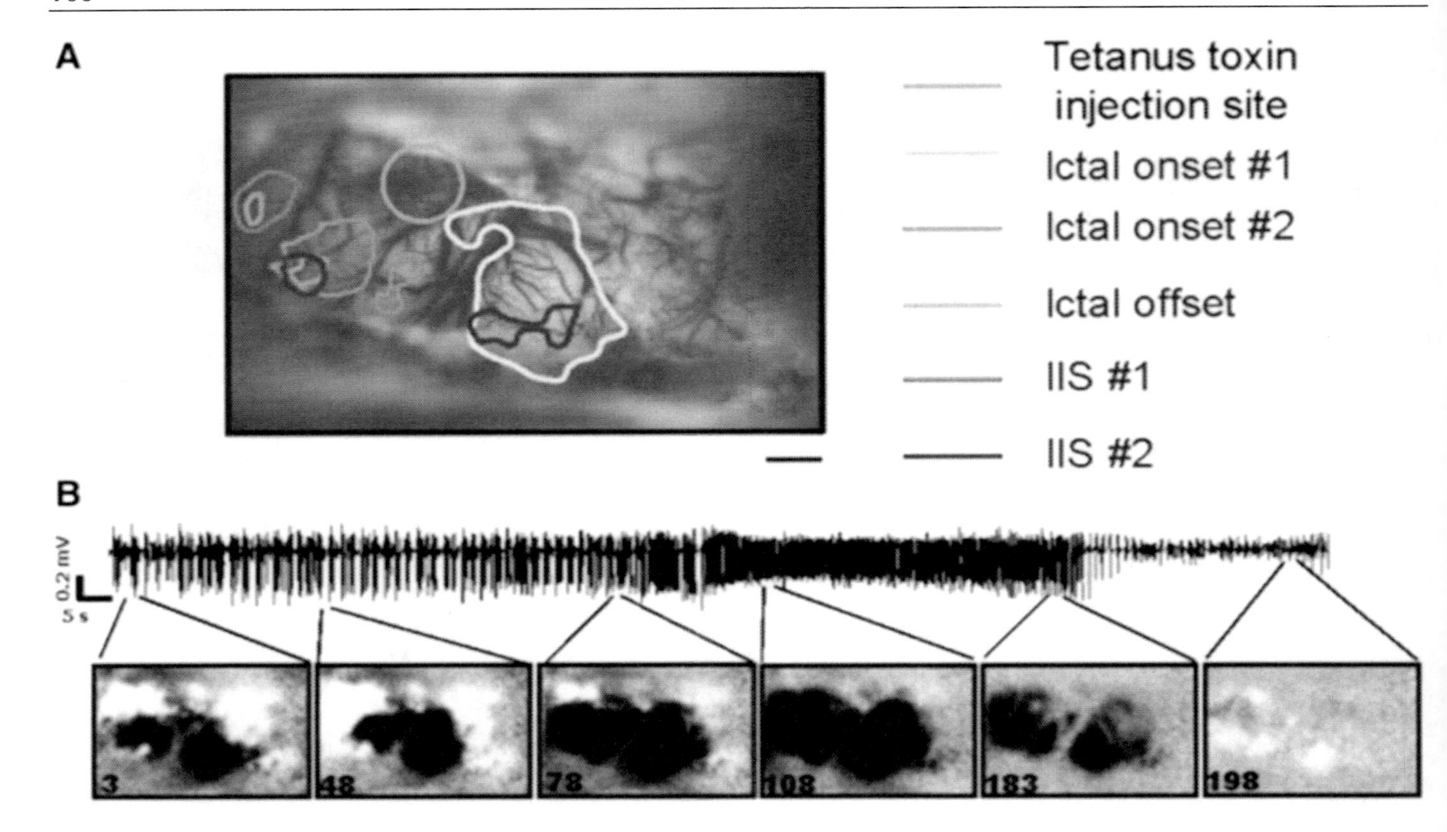

Fig. 15. (**A**) IOS imaging map of separate ictal onsets, offsets and IISs from the same animal as that illustrated in Fig. 14. Thresholds for area measurements are derived from pixel values 15% above the median of all pixel values. Scale bar, 1 mm. (**B**) The transition from periodic spiking activity to faster spiking in the f.p. is correlated with an increase in the size of the excitatory optical signal and a disappearance of the inverted optical signal in the imaging. The numbers in the images refer to the time(s) when the images were acquired after the onset of the ictal event. *See* color version on Companion CD.

epileptiform afterdischarges and cognitively evoked functional activity such as activation of Wernicke's and Broca's areas during language tasks. A group at UCLA lead by Arthur Toga has published several articles intraoperative human imaging of somatosensory and language cortex *(16,184–189)*. Other groups have also imaged somatosensory cortex, including Shoham and Grinvald *(38)* and Sato et al. *(190)*, who described the ability to image both primary and secondary somatosensory areas.

3.2. TECHNICAL CHALLENGES

Although IOS of human cortex in the operating room is quite feasible, the signal is not as robust as in the laboratory and the spatial resolution is lower due to several technical challenges and large sources of noise *(16,38,102,184,185,189–191)*. The major sources of noise include motion of the cortex induced by heartbeat and respiration, as well as a 0.1 Hz vasomotor signal, each of which change the reflected light signal with a different periodicity *(33,189,192)*. In addition to the time constraints of the operating room, the environment is more difficult to control, resulting in larger fluctuations in ambient light *(102,191)* and anesthesia *(191)*.

Various mechanisms have been developed to compensate for these sources of noise. Image acquisition can be synchronized to the cardiac and respiratory cycles *(184,191)*. The cortex can be stabilized with a glass footplate *(38,102,190)*. Various post-hoc algorithms can also be applied to remove noise from the imaging data. These can include warping algorithms *(102,184)*, or, if a sufficiently long series of images can be obtained, an inverse FFT algorithm can be employed to remove periodic fluctuations from the time course of the signal *(10,33,193)*.

Another difficulty in intraoperative human imaging is the large field of view and the natural curvature of the surface of the human brain. The typical craniotomy is 10–14 cm in diameter. Given the curvature of the brain, keeping the entire surface homogeneously illuminated and in focus would require multiple sources of light and a very large depth of field, which in turn would increase blood vessel artifact, since in the laboratory setting a narrow depth of field permits imaging from a plane below the pial surface and vasculature *(130)*. Various solutions have been implemented such as using a dedicated zoom lens with separate sources of epi-illumination, using the operating microscope lenses and built-in white light illumination with absorption filters or a ring illuminator with filtered incident light. Suspending the camera over the patient's head is also a challenge and investigators have used rigid stands on the floor, attachments to the operating table or the operating microscope itself.

We recorded fluctuations in light reflectance in one pixel over time during an imaging session. Data was sampled at a rate of 1 frame/300 ms during a period of 105 s, an FT was performed and the power spectrum calculated (Fig 17B). The peak indicated by the arrow at left shows the peak for vasomotor noise (~0.1 Hz); the arrow at the right shows the peak corresponding to the patient's heartbeat (~1 Hz). Methods for eliminating these sources of noise will be discussed.

3.3. METHODS

Patients suitable for intraoperative imaging are undergoing craniotomy for resection of neocortical pathology such as epilepsy, tumors or vascular malformations adjacent to functional cortex. Once in the operating room, their heads are fixed rigidly to the table with a Mayfield headholder (Ohio Medical). General anesthesia is induced with i.v. thiopental, 3-5 mg/kg and i.v. fentanyl, 50–150 µg, in an average adult and maintained with N_2O/O_2 and isoflurane 1%. A nondepolarizing muscle relaxant, Pavulon, 0.1–0.15 mg/kg is given i.v. to facilitate endotracheal intubation. The line of incision is infiltrated with a mixture of equal volumes 1% Lidocaine (with epinephrine 1;100,000). Craniotomy is performed using standard neurosurgical techniques. Once the cortex is exposed, the N_2O/O_2 is discontinued and the patient is maintained on isoflurane 0.2% and supplemental fentanyl 50–150 µg as necessary for the electrical and optical recordings. This combination of anesthetics has been shown to have a minimal effect on the ECoG *(194)*.

In our first experiments (Yale University), we used we reproduced the laboratory setting and imaged a small field of view using a tandem lens arrangement (refs. *130,190*). The camera was held in place by modifying the Mayfield (Ohio Medical) U-bar that attached to the table and sits over the patient's chest (Fig. 16A). The camera was held by an X,Y,Z-manipulator (Narishige, Japan) and the entire apparatus was placed in a sterile drape. Cortical illumination was achieved with a ring illuminator attached to a DC-regulated power supply and a broadband filter 650 ± 50 nm. A second locking retractor provided another point of fixation between the Mayfield head-holder and the ring-illuminator, which was fixed to the camera lens. In this way, there was minimal relative movement between the cortical surface and the camera lens. Cortical stabilization was achieved with a glass footplate held with the Greenberg retractor system (Fig. 16B). In our later experiments (Weill-Cornell Medical College of Cornell University), we built a camera holder that sat on the floor at the head of the bed and locked to the headholder while suspending the camera over the head on a gross and fine X,Y,Z-manipulator (Fig. 17A). The cortex was illuminated with a ring illuminator on a retractable arm extending down from the camera lens (single 50 mm lens). By lowering the ring illuminator closer to the cortex we increased the intensity of the light and were able to use narrow band filters (± 10 nm). The 50-mm lens increased the field of view to 8 × 8 cm. The cortex was stabilized with a glass footplate.

3.4. INTRAOPERATIVE IOS OF SOMATOSENSORY ARCHITECTURE

We used IOS imaging to investigate somatotopy in the human face area *(195)*. In particular, we wanted to determine the relative cortical location of peri-orbital skin versus skin of the lateral face. Animal studies using microelectrode recordings from Macaque and Cebus monkeys have shown that cortical representation of the peri-orbital skin in Brodmann Area 1 is both rostral and medial to peribuccal skin *(196–198)*. Maps of human face somatotopy generally show peri-orbital skin medial, but not rostral, to lateral face, but these have been generated using cortical stimulation mapping which has a limited spatial resolution *(199–201)*. The field of view using the tandem lens was 14 × 8.7 mm. Three stimulus conditions were used: upper face stimulation, lower face stimulation, and blank. Stimuli consisted of four constant current electrical pulses of 1.5 msec duration with a magnitude of 2.2 mA delivered at 2 Hz through ball electrodes placed below and lateral to the left eye and along the left lower cheek parallel to the lips by an S-88 stimulator (Grass Instruments, Quincy, MA). Each stimulus condition was presented in pseudo-randomized order with a 10- to 15-s interstimulus interval. For each stimulus condition, we collected 10 consecutive 300 ms image frames after stimulus onset and these were stored for subsequent analysis. We collected five blocks of five trials per stimulus.

Stimulation of the upper face produced a focal change in reflectance in a different location than stimulation of the lower face (Fig. 16). Here we demonstrate with high-resolution optical imaging that indeed, in the human, the lateral face is represented both rostral as well as medial to peri-buccal skin *(195)*. We also examined the cortical magnification factor (CMF), which is defined as the area of cortex dedicated to the representation of an area of skin *(202)*. The CMF was calculated from the distance between the electrode contacts on the face (~7 cm) and the distance between the centers of the change in reflectance on the cortical surface (~2.5 mm). The calculated CMF was 0.36 mm per cm (2.5 mm/7 cm) of facial skin. A comparable measurement from the face area of the cynomolgus Macaque is 0.75 mm of cortex per cm of facial skin (cf. Nelson et al., 1980 *[196]*, Fig. 11, distance between center of cortical representation of orbital skin and penetration 1 is 3 mm; distance between orbital skin and upper lip is about 4 cm). Therefore the CMF for this area of skin in the human may be smaller than in the macaque by a factor of 2 *(195)*.

3.5. INTRAOPERATIVE IOS IMAGING OF CORTICAL STIMULATION AT MULTIPLE WAVELENGTHS

Currently several groups are performing human IOS imaging of functional architecture and drawing conclusions about human physiology. However, little is known about the IOS response in humans at different wavelengths. We are currently investigating the IOS characteristics following a reproducible, focal cortical stimulus recorded at multiple wavelengths in the human. A two-contact ECoG strip is placed on the cortex underneath a 5 × 5-cm glass footplate. The operating room is darkened, and the cortex illuminated with a ring illuminator at 546 ± 10 nm to record the surface blood vessel pattern and then at 546 ± 10, 605 ± 10, and 700 ± 10 nm for IOS imaging. The optical reflectance signal is recorded a 10-bit camera (Imager 3001, Optical Imaging Inc., Germantown, NY) and digitized onto a PC at 33 frames per second, and integrated to variable frame rates from 10-2 frames/s. Constant current stimulation (Ojemann Cortical Stimulator, Radionics) was applied (3 s, 60 Hz, biphasic square waves of 0.5 ms duration each) at 4 mA.

We find that the optical signal recorded at 546 nm, corresponding with cerebral blood volume, is larger, both in magnitude (8%) and area, than the signal recorded at higher wavelengths (605 nm, 1.1%; 700 nm, 0.7%; ANOVA $p < 0.01$; SNK post-hoc test; Fig. 17). The signal at 546 spreads along the

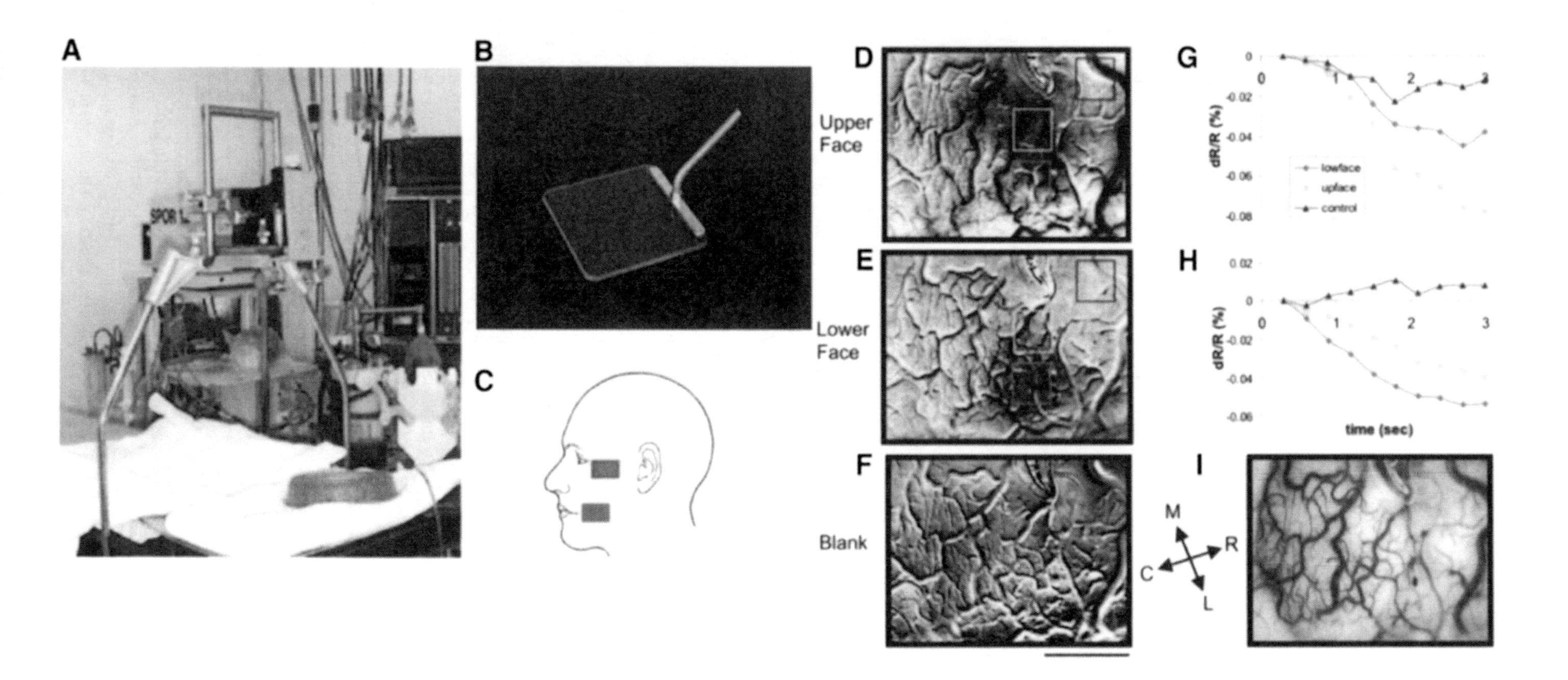

Fig. 16. (A) The camera is suspended above the head by attaching it to a U-bar, which is attached to the operating table on either side of the patient's chest. An X, Y, Z-manipulator holds the camera. **(B)** A glass footplate held by a retractor arm is gently placed on the cortical surface to dampen cortical pulsations. **(C)** Location of stimulating electrodes on the face. **(D)** Activation to stimulation of upper face. Sum of five trials. **(E)** Activation to stimulation of lower face. Sum of five trials. **(F)** No stimulation condition. (D) Time course of signals obtained during stimulation of upper face. Three locations were evaluated: red box is located over the lower face site, yellow box is located over the upper face site, and blue box is located over a site away from the activated sites. **(G)** Activation of upper face produces large reflectance change over the upper face location (yellow square), smaller reflectance changes over the lower face location (red box), and little change over the control location (blue triangle). **(H)** Time course of signals obtained during stimulation of lower face. Activation of lower face produces large reflectance change over lower face location, smaller reflectance change over upper face location, and little change over control location. Scale bar,5 mm. **(I)** Blood vessel map of imaged field of view. R, rostral; C, caudal; M, medial; L, lateral. *See* color version on Companion CD.

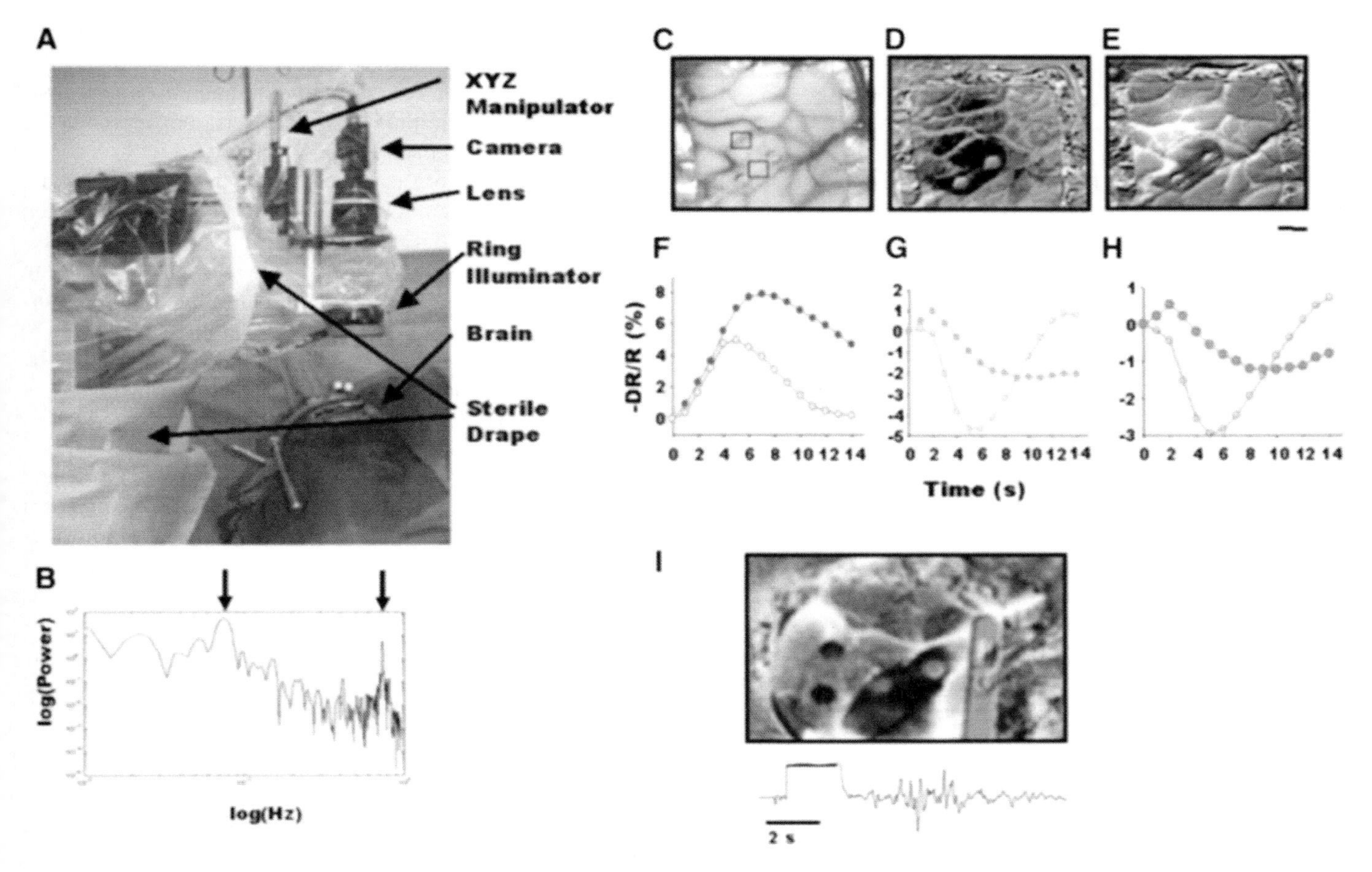

Fig. 17. **(A)** The camera holder is draped and the camera and ring illuminator suspended over the cortex. Using a glass footplate for cortical stabilization, we can image bipolar cortical stimulation (3 s, 60 Hz, biphasic square waves of 0.5 msec duration each and amplitude 4 mA) averaged over five trials, and **(I)** afterdischarges at 546 ± 10 nm **(D)** and 605 ± 10 nm **(E)**. **(C)** Location of the bipolar electrode on the cortical surface and the glass footplate. **(F–H)** Change in reflectance of light from two separate regions of interest. The filled circles show reflectance change in the region of interest located between the stimulating electrodes (lower box in [C]), and the open circles are from the region enclosed by the upper box in (C). Imaging was performed at 546 nm (F), 605 nm (G), and 700 nm (H). **(B)** Fourier transform of a single pixel shows that the noise associated with human imaging is mainly from vasomotor noise (left arrow) at 0.1 Hz and heart beat (right arrow) at 1 Hz. **(I)** Stimulation at a higher amplitude (10 mA) produces afterdischarges seen in ECoG recording below and a large change in reflectance in the area of the stimulating electrodes. *See* color version on Companion CD.

brain parenchyma and does not involve the blood vessels themselves. At 605 nm and 700 nm, we see an early focal decrease in reflectance in the brain parenchyma between the electrodes, consistent with the "initial dip" in oxygenation associated with metabolism that peaks at 2 s after the start of the stimulus (closed circles). A second, later and larger inverted IOS (open circles), peaking at 5 s, arises from both parenchyma and vessels, extending over approximately the same larger area as the darkening signal at 546 nm. The significance of these results will be discussed below.

3.6. INTRAOPERATIVE IOS OF EPILEPTIFORM EVENTS

Haglund et al. *(102)* first imaged afterdischarges triggered with bipolar stimulation in 1992. We have successfully repeated these experiments (Fig. 17I). However, there have been no reports of IOS imaging of spontaneous epileptiform events recorded from the human and correlated with intraoperative ECoG. We have attempted to record spontaneous IIS from human cortex and have found that the noise is a more significant factor than in studies involving sensory mapping or bipolar stimulation. We have tried to average among multiple IIS to improve our signal-to-noise ratio; however, if there is a fluctuation in the location of each IIS, the signals will cancel out. It is likely the amplitude of the IOS change associated with each IIS in chronic human epilepsy is much smaller than in the acute animal models described earlier. Other post-hoc methods to eliminate contamination from noise will have to be implemented, such as FT and PCA (discussed below).

4. CONCLUSIONS

IOS imaging is clearly a valuable technique in the study of epileptogenesis. As we have shown, the data are complementary to electrographic data and provide quantitative spatial information about blood volume, flow, and oxygenation of hemoglobin associated with the neuronal population activity underlying a variety of epileptiform events. Using several experimental models, as well as the human, we have explored the relationship between the intrinsic signal and the electrophysiology of interictal and ictal events. Although much of the data is consistent, some is contradictory and there are many unanswered questions.

4.1. 546 NM

The signal recorded at 546 nm, thought to show changes in blood volume, is clearly the highest in amplitude. In all models, this signal peaks latest, rises more slowly, and appears to arise from the brain parenchyma and not the blood vessels. However, with 10-Hz temporal resolution, we show that as early as 100 ms after the acute disinhibition IIS, there is a focal change in reflectance at 546 nm in many of the animals we studied, a finding that is surprising because blood volume is not supposed to respond so rapidly to neuronal activity. We hypothesize that the acute disinhibited focus is a special case in which a large population of neurons is all firing simultaneously and the demand for increased blood volume is extreme. Alternatively, blood volume may in fact respond more quickly than previously thought even under physiologic circumstances, and it requires a model with an enormous focal metabolic demand with a large signal amplitude to record it.

The signal at 546 is also thought to be less focal and, hence, less sensitive to the population of active neurons. Our data are consistent with this hypothesis particularly in the human imaging. The spread of bipolar stimulation recorded at 546 nm went far beyond the bipolar stimulating electrode to the adjacent 2 or 3 gyri in the absence of afterdischarges. At such low amplitudes (4 mA), focal cortical stimulation is known to disrupt or activate only local neuronal populations, which is why it is a useful technique or brain mapping in the operating room. Hence, imaging at 546 nm does appear to overestimate the population of active neurons, unless there is a large region of subthreshold excitation which is revealed at 546 nm and not other wavelengths.

Our data in the rat, during IIS and ictal events, however, is not in agreement with the human data. Although the area of spread was larger at 546 nm in some animals, on average, there was no significant difference compared with higher wavelengths. Once again, this result may be unique to acute pharmacologically induced epileptic events in the brain, in which metabolic demands are so high, in such a large area of cortex, that the oxy-deoxy signal is as widespread as the blood volume signal.

At 546 nm, the inverted optical signal from the "surround" is less intense than at higher wavelengths in the ictal model, but not in the interictal model. In contrast, in the human, following bipolar stimulation, we were not able to record any negative signal whatsoever at 546 nm. It is not clear what the significance is of the inverted signal at 546 nm. Whether it represents shunting of blood volume to the interictal focus or an indirect marker for neuronal inhibition (or both) is unclear and will require further investigation. However, in the chronic model, the negative signal at 546 nm is clearly present, particularly during spike-and-wave activity and its disappearance during faster activity appears to correlate with horizontal spread of the seizure.

4.2. 605 AND 630 NM

IOS imaging between 600 and 650 is thought to represent changes in oxy/deoxyhemoglobin. In all experiments at these wavelengths, we observe a focal decrease in light reflectance (darkening) in the brain parenchyma that occurs within 100 ms of the event. The signal rises more rapidly than at 546 nm, peaks earlier and, in the human, and some animal experiments, is more focal. We suggest this represents the "initial dip", or decrease in oxygenation resulting from an increase in metabolism. Following this initial dip we then see an inversion in the signal, particularly in the draining veins that is less well-localized, which likely represents an increase in blood flow and a rise in oxygenated hemoglobin as found in the blood oxygen level dependent signal. However, there is another inverted signal that occurs early in the brain parenchyma in the surrounding cortex. In our fast imaging experiments, this signal begins within 100 ms. We propose that this signal may correspond to surround inhibition; the relative timings of the signals may be critical in distinguishing these two distinct inverted signals.

4.3. 700 NM

Our results at 700 nm are quite similar to our results at 605 and 630 nm, although the signal is smaller in amplitude and more focal. Reflected light at 700 nm is believed to represent cell swelling; however, we do not employ an absorption filter and there may be significant contamination from the oxy/deoxyhemoglobin signal. We anticipated that the signal at this wavelength would be more rapid than at 605 or 630 nm, but this was not observed. The focality of the change in reflection may imply that it is an even better signal for localizing the population of epileptic neurons. Correlation with multifocal single-unit and f.p. recordings may answer this question.

4.4. INVERTED IOS

Whether the inverted IOS recorded from the surrounding brain parenchyma as early as 100 ms after the IIS represents shunting of blood volume, oxygenation, or, indirectly, neuronal inhibition, is unclear. Single unit recordings from the inverted optical signal region adjacent to a BMI-induced interictal focus in the ferret found decreased neuronal activity *(100)*. Das and Gilbert *(20)* observed an increase in reflectance in the cat primary visual cortex that they hypothesize may be an "inhibitory moat surrounding the excitatory center." Similarly, single-unit recordings from a ring of inverted optical signal in mouse visual cortex, in response to retinotopic photic stimulation, also revealed neuronal inhibition *(19)*.

However, Haglund et al. *(102)* raised the possibility that the increase in light reflectance may result from a decrease in blood flow, resulting from a shunting of blood toward the focus, rather than from a decrease in electrophysiologic activity. We found, in our interictal model, an early, wavelength-independent, inverted signal, and in our ictal and human studies, a late inverted wavelength-dependent signal. We hypothesize that the early wavelength-independent inverted signal is more likely to correlate with neuronal inhibition than the later signal. Our future experiments will explore this question using muticontact electrodes, voltage-sensitive dyes and other methods for measuring blood volume (Texas Red Dextran) in our epilepsy models.

4.5. ANALYSIS TECHNIQUES

The classic technique for IOS image analysis has been to divide images acquired during an activated state by images acquired during an inactive state. Image acquisition is triggered by stimulus onset. Fluctuations in the signal that are not related to the stimulus are removed by trial repetition and signal averaging. Recently, there has been rising interest in applying other methods of analysis to the intrinsic signal. There are several

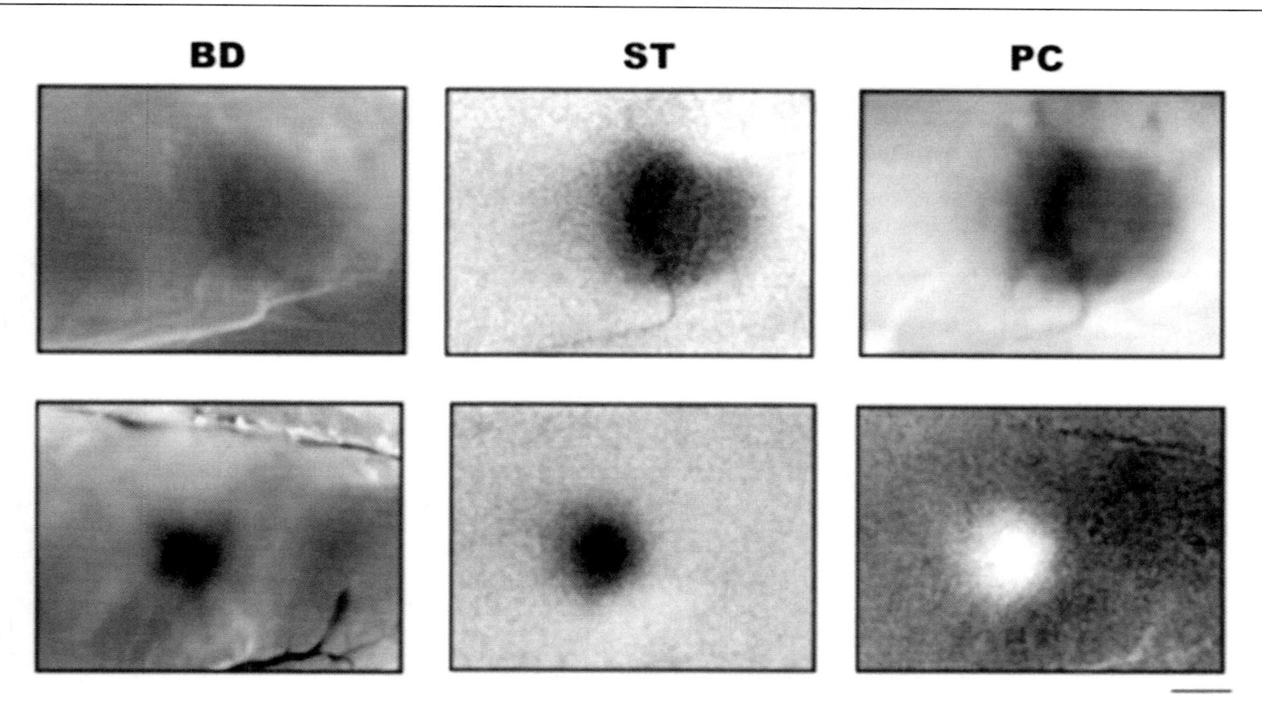

Fig. 18. Image analysis is compared using blank divided (BD), spike-triggered (ST), and PCA (PC). All three maps are comparable. The BD and ST maps are calculated with image division, whereas the PCA maps are calculated from the raw signal. There is more vascular noise in the BD maps because several minutes separated the denominator frame from the numerator frame. Scale bar, 1 mm.

limitations to requiring the use of a "blank" state for image division. The cortex is always active, so a true "blank" is impossible to achieve. Any signal within the blank will introduce signal into the final image (e.g., Fig. 10). The experiments take twice as long since an equal number of blank images must be acquired. Lengthy experimentation is also a necessity if trial averaging is used to reduce noise. For epilepsy imaging, these two issues are particularly bothersome. In chronic epilepsy, the focus is constantly spiking and so there is no "blank" state when nothing is happening. Likewise, as discussed earlier, each epileptiform event is different and the dynamic fluctuations in the epileptic pool of neurons make averaging between events counterproductive. Two interesting methods for addressing these problems are FT and PCA *(33,34)*. We have begun exploring both methods as a means of eliminating noise with a known periodicity. However, to extract epileptic data, one must know the periodicity of the signal of interest. For acute interictal spikes that occur with a regular periodicity, the epileptic intrinsic signal can be easily extracted (Fig. 18). However, in chronic epilepsy, the periodicity of the epileptiform events is irregular and signal extraction is much more difficult. Nevertheless, as technology and computational sophistication improve, many of our current obstacles will be overcome. The ultimate role for IOS imaging in the treatment of epilepsy has yet to be determined, but the outlook is promising.

ACKNOWLEDGMENTS

This work was supported by NINDS (K08, R21), as well as grants from the Alexander von Humboldt Fellowship, Van Wagenen Fellowship of the American Association for Neurological Surgeons (AANS), Research Clinical Training Grant from the Epilepsy Foundation of America, DANA Foundation, CURE Foundation, and Junior Clinical Investigator Award from the AANS. We thank our collaborators, including Tobias Bonhoeffer, Mark Hubener, Frank Sengpiel, Sven Shuett, Anna Roe, Dennis Spencer, Robert Friedman, Li-Min Chen and Amiram Grinvald, as well as Koon-Ho Danny Wong for technical assistance.

REFERENCES

1. Hill DK, Keynes RD. Opacity changes in stimulated nerve. J Physiol 1949;108:278–281.
2. Chance BC, P., Jobsis F, Schoener B. Intracellular oxidation-reduction states in vivo. Science 1962;137:499–508.
3. Jobsis FF. Noninvasive, infrared monitoring of cerebral and myocardial oxygen sufficiency and circulatory parameters. Science 1977;198:1264–1266.
4. Cohen LB, Keynes RD. Changes in light scattering associated with the action potential in crab nerves. J Physiol 1971;212:259–275.
5. Cohen L. Optical approaches to neuronal function. In: Hoffman JF, De Weer P, eds. Annual Review of Physiology. Palo Alto, CA: Annual Review Inc.; 1989:487–582.
6. Grinvald A, Lieke EE, Frostig RD, Gilbert CD, Wiesel TN. Functional architecture of cortex revealed by optical imaging of intrinsic signals. Nature 1986;324:361–364.
7. Bonhoeffer T, Grinvald A. Iso-orientation domains in cat visual cortex are arranged in pinwheel-like patterns. Nature 1991;353:429–431.
8. Bonhoeffer T, Grinvald A. The layout of iso-orientation domains in area 18 of cat visual cortex: optical imaging reveals a pinwheel-like organization. J Neurosci 1993;13:4157–4180.
9. Rubin BD, Katz LC. Optical imaging of odorant representations in the mammalian olfactory bulb. Neuron 1999;23:499–511.

10. Mrsic-Flogel T, Hübener M, Bonhoeffer T. Brain mapping: new wave optical imaging. Curr Biol 2003;13.:R778–R780.
11. Bonhoeffer T, Grinvald A. Optical imaging based on intrinsic signals. The methodology. In: Toga AW, Mazziota JC, eds. Brain Mapping The Methods. San Diego: Academic Press; 1996:55–99.
12. Holthoff K, Witte OW. Intrinsic optical signals in rat neocortical slices measured with near-infrared dark-field microscopy reveal changes in extracellular space. J Neurosci 1996;16:2740–2749.
13. Frostig RD, Lieke EE, Ts'o DY, Grinvald A. Cortical functional architecture and local coupling between neuronal activity and the microcirculation revealed by in vivo high-resolution optical imaging of intrinsic signals. Proc Natl Acad Sci 1990;87:6082–6086.
14. Malonek D, Grinvald A. Interactions between electrical activity and cortical microcirculation revealed by imaging spectroscopy: implications for functional brain mapping. Science 1996;272:551–554.
15. Vanzetta I, Grinvald A. Increased cortical oxidative metabolism due to sensory stimulation: implications for functional brain imaging. Science 1999;286:1555–1558.
16. Pouratian N, Sicotte N, Rex D, et al. Spatial/temporal correlation of BOLD and optical intrinsic signals in human. Magn Res Med 2003;47:766–776.
17. Fox PT, Raichle ME. Focal physiological uncoupling of cerebral blood flow and oxidative metabolism during somatosensory stimulation in human subjects. Proc Natl Acad Sci USA 1986;83:1140–1144.
18. Kwong KK, Belliveau JW, Chesler DA, et al. Dynamic magnetic resonance imaging of human brain activity during primary sensory stimulation. Proc Natl Acad Sci USA 1992;89:5675–5679.
19. Schuett S, Bonhoeffer T, Hubener M. Mapping retinotopic structure in mouse visual cortex with optical imaging. J Neurosci 2002;22: 6549–6559.
20. Das A, Gilbert CD. Long-range horizontal connections and their role in cortical reorganization revealed by optical recording of cat primary visual cortex. Nature 1995;375:780–784.
21. Toth LJ, Rao SC, Kim DS, Sur M. Subthreshold facilitation and suppression in primary visual cortex revealed by intrinsic signal imaging. Proc Natl Acad Sci USA 1996;93:9869–9874.
22. Hübener M, Shoham D, Grinvald A, Bonhoeffer T. Spatial relationships among three columnar systems in cat area 17. J Neurosci 1997; 17:9270–9284.
23. Sengpiel F, Stawinski P, Bonhoeffer T. Influence of experience on orientation maps in cat visual cortex. Nature Neurosci 1999;2:727–732.
24. Issa NP, Trachtenberg JT, Chapman B, Zahs KR, Stryker MP. The critical period for ocular dominance plasticity in the ferret's visual cortex. J Neurosci 1999;19:6965–6978.
25. Chapman B, Bonhoeffer T. Overrepresentation of horizontal and vertical orientation preferences in developing ferret area 17. Proc Natl Acad Sci USA 1998;95:2609–2614.
26. White LE, Bosking WH, Williams SM, Fitzpatrick D. Maps of central visual space in Ferret V1 and V2 lack matching inputs from the two eyes. J Neurosci 1999;19:7089–7099.
27. Schwartz TH. Optical imaging of epileptiform events in visual cortex in response to patterned photic stimulation. Cereb Cortex 2003; 13:1287–1298.
28. Fitzpatrick D. The functional organization of local circuits in visual cortex: insights from the study of tree shrew cortex. Cereb Cortex 1996;6:329–341.
29. Weliky M, Bosking WH, Fitzpatrick D. A systematic map of direction preference in primary visual cortex. Nature 1996;379:725–728.
30. Grinvald A, Frostig RD, Siege RM, Bartfeld E. High-resolution optical imaging of functional brain architecture in the awake monkey. Proc Natl Acad Sci U S A 1991;88:11559–11563.
31. Roe AW, Ts'o DY. Specificity of color connectivity between primate V1 and V2. J Neurophysiol 1999;82:2719–2730.
32. Bonhoeffer T, Kim D-S, Maloniek, Shoham D, Grinvald A. Optical imaging of the layout of functional domains in area 17 and across the area 17/18 border in cat visual cortex. Eur J Neurosci 1995;7:1973–1988.
33. Kalatsky VA, Stryker MP. New paradigm for optical imaging: temporally encoded maps of intrinsic signal. Neuron 2003;38:529–545.
34. Sornborger A, Sailstad C, Kaplan E, Sirovich L. Spatiotemporal analysis of optical imaging data. NeuroImage 2003;18:610–621.
35. Brett-Green BA, Chen-Bee CH, Frostig RD. Comparing the functional representations of central and border whiskers in rat primary somatosensory cortex. J Neurosci 2001;21:9944–9954.
36. Takashima I, Kajiwara R, Iijima T. Voltage-sensitive dye versus intrinsic signal optical imaging: comparison of optically determined functional maps from rat barrel cortex. Neuroreport 2001;12:2889–2894.
37. Sheth SA, Nemoto M, Guiuo M, Walker M, Pouratian N, Toga AW. Evaluation of coupling between optical intrinsic signals and neuronal activity in rat somatosensory cortex. NeuroImage 2003;19: 884–894.
38. Shoham D, Grinvald A. The cortical representation of the hand in macaque and human area S-1: high resolution optical imaging. J Neurosci 2001;21:6820–6835.
39. Chen LM, Friedman RM, Ramsden BM, LaMotte RH, Roe AW. Fine-scale organization of S1 (Area 3b) in the squirrel monkey revealed with intrinsic optical imaging. J Neurophysiol 2001;86:3011–3029.
40. Chen LM, Friedman RM, Ramsden BM, Roe AW. Organization of the somatosensory cortex revealed with intrinsic optical imaging in the squirrel monkey. Soc Neurosci Abstr 1999;25:1167.
41. Arieli A, Grinvald A, Slovin H. Dural substitute for long-term imaging of cortical activity in behaving monkeys and its clinical implications. J Neurosci Methods 2002;114:119–133.
42. Leão AAP. Pial circulation and spreading depression of activity in the cerbral cortex. J Neurophysiol 1944;7:391–396.
43. Leão AAP. Spreading depression of activity in the cerebral cortex. J Neurophysiol 1944;7:259–390.
44. Martins-Ferreira H, Nedergaard M, Nicholson C. Perspectives on spreading depression. Brain Res Rev 2000;32:215–234.
45. Richter F, Lemenkühler A. Spreading depression can be restricted to distinct depths of the rat cerebral cortex. Neurosci Lett 1993;152: 65–68.
46. Ba AM, Guiou G, Pouratian N, et al. Multiwavelength optical intrinsic signal imaging of cortical spreading depression. J Neurophysiol 2002;10:2726–2735.
47. Hossmann KA. Glutamate-mediated injury in focal cerebral ischemia: the excitotoxin hypothesis revised. Brain Pathol 1994; 4:23–36.
48. Koroleva VI, Bures J. Blockade of cortical spreading depression in electrically and chemically stimulated areas of cerebral cortex in rats. EEG Clin Neurophysiol 1980;48:1–15.
49. Aitken PG, Tombaugh GC, Turner DA, Somjen GG. Similar propagation of SD and hypoxic SD-like depolarization in rat hippocampus recorded optically and electrically. J Neurophysiol 1998;80:1514–1521.
50. Peixonto NL, Fernandes de Lima VM, Hanke W. Correlation of the electrical and intrinsic optical signals in the chicken spreading depression phenomenon. Neurosci Lett 2001;299:89–92.
51. Müller M, Somjen GG. Intrinsic optical signals in rat hippocampal slices during hypoxia induced spreading depression-like depolarization. J Neurophysiol 1999;82:1818–1831.
52. Bahar S, Fayuk D, Somjen GG, Aitken PG, Turner DA. Mitochondrial depolarization and intrinsic optical signal Imaged during hypoxia and spreading depression in rat hippocampal slices. J Neurophysiol 2000;84:311–324.
53. O'Farrell AM, Rex DE, Jmutialu A, et al. Characterization of optical intrinsic signals and blood volume during cortical spreading depression. Neuroreport 2000;11:2121–2125.
54. Hauser WA, Hesdorfer DC. Epilepsy: frequency, causes and consequences. New York: Demos; 1990.
55. Fisher RS, Weber WR, Lesser RP, Aroyo S, Uematsu S. High-frequency EEG activity at the start of seizures. J Clin Neurophsyiol 1992;9:441–448.
56. Bragin A, Mody I, Wilson CL, Engel JJ. Local generation of fast ripples in epileptic brain. J Neurosci 2002;22:2012–2021.
57. Staba RJ, Wilson C, Bragin A, Fried I, Engel JJ. Quantitative analysis of high-frequency oscillations(80–500 Hz) recorded in human epileptic hippocampus and entorhinal cortex. J Neurophysiol 2002;88:1743–1752.

58. Ylinen A, Bragin A, Nádasdy Z, et al. Sharp wave associated high frequency oscillation (200 Hz) in the intact hippocampus: network and intracellular mechanisms. J Neurosci 1995;14:30–46.
59. Grenier F, Timofeev I, Steriade M. Neocortical very fast oscillations (ripples, 80–200 Hz) during seizures: intracellular correlates. J Neurophysiol 2003;89:841–852.
60. Traub RD, Whittington MA, Buhl EH, et al. A possible role of gap junctions in generating very fast EEG oscillations preceding the onset of, and perhaps initiating, seizures. Epilepsia 2001;42:153–170.
61. Schwartzkroin PA. Epilepsy. Models, Mechanisms and Concepts. Cambridge: Cambridge University Press; 1993.
62. Matsumoto H, Ajmone-Marsan C. Cortical cellular phenomena in experimental epilepsy: interictal manifestations. Exp Neurol 1964;9: 286–304.
63. Matsumoto H, Ajmone-Marsan C. Cortical cellular phenomena in experimental epilepsy: ictal manifestations. Exp Neurol 1964;9:305–326.
64. Prince DA. The depolarizing shift in "epileptic" neurons. Exp Neurol 1968;21:467–485.
65. Chagnac-Amitai Y, Connors BW. Synchronized excitation and inhibition driven by intrinsically bursting neurons in neocortex. J Neurophysiol 1989;62:1149–1162.
66. Chagnac-Amitai Y, Connors BW. Horizontal spread of synchronized activity in neocortex and its control by GABA mediated inhibition. J Neurophysiol 1989;61:747–758.
67. Chervin RD, Pierce PA, Connors BW. Periodicity and directionality in the propagation of epileptiform discharges across neocortex. J Neurophysiol 1988;60:1695–1713.
68. Connors BW. Initiation of synchronized bursting in neocortex. Nature 1984;310:685–687.
69. Telfian AE, Connors BW. Layer-specific pathways for the horizontal propagation of epileptiform discharges in neocortex. Epilepsia 1998;39:700–708.
70. Gutnick MJ, Connors BW, Prince DA. Mechanisms of neocortical epileptogenesis in vitro. J Neurophysiol 1982;48:1321–1335.
71. de Lanerolle NC, Kim JH, Robbins RJ, Spencer DD. Hippocampal interneuron loss and plasticiy in human temporal lobe epilepsy. Brain Res 1989;495:387–395.
72. Prince DA, Jacobs KM, Salin PA, Hoffman S, Parada I. Chronic focal neocortical epileptogenesis: does disinhibition play a role? Can J Physiol Pharmacol 1997;75:500–507.
73. Schwartzkroin PA, Haglund MM. Spontaneous rhythmic activity in epileptic human and normal monkey temporal lobe. J Neurophysiol 1986;27:523–533.
74. Davenport CJ, Brown WJ, Babb TL. Sprouting of GABAergic and mossy fiber axons in dentate gyrus following intrahippocampal kainate injections in the rat. Exp Neurol 1990;109:180–190.
75. Nusser Z, Hajos N, Somogyi P, Mody I. Increased numbers of synaptic GABA(A) receptors underlies poteniation at hippocampal inhibitory synapses. Nature 1998;395:172–177.
76. Esclapez M, Hirsch JC, Khazipov R, Ben Ari Y, Bernard C. Operative GABAergic inhibition in hippocampal CA1 pyramidal neurons in experimental epilepsy. Proc Natl Acad Sci USA 1997;94:12151–12156.
77. Isokawa-Akeson M, Wilson CL, Babb TL. Inhibition in synchronously firing human hippocampal neurons. Epilepsy Res 1989;3:236–247.
78. Colder BW, Frysinger RC, Wilson CL, Harper M, Engel JJ. Decreased neuronal burst discharge near site of seizure onset in epileptic human temporal lobes. Epilepsia 1996;37:113–121.
79. Kisvárday ZF. GABAergic networks of basket cells in the visual cortex, In: Mize RR, Marc R, Sillito AM, eds. Progress in Brain Research, Vol 90, Mechanisms of GABA in the Visual System. Amsterdam: Elsevier; 1992:385–405.
80. Andersen P, Eccles JC, Løyning Y. Pathway of postsynaptic inhibition in the hippocampus. J Neurophysiol 1964;27:608–619.
81. Dichter M, Spencer WA. Penicillin-induced interictal discharges from cat hippocampus. I. Characteristics and topographical features. J Neurophysiol 1969;32:649–662.
82. Prince DA, Wilder J. Control mechanisms in cortical epileptogenic foci. "Surround" inhibition. Arch Neurol 1967;16:194–202.
83. Goldensohn ES, Salazar AM. Temporal and spatial distribution of intracellular potentials during generation and spread of epileptogenic discharges. Adv Neurol 1986;44:559–582.
84. Tucker TR, Katz LC. Recruitment of local inhibitory networks by horizontal connections in layer 2/3 of ferret visual cortex. J Neurophysiol 2003;89:501–512.
85. Tucker TR, Katz LC. Spatiotemporal patterns of excitation and inhibition evoked by the horizontal network in layer 2/3 of ferret visual cortex. J Neurophysiol 2003;89:488–500.
86. Duncan JS. Imaging and epilepsy. Brain 1997;120:339–377.
87. Pedley TA. Interictal epileptiform discharges: discriminating characteristics and clinical correlations. AM J EEG Technol 1980;20:101–119.
88. Dichter MA, Ayala GF. Cellular mechanisms of epilepsy: a status report. Science 1987;237:157–164.
89. McNamara JO. Cellular and molecular basis of epilepsy. J Neurosci 1994;14:3413–3425.
90. Wyler AR, Burchiel KJ, Ward AAJ. Chronic epileptic foci in monkeys: correlation between seizure frequency and proportion of pacemaker neurons. Epilepsia 1978;19:475–483.
91. Prince DA, Futamachi KJ. Intracellular recordings from chronic epileptogenic foci in the monkey. Electroencephalogr Clin Neurophysiol 1970;29:496–510.
92. Cohen I, Navaro V, Clemenceau S, Baulac M, Miles R. On the origin of interictal activity in human temporal lobe epilepsy in vitro. Science 2002;298:1418–1421.
93. Penfield W. The evidence for cerebral vascular mechanism in epilepsy. Ann Intern Med 1933;7:303–310.
94. Hochman DW, Baraban SC, Owens JWM, Schwartzkroin PA. Dissociation of synchronization and excitability in furosemide blockade of epileptiform activity. Science 1995;270:99–102.
95. Buchheim K, Schuchmann S, Siegmund H, Weissinger F, Heinemann U, Meierkord H. Comparison of intrinsic optical signals associated with low Mg2+-and 4-aminopyridine-induced seizure-like events reveals characteristic features in the adult rat limbic system. Epilepsia 2000;41:635–641.
96. Weissinger F, Buchheim K, Siegmund H, Heinemann U, Meierkord H. Optical imaging reveals characteristic seizure onsets, spread patterns, and propagation velocities in hippocampal-entorhinal cortex slices of juvenile rats. Neurobiol Dis 2000;7:286–298.
97. D'Arcangelo G, Tancredi V, Avoli M. Intrinsic optical signals and electrographic seizures in the rat limbic system. Neurobiol Dis 2001; 8:993–1005.
98. Federico P, Borg SG, Salkauskus AG, MacVicar BA. Mapping patterns of neuronal activity and seizure propagation in the isolated whole brain of the guinea-pig. Neuroscience 1994;58:461–480.
99. Chen JWY, O'Farrell AM, Toga AW. Optical intrinsic signal imaging in a rodent seizure model. Neurology 2000;55:312–315.
100. Schwartz TH, Bonhoeffer T. In vivo optical mapping of epileptic foci and surround inhibition in ferret cerebral cortex. Nat. Med 2001; 7:1063–1067.
101. Haglund MM, Blasdel GG. Optical imaging of acute epileptic foci in monkey visual cortex. Epilepsia 1993;34:21.
102. Haglund MM, Ojemann GA, Hochman DW. Optical imaging of epileptiform and functional activity in human cerebral cortex. Nature 1992;358:668–671.
103. Engel JJ, Shewmon DA. Who should be considered a surgical candidate?, In: Engel JJ, ed. Surgical Treatment of the Epilepsies. New York: Raven Press; 1993:23–34.
104. Buzsáki G, Traub RD. Physiological basis of EEG activity, In: Engel TA, Pedley TA, ed. Epilepsy: A Comprehensive Textbook. Philadelphia: Lippincott-Raven Publishers; 1997:819–830.
105. Gloor P. Neuronal generators and the problem of localization in electroencephalography: application of volume conduction theory to electroencephalography. J Clin Neurophysiol 1985;2:327–354.
106. Alarcon G, Guy CN, Binnie CD, Walker SR, Owes R, Polkey CE. Intracerebral propagation of interictal spikes in partial epilepsy: implications for source localization. J Neurol Neurosurg Psychiatry 1994;57:435–449.
107. Uematsu D, Araki N, Greenberg JH, Reivich M. Alterations in cytosolic free calcium in the cat cortex during bicuculline-induced epilepsy. Brain Res Bull 1990;24:285–288.

108. Purpura DP, Pernry JK, Tower D, Woodbury DM, Walter R. Experimental Models of Epilepsy - A Manual for the Laboratory Worker. New York: Raven Press; 1972.
109. Fisher RS. Animal models of the epilepsies. Brain Res Rev 1989; 14:245–278.
110. Prince DA. Topical convulsant drugs and metabolic antagonists, In: Purpura DP, Penry JK, Tower DB, Woodbury DM, Walter RD, eds. Experimental Models of Epilepsy—A Manual for the Laboratory Worker. New York: Raven Press; 1972:52–83.
111. Szente M, Pongracz F. Aminopyridine-induced seizure activity. Electroencephalogr Clin Neurophysiol 1979;46:605–608.
112. Szente BM, Baranyi A. Mechanism of aminopyridine-induced ictal seizure activity in the cat neocortex. Brain Res 1987;41:386–373.
113. Ajmone Marsan C. Focal electrical stimulation, In: Purpura DP, Penry JK, Tower DB, Woodbury DM, Walter RD, ed. Experimental Models of Epilepsy—A Manual for the Laboratory Worker. New York: Raven Press; 1972:148–169.
114. Bashir ZI, Holmes O. Phases in the development of a penicillin epileptiform focus in rat neocortex. Exp Brain Res 1993;96:319–327.
115. Campbell A, Homes O. Bicuculline epileptogenesis in the rat. Brain Res 1984;323:239–246.
116. Petsche H, Prohaska O, Rappelsburger P, Vollmer R, Kaiser A. Cortical seizure patterns in a multidimensional view: the information content of equipotential maps. Epilepsia 1974;15:439–463.
117. Goldensohn ES, Zablow L, Salazar A. The penicillin focus. I. Distribution of potential at the cortical surface. Electroencephalogr Clin Neurophysiol 1977;42:480–492.
118. Wyler AR, Ward AAJ. Epileptic neurons, In: Lockard JSS, Ward AAJ, eds. Epilepsy: A Window to Brain Mechanisms. New York: Raven; 1980: 51–68.
119. Wyler AR, Ojemann GA, Ward AA, Jr. Neurons in human epileptic cortex: correlation between unit and EEG activity. Ann Neurol 1982;11:301–308.
120. Ishijima B, Hori T, Yoshimasu N, Fukushima T, Hirakawa K, Seikino H. Neuronal activities in human epileptic foci and surrounding areas. EEG Clin Neurophysiol 1975;39:643–650.
121. Mattia D, Haw GG, Avoli M. Epileptiform activity induced by 4-aminopyridine in guinea-pig and rat neocortices. Neurosci Lett 1993;154:157–160.
122. Stansfeld CE, Marsh SJ, Halliwell JV, Brown D. 4-Aminopyridine and dendrotoxin induce repetitive firing in rat visceral sensory neurons by slowly inactivating outward current. Neurosci Lett 1986;64: 299–304.
123. Barkai E, Friedman A, Grossman Y, Gutnick MJ. Laminar pattern of synaptic inhibition during convulsive activity induced by 4-aminopyridine in neocortical slices. J Neurophysiol 1995;73:1462–1467.
124. Benardo LS. Recruitment of GABAergic inhibition and synchronization of inhibitory interneurons in rat neocortex. J Neurophysiol 1997;77:3134–3144.
125. Rogawski MA, Barker JA. Effects of 4-aminopyridine on calcium action potentials and calcium current under voltage clamp in spinal neurons. Brain Res 1983;280:180–185.
126. Szente BM, Baranyi A. Properties of depolarizing plateau potentials in aminopyridine-induced ictal seizure foci of cat motor cortex. Brain Res 1989;495:261–270.
127. Yang XF, Rothman SM. Focal cooling rapidly terminates experimental neocortical seizures. Ann Neurol 2001;49:721–726.
128. Yang XF, Chang JH, Rothman SM. Intracerebral temperature alterations associated with focal seizures. Epilepsy Res 2002;52:97–105.
129. Wong BY, Prince DA. The lateral spread of ictal discharges in neocortical brain slices. Epilepsy Res 1990;7:29–39.
130. Ratzlaff EH, Grinvald A. A tandem-lens epifluorescence microscope: hundred-fold brightness advantage for wide field imaging. J Neurosci Methods 1991;36:127–137.
131. Chen-Bee CH, Kwon MC, Masino SA, Frostig RD. Areal extent quantification of functional representations using intrinsic signal optical imaging. J Neurosci Meth 1996;68:27–37.
132. Masino SA, Kwon MC, Dory Y, Frostig R. Characterization of functional organization within rat barrel cortex using intrinsic signal optical imaging through a thinned skull. Proc Natl Acad Sci USA 1993;90:9998–10002.
133. Szente MB, Boda B. Cellular mechanisms of neocortical secondary epileptogenesis. Brain Res 1994;648:203–214.
134. Schwartzkroin PA, Futamachi KJ, Noebels JL, Prince DA. Transcallosal effects of a cortical epileptiform focus. Brain Res 1975;99:59–68.
135. Engel JJ. Surgery for seizures. N Engl J Med 1996;334:647–652.
136. Babb TL, Crandall PH. Epileptogenesis of human limbic neurons in psychomotor epileptics. EEG Clin Neurophysiol 1976;40:225–243.
137. Engel JJ. Functional explorations of the human epileptic brain and their therapeutic implications. EEG Clin Neurophysiol 1990;76:296–316.
138. Engel JJ. Intracerebral recordings: organization of the human epileptogenic region. J Clin Neurophysiol 1993;10:90–98.
139. Kutsy RL, Farrell DF, Ojemann GA. Ictal patterns of neocortical seizures monitored with intracranial electrodes: correlation with surgical outcome. Epilepsia 1999;30:257–266.
140. Spencer SS, Guimaraes P, Kim J, Spencer DD. Morphological patterns of seizures recorded intracranially. Epilepsia 1992;33:537–545.
141. Williamson A, Spencer SS, Spencer DD. Depth electrode studies and intracellular dentate granule cell recordings in temporal lobe epilepsy. Ann Neurol 1995;38:778–787.
142. Spencer SS, Spencer DD. Implications of seizure termination location in temporal lobe epilepsy. Epilepsia 1996;37:455–458.
143. Netoff TI, Schiff SS. Decreased neuronal synchronization during experimental seizures. J Neurosci 2002;22:7297–7307.
144. Ayala GF, Matsumoto H, Gumnit RJ. Excitability changes and inhibitory mechanisms in neocortical neurons during seizures. J Neurophysiol 1970;33:73–85.
145. Haycock JW, Levy WB, Cotman CW. Stimulation-dependent depression of neurotransmitter release in the brain: [Ca] dependence. Brain Res 1987;155:192–195.
146. Bragin A, Penttonen M, Buzsáki G. Termination of epileptic afterdischarge in the hippocampus. J Neurosci 1997;17:2567–2579.
147. Binnie CD, Wilkins AJ. Visually induced seizures not caused by flicker (intermittent right stimulation), In: Zifkin BG, Andermann F, Beaumanoir A, Rowan AJ, ed. Advances in Neurology. Philadelphia: Lippincott-Raven; 1998:123–138.
148. Dorothée GA, Trenité K-N. Reflex seizures induced by intermittent light stimulation, In: Zifkin BG, Andermann F, Beaumanoir A, Rowan AJ, eds. Advances in Neurology. Philadelphia: Lippincott-Raven; 1998:99–121.
149. Ribak CE, Reiffenstein RJ. Selective inhibitory synapse loss in chronic cortical slabs: a morphological basis for epileptic susceptibility. Can J Physiol Pharmacol 1982;60:864–870.
150. Prince DA, Tseng G-F. Epileptogenesis in chronically injured cortex: in vitro studies. J Neurophysiol 1993;69:1276–1291.
151. Salin P, Tseng G-F, Hoffman S, Parada I, Prince DA. Axonal sprouting in layer V pyramidal neurons of chronically injured cerebral cortex. J Neurosci 1995;15:8234–8245.
152. Lewis DV, Mutsuga N, Schuette WH. Potassium clearance and reactive gliosis in the alumna cream model. Epilepsia 1977;18:499–506.
153. Haglund MM, Berger MS, Kunkel DD, Franck JE, Ghatan S, Ojemann GA. Changes in γ-aminobutyric acid and somatostatin in epileptic cortex associated with low-grade gliomas. J Neurosurg 1992;77:209–216.
154. Ward AA. Topical convulsant metals, In: Purpura DP, Penry JK, Woodbury DM, Tower DB, Walter RD, eds. Experimental Models of Epilepsy—A Manual for the Laboratory Worker. New York: Raven; 1972:13–35.
155. Prince DA, Futamachi KJ. Intracellular recordings in chronic focal epilepsy. Brain Res 1968;11:681–684.
156. Atkinson JR, Ward AAJ. Intracellular studies of cortical neurons in chronic epileptogenic foci in the monkey. Exp Neurol 1964;10:285–295.
157. Ward AAJ. The epileptic neurone. Epilepsia 1961;2:70–80.
158. Willmore LJ, Sypert GW, Munson JB, Hurd RW. Chronic focal epileptiform discharges induced by injection of iron into rat and cat cortex. Science 1978;200:1501–1503.
159. Willmore LJ, Sypert GW, Munson JB. Recurrent seizures induced by cortical iron injection: a model of posttraumatic epilepsy. Ann Neurol 1978;4:329–336.

160. Moriwaki A, Hattori Y, Nishida N, Hori Y. Electrographic characterization of chronic iron-induced epilepsy in rats. Neurosci Lett 1990;110:72–76.
161. Moriwaki A, Hattori Y, Hayashi Y, Hori Y. Development of epileptic activity induced by iron injection into rat cerebral lcortex: electrographic and behavioral characteristics. EEG Clin Neurophysiol 1992;83:281–288.
162. Willmore LJ, Rubin JJ. Antiperoxidant pretreatment and iron-induced epileptiform discharges in the rat: EEG and histopathologic studies. Neurology 1981;31:63–69.
163. Engstrom R, Hillered L, Flink R, Kihlstrom L, Lindquist C, Nie J-X, Olsson Y, Silander HC. Extracellular amino acid levels measured with intracerebral microdialysis in the model of posttraumatic epilepsy induced by intracortical iron injection. Epilepsy Research 2001;43:135–144.
164. Reid SA, Sypert GW, Boggs WM, Williams LJ. Histopathology of the ferric-induced chronic epileptic focus in the cat: a golgi study. Exp Neurol 1979;66:205–219.
165. Westrum LE, White LE, Ward AAJ. Morphology of the experimental epileptic focus. J Neurosurg 1964;21:1033–1046.
166. Schiebel ME, Crandall PH, Schiebel AB. The hippocampal-dentate complex in temporal lobe epilepsy-a golgi study. Epilepsia 1974; 15:55–80.
167. Brooks VB, Asunama H. Action of tetanus toxin in the cerebral cortex. Science 1962;137:674–676.
168. Carrea R, Lanari A. Chronic effects of tetanus toxin applied locally in the cerebral cortex of the dog. Science 1962;137:342–343.
169. Mellanby J, George G, RObinson A, Thompson PA. Epileptiform syndrome in rats produced by injecting tetanus toxin into the hippocampus. J Neurol Neurosurg Psychiatry 1977;40:404–414.
170. Brener K, Amitai Y, Jeffreys JGR, Gutnick MJ. Chronic epileptic foci in neocotex: In vivo and in vitro efects of tetanus toxin. Eur J Neurosci 1990;3:47–54.
171. Finnerty GT, Jeffreys JGR. Investigations of the neuronal aggregate generating seizures in the rat tetanus toxin model of epilepsy. J Neurophysiol 2002;88:2919–2927.
172. Louis ED, Williamson PD, Darcey TM. Chronic focal epilepsy induced by microinjection of tetanus toxin ino the cat motor cortex. EEG Clin Neurophysiol 1990;75:548–557.
173. Hagemann G, Bruehl C, Lutzenburg M, Wite OW. Brain hypometabolism in a rat model of chronic focal epilepsy in rat neocortex. Epilepsia 1998;39:339–346.
174. Penner R, Heher E, Dreyer F. Intracellularly injected tetanus toxin inhibits exocytosis in bovine adrenal chromafin cells. Nature Med 1986;324:76–78.
175. Calabresi P, Benedeti M, Mercuri NB, Bernardi G. Selective depression of synaptic transmission by tetanus toxin. A comparative study on hippocampal and neostriatal slices. Neurosci 1989;30:663–670.
176. Bergey GK, Bigalke H, Nelson PG. Differential effects of tetanus toxin on inhibitory and excitatory synaptic transmission in mammalian spinal cord neurons in culture; a presynaptic locus of action for tetanus toxin. J Neurophysiol 1987;57:121–131.
177. Empson RM, Jeffreys JGR. Synaptic inhibition in primary and secondary chronic epileptic foci induced by intrahippocampal tetanus toxin in the rat. J Physiol 1993;465:595–614.
178. Habermann E, Erdmann G. Pharmacokinetic and histoautoradiographic evidence for the intraaxonal movement of toxin in the pathogenesis of tetanus. Toxicon 1974;16:611–623.
179. Schwab ME, Suda K, Thoenen H. Selective retrograde synaptic transfer of a protein, tetanus toxin, subsequent to its retrotransport. J Cell Biol 1979;82:798–810.
180. Liang F, Jones EG. Differential and time-dependent changes in gene expression or type II calcium/calmodulin-dependent protein kinase, 67 kDa glutamic acid decarboxylase, and glutamate receptor subunits in tetanus toxin-induced focal epilepsy. J Neuroscience 1997; 17:2168–2180.
181. Bergey GK, Macdonald RL, Habig WH, Hardegree MC, Nelson PG. Tetanus toxin convulsant action in spinal cord neurons in culture. J Neurosci 1983;3:2310–2323.
182. Jeffreys JGR, Evans BJ, Hughes SA, Williams SF. Neuropathology of the chronic epileptic syndrome induced by intrahippocampal tetanus toxin in the rat: preservation of pyramidal cells and incidence of dark cells. Neuropath Appl Neurobiol 1992;18:53–70.
183. MacVicar BA, Hochman D, LeBlanc FE, Watson TW. Stimulation evoked changes in intrinsic optical signals the human brain. Soc Neurosci Abstr 1990;16:309.
184. Cannestra AF, Black KL, Martin NA, et al. Topographical and temporal specificity of human intraoperative optical intrinsic signals. NeuroReport 1998;9:2557–2563.
185. Cannestra AF, Blood AJ, Black KL, Toga AW. The evolution of optical signals in human and rodent cortex. NeuroImage 1996;3: 202–208.
186. Cannestra AF, Bookheimer SY, O'Farrell A, et al. Temporal and topographical characterization of language cortices utilizing intraoperative optical intrinsic signals. NeuroImage 2000;12:41–54.
187. Cannestra AF, Pouratian N, Bookheimer SY, Martin NA, Becker DP, Toga AW. Temporal spatial differences observed by functional MRI and human intraoperative optical imaging. Cerebral Cortex 2001;11:773–782.
188. Pouratian N, Bookheimer SY, O'Farrell AM, et al. Optical imaging of bilingual cortical representations. Case report. J Neurosurg 2000; 93:676–681.
189. Pouratian N, Sheth SA, Martin NA, Toga AW. Shedding light on brain mapping: advances in human optical imaging. Trends Neurosci 2003;26:277–282.
190. Sato K, Nariai T, Sasaki S, et al. Intraoperative intrinsic signal imaging of neuronal activity from subdivisions of the human primary somatosensory cortex. Cerebral Cortex 2002;12:269–280.
191. Toga AW, Cannestra AF, Black KW. The temporal/spatial evolution of optical signals in human cortex. Cerebral Cortex 1995;5:561–565.
192. Mayhew JEW, Askew S, Zheng Y, et al. Cerebral vasomotion: a 0.1-Hz oscillation in reflected light imaging of neural activity. NeuroImage 1996;4:183–193.
193. Mitra PP, Pesaran B. Analysis of dynamic brain imaging data. Biophys J 1999;76:691–708.
194. Kraemer DL, Spencer DD. Anesthesia in epilepsy surgery, In: Engel JJ, ed. Surgical Treatment of the Epilepsies. New York: Raven Press, Ltd.; 1993:527–538.
195. Schwartz TH, Chen L-M, Friedman RM, Spencer DD, Roe AW. Intraoperative optical imaging of human face cortical topography: a case study. Neuroreport 2004;15:1527–1531.
196. Nelson RJ, Sur M, Felleman DJ, Kaas JH. Representation of the body surface in postcentral parietal cortex of Macaca fascicularis. J Comp Neurol 1980;192:611–643.
197. Felleman DJ, Nelson RJ, Sur M, Kaas JH. Representations of the body surface in areas 3b and 1 of postcentral parietal cortex of cebus monkeys. Brain Res 1983;268:15–26.
198. Jain N, Qi H-X, Catania KC, Kaas JH. Anatomic correlates of the face and oral cavity representations in the somatosensory cortical area 3b of monkeys. J Comp Neurol 2001;429:455–468.
199. Penfield W, Jasper H. Epilepsy and the functional anatomy of the human brain. Boston, Little Brown, 1954.
200. Van Buren JM. Sensory responses from stimulation of the inferior Rolandic and Sylvian regions in man. J Neurosurg 1983;59:119–130.
201. Uematsu S, Lesser R, Fisher RS, et al. Motor and sensory cortex in human: topography studied with chronic subdural stimulation. Neurosurgery 1992;31:59–72.
202. Sur M, Merzenich MM, Kaas JH. Magnification, receptive-field area, and "hypercolumn" size in areas 3b and 1 of somatosensory cortex in owl monkeys. J Neurophysiol 1980;44:295–311.

15 Intraoperative Magnetic Resonance Imaging in the Surgical Treatment of Epilepsy

Theodore H. Schwartz, MD

SUMMARY

Epilepsy is the only chronic neurodegenerative disorder that can be cured with surgery. Epilepsy surgery consists of both curative and palliative procedures. Curative surgeries are based on the removal of a focal area of the brain from which the epilepsy arises. Often, this region can be visualized on magnetic resonance imaging (MRI) but is not apparent macroscopically in the operating room. Stereotactic navigation in the operating room that is based on a preoperative acquired dataset does not account for intraoperative "brain shift" because of gravity, loss of cerebrospinal fluid, and retraction. The ability to perform intraoperative MRI can be a useful adjunct for intraoperative neuronavigation in these situations to assure complete removal of the lesion. Surgeries such as amygdalohippocampectomy, functional hemispherectomy, and corpus callosotomy require a complete removal or disconnection of an anatomic structure in the brain. Intraoperative judgment, although often adequate, can sometimes be inaccurate in assuring that all operative goals are met. The ability to perform intraoperative MRI scans combined with neuronavigation may be helpful in confirming, or directing, the surgeon's impressions of various anatomic margins. Whether the implementation of intraoperative MRI is cost effective or significantly alters patient outcome is unknown.

Key Words: Epilepsy; magnetic resonance imaging; temporal lobe; corpus callosotomy; seizure; surgery.

1. INTRODUCTION

Epilepsy is one of the few chronic neurodegenerative diseases of the brain that can be cured with surgery. Although surgical treatments exist for the symptoms of other diseases, such as Parkinson's disease, multiple sclerosis, or Alzheimer's disease, these are merely palliative at best. In contrast, surgical resection of a well-defined epileptic focus may be curative in as many as 70–80% of cases *(1)*. More impressive is the fact that the progressive cognitive decline associated with persistent epilepsy can actually be reversed after the cessation of seizures and that these neuropsychologic gains persist as long as 10 yr after surgery *(2)*. Our ability to cure epilepsy is directly proportional to the focality of the epileptic zone in the brain. Patients with localized pathology have a much higher rate of cure than those with diffuse brain abnormalities *(3)*. Similarly, patients with focal abnormalities on magnetic resonance imaging (MRI) scan also have a higher rate of cure than patients with diffuse findings or even normal MRI scans *(4)*. Although epilepsy is primarily a disease of brain function and not structure, association with a structural abnormality is quite common and precise localization of the structural abnormality is often critical to successful epilepsy surgery. Not uncommonly, structural abnormalities that can be seen on MRI scan are not apparent macroscopically to the surgeon in the operating room (OR). The ability to identify abnormal MRI signals in real-time and correlate these signals with visualized brain structures in the OR is a powerful advance in our ability to surgically treat chronic seizure disorders that is now offered by intraoperative MRI (iMRI).

Despite the advantage provided by the ability to localize abnormal MRI findings in "surgical space," epilepsy is first and foremost a disease of brain function and not structure. Accordingly, surgeons often define the margins of their resections based on electrographic data, which has no MRI correlate and not uncommonly the MRI is normal. What use, then, is iMRI, in this situation? Because epilepsy neurosurgeons, unlike vascular, oncologic, or spine neurosurgeons, often find themselves operating on what appears grossly to be a completely normal anatomy, with no pathologic substrate, resections are frequently based on known anatomical boundaries. As will be discussed below, surgeries such as selective amygdalohippocampectomy, corpus callosotomy, and functional hemispherectomy are based on aggressive removal or disconnections of anatomical structures, such as deep grey matter nuclei or white matter tracts. The margins of these landmarks can be quite difficult to differentiate from neighboring structures in the OR, even under

From: *Bioimaging in Neurodegeneration*
Edited by P. A. Broderick, D. N. Rahni, and E. H. Kolodny

microscopic illumination and magnification. Real-time anatomical feedback from an iMRI can be extremely useful in anatomic-based resections for epilepsy.

Finally, we will discuss other uses for iMRI in epilepsy surgery, such as stereotactic placement of depth electrodes, real-time functional MRI (fMRI) to identify functional cortex, and more experimental ideas, such as real-time monitoring of tissue response to photoablative therapy. Ultimately, however, iMRI is an extremely expensive technology that has never been shown to actually affect patient outcome. Although this chapter will present theoretical advantages of iMRI in epilepsy surgery, it is unclear whether the actual implementation of the technology makes economic or even medical sense or whether it offers any real advantages over current, more affordable technology. Hence, it is not the author's intention to imply that iMRI must be used for successful epilepsy surgery but rather that its use may improve outcome in certain situations.

2. INTRAOPERATIVE MRI

2.1. JUSTIFICATION FOR THE TECHNOLOGY

The implementation of iMRI into neurosurgical practice and its combination with "frameless stereotaxy" is the end result of decades of progress in the fields of "image-guided neurosurgery," "computer-guided neurosurgery," and "stereotactic neurosurgery." The goal of these fields is to translate a point in space defined by an imaging modality such as x-rays, computed tomography (CT),or MRI, into 3D space within a patient's head in the operating room to guide a neurosurgical procedure. Another term for this technology is "neuronavigation." The use of computers with increasing processor speeds and developments in graphical interfaces, have been instrumental in the growth of this field. The motivation for this technology is based on the following facts: (1) Certain anatomical boundaries in the brain that can be viewed on MRI or CT may not be appreciable with the naked eye macroscopically in the OR. (2) Certain brain pathology that can be visualized on MRI or CT may not be distinguishable from normal brain by the naked eye macroscopically in the OR. (3) The location and trajectory to a focal lesion in the brain that does not present to the cortical surface is unknown when the brain is exposed in the OR. (4) The precise location and boundaries of a focal lesion in the brain are unknown when preparations are made to incise the skin and open the bone to expose these pathologies. (5) The ability to see around corners and beyond anatomical surfaces is impossible with a conventional operating microscope but critical to successful neurosurgical procedures. (6) Surgeons are notoriously poor at determining intraoperatively the actual extent of resection of a given structure, and postoperative imaging generally reveals residual tissue that was believed to resected in the OR. (7) Postoperative hematomas can develop between the time that hemostasis is obtained in the brain and the skin is closed. Hence, an accurate system for image-guided stereotactic neuronavigation is essential for determining (1) where to make a skin incision; (2) the extent of bone removal required; (3) where to enter the cortex to reach a lesion without damaging a normal structure; (4) the trajectory required to reach the lesion; (5) the boundaries required for a successful biopsy or complete removal of a structure; (6) the absence residual tissue that should have been resected; and (7) the absence of an expanding hematoma prior to extubating the patient.

Before the introduction of iMRI, most surgeons relied on frame-based or frameless stereotactic systems, which were based on preoperatively acquired data sets. These systems are excellent if there are no changes in the structure of the brain between the time the data are acquired and the time they are needed for intraoperative decisionmaking. However, once the brain is exposed and cerebrospinal fluid is lost, or some tissue is resected, certain brain structures can shift by as much as 1–1.5 cm *(5–9)*. Hence, a truly accurate image-guided system for neuronavigation requires real-time imaging to compensate for this known inaccuracy. One such modality is intraoperative ultrasound; however, the image quality is generally poor for differentiating fine anatomic detail *(10)*. Intraoperative CT is another solution; however, soft tissue resolution is not comparable with MRI, and the posterior fossa is degraded by bone artifact *(11)*. The best solution was to integrate MRI technology with the OR environment. Although we refer to iMRI as "real-time," in actuality, it usually takes anywhere from several seconds to minutes to obtain a scan, during which time the progress of the operation must cease.

2.2. TECHNOLOGY

iMRI was first introduced into clinical practice at the Brigham and Women's Hospital in Boston by Black and colleagues *(12)*. This system used a 0.5-T "double donut" magnet (GE Medical Systems, Milwaukee WI) with two vertical magnets in which the surgeons operated within the magnet using nonferromagnetic equipment. Although most clinical magnets are 1.5 T, adequate image resolution can be obtained with lower tesla magnets for intraoperative decisionmaking. This first system had no method for neuronavigation and was based on the concept of bringing the OR into the MR environment. Several other "low-field" systems are currently operational based on 0.12-, 0.2-, and 0.3-T magnets *(13–17)*. Surgeries are either performed within the magnet using nonferromagnetic equipment in adjacent rooms using standard instrumentation or near the magnet in the "fringe-fields," where standard instrumentation may be used. The advantage of this latter approach is to minimize the distance that the patient must be moved for imaging and the associated increased time and risk. Magnets are either horizontal or vertical and integrated neuronavigation systems were incorporated. High-field (1.5-T) magnets, which provide a much higher image resolution, have also been adapted for use in the OR *(18–20)*. These systems have been designed either to move to the patient on ceiling-mounted tracks or sit in the operating room. The strength of the field requires the use of nonferromagnetic instruments for surgeries performed in proximity to the magnet but the imaging capabilities of the system extend to include MR spectroscopy, fMRI, and MR angiography.

The lowest field magnet available (0.12 T) is the Polestar N-10 (Odin Technologies, Yokneam, Israel), which was first installed in Chaim Sheba Medical Center in Tel Hashomer, Israel, under the direction of Hadani and colleagues *(21)*. The concept was to alter the MR scanner to fit into a standard operating room environment. Surgeries can be performed with standard instrumentation with the magnets lowered under the operating table or, alternatively, with nonferromagnetic equip-

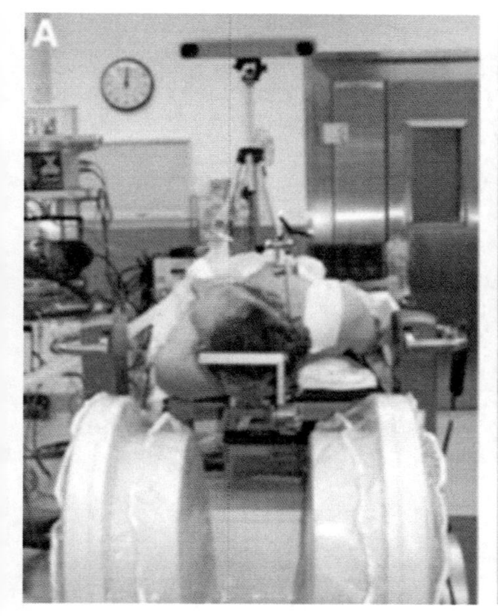

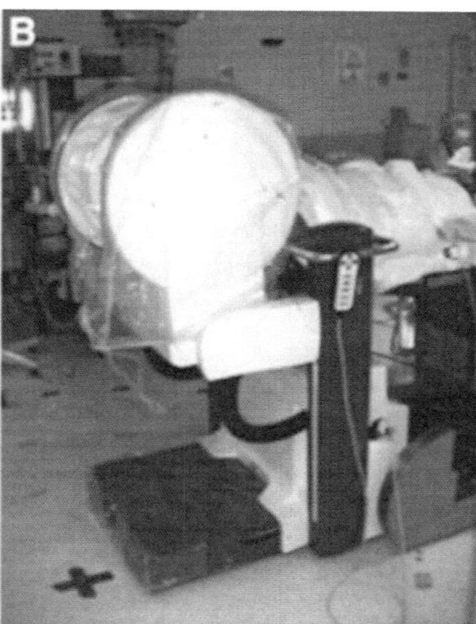

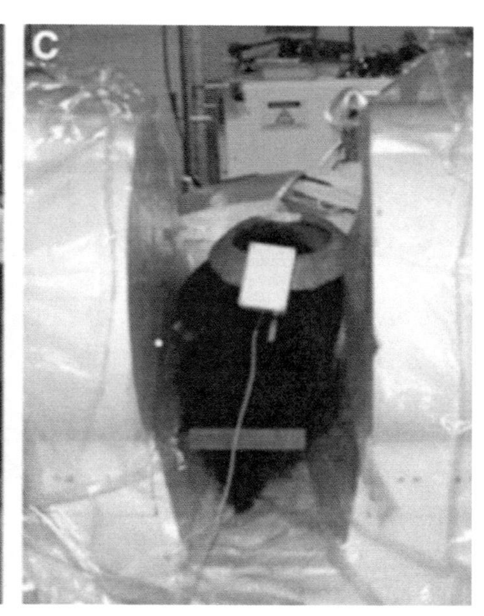

Fig. 1. The ODIN Polestar N-10 0.12T intraoperative magnet with integrated neuronavigation. **(A)** The two poles of the magnet sit on the floor under the patient's head. The head is fixed in a MR-compatible head-holder. There is a reference frame attached to the head-holder on the right side of the patient's neck with three reflective spheres. At the foot of the bed is an infrared camera, which notes the position of the patient, the magnet and a reflecting wand for neuronavigation (not shown). **(B)** The magnets can be robotically elevated with a push of a button into "scanning position," and the position is remembered by the system for later comparison images taken during the surgery. **C,** A surface coil (blue) must be placed in the center of the magnet over the surface of the head for each image. This can be placed in a sterile bag for use during surgery. *See* color version on Companion CD.

ment when the magnets are in place for imaging on either side of the head (Fig. 1A and B). The placement of surface coils directly on the head increases the signal-to-noise ratio (Fig 1C). The magnet also can be lowered slightly from the "scanning position" and used as an armrest for the surgeon. This permits the use of normal instrumentation and avoids lowering the magnet all the way to the floor, which may contaminate the sterile drapes (Fig. 2A and B). Given the low field strength of the magnet, imaging of the entire head is impossible and the field of view is generally limited to the operative field (11 × 16 cm). However, because the magnet is small, it is easily stored away at the end of the case so the OR can be used as a regular OR for other cases (Fig. 2C). This system is currently installed at the author's institution. A newer model, the Polestar N-20 (Medtronic, Louisville, CO), will have a slightly stronger magnet (0.15 T; Fig. 3). The strength of the magnet permits a larger field of view (16 × 20 cm; Fig. 4). At the time this chapter was written, this system was not yet commercially available. Both systems have integrated neuronavigation (Fig. 1A).

The purchase and installation of an iMRI can be extremely expensive and, in addition to the cost of the magnet, the OR must have radiofrequency shielding and dedicated anesthesia equipment, including ventilator, infusion pumps, and monitoring screens (Fig. 5). An alternative to shielding the entire OR is the portable shield, which can be pulled over the patient during scanning *(22)*. Several modifications in anesthesia technique must be made to accommodate the high magnetic field and limit radiofrequency noise *(23)*. Cables cannot come in contact with either the scanner or the patient's skin to prevent burns, and the pressure transducers must be kept outside the scanner to prevent artifacts. The increase in OR time has been reported to be approx 30 min; however, there is clearly a learning curve *(18,23)*.

3. EPILEPSY SURGERY

Epilepsy is a clinical term referring to a disease that effects between 1% and 2% of the population of the United States involving recurrent seizures *(24)*. Epilepsies can be divided into "partial," which start in a focal area of the brain, or "generalized," which involve large areas of the brain simultaneously. Initial therapy is always medical, which is variably successful at reducing the frequency and severity of seizures, depending on the etiology. Patients who are refractory to medical treatment are considered for surgical therapy. Surgical therapy for epilepsy can be divided into "curative" or "palliative." Curative surgeries aim at removing the region of brain that is responsible for the onset of the seizure, the so-called "ictal onset zone." Palliative surgeries either prevent the spread of epileptic events by disrupting pathways of propagation or partially interrupt seizure initiation by subtotally removing or disabling the ictal onset zone.

The most common partial-onset epilepsy treated surgically is medial temporal lobe epilepsy. Seizures arise from a local network involving the hippocampus, amygdala, entorhinal cortex, and parahippocampal gyrus. The removal of these structures is curative in as many as 70–80% of well-selected cases *(1)*. Neocortical epilepsy, in which the seizures arise from a specific abnormal region in the cerebral convexity, is also potentially curable with surgery. This disease can be divided into lesional epilepsy, in which there is an abnormality found

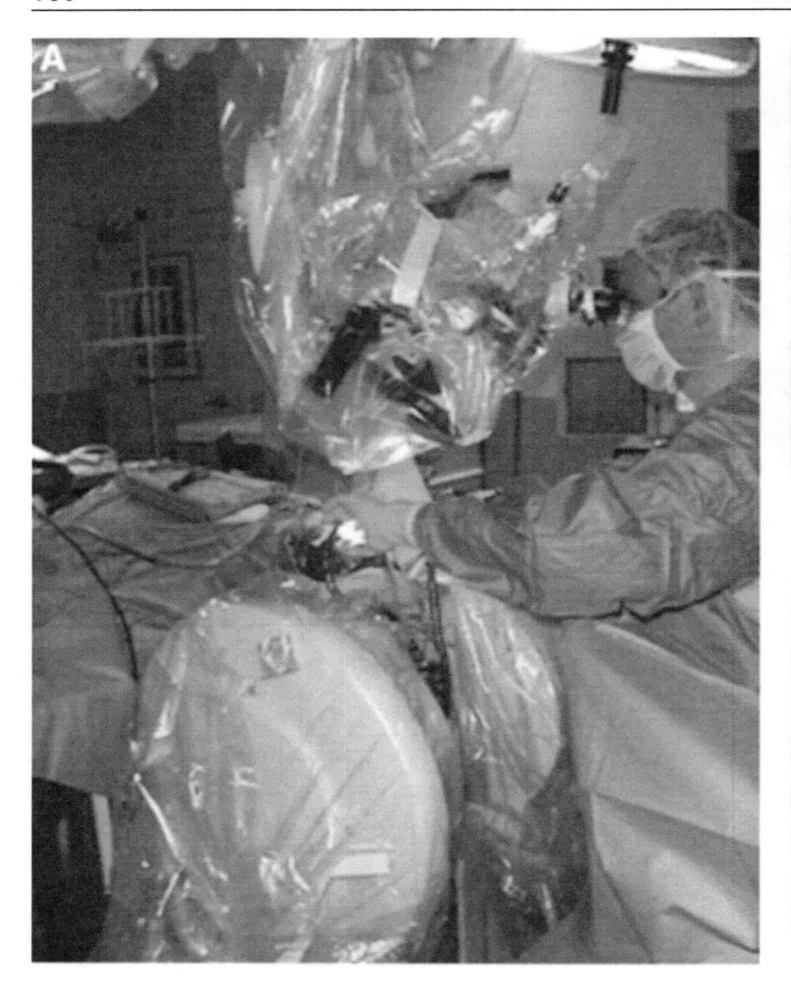

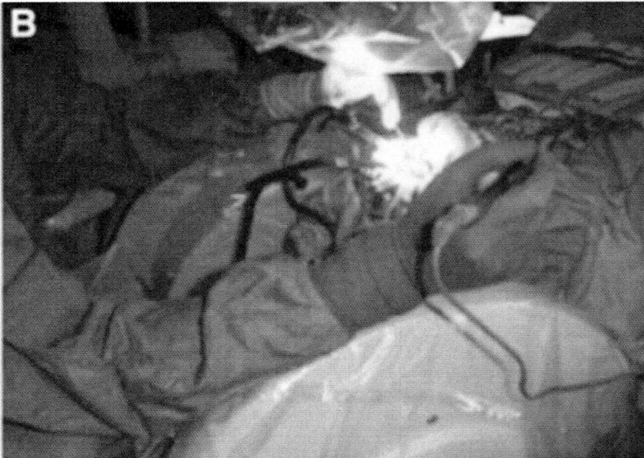

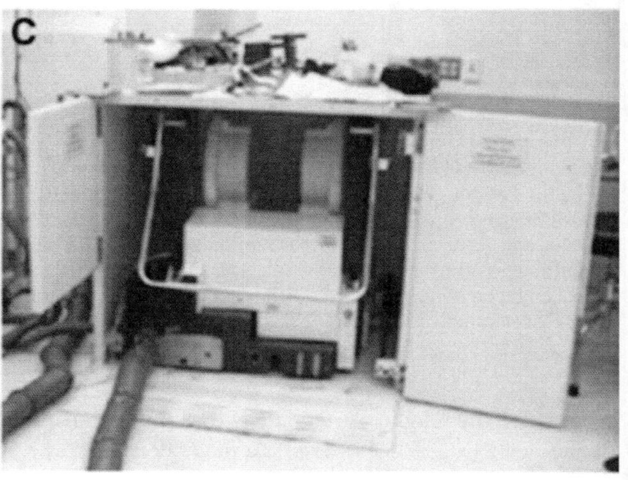

Fig. 2. The magnets can be placed in an intermediate position during surgery. (**A**) Clear sterile drapes are placed over the magnets and normal which remain sterile with the magnet lowered slightly. (**B**) In this position, the surgeon can use the magnets are armrests and normal instruments can still be used. (**C**) When not in use, the magnets are stored in a lead-shielded box and the OR can be used for other non-iMRI cases. *See* color version on Companion CD.

on MRI scan, and nonlesional epilepsy, in which the MRI scan appears normal. Lesional epilepsy may be caused either by a macrostructural abnormality, such as a low-grade tumor, vascular malformation, or gliotic scar, or a microstructural abnormality, such as focal cortical dysplasia. If the epileptogenic area is focal, then surgical cure may be expected in a high percentage of patients (50–95%; refs. *25* and *26*), whereas a more diffuse epileptic area may be more difficult to remove and the rate of cure will be accordingly lower. Nonlesional neocortical epilepsy in a patient with a normal MRI scan usually is caused by more diffuse, subtle cortical dysplasia or an idiopathic etiology, and the cure rates with resective surgery are lower (25–40%), again depending on the focality of the epileptic onset zone *(3,27,28)*.

Palliative surgical procedures are also extremely effective in preventing the initiation and spread of seizures, which can decrease the severity and frequency of the ictal events. Corpus callosotomy is one such procedure, in which the fibers connecting the hemispheres are disrupted to prevent secondary generalization of partial onset seizures. Likewise, partial resection of an epileptic focus that overlaps with functional brain, such as motor or language cortex, may provide palliation rather than cure. Novel techniques such as multiple subpial transections, vagal nerve stimulation, deep brain stimulation, direct cortical stimulation, chronic drug infusion, and gene therapy are currently being investigated as further palliative therapy for patients who are not suitable for resective surgery of their epileptic focus.

4. INTRAOPERATIVE MRI AND EPILEPSY SURGERY

4.1. HISTORY

At the time of the writing of this chapter, a total of 94 epilepsy surgeries using iMRI have been reported in the literature (Table 1). The first report of epilepsy surgery performed in an IMRI was by Steinmeier et al. in 1998 *(16)*. They reported six patients who underwent temporal lobe resections for pharmacoresistent epilepsy in the 0.2-T Magnetom Open Scanner as part of a larger series of tumor patients. Utility in tailoring the extent of mesial and neocortical resection to fit the "preoperative findings of morphological and electrophysiological alterations" was described. Seven additional temporal lobectomies were mentioned in a larger series of tumor patients presented by Kaibara et al. in 2000 *(18)*; however,

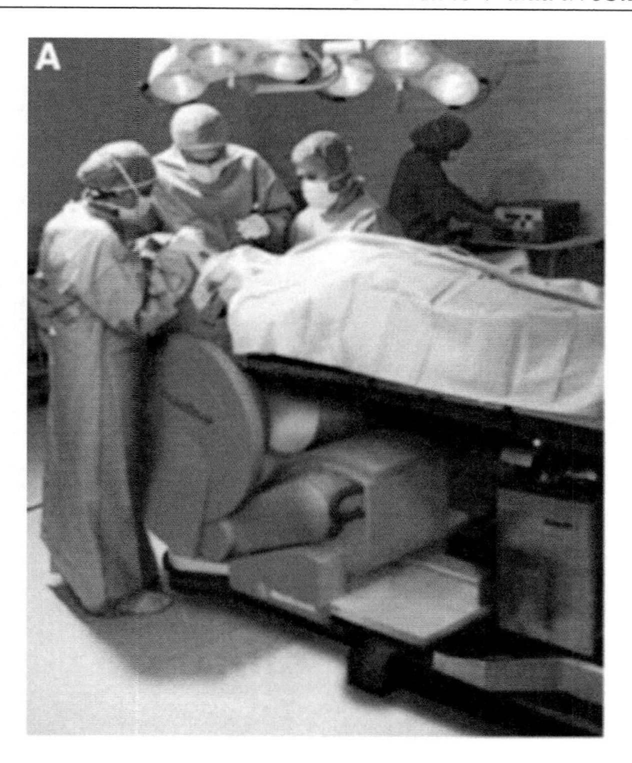

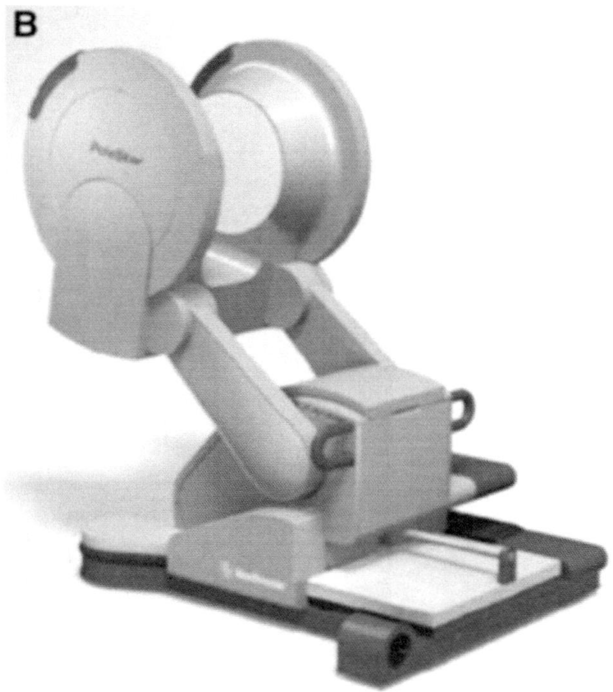

Fig. 3. The Medtronic Polestar N-20 is a slightly stronger magnet (0.15 T). (**A**) The magnet, although slightly larger than the N-10 also sits under the patient's head during the operation at a distance safe enough for using normal instruments. (**B**) The magnets are robotically elevated into position for imaging. Photos courtesy of Medtronic. *See* color version on Companion CD.

there was no specific discussion of these cases. Buchfelder et al. then reported an additional 61 cases in 2000 involving temporal lobectomy, corpus callosotomy, and lesionectomy *(29)*. Since then, an additional 33 cases have been reported (Table 1).

4.2. TEMPORAL LOBE EPILEPSY

Advances in operative techniques and imaging technologies have made a dramatic impact on our understanding of the etiology and surgical treatment for medial temporal lobe epilepsy. Although still a controversial topic *(30– 33)*, perhaps the most significant recent innovation has arisen from the mounting evidence to support a radical resection of the medial structures to optimize outcome. As a result of depth electrode studies, investigators have demonstrated ictal onsets arising not only from the anterior hippocampus but also the posterior hippocampus, amygdala, and parahippocampal gyrus *(34–37)*. Studies using high-resolution postoperative MRI to correlate residual medial temporal lobe structures with outcome have demonstrated the important role of residual hippocampus and parahippocampal gyrus in causing recurrent seizures after medial temporal lobe resections *(38–41)*. Several retrospective and prospective studies have also confirmed a higher seizure-free rate after a more extensive hippocampal resection, and many surgeons now perform a radical amygdalohippocampectomy as an integral part of their temporal lobe resections *(42–45)*. Whether this approach actually confers a higher rate of seizure freedom or a greater amount of visual field or neuropsychological impairment than electrographically tailored resections has never been tested in a randomized prospective fashion. Although there are clearly a subset of patients who will become seizure free with a subtotal hippocampectomy *(30,31)*, it is not always apparent how to reliably identify these patients and whether partial hippocampectomy minimizes postoperative neuropsychologic deficits. Hence, in many epilepsy centers, the goal of surgery in the treatment of medial temporal lobe epilepsy is an extensive resection of the hippocampus, amygdala and parahippocampal gyrus.

Despite the increased visibility provided by the operating microscope, careful evaluation of postoperative MRIs has indicated that the extent of the hippocampal resection is often overestimated by the operating surgeon *(43,45–48)*. Initially, surgeons integrated frame-based and frameless stereotaxy into their surgical approach to the medial temporal lobe *(49–54)*. These authors describe the utility of computer-assisted surgical navigation in planning a smaller craniotomy, locating the temporal horn of the lateral ventricle, and assisting in maximizing the removal of the hippocampus and amygdala *(49–53)*. The major flaw with this technique, as previously described, is its reliance on preoperative data sets. Brain shift caused by loss of cerebrospinal fluid, gravity, retraction, or resection can render inaccurate the stereotactic coordinates and the surgical plans derived from preoperative images *(6,8,9,55)*. Intraoperative MRI, combined with neuronavigation, is an ideal technique for ensuring a complete resection of the medial temporal lobe structures and optimizing outcome in the surgical treatment of temporal lobe epilepsy (Figs. 6 and 7).

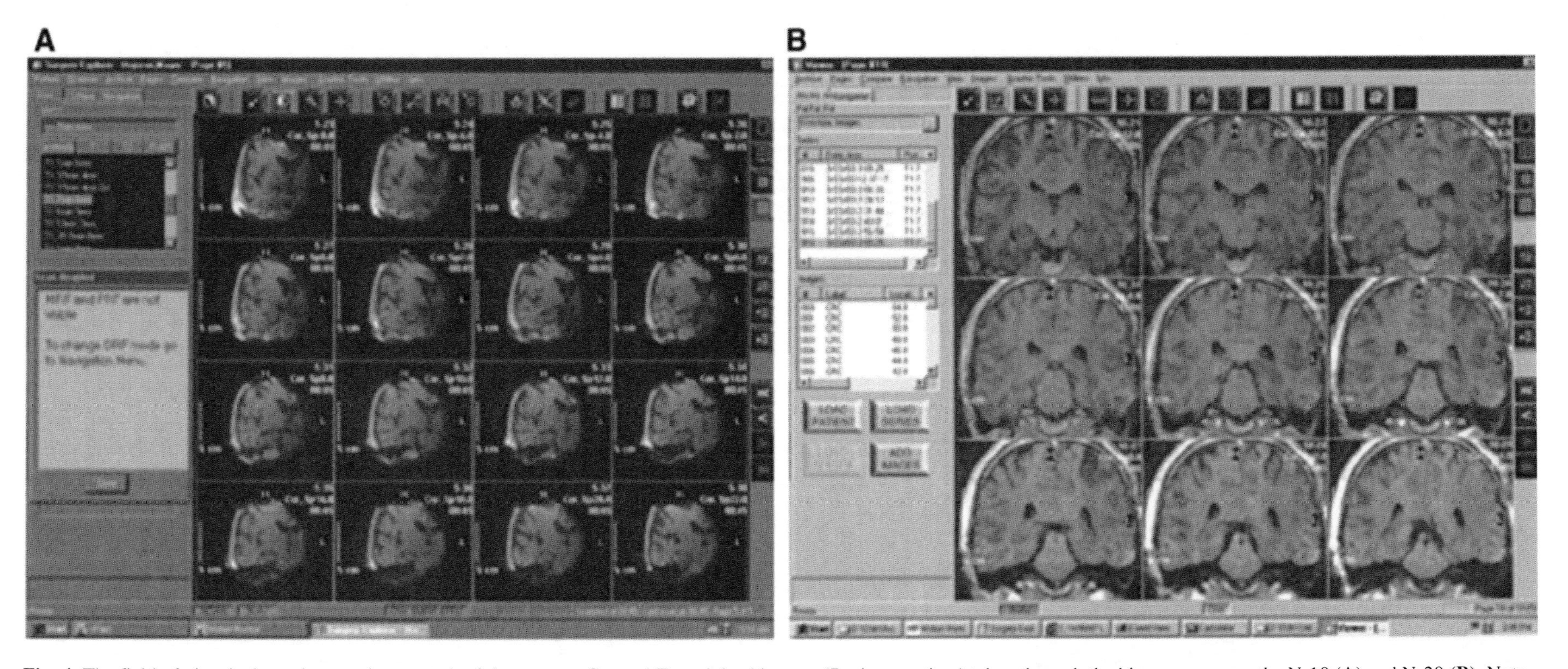

Fig. 4. The field of view is dependent on the strength of the magnet. Coronal T_1-weighted images (7 min scan time) taken through the hippocampus on the N-10 (**A**) and N-20 (**B**). Note the larger field of view using the magnet with greater strength (B). Also note hat although the field of view of the N-10 is limited, it is adequate for imaging the entire ipsilateral amygdala, hippocampus, and temporal lobe neocortex (A). *See* color version on Companion CD.

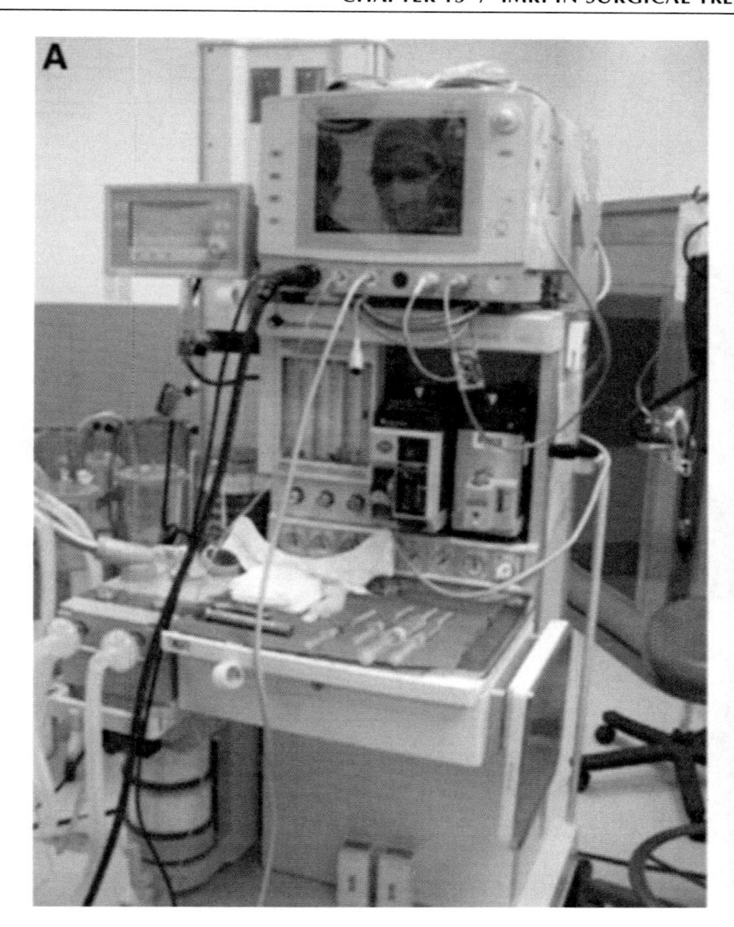

Fig. 5. iMRI requires extensive radiofrequency shielding. In addition to shielding the OR walls, anesthesia equipment (**A**) must be MR compatible because these machines cannot be switched off during imaging. All other electronic equipment must be shielded, such as phones and computers (**B**). Even the monitor and controls for the iMRI (**C**) must be shielded during imaging. *See* color version on Companion CD.

Table 1
Reported Epilepsy Surgeries Using iMRI

Author	*Year*	*No. of Patients*	*Surgery*	*Scanner*
Steinmeyer et al.	1998	6	Temporal lobectomy	0.2 T
Kaibara et al.	2000	7	Temporal lobectomy	1.5 T
Buchfelder et al.	2000	57	Temporal lobectomy Temporal lesionectomy	
		4	Corpus callosotomy	0.2 T
Nimsky et al.	2001	1	Corpus callosotomy	0.2 T
Schwartz et al.	2002	5	Amygdalohippocampectomy	0.12 T
Kaibara et al.	2002	14	Amygdalohippocampectomy	1.5 T
Buchfelder et al.	2002	N/A*	Temporal lobectomy	0.2 T
Walker et al.	2002	13	Neocortical lesionectomy	0.5 T

*Same patients reported earlier.

Eighty-nine temporal lobe resections for medically intractable epilepsy have been reported thus far (Table 1). However, in the 57 cases reported by Buchfelder et al. *(29)*, 29 cases were lesional, and all nonlesional resections were tailored to intraoperative electrocorticography (ECoG) and ranged from lesionectomies, neocortical resections sparing the hippocampus, to selective amygdalohippocamectomies. Advantages afforded by the use of iMRI include (1) smaller craniotomies; (2) ease of finding the temporal horn of the lateral ventricle; (3) assuring complete and safe resection of the amygdala, hippocampus, and parahippocampal gyrus; 4) assuring complete removal of any associated lesions; 5) visualizing postoperative hematomas prior to extubating the patient; and 6) minimizing morbidity during reoperations when anatomy is distorted. There are few outcome data available to compare patients operated upon using iMRI vs standard image guidance and certainly no randomized studies. Of the 61 reported cases of nonlesional temporal lobe resections, there are follow-up data on only 48

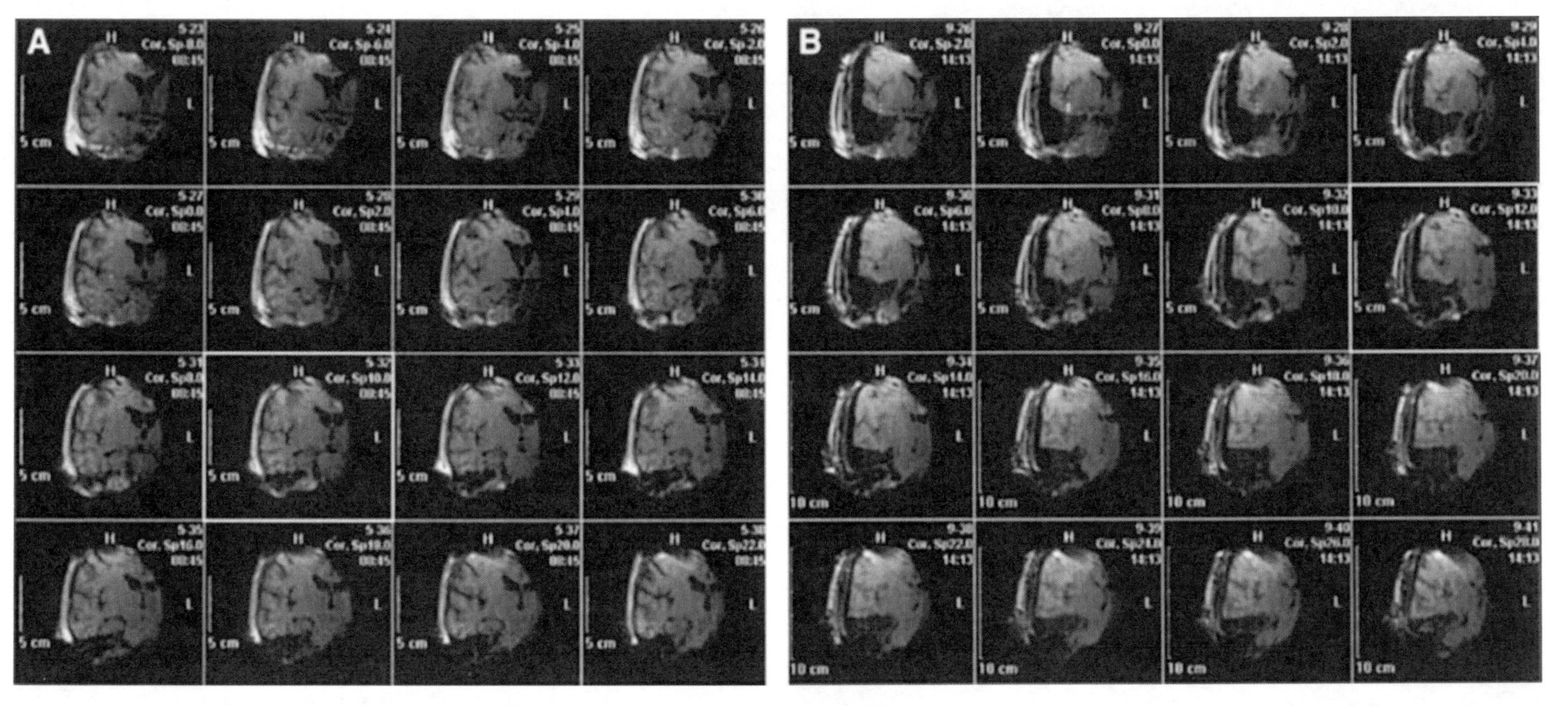

Fig. 6. Coronal T_1-weighted scans through the temporal lobe, amygdala, and hippocampus obtained with the Polestar N-10 before resection **(A)** and after resection **(B)** while craniotomy is still open. If residual structures are seen, they can be removed prior to closing the skin.

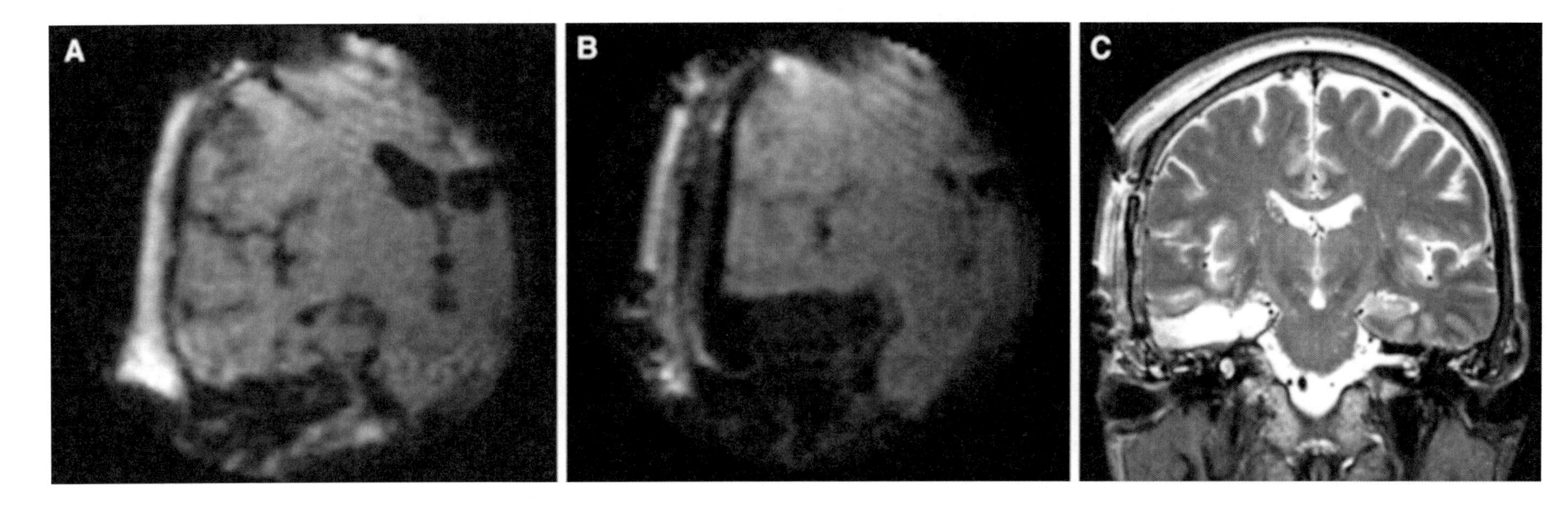

Fig. 7. Enlarged coronal T_1-weighted scan taken of the temporal lobe structures with Polestar N-10 iMRI. **(A)** Image acquired in operating room prior to resection. **(B)** Image acquired after resection felt to be complete by the surgeon but the craniotomy is still open. **(C)** Postoperative 1.5-T coronal T_2-weighted image confirms complete resection of mesial structures as seen with iMRI.

patients. In the series by Schwartz et al. *(56)*, postoperative MRI scans revealed complete hippocampal resections in all five patients, and all patients were seizure-free after a mean follow-up of 10 mo. In the study by Kaibara et al. *(57)*, 7 of 14 patients had inadequately resected medial temporal lobe structures demonstrated by the iMRI requiring reexploration and further surgery. At a mean follow-up of 17 mo, 93% of patients were seizure-free. In the series by Buchfelder et al. *(58)*, only 29 of the cases were actually nonlesional and, after a mean follow-up of 10.7 mo, 76% were seizure-free. However, as mentioned above, in this series standard resections were not performed and intraoperative ECoG was used to determine the extent of mesial and lateral resections.

4.3. NEOCORTICAL EPILEPSY

Although the mesial temporal structures are the most common location for partial onsets seizures, epilepsy can potentially arise from any neocortical location *(28,59,60)*. In contrast to mesial temporal lobe epilepsy, in which resections can be guided uniquely by anatomical boundaries, neocortical epilepsy can be lesional or nonlesional. Because nonlesional epilepsy is characterized by a normal MRI scan, the limits of the resection are generally defined by electrophysiology, that is, the boundaries of ictal and interictal events, which can be recorded acutely in the operating room or with chronically implanted electrodes. Lesional neocortical epilepsy, however, can be treated with a lesionectomy, although improved outcomes are generally achieved from additional electrophysiological recordings to identify adjacent epileptogenic brain from which seizures arise *(26,61,62)*.

The vast majority of lesions causing chronic epilepsy are extremely difficult to visualize in the OR with the naked eye. Examples include gangliogliomas, dysembryoplastic neuroepithelial tumors, and cortical dysplasia *(26,27)*. Focal cortical dysplasia, first described by Taylor et al. in 1971 *(63)*, is divided into two subcategories: type 1 and type 2 *(64)*. Type 1 is a milder form without balloon cells, and type 2 is more severe and balloon cells are present. Although invisible to the naked eye on the cortical surface, several distinct MR signal abnormalities are associated with these lesions *(64)*. The advantage of iMRI is the ability to localize the anatomical boundaries of the underlying lesion during real time in the OR. This information is critical not only if a lesionectomy is being performed but also in identifying the area where electrode grids must be placed for intraoperative ECoG or chronic recording (Fig. 8).

There are limited data on the use of iMRI for neocortical epilepsy. Buchfelder et al. *(58)* reported a series of 29 cases of lesional temporal lobe epilepsy in which iMRI (0.2 T) increased the completeness of their resection from 73% to 87%. After 15.5 mo, 24 of 29 (83%) patients were free of seizures. Of note, intraoperative ECoG was also performed. In contrast, Walker et al. *(65)* reported 12 patients, of whom 6 had extratemporal neocortical lesions, who underwent a lesionectomy without intraoperative ECoG in a 0.5-T double doughnut magnet. The authors report that electrical recordings were not reliable in such close proximity to the magnet. Residual tissue was identified by the iMRI in five patients who required further resection, after which all patients had complete radiographic removal of the lesion. Patients were followed for a mean of 22.1 mo, and only 5/12 (41%) were free of seizures *(65)*. These results indicate that although iMRI may be helpful, and in some cases even critical in identifying epileptogenic lesions, resections guided only by anatomy may not be sufficient to maximize results with respect to seizure outcome. Hence, a magnet that is fixed in close approximation to the patient, as is the 0.5-T double doughnut, may not be ideal for epilepsy surgery. In contrast, successful ECoG has been reported in other situations where the magnet or patient are mobile with varying strengths from 0.12 T *(56)*, 0.2 T *(58)*, to 1.5 T *(57)*.

4.4. CORPUS CALLOSOTOMY

Division of the corpus callosum for the treatment of patients with seizure disorders dates back to the observations of Van Wagenan and Herren in the 1930s *(66)*. They noticed that patients with strokes affecting the corpus callosum often had improvement in the frequency of their attacks. Experimental evidence for the importance of commissural fibers for the spread of epilepsy was demonstrated in the primate by Erickson, lending further support for this therapeutic approach *(67)*. Additional reports by Bogan and Vogel in adults *(68)* and Luessenhop in children *(69)* established corpus callosotomy (CC) as a standard technique in the surgical treatment of epilepsy.

Candidates for CC include patients with medically intractable primary generalized epilepsy or partial epilepsy with rapid secondary generalization that is either unlocalized or localized to unresectable cortex. The most common generalized epilepsies treated with CC are characterized by atonic or akinetic seizures, often involving sudden drop attacks *(70)*. In particular, children with Lennox–Gastaut show a significant improvement in seizure control and quality of life *(71,72)*. Complete section is approximately twice as effective as partial section in controlling seizure frequency *(73,74)*. Nevertheless, most centers currently perform a partial section initially to avoid a potential acute disconnection syndrome, although even partial CC may disrupt cortical function, particularly in patients with mixed dominance. Completion of CC is then performed as a second operation if seizure control is inadequate. For all patients, secondarily generalized seizures are controlled in 70–80% of patients whereas only 25–50% of patients find relief from complex-partial seizures *(70)*. Although CC is not intended as a curative procedure, reports of complete cessation of seizures can occur in 5–7% of patients *(75)*. In one series a more than 80% reduction in the frequency of patients' seizures was found for major seizures in 65%, focal motor seizures in 38%, atonic seizures in 76%, and absence seizures in 68% *(74)*. For all seizures together, the percentage of patients obtaining a greater than 80% reduction in generalized seizures improves from 29% after anterior section to 62% after completion of callosal section *(74)*.

Image guidance and stereotaxy have been shown to be useful in performing an anterior two thirds callosotomy *(50,53,76)*. Although most surgeons prefer to approach the brain from the right side in right-handed patients to minimize retraction injury to the dominant hemisphere, the identification of draining veins that might limit exposure may lead the surgeon to approach the callosum from the opposite side of the head. During the approach, image guidance may be helpful in distinguishing the cingulate gyrus from the callosum to avoid cortical injury. Finally, the exact extent of the callosum section can be determined intraoperatively to ensure that only the anterior two thirds

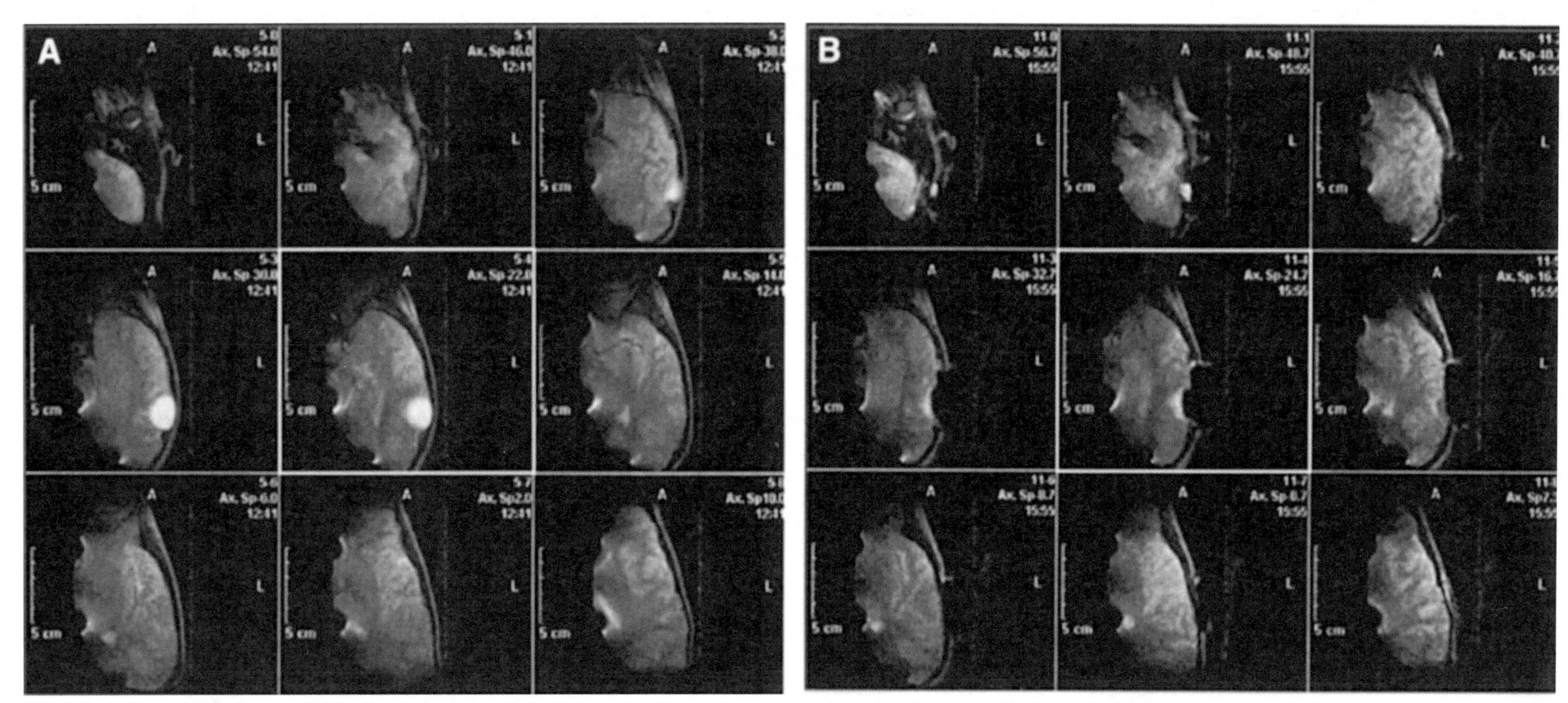

Fig. 8. Axial T_2-weighted images (3.75-min scan) taken with the Polestar N-10 reveal a neocortical ganglioglioma (**A**). The lesion is clearly seen at the cortical surface and neuronavigation can be used to locate the skin incision as well as the margins of resection at the cortical. (**B**) Postresection scan obtained before the bone flap has been replaced reveals that the lesion has been removed. Images courtesy of Dr. Mark Souweidane.

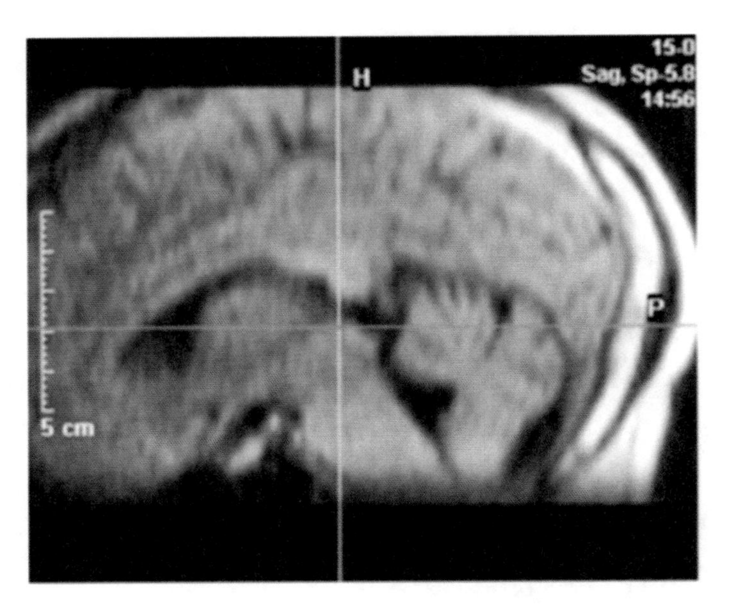

Fig. 9. Sagittal T_1-weighted image taken with the Polestar N-20 at the midline reveals the extent of the corpus callosum. Image courtesy of Medtronic.

is divided (Fig. 9). However, because entry into the ventricle often occurs in this surgery, the degree of brain shift can be significant. Hence, achieving the stated objectives would be aided by iMRI. Buchfelder et al. have reported four cases in which the 0.2 T Magnetom Open MRI was used for corpus callosotomy with good results *(29)*.

4.5. HEMISPHERECTOMY

Complete removal, or disconnection of one cerebral hemisphere, is indicated in patients with unilateral ictal onsets that are either multifocal or nonlocalizable to a single lobe associated with severe hemispheric dysfunction involving movement. Ideal candidates have already lost any usable function of the hand, particularly fine finger movements, and many are almost hemiplegic. The absence of a visual field deficit is not a contraindication to hemispherectomy as long as the patient and/or primary caregivers are well-informed of the inevitable deficit and language function will relocate to the opposite hemisphere if the surgery is performed before the age of 10 and ideally before the age of 6. Outcomes can be excellent, with seizure-free rates of 60–70% in well selected candidates *(77,78)*. Surgical procedures range from complete removal of the entire hemisphere to decortication or functional hemispherectomy. None of these procedures have been performed within an iMRI; however, the latter procedure might be aided by iMRI-guided neuronavigation.

Functional hemispherectomy involves anatomic disconnection of the grey matter of the hemisphere without removal of significant amounts of tissue *(79)*. The surgeon enters the ventricle circumferentially around the Sylvian fissure, performs a CC from within the ventricle, removes the medial temporal lobe structures, and then disconnects the basal frontal lobe and insula. At each of these steps, complete disconnection must be achieved otherwise the procedure will fail. Assessment of disconnection is difficult intraoperatively because significant tissue remains to obstruct the surgeon's view and intraoperative electrophysiologic recordings are not helpful because the residual tissue often exhibits increases in interictal activity *(80,81)*. Reports of ultrasound-guided hemispherectomy indicate the real-time intraoperative imaging and neuronavigation are helpful and theoretically iMRI would provide additional information that might increase the efficacy of the operation and ensure a complete disconnection *(82)*.

4.6. HYPOTHALAMIC HAMARTOMAS

Hamartomas arising from the hypothalamus are a rare case of chronic epilepsy in children characterized by precocious puberty and gelastic seizures (paroxysmal laughter). Confirmation of the etiology of the seizures has been demonstrated both with direct electrical recordings as well as functional imaging *(83,84)*. Reasonable relief of seizures can be obtained either with direct surgical resection of the hamartoma or stereotactic radiosurgery, if the lesion is of the appropriate size *(85 86)*. The surgery can be quite risky based on the deep location of the lesion and the indistinct margin between the hamartoma and normal hypothalamus. There is only one patient who has had a hypothalamic hamartoma removed using iMRI and although the patient had "worthwhile" improvement in the frequency of seizures, cure was not achieved *(65)*. In theory, iMRI should be quite helpful in this situation because the margin between the hamartoma and the normal hypothalamus might be apparent if the image quality were sufficiently high.

4.7. DEPTH ELECTRODES

Localization of epileptogenic foci often requires placement of chronically implanted electrodes to record seizure onsets. Subdural grid and strip electrodes provide a means for localizing foci on the cortical surface. However, potential foci also include deep structures such as the hippocampus, amygdala, and subcortical heterotopias *(87–89)*. Volume conduction of the electric fields generated by deep foci interferes with accurate localization by recordings at the surface of the brain *(90,91)*. Thus, accurate assessment of deep foci often requires placement of depth recording electrodes.

Depth electrodes were originally placed using a double-grid system in conjunction with angiography *(92)*. This technique, although accurate and safe, was time consuming and imposed limitations upon the working space at the implantation site. The development of the Leksell frame enabled the use of multiple entry sites and more working space *(93)*. Until recently, depth electrodes have been placed using frame-based systems in conjunction with computed tomography, angiography or MRI *(94–100)*. Although frame-based methodologies are highly accurate, they suffer from a number of drawbacks, including potential patient discomfort and time involved in frame placement, restricted access to the surgical field with limitations on craniotomy size, and most importantly a limited ability to rapidly define new targets and trajectories in real-time during surgery. Although frameless stereotaxy for placing depth electrodes has recently been reported *(101,102;* Mehta AM, Labar D, Dean A, Harden C, Hosain S, Pak J, Marks D, Schwartz TH, submitted for publication*)*, the problem of brain shift, field inhomogeneities, warping, and reliance on a preoperatively acquired data set, may render the placement inaccurate. The lateral temporal approach is often used when larger craniotomies are performed and depth electrodes are placed in conjunction with large arrays of subdural grids and strip electrodes. In these cases, the mesial structures may shift away from the tentorium *(8,9,55)*, causing the target to migrate in a superior direction.

Although there are no published reports of depth electrodes being placed with iMRI guidance, the technology is available. iMRI has been used successfully to perform stereotactic brain biopsies with an accuracy of 0.2 mm and a 100% success rate *(103,104)*. If the iMRI system is equipped with an optical tracking device, all that is required is a titanium cannula that can be tracked in real time. Another potential advantage of iMRI is the ability to identify intraoperative hemorrhages before extubating a patient *(103)*. This is not insignificant because the risk of hemorrhage after placing depth electrodes with frame-based angiography, CT, and MRI-based stereotaxy has been reported to be 1–4% *(88,105,106)*. Schwartz et al. *(56)* and Buchfelder et al. *(29)* have shown that commercially available depth electrodes can be visualized with iMRI even when field strength is relatively low (Fig. 10).

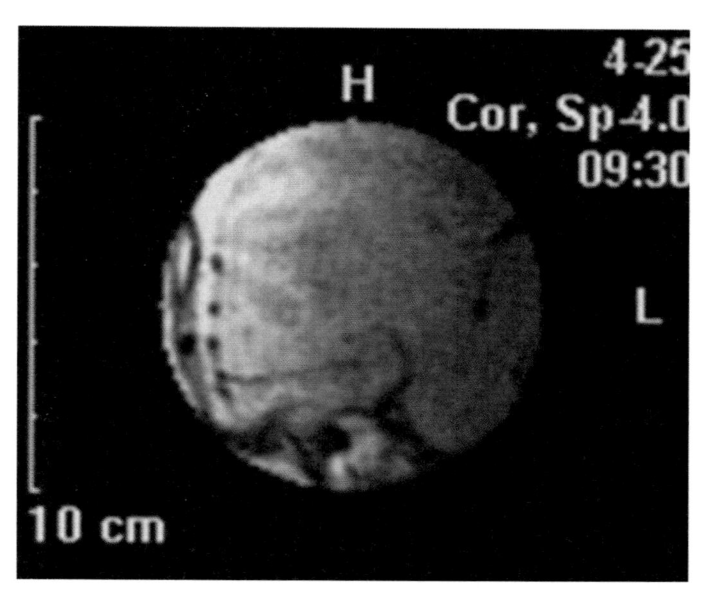

Fig. 10. T_1-weighted coronal scan taken with the Polestar N-10 before surgery (7-min acquisition time) demonstrates a surface grid and depth stainless steel electrodes in the right temporal lobe. From ref. *56* with permission from Blackwell Publishing.

4.8. ELECTROCORTICOGRAPHY, AWAKE MAPPING

Although high field strength magnets will clearly interfere with intraoperative ECoG and the electrical equipment, if turned on, will interfere with imaging, if the two technologies are not performed simultaneously, there should be no interference. Thus, although Walker et al. *(65)* report difficulty performing ECoG within a 0.5-T double donut MR, Buchfelder et al. *(58)* had no difficulties adjacent to a 0.2-T scanner. Our personal experience is that using the 0.12-T magnet at a distance of 2–3 feet from the brain and 5 feet from the electroencephalography (EEG) machine, ECoG can be performed successfully, even in patients who are awake. Once the EEG machine is powered off, the imaging can proceed smoothly without artifact. If a local shield is used, then the machine can remain powered on. The use of platinum electrodes is preferred because they do not heat up and do not emit any artifact on the MRI scan. The ability to perform on patients while they are awake is merely a question of patient positioning and comfort and is not, in and of itself, contraindicated with iMRI. Surgeries can be performed either with the patient experiencing rigid fixation and receiving generous dose of local anesthesia or with the patient resting on a soft headrest. The author has successfully performed language mapping in the Polestar N-10 without any significant problems (Fig. 11).

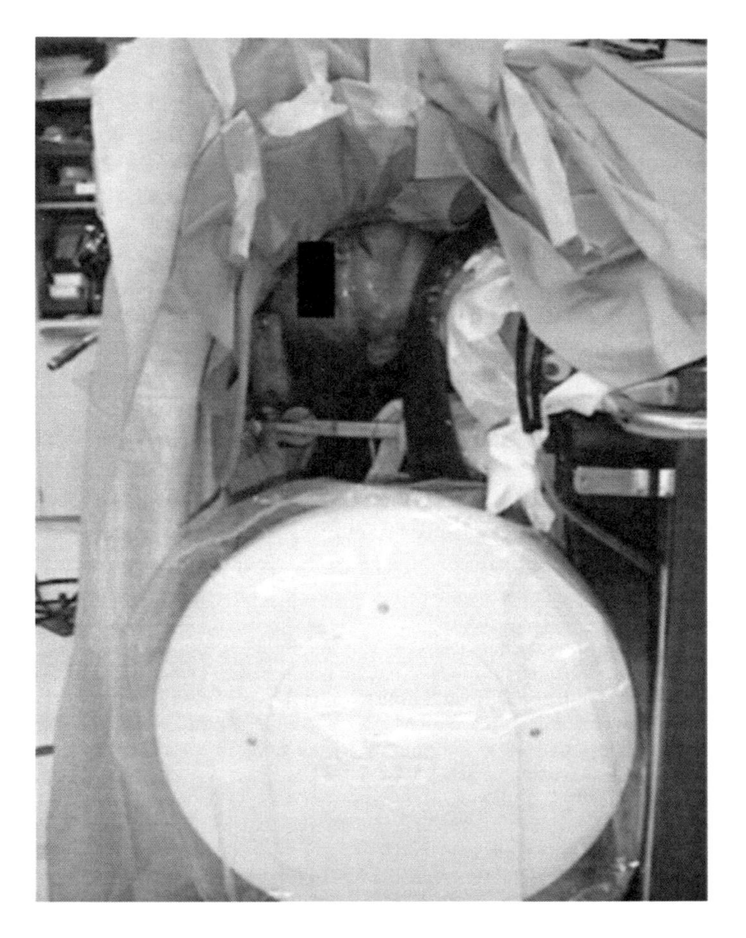

Fig. 11. Awake surgery for language mapping can be performed in the Polestar N-10 iMRI. The ECoG machine should be turned off during imaging to reduce radiofrequency noise, and the magnet should be on the floor during stimulation mapping. *See* color version on Companion CD.

4.9. FMRI AND PHOTOABLATIVE THERAPY

fMRI is a technique that takes advantage of the uncoupling of blood flow and oxygen use in activated cortical tissue *(107)* resulting in a decrease in paramagnetic deoxyhemoglobin in local capillaries and veins. Gradient-echo T_2* and T_2-weighted MR images demonstrate this change as an increase in signal, commonly known as the blood oxygen level-dependent (BOLD) signal *(108)*. FMRI permits mapping of functional cortex such as motor, sensory, visual, and language cortex with MRI and is useful in neurosurgical operative planning, particularly when resections are planned near eloquent cortex as is often the case in epilepsy surgery *(109,110)*. The ability to perform fMRI in an intraoperative setting requires a magnet with sufficient strength to image the BOLD signal. Liu et al.

(111) have reported successful fMRI in their 1.5-T iMRI on patients before the induction of anesthesia and have used the results for operative planning. However, operative planning is still based on a preoperatively acquired data set. Eventually, iMRI may be able to map functional cortex during an operative procedure while the patient's brain is exposed and after brain shift. Alternatively, one cold coregister a peroperatively acquired fMRI with an iMRI image and warp the fMRI to account for the brain shift.

The use of iMRI to monitor photodynamic therapy of tumors was recently described by Gross et al. *(112)*. This technique involves injecting a circulating drug that generates rapid vascular occlusion and necrosis in reaction to local illumination (photoexcitation). One can easily imagine the use of an intraventricular endoscope to illuminate the hippocampus and amygdala to selectively destroy the medial temporal lobe structures with photoablative therapy as an alterative to surgical resection. The results could then be followed in real time using BOLD-contrast iMRI to ensure adequate treatment before the patient leaves the OR.

5. COST–BENEFIT ANALYSIS

An iMRI is an extremely expensive piece of equipment for a hospital to purchase. In addition to the cost of the hardware and software are costs of shielding the OR, training personnel, and lost revenue because of increased length of the surgery. Despite all of these expenses, there is little evidence that the iMRI has a significant impact on patient outcome *(113)*. However, we do know that increasing the extent of resection of a lesion or the mesial structures will improve seizure-freedom and a more complete disconnecting of one hemisphere from another will lead to fewer drop attacks. We also know that, in certain circumstances, the iMRI will cause surgeons to remove tissue that had been left behind. Yet, once the iMRI is in the room, isn't it possible that it affects the surgeon's behavior and perhaps he is less aggressive initially because he knows an image will be obtained to confirm the resection margin? Until randomized controlled studies are performed, which likely will never be done, we may never know the answer to this question. Nevertheless, iMRI is the current state-of-the-art in intraoperative neuronavigation and although it may be only a very expensive stepping stone to a better technology, it is likely a necessary stage in the evolution of neurosurgical technology.

REFERENCES

1. Doyle WK, Spencer DD. Anterior temporal resections, In: Engel JJ, Pedley TA, ed. Epilepsy. A Comprehensive Textbook. Philadelphia: Lippincott-Raven Publishers, 1997: 1807–1817.
2. Helmstaedter C, Kurthen M, Lux S, Reuber M, Elger CE. Chronic epilepsy and cognition: a longitudinal study in temporal lobe epilepsy. Ann Neurol 2003;54:425–432.
3. Olivier A. Extratemporal resections in the surgical treatment of epilepsy, In: Spencer SS, Spencer DS, eds. Contemporary Issues in Neurological Surgery: Surgery for Epilepsy. Boston: Blackwell Scientific Publications; 1991:150–167.
4. Berkovic SF, McIntosh AM, Kalnins RM. Pre-operative MRI predicts outcome of temporal lobectomy: an actuarial analysis. Neurology 1995;45:1358–1363.
5. Nimsky C, Ganslandt O, Cerny S, Hastreiter P, Greiner G, Fahlbusch R. Quantiication of, visualization of, and compensation for brain shift using intraoperative magnetic resonance imaging. Neurosurgery 2000;47:1070–1080.
6. Dorward NL, Alberti O, Velanti B, Gerritsen FA, Harkness WFJ, Kitchen ND, Thomas DGT. Postimaging brain distortion: magnitude, correlates, and impact on neuronavigation. J Neurosurg 1998;88:656–662.
7. Hill DLJ, Maurer CR, Maciunas RJ, Barwise JA, Fitzpatrick JM, Wang MY. Measurement of intraoperative brain surface deformation under a craniotomy. Neurosurgery 1998;43:514–528.
8. Nabavi A, Black PM, Gering DT, et al. Serial intraoperative magnetic resonance imaging of brain shift. Neurosurgery 2001;48:787–798.
9. Roberts DW, Hartov A, Kennedy FE, Miga MI, Paulsen KD. Intraoperative brain shift and deformation: a quantitative analysis of cortical displacement in 28 cases. Neurosurgery 1998;43:749–760.
10. Chandler W, Knake J, McGillicudy J, Lillehei K, Silver T. Intraoperative use of real-time ultrasonography in neurosurgery. J Neurosurg 1982;57:157–163.
11. Lunsford LD, Parrish R, Allbight L. Intraoperative imaging with a therapeutic computed tomographic scanner. Neurosurgery 1984;15: 559–61.
12. Black PM, Moriarty T, Alexander EI, et al. Development and implementation of intraoperative magnetic resonance imaging and its neurosurgical applications. Neurosurgery 1997;41:831–845.
13. Rubino GJ, Farahani K, McGill D, Van de Wiele B, Villablanca JP, Wang-Mathieson A. Magnetic resonance image-guided neurosurgery in the magnetic fringe fields: the next step in neuronaviagation. Neurosurgery 2000;46:643–654.
14. Nimsky C, Ganslandt O, Kober H, Buchfelder M, Fahlbush R. Intraoperative magnetic resonance imaging combined with neuronavigation: a new concept. Neurosurgery 2001;48:1082–1091.
15. Wirtz CR, Knauth M, Staubert A, Bonsanto MM, Sartor K, Kunze S, Tronnier VM. Clinical evaluation and follow-up results or intraoperative magnetic resonance imaging in neurosurgery. Neurosurgery 2000;46:1112–1122.
16. Steinmeier R, Fahlbusch R, Ganslandt O, Nimsky C, Buchfelder M, Kaus M, Heigl T, Lenz G, Kuth R, Huk W. Intraoperative magnectic resonance imaging with the Magnetom open scanner: concepts, neurosurgical indications, and procedures: a preliminary report. Neurosurgery 1998;43:739–748.
17. Bohinski RJ, Kokkino AK, Warnick RE, Gaskill-Shipley MF, Kormos DW, Lukin RR, Tew JMJ. Glioma resection in a shared-resource magnetic resonance operating room after optical image-guided frameless stereotactic resection. Neurosurgery 2001;48: 731–744.
18. Kaibara T, Saunders JK, Sutherland GR. Advances in mobile intraoperative magnetic resonance imaging. Neurosurgery 2000;47:131–138.
19. Sutherland GR, Kaibara T, Louw D, Hoult DI, Tomanek B, Saunders J. A mobile high-field magnetic resonance system for neurosurgery. J Neurosurg 1999;91:804–813.
20. Hall WA, Liu H, Martin AJ, Pozza CH, Maxwell RE, Truwit CL. Safety, efficacy, and functionality of high-field strength interventional magnetic resonance imaging for neurosurgery. Neurosurgery 2000;46:632–642.
21. Hadani M, Spiegelman R, Feldman Z, Berkenstadt H, Ram Z. Novel, compact, intraoperative magnetic resonance imaging-guided system for conventional neurosurgical operating rooms. Neurosurgery 2001;48:799–809.
22. Levivier M, Wikler D, De Witte O, Van de Steene A, Baleriaux D, Brotchi J. PoleStar N-10 low-field compact intraoperative magnetic resonance imaging system with mobile radiofrequency shielding. Neurosurgery 2003;53:1001–1006.
23. Schmitz B, Nimsky C, Wendel G, - et al. Anesthesia during high-field intraoperative magnetic resonance imaging experience with 80 consecutive cases. J Neurosurg Anes 2003;15:255–262.
24. Hauser WA, Hesdorfer DC. Epilepsy: frequency, causes and consequences. New York: Demos; 1990.
25. Berger MS, Ghatan S, Geyer JR, Keles GE, Ojemann GA. Seizure outcome in children with hemispheric tumors and associated intractable epilepsy: the role of tumor removal combined with seizure foci resection. Pediatr Neurosurg 1991;17:185–191.

26. Berger MS, Ghatan S, Haglund MM, Gartner M. Low-grade gliomas associated with intractable epilepsy: seizure outcome utilizing electrocorticography during tumor resection. J Neurosurg 1993;79: 62–69.
27. Bautista RED, Cobbs MA, Spencer DD, Spencer SS. Prediction of surgical outcome by interictal epileptiform abnormalities during intracranial EEG monitoring in patients with extrahippocampal seizures. Epilepsia 1999;40:880–890.
28. Rasmussen T. Tailoring of cortical excisions for frontal lobe epilepsy. Can J Neurol Sci 1991;8:606–610.
29. Buchfelder M, Ganslandt O, Fahlbusch R, Nimsky C. Intraoperative magnetic resonance imaging in epilepsy surgery. J Magn Reson Imaging 2000;12:547–555.
30. McKhann GMI, Schonfield-McNeill J, Born DE, Haglund MM, Ojemann GA. Intraoperative hippocampal electrocorticography to predict the extent of hippocampal resection in temporal lobe epilepsy surgery. J Neurosurg 2000;93:44–52.
31. Kanner AM, Kaydanova Y, deToledo-Morrell L, Morrell F, Smith MC, Bergen D, Pierre-Louis JC, Ristanovic R. Tailored anterior tempora lobectomy. Relation between extent of mesial structures and postsurgical outcome. Arch Neurol 1995;52:173–178.
32. Goldring S, Edwards I, Harding GW, Bernardo KL. Temporal lobectomy that spares the amygdala for temporal lobe epilepsy, In: Silbergeld DL, Ojemann GA, ed. Neurosurgery Clinics of North America. Philadelphia: W.B. Saunders Co., 1993: 263–272.
33. Keogan M, McMackin D, Peng S, Phillips J, Burke T, Murphy S, Farrell M, Staunton H. Temporal neocorticectomy in management of intractable epilepsy: Long-term outcome and predictive factors. Epilepsia 1992;33:852–861.
34. Spencer DD, Spencer SS, Mattson RH, Novelly RA, Williamson PD. Access to to the posterior temporal lobe structures in the surgical treatment of temporal lobe epilepsy. Neurosurgery 1984;15:667–671.
35. Jooma R, Yeh S-H, Privitera MD, Rigrish D, Gartner M. Seizure control and extent of mesial temporal resection. Acta Neurochir (Wein) 1995;133:44–49.
36. Hudson LP, Munoz DG, Miller L, Maclachlan RS, Girvin JP, Reichman H, Blume WL. Amygdalar sclerosis in temporal lobe epilepsy. Ann Neurol 1993;33:622–631.
37. Weiser HG. Electroclinical features of psychomotor seizures. London: Butterworths; 1983.
38. Holmes MD, Wilensky AJ, Ojemann LM, Ojemann GA. Predicting outcome following reoperation for medically intractable epilepsy. Seizure 1999;8:103–106.
39. Germano IM, Poulin N, Olivier A. Reoperation for recurrent temporal lobe epilepsy. J Neurosurg 1994;81:31–36.
40. Awad IA, Nayel MH, Lüders H. Second operation after the failure of previous resection for epilepsy. Neurosurgery 1991;28:510–518.
41. Wyler AR, Hermann BP, Richey ET. Results of reoperation for failed epilepsy surgery. J Neurosurg 1989;71:815–819.
42. Wyler AR, Hermann BP, Somes G. Extent of medial temporal resection on outcome from anterior temporal lobectomy: a randomized prospective study [see comments]. Neurosurgery 1995;37:982–990; discussion 990–991.
43. Awad IA, Katz A, Hahn JF, Kong JF, Ahl J, Lüders H. Extent of resection in temporal lobectomy or epilepsy. I. Interobserver analysis and correlation with seizure outcome. Epilepsia 1989;30:756–762.
44. Bengzon ARA, Rasmussen T, Gloor P, Dussault J, Stephens M. Prognostic factors in the surgical treatment of temporal lobe epileptics. Neurology 1968;18:717–731.
45. Nayel MH, Awad IA, Luders H. Extent of mesiobasal resection determines outcome after temporal lobectomy for intractable complex partial seizures. Neurosurgery 1991;29:55–61.
46. Van Roost D, Schaller C, Meyer B, Schramm J. Can neuronavigation contribute to standardization of selective amygdalohippocampectomy. Stereotac Funct Neurosurg 1997;69:239–242.
47. Jack CR, Sharbrough FW, Marsh WR. Use of MR imaging for quantitive evaluation of resection for temporal lobe epilepsy. Radiology 1988;169:463–468.
48. Siegel AM, Wieser HG, Wichmann W, Yasargil GM. Relationships between MR-imaged total amount of tissue removed, resection scores of specific mediobasal limbic subcompartments and clinical outcome following selective amygdalohippocampectomy. Epilepsy Res 1990;6:56–65.
49. Hirabayashi H, Chitoku S, Hoshida T, Sakaki T. Accuracy and availability of the computed assisted neurosurgery navigation system during epilepsy surger. Stereotact Funct Neurosurg 1999;72:117–124.
50. Olivier A, Germano IM, Cukiert A, Peters T. Frameless stereotaxy for surgery of the epilepsies: preliminary experience. Technical note. J Neurosurg 1994;81:629–633.
51. Roberts DW, Darcey TM. The evaluation and image-guided surgical treatment of the patient with a medically intractable seizure disorder. Neurosurg Clin N Am 1996;7:215–225.
52. Wurm G, Wies W, Schnizer M, Trenkler J, Holl K. Advanced surgical approach for selective amygdalohippocampectomy through neuronavigation. Neurosurgery 2000;46:1377–1383.
53. Olivier A, Alonso-Vanega M, Comeau R, Peters TM. Image-guided surgery of epilepsy. Neurosurg Clin N Am 1996;7:229–243.
54. Kelly PJ, Sharbrough FW, Kall BA, Goerss SJ. Magnetic resonance imaging-based computer-assisted stereotactic resection of the hippocampus and amygdala in patients ith temporal lobe epilepsy. Mayo Clin Proc 1987;62:103–108.
55. Hill DLG, Maurer CRJ, Maciunas RJ, Barwise JA, Fitzpatrick JM, Wang MY. Measurement of intraoperative brain surface deformation under a craniotomy. Neurosurgery 1998;43:514–528.
56. Schwartz TH, Marks D, Pak J, Hill J, Mandelbaum DE, Holodny AI, Schulder M. Standardization of amygdalohippocampectomy with intraoperative magnetic resonance imaging: preliminary experience. Epilepsia 2002;43:430–436.
57. Kaibara T, Myles ST, Lee MA, Sutherland GR. Optimizing epilepsy surgery with intraoperative MR imaging. Epilepsia 2002;43:425–429.
58. Buchfelder M, Fahlbusch R, Ganslandt O, Stefan H, Nimsky C. Use of intraoperative magnetic resonance imaging in tailored temporal lobe surgeries or epilepsy. Epilepsia 2002;43:864–873.
59. Williamson PD, Van ness PC, Wieser HG, Quesney LF. Surgically remediable extratemporal syndromes, In: Engel JJ, ed. Surgical Treatment of the Epilepsies. New York: Raven Press, 1993: 65–76.
60. Engel JJ, Shewmon DA. Who should be considered a surgical candidate?, In: Engel JJ, ed. Surgical Treatment of the Epilepsies. New York: Raven Press; 1993:23–34.
61. Awad IA, Rosenfeld J, Ahl J, Hahn JF, Lüders H. Intractable epilepsy and structural lesions of the brain: mapping, resection strategies, and seizure outcome. Epilepsia 1991;37:179–186.
62. Berger MS, Ojemann GA. Intraoperative brain mapping techniques in neuro-oncology. Stereotact Funct Neurosurg 1992;58:153–161.
63. Taylor DC, Falconer MA, Bruton CJ, Corsellis JA. Focal dysplasia of the cerebral cortex in epilepsy. J Neurol Neurosurg Psychiatry 1971;34:369–387.
64. Kuzniecky R, Garcia JH, Faught E, Morawetz RB. Cortical dysplasia in temporal lobe eplepsy: magnetic resonance imaging correlations. Ann Neurol 1991;29:293–298.
65. Walker DG, Talos F, Bromfield EB, Black PM. Intraoperative magnetic resonance for the surgical treatment of lesions producing seizures. J Clin Neursci 2002;9:515–520.
66. Van Wagenan WP, Herren RY. Surgical division of the commissural pathways in the corpus callosum: Relation to spread of an epileptic attack. Arch Neurol Psych 1940;44:740–759.
67. Erickson TE. Spread of the epileptic discharge; An experimental study of the afterdischarge induced by electrical stimulation of the cerebral cortex. Arch Neurol Psychiat 1940;43:429–452.
68. Bogan JE, Vogel PJ. Cerebral commissurotomy in man: preliminary case report. Bull LA Neurol Soc 1962;27:169–172.
69. Luessenhop AJ. Interhemispheric commissurotomy: (the split brain operation) as an alternative to hemispherectomy for control of intractable seizures. Am Surg 1970;36:265–268.
70. Schwartz TH. Corpus callosotomy: indications and techniques, In: Schulder M, ed. Handbook of Stereotactic and Functional Neurosurgery. New York: Marcel Dekker, Inc.; 2003:529–538.

71. Collins S, Walker J, Barbaro N, Laxer K. Corpus callosotomy in the treatment of Lennox-Gastaut syndrome. Epilepsia 1989;30:670–689.
72. Sorenson JM, Wheless JW, Baumgartner JE, Clifton GL, et al. Corpus callosotomy for medically intractable seizures. Pediatr Neurosurg 1997;27:260–267.
73. Spencer SS, Spencer DD, Williamson PD, Sass K, Novelly RA, Mattson RH. Corpus callosotomy for epilepsy: I. Seizure effects. Neurology 1988;38:19–24.
74. Roberts DW. Section of the corpus callosum for epilepsy, In: Schmidek HH, Sweet WH, eds. Operative Neurosurgical Techniques. Philadelphia: W. B. Saunders Co.; 1995:1351–1358.
75. Engel JJ, Van Ness PC, Rasmussen TB.Outcome with respect to epileptic seizures, In: Engel JJ, ed. Surgical Treatment of the Epilepsies. New York: Raven Press; 1993:609–621.
76. Hodaie M, Musharbash A, Otsubo H, et al. Image-guided, frameless stereotactic sectioning of the corpus callosum in children with intractable epilepsy. Pediatr Neurosurg 2001;34:286–294.
77. Kosoff EH, Vining EPG, Pillas DJ, et al. Hemispherectomy for intractable unihemispheric epilepsy. Etiology vs outcome. Neurology 2003;61:887–890.
78. Vining EPG, Freeman JM, Pillas DJ, et al. Why would you remove half a brain? The outcome of 58 children after hemispherectomy—the Johns Hopkins experience: 1968 to 1996. Pediatrics 1997;100:163–171.
79. Villemure J-G, Mascott CR. Per-insular hemispherectomy: surgical principles and anatomy. Neurosurgery 1995;37:975–981.
80. Wennberg RA, Quesney LF, Villemure J-G. Epileptiform and non-epileptiform paroxysmal activity from isolated cortex after functional hemispherectomy. Electroenceph Clin Neurophys 1997;102: 437–442.
81. Smith SJM, Andermann F, Villemure J-G, Rasmussen TB, Quesney LF. Functional hemispherectomy: EEG findings, spiking from isolated brain post-operatively, and prediction of outcome. Neurology 1991;41:1790–1794.
82. Kanev PM. Ultrasound-tailored functional hemispherectomy for surgical control of seizures in children. J Neurosurg 1997;86:762–767.
83. Munari C, Kahane P, Francione S, et al. Role of the hypothalamic hamartoma in the genesis of gelastic fits (a video-stereo-EEG study). EEG Clin Neurophysiol 1995;95:154–160.
84. Tasch E, Cendes F, Li LM, Dubeau F, Montes J, Rosenblatt B, Andermann F, Arnold D. Hypothalamic hamartoma and gelastic epilepsy: a spectostopic study. Neurology 1998;51:1046–1050.
85. Regis J, Bartolomei F, de Toffol B, et al. Gamma knife surgery for epilepsy related to hypothalamic hamartomas. Neurosurgery 2000;47:1343–1352.
86. Rosenfeld JV, Harvey A, Wrennall J, Zaharin M, Berkovic SF. Transcallosal resection of hypothalamic hamartomas, with control of seizures, in children with gelastic epilepsy. Neurosurgery 2001;48: 108–118.
87. Smith JR, Flanigin HF, King DW, et al. An 8-year experience with depth electrodes in the evaluation of ablative seizure surgery candidates. Stereotactic Funct Neurosurg1990;54:60–66.
88. Ross DA, Brunberg JA, Drury I, Henry TR. Intracerebral depth electrode monitoring in partial epilepsy: the morbidity and efficacy of placement using magnetic resonance image-guided stereotactic surgery. Neurosurgery 1996;39:327–334.
89. Brekelman GJ, van Emde Boas W, Velis DN, Lopes de Silva FH, van Rijen PC, van Weelen CW. Comparison of combined versus subdural or intraerebral electrodes alone in presurgical focus lateralization. Epilepsia 1998;39:1290–1301.
90. Ajmone-Marsan C. Chronic intracranial recording and electrocorticography, In: Daly DD, Pedley TA, ed. Current Practice of Clinical Electroencephalography. New York: Raven; 1990:535–560.
91. Spencer SS, So NK, Engel JJ, Williamson PD, Levesque MF, Spencer DD. Depth electrodes, In: Engel JJ, ed. Surgical Treatment of the Epilepsies. New York: Raven; 1993:359–376.
92. Tailarach J, David M, Tournoux R. Atlas d'Anatomie Stereotaxique. Paris: Masson; 1957.
93. Olivier A, Bertrand G. Stereotaxic device for percutaneous twist-drill insertion of depth electrodes and for brain biopsy. J Neurosurg 1982;56:307–308.
94. Olivier A, Marchand E, Peters T, Tyler J. Depth electrode implantation at the Montreal Neurological Institute and Hospital, In: Engel JJ, ed. Surgical Treatment of the Epilepsies. New York: Raven Press; 1987:595–605.
95. So NK. Depth electrode studies in mesial temporal epilepsy, In: Luders H, ed. Epilepsy Surgery. New York: Raven Press; 1992:371–384.
96. Van Roost D, Solymosi L, Schramm J, van Oosterwyck B, Elger CE. Depth electrode implantation in the length axis of the hippocampus for the presurgical evaluation of medial temporal lobe epilepsy: a computed tomography-based stereotactic insertion technique and its accuracy. Neurosurgery 1998;43:819–827.
97. McCarthy G, Spencer DD, Riker RJ. The stereotaxic placement of depth electrodes in epilepsy, In: Luders H, ed. Epilepsy Surgery. New York: Raven Press; 1992:385–394.
98. Krateminos GP, Thomas DGT, Shorvon SD, Fish DR. Stereotactic insertion of intracerebral electrodes in the investigation of epilepsy. Br J Neurosurg 1993;7:45–52.
99. Roach MR. A model study of why some intracranial aneurysms thrombose but others rupture. Stroke 1978;9:583–587.
100. Blatt DR, Roper SN, Friedman WA. Invasive monitoring of limbic epilepsy using stereotactic depth and subdural strip electrodes: surgical technique. Surg Neurol 1997;48:74–79.
101. Yeh H, Taha JM, Tobler WD. Implantation of intracerebral depth electrodes for monitoring seizures using the Pelorus stereotactic system guided by magnetic resonance imaging. J Neurosurg 1993;78: 138–141.
102. Murphy MA, Obrien TJ, Cook MJ. Insertion of depth electrodes with or without subdural grids using frameless stereotactic guidance systems - technique and outcome. Br J Neurosurg 2002;16:119–125.
103. Moriarty TM, Quinones-Hinojosa A, Larson PS, Alexander E 3rd, et al.. Frameless stereotactic neurosurgery using intraoperative magnetic resonance imaging: stereotactic brain biopsy. Neursurgery 2000;47:1138–45; discussion 1145–6.
104. Hall WA, Martin AJ, Liu H, Nussbaum ES, Maxwell RE, Truwit CL. Brain biopsy using high-field strength interventional magnetic resonance imaging. Neurosurgery 1999;44:807–814.
105. Van Buren JM. Complications of surgical procedures in the diagnosis and treatment of epilepsy, In: Engel JJ, ed. Surgical Treatment of the Epilepsies. New York: Raven Press; 1987:465–475.
106. Merriam MA, Bronen RA, Spencer DD, McCarthy G. MR findings after depth Electrode implantation for medically refractory epilepsy. AJNR Am J Neuroradial 1993;14:1343–1346.
107. Fox PT, Raichle ME. Focal physiological uncoupling of cerebral blood flow and oxidative metabolism during somatosensory stimulation in human subjects. Proc Natl Acad Sci USA 1986;83:1140–1144.
108. Ogawa S, Lee TM, Kay AR, Tank DW. Brain magnetic resonance imaging with contrast dependent on blood oxygenation. Proc Natl Acad Sci USA 1990;87:9868–9872.
109. Duncan JS. Imaging and epilepsy. Brain 1997;120:339–77.
110. Morris GL, Mueller WM, Yetkin FZ, et al. Functional magnetic resonance imaging in partial epilepsy. Epilepsia 1994;35:1194–1198.
111. Liu H, Hall WA, Truwit CL. The roles of functional MRI in MR-guided neurosurgery in a combined 1.5 Tesla MR-operating room. Acta Neurochir Suppl 2003;85:127–135.
112. Gross S, Gilead A, Scherz A, Neeman M, Salomon Y. Monitoring photodynamic therapy of solid tumors online by BOLD-contrast MRI. Nature Med 2003;9:1327–1331.
113. Kelly PJ. Neuronavigation and surgical neurology. Surg Neurol 1999;52:7–12.

16 Periodic Epileptiform Discharges Associated With Increased Cerebral Blood Flow

Role of Single-Photon Emission Tomography Imaging

IMRAN I. ALI, MD AND NOOR A. PIRZADA, MD

SUMMARY

Periodic lateralized epileptiform discharges are associated with a variety of acute neurological disorders such as stroke, encephalitis and intracerebral hemorrhage. This particular EEG pattern may also be seen in association with status epilepticus. There is a debate in the neurology literature whether this pattern represents ongoing seizure activity or is an interictal pattern. Researchers are now beginning to utilize functional imaging studies such as SPECT to help differentiate whether this EEG pattern is ictal or interictal. This chapter provides an overview of the current state of literature regarding this topic.

Key Words: Periodic lateralized epileptiform discharges; EEG; status epilepticus; ictal patterns; SPECT.

Periodic lateralized epileptiform discharges (PLEDs) were first described as an electroencephalographic pattern more than 40 yr ago as a marker of neuronal injury *(1)*. Since then, PLEDs have been described in a variety of neurologic disorders, including cerebral infarct, intracerebral hemorrhage, encephalitis, cerebral neoplasm, and in association with seizures *(2–5)*. On electroencephalography (EEG), PLEDs are defined as focal, repetitive spike or sharp wave discharges that occur at a regular or irregular interval (1–4 s), either continuous or persisting for longer than 10 min *(6)*.

There is a considerable amount of debate in neurologic literature about the significance of PLEDs in a patient with seizure disorder *(7–13)* primarily because PLEDs in such a patient usually occur in association with status epilepticus *(7,9)*. Treiman *(10)* described various stages of status epilepticus and, in his original description, PLEDs appear in the later stages of status epilepticus, usually at a time when clinical activity may have ended. This EEG pattern in that situation appears to signify ongoing epileptiform activity without any clear clinical manifestations. In a patient who appears to be improving, one could carefully observe the presence of these discharges without attempting to intervene, but in an unresponsive or encephalopathic patient, it is unclear in the presence of PLEDs whether there is ongoing seizure activity that is responsible for the patient's level of mental dysfunction. Attempts to aggressively treat PLEDs have led to mixed results, with some patients showing remarkable improvements and others doing poorly *(14)*, causing further debate about the significance of PLEDs, especially in association with status epilepticus.

Recent cases, including from our institution *(15)* and some others *(16)*, have shown that PLEDs in association with status epilepticus are associated with increased cerebral blood flow. The increased cerebral blood flow has been documented by single-photon emission tomography (SPECT). SPECT appears to be a useful tool to assess cerebral blood flow and has been routinely used in epilepsy surgery evaluation to identify the seizure focus by injecting during a seizure *(17–19)*. Either ^{99}Tc hexamethyl-propylenamine oxime or ethylcysteinate dimer is used as the radioactive compound. Ictal SPECT in performed in most epilepsy centers by continuously monitoring patients with video-EEG recording and injecting the radioactive compound at the onset of seizure or as soon as possible after the onset. In approx 69–90% of cases *(14)* the SPECT scan correctly identifies the seizure focus by showing hyperperfusion that correlates with the ictal onset on EEG. Some hyperperfusion also is seen in the ipsilateral thalamus and contralateral cerebellum that correlates with the activation of the cortical network. In both interictal as well as postictal state there is no significant increase in blood flow, but rather a decrease or hypoperfusion is noted.

This observation, then, has quite a lot of relevance in evaluation of PLEDs, as increased cerebral blood flow correlating with PLEDs would distinguish between an ictal and an interictal abnormality. The former would be associated with increased cerebral blood flow and the latter would either not show any significant change or show a decrease *(16)*.

The following case illustrates this finding quite well: A 51-yr-old right-handed woman who had metastatic lung adenocarcinoma to the left parietal lobe resected six months previously

From: *Bioimaging in Neurodegeneration*
Edited by P. A. Broderick, D. N. Rahni, and E. H. Kolodny

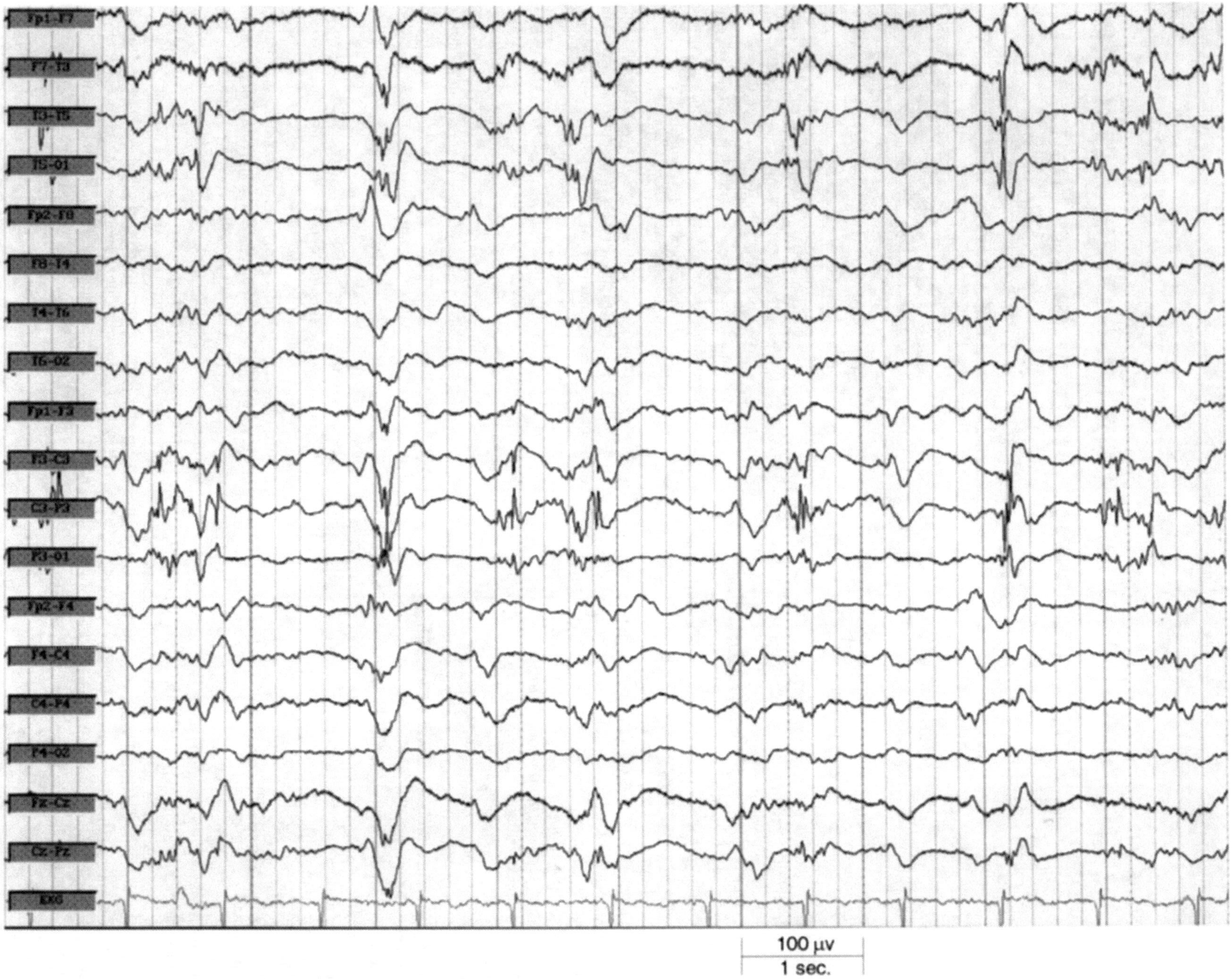

Fig 1. EEG showing periodic lateralized epileptiform discharges.

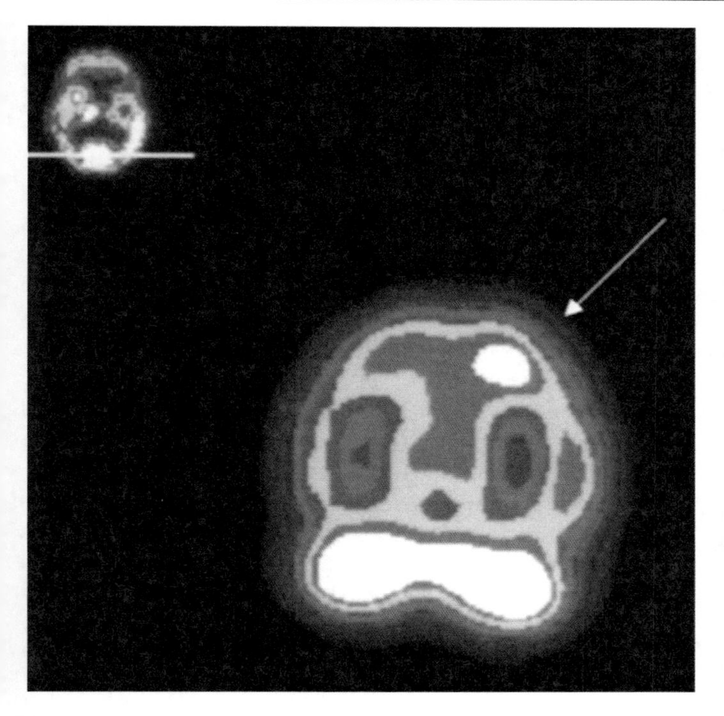

Fig. 2. Coronal section of ^{99}Tc SPECT showing increased blood flow in the left temporal and parietal region. *See* color version on Companion CD.

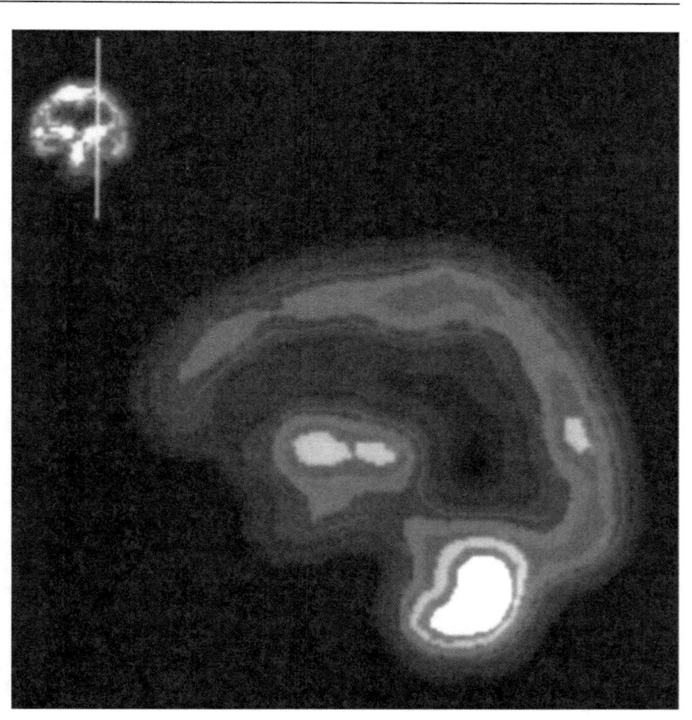

Fig. 4. Postictal SPECT showing resolution of increased blood flow. *See* color version on Companion CD.

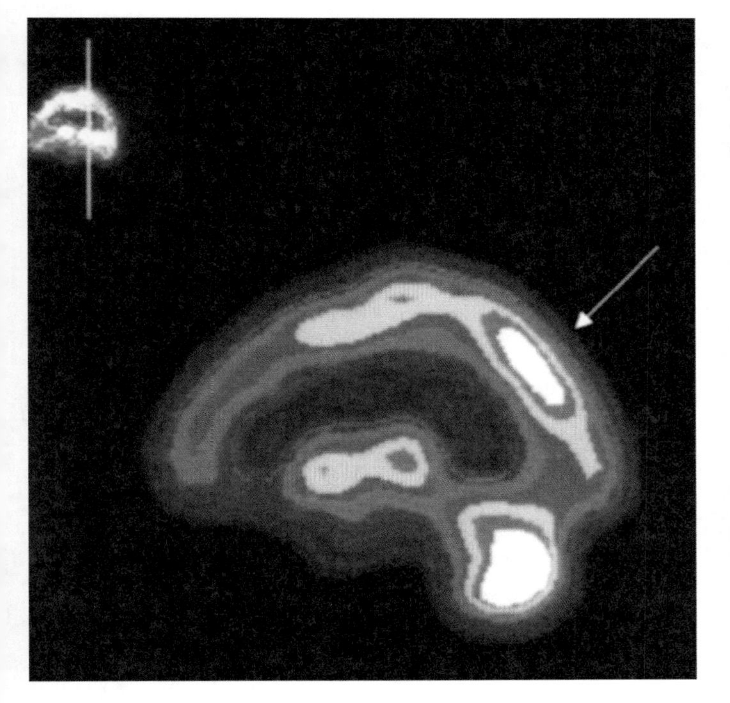

Fig. 3. Sagittal section of ^{99}Tc SPECT scan showing increased left parietal blood flow. *See* color version on Companion CD.

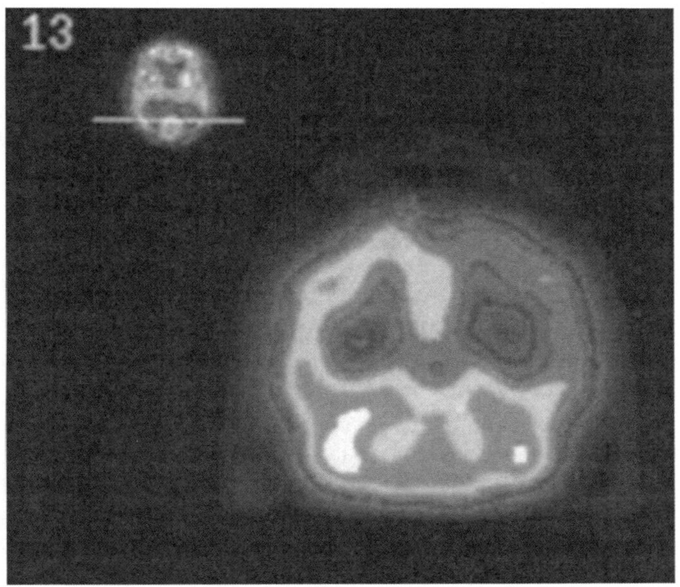

Fig. 5. ^{99}Tc SPECT in a 73-yr-old man with CJD showing increased cerebral blood flow in the right parietal region that did correspond to the location of the PLEDs. *See* color version on Companion CD.

presented to the Medical College of Ohio Emergency Department with generalized tonic-clonic seizures. She was prescribed phenytoin 100 mg by mouth three times a day and discharged. The next day she presented again to the emergency room with confusion and aphasia. Before her arrival at the emergency department, she had complained of right upper extremity paresthesiae and had involuntary clonic activity of the right arm. Initial examination revealed evidence of partial expressive and receptive aphasia and mild right-sided weakness. She was given an additional 10 mg/kg of phenytoin intravenously without significant improvement. EEG showed continuous rhythmic left temporal and parietal spike and wave activity consistent with complex partial status epilepticus. The patient was then given 15 mg/kg of intravenous valproic acid followed by 1 mg/kg/h infusion for 6 h with partial resolution of the aphasia. EEG monitoring continued to show frequent electrographic seizures, and the patient's neurologic examination fluctuated. She was transferred to the intensive care unit, intubated, and loaded with 20 mg/kg of phenobarbital. The ictal EEG pattern persisted, and another 10 mg/kg of phenobarbital was administered intravenously. This EEG pattern remained refractory to phenobarbital, midazolam, and propofol infusions at optimal doses. On the 11th day after presentation, PLEDs were noted over the left temporal and parietal region (Fig. 1) with no electrographic or clinical seizure activity. A computerized axial tomography (CT) and magnetic resonance imaging scan of the brain with contrast did not show evidence of tumor recurrence. A Technetium 99-SPECT scan was performed and showed focal increased blood flow over the left parietal and temporal region (Figs. 2 and 3). Pentobarbital infusion at 5.5 mg/kg/h (loading dose 12 mg/kg) resulted in resolution of this pattern. The patient subsequently recovered with some residual memory and attention deficits. A repeat SPECT scan after termination of the PLEDs showed resolution of focal increased cerebral blood flow (Fig. 4). The patient remained stable and has not had any recurrence of the tumor in the brain.

This case illustrates a number of important principles. First, PLEDs can occur after complex partial status epilepticus, although most often they have been associated with focal motor seizures. Secondly, PLEDs in association with status epilepticus are likely to be ictal especially if they are accompanied by persistent alteration of neurological status, as was the case in our patient. Our patient remained comatose despite appropriate therapy, and we were unable to differentiate between medication-induced coma and persistent ictal activity as a cause for this state of unresponsiveness until the SPECT scan was obtained.

A larger study from Switzerland published recently looked at a larger group of patients and found a similar pattern. Eighteen patients with PLEDs and seizures were studied by SPECT using intravenous injection of either ^{99m}Tc hexamethylpropylenamine oxime ^{99}Tc hexamethyl-propylenamine oxime or ethylcysteinate dimer. There was increased cerebral blood flow in all patients and correlated either completely or partially with the location of PLEDs in 94%. Decreased cerebral blood flow corresponded to either an intracerebral infarct or hemorrhage and occurred in patients with areas of hyperperfusion correlating with PLEDs. In three cases when the SPECT was repeated after the resolution of PLEDs, normalization of blood flow pattern was noted.

At our institution, we routinely use SPECT scans to differentiate between an ictal and an interictal event. We also find it useful in other degenerative or unexplained conditions where PLEDs are observed in association with altered level of consciousness. We recently studied a 73-yr-old man who developed nonconvulsive status epilepticus with seizures arising from the right frontal and parietal cortex. After treatment with intravenous phenytoin, his seizures stopped, but periodic sharp waves were noted over the right parietal region. Because of lack of improvement, a SPECT scan was obtained and showed evidence of focal right parietal hyperperfusion (Fig. 5). EEG showed corresponding typical periodic sharp wave discharges. Unfortunately, he did not respond to antiepileptic treatment and subsequently died after developing aspiration pneumonia. Diagnosis of Creutzfeldt-Jakob disease (CJD) was confirmed by cerebrospinal findings of 14-3-3 protein and pathologic evidence of prion disease in the brain. This case also illustrates an important and interesting fact that focal periodic discharges observed in other neurologic disorders such as CJD *(20–22)* also may be associated with associated hyperperfusion on SPECT imaging. The clinical significance of this finding is presently unknown but may imply that the PLEDs in CJD are possibly an epileptic event and may respond to antiepileptic therapy. However, patients treated aggressively do not respond well to such therapy, most likely because of the degenerative nature of this condition.

In conclusion, the presence of cerebral hyperperfusion with persistent altered neurologic function after status epilepticus or prolonged seizures may indicate that PLEDs are ictal and need to be treated aggressively. Alteration of neurological function that occurs in all patients with status epilepticus usually resolves with appropriate treatment. However, a persistent state of altered neurologic function (such as coma or encephalopathy) that does not resolve during an expected time frame in the presence of PLEDs, especially in association with therapy that may impair neurologic function, requires further evaluation. This approach is, therefore, most useful in patients in whom periodic lateralized epileptiform discharges develop after status epilepticus and expected clinical improvement does not occur. This is of great clinical significance as ongoing seizure activity, even in the absence of clinical changes may result in neuronal damage *(14,22)*. Larger, prospective and blinded studies are needed to further understand the significance of periodic discharges on EEG and hyperperfusion noted on SPECT imaging.

REFERENCES

1. Chatrian GE, Shaw CM, Leffmen H. The significance of PLEDs in EEG: an electrographic, clinical and pathological study. Electrencephalogr Clin Neurophysiol 1964;17:177–193.
2. Garcia-Morales I, Garcia MT, Galan-Davila L, Gomez-Escalonilla C, Saiz-Diaz R, Martinez-Salio A, et al Periodic Epileptiform Discharges; Etiology, Clinical Aspects, Seizures and Evolution in 130 patients. J Clin Neurophys 2002;19:172–177.
3. Pohlmann-Eden B, Hoch D, Cochius J, Chiappa K. Periodic lateralized epileptiform discharges—a critical review. J Clin Neurophysiol 1996;13:519–530.
4. Brenner RP, Schaul N. Periodic EEG patterns Classification, clinical correlation, and pathophysiology. J Clin Neurophysiol 1990; 7:249–267.
5. Walsh JM, Brenner RP. Periodic lateralized epileptiform discharges—long term outcome in adults. Epilepsia 1987;28:533–536.
6. Assal F, Papazyan JP, Slosman DO, Jallon P, Goerres GW. SPECT in periodic epileptiform discharges; a form of partial status epilepticus. Seizure 2001;10:260–264.
7. Reiher J, Rivest J, Grand'Maison F, Leduc CP. Periodic lateralized epileptiform discharges with transitional rhythmic discharges: association with seizures. Electroencephalogr Clin Neurophysiol 1991; 78:12–17.
8. Snodgrass SM, Tsuburaya K, Ajmone-Marsan C. Clinical significance of periodic lateralized epileptiform discharges: relationship with status epilepticus. J Clin Neurophysiol 1989;6:159–172.
9. Terzano MG, Parrino L, Mazzucchi A, Moretti G. Confusional states with periodic lateralized epileptiform discharges (PLEDs):

a peculiar epileptic syndrome in the elderly. Epilepsia 1986:27: 446–457.

10. Treiman DM. Electroclinical features of status epilepticus. J Clin Neurophysiol 1995;12:343–362.
11. Handforth A, Treiman DM. Functional mapping of the late stages of status epilepticus in the lithium-pilocarpine model in rat: a 14C-2-deoxyglucose study. Neuroscience 1995;64:1075–1089.
12. Handforth A, Cheng JT, Mandelkern MA, Treiman DM. Markedly increased mesiotemporal lobe metabolism in a case with PLEDs: further evidence that PLEDs are a manifestation of status epilepticus. Epilepsia 1994;35:876–881.
13. KuroiwaY, Celesia GG. Clinical significance of periodic EEG patterns. Arch Neurol 1980;37:15–20.
14. Krumholz A, Sung GY, Fisher RS, Barry E, Bergey GK, Gratten LM. Complex partial status epilepticus accompanied by serious morbidity and mortality. Neurology 1995;45:1499–1504.
15. Ali I, Pirzada NA, Vaughn BV. Periodic lateralized epileptiform discharges after complex partial status epilepticus associated with increased cerebral blood flow. J Clin Neurophys 2001;18:565–569.
16. Duncan JS. Imaging and epilepsy. Brain 1997;120:339–377.
17. So, Elson L. Role of neuroimaging in management of seizure disorders. Mayo Clin Proc 2002;77:1251–1264.
18. Duncan R, Patterson J, Roberts R, Hadley DM, Bone I. Ictal/postictal SPECT in the presurgical localization of complex partial seizures. J Neurol Neurosurg Psychiatry 1993;56:141–48.
19. Kuzneicky R, Knowlton RC. Neuroimaging in epilepsy. Sem Neurol 2002:22:279–289.
20. Au W, Gabor AJ, Vijayan N, Markand O. Periodic lateralized epileptiform discharges in Creutzfeldt-Jakob Disease. Neurology 1980;30:611–617.
21. Fushimi M, Sato K, Shimuzo T, Hadeishi H. PLEDs in Creutzfeldt-Jakob Disease following a cadaveric dural graft. Clin Neurophys 2002;113:1030–1035.
22. Lothman E. The biochemical basis and pathophysiology of status epilepticus. Neurology 1990;40:13–23.

17 Imaging White Matter Signals in Epilepsy Patients

A Unique Sensor Technology

PATRICIA A. BRODERICK, PhD, AND STEVEN V. PACIA, MD

SUMMARY

We studied 15 temporal lobe epilepsy patients who presented with either neocortical or mesial temporal lobe epilepsy. Resected tissue from these patients was studied with neuromolecular imaging (NMI) using miniature carbon-based BRODERICK PROBE® stearic and lauric acid sensors. In separate studies, a sensor was inserted into neocortex or hippocampus, specifically into individual layers of neocortical temporal gyrus as well as temporal stem and hippocampal pyramidal layer, dentate gyrus, alveus, and subiculum. The catecholamines, dopamine and norepinephrine, and the indoleamine, serotonin, as well as ascorbic acid, an enzyme catalyst in the dopamine metabolic pathway, were detected in separate electroactive signals by these sensors in real time. Results showed that gray matter and white matter were readily distinguished both by characteristic microvoltammetric waveforms and by concentration differences, that is, (1) electroactive species for catecholamine signals in white matter were diffusion waveforms whereas the indoleamine waveform was an adsorption waveform and (2) concentrations of neurotransmitters were significantly lower in white matter than in gray matter. White matter signals were detected and distinguished from gray matter signals in both neocortical and hippocampal neuroanatomic substrates. Results showed that an average of 80% of patient electroactive signals for white matter in the three white matter structures were positive both for waveforms and for concentration characteristics. Another 21.4% and 10% were partially positive for white matter in alveus and subiculum, respectively, that is, positive for concentration characteristics but not positive for both catecholamine and indoleamine waveforms. An average of 90% of patient electroactive signals for gray matter were positive both for waveforms and for concentrations. Thus, NMI with these specialized sensors provide a unique technology for detecting, monitoring, and measuring neurotransmitters and related neurochemicals in white matter vis-à-vis gray matter in temporal lobe epilepsy patients.

Key Words: Temporal lobe epilepsy; neocortex; hippocampus; monoamines; in vivo microvoltammetry; monoamines; nanotechnology; neuromolecular imaging; NMI.

1. INTRODUCTION

White matter demyelination with and/or without the destruction of white matter causes devastating consequences to neuronal conduction. White matter diseases and inherited metabolic disorders include multiple sclerosis, cerebral ischemia, leukodystrophies, adrenoleukodystrophies, leukoaraioses, leukoencephalopathies, as well as mitochondrial disorders. Since the discovery that white matter can be imaged on computed tomography, correlations between white matter and neuronal pathologies have been under intense examination. One study has found that age-related changes in white matter in frontal lobe were significantly associated with decreased cognitive status whereas leukoaraiosis was associated with loss of motivation *(1)*. In addition, using diffusion tensor magnetic resonance imaging (MRI), O'Sullivan et al. *(2)* confirmed a role for white matter damage and disruption of white matter connections in the pathogenesis of cognitive impairment in cerebral small vessel disease. van der Knapp et al. have reported a distinct encephalopathic nosologic entity in children called H-ABC. This entity involves white matter atrophy in basal ganglia and the cerebellum as shown by MRI and spectroscopy; H-ABC is associated with extrapyramidal symptoms, movement dysfunction likely related to basal ganglia nigrostriatal neuronal systems *(3)*. This same group also has discovered genetic links to vanishing white matter encephalopathy *(4)*.

White matter abnormalities have been reported in temporal lobe epilepsy (TLE) patients by using glial fibrillary acidic protein analysis.The results from this study showed that tissue from hippocampal sclerotic patients and tumor/TLE patients exhibited significantly reduced densities for cells immunopositive for glial fibrillary acidic protein when com-

From: *Bioimaging in Neurodegeneration*
Edited by P. A. Broderick, D. N. Rahni, and E. H. Kolodny

pared with control tissue as well as tissue from idiopathic epilepsy patients, *(5)*. Another study, an immunohistochemical study, showed that temporal lobe white matter in normal tissue existed in higher levels than white matter that was seen in white matter resected from epilepsy patients *(6)*. Positron emission tomography imaging with [11C] flumazenil, a γ–aminobenzoic acid antagonist, also has been used to show discrepancies in white matter in epilepsy patients, that is, increased flumazenil binding in temporal lobe white matter was observed in 11 of 18 epilepsy patients *(7)*.

Using single-voxel MRI technology, temporal lobe white matter of patients with hippocampal sclerosis was found to be reduced by using *N*-acetyl-aspartate/choline ratios as a marker. The data implicate a decreased number of axons and a reduction in myelin density *(8)*. Interestingly, coronal spin-echo MRI sequence was used to study a possible differentiation in white matter function between patients with TLE and those with chronic interictal psychosis; the results showed that only TLE patients had white matter deficits *(9)*. Volumetric studies further showed a corpus callosum involvement in childhood onset of TLE *(10)*.

The importance of neurotransmitters in white matter is becoming more and more critical to our understanding of neurologic disorders and white matter abnormalities. Indeed, white matter glia facilitates both the redox-cycling of dopamine (DA) to free radicals and the reverse transport of these free radicals to DA in neuronal gray matter. This oxidation/reduction reaction helps to suppress the yield of toxic free radicals in gray matter *(11)*. This laboratory has been engaged in detecting neurotransmitters in white matter and in gray matter in neocortical and hippocampal tissue from epilepsy patients by using NMI. With such devices, based on the presence or absence of specific neurotransmitters and/or concentration differences among neurotransmitters, TLE was neurochemically subtyped *(12,13)*. In the present edition of this book, correlations between serotonin (5-HT) and its precursor, L-tryptophan (L-Trp) are reported that further differentiate subtypes of TLE in human epilepsy both neocortically and mesially *(14)*.

The purpose of this chapter is (1) to present a unique technology to bioimage white matter distinct from gray matter in TLE patients, and (2) to compare NMI with magnetic resonance imaging (MRI).

2. MATERIALS AND METHODS

2.1. OVERVIEW

We used NMI with BRODERICK PROBE®stearic acid and lauric acid sensors to detect ascorbic acid (AA), DA, norepinephrine (NE), and 5-HT in resected temporal lobe tissue (n = 15) from patients with TLE. Neurotransmitters were detected in separate signals within the same recording cycle in seconds, on line, and in real time. Three gray matter structures (neocortical temporal gyrus, hippocampal pyramidal layer and hippocampal dentate gyrus, [granular layer] tissue) and three white matter structures (neocortical temporal stem, hippocampal alveus, and hippocampal subiculum) were investigated. Experimentally derived oxidative potentials were determined in Ringer's Lactate and/or phosphate buffer. Sensors were placed as neuroanatomically appropriate. Methods are published in detail elsewhere *(15–23)*. Criteria for distinguishing electroactive species for white matter vs gray matter were (1) characteristic waveform differences and (2) characteristic concentration differences.

2.2. CONVENTIONAL TECHNOLOGY

In vivo voltammetry, also known as in vivo electrochemistry, has consisted of placing an indicator electrode into brain, and after triggering a potential difference, which causes a redox reaction, a current is produced. One can derive information about an analyte, a neurotransmitter, or its metabolite, from an electrochemical current which is a function of potential difference as this activates the surface of an electrochemical electrode. Therefore,

- Voltammetry involves the measurement of current in an electrochemical cell as a function of applied potential.
- Electroactive species undergo oxidation or reduction at a characteristic half wave potential, according to the formula: O + $ne^- \rightarrow$R, wherein, O = oxidation, ne = number of electrons R = reduction.
- The amount of current is proportional to concentration according to the Cottrell equation. (Fig 1).
- Chemically modified conventional electrodes utilize more specific chemical reactions.

2.3. NOVEL TECHNOLOGY

BRODERICK PROBE® inventions (depicted schematically in Fig. 2) relate to a variety of unique, patented, and trademarked carbon sensors that are composed selectively of a series of compounds that include, among others, classes of molecules in the biochemical categories of lipid, glycolipid, lipoprotein, saturated, and unsaturated fatty acid. These inventions are able to detect electrochemical signals for a vast number of neurotransmitters, neuromodulators, and metabolites, including neuropeptides, hormones and vitamins *(15–23,28)*. In this laboratory, we routinely and selectively detect, in discrete neuroanatomic substrates of living human and animal brain, the monoamines, DA, NE, and 5-HT, in addition to L-Trp (the precursor to 5-HT), AA, and uric acid *(15–23)*. This laboratory also has differentiated the catecholamines, DA and NE, electrochemically *(12,13)*. The electrochemical detection of somatostatin and dynorphin A are among our latest discoveries. Nitrous oxide detection during intraoperative brain surgery for treatment of TLE epilepsy is another recent discovery from this laboratory (see present edition of this book; ref. *24*).

Each new formulation of the sensor reliably detects selective properties for detection of specific analytes, neurotransmitters and metabolites. Moreover, the size of each new formulation is directly related to capacitance at the surface of this sensor, this nanotechnology (as low as 1 nm in surface *[19–21]*) reliably reduces excess charging current.

These nano- and microdevices detect basal (normal or steady state) concentrations in vivo, *in situ*, and in vitro and any alterations in neurochemistry in brain, body, beaker, or body fluids before and after pharmacologic manipulation with drugs and other compounds. Neurochemicals during actual, induced, or mimicked brain diseases can be detected. Neurochemicals in the brain and body of animals and humans can be detected.

2.4. RECORDINGS AND CIRCUIT

One recording selectively recorded several analytes, each at distinct and separate signature oxidation potentials. Each

$$i_t = nFAC_0D_0^{1/2}/3.14^{1/2}\, t^{1/2}$$

where i = *current at time t*
n = *number of electrons, eq/mol*
F = *Faraday's constant, 96,486C/eq*
A = *electrode area,* cm^2
C = *concentration of O,* mol/cm^3
D = *diffusion coefficient of O,* cm^2/s

Fig. 1. The Cottrell equation: the proportionality, between charge and mass of an electrochemical reaction, describes the relationship between the charge of each monoamine and AA being oxidized and/or reduced and the concentration of each monoamine and AA, respectively.

analyte was detected on line within 10 seconds. Repetitive scanning with this nano-and microtechnology is also reliable. Charging (background) current was recorded and eliminated in the first 20 s of each recording. Scans were recorded for at least half an hour in each neocortical and hippocampal subparcellation. For these studies, a semiderivative circuit was used; this circuit provides the first half derivative of the linear analog signal. A semiderivative voltammetric circuit combines an additional series of resistors and capacitors called a "ladder network" *(25)* with the traditional linear scanning technology. The addition of the "ladder network" to the original linear scanning circuit allows for sharper and more clearly defined peaks than those previously obtained from the linear scanning methodology. Early on, the detection of electrochemical signals was computerized and operated with remote control in this laboratory. Recently, we have updated our computerized remote control detection and measurement of electrochemical signals with an Autolab (manufactured in the Netherlands and distributed by Brinkmann, Long Island, NY).

2.5. STUDY PROTOCOL OF THE PATIENTS

In separate experiments, each layer of neocortex and each subparcellation of hippocampus was studied. As recordings were taken from each neocortical or hippocampal specimen O_2 was intermittently bubbled through the Ringer's lactate medium to diminish possible effects of hypoxia; electrochemical signals for monoamines have been studied and are easily recognizable when hypoxia is present *(26)*.

After the indicator sensor was placed in specific neuroanatomic substrates, scanning potentials (E_{app}) in mV were applied to the indicator, versus Ag/AgCl referenced stainless steel auxillary electrodes, via a CV37 potentiostat (BAS) at a scan rate of 10 mV/s, with an initial applied potential of 0.2 V and an ending applied potential of 0.4 V. Sensitivity was set at 5 nA/V; time constant was set at a 5- and/or 1-s time constant. The CV37 detector was electrically connected to a Minigard surge suppressor (Jefferson Electric, Magnetek, NY), which was then connected to an electrical ground in isolation.

2.6. PRESURGICAL EVALUATION OF PATIENTS

After intracranial EEG seizure localization and functional mapping of language for dominant temporal lobe patients, patients underwent anterior and mesial temporal lobe resections. There was a variable extent to lateral resection, which depended on the location of ictal onset and its proximity to language function. Tissue resection limitations were established by chronic and intraoperative electrocorticography, and these included the epileptogenic zone. The amygdala and hippocampus were removed in both neocortical TLE patients (with the exception of one) and mesial TLE patients. This procedure was performed in neocortical patients despite the lack of involvement of the mesiobasal temporal regions at ictal onset because of the known rapid spread of seizures into mesial temporal regions in many patients with neocortical foci. Data from animal studies have revealed that the hippocampus is a pacemaker or an amplifier of seizures beginning in temporal lobe *(27)*. Additionally, neocortical TLE patients demonstrated memory dysfunction on presurgical WADA testing.

2.7. CLASSIFICATION OF PATIENTS

Fifteen consecutive patients who had temporal lobectomies for intractable seizures were studied. Neocortical tissue was resected in all 15 patients and hippocampal tissue was resected in 14 of the 15 patients. Five patients were diagnosed with neocortical TLE, and ten patients were diagnosed with mesial TLE. Neocortical TLE patients were classified as such based on MRI scans that showed no atrophy. The classification was further based on tissue examination pathologically as well as seizure onset seen in temporal neocortex during chronic intracranial EEG studies. These studies were performed by using lateral temporal subdural grid electrodes and multiple basomesial temporal subdural strip electrodes. ten patients were classified as having mesial TLE if pathologic examination of the resected temporal lobe revealed severe hippocampal neuronal loss and gliosis and if examination of the neocortical tissue revealed no other etiology for the patient's epilepsy.

3. RESULTS

Catecholamine (DA,NE) as well as AA electroactive signals in white matter were comprised of (1) representative waveforms known as diffusion (broad) waveforms and (2) lower concentrations when compared with gray matter.

Indoleamine (5-HT) electroactive signals in white matter were (a) representative waveforms known as adsorption (sharp) waveforms and (b) lower concentrations when compared with gray matter.

There were 79 tissues studied. Of the 41 temporal lobe gray matter tissues sampled, 37 gray matter tissues were positive for gray matter, that is, these tissues exhibited electroactive signals consistent with the above-defined criteria for identification of gray matter. Only four of these tissues did not meet criteria. Of the 38 white matter tissues sampled, 30 white matter tissues were positive for white matter, that is, these tissues exhibited electroactive signals consistent with the above-defined criteria for identification of white matter. Moreover, there were three tissues in alveus and one in subiculum that exhibited partial positive identification, that is, these tissues were positive for low concentrations of catecholamines and indoleamines but did not reveal characteristic white matter waveforms for the catecholamines, although they did exhibit characteristic white matter waveforms for the indoleamine, 5-HT.

White matter signals were detected and distinguished from gray matter in both neocortical and hippocampal neuroanatomic structures. An average of 80% of patient electroactive signals

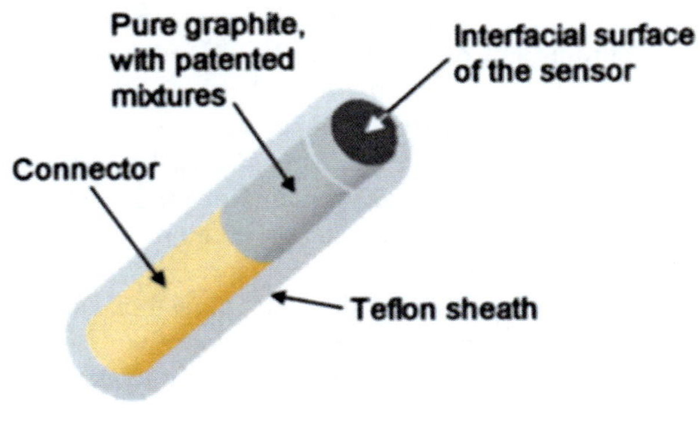

Fig. 2. The BRODERICK PROBE®: A schematic diagram depicting a medical nanotechnology and microtechnology device.

for white matter in the three white matter structures were positive for both catecholamine and indoleamine waveforms and characteristic concentrations as well. Another 21.4% and 10% were partially positive for white matter in alveus and subiculum, respectively, that is, positive for concentration characteristics but not positive for both catecholamine and indoleamine waverforms. An average of 90% of patient electroactive signals for gray matter were positive for both characteristic catecholamine and indoleamine waveforms and characteristic concentration values.

White matter exhibited substantially lower concentrations of catecholamines and the indoleamine, 5-HT (approximately three orders of magnitude less in white matter as compared with gray matter, that is, p*M* concentrations for white matter as opposed to n*M* concentrations for gray matter; unpublished data). Recordings drawn from raw data, from human TLE patients, are shown in Figs. 3–5. Figure 6 shows results from our white versus gray matter studies in histogram form.

There appeared to be no significant difference in the distinction between subtypes of TLE as to characterization of white matter vs gray matter electroactive signals, although this aspect remains under study. The possible exception to this may only be in concentrations differences, that is, that catecholamines in temporal stem white matter exhibit higher amplitude waveforms associated with higher concentrations than do hippocampal white matter.

4. DISCUSSION

White matter has inherent differences in chemical composition from gray matter. Furthermore, white matter has inherent neuroanatomic, physiologic, neuronal, structural, and functional differences from gray matter. White and gray matter of the central nervous system significantly differ in gross morphology, in water content, and macromolecular components, for example, membrane lipids. White matter is composed of myelinated bundles of axons, whereas gray matter primarily is composed of neurons and neuronal processes.

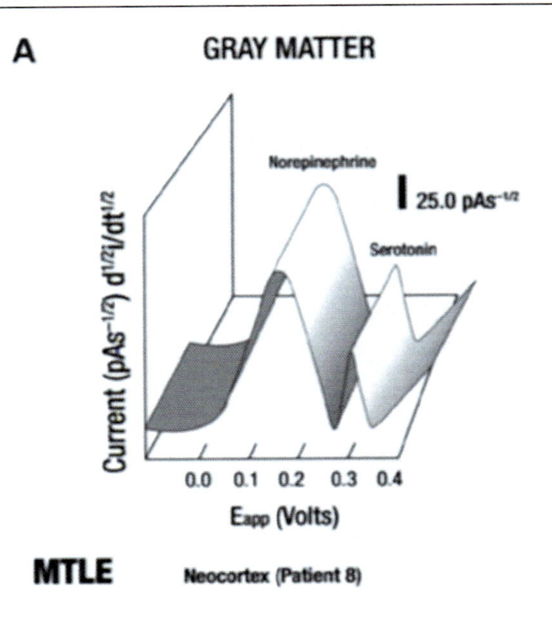

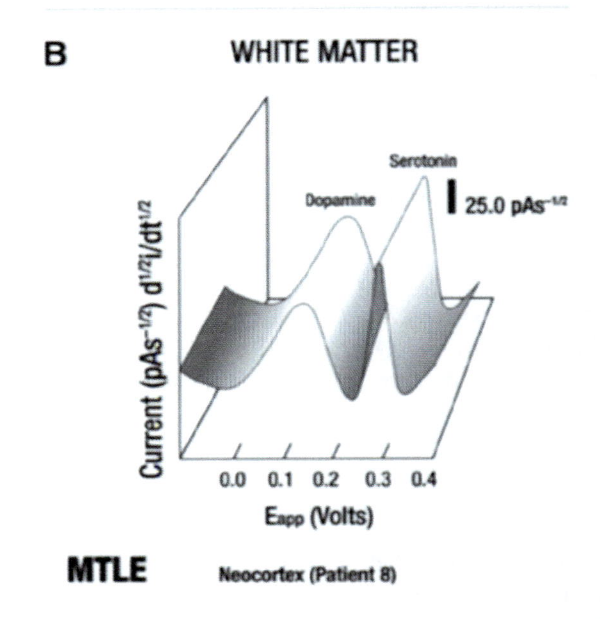

Fig. 3. (A) Gray matter recording from resected neocortical tissue from mesial temporal lobe epilepsy (MTLE) patient 8. Recording, taken with NMI and BRODERICK PROBE® sensors, show the real-time electrochemical detection of norepinephrine and serotonin at experimentally derived oxidation potentials. *x-axis*: oxidation potentials in millivolts. *y-axis*: current in picoamperes (pA) per semidifferentiation of the second. Temporal gyrus gray matter signals were derived from separate studies in each of the neocortical layers 1 through 6 with the exception of layer 4, the Baillarger's band. Electroactive signals from Baillarger's band, which is composed of white matter, were previously published *(13)*. It is noteworthy that the terms semidifferentiation and semiderivative are used interchangeably, although they are not electronically equal. The difference resides in the feedback circuit to operational amplifiers *(25)*. **(B)** White matter recording from resected neocortical tissue from mesial temporal lobe epilepsy (MTLE) patient 8. Recording, taken with NMI and BRODERICK PROBE ® sensors, shows the real-time electrochemical detection of dopamine and serotonin at experimentally derived oxidation potentials. *x-axis*: oxidation potentials in millivolts. *y-axis*: current in picoamperes (pA) per semidifferentiation of the second. White matter signals in patients were studies in the neuroanatomic substrate, temporal stem.

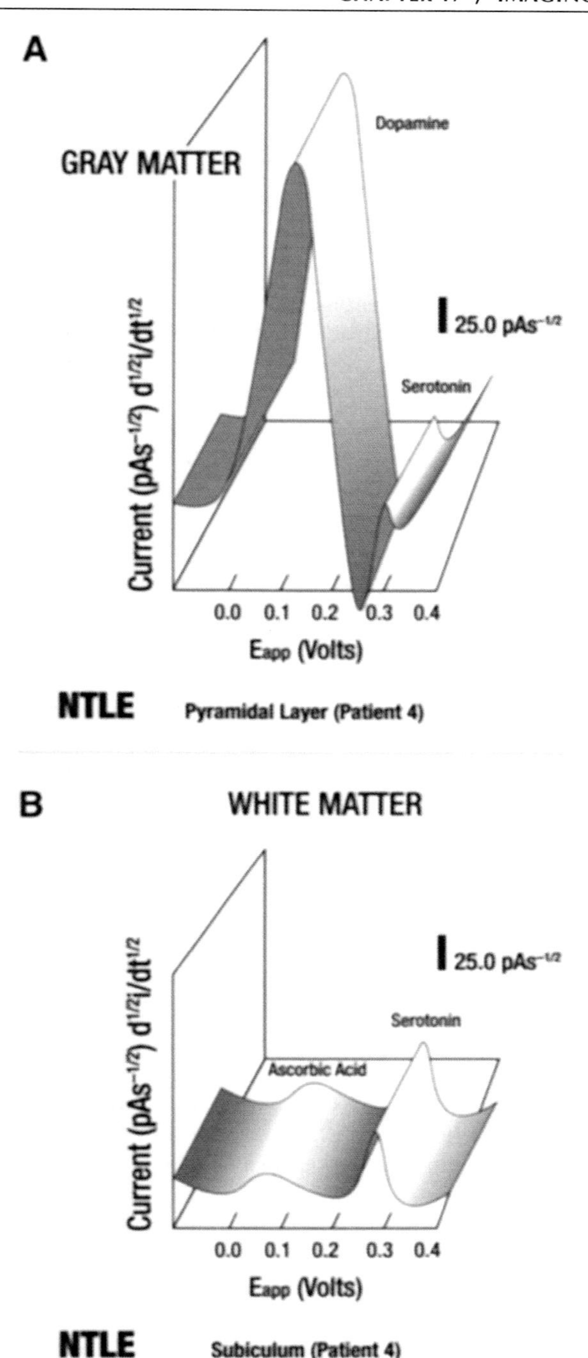

Fig. 4. **(A)** Gray matter recording from resected hippocampal tissue from neocortical temporal lobe epilepsy (NTLE) patient 4. Recording, taken with NMI and BRODERICK PROBE® sensors, shows the real-time electrochemical detection of dopamine and serotonin at experimentally derived oxidation potentials. *x-axis*: oxidation potenitals in millivolts. *y-axis*: current in picoamperes (pA) per semidifferentiation of the second. Gray matter signals in patients were studied in the neuroanatomic substrate, pyramidal layer. **(B)** White matter recording from resected hippocampal tissue of neocortical temporal lobe epilepsy (NTLE) patient 4. Recording, taken with NMI and BRODERICK PROBE ® sensors, shows the real-time electrochemical detection of asorbic acid and serotonin at experimentally derived oxidation potentials. *x-axis*: oxidation potentials in millivolts. *y-axis*: current in picoamperes (pA) per semidifferentiation of the second. White matter signals in patients were studies in the neuroanatomic substrate, subiculum.

Several lipids are found in the lipid bilayer of the membrane of white matter, for example, phospholipid, sphingomyelin, galactocerebroside, and cholesterol. These are lipids that account for myelin membrane strength *(29,30)*, and demyelination of the lipid bilayer may occur from any number of missteps such as trauma, injury, and/or chemical denaturation *(31)*. Indeed, it has been reported that the ionic environment of glial cells is a significant factor in the regulation of white matter *(32)*. Importantly, a change in the ionic environment of glial cells has recently been associated with the pathophysiology of epileptic seizures *(33,44)*.

These differences in neuroanatomy, physiology, as well as structural, functional, and chemical composition differences, lend explanation for the representative electroactive waveforms observed here for white matter vs gray matter. These differences may also lend explanatory notes for the observed differences in concentrations of neurotransmitters in white matter vs gray matter in TLE.

Simply stated, white matter is composed of fewer neuronal processes than is gray matter, and this difference provides a rational explanation for the observed lower concentrations of neurotransmitters and neurochemicals in white matter as compared with gray matter. However, the waveform differences observed in the detection of white matter vs gray matter very likely depend on chemical compositional differences. In fact and a priori, a lipid environment is an important determinant in signal amplification and in the mechanism of action of BRODERICK PROBE® sensors Several studies, in vitro and in vivo have validated these data *(15–23,34)*. Other technologies such as Raman resonance technology and surface-enhanced Raman spectroscopy confirm these results *(34)*.

Moreover, the exact chemical modification of each specific formulation of these miniature sensors, provides a further rational explanation for the ability of these miniature sensors to distinguish electroactive signals for white matter from gray matter in terms of oxidative waveforms. Electron transfer is not only enhanced by lipids and suppressed by protein *(19–23,34)* but electron transfer is affected by the hydrophobicity, hydrophilicity, carbon chain length, and polar end groups of saturated and unsaturated fatty acids, simple lipids, and complex lipids.

Therefore, relevant to the present study is the difference in electron transfer properties between the BRODERICK PROBE® stearic acid and lauric acid sensors in that electron transfer for lauric acid, a shorter chain, 12-carbon saturated fatty acid is greater than that of stearic acid, a longer chain, 18-carbon saturated fatty acid. Thus, differences in electron transfer, depending on the chemical modification of the surface, factor into the specific detection properties of sensors as each sensor interacts with ionic differences in white vs gray matter. Therefore, it is suggested that electron transfer may well be a key mechanism for detection of analytes with this nanomicrotechnology. In addition, neuroanatomic structure, ionic, and chemical composition and such of white matter vs gray matter, as these relate to detection properties on the surface of the microelectrode sensor is another key by which distinct signals for white vs gray matter are provided by this unique sensor technology.

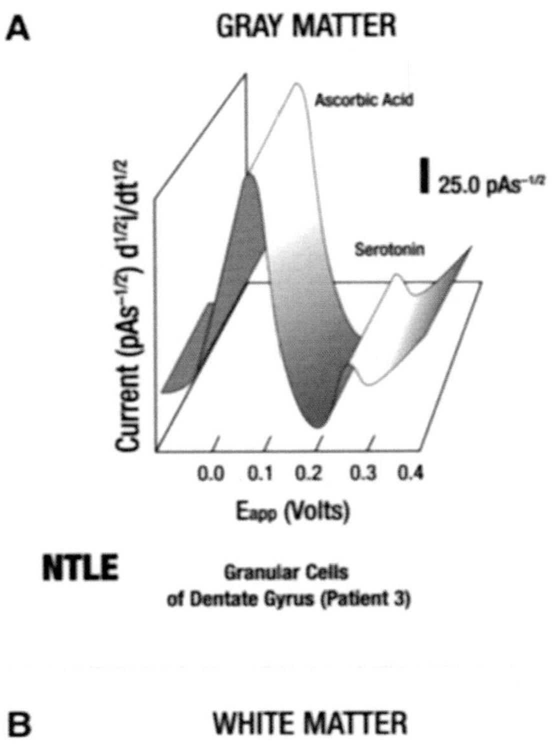

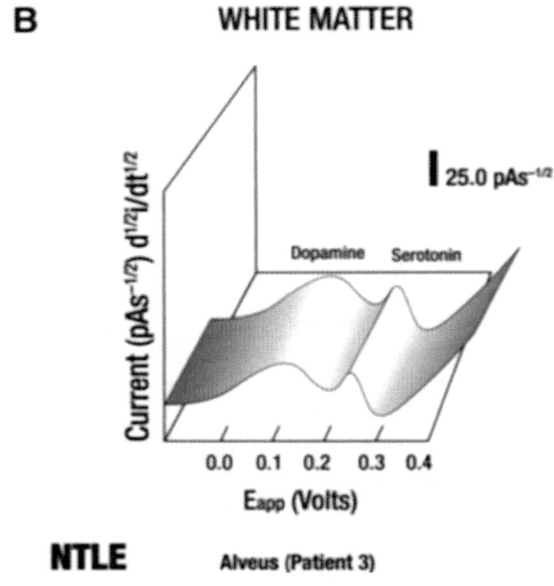

Fig. 5. **(A)** Gray matter recording from resected hippocampal tissue from neocortical temporal lobe epilepsy (NTLE) patient 3. Recording, taken with NMI and BRODERICK PROBE® sensors, shows the real-time electrochemical detection of asorbic acid and serotonin at experimentally derived oxidation potentials. *x-axis*: oxidation potenitals in millivolts. *y-axis*: current in picoamperes (pA) per semidifferentiation of the second. Gray matter signals in patients were studied in the neuroanatomic substrate, granular cells of the dentate gyrus. **(B)** White matter recording from resected hippocampal tissue of neocortical temporal lobe epilepsy (NTLE) patient 3. Recording, taken with NMI and BRODERICK PROBE ® sensors, shows the real-time electrochemical detection of asorbic acid and serotonin at experimentally derived oxidation potentials. *x-axis*: oxidation potentials in millivolts. *y-axis*: current in picoamperes (pA) per semidifferentiation of the second. White matter signals in patients were studies in the neuroanatomic substrate, alveus.

4.1. NMI TECHNOLOGY

In NMI, action potentials are not involved; it is not a depolarization technology. It is important to note that these miniature sensors do not sense membrane potentials. These sensors pass small but finite currents while neurotransmitters close to the surface undergo oxidation and/or reduction *(35)*. The current, which is formed from this flow of electrons, is dependent on voltage according to Ohm's law. Thus, the detection of electrochemical or electroactive signals from neuroanatomic brain sites is termed *faradaic* because the amount of the oxidative and/or reductive species detected at the surface of the microelectrode surface is calculated by a derivation of Faraday's law, which is the Cottrell equation *(36)*. Faradaic electrochemistry using conventional electrodes has been described *(37)*.

4.2. MRI TECHNOLOGY

In MRI, action potentials are not involved; MRI is not a depolarization technology. It is important to note that as myelination occurs, there is a loss of water within the myelin sheath which decreases proton density, thus, enhancing the T_1-weighted image and reducing the T_2-weighted image (*see* ref. *43* for overview). Too, signal changes on T_1-weighted MRI patterns parallel increases in lipids that occur during myelin formed from oligodendrocytes *(38–40)*. Signal changes on T_2-weighted MRI patterns may correlate with the period of maturation of the myelin sheath *(31)*. In MRI detection, hydrogen atoms of lipids do not contribute appreciably to the MR signal because hydrogen atoms of lipids are immobile and bound tightly to long chain fatty acids. *(38)*. Instead, lipid protons affect MRI by interacting with the mobile water protons by, for example, chemical exchange with transiently immobilized water protons *(30)*. Chemicals such as myelin-bound cholesterol *(38,39,42)* and galactosecerebroside *(30)* are responsible for shorter T_1/T_2-weighted relaxation times for normal myelinated white matter on MRI (*see* ref. *41* for review).

4.3. SIMILARITIES BETWEEN NMI AND MRI

It is of interest that neither NMI nor the MRI technologies use action potentials. Neither technology is a depolarization technique. Whereas T_1/T_2-weighted images in MRI signals white matter changes and the presence or absence of white matter, it is electron transfer in NMI with our sensors that signals white matter changes and the presence or absence of white matter. In addition, in MRI detection, the hydrogen atoms of lipids do not contribute appreciably to the MR signal. Lipid protons affect the MR signal in MRI indirectly by communicating with water protons. In NMI with our sensors, lipids act to enhance signals indirectly as well, by enhancing electron transfer for electroactive species. Finally, in both technologies, , structural and chemical changes in axons, in addition to length of carbon chain in fatty acids, and properties such as hydrophobicity and hydrophilicity, e.g., are significant factors, affecting and influencing the image and waveform patterns of white matter.

In summary, NMI with BRODERICK PROBE® sensors reliably differentiates temporal lobe gray from white matter in separate subparcellations of neocortex and hippocampus in TLE patients. These results have important implications for enabling direct in vivo measurement of neurotransmitters and critical analytes in diverse neuroanatomic substrates during surgery. Thus, one can envision, with numerous sensors, the pro-

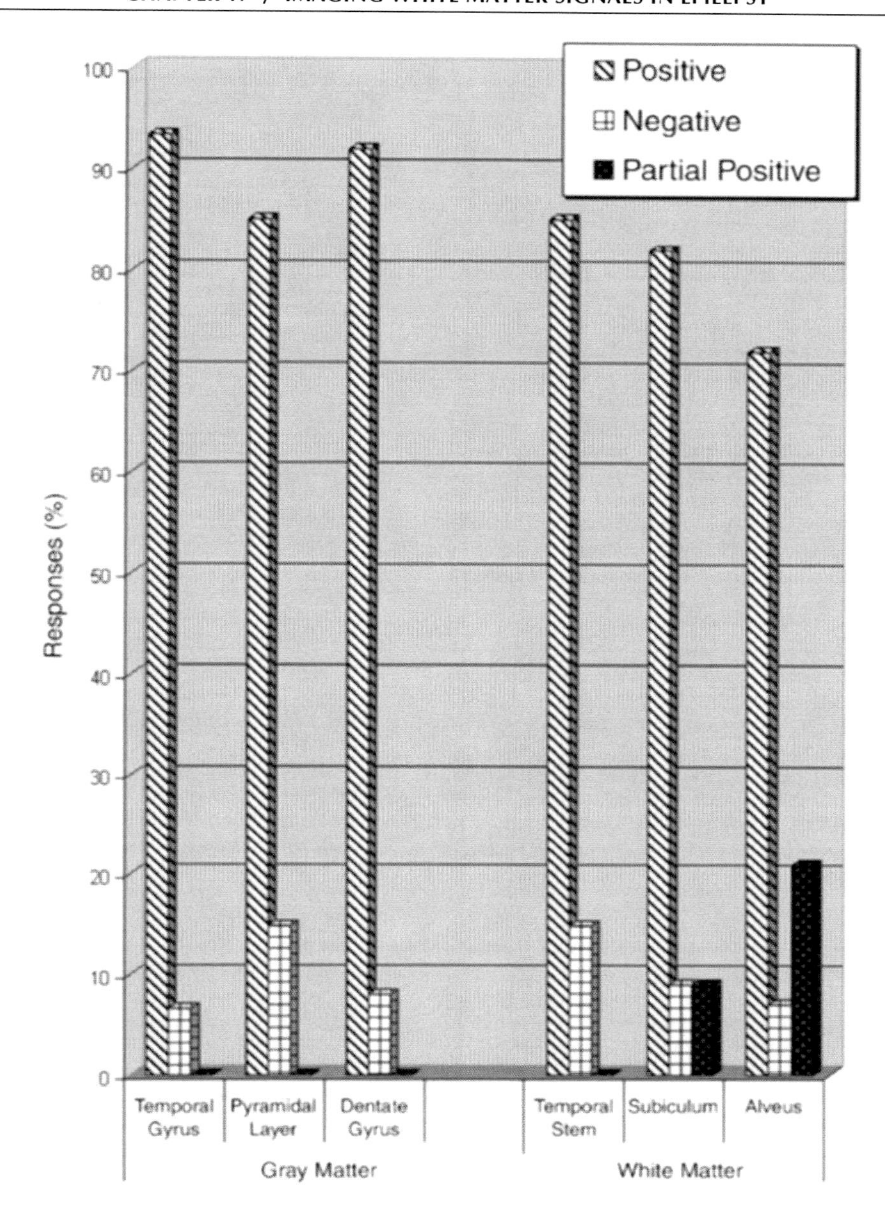

Fig. 6. Patient percentages. On the left is a histogram representation of percentages of patients who met criteria for positive identification of gray matter waveforms; negative percentages are also shown. In the temporal gyrus, pyramidal layer, and dentate gyrus, 14 of 15 patients, 11 of 13 patients, and 12 of /13 patients, respectively, exhibited waveforms positive for gray matter. On the right is a histogram representation of percentages of patients who met criteria for positive identification of white matter waveforms; negative percentages are also shown. In temporal stem, alveus, and subiculum, 11 of 13 patients, 10 of 14 patients and 9 of 11 patients, respectively, exhibited waveforms positive for white matter. In alveus and subiculum, 3 of 14 patients and 1 of 11 patients respectively, met criteria for partially positive identification of white matter waveforms.

duction of a temporospatial image of synaptic changes in TLE neuronal circuitry, which may well enable more precise diagnosis and pharmacotherapy for TLE, in addition to enabling more precise intraoperative interventions. These data hopefully will lead to human trials and be helpful also in the areas of tumor, leukodystrophies, leukoencephalopathies, stroke and spinal cord injury.

ACKNOWLEDGMENTS

Our protocol for human studies was approved by CUNY and NYU Investigational Review Boards. We thank the FACES Campaign, Parents Against Childhood Epilepsy (PACE Foundation) and the National Institute of Health, NIH/NIGMS SCORE AWARD # SO 6 GM 08168 for partial financial support. The authors gratefully acknowledge Bridget T. O'Sullivan, OP, MA; Msgr. Scanlan, HS; and Karen Schulz, Humana Press, for secretarial and artistic assistance, respectively. We thank Patricia A. Burns and Opeyemi A. Olabisi, CCNY students, for participation in this project. Patricia is an M.D. student at the NYU School of Medicine and Opeyemi is an MD PhD student at the Albert Einstein College of Medicine, Yeshiva University. We also thank Fang Zhou and Ratna Medicherla, second year CUNY Medical School students, for their participation.

REFERENCES

1. Thomas P, Hazif-Thomas C, Saccardy F, Vandermarq P. Loss of motivation and frontal dysfunction. Role of the white matter change. Encephale 2004;30:52–59.
2. O'Sullivan M, Morris RG, Huckstep B, Jones DK, Williams SC, Markus HS. Diffusion tensor MRI correlates with executive dysfunction in patients with ischaemic leukoaraiosis J Neurol Neurosurg Psychiatry 2004;75:441–447.
3. van der Knaap MS, Naidu S, et al. New syndrome characterized by hypomyelination with atrophy of the basal ganglia and cerebellum. AJNR Am J Neuroradiol 2002;23:1466–1474.
4. Leegwater PA, Pronk JC, van der Knaap MS. Leukoencephalopathy with vanishing white matter: from magnetic resonance imaging pattern to five genes. J Child Neurol 2003;18:639–645.
5. Kendal C, Everall I, Polkey C, Al-Sarraj S. Glial cell changes in the white matter in temporal lobe epilepsy. Epilepsy Res 1999;36:43–51.
6. Emery JA, Roper SN, Rojiani AM. White matter neuronal heterotopia in temporal lobe epilepsy: a morphometric and immunohistochemical study. J Neuropathol Exp Neurol 1997;56:1276–1282.
7. Hammers A, Koepp MJ, Hurlemann R, et al. Abnormalities of grey and white matter [11C] flumazenil binding in temporal lobe epilepsy with normal MRI. Brain 2002;125:2257–2271.
8. Meiners LC, van der Grond J, van Rijen PC, Springorum R, de Kort GA, Jansen GH. Proton magnetic resonance spectroscopy of temporal lobe white matter in patients with histologically proven hippocampal sclerosis. J Magn Reson Imaging 2000;11:25–31.
9. Marsh L, Sullivan EV, Morrell M, Lim KO, Pfefferbaum A. Structural brain abnormalities in patients with schizophrenia, epilepsy, and epilepsy with chronic interictal psychosis. Psychiatry Res 2001;108:1–15.
10. Hermann B, Hansen R, Seidenberg M, Magnotta V, O'Leary D. Neurodevelopmental vulnerability of the corpus callosum to childhood onset localization-related epilepsy. Neuroimage 2003; 18:284–292.
11. Cadet JL, Brannock C. Mechanisms of methamphetamine-induced neurotoxicity. In: Massaro EJ, Broderick PA, Mattsson JL, Schardein JL, Schlaepfer TE, eds. Handbook of Neurotoxicology, vol. 2. Totowa, NJ: Humana Press; 2002:259–268.
12. Broderick PA, Pacia SV, Doyle WK, Devinsky O. Monoamine neurotransmitters in resected hippocampal subparcellations from neocortical and mesial temporal lobe epilepsy patients: in situ microvoltammetric studies. Brain Res 2000;878:49–63.
13. Pacia SV, Doyle WK, Broderick PA. Biogenic amines in the human neocortex in patients with neocortical and mesial temporal lobe epilepsy: identification with in situ micovoltammetry. Brain Res 2001; 899:106–111.
14. Pacia SV, Broderick PA. Bioimaging L-Tryptophan in human hippocampus and neocortex: subtyping temporal lobe epilepsy. In: Broderick PA, Rahni DN, Kolodny EH, eds. Bioimaging in Neurodegeneration, Totowa, NJ: Humana Press; 2005: pp. 141–147.
15. Broderick PA. Distinguishing in vitro electrochemical signatures for norepinephrine and dopamine. Neurosci Lett 1988;95:275–280.
16. Broderick PA. Characterizing stearate probes in vitro for the electrochemical detection of dopamine and serotonin. Brain Res 1989;495: 115–121.
17. Broderick PA. State-of-the-art microelectrodes for in vivo voltammetry. Electroanalysis 1990;2:241–251.
18. Broderick PA. In vivo voltammetric studies on release mechanisms for cocaine with γ-butyrolactone. Pharmacol Biochem Behav 1991; 40:969–975.
19. Broderick PA. Microelectrodes and their use in cathodic electrochemical circuits. 1995;US Patent #5,433,710.
20. Broderick PA. Microelectrodes and their use in cathodic electrochemical arrangement with telemetric application. 1996; European Patent #90914306.7.
21. Broderick PA. Microelectrodes and their use in an electrochemical arrangement with telemetric application. 1999; US Patent #5,938,903.
22. Broderick PA, Pacia SV. Identification, diagnosis and treatment of neuropathologies, neurotoxicities, tumors, and brain and spinal cord injuries using microelectrodes with microvoltammetry. 2002;#PCT/02/11244. Pending.
23. Broderick PA, Pacia SV. Identification, diagnosis and treatment of neuropathologies, neurotoxicities, tumors, and brain and spinal cord injuries using microelectrodes with microvoltammetry. 2002; US Patent Application Serial No. 10/118,571. Pending.
24. Broderick PA., Rahni DN and Pacia SV. Nano- and microimaging surgical anesthesia in epilepsy patients. In: Broderick PA, Rahni DN, Kolodny EH, eds. Bioimaging in Neurodegeneration, Totowa, NJ: Humana Press Inc.; 2005: pp. xiii–xvi.
25. Oldham K. Semi-integral electroanalysis: analog implementation. Anal Chem 1973;45:39–50.
26. Broderick PA, Gibson GE. Dopamine and serotonin in rat striatum during in vivo hypoxic-hypoxia. Metab Br Disease 1989;4:143–153.
27. Wieser H. Electroclinical Features of the Psychomotor Seizure: A Stereo-Encephalographic Study of Ictal Symptoms and Chronotopographical Seizure Patterns Including Clinical Effects of Intracerebral Stimulation. New York: Gustav Fisher Vertag; 1983.
28. Broderick PA. Striatal neurochemistry of dynorphin-(1–13): In vivo electrochemical semidifferential analyses. Neuropeptides 1987;10: 369–386.
29. Van der Knaap MS, Valk J. Magnetic Resonance of Myelin, Myelination, and Myelin Disorders. 2nd ed. Berlin: Springer; 1995:1–17.
30. Kucharczyk W, Macdonald PM, Stanisz GJ, et al. Relaxivity and magnetization transfer of white matter lipids at MR imaging: importance of cerebrosides and pH. Radiology 1994;192:521–529.
31. Barkovich AJ, Lyon G, Evrard P. Formation, maturation, and disorders of white matter. AJNR Am J Neuroradiol 1992;13:447–461.
32. Walz W. Role of glial cells in the regulation of brain microenvironment. Prog Neurobiol 1989;33:309–333.
33. Bordey A, Sontheimer H. Properties of human glial cells associated with epileptic seizure foci. Epilepsy Res 1998;32:286–303.
34. Foucault R, Broderick PA, Rahni DN, Lombardi JR, Birke RL. Neurotransmitter signatures: a correlation between Raman spectroscopy and the microvoltammetric BRODERICK PROBE®. NIH/NIGMS Symposium, New Orleans, LA, Nov. 2002.
35. Adams, RN, Marsden CA. New techniques in psychopharmacology. In: Iversen, LL, Synder SH, eds. Handbook of Psychopharmacology. New York: Plenum Press; 1982.
36. Kissinger PT, Preddy CR, Shoup RE, Heineman WR. Fundamental concepts of analytical electrochemistry. In: Kissinger PT, Heineman, WR. eds. Laboratory Techniques in Electroanalytical Chemistry, New York: Marcell Dekker Inc.; 1996:11–50.
37. Dayton MA, Brown JC, Stutts KJ, Wightman RM. Faradaic electrochemistry at microvoltammetric electrodes. Anal Chem 1980; 52:948–950.
38. Koenig SH. Cholesterol of myelin is the determinant of gray-white matter contrast in MRI of the brain. Magn Reson Med 1991;20: 285–296.
39. Fralix TA, Ceckler TL, Wolff SD, et al. Lipid bilayer and water proton magnetization transfer: effect of cholesterol. Magn Reson Med 1991;18:214–223.
40. Brody BA, Kinney HC, Kloman AS, et al. Sequence of central nervous system myelination in human infancy. I. An autopsy study of myelination. J Neuropathol Exp Neurol 1987;46:283–301.
41. Nusbaum AO, Fung K, Atlas SW. White matter diseases and inherited metabolic disorders. In: Atlas SW, ed. Magnetic Resonance Imaging of the Brain and Spine. 3rd ed. Philadelphia: Lippincott Williams & Wilkins; 2002:457–563.
42. Koenig SH, Brown RD, Spiller M, Lundbom N. Relaxometry of brain: why white matter appears bright in MRI. Magn Reson Med 1990;14:482–490.
43. Kolodny, EH. Overview of the leukoencephalopies: an MRI point of view. In: Broderick, PA, Rahni DN, Kolodny EH, eds. Bioimaging in Neurodegeneration, Totowa, NJ: Humana Press Inc.; 2005: pp. 209–214.
44. Bordey A, Spencer DD. Distinct electrophysiological alterations in dentate gyrus versus glial cells from epileptic humans with temporal lobe sclerosis. Epilepsy Res 2004;59:107–122.

IV

Leukodystrophy (White Matter) Diseases

18 Overview of the Leukoencephalopathies

An MRI Point of View

EDWIN H. KOLODNY, MD

SUMMARY

Many forms of inherited leukodystrophies are now known, each characterized by specific biochemical and molecular abnormalities. The end result may be hypomyelination, i.e., the failure to form specific myelin proteins; delay in myelination attributable to an inadequate supply of myelin precursors or accumulation of substances toxic to oligodendroglia; demyelination with loss of normally formed myelin; vacuolating myelinopathy, wherein degenerating white matter is replaced by fluid and vacuolization; and secondary demyelination with destruction of both axons and myelin. Clinical signs may develop after a period of normal development. These can include abnormalities in behavior, cognition and memory, long tract signs, optic atrophy, peripheral neuropathy and macro- or microcephaly. Their clinical delineation is facilitated by magnetic resonance imaging (MRI), including the use of diffusion-weighted imaging and MR spectroscopy (MRS). Genetic leukodystrophies are progressive and can produce specific patterns of abnormality on MRI that help to distinguish them. The loss of myelin is accompanied by an increase in water, causing a decrease in the white matter signal on T1-weighted images and an increase on T2-weighted images. The finding of a leukodystrophy by MRI presymptomatically can provide the clinician with a window of opportunity to intervene therapeutically prior to overt clinical signs. This is particularly relevant for metachromatic leukodystrophy, globoid cell leukodystrophy, and X-linked adrenoleukodystrophy, each of which may respond to hematopoietic stem cell transplantation if treated early in the clinical course.

Key Words: Leukodystrophy; hypomyelination; demyelination; vacuolating myelinopathy; diffusion-weighted imaging; MR spectroscopy.

From: *Bioimaging in Neurodegeneration*
Edited by P. A. Broderick, D. N. Rahni, and E. H. Kolodny

Magnetic resonance imaging (MRI) is a highly sensitive technique for detecting pathologic changes in brain white matter. Collectively, these abnormalities are referred to as leukoencephalopathies and may be inherited or acquired. The inherited forms have a genetic basis, a progressive clinical course, and demonstrable biochemical or molecular defects *(1,2)*. The acquired leukoencephalopathies encompass disorders of an inflammatory, autoimmune, vascular or infectious nature, tumors, and injury by neurotoxins *(3)*.

Myelin, a key component of brain white matter, is composed of multiple layers of lipids and proteins representing extensions of oligodendroglial cell processes that are wrapped in a spiral fashion around portions of an axon *(4)*. The major myelin proteins are myelin basic protein and proteolipid protein. The lipid layer is composed of cholesterol, phospholipid, and glycolipid in a ratio of approximately 4:3:2. The glycolipids consist of galactocerebroside and sulfatide, which together with cholesterol are in the outer layer of the membrane and are exposed to the extracellular space. Their functional groups interact with water in contrast to the phospholipids, which are located in the inner, cytoplasmic, side of the membrane and are hydrophobic.

MRI of brain white matter is influenced by two different populations of water molecules. Water within the myelin sheath, forming transient hydrogen bonds with hydroxyl and ketone residues, has relatively short T_1 and T_2 relaxation times. Water molecules outside the myelin sheath, within axons and the interstitial space, are not bound to macromolecules and therefore produce longer T_1 and T_2, relaxation times. As myelination proceeds, there is a loss of water within the myelin sheath, reducing the proton density. The intensity of the white matter signal on T_1-weighted images increases and the signal on the T_2-weighted images decreases. With white matter diseases, there is an increase in water and high-signal intensity on T_2-weighted images.

MRI has expanded greatly our awareness of the leukoencephalopathies. Pattern recognition together with clinical information and laboratory data enable us to categorize many of these disorders and to define new forms of white matter disease. Yet as many as half of the patients we see with MRI evidence of a leukoencephalopathy remain undiagnosed *(5)*. We have learned many lessons using MRI, especially in studying the inherited diseases. Often abnormal MRI signals may precede clinical signs of the disease *(6)*. For example, a 15-mo-old child with infantile metachromatic leukodystrophy may present to the hospital with lethargy during a febrile illness, then recover with a presumably intact nervous system. However, should an MRI be performed, symmetrical periventricular signal abnormalities may already be observed. Several months later, motor and sensory signs of metachromatic leukodystrophy become overt. In a family with cerebral autosomal-dominant arteriopathy with subcortical infarcts and leukoencephalopathy (CADASIL), we have observed white matter abnormalities in the MRIs of asymptomatic children whose parents were clinically affected.

The location of the myelin disturbance is critically important to disease expression. Frontal lobe involvement and loss of myelin in the genu of the corpus callosum produce greater cognitive loss and neurobehavioral symptoms than lesions in white matter of the occipital lobe and splenium of the corpus callosum *(7)*. The early stages of an illness such as juvenile onset globoid cell leukodystrophy may involve primarily the corticospinal tracts, causing gait difficulty and not much else. In those patients with diffuse symmetrical involvement of subcortical white matter, receptive language may be relatively preserved so that simple commands are executed but usually in slow motion and after some delay in processing. Finally, environmental precipitants such as fever or trauma can quickly and dramatically exacerbate an underlying metabolic disorder of myelin such as adrenoleukodystrophy or vanishing white matter disease/childhood ataxia with diffuse central nervous system hypomyelination *(8)*.

Van der Knaap and associates have identified seven patterns of abnormality in MRIs among patients with leukoencephalopathies *(5)*. Patients in category A with severely deficient myelination include Pelizaeus-Merzbacher disease, Cockayne syndrome type II, and Menkes Syndrome *(9)*. Category B patients have global cerebral white matter involvement, of which megaloencephalic leukodystrophy *(10)* is an example. In category C are patients with extensive cerebral white matter abnormalities with a fronto-occipital gradient and relative sparing of the occipital lobes. Alexander disease is one example *(11)*. Approximately 10% of patients with X-linked adrenoleukodystrophy also show predominately a frontal lobe involvement. Periventricular white matter abnormalities are a feature of category D patients. This pattern, which includes relative sparing of the arcuate fibers, is found in metachromatic leukodystrophy, Krabbe disease, and X-linked adrenoleukodystrophy. The MRI scans of category E patients show multifocal white matter abnormalities with a predominately lobar location and relative sparing of arcuate fibers and periventricular white matter. These individuals have acquired diseases, especially of inflammatory or infectious etiology. Examples are multiple sclerosis and congenital cytomegalovirus infection. Category F includes diseases in which the white matter changes are primarily subcortical and involve arcuate fibers, such as L-2-hydroxyglutaric aciduria and Kearns-Sayre syndrome. The white matter lesions in category G patients are predominately in the posterior fossa. Cerebrotendinous xanthomatosis and Refsum disease fit this pattern.

To these, one may add disorders in which the white matter abnormality may first or primarily appear in the posterior aspect of brain, such as X-linked adrenoleukodystrophy and the posterior reversible leukoencephalopathy syndrome *(12)*. Also, other aspects to consider are

1. The presence of radially oriented stripes (Pelizaeus-Merzbacher disease, metachromatic leukodystrophy).
2. Enlarged Virchow-Robin spaces (mucopolysaccharidosis I, congenital muscular dystrophy, Lowe syndrome).
3. Contrast enhancement (X-linked adrenoleukodystrophy, Alexander disease, Krabbe disease, methylmalonic acidemia, mitochondrial complex I oxidative phosphorylation defect; ref. *13*).
4. Infarction (CADASIL; mitochondrial myopathy, encephalopathy, lactic acidosis, and stroke-like episodes [MELAS]; Fabry disease; amyloid angiopathy).
5. Cortical dysplasia (lissencephaly, congenital muscular dystrophy, glutaric aciduria, Zellweger syndrome).
6. Calcifications (Cockayne; Krabbe disease; X-linked adreno-leukodystrophy; Kearns-Sayre syndrome; MELAS, myoclonic epilepsy and ragged red fiber disease [MERRF]; toxoplasmosis, other, rubella, cytomegalovirus, herpes simplex [TORCH] infections).

The subcortical arcuate fibers are lost early in Canavan disease and persist or are lost late in the course of other leukoencephalopathies.

Serial studies will disclose whether the process of myelination is proceeding, if there is a delay or, conversely, whether the myelin abnormality is spreading and, if so, in which direction (centrifugal, centripetal, ventrodorsal, or dorsoventral). Repeat MRI scans are also useful to gauge the effects of treatments, such as bone marrow transplantation, and the reversibility of lesions, which are seen in MELAS and the posterior reversible leukoencephalopathy syndrome. The integrity of brain white matter may be altered by an increase or decrease in essential lipid or protein components, in abnormal carbon chain lengths of the fatty acid residues, by the loss of oligodendroglia cells as the result of specific toxins, or by the death of neurons with degeneration of their axons. If, as a result, normally formed myelin becomes unstable, demyelination occurs. When there is a failure in myelin production, hypomyelination is the result. The term leukodystrophy is used to encompass both processes. Most leukodystrophies have a genetic basis, are progressive, and produce symmetrical changes on brain MRI scans *(1,2,9)*.

Helpful clinical features in assessing patients with leukodystrophies are shown Table 1. Evoked potential studies, especially brainstem auditory-evoked potentials, that demonstrate delays in central conduction may provide ancillary information. Seizures, if they do occur, are a late phenomenon and

Table 1
Leukodystrophy Test Panel

Routine hematology
Electrolytes
Liver chemistries
Chromosomes
Ammonia, uric acids
Amino acids, organic acids
Lactate, pyruvate
Very long chain fatty acids
Blood and urine copper
Cholesterol and cholestanol
Lysosomal enzymes
Mitochondrial deoxyribonucleic acid (DNA) studies
DNA repair studies
Muscle, nerve and skin biopsy
TORCH Titers
DNA sequencing (e.g., *PLP, GFAP, Notch 3*)

Table 2
Clinical Signs of Leukodystrophy

Early period of normal development
Abnormalities in behavior, cognition and memory
Long tract signs (usually bilateral and symmetrical)
Hypertonia, spasticity
Hyperreflexia, weakness
Dysmetria, ataxia
Optic atrophy occurs late
Peripheral neuropathy in some diseases
Head size may be smaller or larger than normal

therefore the EEG often remains normal. An exception is infantile Krabbe disease. Nerve conduction studies assist in the evaluation of peripheral neuropathy. Other useful laboratory tests are noted in Table 2.

A classification of the genetic leukodystrophies is presented in Table 3. For each condition, the accumulating substrate, abnormal protein, or mutated gene are listed and references provided that give descriptions of their MRI findings.

They are subdivided as follows:

1. Hypomyelination: disorders resulting from failure of synthesis of a specific myelin protein.
2. Delayed myelination: disorders resulting from an inadequate supply of myelin precursors or accumulation of substances toxic to the oligodendroglia and formation of myelin.
3. Primary demyelination: loss of normally formed myelin with relative preservation of axons.
4. Vacuolating myelinopathies: a subcategory of primary demyelination in which the degenerating white matter is replaced by fluid and vacuolization.
5. Secondary demyelination: destruction of both axons and myelin by a more diffuse process.

Detailed descriptions of the genetic leukodystrophies may be found in refs. *1, 2,* and *9*.

One epidemiological study estimates that progressive childhood encephalopathies occur with a frequency of approx 0.6 in 1000 live births *(14)*. A major proportion of these abnormalities involve white matter. In another study, 215 of 7784 MRI scans performed at a children's hospital during a 1-yr period showed some abnormality within the white matter *(15)*. From these and other studies, it is evident that white matter abnormalities are a significant cause of degenerative disease. Different MRI techniques, including diffusion-weighted imaging *(16,17)*, MR spectroscopy, and other forms of neuroimaging, are vital tools not only for the elucidation of causality in these diseases but also for the discovery of new disease entities *(18–21)*.

REFERENCES

1. Kolodny EH. Genetic and metabolic aspects of leukodystrophies. In: Dangond F (ed) Disorders of Myelin in the Central and Peripheral Nervous Systems. Woburn, MA: Butterworth, 2002:83–101.
2. Nusbaum AO, Fung K, Atlas SW. White matter diseases and inherited metabolic disorders. In: Atlas SW. Magnetic Resonance Imaging of the Brain and Spine. 3rd ed. Philadelphia: Williams & Wilkins, 2002:457–563.
3. Filley CM, Kleinschmidt-DeMaster BK. Toxic leukoencephalopathy. N Engl J Med 2001;345:425–432.
4. Barkovich, AJ. Concepts of myelin and myelination in neuroradiology. Am J Neuroradiol 2000;21:1099–1109.
5. van der Knapp MS, Breiter SN, Naidu S, Hart AAM, Valk J. Defining and categorizing leukoencephalopathies of unknown origin: MR imaging approach. Radiology 1999;213:121–133.
6. Aubourg P, Sellier N, Chaussain JL, Kalifa G. MRI detects cerebral involvement in neurologically asymptomatic patients with adrenoleukodystrophy. Neurology 1989;39:1619–1621.
7. Filley C. The behavioral neurology of cerebral white matter. Neurology 1998;50:1535–1540.
8. van der Knapp MS, Barth PG, Gabreëls FJM, et al. A new leukoencephalopathy with vanishing white matter. Neurology 1997;48: 845–855.
9. Faerber EN, Poussaint TY. Magnetic resonance of metabolic and degenerative diseases in children. Topics Magn Reson Imaging 2002;13:3–22.
10. Topcu M, Gartioux C, Ribierre F, et al. Vacuolating megaloencephalic leukoencephalopathy with subcortical cysts, mapped to chromosome 22q. Am J Hum Genet 2000;66:733–739.
11. van der Knaap MS, Naidu S, Breiter SN, et al. Alexander disease: diagnosis with MR imaging. Am J Neuroradiol 2001;22:541–522.
12. Kinoshita T, Moritani T, Shrier DA, et al. Diffusion-weighted MR imaging of posterior reversible leukoencephalopathy syndrome a pictorial essay. J Clin Imaging 2003;27:307–315.
13. Moroni I, Bugiani M, Bizzi A, Castelli G, Lamantea E, Uziel G. Cerebral white matter involvement in children with mitochondrial encephalopathies. Neuropediatrics 2002;33:79–85.
14. Uvebrant P, Lanneskog K, Hagberg B. The epidemiology of progressive encephalopathies in childhood: live birth prevalence in West Sweden. Neuropediatrics 1992;23:209–211.
15. Lasbury N, Garg B, Edwards-Brown M, Cowan LD, Dimassi H, Bodensteiner JB. Clinical correlates of white-matter abnormalities on head magnetic resonance imaging. J Child Neurol 2001;16:668–672.
16. Engelbrecht V, Scherer A, Rassek M, Witsack HJ, Modder U. Diffusion-weighted MR imaging in the brain in children: findings in the normal brain and in the brain with white matter diseases. Radiology 2002;222:410-418.
17. Sener RN. Diffusion magnetic resonance imaging patterns in metabolic and toxic brain disorders. Acta Radiol 2004;45:561–570.
18. Wolf NI, Willemsen MA, Engelke UF, et al. Severe hypomyelination associated with increased levels of *N*-acetylaspartylglutamate in CSF. Neurology 2004;62:1503–1508.

Table 3
Genetic Leukodystrophies

Disorder	Metabolite	Clinical Features	Refs.
Hypomyelination			
Pelizaeus-Merzbacher	Proteolipid protein	Rotatory nystagmus, spastic quadriparesis, ataxia, choreoathetosis, head titubation, absent speech	*16,22*
18q- Syndrome	Myelin basic protein	Short stature, microcephaly, stenotic ear canals, dysmorphic face and hands, nystagmus, MR	*23*
Delayed myelination			
Amino acidopathies Organic acidopathies	Example: phenylketonuria Example: maple syrup urine disease	Developmental delay, abnormal muscle tone, psychomotor regression, seizures, ataxia, deterioration secondary to metabolic stress	*17,24,25*
Mitochondrial disorders	Example: complex I-IV deficiency, Kearns-Sayre syndrome, MELAS, MERFF, MNGIE, Leber's hereditary optic neuropathy	Progressive external ophthalmoplegia, short stature, exercise intolerance, neuropathy, hearing loss, seizures (MERFF), retinitis pigmentosa (Kearns-Sayre), GI complaints (MNGIE)	*13,26–29*
Primary demyelination			
Globoid cell leukodystophy (including Krabbe disease)	Galactosylceramide	Irritability, rigidity, polyneuropathy, microcephaly, blindness	*30–32*
Metachromatic leukodystrophy	Sulfatide	Ataxia, spastic quadriparesis, neuropathy	*33–34*
Multiple sulfatase deficiency	Sulfatide, mucopolysaccharide	Above, plus ichthyosis, dysmorphism, enlarged liver	*35*
Adrenoleukodystrophy (X-linked)	Very long chain fatty acids	Psychomotor regression, spastic quardriparesis, blindness, adrenal insufficiency	*36,37*
Alexander disease	GFAP	Megalencephaly, ataxia, spastic paraparesis, seizures, relatively spared cognition	*11,38*
Vacuolating myelinopathies			
Canavan disease	*N*-acetylaspartate	Lethargy, hypotonia, psychomotor delay, macrocephaly, seizures, blindness	*39*
Leukodystrophy with vanishing white matter (CACH)	Subunits of eIF2B	Onset after fever or minor head trauma, ataxia, spasticity, seizures, optic atrophy, relatively preserved intellect	*8,40*
Vacuolating megalencephalic leukoencephalopathy	MLC1 protein, unknown second gene	Macrocephaly, ataxia, spastic paraparesis, seizures, cognition relatively spared	*10,41–43*

Secondary demyelination			
Lysomal storage disease	Example: G_{M1}- and G_{M2}-gangliosidoses, fucosidosis, Neimann-Pick disease, silalic acid storage disease	Seizures, megalencephaly (G_{M1} and G_{M2}), psychomotor regression, visceromegaly	*44,45*
Merosin-deficient congenial muscular dystrophy	Merosin (laminin-α2)	Slowly progressive weakness, seizurres, high CK	*46*
Sjögren-Larson syndrome	Fatty alcohols	Developmental delay spasticity, ichthyosis, pseudobulbar dysarthia, seizures, eye findings	*47*
Cockayne syndrome	Defective nucleotide excision repair	Chachectic dwarfism, microcephaly, facial dysmorphism, photosensitivity, retinal pigmentary degeneration, nystagmus, neuropathy, behavioral changes	*48–50*
CADASIL	*Notch3* transmembrane protein, GOM's	Migraine, lacunar strokes, dementia	*51,52*
Cerebrotendinous xanthomatosis	Cholestanol	Mental retardation, tuberous xanthomata, cerebellar and pyramidal tract-signs, neuropathy, seizures	*53,54*
Proximal myotonic myopathy (PROMM)	CCTG expansion in *ZNF9*	Myotonia, proximal muscle weakness, hypersomnia, Parkinsonianism, cataracts, sensorineural hearing loss, strokes, seizures	*55*
SOX10 deficiency	*SOX10*	Pigmentary defects, intestinal aganglionosis, peripheral neuropathy	*56*
Other Leukodystrophies			
Chorea-Acanthocytosis	Chorein	Chorea, dysarthia, neuropathy, myopathy, seizures, dementia, acanthocytosis	*57,58*
Aicardi-Goutieres syndrome	Interferon-α	Progressive encephalopathy, spastic quadraparesis, unexplained fevers, acrocyanosis	*59*
Leukodystrophy with ovarian dysgenesis	Unknown	Mild mental deficiency, frontal cortical atrophy, ovarian failure	*60*
Hereditary leukoencepalopathy and palmoplantar keratoderma	Collagen	Cognitive impairment, progressive tetraparesis, thickened hyperkeratotic skin	*61*

Adapted from Table 4-2 in ref *1*.
MR, mental retardation; CK, creatine kinase; GOM, granular osmiophilic material.

19. Nagae-Poetscher LM, Bibat G, Philippart M, et al. Leukoencephalopathy, cerebral calcifications, and cysts. Neurology 2004;62: 1206–1209.
20. Huck JHJ, Verhoeven NM, Struys EA, Salomons GS, Jokobs C, van der Knapp MS. Ribose-5-phosphate isomerase deficiency: new inborn error in the pentose phosphate pathway associated with a slowly progressive leukoencephalopathy. Am J Hum Genet 2004;74:745–751.
21. Kurian MA, Ryan S, Besley GTN, Wanders RJA, King MD. Straight-chain acyl-CoA oxidase deficiency presenting with dysmorphia, neurodevelopmental autistic-type regression and a selective pattern of leukodystrophy. J Inherit Metab Dis 2004;27:105–108.
22. Plecko B, Stöckler-Ipsiroglu S, Gruber S, et al. Degree of hypomyelination and magnetic resonance spectroscopy findings in patients with Pelizaeus Merzbacher phenotype. Neuropediatrics 2003;34:127–136.
23. Linnankivi TT, Autti TH, Pihko SH, et al. 18q⁻ syndrome: brain MRI shows poor differentiation of gray and white matter on T_2-weighted images. J Magn Reson Imaging 2003;18:414–419.
24. Neumaier-Probst E, Harting I, Seitz A, Ding C, Kolker S. Neuroradiological findings in glutaric aciduria type I (glutaryl-Co A dehydrogenase deficiency). J Inherit Metab Dis 2004;27:869–876.
25. Schonberger S, Schweiger B, Schwahn B, Schwarz M, Wendel U. Dysmyelination in the brain of adolescents and young adults with maple syrup urine disease. Mol Genet Metab 2004;82:69–75.
26. Bianchi MC, Tosetti M, Battini R, et al. Proton MR spectroscopy of mitochondrial diseases: analysis of brain metabolic abnormalities and their possible diagnostic relevance. Am J Neuroradiol 2003; 24:1958–1966.
27. Parry A, Matthews PM. Roles for imaging in understanding the pathophysiology, clinical evaluation, and management of patients with mitochondrial disease. J Neuroimaging 2003;13:293–302.
28. Abe K, Yoshimura H, Tanaka H, Fujita N, Hikita T, Sakoda S. Comparison of conventional and diffusion-weighted MRI and proton MR spectroscopy in patients with mitochondrial encephalomyopathy, lactic acidosis, and stroke-like events. Neuroradiology 2004;46:113–117.
29. Millar WS, Lignelli A, Hirano M. MRI of five patients with mitochondrial neurogastrointestinal encephalomyopathy. Am J Roentgenol 2004;182:1537–1541.
30. Zarifi MK, Tzika AA, Astrakas LG, Poussaint TY, Anthony DC, Darras BT. Magnetic resonance spectroscopy and magnetic resonance imaging findings in Krabbe's disease. J Child Neurol 2001;16:522–526.
31. Brockmann K, Dechent P, Wilken B, Rusch O, Frahm J, Hanefeld F. Proton MRS profile of cerebral metabolic abnormalities in Krabbe disease. Neurology 2003;60:819–825.
32. Husain AM, Altuwaijri M, Aldosari M. Krabbe disease: neurophysiologic studies and MRI correlation. Neurology 2004;63:617–620.
33. Kim TS, Kim IO, Kim WS, Choi YS, Lee JY, Kim OW et al. MR of childhood metachromatic leukodystrophy. Am J Neuroradiol 1997; 18:733–738.
34. Faerber EN, Melvin J, Sinergel EM. MRI appearances of metachromatic leukodystrophy. Pediatr Radiol 1999;29:669–672.
35. Mancini GMS, van Diggelen OP, Huijmans JG, Stroink H, de Coo RF. Pitfalls in the diagnosis of multiple sulfatase deficiency. Neuropediatrics 2001;32:38–40.
36. Eichler FS, Itoh R, Barker PB, Mori S, Garritt ES, Van Ziji PCM, et al. Proton MR spectroscopic and diffusion tensor brain MR imaging in X-linked adrenoleukodystrophy: initial experience. Radiology 2002;225:245–252.
37. Loes DJ, Fatemi A, Melhem ER, Gupte N, Bezman L, Moser HW, Raymond GV. Analysis of MRI patterns aids prediction of progression in X-linked adrenoleukodystrophy. Neurology 2003;61:369–374.
38. Johnson AB, Brenner M. Alexander's disease: clinical, pathologic, and genetic features. J Child Neurol 2003;18:625–632.
39. Haselgrove J, Moore J, Wang Z, Traipe E, Bilaniuk L. A method for fast multislice T_1 measurement: feasibility studies on phantoms, young children and children with Canavan's disease. J Magn Reson Imaging 2000;11:360–367.
40. Fogli A, Schiffmann R, Bertini E, Ughetto S, Combes P, Eymard-Pierre E, et al. The effect of genotype on the natural history of eIF2B-related leukodystrophies. Neurology 2004;62:1509–1517.
41. Van der Knaap MS, Valk J, Barth PG, Smit LM, Van Engelen BG, Tortori DP. Leukoencephalopathy with swelling in children and adolescents: MRI patterns and differential diagnosis. Neuroradiology 1995;37:679–686.
42. Teijido O, Martinez A, Pusch M, Zorzano A, Soriano E, del Rio JA, et al. Localization and functional analysis of the MLC1 protein involved in megalencephalic leukoencephalopathy with subcortical cysts. Hum Molec Genet 2004;13:2581–2594.
43. Brockmann K, Finsterbusch J, Terwey B, Frahm J, Hanefeld F. Megalencephalic leukoencephalopathy with subcortical cysts in an adult: quantitative proton MR spectroscopy and diffusion tensor MRI. Neuroradiology 2003;45:137–142.
44. Folkerth RD Abnormalities of developing white matter in lysosomal storage diseases. J Neuropathol Experim Neurol 1999;58:887–902.
45. Barone R, Farano E, Trifiletti RR, Fiumara A, Pavone P. White matter changes mimicking a leukodystrophy in a patient with Mucopolysaccharidosis: characterization by MRI. J Neurol Sci 2002;195:171–175.
46. Philpot J, Pennock J, Cowen F, Sewry CA, Dubowitz V, Bydder G, Muntoni F. Brain magnetic resonance imaging abnormalities in merosin-positive congenital muscular dystrophy. Eur J Paediatr Neurol 2000;4:109–114.
47. Willemsen MA, Van Der Graaf M, Van Der Knapp MS, Heerschap A, Van Domburg PH, Gabreels FJ, Rotteveel JJ. MR imaging and proton MR spectroscopic studies in Sjögren-Larsson syndrome: characterization of the leukodystrophy. Am J Neuroradiol 2004; 25:649–657.
48. Demaerel P, Kendall BE, Kingsley D. Cranial CT and MRI in diseases with DNA repair defects. Neuroradiology 1992;34:117–121.
49. Sugita K, Takanashi J, Ishii M, Niimi H. Comparison of MRI white matter changes with neuropsychologic impairment in Cockayne syndrome. Pediatr Neurol 1992;8:295–298.
50. Spivak G. The many faces of Cockayne syndrome. Proc. Natl Acad Sci USA 2004;101:15273–15274.
51. Tuominen S, Miao Q, Kurki T, Tuisku S, Poyhonen M, Kalimo H, et al. Position emission tomography examination of cerebral blood flow and glucose metabolism in young CADASIL patients. Stroke 2004;35:1063–1067.
52. Vahedi K, Chabriat H, Levy C, Joutel A, Tournier-Lasserve E, Bousser M-G. Migraine with aura and brain magnetic resonance imaging abnormalities in patients with CADASIL. Arch Neurol 2004;61:1237–1240.
53. Frederico A, Dotti M. Cerebrotendinous xanthomatosis: clinical manifestations, diagnostic criteria, pathogenesis, and therapy. J Child Neurol 2003;18:633–638.
54. Inglese M, DeStefano N, Pagani E, Dotti MT, Comi G, Frederico A, Filippi M. Quantification of brain damage in cerebrotendinous xanthomatosis with magnetization transfer MR imaging. Am J Neuroradiol 2003;24:495–500.
55. Kornblum C, Reul J, Kress W, Grothe C, Amanatidis N, Klockgether T, Scröder R. Cranial magnetic resonance imaging in genetically proven myotonic dystrophy type 1 and 2. J Neurol 2004;251:710–714.
56. Inoue K, Tanabe Y, Lupski JR. Myelin deficiencies in both the central and the peripheral nervous systems associated with a SOX10 mutation. Ann Neurol 1999;46:313–318.
57. Dobson-Stone C, Velayos-Baeza A, Filippone LA, Westbury S, Storch A, Erdmann T, et al. Chorein detection for the diagnosis of chorea-acanthocytosis. Ann Neurol 2004;56:299–302.
58. Nicholl DJ, Sutton I, Dotti MT, Supple SG, Danek A, Lowden M. White matter abnormalities on MRI in neuroacanthocytosis. J Neurol Neurosurg Psychiatry 2004;75:1200–1201.
59. Abdel-Salam GM, Zaki MS, Lebon P, Mequid NA. Aicardi-Goutieres syndrome: clinical and neuroradiological findings of 10 new cases. Acta Paediatr 2004;93:929–936.
60. Fogli A, Rodriquez D, Eymard-Pierre E. Ovarian failure related to eukaryotic initiation factor 2B mutations. Am J Hum Genet 2003; 72:1544–1550.
61. Lossos A, Cooperman H, Soffer D, Ben-Nariah Z, Sagi E, Gomori M, et al. Hereditary leukoencephalopathy and palmoplantar keratoderma: A new disorder with increased skin collagen content. Neurology 1995;45:331–337.

19 Pyramidal Tract Involvement in Adult Krabbe's Disease

Magnetic Resonance Imaging and Proton Magnetic Resonance Spectroscopy Abnormalities

LAURA FARINA, MD, ALBERTO BIZZI, MD, AND MARIO SAVOIARDO, MD

SUMMARY

In the last few years, the adult onset of globoid cell leukodystrophy or Krabbe disease has been recognized. A few authors have described an association of progressive spastic paraparesis and selective involvement of the corticospinal tracts in adult patients with deficiency of galactocerebrosidase, the Krabbe disease-related enzyme. This phenotypic expression of the disease should be considered in the differential diagnosis of spastic paraparesis. Magnetic resonance imaging may offer the clues for the recognition of this disorder. Proton MR spectroscopy may be useful in quantifying the damage to the corticospinal tract.

Key Words: Adult Krabbe's disease; MRI; H-MRS; leukodystrophy; pyramidal tract involvement.

1. INTRODUCTION

Krabbe's disease (KD), or globoid cell leukodystrophy, is a dysmyelinating or demyelinating disorder caused by an inherited deficiency of lysosomal enzyme galactocerebrosidase (GALC), a key component in metabolic pathways of myelin turnover and breakdown. The GALC gene maps to chromosome 14q31, and a variety of KD-causing mutations have been identified *(1)*. GALC deficiency results in the accumulation of galactosylsphingosine, which is considered neurotoxic to both the central and peripheral nervous systems, thus leading to extensive demyelination, gliosis, and perivascular infiltration of multinucleated macrophages termed globoid cells. The disease can be diagnosed if severe deficiency of GALC activity in leukocytes or fibroblasts is detected.

Like other leukodystrophies, KD is most commonly seen in children and is essentially considered a pediatric disorder. Although reports of cases with onset even in advanced adulthood have been previously described, the work by Satoh et al. in the journal *Neurology (2)* and the accompanying editorial by Percy *(3)* were the first to highlight the occurrence of an adult form of KD characterized by spastic paraparesis as the main clinical finding and pyramidal tract involvement on magnetic resonance imaging (MRI).

Spastic paraparesis may be the only or most prominent sign in several metabolic or degenerative disorders. A review of the genetically determined spastic paraplegias was published by Fink in 2002 *(4)*; the number of genes causing these disorders is constantly growing. We do not discuss the whole topic of spastic paraplegias but only consider the disorders in which brain MRI or proton MR spectroscopy (H-MRS) show involvement of the pyramidal tracts; we must admit, however, that the reports of brain or spine MRI in patients with the various forms of spastic paraplegia are rare and often incomplete. In the differential diagnosis of adult KD, we shall only consider adrenoleukodystrophy (ALD), adrenomyeloneuropathy (AMN), X-linked recessive spastic paraplegia type 2, and amyotrophic lateral sclerosis (ALS).

2. CLINICAL ASPECTS

KD is traditionally classified into three major forms on the basis of the age of onset: (1) early-infantile, (2) late-infantile or juvenile, and (3) adult form. These forms also differ in their clinical severity, and the late-onset forms are characterized by a milder and more protracted course. The emergence of molecular genetic advances in the metabolic diseases has blurred these discrete boundaries between categories and consolidated the concept of a continuum of disease expression from infancy to adulthood based on the different impact of the specific mutational events *(3)*.

The most frequent and common form of KD is the infantile form, which begins in the first 6 mo of life and rapidly progresses leading to death before the child reaches the age of 2. Children with KD present rapid psychomotor regression, generalized rigidity, and peripheral neuropathy; they subsequently develop optic atrophy, deafness, and cachexia. Increased proteins in the cerebrospinal fluid is a constant finding; cell count is usually normal. Nerve conduction velocity is always abnormal.

From: *Bioimaging in Neurodegeneration*
Edited by P. A. Broderick, D. N. Rahni, and E. H. Kolodny

The late-onset forms of KD are clinically more heterogeneous, progress more slowly and have a milder clinical picture. Various clinical signs are observed including hemiparesis, spastic paraparesis, intellectual impairment, cerebellar ataxia, visual failure, peripheral polyneuropathy and talipes cavus. The increase in cerebrospinal fluid proteins and the reduction in nerve conduction velocity are less constant *(5)*. An adult-onset form can be reasonably, although arbitrarily, isolated from this group.

Adult KD presenting as a pure spastic paraparesis, which may reach variable degrees of severity, is a rare phenotype that we shall now discuss.

These patients present an indolent course that may start even in their 40s; a few patients were diagnosed in their 50s or 60s. However, clinical onset sometimes goes back to childhood or adolescence if we consider the long-standing history of the presence of pes cavus. The deformity of talipes cavus is caused by an increase in the tone of the plantar flexor muscles, which probably indicates a very subtle involvement of the corticospinal tract *(6)*. Most of the patients reported in the literature *(2,7–11)* as well as four of the five patients diagnosed at our Institute also had minimal neurological signs consisting of pes cavus (three of our five patients), subtle signs of peripheral neuropathy (three of five), slight dysarthria (two of five), and mild cognitive impairment (two of five). In addition to spastic paraparesis, other mild neurologic signs occasionally reported in the literature include visual impairment, pallor of the optic disc, tremor, and ataxia.

3. GENETIC CONSIDERATIONS

Molecular studies have demonstrated the heterogeneous nature of KD. Cloning of the GALC gene *(12,13)* has demonstrated that different mutations within this gene are associated with different severity of the disease *(14)*. One common mutation has been found in 40–50% of the mutant alleles in infantile cases of European or Mexican descent *(14)*. Sixty-five disease-causing mutations and polymorphic changes within the GALC gene have been described *(14)*. Most of the mutations causing the late-onset or adult form are located in the region of the GALC gene coding for the 50-kDa subunit *(2)*. However, some adult cases exhibited mutations located in the region encoding the 30-kDa subunit *(2,7,15)*. The phenotypic variation of the disease may depend on the amount of residual GALC activity associated with differing mutant alleles *(14)*. On the basis of in vitro studies, the correlation between residual enzymatic activity and clinical expression also has been proposed by Percy *(3)*, who pointed out, however, that the available enzymatic data did not fully support this hypothesis. A different expression or stability of the mutant enzyme within various brain regions also might play a role *(3)*. Other genetic factors not related to the GALC gene and, in particular, differing rates of psychosine turnover may also have an influence on the variable severity of different forms of KD *(3)*.

4. RADIOLOGICAL FINDINGS

In the classic infantile form of KD, MRI scan shows diffuse abnormalities in the cerebral white matter and progressive atrophy. The white matter lesions may be difficult to recognize at a very early age when the white matter is still normally hyperintense in T_2-weighted images. At this stage, computed tomography studies may be more helpful because they may show hyperdense areas in the thalami or in the posterior periventricular regions that probably correspond to the clusters of globoid cells in which the galactocerebroside and psychosine accumulate; calcium deposits may contribute to the hyperdensity *(16)*. In children in whom the white matter has become sufficiently hypointense in T_2-weighted images, the MRI changes are more easily recognizable; they mostly involve the white matter of the parieto-occipital regions, with sparing of the subcortical arcuate fibers. An almost constant involvement of the pyramidal tracts (one of the first white matter pathways to myelinate) may be recognizable even within a diffusely unmyelinated white matter. Similarly, the cerebellar white matter is often affected. A frequent, although subtle, involvement of the basal ganglia and thalami is present, whereas the dentate nuclei appear more markedly abnormal *(9)*. The brains of these children become progressively atrophic, and microcephaly ensues.

Children who are diagnosed at age 3 or 4 have more restricted and clear MRI abnormalities. They present hyperintense signal in T_2-weighted images mostly in the posterior periventricular regions with involvement of the splenium of the corpus callosum, which may shrink.

Correlative MRI and neuropathologic studies have shown that the areas of hyperintensity on T_2-weighted images correspond to the areas of demyelination with globoid cell infiltration *(16)*. In long surviving patients, the white matter may be totally gliotic and devoid of macrophages *(17)*.

In the patients with late onset of the disease reported in the literature, the radiological data are often incomplete because they are usually described in articles prevalently dealing with the clinical, biochemical, and genetic aspects of the disease *(18,19)*. These patients show well-circumscribed, posterior, periventricular white matter abnormalities with frequent involvement of the corpus callosum. All these white matter areas are shrunk; the posterior parts of the lateral ventricles, therefore, are enlarged. The corticospinal tracts are often abnormal symmetrically or asymmetrically *(2,6–11,20–32*; Fig. 1). Loes et al. *(9)* in their series showed that pyramidal tract involvement is the most characteristic finding in both early- and late-onset KD. Mild cerebral atrophy may be present. However, in late-onset KD patients, MRI studies may show very subtle abnormalities *(11)* or be entirely normal *(14)*.

In late-onset KD, MR-hyperintense lesions suggest demyelination *(17)*. Neuropathological reports of late-onset cases are limited. Choi et al. *(33)* described the neuropathology of 18-yr-old twins who died of graft-vs-host disease 2 mo after allogeneic bone-marrow transplantation. Their brains showed degeneration of the optic radiations, frontoparietal white matter, and corticospinal tracts. Multiple necrotic foci with calcium deposits were found within the lesions. Globoid cell infiltration was present in actively degenerating white matter. In the peripheral nerves of adult KD, loss of myelinated fibers, disproportionately thin myelin sheaths, and inclusions in Schwann cells have been described *(17)*.

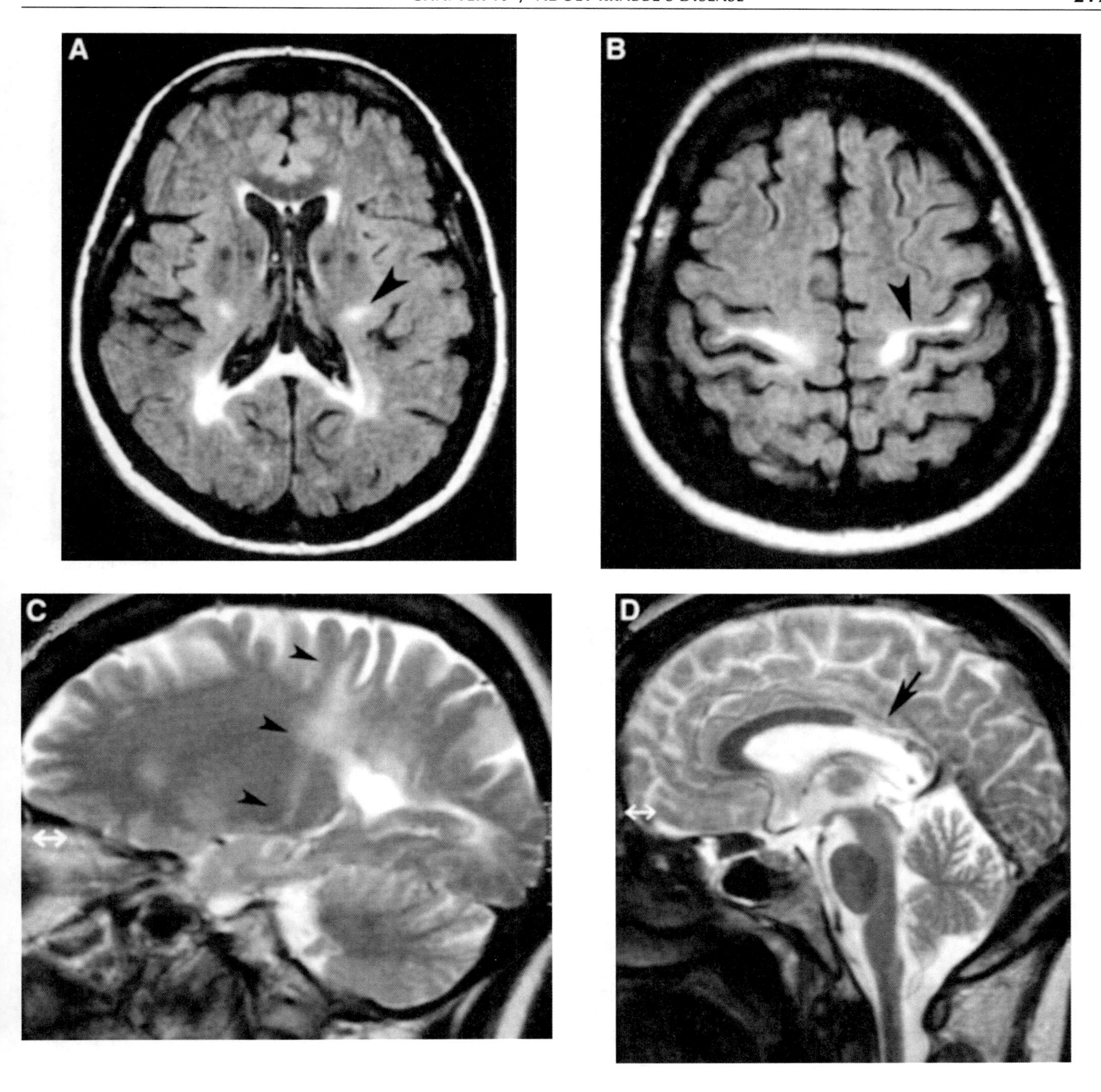

Fig. 1. Shown is a 53-yr-old female patient with onset of the disease at age 30. **(A,B)** Axial fluid-attenuated inversion recovery sections. **(C,D)** Sagittal SE T_2-weighted sections. In A, hyperintense signal in the posterior periventricular regions and splenium of the corpus callosum is demonstrated. The corticospinal tracts are involved from the subcortical white matter within the precentral gyrus through the corona radiata down to the posterior limb of the internal capsule (arrowheads, A–C). On the midline sagittal section (D), involvement of the isthmus of the corpus callosum is also visible (arrow).

Recently, patients with adult-onset KD presenting with pure progressive spastic paraparesis and selective involvement of the corticospinal tracts, which are associated with segmental atrophy of the corpus callosum and occasional slight abnormalities in the parieto-occipital white matter, have been described *(2,7–11)*. In more detail, the patients we observed presented signal abnormalities along the pyramidal tracts visible from the axis of the precentral gyrus down to the corona radiata, posterior limb of the internal capsule, and to variable levels of the brainstem. The spinal cords of two patients that we examined appeared slightly atrophic, but we could not detect any signal abnormalities, perhaps because of the suboptimal quality of the examinations. The neat involvement of the corticospinal tracts was very well appreciated in T_2-weighted sagittal paramedian sections (Fig. 1C); coronal sections demonstrated that the involvement was more marked in the

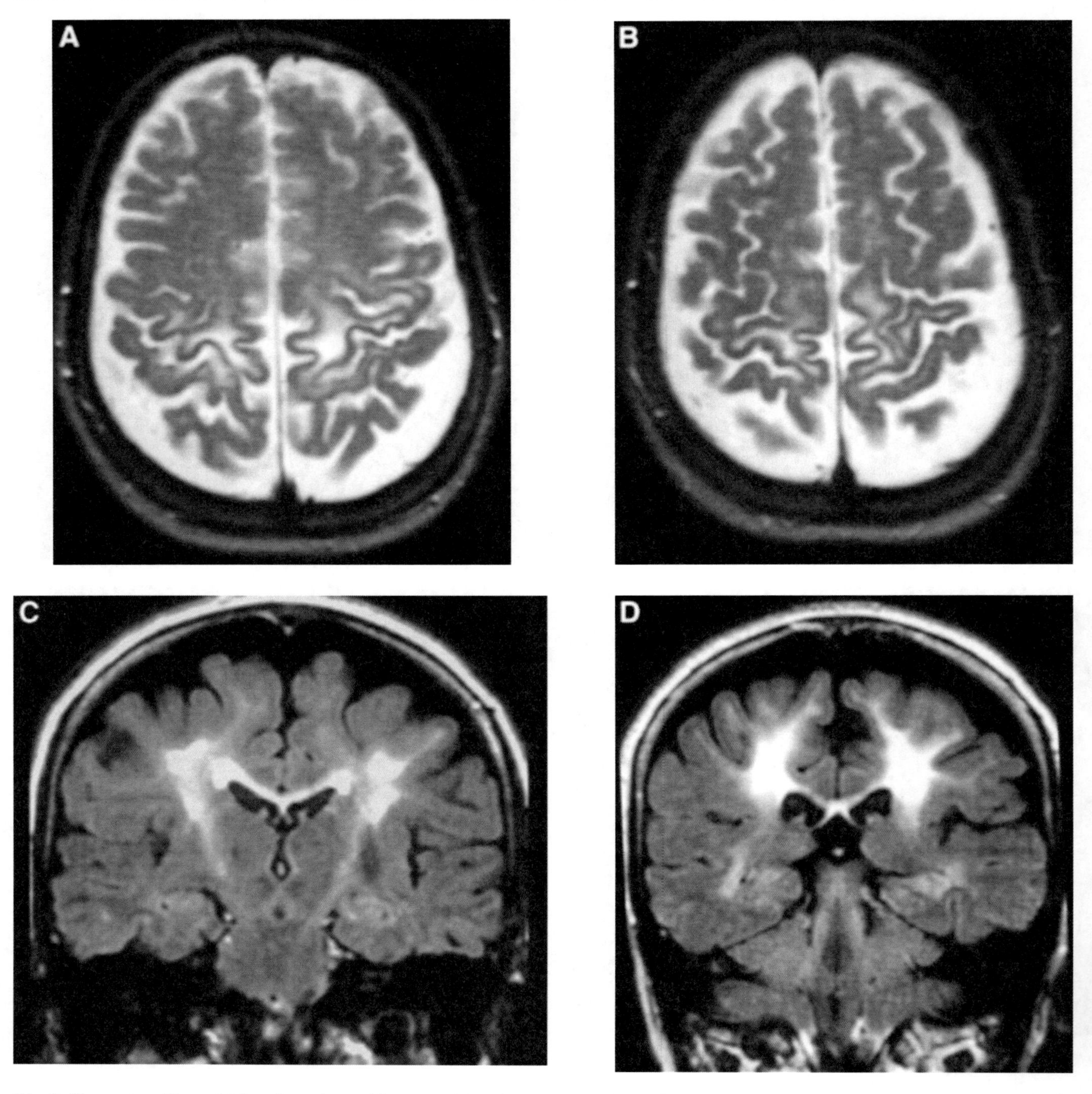

Fig. 2. Shown is a 62-yr-old female patient with onset of the disease at age 32. **(A,B)** Axial T_2-weighted sections show hyperintensity of the white matter in the motor strip, which is moderately atrophic on both sides. **(C,D)** Coronal fluid-attenuated inversion recovery images 5-mm apart demonstrate involvement of the corticospinal tracts (C) and corpus callosum. The opercular areas are spared.

upper part of the motor strip, thus explaining the involvement of the lower extremities (Fig. 2). The lower part of the motor strip, corresponding to the areas of representation of the hand and face, was affected to a lesser extent or not at all. The white matter at the opercula was always spared (Fig. 2C,D). The upper part of the precentral gyrus was always clearly atrophic and became as thin as the postcentral gyrus. The abnormal signal intensity extended across the corpus callosum at the isthmus, which appeared focally thinned (Figs. 1D and 2C). This finding probably corresponds to the degeneration of the association fibers connecting the motor cortex of both sides. The intensity of the signal abnormalities is variable in different cases, ranging from marked T_2 hyperintensity to minimal alterations only in fluid-attenuated inversion recovery images.

It is difficult to understand why the corticospinal tracts are selectively involved in this disorder. It is noteworthy that the corticospinal tracts have some characteristics that are different from other white matter tracts *(34)*. Van der Knaap postulated that in patients with dysmyelinating diseases the oldest myelin breaks down first (M. van der Knaap, Myelination and myelin disorders, Utrecht, the Netherlands: University of Utrecht; 1991, thesis). Because in KD the pathologic process does not involve the formation of myelin but its turnover, the myelin that is formed first, like in the corticospinal tract, will be the first to be affected *(5)*.

Satoh et al. proposed that in patients with late-onset KD the selective vulnerability of the corticospinal tracts might depend on their more active myelin metabolism, compared with the white matter of other parts of the brain where the residual GALC activity is sufficient to maintain a normal or nearly normal myelin. Perhaps the more marked enzymatic defect justifies the more severe white matter involvement of infantile KD. So far, however, the reason for the selective vulnerability of the corticospinal tract still remains unknown. It must be pointed out that the corticospinal tract has some peculiarity compared with the rest of the white matter: in normal subjects, its signal may be slightly hyperintense in T_2-weighted images in the posterior limb of the internal capsule, with some variability between subjects *(35)*. In addition, within the corticospinal tract there are segments that can be affected for a certain length, whereas the adjacent segments are spared as seen in some cases of ALD.

4.1. MRI DIFFERENTIAL DIAGNOSIS

The leukodystrophy that most frequently enters into the differential diagnosis with KD is ALD, both in its typical, severely progressive childhood form and in its adult form of AMN, which is characterized by the elevation of very long chain fatty acids in plasma and skin fibroblasts.

It is worth noting that children with KD who present at age 2 or 3 may show MRI features that are indistinguishable from those of the childhood form of ALD. In both disorders, the lesions are mainly located in the posterior periventricular regions and involve the splenium of the corpus callosum and corticospinal tracts. One feature that may help to differentiate these cases of KD from ALD is the lack of postcontrast ring-enhancement of the lesions that is, on the contrary, very frequent in ALD. This fact may reflect a less marked or absent inflammatory reaction at the margins of the lesions in KD. ALD may typically involve different tracts of the white matter, particularly in the brainstem, including the corticospinal tracts. The pyramidal tracts may be segmentally involved usually below the internal capsule, often only in the pontomedullary region *(34)*. This segmental involvement of the corticospinal tracts may be the first MR imaging manifestation of central nervous system abnormality in asymptomatic ALD children. In these cases, however, the disease may progress to the typical childhood cerebral ALD with rapid appearance of the extensive brain changes seen in this form.

AMN has a very slow course and its initial manifestations, that is, spastic paraparesis, occur in adulthood, normally around age 30. In AMN, MRI of the spinal cord shows atrophy. An MRI of the brain may be entirely normal or show mild cerebellar atrophy or very limited signal changes, similar to those of childhood cerebral ALD, but confined to focal areas such as the splenium of the corpus callosum, cerebellar white matter, or corticospinal tracts *(36)*. In AMN, the MRI abnormalities show very little changes through the years and no postcontrast enhancement; these features reflect the absence of inflammatory changes that characterize the progressive form of the disease.

It must be noted that another disease that usually presents in infancy or childhood such as an X-linked hypomyelinating disorder or leukodystrophy, that is, Pelizaeus–Merzbacher disease, is related to mutations of the proteolipid protein gene, which is also linked to a late onset or adult form of spastic paraplegia (X-linked recessive spastic paraplegia type 2). The neuroradiologic findings in this form of spastic paraplegia are, however, poorly described *(37,38)*.

The chief MRI differential diagnosis of the adult form of KD with involvement restricted to the corticospinal tracts is ALS, a degenerative disorder characterized by the degeneration of upper and lower motor neurons. In ALS, the MR signal changes may extend from the subcortical white matter of the precentral gyrus to the upper part of the brainstem but are usually more evident in the posterior limb of the internal capsule and the corona radiata *(39)*. The signal abnormalities, however, are very faint *(40)* and do not usually reach the intensity seen in the adult-onset KD patients we observed and in those reported in the literature *(2,7–11)*.

In ALS, a hypointense signal in T_2-weighted images in the cortex of the precentral gyrus also has been reported, attributed to iron deposits *(39)*, although the specificity of this sign has been questioned by several authors. Recently, Hecht et al. *(41)* found a significant increase in the quantified hyperintense signals at the subcortical precentral gyrus in fluid-attenuated inversion recovery images and a significant increase over the course of time of hypointense signals in the cortex of the precentral gyrus in T_2-weighted images. This hypointensity may be attributed not only to iron but also to accumulation of paramagnetic substances such as oxygen free radicals *(39,41)*. In T_1-weighted images, hyperintensity along the corticospinal tracts has also been described and considered specific for ALS *(40)*. It might be related to the presence of lipid-laden macrophages or accumulation of intra-axonal neurofilaments *(40)*.

In conclusion, the spectrum of MRI findings in KD varies from diffuse abnormalities involving the whole brain and leading to atrophy in early-infantile cases to a more limited involvement predominantly affecting the posterior periventricular white matter, the splenium of the corpus callosum and the corticospinal tracts in late-onset cases. At the end of the spectrum there is a more selective involvement nearly limited to the corticospinal tracts. This progressive restriction of white matter involvement is not limited to KD as may also be observed in ALD-AMN.

Pyramidal tract involvement is the most characteristic finding in both early- and late-onset KD *(9)*.

The radiologic differential diagnosis of lesions involving the corticospinal tracts is made on MRI.

When quantification of the pyramidal tracts abnormalities is necessary, other techniques such as H-MRS or diffusion tensor MRI should be used *(32,42)*. In adult KD, however, H-MRS has been very rarely applied *(10,11,32)*, whereas, to our knowledge, diffusion tensor imaging has never been employed.

Recently, diffusion tensor-derived anisotropy maps have been shown to provide a quantitative measure of abnormal white matter in patients with the infantile form of KD. They are

A

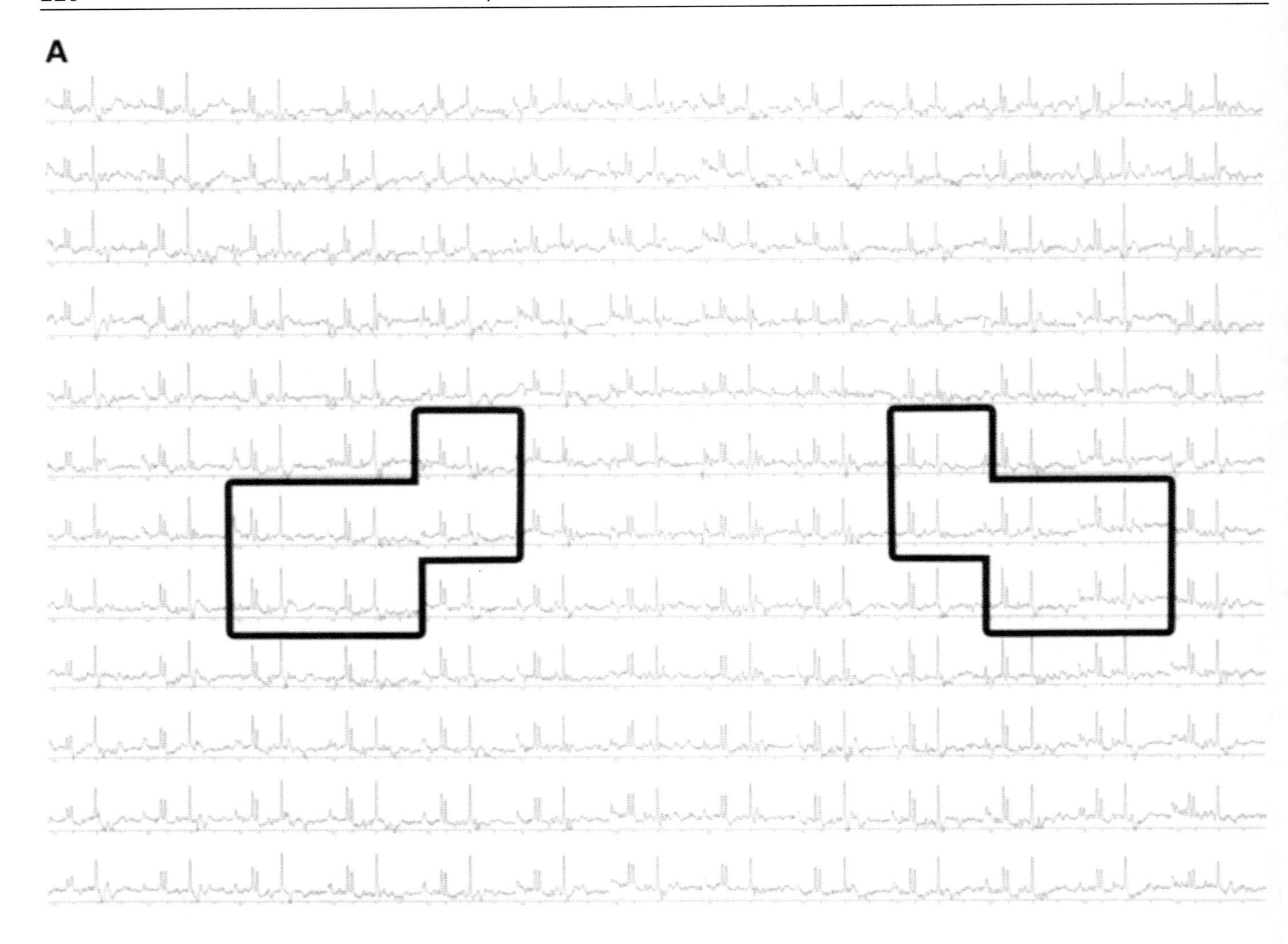

B

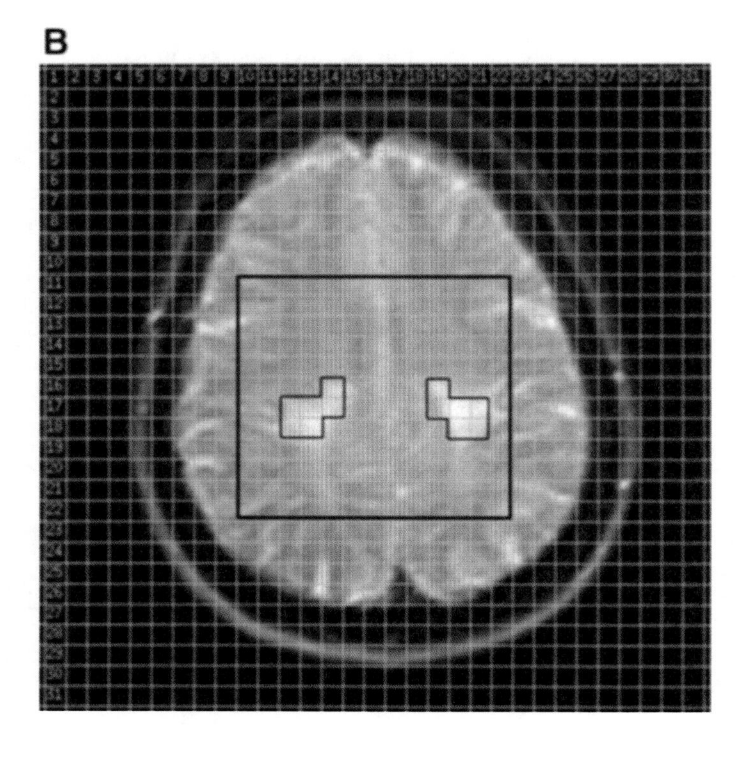

Fig. 3. Same patient as in Fig. 1. **(A)** H-MRSI (multivoxel 2DSE: TR/TE = 1200/136 ms; 32 × 32 phase encoding steps; 20-mm slice thickness) at the level of the centra semiovalia shows symmetrical mild elevation of choline and mild loss of NAA signal intensities within the voxels with T_2 prolongation. Note that choline signal is mildly elevated in several other voxels within the white matter of the centra semiovalia. **(B)** The position of the selected voxels is indicated in the conventional SE T_2-weighted MRI. **(C)** One abnormal spectrum from the left hyperintense area corresponding to the left corticospinal tract is shown (1), compared with a normal spectrum from the parietal interhemispheric cortex (2). The resonance frequencies of the main metabolites are indicated in ppm: choline (3.2), creatine (3.0), and NAA (2.0).

more sensitive than T_2-weighted images in detecting white matter abnormalities and may be a marker of treatment response *(42)*.

Up to a few years ago, KD was considered inevitably fatal. However, hematopoietic stem cell (HSC) transplantation has been recently reported to halt disease progression as observed in a few patients on brain MR scans obtained after HSC transplantation *(43)*. In particular, in four patients with late-infantile onset disease, neurologic deterioration was reversed, and in the

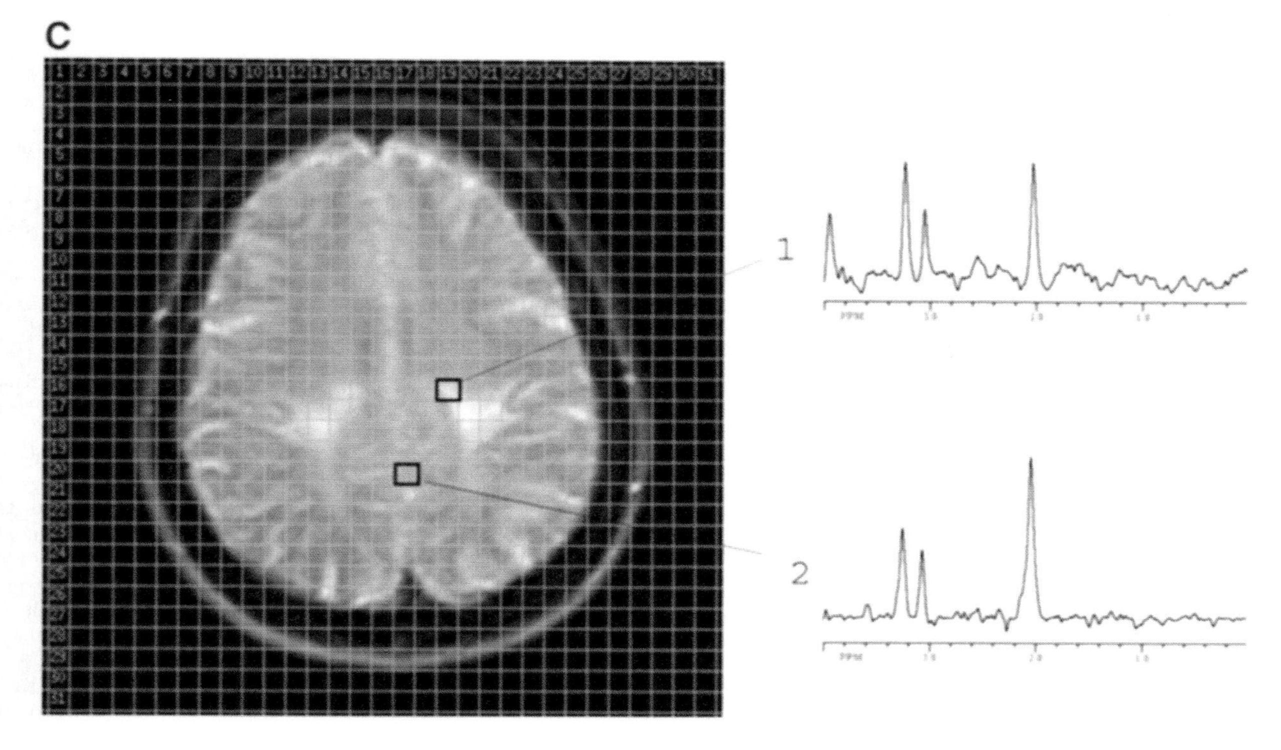

Fig. 3C.

patient with the infantile form of the disease, symptoms and signs did not appear. MRI showed a decrease in signal intensity in the three patients with late-onset disease who were studied both before and after transplantation. These results demonstrate that, like some other leukodystrophies, KD can be treated with HSC transplantation. This procedure, however, carries a high risk of even lethal complications.

4.2. PROTON MR SPECTROSCOPY

To our knowledge, only three reports on H-MRS in adult KD have been published. In 2000, De Stefano et al. *(10)* described symmetrical moderate choline/creatine elevation in the voxels located inside, at the margin, and outside the T_2-weighted MRI signal abnormality in the subcortical white matter of both motor strips. In this 44-yr-old female with slowly progressive gait disturbance since the age of 42, *N*-acetylaspartate (NAA) was still the dominant peak. The NAA/creatine ratio was within normal values. In 2000, Farina et al. *(11)* described similar diffuse H-MRSI abnormalities in the white matter of the centra semiovalia in two siblings with adult KD. These two authors used the same multivoxel point-resolved spectroscopy (PRESS) technique with a nominal voxel resolution of 2 mL. The choline/creatine ratio was elevated, and NAA was mildly decreased compared with the adjacent cortical gray matter. The choline/NAA ratio was close to one not only in the voxel within the T_2-signal abnormality but also in most voxels of the centra semiovalia without signal abnormality on T_2-weighted MRI. In the gray matter of the precentral and postcentral gyri, immediately adjacent to the white matter signal abnormality, the spectra appeared normal. Since that report, we have studied two additional patients with confirmed adult KD (unpublished data) with H-MRSI. In both patients, we confirmed diffuse symmetrical choline/creatine elevation in the centra semiovalia that extended beyond the T_2-weighted signal abnormality. Although the elevation in choline was mild to moderate throughout the white matter, the loss in NAA was more pronounced in the proximal pyramidal tracts at the level of the corona radiata (Fig. 3). Choline was the dominant peak of the spectrum in these affected areas. NAA and choline had near-normal values in the gray matter of the precentral and postcentral gyri. In only one of these two patients was lactate elevated in the affected white matter of both sides (more markedly on the right). More recently, Brockmann et al. described single-voxel H-MRS findings in a 47-yr-old female as part of a larger series of patients with KD *(32)*. They reported near normal H-MRS quantitative metabolite value in their adult patient. However, they used a single-voxel STEAM technique with a larger voxel size and positioned the volume-of-interest in the right posterior paraventricular white matter that is usually less affected in the adult form of this disease.

We have studied five additional KD patients with H-MRSI: two had the infantile form; three had the juvenile form (unpublished data). In all these patients H-MRSI showed diffuse moderate choline elevation in the white matter of the centra semiovalia, within and beyond the signal abnormalities on T_2-weighted MR images. NAA was moderately to markedly reduced in the T_2-weighted affected areas and only slightly decreased in areas without MRI abnormalities. Mild lactate elevation was found in all patients. The spectral profile was near normal in the adjacent cortical gray matter. These spectroscopic findings are in agreement with those reported by Brockmann *(32)*, Frahm and Hanefeld *(44)*, and Zarifi et al. *(45)* in the infantile and juvenile forms of KD. The spectroscopic abnormalities are not specific and are similar to those found in other leukoencephalopathies with demyelinating features. The variable elevation in choline signal shown in KD patients, with onset at different age, most likely reflects differences in degree and extension of demyelination. The choline signal is mainly contributed by free choline, phosphorylcholine, and phosphorylethanolamine that are precursor molecules for membrane synthesis, as well as glycerophosphorylcholine and glycerophosphorylethanolamine that are the corresponding membrane degradation products. The greater signal abnormalities detected in the infantile and juvenile forms might reflect a greater severity of the disease or a greater pool or instability of membrane synthesis in younger patients. Loss of NAA also was found to be more severe in younger patients, both in our experience and in the few reported cases of the literature. This finding most likely reflects the greater and more extensive neuroaxonal damage occurring in these KD patients. This hypothesis is confirmed by the higher incidence of lactate elevation in this class of age.

In conclusion, H-MRS findings in KD are variable and nonspecific but clearly reflect the severity and extent of the disease. Therefore, they are useful to evaluate the degree and progression of the brain damage. H-MRS might be of great value in the follow-up of patients undergoing therapy with HSC transplantation *(43)* or substrate reduction *(46)*.

ACKNOWLEDGMENTS

We thank Dr. Gaetano Finocchiaro and Dr. Davide Pareyson for reviewing the manuscript and Dr. Ugo Danesi for technical support.

REFERENCES

1. Wenger DA, Rafi MA, Luzi P. Molecular genetics of Krabbe disease (Globoid cell leukodystrophy): diagnostic and clinical implications. Hum Mutat 1997;10:268–279.
2. Satoh J-I, Tokumoto H, Kurohara K, et al. Adult-onset Krabbe disease with homozygous T1853C mutation in the galactocerebrosidase gene. Unusual MRI findings of corticospinal tract demyelination. Neurology 1997;49:1392–1399.
3. Percy AK. Krabbe continuum or clinical conundrum? Neurology 1997;49:1203–1204.
4. Fink JK. Hereditary spastic paraplegia: the pace quickens. Ann Neurol 2002:51;669–672.
5. Van der Knaap MS, Valk J. Magnetic Resonance of Myelin, Myelination, and Myelin Disorders. 2nd ed. Berlin, Heidelberg, New York: Springer-Verlag; 1995:68–75.
6. Kolodny EH, Raghavan S, Krivit W, et al. Late-onset Krabbe disease (globoid cell leukodystrophy); clinical and biochemical features of 15 cases. Dev Neurosci 1991;13:232–239.
7. Furuya H, Kukita Y, Nagano S, et al. Adult-onset globoid cell leukodystrophy (Krabbe disease): analysis of galactosylceramidase cDNA from four Japanese patients. Hum Genet 1997;100: 450–456.
8. Turazzini M, Beltramello A, Bassi R, Del Colle R, Silvestri M. Adult onset Krabbe's leukodystrophy: a report of 2 cases. Acta Neurol Scand 1997;96:413–415.
9. Loes DJ, Peters C, Krivit W. Globoid cell leukodystrophy: distinguishing early-onset from late-onset disease using a brain MR imaging scoring method. AJNR Am J Neuroradiol 1999;20:316–323.
10. De Stefano N, Dotti MT, Mortilla M, et al. Evidence of diffuse brain pathology and unspecific genetic characterization in a patient with an atypical form of adult-onset Krabbe disease. J Neurol 2000;247:226–228.
11. Farina L, Bizzi A, Finocchiaro G, et al. MR imaging and proton MR spectroscopy in adult Krabbe disease. AJNR Am J Neuroradiol 2000;21:1478–1482.
12. Chen YQ, Rafi MA, de Gala G, et al. Cloning and expression of cDNA encoding human galactocerebrosidase, the enzyme deficient in globoid cell leukodystrophy. Hum Mol Genet 1993;2: 1841–1845.
13. Sakai N, Inui K, Fujii N, et al. Krabbe disease: isolation and characterization of a full-length cDNA for human galactocerebrosidase. Biochem Biophys Res Commun 1994;198:485–491.
14. Bajaj NPS, Waldman A, Orrel R, Wood NW, Bhatia KP. Familial adult onset of Krabbe's disease resembling hereditary spastic paraplegia with normal neuroimaging. J Neurol Neurosurg Psychiatry 2002;72:635–638.
15. Jardim LB, Giuliani R, Pires RF, et al. Protracted course of Krabbe disease in an adult patient bearing a novel mutation. Arch Neurol 1999;56:1014–1017.
16. Percy AK, Odrezin GT, Knowles PD, Rouah E, Armstrong DD. Globoid cell leukodystrophy: comparison of neuropathology with magnetic resonance imaging. Acta Neuropathol 1994;88:26–32.
17. Suzuki K, Suzuki K. Lysosomal diseases. In: Graham DI, Lantos PL, eds. Greenfield's Neuropathology. 7th ed, vol. 1. London: Arnold; 2002:653–735.
18. Sabatelli M, Quaranta L, Madia F, et al. Peripheral neuropathy with hypomyelinating features in adult-onset Krabbe's disease. Neuromuscul Disord 2002;12:386–391.
19. Henderson RD, MacMillan JC, Bradfield JM. Adult onset Krabbe disease may mimic motor neurone disease. J Clin Neurosci 2003; 10:638–639.
20. Hedley-Whyte ET, Boustany RM, Riskind P, Raghavan S, Zuniga G, Kolodny EH. Peripheral neuropathy due to galactosylceramide β-galactosidase deficiency (Krabbe's disease) in a 73-year old woman. Neuropathol Appl Neurobiol 1988;14:515–516.
21. Demaerel P, Wilms G, Vendru P, Carton H, Baert AL. MR findings in globoid cell leukodystrophy. Neuroradiology 1990;32:520–522.
22. Vendru P, Lammens M, Dom R, et al. Globoid cell leukodystrophy: a family with both late-infantile and adult type. Neurology 1991; 41:1382–1384.
23. Grewal RP, Petronas N, Barton NW. Late onset globoid cell leukodystrophy. J Neurol Neurosurg Psychiatry 1991;54:1011–1012.
24. Kapoor R, McDonald WI, Crockard A, et al. Clinical onset and MRI features of Krabbe disease in adolescence. J Neurol Neurosurg Psychiatry 1992;55:331–332.
25. Inatomi Y, Tomoda H, Itoh Y, Fujii N, Kobayashi T, Ohnishi A. An adult patient with Krabbe's disease. The first case reported in Japan. Clin Neurol 1993;33:1188–1194.
26. De Gasperi R, Gama Sosa MA, Sartorato EL, et al. Molecular heterogeneity of late-onset forms of globoid-cell leukodystrophy. Am J Hum Genet 1996;59:1233–1242.
27. Matsumoto R, Oka N, Nagahama Y, et al. Peripheral neuropathy in late-onset Krabbe's disease: histochemical and ultrastructural findings. Acta Neuropathol 1996;92:635–639.
28. Luzi P, Rafi MA, Wenger DA. Multiple mutations in the GALC gene in a patient with adult-onset Krabbe disease. Ann Neurol 1996;40:116–119.
29. Bernardini GL, Herrera DG, Carson D, et al. Adult-onset Krabbe's disease in siblings with novel mutations in the galactocerebrosidase gene. Ann Neurol 1997;41:111–114.
30. Bataillard M, Richard P, Rumbach L, Vanier MT, Truttmann M. Paraparésie spastique isolée révélant une maladie de Krabbe à l'âge adulte. Rev Neurol 1997;153:347–350.
31. Selleri S, Torchiana E, Pareyson D, et al. Deletion of exons 11–17 and novel mutations of the galactocerebrosidase gene in adult- and early-onset patients with Krabbe disease. J Neurol 2000;415:875–877.
32. Brockmann K, Dechent P, Wilken B, Rusch O, Frahm J, Hanefeld F. Proton MRS profile of cerebral metabolic abnormalities in Krabbe disease. Neurology 2003;60:819–825.
33. Choi KG, Sung JH, Clark HB, et al. Pathology of adult-onset globoid cell leukodystrophy (GLD) (Abstract). J Neuropathol Exp Neurol 1991;50:335.

34. Barkovich AJ, Ferriero DM, Bass N, Boyer R. Involvement of the pontomedullary corticospinal tracts: a useful finding in the diagnosis of x-linked adrenoleukodystrophy. AJNR Am J Neuroradiol 1997;18:95–100.
35. Peretti-Viton P, Azulay JP, Trefouret S, et al. MRI of the intracranial corticospinal tracts in amyotrophic and primary lateral sclerosis. Neuroradiology 1999;41:744–749.
36. Van der Knaap MS, Valk J. Magnetic Resonance of Myelin, Myelination, and Myelin Disorders. 2nd ed. Berlin, Heidelberg, New York: Springer-Verlag; 1995:129–139.
37. Naidu S, Dlouhy SR, Geraghty MT, Hodes ME. A male child with the rumpshaker mutation, x-linked spastic paraplegia/Pelizaeus-Merzbacher disease and lysinuria. J Inher Metab Dis 1997;20:811–816.
38. Hodes ME, Hadjisavvas A, Butler IJ, Aydanian A, Dlouhy SR. X-linked spastic paraplegia due to a mutation (C506T; Ser169Phe) in exon 4 of the proteolipid protein gene (PLP). Am J Med Gen 1998; 75:516–517.
39. Oba H, Araki T, Ohtomo K, et al. Amyotrophic lateral sclerosis: T2 shortening in motor cortex at MR imaging. Radiology 1993;189: 843–846.
40. Waragai M. MRI and clinical features in amyotrophic lateral sclerosis. Neuroradiology 1997;39:847–851.
41. Hecht MJ, Fellner F, Fellner C, Hilz MJ, Neundorfer B, Heuss D. Hyperintense and hypointense MRI signals of the precentral gyrus and corticospinal tracts in ALS: a follow-up examination including FLAIR images. J Neurol Sci 2002;199:59–65.
42. Guo AC, Petrella JR, Kurtzberg J, Provenzale JM. Evaluation of white matter anisotropy in Krabbe disease with diffusion tensor MR imaging: initial experience. Radiology 2001;218:809–815.
43. Krivit W, Shapiro EG, Peters C, et al. Hematopoietic stem-cell transplantation in globoid-cell leukodystrophy. N Engl J Med 1998;338: 1119–1126.
44. Frahm J, Hanefeld F. Localized proton Magnetic Resonance Spectroscopy of brain disorders in childhood. In: Bachelard H, ed: Magnetic Resonance Spectroscopy and Imaging in neurochemistry, Advances in Neurochemistry, vol 8, New York: Plenum Press; 1997:329–402.
45. Zarifi MK, Tzika AA, Astrakas LG, Poussaint TY, Anthony DC, Darras BT. Magnetic resonance spectroscopy and magnetic resonance imaging findings in Krabbe disease. J Child Neurol 2001;16: 522–526.
46. Biswas S, Le Vine SM. Substrate-reduction therapy enhances the benefits of bone marrow transplantation in young mice with globoid cell leukodystrophy. Pediatr Res 2002;51:40–47.

20 Imaging Leukodystrophies

Focus on Lysosomal, Peroxisomal, and Non-Organelle Pathology

Annette O. Nusbaum, MD

SUMMARY

Magnetic resonance imaging (MRI) plays a critical role in the early identification of inherited metabolic disorders. The majority of metabolic disorders present in early childhood and only a few will present in early infancy or adulthood. When metabolic disorders present in infancy, the typical finding on MRI is a delay in normal myelination. In the older child, progressive white matter signal abnormalities are common, resulting in diffuse cerebral atrophy. Although the MRI findings are similar for the majority of these disorders in the later stages, this chapter will discuss the imaging findings of inherited disorders based on: (1) pathology of the subcellular organelles (lysosomes and peroxisomes)-based and (2) non-organelle (i.e., amino or organic-acid)-based pathology, and (3) leukodystrophies presenting with macrocrania.

Key Words: Adrenoleukodystrophy; Alexander's disease; Canavan's disease; Cockayne's syndrome; Fabry's disease; gangliosidosis; Gaucher's disease; Krabbe's disease; lysosomal disorders; metachromatic leukodystrophy; MLD; mucopolysaccharidosis; Pelizaeus-Merzbacher disease; peroxisomal disorders; Sandoff disease; sphingolipidosis; Tay Sachs.

1. INTRODUCTION

Magnetic resonance imaging (MRI) plays a critical role in the early identification of inherited metabolic disorders, which frequently manifest as signal abnormalities of the white matter with or without involvement of the gray matter or primarily as signal abnormalities of the gray matter. Inherited metabolic disorders include dysmyelinating diseases in which there is abnormal development or maintenance of myelin *(1)* and also disorders in which the abnormal accumulation of a biochemical or absence of a specific enzyme primarily affects gray matter structures. The clinical and histopathologic criteria for a leukodystrophy include metabolic disorders in which there is an inherited defect affecting the oligodendroglial cells or myelin that causes progressive neurologic deterioration *(2)*. The terms leukodystrophy and dysmyelination are commonly used interchangeably and several of these disorders (e.g., adrenoleukodystrophy, metachromatic leukodystrophy [MLD], and Krabbe's disease) demonstrate evidence of both demyelination and dysmyelination on pathology that cannot be distinguished by MRI. The majority of metabolic disorders present in early childhood and only a few will present in early infancy or adulthood.

In the majority of the metabolic disorders presenting in infancy, the typical finding on MRI is a delay in the normal myelination. In the older child, progressive white matter signal abnormalities are common, resulting in diffuse cerebral atrophy. Although the MRI findings are similar for the majority of these disorders in the later stages, there are some distinguishing features early in the disease course (Table 1; refs. *3–5*). Abnormalities found on MRI do not always correlate with clinical disability, even when extreme, and the clinical assessment remains the most important measure of disease *(6)*. Studies have shown that ^{1}H MR spectroscopy (MRS) may be more sensitive to early changes in the brain than conventional MRI *(7)* and may contribute additional insight into the in vivo metabolism of metabolic disorders *(8)*. This chapter will discuss inherited disorders based on: (1) pathology of the subcellular organelles (lysosomes and peroxisomes)-based and (2) non-organelle (i.e. amino or organic acid)-based pathology and (3) leukodystrophies presenting with macrocrania *(9)*.

2. LYSOSOMAL DISORDERS

Lysosomes are membrane-bound intracellular hydrolytic enzymes that function as the cell's digestive system and are found primarily within phagocytic cells *(10)*. In lysosomal disorders, specific catabolic enzymes are deficient, resulting in the accumulation of products (such as lipid, carbohydrate, or mucopolysaccharide) that interfere with cell function and eventually may lead to cell death *(11)*. Lysosomal disorders are classified by the nature of the material that abnormally accu-

From: *Bioimaging in Neurodegeneration*
Edited by P. A. Broderick, D. N. Rahni, and E. H. Kolodny

Table 1
Discriminating Features of Common Metabolic Disorders

Deep GM involvement (± white matter)	Mitochondrial disorders
Arrested or highly delayed myelination	Pelizaeus–Merzbacher disease
	Canavan's disease
	Infantile GM1 or GM2 gangliosidosis
Occipital white matter predominant	X-linked ALD
Frontal white matter predominant	Alexander's disease
Enlarged Virchow–Robin spaces	Mucopolysaccharidoses
Macrocephaly	Alexander's disease
	Canavan's disease
Subcortical U fibers involved early	Canavan's disease
Infarcts	mitochondrial encephalomyopathy, lactic acidosis, and stroke-like episodes
Cortical dysplasia	Zellweger's
Hyperdense basal ganglia and thalami on CT	Krabbe's
	GM2 gangliosidosis

mulates (sphingolipidoses, mucopolysaccharidoses, and the mucolipidoses). These disorders often have a relentless, progressive course and vary only in the rate of intellectual and visual deterioration. In many cases, there are no abnormalities on MRI until late in the course.

3. SPHINGOLIPIDOSES

3.1. MLD

3.1.1. Clinical Features

MLD is characterized by a deficiency of the lysosomal enzyme arylsulfatase A (cerebroside sulfatase). It is inherited by an autosomal-recessive pattern, encoded on chromosome 22Q, and primarily affects the central and peripheral nervous system *(12)*. The disorder usually does not begin until after 1 yr of age; however, neonatal onset has been described *(13–15)*. Early clinical features include hypotonia and then ataxia, with progression to spasticity, intellectual failure, and death at age 3 to 6 yr. Cerebroside sulfate (galactosyl sulfatide) abnormally accumulates within the white matter (resulting in breakdown of the membrane of the myelin sheath), kidneys, gallbladder, and other viscera; refs, *16* and *17*). The diagnosis of MLD is based on the finding of marked reduction of arylsulfatase A in the urine and peripheral leukocytes *(18)*.

3.1.2. Pathologic Findings

On gross examination of the brain, the white matter is chalky in appearance, which progresses from patchy to diffuse involvement, often in a symmetric fashion, resulting in a butterfly configuration *(19)*. There is relative preservation of the subcortical arcuate fibers. In cases of severe or long-standing MLD, there is severe atrophy of the white matter with compensatory enlargement of the ventricles. Demyelination is present in the cerebral hemispheres and to a lesser extent in the cerebellum, brainstem, and spinal cord. The hallmark of the disease is metachromatic granules (20–30 mm in diameter), presumably derivatives of cerebroside sulfate, which are found in the brain, liver, kidney, peripheral nerves, and other organs. The areas of demyelination are infiltrated by macrophages that contain metachromatic granules, and gliosis also can be observed. In areas in which myelin is preserved, metachromatic granules may be found within oligodendrocytes. Although axons are relatively spared, some of them are fragmented. Oligodendroglias are absent in areas of demyelination and arereduced in number even in areas where the myelin is still intact. There are no inflammatory cells within areas of demyelination.

3.1.3. MRI Findings in MLD

The characteristic appearance of MLD on MRI is symmetric, confluent areas of high signal intensity on the T_2-weighted images within the periventricular and cerebellar white matter, with sparing of the subcortical U fibers until late in the disease (Figs. 1A, B and 2). In late-onset cases (juvenile and adult forms), the frontal white matter is predominantly involved (Figs. 1A, B), and the signal abnormality progresses from the anterior to the posterior direction *(20)*. In the late-infantile form of MLD *(21)*, the most common type, the signal abnormality is predominantly observed in the occipital lobe and dorsofrontal progression of disease has been reported (Fig. 2). Involvement of the corticospinal tracts may also be seen in the late-infantile form of MLD as abnormal high signal intensity on the T_2-weighted images along the path of the corticospinal tracts in the posterior limbs of the internal capsules and brainstem *(9,21)*. As the disease progresses, the high signal intensity becomes more extensive and confluent with associated atrophy. The corpus callosum is typically affected, and hypointensity within the thalami on T_2-weighted images may be observed. Lesions in the deep gray matter are rarely seen *(9)*. The "tigroid" pattern of demyelination (alternating areas of normal white matter within areas of demyelination), typically described in Pelizaeus–Merzbacher disease, may also be seen in the late-infantile form of MLD. The distinguishing features of MLD include frequent involvement of cerebellar white matter, no involvement of the deep gray matter, and absence of contrast enhancement. Proton MRS frequently demonstrates abnormality in the metabolic peaks before conventional MRI.

3.2. GM1 AND GM2 GANGLIOSIDOSES

3.2.1. Clinical Features

GM1 gangliosidosis is a rare lysosomal storage disorder that occurs as a result of deficiency of the lysosomal enzyme β-galactosidase. The clinical presentation reflects the underlying severity in pathology. If there is severe accumulation, the disease presents in infancy (pseudo-Hurler's); moderate accumulation presents in early childhood or focal accumulation presents in young adulthood *(9)*. GM2 gangliosidosis is the

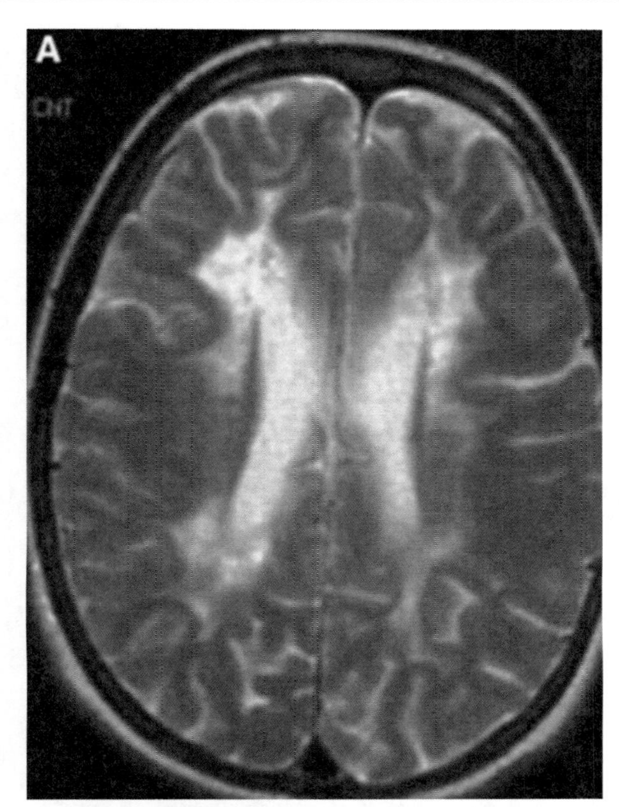

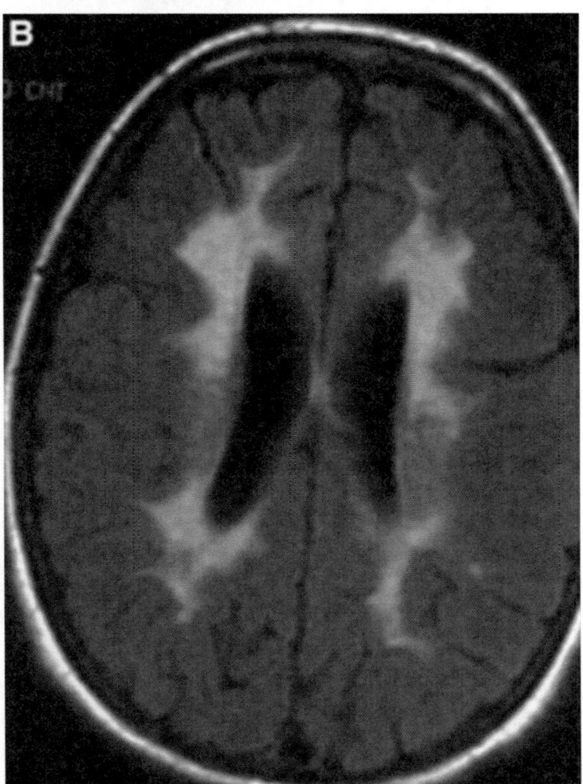

Fig. 1. MLD, juvenile form. Shown is a 14-yr-old female with characteristic involvement of the frontal white matter in the late onset or juvenile form. In (**A** and **B**) note the partial preservation of the subcortical U fibers. Axial T_2 and FLAIR images demonstrate increased signal intensity within the periventricular white matter (frontal white matter more severely affected than the parietal. (Courtesy of Dr. Gary L. Hedlund, Salt Lake City, UT.)

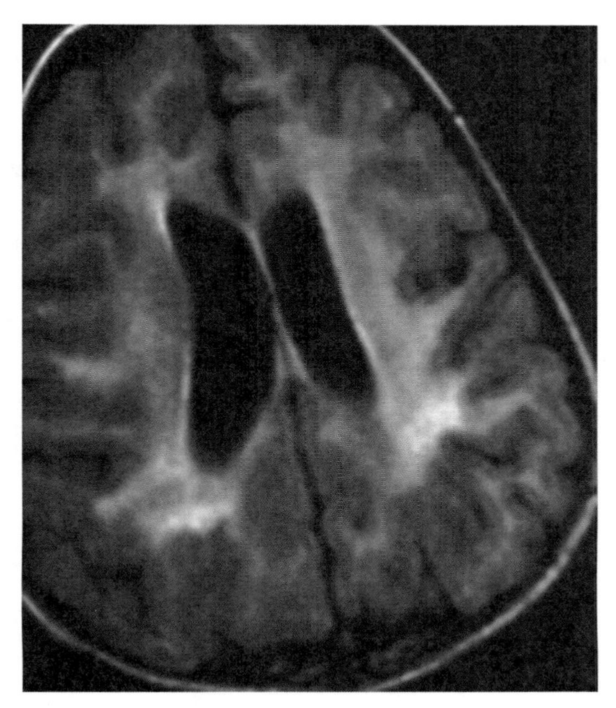

Fig. 2. MLD, late-infantile form. Shown is a 2-yr-old female with characteristic involvement of the posterior white matter. Axial FLAIR image demonstrates predominant high signal intensity within the posterior periventricular white matter with (dorsofrontal) progression to involve the frontal white matter.

general term for two autosomal-recessive lysosomal storage disorders, Tay Sachs disease and Sandhoff disease, which are both characterized by the accumulation of lipids, GM2 gangliosides, primarily in the central nervous system. GM2 gangliosidosis occurs secondary to a deficiency of hexosaminidase A activity. Tay Sachs is a form mainly found in Ashkenazi Jews and is caused by a deficiency in the B-*N*-acetyl hexosaminidase-A isozyme. Sandhoff disease (deficiency of A and B isozymes of hexosaminidase) has a clinical course similar to Tay Sachs but also has visceral involvement *(9)*. There are three clinical variants of the disease based on age of onset: infancy (type I), age 2–6 yr (type II), and adulthood (type III); the earlier the age of onset, the more severe the disease. This presumably reflects the underlying severity of pathology. Late-onset GM2 gangliosidosis occurs during childhood or adolescence and is characterized by poor coordination, tremor, and/or slurred speech. With advancing age, patients develop neurologic symptoms, including ataxia (inability to coordinate voluntary muscle movement), unsteady gait, muscle weakness, and slurred speech. In the fourth decade of life, mental and behavioral involvement may become evident *(22)*.

3.2.2. MRI Findings in GM1 and GM2 Gangliosidosis

The MRI findings in GM1 gangliosidosis differ depending on the age of presentation and thereby reflect the extent of underlying ganglioside accumulation. If the disease presents in infancy, there is extensive demyelination and gliosis in the white matter that will appear as diffuse high signal intensity on the T_2-weighted images on MRI. In the childhood form, diffuse generalized cerebral and cerebellar atrophy has been described.

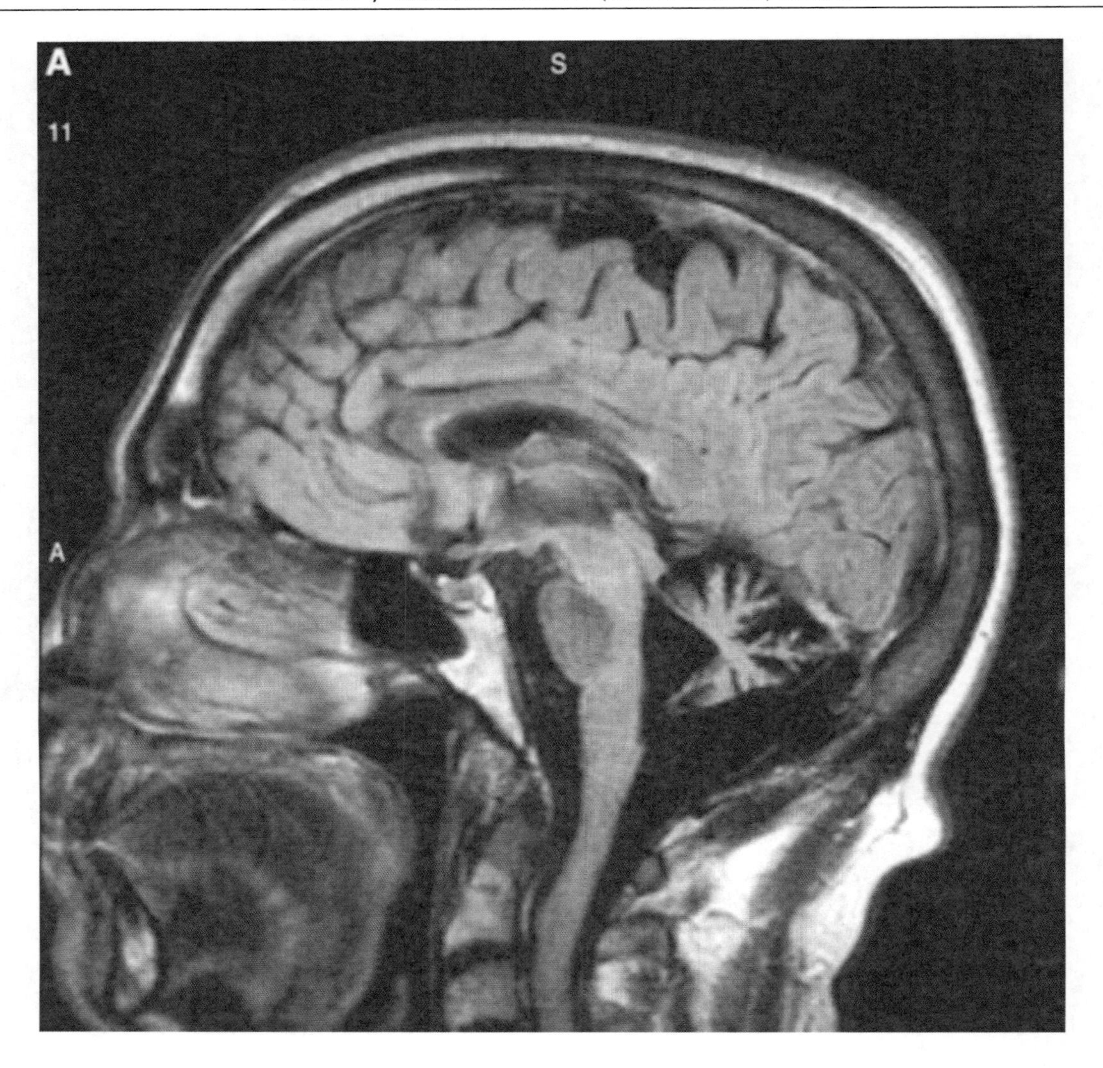

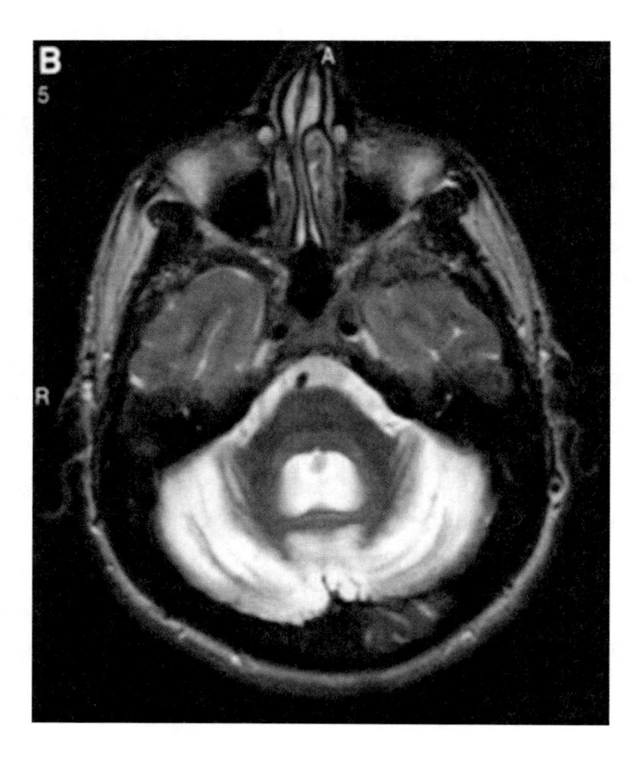

Fig. 3. Tay-Sachs disease. (**A** and **B**) sA 28-yr-old male with characteristic cerebellar atrophy on sagittal FLAIR and axial T_2-weighted images show severe atrophy of the cerebellum.

In young adults, cerebral atrophy and high signal intensity on the T_2-weighted images in the bilateral caudate nucleus and putamen has been reported.

In GM2 gangliosidosis, MRI may demonstrate high signal intensity within the bilateral basal ganglia on the T_2-weighted images early in the disease course. On computed tomography (CT) scan, calcification frequently is observed in the basal ganglia, which will appear as high signal intensity on the T_1-weighted images and low signal intensity on the T_2-weighted images on MRI. Enlargement of the caudate nuclei has been described. Diffuse atrophy may be seen early on and, as the disease progresses, diffuse high signal intensity is seen within the white matter reflecting gliosis, and demyelination and cavitation may be seen. Severe atrophy is most prominent in the cerebellum (Fig. 3A, B; refs. *23–25*).

3.3. FABRY'S DISEASE

3.3.1. Clinical Features

Fabry's disease is an X-linked lysosomal storage disorder caused by deficiency of the enzyme α-galactosidase A *(26–28)*. Glycosphingolipids accumulate in the vascular endothelium, smooth muscles, and neurons. The disease usually presents late in childhood but may not be recognized until the third or fourth decade of life. Clinical features typically consist of acroparesthesias, neuropathic pain, hypohidrosis, corneal inclusions, cataracts, and cutaneous angiokeratoma, followed by progressive renal, cerebrovascular, and cardiac disease *(29)*. The diagnosis of Fabry disease is usually made by recognition of the typical clinical symptoms with or without a family history or by recognition of the ocular or cutaneous manifestations of the disease. Cardiac disease develops with age and is typically worsened by systemic hypertension secondary to renal vascular disease *(9)*. Patients may eventually present with transient ischemic attacks, or strokes secondary to small vessel ischemia and focal infarcts.

3.3.2. MRI Findings in Fabry's Disease

Early in the disease, small areas of high signal intensity are seen on the T_2/fluid-attenuated inversion recovery (FLAIR) sequences, most commonly in the basal ganglia and periventricular white matter (Fig. 4A, B). The periventricular disease becomes more extensive and confluent with time with associated generalized cerebral volume loss. Cerebral hemorrhage has also been reported. Dolichoectasia of the cerebral vasculature was noted by Mitsias and Levine *(30)* in a descriptive meta-analysis of 53 cases of Fabry disease; the vertebrobasilar system was particularly affected, but the carotid circulation was also involved. Recent literature has described hyperintensity in the pulvinar on T_1-weighted images as a common finding in Fabry disease (Fig. 5), likely reflecting the presence of calcification *(31)*. Although other disease entities may result in calcification of deep gray nuclei, exclusive involvement of the pulvinar is thought to be distinctly characteristic of Fabry disease. Increased cerebral blood flow on positron emission tomography studies in the posterior circulation, particularly the thalamus, suggests that the dystrophic calcification is secondary to cerebral hyperperfusion and selective vulnerability of the pulvinar and adjacent thalamic nuclei *(32)*. The finding of isolated pulvinar hyperintensity on T_1-weighted images should suggest Fabry disease, particularly when seen in conjunction with other nonspecific neuroradiologic manifestations of the disease.

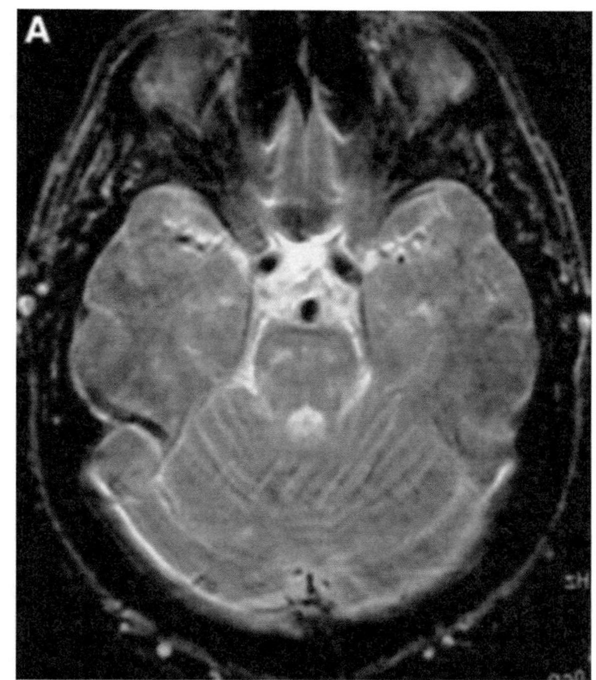

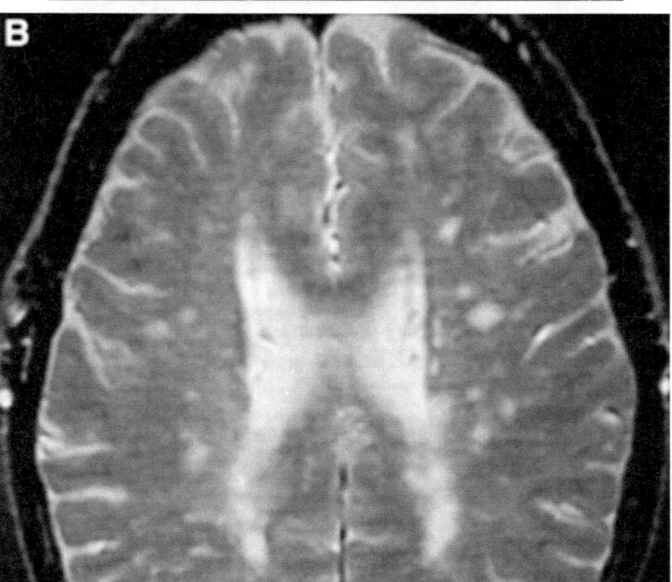

Fig. 4. Fabry's disease. Scattered focal areas of high signal intensity on the T_2-weighted images within the brainstem (**A**) and periventricular white matter (**B**).

3.4. KRABBE'S DISEASE

3.4.1. Clinical Features

Krabbe's disease, or globoid cell leukodystrophy, is an autosomal-recessive inherited disorder that commonly presents within the first 6 mo of life *(33)*. Galactocerebroside accumulates in macrophages as a result of deficiency of the lysosomal enzyme galactocerebroside β-galactosidase (encoded by chromosome 14; refs. *34* and *35*). Because galactocerebroside is an important component of mature myelin, symptoms usually begin during the period of active myelin synthesis. Several clinical types have been distinguished based on the age of onset and disease course (early-infantile, congenital, late-infantile, juvenile, and adolescent-adult; refs. *33* and *36*).

3.4.2. Pathologic Findings

The brain is small and the white matter is rubbery to firm but the cortex is relatively unaffected *(19)*. The pathologic hallmark of Krabbe's disease is a massive accumulation of large multinucleated cells containing periodic acid Schiff-positive material (globoid cells). Demyelination and dysmyelination are seen.

3.4.3. MRI Findings in Krabbe's Disease

Characteristic patterns of involvement in GLD have been described on MRI based on the age of onset of clinical symptoms: (1) early-onset disease (clinical presentation at or before age and (2) late-onset disease (clinical presentation after age 2; refs. *36–38*). The most typical finding on MRI in both the

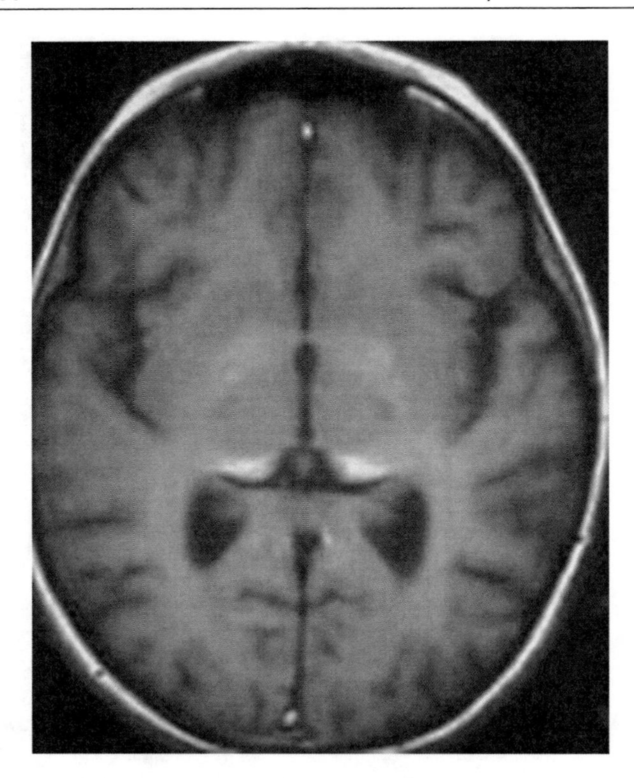

Fig. 5. Fabry's disease. High signal intensity in the pulvinar of the thalamus on the noncontrast T_1-weighted images, which is characteristic of the disease.

infantile- and late-onset forms of GLD is high signal intensity on the T_2-weighted images along the lengths of the corticospinal tracts. In the early-onset form, additional findings include abnormal signal intensity within the cerebellar white matter, deep gray nuclei (dentate, basal ganglia, thalamus), with progressive involvement of the parieto-occipital white matter and posterior portion of the corpus callosum. In the late-onset form, there is also abnormal signal in the posterior portion of the corpus callosum and bilateral symmetric parieto-occipital white matter. The cerebellar white matter and deep gray nuclei are not involved in the late-onset form.

Classically, the subcortical U fibers are spared until late in the disease. Enhancement at the border between the white matter and arcuate fibers has been described *(33)* but is not usually seen. Rare findings that have been described include optic nerve atrophy and bilateral symmetric optic nerve hypertrophy *(39,40)*. Characteristic findings on CT scan include hyperdensity in the thalami, caudate nuclei, and corona radiata and have been shown to correspond to fine calcifications at autopsy *(41)*. Atrophy is common late in the course of the disease resulting from progressive loss of white matter.

3.5. GAUCHER'S DISEASE

3.5.1. Clinical Features

Gaucher's disease includes several autosomal-recessive lipid-storage diseases in which there is a deficiency of a lysosomal enzyme that results in accumulation of the major substrate, glucosylceramide, in cells of monocyte/macrophage lineage and leads to hepatosplenomegaly, destructive skeletal disease, and bone-marrow compromise *(42)*. Neurologic symptoms may include seizures, developmental regression, spasticity, mental deficiency, incoordination, and tics. The diagnosis is made by clinical criteria, by the presence of Gaucher's cells in the bone marrow, and by finding reduced glucocerebroside β-glucosidase in the cultured skin, fibroblasts, or blood leukocytes *(19)*.

3.5.2. MRI Findings

MRI and CT findings are similar to those of Fabry's disease, with atrophy, infarction, and occasional hemorrhage. Chang et al. *(43)* reported a case in which there was unilateral dural thickening that was enhanced on postcontrast images and presumably reflected glucosylceramide infiltration.

4. MUCOPOLYSACCHARIDOSES

4.1. CLINICAL FEATURES

The mucopolysaccharidoses include a group of disorders in which a lysosomal enzyme deficiency results in the inability to degrade the mucopolysaccharides (glycosaminoglycans) heparan sulfate, keratan sulfate, and/or dermatan sulfate *(44)*. There are a number of characteristic clinical features shared by these disorders, including the typical gargoyle features, numerous skeletal abnormalities (dwarfism), cardiac anomalies, and central nervous system abnormalities, and frequent involvement of the visual and auditory systems. The diagnosis of mucopolysaccharidoses is made based on the clinical presentation, family history, and enzymatic assays of cultured skin fibroblasts or peripheral leukocytes. The skeletal features of these disorders are far more characteristic than the MRI findings in the brain.

4.2. MRI FINDINGS IN MUCOPOLYSACCHARIDOSES

Prominent cystic or dilated perivascular spaces may be seen, which represent vacuolated cells distended with mucopolysaccharide. Dilated perivascular spaces may be seen normally in children but have only been described in the corpus callosum (Figs. 6 and 7) in the mucopolysaccharidoses *(44)*. As the disease progresses, the lesions become more widespread and extensive, reflecting the development of infarcts in demyelination *(45)*. In addition, multiple patchy areas of high signal intensity can be seen on the T_2-weighted images within the periventricular white matter (Fig.8). Ventricular enlargement is common and is likely due to a combination of communicating hydrocephalus and white matter volume loss.

Thickening of the skull and meninges is seen on imaging. Spinal cord compression is common, especially at the foramen magnum and upper cervical level, resulting from dural thickening secondary to mucopolysaccharide deposits (Fig. 9). Spinal cord compression may result from atlanto-axial subluxation or thoracic gibbus. Nerve root compression has also been described.

5. PEROXISOMAL DISORDERS

Peroxisomes are membrane-bound subcellular organelles involved in lipid metabolism *(11)*. Peroxisomes are present in all cells but are more prevalent within oligodendrocytes that specialize in lipid metabolism and myelin production and maintenance. The genetic defects are found in the peroxisomes or in one of the enzymes normally located in the peroxisome and lead to abnormal accumulation of biochemicals (i.e., very long chain fatty acids, pipecolic acid, and dicarboxylic acids).

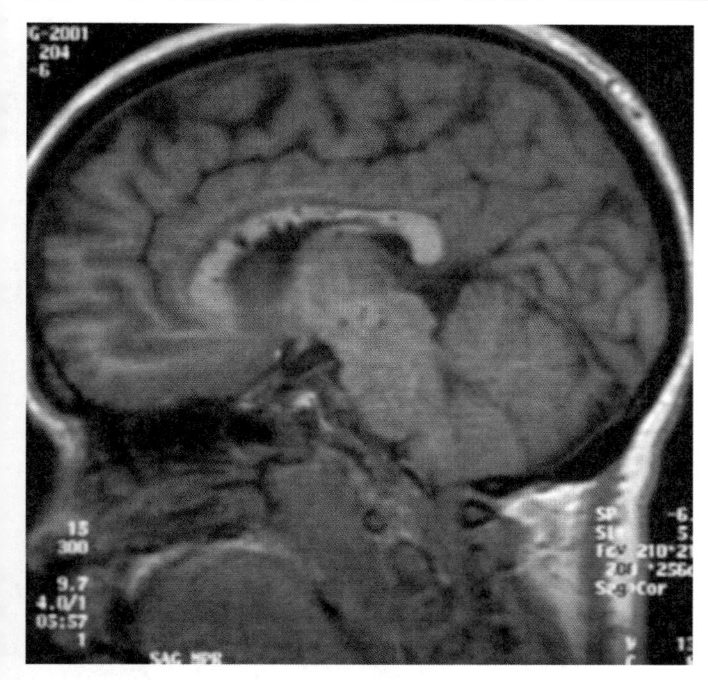

Fig. 6. Mucopolysaccharidosis VI. Prominent cystic spaces are seen within the corpus callosum, which represent vacuolated cells distended with mucopolysaccharide.

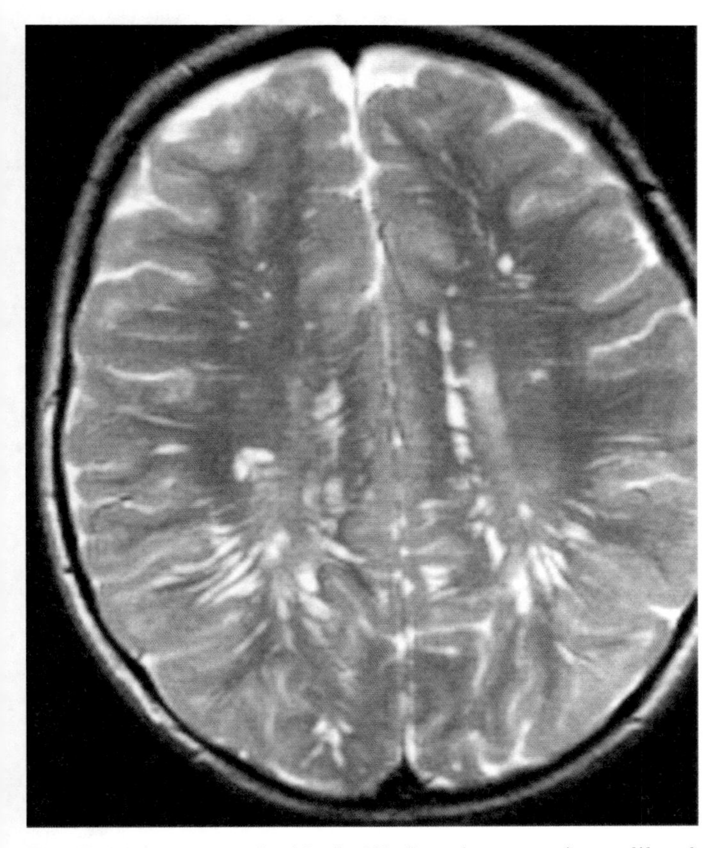

Fig. 7. Mucopolysaccharidosis VI. Prominent cystic or dilated perivascular spaces are seen within the white matter on the axial T_2-weighted sequence which represent vacuolated cells distended with mucopolysaccharide.

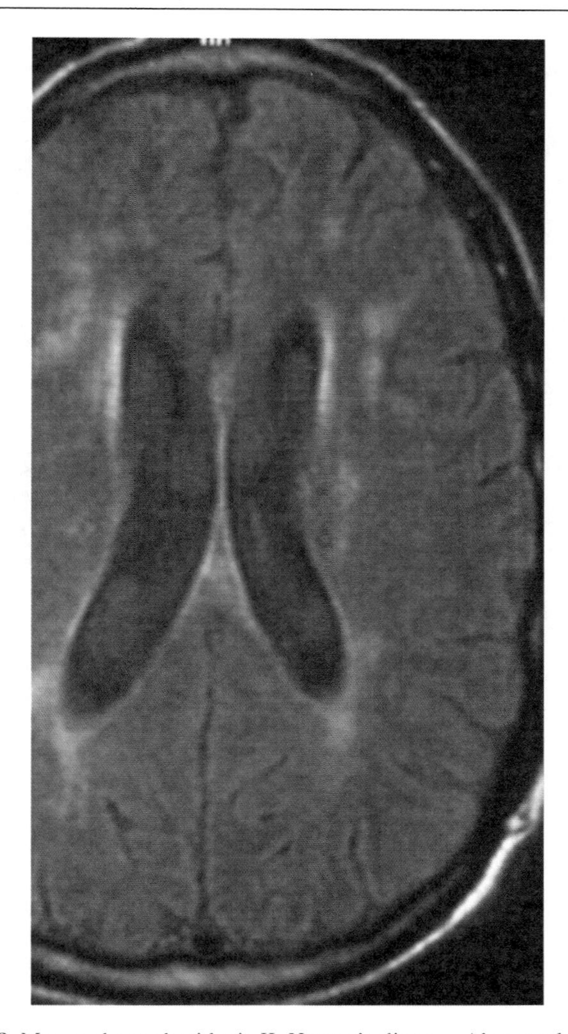

Fig. 8. Mucopolysaccharidosis II, Hunter's disease. Abnormal focal areas of high signal in the periventricular region on the axial FLAIR image reflects the perivascular involvement.

Neuropathologic lesions in the peroxisomal disorders can be divided into three major classes. The first group is characterized by defects in the formation and maintenance of white matter and X-linked adrenoleukodystrophy (ALD) is the prototype. The second group is associated with migrational disorders and Zellweger's syndrome is the classic example. The third group is associated with postdevelopmental neuronal degenerations such as cerebellar atrophy seen in rhizomelic chondrodysplasia punctata *(46)*.

5.1. ALD/ADRENOMYELONEUROPATHY (AMN)

5.1.1. Clinical Features

ALD is an X-linked recessive peroxisomal disorder involving the white matter of the brain and spinal cord and also the adrenal cortex. Biochemically, there is an abnormal accumulation of very long chain fatty acids that become incorporated into myelin. This leads to instability and dysmyelination, with a possible direct cytotoxic effect on the oligodendrocytes *(46)*. The clinical presentation is quite variable. Greater than 50% of patients present with progressive childhood onset, approx 25% have a late-onset presentation with AMN, and 10% have isolated Addison disease. Development in the first few years of life is usually normal. Neurologic symptoms appear later in childhood, between the ages of 5 and 9, with behavior prob-

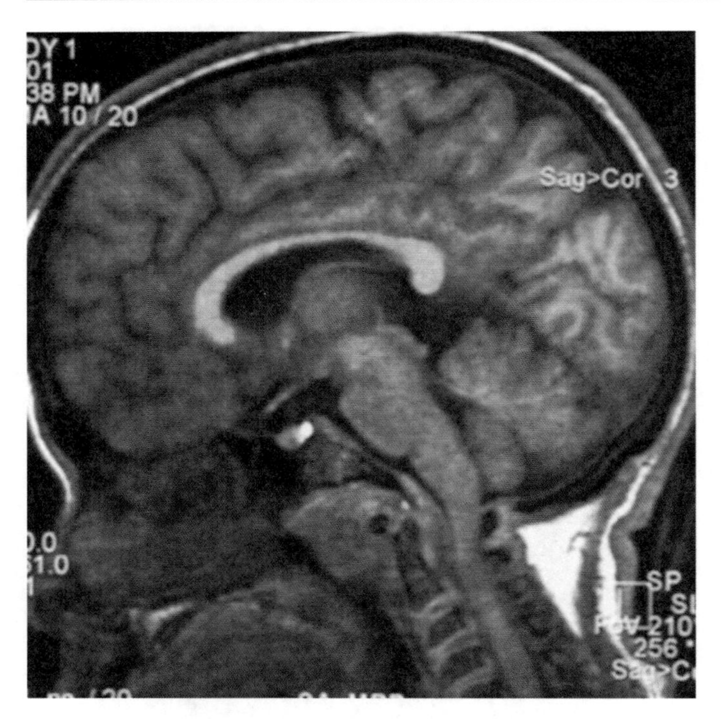

Fig. 9. Mucopolysaccharidosis VI. Sagittal T_1-weighted image demonstrates thickening of the meninges at the foramen magnum and upper cervical level, with mild indentation of the spinal cord. This dural thickening is secondary to mucopolysaccharide deposits.

lems, decreasing mental function, and visual and hearing disorders *(47)* progressing to motor signs and ataxia. Symptoms of Addison's disease commonly appear before the neurologic symptoms but may follow mental deterioration and occasionally even never present. The disease progresses to include seizures, spastic quadriplegia, and decorticate posturing, with death ensuing within the first few years of onset.

5.1.2. Pathologic Findings

Gross examination of the brain in ALD patients reveals that the white matter is gray and firm, with atrophy and cystic cavitation. Histologically, there are confluent areas of demyelination in a symmetric fashion, usually in the bilateral occipital white matter, with extension across the splenium of the corpus callosum. The occipital, parietal, and the temporal lobes are more severely affected than the frontal lobe. The demyelination tends to have a caudorostral progression with relative sparing of the subcortical arcuate fibers.

The histologic findings reflect the zones of activity seen on imaging studies. Three zones of demyelination are characteristically noted. The central portion of the lesion reveals absent myelin sheaths and oligodendroglia. Glial stranding and scattered astrocytes with no evidence of active disease are present. The next zone of involvement shows evidence of active inflammation with many macrophages filled with lipid. Intact axons are identified both with and without myelin sheaths. The outer zone is characterized by active myelin break down with some lipid-laden macrophages but no inflammatory changes. Characteristic lipid lamellae, best demonstrated by electron microscopy, are observed in the brain, testis, adrenal gland, skin, conjunctiva, and Schwann cells *(19)*.

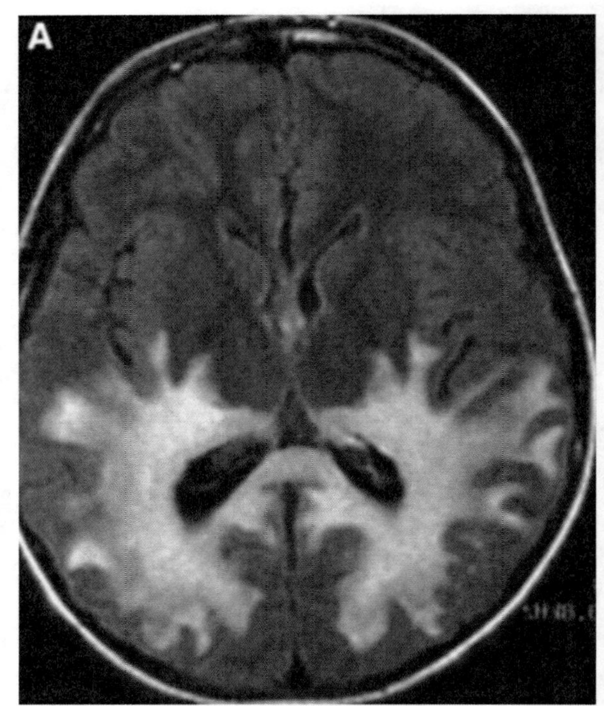

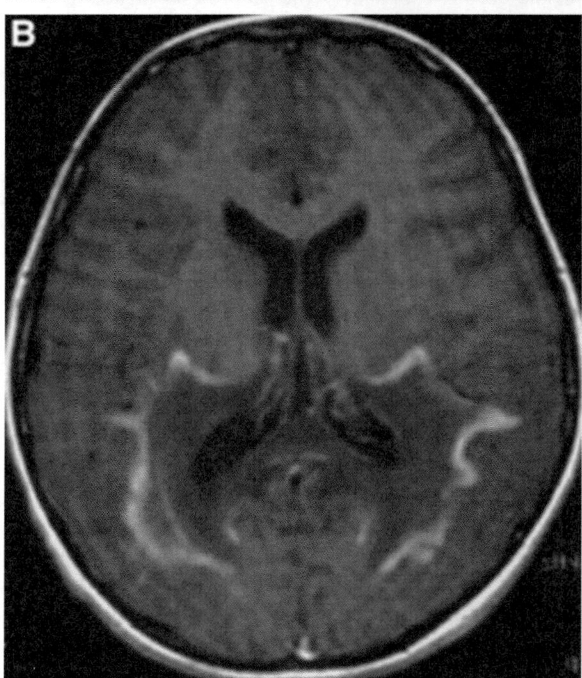

Fig. 10. X-linked ALD. (A) Axial T_2-weighted image. **(B)** Axial contrast-enhanced T_1-weighted image. Note extensive high intensity on the T_2-weighted image (A) in bilateral parietal and occipital white matter through the corpus callosum. Postcontrast image (B) shows enhancement of the leading edges of these areas.

5.1.3. MRI Findings in ALD/AMN

The CT and MRI appearance of ALD is somewhat specific, with symmetric areas of high signal intensity on the T_2-weighted images within the white matter surrounding the atria of the lateral ventricles, extending across the splenium of the corpus callosum (Fig. 10A, B and 11A–C; refs. *46–49*). The parietal and occipital white matter is most commonly involved;

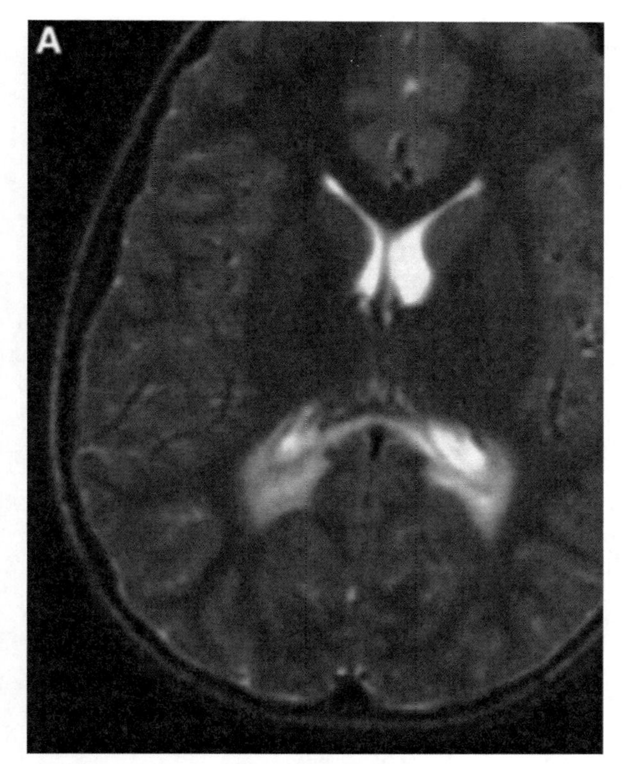

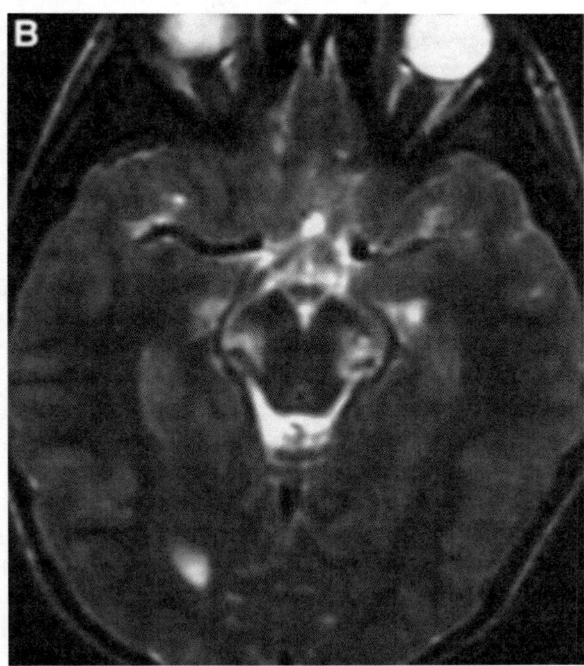

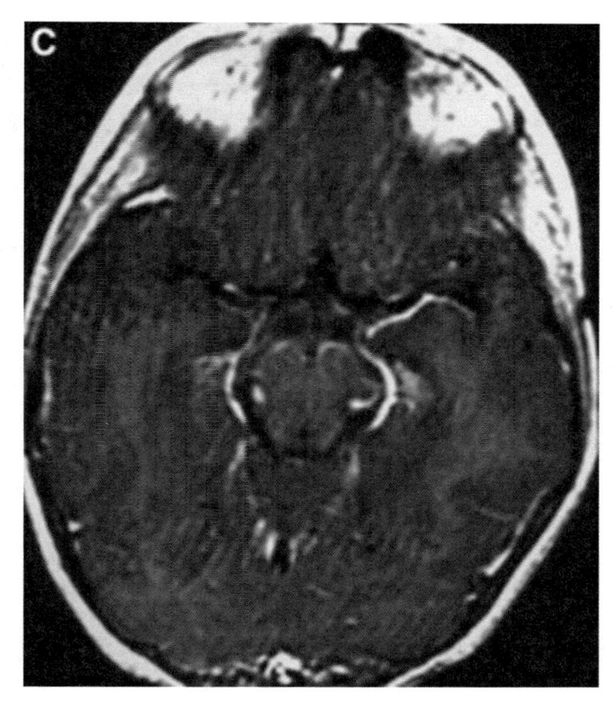

Fig. 11. X-linked ALD. FLAIR images. The characteristic high signal intensity is noted in (**A**) the splenium of the corpus callosum, posterior periatrial white matter, and (**B**) further inferiorly into the brainstem. Note enhancement in the brainstem on the postcontrast T_1-weighted image (**C**).

however, frontal predominance and holohemispheric patterns have been described (Fig. 12; ref. *9*). At the lateral margin of the zones of demyelination, contrast enhancement appears, corresponding to areas of active demyelination accompanied by inflammation (Schaumberg's zones 1 and 2).

MRI of the brain in AMN may be normal, with neurologic involvement confined to the spinal cord and peripheral nerves. In some patients with AMN, the MRI findings may be similar to ALD, and the clinical progression may be as rapid *(46)*.

6. AMINO AND ORGANIC ACID DISORDERS

Disorders of amino acid or organic acid metabolism are very rare. They typically involve an inherited deficiency or altered function of an enzyme or transport system that mediates the disposition of a particular amino or organic acid (Table 2). The oxidation of amino acids gives rise to ammonia, which is neurotoxic in high concentrations *(50)*. Because the urea cycle functions in the disposal of ammonia, congenital deficiencies of the urea cycle cause hyperammonemia or elevated plasma glutamine (formed from ammonia; Fig. 13). The urea cycle defects include carbamyl phosphate synthetase deficiency, ornithine transcarbamylase deficiency citrullinemia, and argininosuccinic aciduria. The severity of presentation is determined by the particular amino or organic acid abnormality, the duration of the accumulation, and the presence of other metabolic alterations (e.g., hypoglycemia). Dysmyelination, neuronal degeneration, and reactive gliosis are common in patients who die in the first few days of life (i.e. maple syrup urine disease).

6.1. LEUKODYSTROPHIES WITH MACROCRANIA

6.1.1. Canavan's Disease

6.1.1.1. Clinical Features

Canavan's disease (Canavan-van Bogaert-Bertrand disease) or spongy degeneration of the brain is an autosomal-recessive disorder of amino acid metabolism found predominantly in children of Ashkenazi Jewish decent. It is caused by a deficiency of *N*-acetylaspartate acylase, which results in excessive accumulation of *N*-acetyl aspartate (NAA; refs. *51* and *52*).

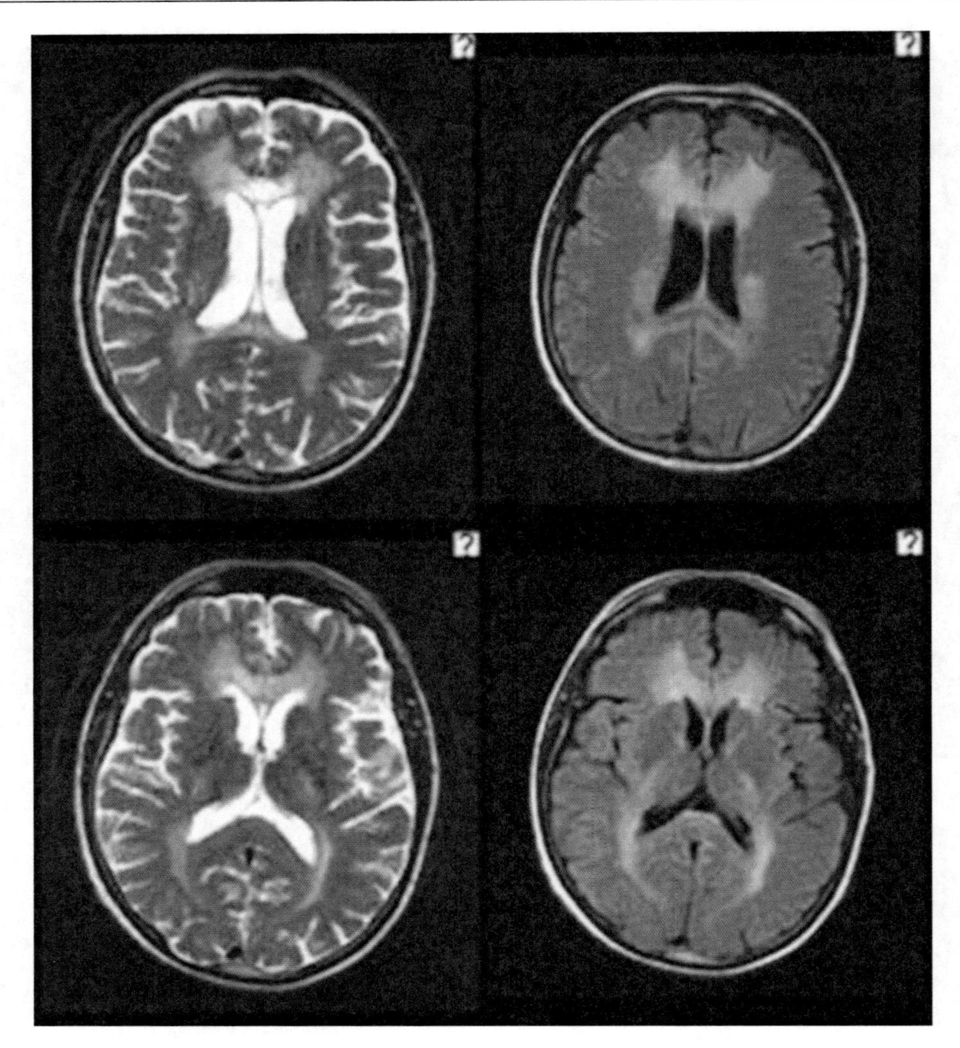

Fig. 12. ALD with unusual feature of predominant frontal involvement. High signal intensity is seen spanning the genu of the corpus callosum on axial FLAIR and T_2-weighted images also involving the bifrontal periventricular white matter. In addition, the posterior periventricular white matter demonstrates abnormal high signal intensity with a lesser degree of abnormal signal in the splenium of the corpus callosum.

Clinical signs and symptoms become manifest within the first few months of life marked by hypotonia, head lag, increased head circumference (>98th percentile), seizures, and failure to achieve motor milestones. Death typically ensues by age 4. Diagnosis can be made by quantitative study of acetylaspartic acid in urine and aspartoacylase level (accumulation of abnormally high level of nonfunctional enzyme) in cultured fibroblasts.

6.1.1.2. Pathologic Findings

The brain may be heavier or of normal weight and the white matter is soft and gelatinous. Histologic features include vacuolization of both gray and white matter and proliferation of Alzheimer's type II astrocytes. The distribution of spongiotic changes is most prominent in the deeper cortex and subcortical white matter, with relative sparing of the deeper white matter and internal capsule. As the disease progress, a more diffuse pattern of demyelination will develop. During the first several years of life, the ventricles are usually narrowed and gradually increase in size as a result of loss of tissue.

6.1.1.3. MRI Findings

In Canavan's disease, the signal abnormality has a centripetal distribution beginning in the subcortical white matter of the cerebrum and cerebellum (Fig. 14A, B). The subcortical white matter may appear swollen *(9)*. Typical findings include diffuse symmetric increased signal intensity on the T_2-weighted images throughout the white matter with relative sparing of the internal and external capsules and corpus callosum. The central white matter becomes involved with disease progression. High signal intensity on the T_2-weighted images is always seen within the globus pallidus, with frequent involvement of the thalamus and relative sparing of the putamen and caudate nucleus. Cerebral and cerebellar atrophy is a late finding (Fig. 14B). The underlying pathology of excessive accumulation of NAA in the brain is readily demonstrated by proton MRS with a characteristic increase in the NAA peak *(53–55)*.

6.1.2. Alexander's Disease

6.1.2.1. Clinical Features

Alexander's disease occurs sporadically with no known pattern of inheritance. There are three forms of the disease: infantile, juvenile, and adult. The infantile form is the most common. The diagnosis is usually made within the first year of life when the infant presents with developmental delay, macrocephaly, spasticity, and seizures *(56–58)*. Progressive deterio-

Table 2
Disorders of Amino Acid Metabolism

Disease	*Imaging Findings*	*Clinical Findings*
Maple syrup urine disease	Swelling and Increased signal T_2-WI: Brainstem, GP, cerebellum postlimb internal capsule	Neonate-coma Seizures Respiratory failure
Glutaric aciduria type I	Prominent extra-axial spaces Increased signal T_2-WI: bilateral caudate, putamen Delay in myelination	Infancy/childhood Encephalopathy Macrocephaly
Phenylketonuria	Delay in myelination Increased signal T_2-WI: Periatrial WM	Normal at birth Developmental delay Varying severity
Nonketotic hyperglycinemia	Increased signal T_2-WI: Periventricular WM Delay in myelination	Early infancy Vomiting, seizures Apnea
Urea cycle defects	Neonates: brain swelling Acute stages: Multifocal swollen areas Cortex and subcortical WM Chronic stages: Diffuse atrophy Delay in myelination	Infancy: lethargy, Coma Childhood: episodic lethargy, confusion, ataxia, dysarthria Coma

GP, globus pallidus; WI, weighted image; WM, white matter.

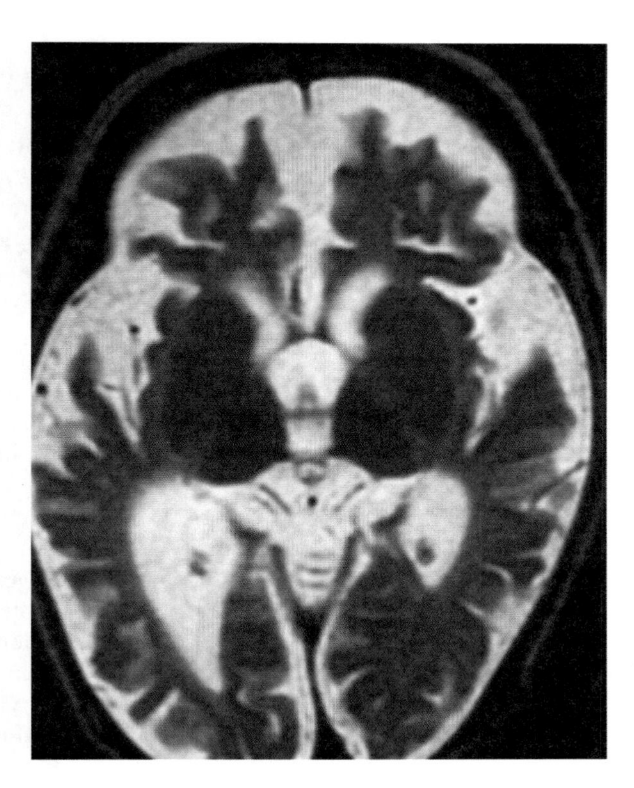

Fig. 13. Ornithine transcarbamylase deficiency (urea cycle defect). Axial T_2-weighted image shows the chronic phase with diffuse severe atrophy. In the presence of an acute metabolic derangement, swelling of the cortex and underlying white matter may be seen in this disorder.

ration of intellectual functioning and spasticity are followed by death in infancy or early childhood.

6.1.2.2. Pathologic Findings

In Alexander's disease, the brain is abnormally enlarged, and the white matter is jelly-like and collapsed. Histologic examination reveals extensive demyelination and rarefaction of the white matter. The salient feature of Alexander's disease is the accumulation of Rosenthal fibers that are found in the perivascular spaces, subpial regions, and subependymal white matter. The basal ganglia and cortex are usually relatively preserved. Cavitation is common, and there is no sparing of the subcortical U fibers. The cerebellum is less affected than in other leukodystrophies.

6.1.2.3. MRI Findings

MR findings usually demonstrate increased signal intensity on the T_2-weighted images in the frontal white matter with extension into the temporal and parietal white matter and external capsules. CT scan demonstrates low density in these regions (Fig. 15). The subcortical arcuate fibers may be involved, and the white matter may appear swollen *(9)*. Signal abnormality may also involve the basal ganglia and brainstem. The occipital white matter and cerebellum are usually spared. As the disease progresses, cavitation and atrophy of the white matter may be seen. In cases where there is diffuse white matter signal abnormality and swelling in a patient with macrocephaly, the differential diagnosis is mainly between Canavan's and Alexander's diseases. Brain biopsy is usually necessary to establish the diagnosis.

7. SUDANOPHILIC LEUKODYSTROPHIES

7.1. CLINICAL FEATURES

The sudanophilic leukodystrophies include several poorly defined diseases that are categorized based on the histopathologic findings of accumulation of sudanophilic droplets con-

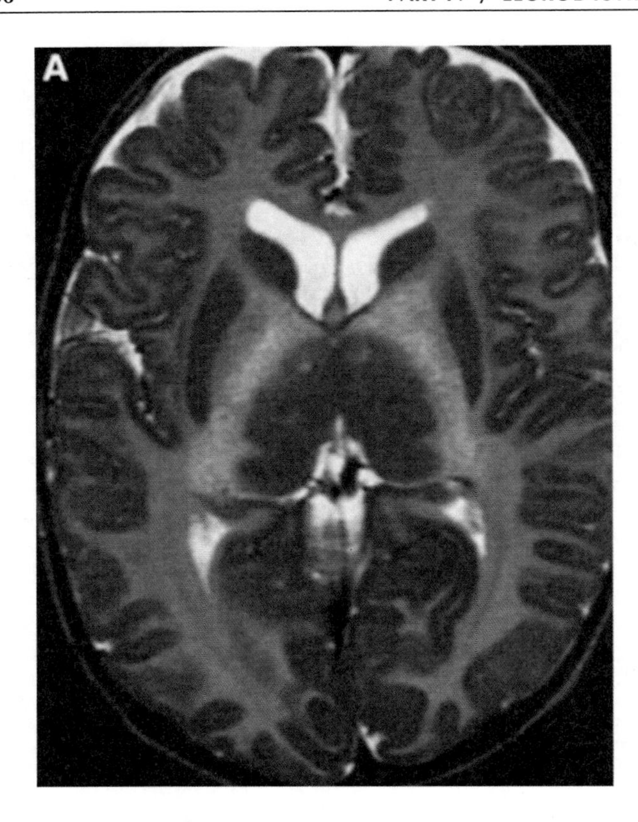

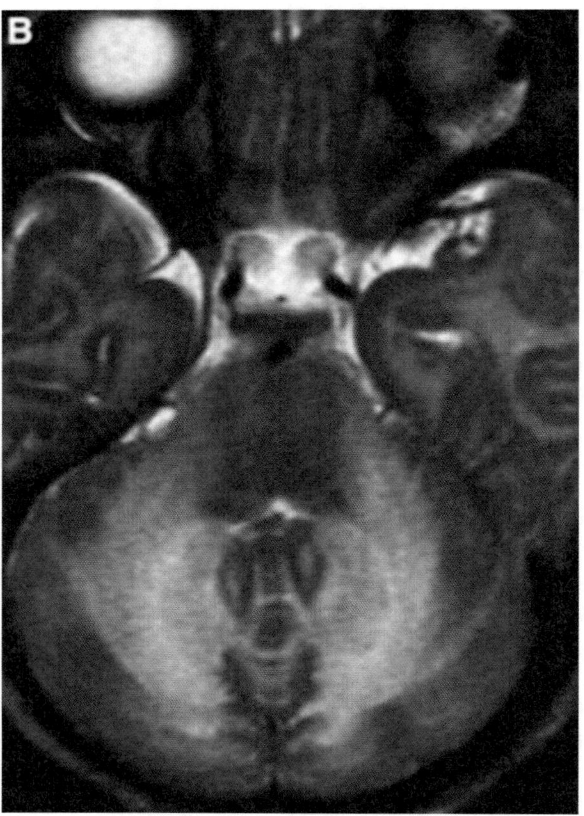

Fig. 14. Canavan's disease. T_2-weighted images in a child with macrocephaly show **(A)** near-complete high-signal intensity in supratentorial white matter, which appears swollen. Note high signal within the globus pallidi (almost always involved) and sparing of the corpus striatum. **(B)** Image of the infratentorium demonstrates high signal and swelling within the cerebellar white matter.

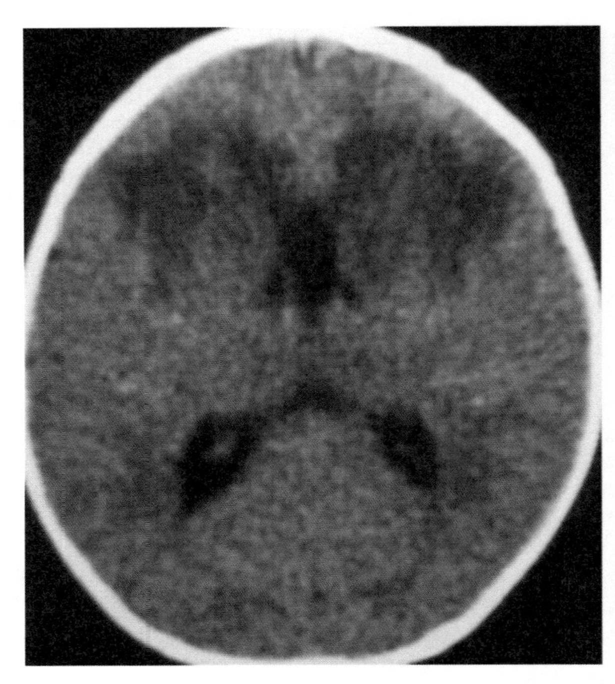

Fig. 15. Alexander's disease in a 2-mo-old infant with macrocephaly. Axial CT scan demonstrates bilateral symmetric low density within the frontal white matter, genu of the corpus callosum and basal ganglia. (Courtesy of Dr. Jill Hunter, Philadelphia, PA.) A brain biopsy was performed to confirm the diagnosis.

taining cholesterol and triglycerides in the white matter. Pelizaeus–Merzbacher disease and Cockayne's syndrome are the two well-documented sudanophilic leukodystrophies *(59–61)*. In Pelizaeus–Merzbacher disease, neurologic dysfunction usually is observed within the neonatal period; in Cockayne's syndrome, children may be normal until late in infancy. Typically, abnormal eye movements are observed, and patients exhibit head shaking, ataxia, and slow development. Progressive dysfunction evolves rapidly to spasticity and encephalopathy. Children with Pelizaeus–Merzbacher usually die in childhood, but those children with Cockayne's syndrome may live long enough to develop dwarfism.

Pelizaeus–Merzbacher disease is an X-linked disease affecting males; however, female patients are occasionally found. It results from a mutation of the proteolipid protein gene on Xq21.33-22 *(61)* that leads to abnormal proteolipid protein and DM20 proteins, which are the two most abundant proteins in the myelin sheath. The clinical presentation varies from mild to severe. The most consistent features include spasticity, a lack of evidence of male-to-male transmission, and diffuse white matter signal abnormality on MRI.

7.2. PATHOLOGIC FINDINGS

The brain usually is atrophic, which is particularly severe in the posterior fossa (cerebellum and brainstem). Affected infants who die during the first year of life may have a normal brain weight. The white matter appears gray and ranges from gelatinous to firm in consistency and the gray white junction is not well demarcated. Cerebellar cortical degeneration is found in most cases. Histologically, there is a profound lack of myelin.

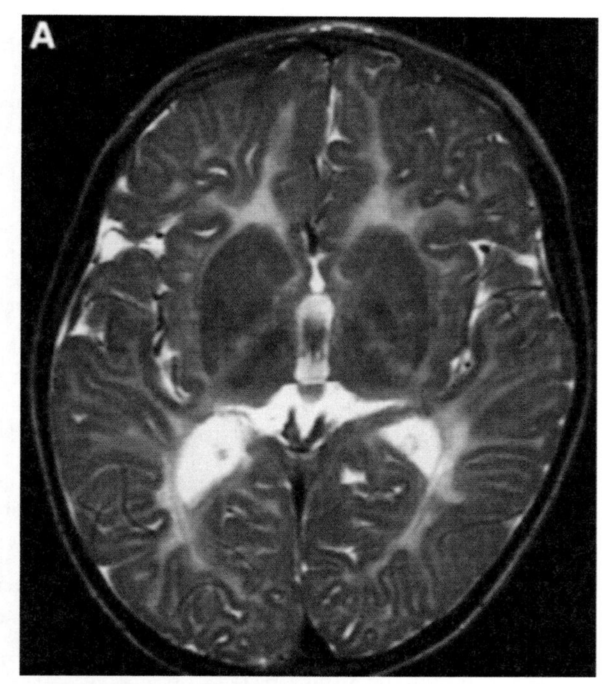

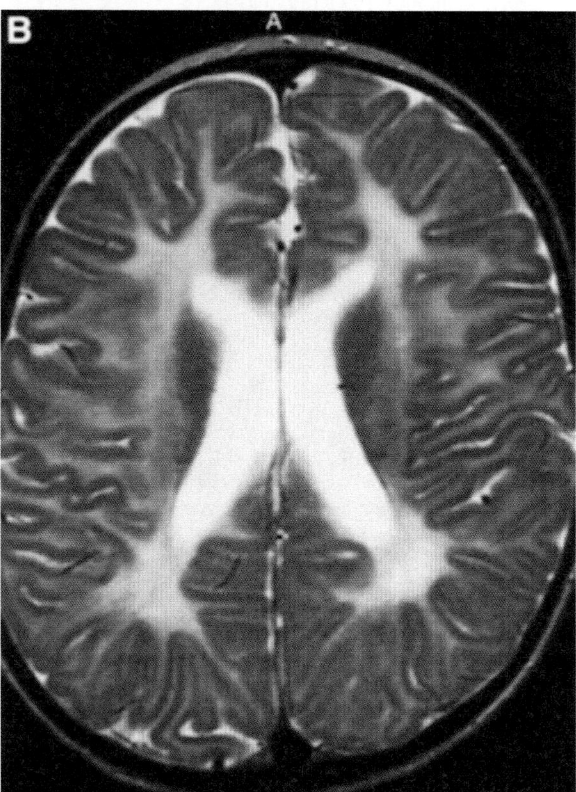

Fig. 16. Pelizaeus–Merzbacher disease. **(A,B)** Axial T_2-weighted images in a 3-yr-old child demonstrate that there is diffuse high signal intensity within the white matter, including the internal capsules, which should be myelinated at birth. Linear low signal intensity regions within the corona radiata characterizes the tigroid pattern. Low signal intensity in the deep gray matter and thalami is presumably due to abnormal iron deposition.

The residual myelin preferentially remains around blood vessels and demonstrates a "tigroid" pattern. There is astrocytosis and a lack of oligodendroglia. As in other types of myelin diseases, the axis cylinders tend to be preserved. There are rare macrophages that contain sudanophilic substances.

Cockayne's syndrome also is characterized by patchy preservation of myelin without sparing of the U fibers. There is granular mineralization of the capillaries, capillary neural parenchyma, cerebral cortex, and basal ganglia *(62)*.

7.3. MRI FINDINGS

The MRI scan typically demonstrates a delay in the pattern of myelination or an arrested pattern of myelination (Fig. 16A and B). In some instances, there is no evidence of myelination. Findings include diffuse white matter signal abnormality (low signal on T_1- and high signal on T_2-weighted images) and atrophy in the late phase of the disease. The tigroid appearance seen histologically is only rarely demonstrated on MRI. The appearance of both Pelizaeus–Merzbacher Disease and Cockayne's syndrome is similar in that the white matter is diffusely abnormal *(63)*. However, in Cockayne's syndrome, calcification may be seen in the basal ganglia and cerebellum on CT scan.

REFERENCES

1. Poser CM. Dysmyelination revisited. Arch Neurol 1978;35:401–408.
2. van der Knaap MS, Valk J. Non-leukodystrophic white matter changes in inherited disorders. Int J Neuroradiol 1995;1:56–66.
3. Menkes JH. Metabolic diseases of the central nervous system. In: Menkes JH, ed. Textbook of Child Neurology. 3rd ed. Philadelphia: Lea & Febiger; 1985:1–122.
4. Friede RI. Developmental Neuropathology. 2nd ed. Berlin: Springer-Verlag; 1989.
5. Mirowitz SA, Sartor K, Prensky AJ, Gado M, Hodges FJ III. Neurodegenerative diseases of childhood: MR and CT evaluation. J Comput Assist Tomogr 1991;15:210–222.
6. Dietrich RB, Vining EP, Taira RK, Hall TR, Phillipart M. Myelin disorders of childhood: correlation of MR findings and severity of neurological impairment. J Comput Assist Tomogr 1990t;14: 693–698.
7. Van der Knaap MS, van der Grond J, van Rijen PC, et al. Age dependent changes in localized proton and phosphorus spectroscopy of the brain. Radiology 1990;176: 509–515.
8. Tzika A, Ball WS Jr, Vigneron D, Dunn RS, Nelson SJ, Kirks D. Childhood adrenoleukodystrophy: assessment with proton MR spectroscopy. Radiology 1993;189:467–480.
9. Van der Knaap MS, Valk J. Magnetic Resonance of Myelin, Myelination, and Myelin Disorders. 2nd ed. Berlin: Springer-Verlag; 1995.
10. Kendall BE. Disorders of lysosomes, peroxisomes, and mitochondria. AJNR Am J Neuroradiol 1992;13:621–653.
11. Suzuki K. Genetic disorders of lipid, glycoprotein, and mucopolysaccharide metabolism. In: Siegel GJ, Agranoff BW, Albers RW, Molinoff PB, eds. Basic Neurochemistry. 5th ed. New York: Raven Press, Ltd.; 1994:793–812.
12. Gieselmann V, Polten A, Kreysing J, Kappler J, Fluharty A, von Figura K. Molecular genetics of metachromatic dystrophy. Dev Neurosci 1991;13:222–227.
13. Feigen I. Diffuse cerebral sclerosis (metachromatic leukoencephalpathy) Am J Pathol 1954;20:715.
14. Bubis, J.J.Adlesberg, L. Congenital metachromatic leukodystrophy. Report of a case. Acta Neuropathol (Berl) 1966;6:298–302.
15. Kolodny EH. Metachromatic leukodystrophy and multiple sulfatase deficiency: Sulfatide lipidosis. In: Scriver CR, Beaudet AL, Sly WS, et al., eds. The Metabolic Basis of Inherited Disease. 6th ed. New York: McGraw-Hill; 1989.
16. Kappler J, Sommerlade HJ, von Figura K, Gieselmann V. Complex arylsulfatase A alleles causing metachromatic leukodystrophy. Hum Mutat 1994;4:119–127.
17. Wolfe HJ, Pietra GG. The visceral lesions of metachromatic leukodystrophy. Am J Pathol 1064; 44:921–930.

18. Jatkewitz H., Mehl E.: Cerbroside-sulphatase and arylsulfatase A deficiency in metachromatic leukodystrophy (ML). J Neurchem. 1969;16:19–28.
19. Lake BD. Lysosomal and peroxisomal disorders. In: Graham DI, Lantos PL, eds. Greenfields's Neuropathology. 6th ed. United Kingdom: Arnold; 1997:657–753.
20. Schipper HI, Seidel D. Computed tomography in late-onset metachromatic leukodystrophy. Neuroradiology 1984;26:39–44.
21. Kim TS, Kim IO, Kim WS et al. MR of childhood metachromatic leukodystrophy. AJNR Am J Neuroradiol 1997;18:733–738.
22. Okada S, McCrea M, O'Brien JS. Sandhoff's disease (GM2-gangliosidosis type 2): clinical, chemical and enzyme studies in five patients. Pediatr Res 1972;6:606.
23. Mugikura S, Takahashi S, Higano S, Kurihara N, Kon K, Sakamoto K. MR findings in Tay-Sachs disease. J Comput Assist Tomogr 1996;20:551–555.
24. Fukumizu M, Yoshikawa H, Takashima S, Sakuragawa N, Kurokawa T. Tay-Sachs disease: progression of changes on neuroimaging in four cases. Neuroradiology 1992;34:483–486.
25. Chen CY, Zimmerman RA, Lee CC, Chen FH, Yuh YS, Hsiao HS. Neuroimaging findings in late infantile GM1 gangliosidosis. AJNR Am J Neuroradiol 1998;19:1628–1630.
26. Anderson W. A case of angiokeratoma. Br J Dermatol 1898;10:113.
27. Fabry J. Beitrag zur kenntnis de purpura hemorrhagica nodularis. Arch Dermatol Syph 1898;43:187.
28. Brady R, Gal AE, Bradley RM, Martensson E, Warshaw AL, Laster L. Enzymatic defect in Fabry disease: ceramide trihexosidase deficiency. N Engl J Med 1967;276:1163–1167.
29. Desnick R, Ioannou YA, Eng CM. α-galactosidase A deficiency: Fabry disease. In: Scriver CR, Beaudet AL, Sly WS, Valle D, eds. The Metabolic and Molecular Bases of Inherited Disease. 8th ed. New York: McGraw-Hill; 2001:3733 –3774
30. Mitsias P, Levine SR. Cerebrovascular complications of Fabry's disease. Ann Neurol 1996;40:8 –17.
31. Moore DF, Ye F, Schiffmann R, Butman JA. Increased signal intensity in the pulvinar on T1-weighted images: a pathognomonic MR imaging sign of Fabry disease. AJNR Am J Neuroradiol. 20031;24: 1096–1101.
32. Moore DF, Scott LJC, Gladwin MT, et al. Regional cerebral hyperperfusion and nitric oxide pathway dysregulation in Fabry disease: reversal by enzyme replacement therapy. Circulation 2001;104: 1506–1512.
33. Van der Knaap MS, Valk J. Globoid cell leukodystrophy: Krabbe's disease. In: Valk J, van der Knapp MS, eds. Magnetic Resonance of Myelin, Myelination, and Myelin Disorders. 2nd ed. Berlin: Springer-Verlag; 1995:68–80.
34. Suzuki K, Suzuki Y. Globoid cell leukodystrophy (Krabbe's disease): deficiency of galactocerebroside beta-galactosidase. Proc Natl Acad Sci USA 1970;66:302.
35. Zlotogora J, Chakraborty S, Knowlton R, Wenger D. Krabbe's disease locus mapped of chromosome 14 by genetic linkage. Am J Hum Genet 1990;47:37.
36. Kolodny EH, Raghavan S, Krivit W. Late-onset Krabbe disease (globoid cell leukodystrophy): clinical and biochemical features of 15 cases. Dev Neurosci. 1991;13:232–239.
37. Loes DL, Peters C, Krivit W. Globoid cell leukodystrophy: distinguishing early-onset form late-onset disease using a brain MR imaging scoring method. AJNR Am J Neuroradiol 1999;20:316–323.
38. Barone R, Bruhl K, Stoeter P, et al. Clinical and neuroradiological findings in classic infantile and late-onset globoid-cell leukodystrophy (Krabbe's disease). AM J Med Genet 1996;63:209–217.
39. Jones BV, Barron TF, Towfighi J. Optic nerve enlargement in Krabbe's disease. AJNR Am J Neuroradiol 1999;20:1228–1231.
40. Baker RH, Trautmann JC, Younge BR, Nelson KD, Zimmerman D. Late juvenile-onset Krabbe's disease. Opthalmology 1990;97: 1176–1180.
41. Feanny SJ, Chuang SH, Becker LE, Clarke TR. Intracranial periventricular hyperdensity: a new CT sign in Krabbe globoid cell leukodystrophy. J Inherit Metab Dis 1987;10:24.
42. Kolodny EH. Clinical and biochemical genetics of the lipidoses. Semin Hematol 1972;9:251–271.
43. Chang YC, Huang CC, Chen CY, Zimmerman RA. MRI in acute neuropathic Gaucher's disease. Neuroradiology. 2000;42:48–50
44. Van der Knaap MS, Valk J. Mucopolysaccharidoses. In: Magnetic Resonance of Myelin, Myelination, and Myelin Disorders. 2nd ed/ Berlin: Springer-Verlag; 1995:97–109.
45. Lee BCP. Magnetic resonance imaging of metabolic and primary white matter disorders in children. Clin Neuroimaging 1993;3:267.
46. Powers JM, Moser HW. Peroxisomal disorders: genotype, phenotype, major neuropathologic lesions, and pathogenesis. Brain Pathol 1998;8:101–120.
47. Moser HW, Moser AE, Singh I, et al. Adrenoleukodystrophy: survey of 303 cases: biochemistry, diagnosis, and therapy. Ann Neurol 1984;16: 628–641.
48. Hong-Mango ET, Muraki AS, Huttenlocher PR. Atypical CT scans in adrenoleukodystrophy. J Comput Assist Tomogr 1987;11: 333–336.
49. Tzika AA, Ball WS Jr, Vigneron DB, Dunn RS, Nelson SJ, Kirks DR. Childhood adrenoleukodystrophy: assessment with proton MR spectroscopy. Radiology 1993;189:467–480.
50. Yudkoff M. Disorders of amino acid metabolism. In: Siegel GJ, Agranoff BW, Albers RW, Molinoff PB, eds. Basic Neurochemistry. 5th ed. New York: Raven Press, Ltd.; 1994:813–839.
51. Matalon R, Kaul R, Michals K. Canavan disease: biochemical and molecular studies. J Inherited Metab Dis 1993;16:744–752.
52. Matalon R, Kaul R, Casanova J, et al. SSIEM Award. Aspartocyclase deficiency: the enzyme defect in Canavan disease. J Inherited Metab Dis 1989;12:329–331.
53. Grodd W, Krageloh-Mann I, Peterson D, Trefz FK, Harzer K. In vivo assessment of *N*-acetylaspartate in brain in spongy degeneration (Canavan's disease) by proton spectroscopy (letter). Lancet 1990;2:437–438.
54. Wang ZJ, Zimmerman RA. Proton MR spectroscopy of pediatric brain metabolic disorders. Neuroimaging Clin N Am 1998;8: 781–807.
55. Schuster V, Horwitz AE, Kreth HW. Alexander's disease: cranial MRI and ultrasound findings. Pediatr Radiol 1991;21:133–134.
56. Wohlwill FJ, Bernstein J, Yakovlev PI. Dysmyelinogenic leukodystrophy. J Neuropathol Exp Neurol 1959;18:359.
57. Neal JW, Cave EM, Singhrao SK et al.: Alexander's disease in infancy and childhood—a report of 2 cases. Acta Neuropathol 1992;84:322–327.
58. Pridmore CL, Baraitser M, Harding B et al: Alexander's disease: Clues to diagnosis. J Child Neurol 1993;8:134–144.
59. Seitelberger F. Pelizaeus–Merzbacher disease. In: Vonken PJ, Bruyn GW, eds. Handbook of Clinical Neurology. Amsterdam: Elsevier North-Holland; 1970.
60. Nance MA, Berry SA. Cockayne syndrome: review of 140 cases. Am J Med Genet 1992;42:68–84.
61. Griffiths IR, Montague P, Dickinson P. The proteolipid protein gene. Neuropathol Appl Neurobiol 1995 21:85–96.
62. Freide RI. Developmental Neuropathology. 2nd ed. Berlin: Springer Verlag; 1989.
63. Mirowitz SA, Sartor K, Prensky AJ, et al. Neurodegenerative diseases of childhood: MR and CT evaluation. J Comput Assist Tomogr 1991;15:210–222.

21 Advanced Magnetic Resonance Imaging in Leukodystrophies

Edwin Y. Wang, MD and Meng Law, MD, FRACR

SUMMARY

Advanced magnetic resonance (MR) techniques, namely MR spectroscopy and perfusion MR imaging, have provided the researcher with new tools for evaluating leukodystrophies. They have allowed for better characterization of diseases, improved sensitivity of detection, and a means of tracking brain metabolites and perfusion in vivo to monitor therapy or disease progression. In some cases, these techniques also allow for the specific diagnosis of certain abnormalities. Evaluation of leukodystrophies with MR spectroscopy has advanced from the simple observation of metabolic derangements and correlation to pathophysiology to the prognostication of patient clinical outcome and the in vivo tracking of metabolic abnormalities after therapeutic intervention. We aim to describe how these technologies can aid the clinician in the preventative, diagnostic, and treatment schemes in these diseases.

Key Words: Magnetic resonance spectroscopy; leukodystrophies; spectrum analysis; magnetic resonance imaging.

1. INTRODUCTION

During the past decade or so, clinical magnetic resonance imaging (MRI) has allowed for significant improvement in the evaluation of patients with generalized white matter disease, providing a visual anatomic correlate of disease involvement and, in some cases, allowing for more specific diagnosis or, at least, a narrowing of differential considerations.

However, it has become apparent that a nonspecific appearance of white matter signal abnormality can be seen on conventional MRI in many white matter diseases, both metabolic and nonmetabolic in origin. Advanced MR techniques, namely MR spectroscopy (MRS), perfusion MRI, and diffusion MRI, have provided researchers with new tools for evaluating these diseases. They have allowed for better characterization of diseases, improved sensitivity of detection, and a means of tracking brain metabolites and perfusion in vivo to monitor therapy or disease progression. In some cases, these techniques also allow for the specific diagnosis of certain abnormalities. Additional advantages of these techniques are the ability to perform these experiments during the same scanning session for routine, conventional MRI without ionizing radiation at minimal cost compared with other techniques such as ^{18}F-deoxyglucose positron emission tomography (PET), single-photon emission computed tomography (SPECT), and perfusion Xenon-CT scanning.

This chapter begins with a discussion of MR perfusion, with subsequent concentration on the role of MRS in the evaluation of white matter diseases and leukodystrophies, with some discussion of developments in this field. We aim to describe how these technologies can aid the clinician in the preventative, diagnostic, and treatment schemes in these diseases. For a more detailed discussion on related topics, the reader should also refer to other chapters in this text, particularly chapters on MRI of leukodystrophies and MRS.

2. PERFUSION IMAGING

2.1. PHYSIOLOGIC BASIS OF PERFUSION IMAGING

Cerebral hemodynamic parameters of clinical interest include cerebral blood flow (CBF), cerebral blood volume (CBV), and mean transit time (MTT). The CBV is the volume of blood in a given amount of brain tissue (units are usually mL/100 mL), whereas CBF reflects the volume of blood that flows through a given volume of tissue in a certain amount of time (mL/100 g/min). Often, CBV and CBF correlate to some degree. The parameter of CBV is often measured in MR perfusion imaging, as it is the most easily determined with routine dynamic susceptibility contrast-enhanced MRI (DSC-MRI; discussed in Section 2.2.). The equation MTT = CBV/CBF defines the relationship between these three parameters. MTT is a measure of the transit of blood through tissue, measured in seconds. Hence a prolongation of MTT indicates either ischemic tissue or increased collateral blood flow.

From: *Bioimaging in Neurodegeneration*
Edited by P. A. Broderick, D. N. Rahni, and E. H. Kolodny

2.2. TECHNICAL CONSIDERATIONS (FOCUS ON DSC-MRI)

The most common means of performing perfusion MRI involves the bolus injection of gadolinium with subsequent observation of its passage through cerebral circulation. In DSC-MRI, the $T_2/T_2{}^*$ effects of gadolinium are exploited. As a result of increased concentration of gadolinium and its relative compartmentalization within the vascular space, assuming an intact blood–brain barrier, the injection of contrast results in increased heterogeneity of the magnetic field and signal loss. Using rapid MRI techniques, the passage of contrast can be monitored after bolus intravenous administration. The generated curves are then analyzed mathematically to extract information about regional CBV, using a relationship between the measured signal drop and the amount of contrast agent that is used. Further evaluation and quantitation of absolute perfusion parameters requires the knowledge of an arterial input function.

Table 1 illustrates the routine MRI protocol with typical parameters used at our institution. In some situations, the blood–brain barrier is not preserved, which will lead to an inaccurate estimation of CBV because of the competing T_1 effects of gadolinium leakage. This can be dealt with by administering a small amount of contrast before perfusion imaging, as well as to use algorithms that take this into account. These calculations also allow for the determination of vascular permeability constants, which assesses for relative leakage of contrast or transfer of contrast between two compartments.

Another means of performing perfusion MRI that does not involve the use of an extrinsic contrast agent uses arterial spin labeling (ASL). The spins of flowing blood are labeled to allow for imaging of this flow *(1)*. At present, however, the signal-to-noise ratio (SNR) of ASL imaging remains fairly poor, and ASL studies are often lengthier in acquisition, which makes the degradation of data acquisition a problem with patient motion. Although these techniques are promising in the future evaluation of CBF, particularly at higher field strengths (3 Tesla and above), the techniques have not been as widely applied as DSC-MRI. Difficulties in transit time evaluation also render determination of MTT and CBV difficult with ASL.

One consideration in the performance of DSC-MRI is the selection of a gradient-echo vs spin-echo imaging protocol. Spin-echo echoplanar imaging technique has increased specificity for the microvasculature. This allows for finer distinctions in the amount of blood volume and may improve evaluation of angiogenesis in the setting of tumor. Some disadvantages include some decrement in the SNR and the need for increased amounts of contrast.

2.3. CLINICAL APPLICATIONS

Currently, experience in the application of perfusion-weighted imaging in the evaluation of leukodystrophies is limited. Moore et al. *(2)* describe the results of arterial-spin tagged perfusion experiments performed on three patients with Fabry disease, demonstrating increased CBF in the posterior circulation, which was theorized to result from hyperperfusion (Fig. 1). This correlated well with results from CBF studies performed with positron emission tomography *(2)* and may occur as a result of impaired vasoregulation or as a steal phenomenon, with blood shunted from areas of higher resistance in the middle cerebral artery (MCA) territory.

Evaluation of perfusion also can be helpful in disorders that involve an element of white matter dementia. Although these have been previously investigated with SPECT, MR perfusion would be another means to evaluate perfusion, also allowing routine MR imaging during the same patient visit. Fukutani et al. *(3)* discussed the use of ^{123}I- IMP SPECT in metachromatic leukodystrophy; they found the presence of diffuse cerebral hypoperfusion that was most notable in the frontal lobes. Other investigators *(4)* were able to document a progressive reduction in regional CBF in cerebral cortex in another patient with metachromatic leukodystrophy.

Correlation with FDG-PET metabolic imaging may be another facet in which MR perfusion imaging may have a role. Multiple authors have demonstrated the presence of glucose metabolic abnormalities in leukodystrophies. Salmon et al. *(5)* found a unique pattern of hypometabolism in the thalami, medial, and frontal cortex and in the occipital lobes in a patient with dementia caused by metachromatic leukodystrophy. There have been additional reports of hypometabolism in the frontal matter of a patient with juvenile Alexander disease corresponding to areas of MRI abnormality *(6)* and of decreased cortical uptake and absent uptake in the caudate nuclei in a patient with globoid cell leukodystrophy *(7)*.

Perfusion analysis also may be helpful in other white matter diseases where a significant vascular or inflammatory component to the pathology is known (e.g., cerebral autosomal-dominant arteriopathy with subcortical infarcts and leukoencephalopathy [CADASIL]). One field of possible exploration is to study whether changes in perfusion parameters can be detected in advance of conventional MRI or clinical abnormalities, much in the way metabolic changes can be identified with MRS. Analysis of vascular permeability constants may represent another variable that could be evaluated, perhaps to assess patients at risk of developing changes of cerebral edema.

3. MRS

3.1. AN INTRODUCTION TO MRS

MRS is a technique based on the concept of chemical shift. This refers to the change in the Larmor resonance frequency of a nucleus caused by its chemical environment. After the application of a radiofrequency pulse, the receiving coil detects voltage variation as a free induction decay signal. The time domain components of the free induction decay signal can be transformed by Fourier analysis to provide frequency domain data. After postprocessing, a spectrum is plotted demonstrating signal amplitude in arbitrary units in the vertical axis, measured against frequency (chemical shift) in parts per million along the horizontal axis. For purposes of consistency, metabolites are evaluated by their chemical shift from a standard reference point, measured in parts per million (ppm). Because of its abundance, increased MR sensitivity to protons in contrast to other nuclei and technical experience with proton NMR in general, ^1H proton spectroscopy is most commonly performed *(8)*. The standard zero reference point in proton MRS used is tetramethylsilane.

Other nuclei that are studied include ^{31}P and ^{13}C. ^{31}P MR spectroscopy allows for the study of energy metabolism, demonstrating metabolites such as adenosine triphosphate, phosphocreatine, and inorganic phosphate. Phosphodiester and

Table 1
MRI Protocol With Typical Parameters

Sequence	*TR (ms)*	*TE (ms)*	*Flip angle /TI*	*Acq /NEX*	*Thickness (mm)*	*No. slices*	*Matrix*	*FOV (mm)*	*Acq time (min.sec)*
Scout/localizer	15	6	NA	1	8	3	256	280	0.19
Axial T_1 (1)	600	14	90	2	5	20	256	210	3.36
Axial FLAIR	9000	110	180/2500	1	5	20	256	210	3.56
Axial T_2	3400	119	180	1	5	20	256	210	1.36
Diffusion/ADC	3400	95	NA	3	5	20	128	210	1.15
DTI*	4000	95	NA	4	5	20	128	210	1.56
DSC-MRI	**1000**	**54**	**30 degrees**	**60 (1/s)**	**3 to 8**	**10**	**128**	**210**	**1**
Post Gd T_1	600	14	90	1	5	20	256	210	3.36
MRS	**1500**	**144**	**90 2D CSI**	**3**	**10**	**1**	**16 × 16**	**160**	**6.05**
		30	**90 3D CSI**	**2**	**10**	**8**	**16 × 16**	**160**	**7.53**
MPRAGE*	1100	4.38	15	1	0.9	192	256	230	3.33

Total imaging time is approx 30 min. Sequences with the * are optional sequences. These parameters are manufacturer specific at 1.5 Tesla. TI denotes inversion time. DTI, diffusion tensor imaging; Gd, gadolinium contrast agent; FLAIR, fluid-attenuated inversion recovery; FOV, field of view; MRS, magnetic resonance spectroscopic imaging; MPRAGE, magnetization-prepared rapid gradient echo.

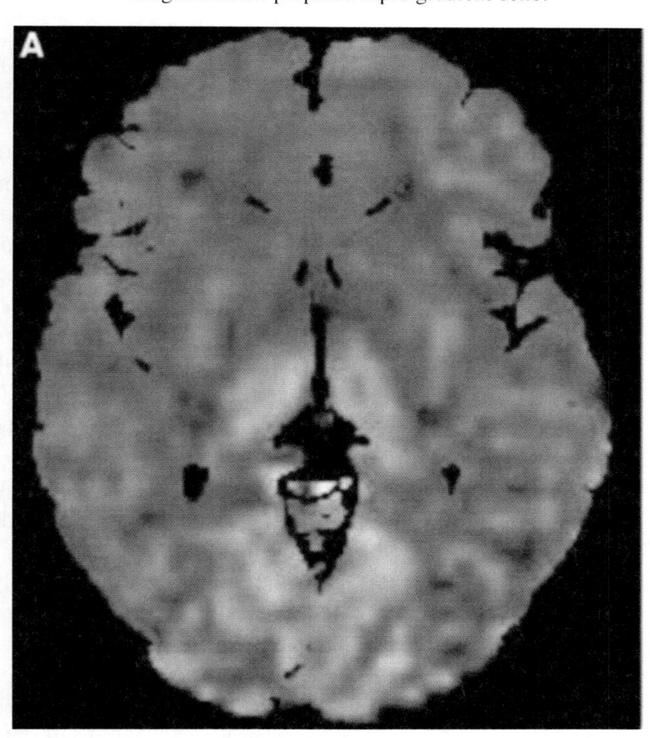

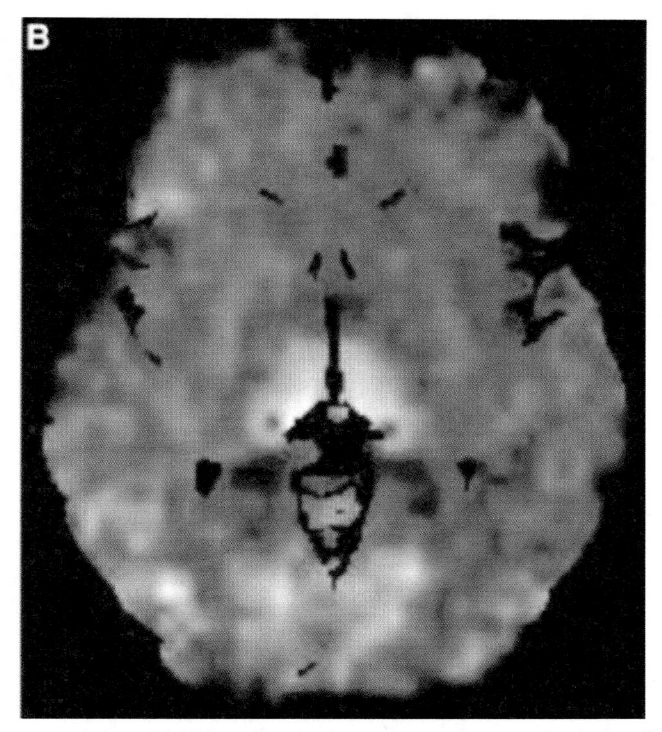

Fig. 1. Axial images from arterial-spin labeled perfusion experiment in two patients with Fabry disease demonstrates increased relative CBF in the posterior circulation and thalami. Used with permission from Moore DF, Frank Ye F, Schiffmann R, and Butman JA. Increased signal intensity in the pulvinar on T1-weighted images: a pathognomonic MR imaging sign of Fabry disease. Am J Neuroradiol 2003;24:1096–1101.

phosphomonoester, which are membrane constituents, can also be tracked *(9)*. Although used in stroke studies, ^{31}P evaluation also has become of interest in the setting of mitochondrial disease, given the inherent link with energy metabolism. A more detailed account of ^{31}P spectroscopy can be found in Chapter **23**. ^{13}C MR spectroscopy is technically challenging given the rarity of this nuclei and the subsequent low signal and requires the use of effective decoupling techniques for imaging. Some work has been performed recently looking at the use of monitoring ^{13}C-labeled glucose to evaluate its metabolism within brain *(10)*. The evaluation of nonproton spectra often involves the use of nuclear Overhauser effect (NOE) enhancement.

3.2. NEUROCHEMISTRY OF PROTON MRS

The following metabolites are those most often demonstrated on routine clinical proton MRS (Fig. 2).

3.2.1. *N*-Acetylaspartate (NAA) 2.02 ppm

NAA is the second most prevalent amino acid in the central nervous system (CNS) after glutamate. NAA is general considered to be a marker of neurons within the adult brain; a free amino acid, it is localized within neuronal tissue *(11)*, with recent investigations *(12)* demonstrating it to be axon-specific in white matter. Histopathologic correlation has demonstrated parallel decreases of NAA and axonal density in demyelinating plaques *(13)*. Gonen et al. demonstrated a means of following

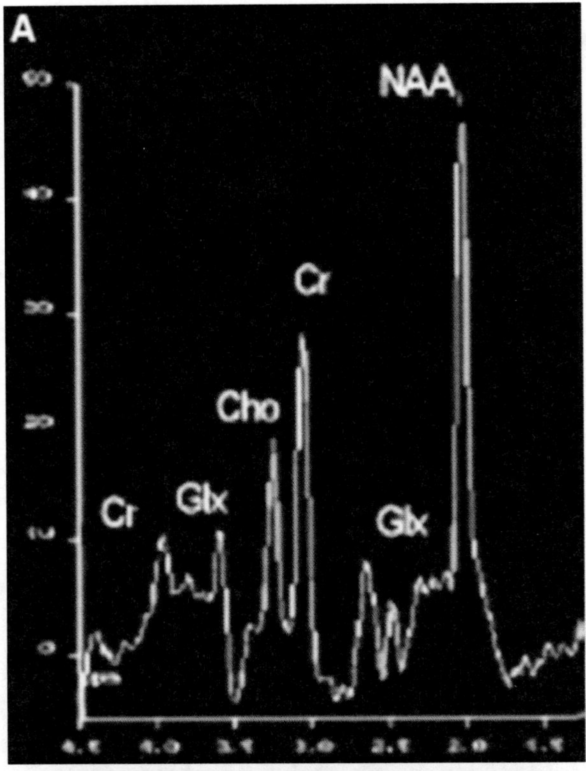

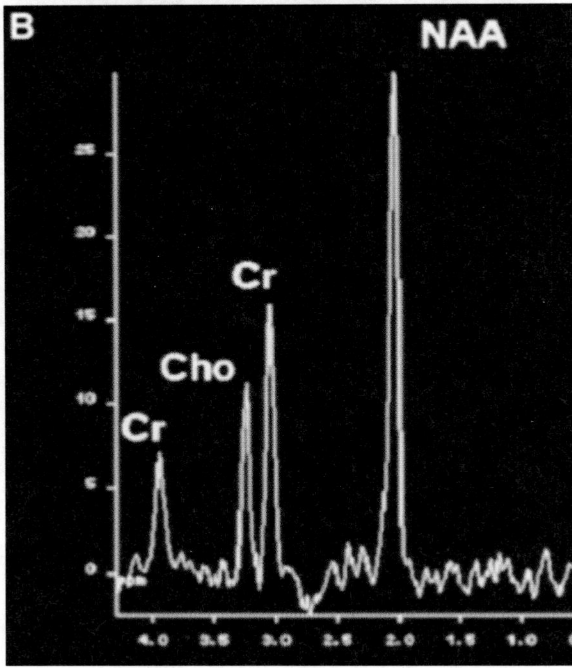

Fig. 2. Spectra acquired from white matter at **(A)** short echo time (TE = 30), demonstrating NAA, choline, creatine, as well as α-glutamine/glutamate, β/γ-glutamine/glutamate, and myo-inositol peaks and **(B)** long echo time (144 ms), demonstrating NAA, choline, and creatine.

total brain NAA as a means of tracking disease load in multiple sclerosis. This could represent an efficient and reproducible means of monitoring patients with leukodystrophy, where in many processes, NAA is seen to correlate with severity of disease *(14)*. However, NAA also has been identified in oligodendroglial precursors cells *(15)* in the developing brain as an acetyl-group source important in fatty acid synthesis *(16)*. Regional elevation of NAA in areas of active myelination during development may reflect this important role.

3.2.2. Glutamine/Glutamate 2.2–2.4 ppm

Glutamate is an excitatory neurotransmitter, precursor of γ-aminobutyric acid, and an intermediate of amino acid metabolism, whereas glutamine aids in the recycling of glutamate. Elevations in glutamine are seen in the setting of hepatic encephalopathy.

3.2.3. Creatine 3.02 ppm

Creatine is a measure of energy in brain tissue, serving as a buffer in the ATP-ADP reservoir and as a reserve of high-energy phosphates *(17,18)*. It is a fairly stable metabolite and is often used in metabolite ratios as an internal reference. However, creatine levels too can drop, often in the setting of severe brain tissue damage.

3.2.4. Choline 3.22 ppm

Choline is identified in elements needed for membrane synthesis and myelination, well as membrane breakdown, and is considered a metabolic marker of membrane turnover. The metabolism of choline is complex and beyond the scope of this text; however, the release of phosphorylcholine and glycerophosphorylcholine from cell membrane destruction and the synthesis of these metabolites during cell membrane synthesis are thought to be the biochemical basis for choline elevation. In proton spectroscopy, an elevation in choline may be the result of either cell membrane synthesis, destruction, or both. Despite a relatively low concentration relative to NAA and creatine within the CNS, signal is relatively high as a $N(Ch_3)^3$ group of choline with nine protons is evaluated *(8)*. Griffin et al. *(19)* reported elevations of phosphocholine on nuclear magnetic resonance spectroscopy performed during cell transfection, suggesting that increased choline seen in vivo on MRS studies is related to cell membrane disruption.

3.2.5. Myo-Inositol 3.56 ppm

Myo-inositol is seen to originate almost exclusively from glial cells *(20)* and is found in the setting of active myelin breakdown as well as in inactive demyelinated lesions. Its predominant function is as an osmolyte and is observedas being deranged in disorders of sodium regulation *(21)*. Largely, MRS myo-inositol signal is felt to be related to astrocytic cell density. Elevated myo-inositol is likely seen in the setting of myelin breakdown products as well as in gliosis or in gliomas.

3.2.6. Lactate 1.33 ppm

A prominent lactate doublet peak is not normally anticipated in normal brain MR spectra. A product of anaerobic metabolism, it is felt to represent a disorder in oxidative metabolism. In the context of leukodystrophies, it is most commonly seen in the setting of mitochondrial abnormalities. However, it also can be seen in other extramitochondrial diseases *(22)*, often in more severe cases, and often portending a poorer prognosis. It should be noted that lactate will invert at a TE of 135 or 144 ms, with an upright doublet at short TE (e.g., 30 ms) as well as higher TEs (e.g., 288 ms). In this way, MRS can be performed to help distinguish lactate signal from lipid signal. Research has been conducted into spectral editing sequences that will allow for more reliable quantification of lactate *(23)*. It also should be noted that some lactate signal can be seen in normal CSF and that points spread function factors (*see* Section 3.3.) may result in spurious lactate signal in voxels adjacent to CSF collections in multivoxel spectroscopy.

3.2.7. Macromolecules 0.5–2.0 ppm

Spectral peaks identified across the 0.5–2.0 ppm range are compatible with a number of species, including lactate (described above); varying portions of lipid, including methyl- and methylene- groups, proteins; and branched chained amino acids. Given substantial peak overlap and low signal, resolution between peaks often is poor. Spatial presaturation of fat-containing regions is of particular importance to allow reproducible and consistent detection of macromolecular resonances *(24)*. Improved detection of macromolecular peaks also can be achieved with metabolite-nulling sequences, although this can add to the required time of MRS evaluation *(25)*.

3.3. LOCALIZATION TECHNIQUES

MRS can be performed using a variety of sequences to allow for a variety of coverage. These include single voxel spectroscopic (SVS) and chemical shift imaging (CSI) protocol. Single voxel techniques most commonly used currently include stimulated echo acquisition mode (STEAM) and point-resolved spectroscopy (PRESS). In theory, PRESS acquisitions should have a SNR higher than that of STEAM and less sensitivity to diffusion. However, J-coupling effects are reduced on short TE STEAM compared with PRESS and allow for shorter TE after visualization of metabolites with short T_2 *(26)*. However, with modern MR scanners and software, PRESS sequences can be used to acquire MRS data at short TE. SVS spectra acquisition is simple; good shimming is easily achieved on a small voxel. Moreover, the reproducibility of position localization with SVS is thought to be superior.

CSI allows for 2D and 3D mutlivoxel spectra acquisition, which is also termed magnetic resonance spectroscopic imaging. Three-dimensional coverage also can be achieved with sequential multisection two-dimensional CSI as well as a Hadamard spectroscopic imaging/2D CSI hybrid *(27–29)*.

CSI imaging protocols have been helpful in increasing the efficiency of spectroscopic interrogation of white matter diseases. Considering the general diffuse nature of these disorders as well as the regional differences in distribution of pathology, two-dimensional and three-dimensional CSI are suited to allow metabolite mapping over a larger region of white matter. However, the greater a volume selected, the increased chance of contamination from adjoining calvarium or scalp soft tissue. This tends to be a greater difficulty in the setting of very peripheral lesions, lesions involving the brain adjacent to the skull base, and the cerebellum. CSI does allow for repositioning voxels after measurement. When more peripheral or focal lesions cannot be adequately addressed with a CSI protocol, the use of outer-volume lipid suppression pulses or multiple SVS sequences interrogating areas of suspected abnormality may be helpful. Obtaining good water suppression and shimming can be difficult over the large volumes used in CSI. CSI also is affected by the point-spread function, which can result in a type of voxel "bleed," with spectra from one voxel receiving contributions from neighboring spectra.

3.4. ADDITIONAL TECHNICAL FACTORS

3.4.1. Field Homogeneity and Water Saturation

MRS is technically challenging given its sensitivity to field homogeneity, as well as the low SNRs involved. Shimming is an important part of all MRS evaluations, although, currently, most manufacturers of MR scanners provide automated shimming in their software. Spielman et al. *(30)* described benefits of higher-order shimming, including increased volume of brain tissue that could be effectively shimmed on CSI studies, as well as improvements in field homogeneity with the brain.

One challenge of MRS is the elimination of water signal because its concentration is approx 1×10^5 times higher than that of other metabolites. This is traditionally reduced with initial chemical shift selective saturation pulses, to suppress water signal. However, these pulses can interfere with metabolite signal because of saturation and magnetization transfer effects. With the use of digital filters and an improved analog-to-digital converter, the performance of ^{1}H MRS without water suppression has been described, with effective editing of water signal on postprocessing *(31)*. This would allow for use of water signal as a reference standard for correction for magnetic field inhomogeneity and for signal variations across image repetitions.

Difficulties in quantification of metabolic information are complicated by a number of factors, including line-fitting algorithms, the use of area ratios vs absolute quantification, difference in metabolite relaxation, spin interaction, and overlap of resonant peaks. Peak parameters include frequency, height, and width at half-height. The width at half-height is proportional to $1/T_2$. MRS is prominently affected by inhomogeneity of the magnetic field, which increases peak width and reduces peak resolution.

3.4.2. Echo Time

Short echo time proton MRS allows for improved demonstration of peaks related to myo-inositol and glutamate/glutamine. It can be helpful in distinguishing lactate from lipid signal and in the evaluation of macromolecular resonances. However, short TE proton MRS can be complicated by more baseline variability, making data processing more difficult. In addition, issues of peak overlap often render quantification efforts impossible for some peaks. Long TE MRS allows for the reliable demonstration of NAA, creatine, choline, and lactate/lipid. At our institution, we routinely use a longer TE sequence unless metabolites identified with short TE imaging (α glutamine/glutamate, β/γ glutamine/glutamate, myo-inositol peaks) are of interest.

3.4.3. Contrast Administration

The effect of gadolinium administration on MRS is controversial. Although Sijens et al. *(32)* determined that spectral resolution and SNR remained stable after contrast administration, they noted a decrease of approx 15% of choline peak area at a TE of 135 ms in 17 patients with brain tumors studied with a CSI protocol. They theorized that T_2 shortening of an extracellular choline might be caused by gadolinium, suggesting decreased loss at short TE and worsened loss at higher TEs. Although possibly more of a difficulty in the MRS evaluation of neoplasm, this should be considered in the evaluation of patients after contrast-enhanced conventional MRI. By the same token there is also literature demonstrating that gadolinium has a negligible affect on metabolite ratios and peak areas *(33–35)*. At our institution, almost all of our MRS scans are performed after contrast administration.

3.5. PRACTICAL CONSIDERATIONS IN MRS ACQUISITION

Voxel sizes typically range from 1 to 8 cm^3; smaller voxels create less signal because of the decreased amount of tissue that is localized. This can increase the number of averages required in spectra acquisition *(17)*, thereby increasing the scan time. Voxels should be placed with care to avoid skull and scalp to avoid lipid contamination. In addition, hemorrhagic, melanotic, or calcified lesions also should be avoided. Hunter and Wang *(36)* make mention of the issue of dental braces, which is encountered with relatively frequency in the pediatric population often scanned for leukodystrophy. In their experience, occipital lobe spectra could still be obtained despite prominent resulting field inhomogeneity.

Ross and Bluml *(37)* recommend the placement of single voxels for gray matter evaluation across the falx above the posterior commissures of the corpus callosum and recommend parietal white matter for white matter evaluations. In the setting of suspected leukodystrophy, evaluation of basal ganglia, normal-appearing white matter, and areas of signal abnormality are recommended.

The processing of MRS data is now fairly easily performed in the clinical setting without the need of a dedicated physicist or spectroscopist. Many manufacturer-based systems will allow for improvements in shimming and automatic postprocessing.

This account was intended to provide a brief synopsis of some of the technical issues involved in MR spectroscopy. A more detailed account of the technical aspects MR spectroscopy can be found in Salibi and Brown *(26)*.

4. CLINICAL APPLICATIONS OF MRS

4.1. MRS IN THE DEVELOPING BRAIN

Aside from differences of metabolite levels in different regions of the brain (decreased NAA in white matter and lower creatine in white matter *(38)*, the evaluation of proton MR spectra in the pediatric patient is complicated by known developmental changes in metabolites. As such, quantified changes in spectra should be compared with known age-matched and regional controls to avoid misinterpretation of what may be normal changes in metabolite content. Reference to published quantitative data of normal individuals may be helpful in these regards *(18,39)*.

Regional NAA increases occur in concordance with neuronal density and myelination, which are seen in higher amounts in the thalami early, with subsequent increase in the occipitoparietal and periventricular white matter afterward *(40)*. Newborn NAA is observed to be almost half that of adult values, with subsequent increase to adult levels. As a marker of cell membrane turnover and lipid metabolism, increased choline and myoinositol is noted in early brain development, likely relating to ongoing myelination and membrane formation. At birth, there are relatively high levels of choline. A gradual decrease in choline is noted during the first 2 yr of life until age 2, when myelination is complete *(41)*. By 2 yr of age, NAA and choline levels should be similar to adult levels. Creatine, as a marker of cellular energy metabolism, has been found to be approximately constant with age. Interestingly, Kreis et al. *(18)* found no difference in the metabolite content in premature vs term infants.

4.2. ^{1}H MRS EVALUATION OF SPECIFIC ENTITIES

Since Grodd et al. *(42)* described the clinical use of ^{1}H-MR spectroscopy in the evaluation of patients with Canavan's disease, investigators have demonstrated the utility of ^{1}H-MRS in patients with inherited metabolic disease *(43–45)*. There have been many publications discussing the use of MRS in metabolic diseases and leukodystrophies. The following is an account of the utility of MR proton spectroscopy in particular, although a synopsis can be found in Table 2 of some of the more general findings in some particular disorders. Some helpful features of MRS the evaluation of specific diagnoses are listed in Table 3. Typically derangements of ^{1}H spectra in patients with white matter disease fall in two groups: those that demonstrate peak derangements related to the specific metabolic abnormality related to the disease or those derangements in more commonly noted peaks that reflect underlying neuropathophysiologic processes. The evaluation of leukodystrophies with MRS has advanced from the simple observation of metabolic derangements and correlation to pathophysiology, to the prognostication of patient clinical outcome and the in vivo tracking of metabolic abnormalities after therapeutic intervention.

The role of MRS in leukodystrophy is well described by Wang and Zimmerman et al., and interested readers should also refer to these references *(46,47)*. In contrast to the organelle-based approach used in these works, we have chosen to categorize the disorders studied in terms of the mechanism of disease proposed by Kolodny (see Chapter 18; ref. *[48]*).

A current issue in the research and clinical realm of spectroscopy is that there is no standardized protocol for acquisition of MRS data, and there also is variation in methods of quantification. This makes for some difficulty in comparing studies, particularly if it involves trials with new therapies. In fact, one may argue that even with the same protocols, metabolite concentrations can be affected by different MR imagers, localization method differences, gain instabilities, regional susceptibility variations, and partial volume effects *(49)*.

4.2.1. Hypomyelinating Disorders

4.2.1.1. Pelizaeus–Merzbacher Disease (PMD)

PMD is a rare disorder of dysmyelination of X-linked recessive inheritance with involvement of the proteolipid protein (PLP) gene. This gene encodes for major structural myelin proteins PLP and DM20. The disease typically presents with mental retardation, choreoathetosis, seizures, nystagmus, and spasticity. MRI generally demonstrates white matter T_2 signal abnormality with variable involvement of hemispheric white matter and the corticospinal tracts, with description of a tigroid appearance of white matter. MRS evaluations have demonstrated variable findings, including normal spectra, variably decreased NAA, increased and decreased choline, and increased myo-inositol (Fig. 3). On pathology, there are areas of demyelination in white matter essentially devoid of myelin *(50)*. It is thought that the presence of normal spectra may correlate to pathologic findings of well-preserved axons and

Table 2
MRS Findings in Selected Leukodystrophies

Disorder (References)	*Metabolites* *NAA*	*Cho*	*Lac*	*Lip*	*Cre*	*mI*	*Glx*	*Additional peaks*
Adrenoleukodystrophy (106,109–113)	**D**	**I**	I			I		
Alexander's disease (115–117)	**D**	I	**I**			I		
CADASIL (145)	D	D			D			
Canavan's disease (42,120,121)	**I**	**D**	I		D			
CTX (142)	D		I					
Galactosemia (146–148)		D		D				3.67/3.74 ppm during exacerbation
GA II (78,79)		**I**	I					
HMG CoA lyase deficiency (70,71)	**D**	**I**	I			**I**		2.42 ppm
Krabbe disease (101,102)	**D**	**I**	I	I	I	**I**		
MSUD (62,63)	D		I					0.9 ppm during exacerbation
MLD (104)	D		I			I		
MMA (65,66)	**D**		**I**					
Mitochondrial abnormalities (84–87,91,93,94,96,97)	**D**	I	**I**		I			2.4 ppm in complex II deficiency
MPS (135,136)	**D**	I						broad-based peak at 3.7 ppm
Neuronal ceroid lipofuscinosis (130,131)	**D**	D	I		D	**I**	I	
Niemann–Pick disease (129)	D	I						
Nonketotic hyperglycinemia (67–69)	D				I		I	3.56 ppm
PMD (51–53)	D					I		
PKU (55,60)								7.37 ppm
Propionic acidemia (82,83)	**D**					D	I	
Sjogren–Larsson syndrome (140,141)	D							lipid peak at 0.9 ppm
VML (126,129)	**D**	D	I		D	D		
Zellweger disease (45,114)	**D**		I			I		lipid peak at 0.9 ppm

Letters in boldface represent the most likely finding.
Cho, choline; Cre, creatine; D, decreased; Glx, glutamine/glutamate; I, increased; mI, myo-inositol.

Table 3
Utility of MRS in Selected White Matter Disorders

Disorder	*Utility of MRS*
PKU	Direct quantification of brain phenylalanine concentration
	Allows tracking of dietary therapy efficacy
	Specific peak on short TE MRS at 7.36 ppm
MSUD	Specific peak identified during decompensation at 0.9–1.0 ppm
	Allows tracking of therapy
MMA	Tracking of NAA loss and lactate elevation with L- carnitine therapy
Nonketotic hyperglycinemia Tr	acking of glycine-associated peak at 3.56 ppm with therapy
HMG-CoA lyase deficiency	Specific peak at 2.42 ppm
GA II	Increased lactate indicates worse prognosis
Mitochondrial disorders	Lactate can be followed to monitor therapy
	Lactate can be seen in normal-appearing white matter during acute exacerbations
	Specific peak at 2.8 ppm in complex II deficiency
	Elevated alanine seen in pyruvate dehydrogenase complex deficiency
ALD	Early choline elevation may predict CNS involvement; this may allow for stratification in selecting
BMT candidates	
Canavan disease	Specific NAA elevation
	NAA can be used to monitor novel gene therapies
MPS	Specific glycosaminoglycan peak in some forms at 3.7 ppm

absence of sclerosis in uninvolved areas; in one group of three patients with PMD without neurologic regression during a 5-yr period, no metabolic abnormalities were identified on proton MRS *(51)*. Similarly, Takanashi et al. *(52)* noted normal metabolite ratios in two patients with genetically defined PMD despite the presence of extensive white matter T_2-signal abnormality. TEs used in these studies were 270 ms and 136 ms, respectively. More variable findings were obtained in a group of four patients with PMD, including decreased NAA in two patients, increased myo-inositol in one patient, and variable choline (both increased and decreased) in one *(53)*. The presence of fairly normal metabolic spectra in the setting of extensive white matter signal abnormality may be a diagnostic clue to the presence of PMD. An evaluation of patients of Pelizaeus-Merzbacher-like disease, a disorder without defects of the PLP gene, also demonstrated variable findings on MRS, including decreased choline, increased myo-inositol, decreased NAA, and the presence of a prominent peak at 3.35 ppm of uncertain

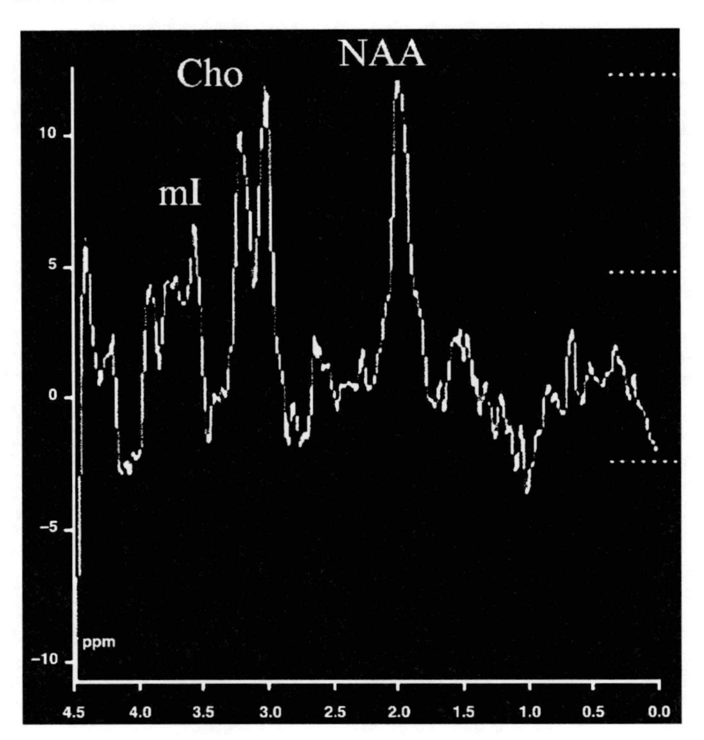

Fig. 3 Proton MRS (TE = 30 ms) in PMD demonstrating decreased NAA, increased choline, and increased myo-inositol.

significance *(53)*. More work is needed to evaluate the significance of these varied findings.

4.2.2. Delayed Myelination

4.2.2.1. Amino Acidopathies

4.2.2.1.1. PHENYLKETONURIA (PKU)

PKU is a disorder of autosomal-recessive inheritance that results in the deficiency of phenylalanine hydroxylase. Although it can be controlled with a phenylalanine-restricted diet, untreated PKU results in abnormal brain development, retardation, behavioral abnormalities, and epilepsy. MRS has been used extensively to evaluate patients with PKU. Some initial results at routine frequencies of chemical shift revealed no abnormality, but modified sequences and the use of subtraction spectra have allowed for the identification of abnormally increased peaks at 7.36 ppm, from phenylalanine, on short TE MRS. Subsequently, MRS has been used to quantify brain phenylalanine concentrations using a variety of models *(54,55)*. Review of the literature suggests that the correlation between brain and serum phenylalanine concentrations is poor and that it is likely that brain phenylalanine, and not plasma phenylalanine, that serves as a better predictor of clinical outcome of these patients. Koch et al. *(56)* described concordance of brain Phe concentration determined by MRS and the degree of abnormality identified on MRI. Weglage et al. *(57)* reported nondetectable brain phenylalanine concentrations in untreated patients with normal intelligence despite comparable serum levels when compared with retarded patients. A group of patients with untreated PKU with normal intelligence was noted who demonstrated prominently increased plasma phenylalanine concentration but nonelevated brain phenylalanine concentrations compared with elevated brain Phe detected in untreated retarded patients and early treated patients with near-normal intelligence *(58)*. These authors argued that MRS can help to establish individual treatment strategies by taking into account interindividual variations in Phe uptake and metabolism. Leuzzi et al. *(59)* studied brain and serum Phe concentrations of patients with varying clinical status before and after dietary therapy discontinuation. Although in most patients there was some correlation of brain and plasma Phe, one individual with moderate retardation and epilepsy demonstrated low plasma Phe and the highest brain Phe in their sample. A patient with normal development after dietary therapy demonstrated high plasma Phe and a low Phe/creatine on MRS. The nonscreened patients evaluated in this study demonstrated clinical prognosis in good correlation to the concentration of brain Phe. In a study of eight patients, normal metabolites on MRS were noted aside from Phe *(60)*.

4.2.2.1.2. MAPLE SYRUP URINE DISEASE (MSUD)

In MSUD there is a disorder of the branched-chain 2-oxoacid dehydrogenase complex, causing accumulation of branched-chain amino acids and related oxoacids in the body. Although dietary changes improve long-term outcome, metabolic decompensation still occurs *(61)*. Two studies evaluating spectroscopic findings during MSUD metabolic decompensation identified peaks at 0.9–1.0 ppm, which were assigned to branched-chain amino acids (BCAA) and branched-chain 2-oxoacids (BCOA) on the basis of in vitro MRS on these substances *(62,63)*. In one case, a note was also made of NAA decrement and lactate elevation, in a patient that had progressed to coma *(62)*. These metabolic abnormalities as well as MRI abnormalities were seen to resolve over a course of 14 days as the patient recovered.

4.2.2.1.3. METHYLMALONIC ACIDEMIA (MMA)

MMA is a disorder resulting from abnormalities of methylmalonyl CoA mutase or an associated coenzyme, adenosyl cobalomine. This complex normally aids in the conversion of methylmalonic acid to succinic acid. High levels of methylmalonic acid result in inhibition of succinate hydrogenase, which are required for mitochondrial glucose oxidation *(64)*. Corresponding, reports of MRS studies on patients with MMA demonstrate decreased NAA and increased lactate without additional abnormality. Specifically, Takeuchi et al. *(65)* discuss a case report where a patient with MMA demonstrated NAA loss and lactate elevation within the basal ganglia (Fig. 4). After L-carnitine therapy, both these metabolic abnormalities improved, in conjunction with improved clinical outcome. They report that this suggests some degree of reversibility in damage sustained in the early stages of MMA. Another report *(66)* found a varied presentation, demonstrating normal brain spectra in one patient with elevated lactate in CSF spaces, and another report demonstrated similar spectral derangement as noted by Takeuchi et al.

4.2.2.1.4. NONKETOTIC HYPERGLYCINEMIA (NKH)

NKH is a disorder of autosomal-recessive inheritance that results from a deficiency of the glycine cleavage system. This deficiency results in glycine accumulation in all tissues, with resulting hypotonia, myoclonus, lethargy, mental retardation, and apnea. One form of treatment consists of sodium benzoate (which conjugates with Gly to form hippurate, allowing urinary excretion) and dextromethorphan administration, as well as

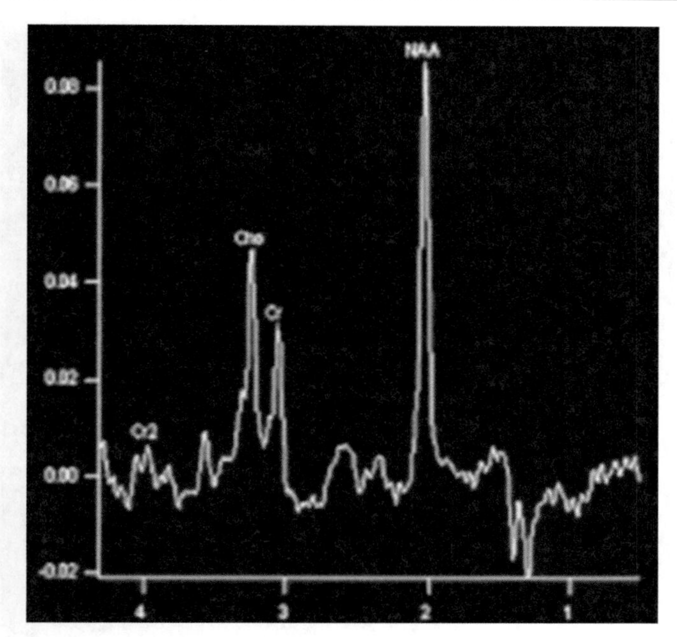

Fig. 4. Proton MRS (TE = 144 ms) in MMA demonstrating increased lactate at 1.33 ppm. See color version on Companion CD.

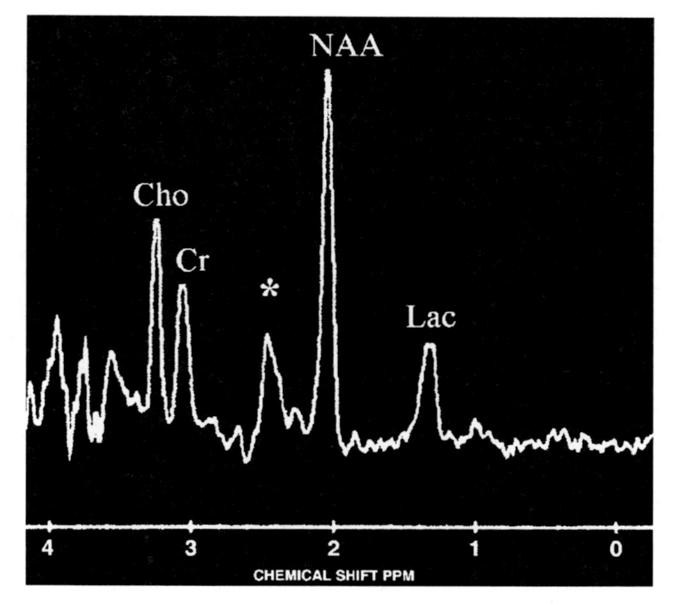

Fig. 5. Long TE MRS in HMG CoA Lyase deficiency. Note is made of an anomalous peak at 2.4 ppm (asterisk), felt to represent 3-OH-isovalerate and/or 3-OH-3-methyglutarate. Cho, choline; Cr, creatine, NAA, *N*-acetyl aspartate; Lac, lactate. Reproduced with permission from Yalcinkaya C, Dincer A, Gunduz E, Ficicioglu C, Kocer N, Aydin A. MRI and MRS in HMG-CoA lyase deficiency. Pediatr Neurol 1999;20:375–380. Copyright 1999, with permission from Elsevier.

dietary restriction of glycine. However, the outcome of these patients remains poor. Both short TE and long TE MRS demonstrate elevation in spectral peaks associated with glycine *(67)*. A decrease in glycine/inositol-associated peak on short TE MRS has been described after treatment, as well as variable NAA loss, which was observed in a patient with more severe presentation *(68)*. A decreased NAA/inositol/glycine ratio may imply a poor prognosis. On short TE MRS, note was also made of increased creatine after therapy, which was felt to result from its production on glycine conversion. Heindel et al. *(69)* documented changes in this glycine associated peak (seen at 3.56 ppm) using long TE MRS to decrease myo-inositol-related signal. It was felt that glycine levels as determined from MRS corresponded more reliably to clinical outcome than did plasma and CSF values; in this study, MRS was also used to track glycine levels after the administration of sodium benzoate therapy.

4.2.2.1.5. 3-Hydroxy-3-Methylglutaryl-Coenzyme A (HMG-CoA) Lyase Deficiency

HMG CoA lyase deficiency is a disease of leucine degradation characterized by episodic hypoglycemia, metabolic acidosis, and hyperammonemia without ketosis, resulting in lethargy, hepatomegaly, and seizures. These crises often occur in the neonatal period. MRS investigation of these patients has found decreased NAA, increased choline, and increased lactate, with variable increases in myo-inositol that was more often seen in more severe cases *(70)*. Yalcinkaya et al. *(71)* also discussed the presence of a peak at 2.42 ppm on long and short TE spectra, which is felt to possibly represent 3-hydroxy-isovalerate and/or 3-hydroxy-3-methylglutarate, thought to be diagnostic for HMG-CoA lyase deficiency on urine spectra (Fig. 5; ref. *72*). Interestingly, this peak is also noted retrospectively in a figure detailing a spectrum from a patient with HMG CoA lyase deficiency in another report *(70)*.

4.2.2.2. Other Disorders

A decrease in NAA, creatine, and choline with elevation of myo-inositol in white matter was described in a case of L-2-hydroxyglutaric acidemia, with worsening of NAA decrement mirroring changes on MRI of volume loss *(73)*. A prominent decrease in nearly all metabolites with prominent lactate and lipid signal was seen in a patient with molybdenum cofactor deficiency, although voxel acquisition was acquired in a largely cystic lesion *(74)*. Engelbrecht et al. describe decreased NAA in a patient with methylenetetrahydrofolate reductase deficiency (a disorder of methionine synthesis) and the presence of lipid peak *(75)*. In a case of infantile cobalamin deficiency, a note was made of decreased NAA (in gray matter), decreased choline, and decreased myo-inositol and the presence of lactate; after treatment, these abnormalities improved upon follow-up *(76)*.

4.2.2.3. Organic Acidopathies

4.2.2.3.1 Glutaric Acidemia type II (GA II)

GA II is a disorder of fatty acid and amino acid metabolism, a consequence of defective electron transfer flavoprotein or electron transfer flavoprotein:ubiquinone oxidoreductase *(77)*. The most severe form of the disorder presents in neonates and infants with hypotonia, metabolic acidosis, and hypoglycemia, in addition to other varied congenital anomalies, with death typically occurring in the first week. Currently diagnosis is often made with urinary organic aid screening and L-carnitine loading tests. Other patients present with a less severe clinical phenotype and even survive into adulthood with varying degrees of myopathy or movement disorder. Pathologic analysis demonstrates gliosis and neuronal loss that is most notable in the basal ganglia. Reports of MRS performed on two separate patients with GA II demonstrated prominent choline elevation on long TE MRS. Of note, one patient demonstrated elevated lactate; this patient presented with more rapid deterioration and died at 4 mo of age in contrast to the other case report of a 3-yr-old child demonstrating clinical improvement after dietary therapy *(78)*. As such, increased lactate in the setting of glutaric acidemia type II may reflect a greater degree of metabolic decompensation and portend a worse prognosis. In addition, Shevell et al. *(79)* described prominent abnormality of ^{31}P MRS with low energy state of phosphate-containing metabolites. It was felt that the combination of proton and phosphorous spectroscopic abnormalities were compatible with combined abnormalities of oxidative metabolism with functional abnormalities of myelination.

4.2.2.3.2, PROPIONIC ACIDEMIA

Proprionic acidemia results from an inborn deficiency of propionyl-CoA carboxylase, resulting in accumulation of organic acids. This results in metabolic acidosis, hyperammonemia, and hyperglycinuria and hyperglycinemia. Clinically, patients present with hypotonia, convulsions, hepatomegaly, respiratory, and feeding difficulties *(80)*. Imaging demonstrates volume loss and lesions within the basal ganglia, which can be seen to be transient after dietary therapy *(81)*. On short TE imaging, a note was made of increased glutamine/glutamate at the basal ganglia in a group of three patients, which was theorized to occur from hyperammonemia with stimulation of glutamine synthesis and inhibition of glutamate re-uptake *(82)*. This may relate to increased excitotoxic damage to the basal ganglia. Decreased NAA and myo-inositol also were noted (Fig. 6). Overall, these changes were not seen to clearly correlate to patient clinical presentation, although the two patients without choreoathetosis were observed to be spared of basal ganglia lesions on MRI. On long TE MRS, lactate was noted in periventricular white matter, even in a patients with properly treated propionic acidemia and minimal abnormalities on routine MR imaging *(83)*. It was felt that this reflected the persistence of a mitochondrial energy metabolic defect despite treatment.

4.2.2.4. Fatty Acid Disorders

4.2.2.4.1. MITOCHONDRIAL DISORDERS

Disorders of mitochondrial metabolism represent a diverse group of disorders that largely present with lactic acidemia with varying degrees of encephalopathy and myopathy. This includes disease such as pyruvate dehydrogenase complex deficiency, respiratory chain complex enzyme deficiencies, Leigh disease, myoclonic epilepsy with ragged red fiber (MERRF), and mitochondrial myopathy, encephalopathy, lactic acidosis, and stroke-like episodes . Several studies have been written regarding the findings on MRS *(84–89)*. The predominant abnormality noted is increased lactate (Fig. 7), which has been shown to correlate to levels of CSF lactate in some patients, although elevations in lactate are seen in many extra-mitochondrial neurodegenerative disorders *(22)*. This is presumably the result of abnormal oxidative phosphorylation related to the mitochondria, resulting in increased reliance on the glycolytic pathway, in which lactate is produced.

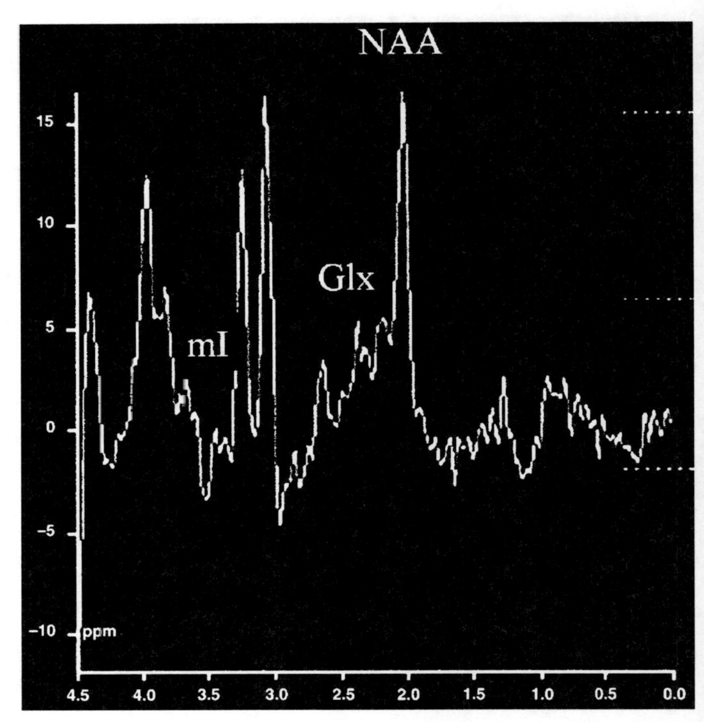

Fig. 6. Proton MRS in propionic acidemia (TE = 30 ms) demonstrating elevation in the glutamine/glutamate complexes (2.05–2.5 ppm range) as well as decrease in NAA and decrease in myo-inositol (3.6 ppm).

Krageloh–Mann et al. *(90)* found in a series of patients with Leigh's disease that lactate elevation on MRS could also be elevated in the absence of peripheral lactate elevation. Multiple studies describe improvement in abnormal lactate signal with treatment, potentially allowing for treatment monitoring. Specifically, Moller et al. *(91)* described significant improvement in lactate and NAA peaks in a patient after creatine treatment, although metabolite peaks did not reach control levels. The authors raise the possibility that this transient NAA recovery may relate to an improvement in decreased NAA production because of abnormalities in the respiratory chain. In a group of nine patients, lactate elevation was not correlated with the degree of NAA loss noted *(92)*. In a case report of Leigh syndrome, NAA remained decreased after sodium dichloroacetate therapy despite a decrease in lactate, and the patient experienced permanent severe psychomotor retardation *(93)*. Lin et al. *(94)* describe some of the difficulties in using lactate in the evaluation of mitochondrial disorders, focusing on the placement of a region of interest and describing cases where abnormal lactate was missed secondary to poor voxel placement. As such, the use of spectroscopic imaging protocols to maximize brain coverage is recommended. It also was found that abnormal lactate identified 63% of patients with diagnoses established by other means in their group of 29 patients and that MRS may be best used as a confirmatory test. The presence of lactate has been noted in normal-appearing white matter in patients experiencing a clinical relapse, whereas the same finding could not be reproduced during quiescence *(95)*. One unique finding is noted in complex II deficiency; Brockmann et al. noted an elevated peak at 2.8 ppm, compatible with succinate (Fig. 8; ref. *96*).

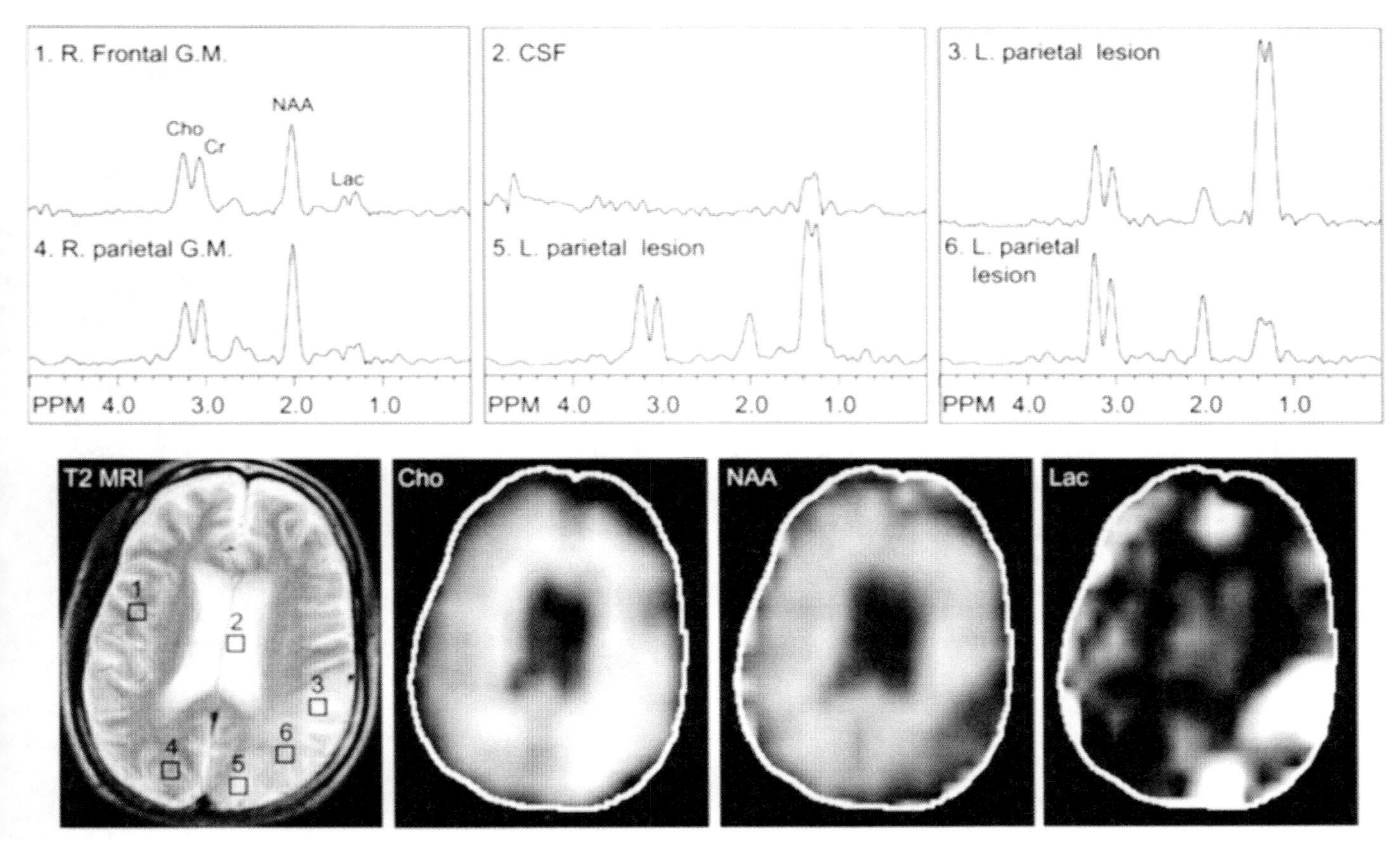

Fig. 7. MRSI in a patient with mitochondrial myopathy, encephalopathy, lactic acidosis, and stroke-like episodes demonstrates increased choline (Chol), decreased NAA, and increase lactate (Lac), most prominently at the left parietal aspect, corresponding to signal abnormality at a site of prior infarction. Reproduced with permission from Lin DDM, Crawford TO, Barker PB. Proton MR Spectroscopy in the diagnostic evaluation of suspected mitochondrial disease. Am J Neuroradiol 2003;24:33–41.

The presence of an alanine peak has also been described in a patient with pyruvate dehydrogenase complex deficiency, which is felt to correspond to increased transamination of pyruvate to alanine. They also described increased creatine and lactate levels (Fig. 9; ref. *97*). One attempt at prenatal MRS demonstrated no elevated lactate in a fetus with subsequently diagnosed pyruvate dehydrogenase deficiency *(98)*. Bianchi et al. describe two cases where elevated lactate was noted in normal appearing brain during episodes of acute exacerbation *(95)*. Decreased NAA was noted in more severe cases; however, it is noted by Bates that decreased NAA can be a result of partial inhibition of the respiratory chain, such that decreased NAA may reflect mitochondrial dysfunction without neuronal loss *(99)*. ^{31}P spectroscopy is of significant interest in mitochondrial disorders, given their role in energy metabolism as whole. For a discussion of this topic, the reader is referred to Chapter **23**.

4.2.3. Primary Demyelination

4.2.3.1. Globoid Cell Leukodystrophy

Globoid cell leukodystrophy, or Krabbe disease, is characterized by deficiency of the lysosomal enzyme galatocerbrosidase, which is involved in myelin-related metabolic pathways. This deficiency causes galactosylsphingosine accumulation, which is neurotoxic and results in demyelination. The most common form of the disease is infantile in onset, with rapid progression to death. MRI demonstrates T_2 signal abnormality of white matter, with variable callosal involvement. Prominent signal abnormality in the corticospinal tracts is another finding seen in Krabbe disease *(100)*. Spectroscopic evaluation has demonstrated increased choline and myo-inositol, with decrease in NAA that is thought to relate to pathologic findings of astrocytic gliosis, widespread demyelination, and secondary axonal degeneration. In a few cases, lactate and lipid peak elevation were noted *101)*. Patients who have late onset demonstrate a varied clinical presentation; Farina et al. *(102)* found a mild decrease in NAA and choline elevation in these patients, with a variable increase in myo-inositol in a group of two patients. choline elevation was noted in areas of white matter without signal abnormality.

4.2.3.2. Metachromatic Leukodystrophy (MLD)

MLD is a lysosomal storage disorder resulting from arylsulfatase A deficiency. A disease of autosomal-recessive inheritance, MLD results in galactosylceramide sulfatide accumulation within white matter of the CNS. Clinical phenotypes of the disease are separated by the age of onset, with infantile, juvenile, and adult forms. Survival and rate of deterioration improve with later onset of disease. Conventional MRI demonstrates symmetric T_2 signal abnormality of white matter, with occasional demonstration of a striate or "tigroid" of parallel lineal hypointensity, which may represent spared periventricular white matter *(103)*. These MRI features can also be seen in PMD. In a series of seven patients, Kruse et al. *(104)* found decreased NAA, prominently elevated myo-inosi-

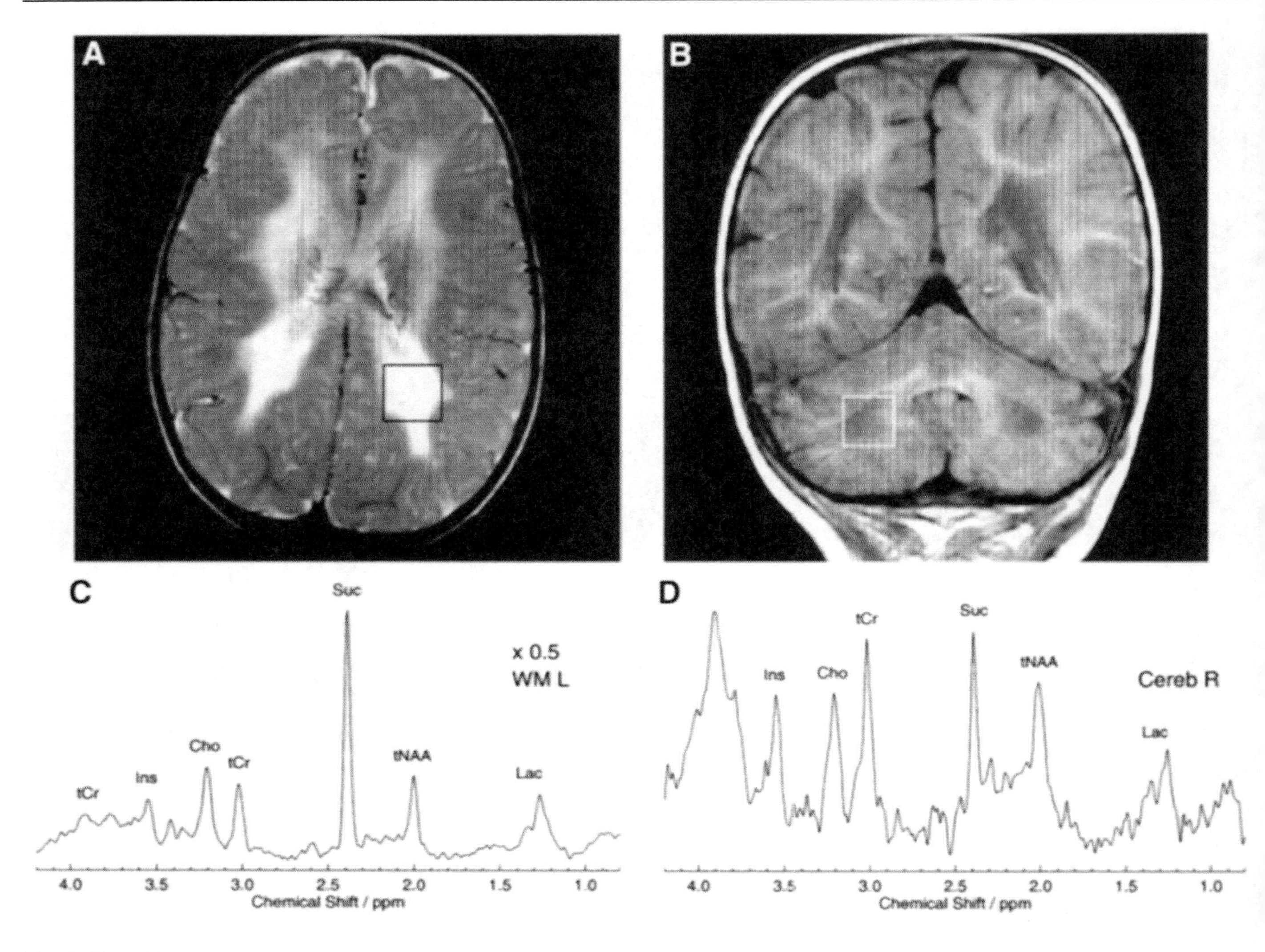

Fig. 8. Short TE spectra acquired from left parietal white matter (**A and C**) and right cerebellum (**B and D**) in a patient with complex II deficiency demonstrates prominent peaks at the 2.4 ppm range, felt to be compatible with succinate. Associated T_2 signal hyperintensity and T_1 hypointensity on routine imaging is noted. Reproduced with permission from Brockmann K, Bjornstad A, Dechent P, et al. Succinate in dystrophic white matter: a proton magnetic resonance spectroscopy finding characteristic for complex II deficiency. Ann Neurol 2002;52:38–46.

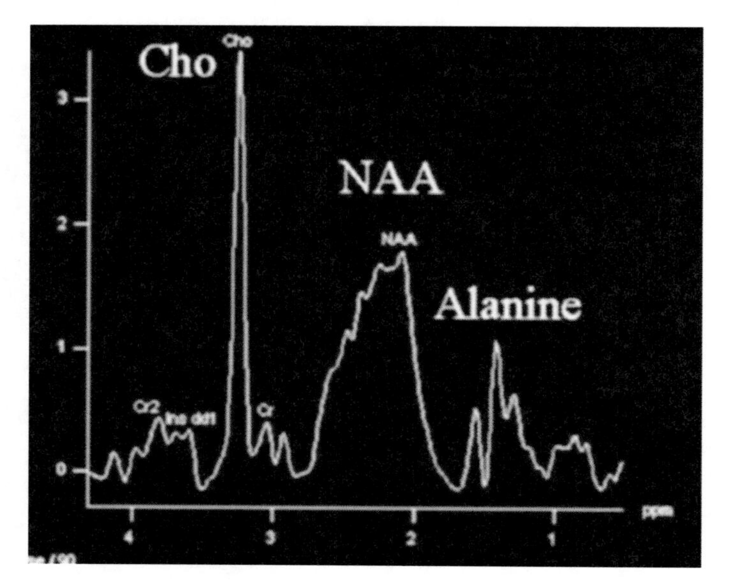

Fig. 9. Proton MRS (TE = 30 ms) in pyruvate dehydrogenase complex deficiency demonstrating an alanine peak at 1.48 ppm (upright doublet) as well as decrease in NAA. Decreased NAA is noted in more severe cases; decreased NAA can be a result of partial inhibition of the respiratory chain, such that decreased NAA may reflect mitochondrial dysfunction without neuronal loss. See color version on Companion CD.

tol, mildly elevated choline, and an occasional lactate peak presence. This may relate to reported autopsy findings of strong elevation of phosphatidylinositol in myelin in patients with MLD. Although most of these findings were not seen in a less severely affected patient in this study, mild myo-inositol elevation was still present.

4.2.3.3. Adrenoleukodystrophy (ALD)

ALD is a peroxisomal disease of significant phenotypic variation. The variety of disorders in this group of diseases results in accumulation of very long chain fatty acids. The most common forms are the childhood-onset form of cerebral adrenoleukodystrophy (cALD) and later-onset adrenomyeloneuropathy (AMN), which is typically less severe. Clinical presentation includes progressive intellectual impairment and, commonly, adrenal insufficiency. Multiple studies have outlined some of the spectroscopic changes noted in ALD, looking at various subtypes of disease, and spectra from different portions of brain *(105)*, typically demonstrating decreased NAA and increased choline (Fig. 10). Notably, Izquierdo et al. *(106)* found no significant difference in metabolites in spectra obtained in normal-appearing white matter, whereas others *(107)* did note some increase in choline even in normal-appearing white matter. This difference may relate to investigation of patients with AMN with decreased impairment described by Izquierdo et al. The early identification of choline in normal-appearing white matter may be the earliest metabolic abnormality identified, and may help in predicting sites of demyelination *(108)*. Similar findings were noted by Kruse and Tzika *(108,109)*. In a group of four patients, Engelbrecht et al. *(110)* noted that the degree of choline elevation and NAA reduction correlated with the extent of demyelination noted on conventional MRI. In one case report, note was made of a decrease in NAA, an increase in choline, and an increased "macromolecular" peak at 1.4 ppm; however, MRS was only performed at a TE of 40 ms *(111)*. Others hav *(112)* described three patients with cALD with an arrest of neurologic deterioration who demonstrated only moderate NAA loss and normal to decreased choline. MRS in such patients may predict an "arrested" cALD or patient who will progress to have the milder AMN form of disease, as opposed to patients with ongoing demyelination, allowing for treatment planning; this is important as bone marrow transplantation is used in treating children with early CNS involvement and, ideally, for patients who will progress to develop more severe disease *(108)*. Rajanayagam et al. *(113)* demonstrated modest improvement in metabolite abnormalities in patients with childhood-onset cerebral ALD who received bone marrow transplantation, although pretreatment MRS was not performed in this group; although for the most part, they found spectroscopic abnormalities to correlate well with clinical status and MRI scores, they did not notice as good a correlation when the clinical status was already poor. Other investigators have documented some reversal in metabolic abnormalities following bone marrow transplant in a patient with cALD *(107)*.

4.2.3.4. Zellweger Syndrome

The Zellweger cerebrohepatorenal syndrome is one of the more severe peroxisomal disorders, resulting in hearing loss, hypotonia, seizures, liver disease, and death. Elevated levels of very long chain fatty acids and pipecolic acid are noted. In a MRS study of two patients *(114)*, a note was made of decreased NAA, increased lactate, and abnormal signal at 0.9 ppm, which were felt to represent mobile lipids; this was felt to relate to abnormal storage of fat in astrocytes and phagocytes in patients with Zellweger syndrome. In another study of peroxisomal disorders, including Zellweger syndrome patients *(45*, a decrease in NAA was noted without lipid or lactate elevation. However, in two patients with concomitant liver dysfunction, a note was made of increased glutamine and decreased myo-inositol.

4.2.3.5. Alexander's Disease

Alexander's disease is a progressive leukodystrophy characterized by prominent Rosenthal fibers in perivascular, subpial, and subependymal portions of the CNS that is associated with dense fibrillary gliosis, and cystic areas of cavitation *(50)*. It presents with developmental delay, macrocephaly, seizures, and spasticity. Findings using MRS include decreased NAA and prominently elevated myo-inositol (felt to relate to intense gliosis), increased choline, and increased lactate *(115)*, which was felt by Brockmann et al. to correlate with active neuroaxonal degeneration, glial proliferation, and active demyelination. It was concluded that a normalized choline level may be seen in areas that had already experienced demyelination. Grodd et al. *(43)* reported lactate elevation and decreased NAA in the frontal regions with normal NAA and no lactate elevation posteriorly, correlating with the expected anterior-to-posterior progression of this disease. Of note, Imamura et al. *(116)* demonstrated the presence of these MRS abnormalities in areas of normal-appearing white matter, possibly allowing for early detection of neuronal change. A case of adolescent Alexander's disease was reported, demonstrating decreased NAA and elevated lactate *(117)*. Decrement in NAA allows for the differentiation of this macrocephalic leukodystrophy from Canavan's disease. There are no accounts in the literature of MRS tracking of progression and therapeutic trials.

4.2.4. Vacuolating Myelinopathies

4.2.4.1. Canavan's Disease

Canavan's disease is an inherited disorder (autosomal recessive) caused by aspartoacylase deficiency, which results in progressive white matter vacuolization; this disorder has been localized to chromosome 17p *(118)*. Patients present with early psychomotor retardation, megalencephaly, blindness, and spasticity. Aspartoacylase aids in the metabolism of NAA into acetate and aspartate NAA buildup ensues, with prominent NAA peaks reported in all published cases of MRS and Canavan's (Fig. 11). A note is also made of decreased choline, which corresponds to pathologic findings of decreased white matter lecithin and sphingomyelin *(119–121)*. In a few cases, lactate elevation and creatine decrease were noted. Recently, advances in gene transfer technology have allowed for the development of novel gene therapies in Canavan's. Leone et al. *(122)* Describe a plasmid-based aspartoacylase gene therapy used in a mammalian model and in two human subjects; MRS in these patients demonstrated some modest decrease in NAA after therapy.

4.2.4.2. Vacuolating Megaloencephalic Leukoencephalopathy (VML)

A recently described leukodystrophy is VML, also known as megaloencephalic leukoencephalopathy with subcortical cysts or van der Knaap syndrome *(123–127)*. Of interest, the characterization of this disorder has occurred at a point at which proton MRS has been widely available as a clinical tool of investigation; van der Knaap et al. described MRS to be helpful in characterizing the disorder as not being a demyelinating disorder, given the absence of changes in choline/creatine ratios, with confirmation of this on histopathologic examination *(128)*. This disorder presents in childhood with macrocephaly, progressive motor dysfunction, and disproportionately mild

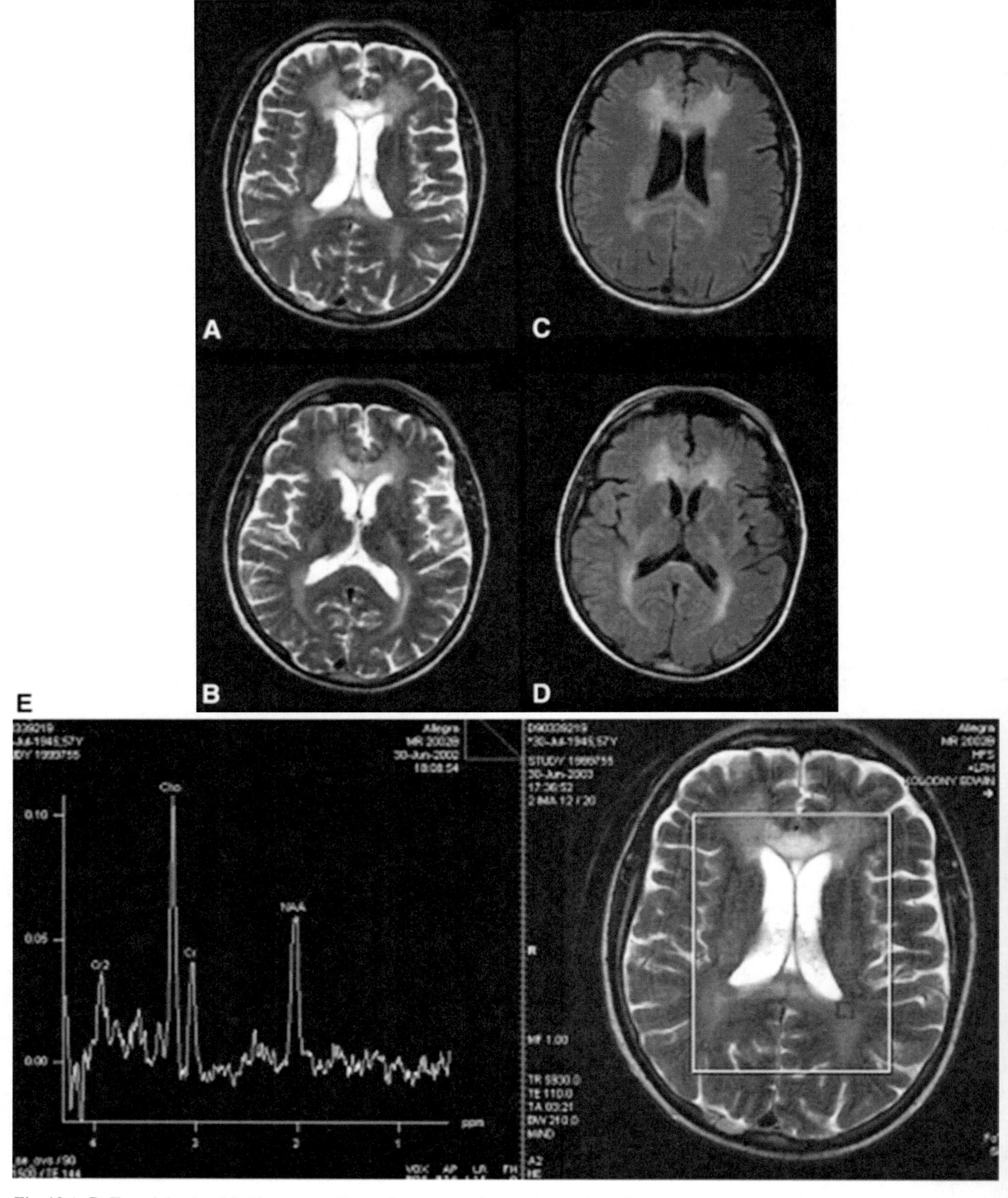

Fig. 10 A–D, T_2-weighted and fluid-attenuated inversion recovery images in a patient with adult-onset AMN variant demonstrating increased signal in the anterior and posterior periventricular white matter. The typical distribution for ALD is in the posterior periventricular white matter. **E,** Proton MRS (TE = 144 ms), in demonstrating decrease in NAA, with an increase in choline acting as a marker of active demyelination in the periventricular white matter.

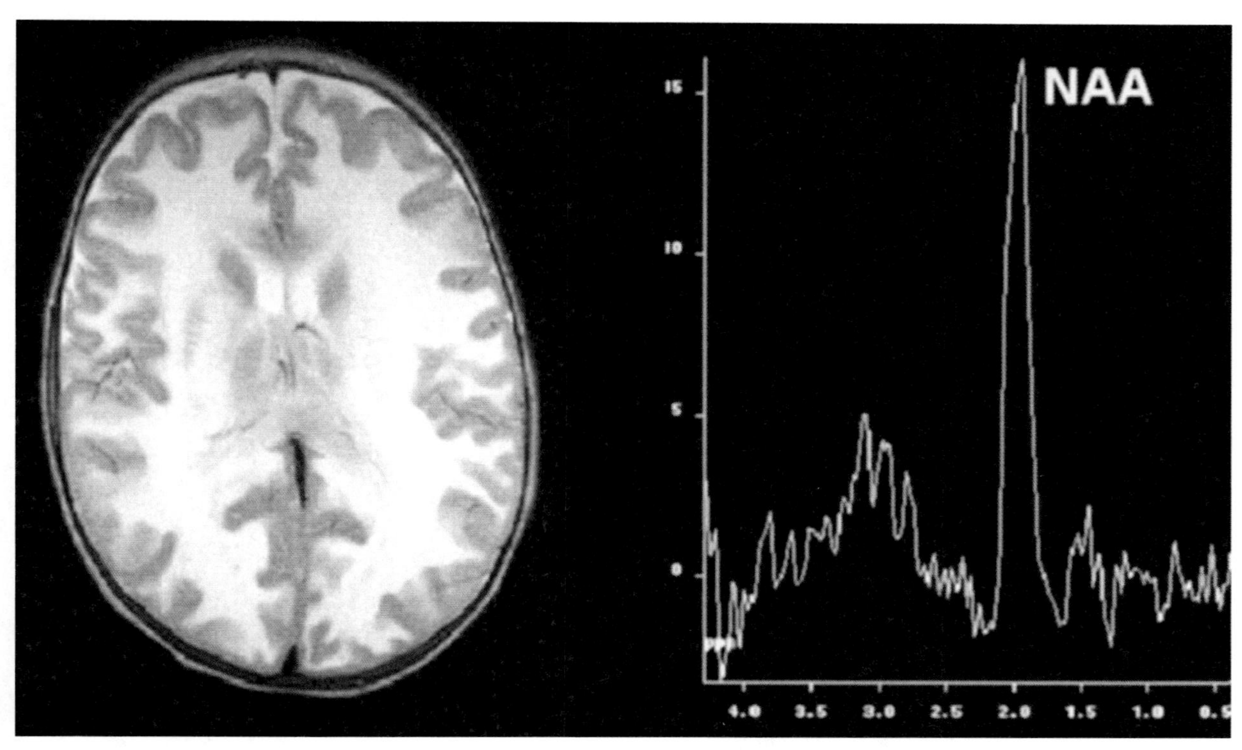

Fig. 11. Proton MRS (TE = 144 ms) in Canavan's disease demonstrating marked elevation in NAA caused by aspartoacylase deficiency. Aspartoacylase aids in the metabolism of NAA into acetate.

cognitive impairment; an autosomal recessive-inheritance pattern is suggested, localized to chromosome 22q. MRI is notable for extensive white matter signal abnormality and the formation of subcortical cysts. In most studies note is made of prominent metabolite loss within white matter, with only mild NAA decrement in gray matter. One study identified relatively preserved myoinositol in white matter in the face of widespread loss of other metabolites, felt to correlate to the astrocytosis noted on histopathology (126).

4.2.5. Secondary Demyelination

4.2.5.1. Lysosomal Storage Diseases

Niemann-Pick disease is a diffuse brain disease of autosomal-recessive inheritance resulting from an error in the handling of cholesterol by lysosomes, resulting in ataxia, developmental delay, and motor dysfunction. A study of 10 patients demonstrated decreased NAA and increased choline, with changes most prominent at the frontoparietal cortex, caudate nucleus, and centrum semiovale *(129)*. Neuronal ceroid lipofuscinosis (NCL) includes a group of disorders that result in lysosomal lipopigment accumulation. Seitz et al. *(130)* describe decrease in NAA, with prominent myo-inositol increase in a group of three patients, becoming the most prominent signal in a case of late infantile NCL with similar findings reported in other studies *(131,132)*; no elevation of lactate was noted. In another series of patients with NCL, patients with the juvenile type of NCL demonstrated normal metabolic profiles *(131)*. The severity of metabolic derangement correlated with the clinical symptoms of the patients and decreasing age of onset.

Fabry disease is a lysosomal storage disease resulting from a deficiency of the enzyme α-galactosidase A. This results in glycosphingolipid intracellular accumulation in heart, brain, kidneys, and endothelium. A new therapeutic option is enzyme replacement therapy. Breunig et al. *(133)* describe the use of ^{31}P spectroscopy of the myocardium to allow for determination of an energetic index in patients with Fabry disease. At our institution, we have begun an initiative in following patients with adult-onset Tay Sach's disease by using SVS and CSI MRS protocols to evaluate spectra through the basal ganglia and cerebellum. Doing so, we can track the status of these patients and metabolic abnormalities identified on MRS.

4.2.5.2. Mucopolysaccharidosis (MPS)

The mucolipidoses are a group of lysosomal storage diseases result in varying degrees of mental retardation, CNS, and systemic involvement. There is a deficiency of enzymes that help degrade glycosaminoglycans. The different diseases result in accumulation of dermatan sulfate, heparan sulfate, or keratan sulfate *(134)*. On MRS, a diffuse decrease in NAA has been noted in mucolipidosis IV *(135)*. Takahashi et al. *(136)* observed a resonance peak at 3.7 ppm that was felt to represent signal from the mucopolysaccharides, as identified on urine spectroscopy (Fig. 12). This region correlated to a ring-proton region peak identified in an in vitro study of urinary glycosaminoglycans *(137)*, also identifying peaks distinguishing heparan sulfate, dermatan sulfate, keratan sulfate, and chondroitin sulfate and, subsequently, different forms of MPS; the use of these spectral profiles in vivo has not been reported. Elevation of this peak was noted within white matter lesions on in vivo studies of seven patients with MPS of mixed variety, with elevated choline/creatine ratios were noted. Although this peak was not increased in white matter without lesions, increased choline/creatine ratios were found. After bone marrow transplantation, the MPS/creatine ratio was

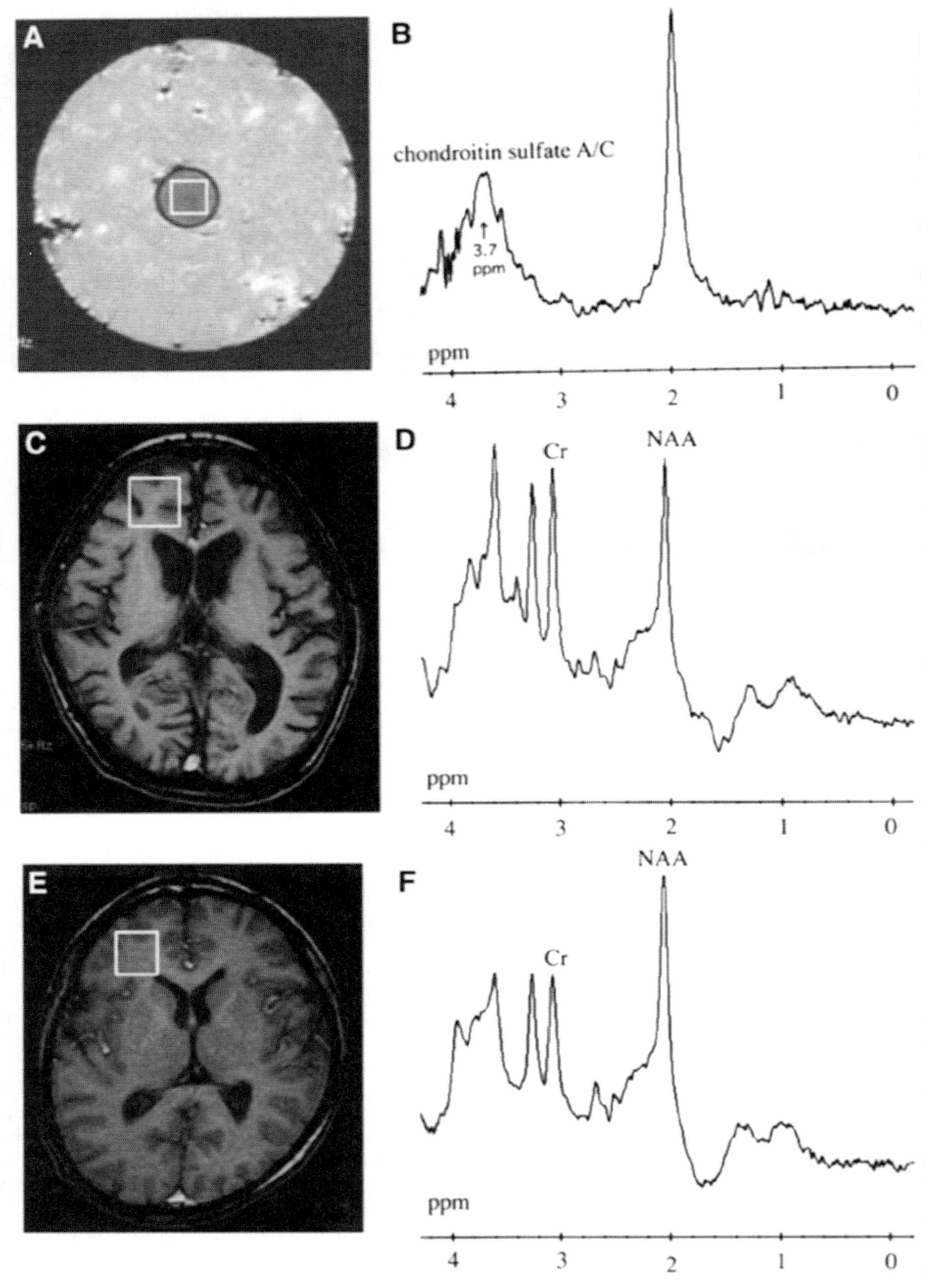

Fig. 12. (**A,B**) MR spectrum acquired from urine sample of chondroitin sulfate, demonstrating a broad peak centered at 3.7 ppm. (**C–F**) Short TE spectra acquired from right frontal white matter in two separate patients with mucolipidoses demonstrates elevation of a peak in this same range. Reproduced with permission from Takahashi Y, Sukegawa K, Aoki M, et al. Evaluation of accumulated mucopolysaccharides in the brain of patients with mucopolysaccharidoses by (1)H-magnetic resonance spectroscopy before and after bone marrow transplantation. Pediatr Res 2001; 49:349–355.

slightly decreased, but never to control ratios, which may relate to an immutable deposition of these materials. Similarly, in white matter without lesions, although choline/creatine ratios were seen to decrease after bone marrow transplant (BMT), they remained higher than control ratio. Within the same study, it was found that patients of higher intelligence quotient tended to have less MPS signal and higher NAA. Similarly elevated peaks was noted in a study of 23 patients with MPS, but the presumed MPS peaks were not quantified *(138)*; of note, significant improvement in white matter abnormalities on conventional MR imaging was noted in two of five patients receiving BMT. In combination with routine MRI, proton MRS may be helpful in evaluating the efficacy and status of individuals with MPS by monitoring levels of NAA, choline, and creatine.

4.2.5.3. Sjogren–Larsson Syndrome

Sjogren–Larsson syndrome is a disorder of autosomal-recessive inheritance that causes long-chain fatty alcohol accumulation as a result of fatty aldehyde dehydrogenase *(139)*. The disease results in retardation, spastic paraplegia, and ichthyosis. A description of MRS in two cases of Sjogren–Larsson's syndrome noted prominence of peaks at 1.3 and 0.9 ppm (seen on short TE; ref. *140*). These were felt to represent methyl and methylene peaks of lipid and possibly represent the peaks of accumulating long-chain fatty alcohols (hexedecanol, octadecanol, and phytol). Another case report of SLS demonstrated this elevated lipid peak, as well as decreased NAA *(141)*, suggesting that periventricular lesions of T_2 signal abnormality with an elevated fatty peak may be pathognomonic for Sjogren–Larsson syndrome.

4.2.5.4. Cerebrotendinous Xanthomatosis (CTX)

CTX results from abnormal activity of the enzyme sterol 27-hyroxylase, which is important in the metabolism of cholesterol. Defective bile acid synthesis and high levels of cholestanol occur, resulting in neurologic impairment, juvenile cataracts, and tendinous xanthomas. De Stefano et al. *(142)* demonstrated increased lactate and decreased NAA in a group of 12 patients with CTX. It was felt that this decrease in the absence of choline derangement correlated to the accumulation of metabolites that were directly neurotoxic, with mitochondrial dysfunction possibly caused by high cholesterol or bile alcohol levels. Notably, there was good correlation between NAA loss and patient disability in this group.

4.2.5.5. CADASIL

CADASIL results from a microangiopathy, which presents with migraine, cognitive dysfunction, and recurrent infarcts, resulting from a mutation in the Notch3 gene *(143)*. An accumulation of a domain of the receptor is seen with deposition in the basal membrane *(144)*. Auer et al. *(145)* evaluated 20 patients with CADASIL and demonstrated choline elevation and NAA loss areas of signal abnormality on conventional MRI, without choline elevation in normal appearing white matter. Elevated normalized myo-inositol concentration was noted that was felt to relate to astrogliosis. Some degree of lactate elevation was also noted, felt by the authors to possible present changes from local hypoperfusion. Worsened NAA decrement was noted in patients with more severe dementia, and note was made of a decrease in all metabolites.

4.2.6. Other White Matter Diseases

White matter disease encompasses a large variety of disorders, which include not only metabolic abnormalities, but inflammatory, hypoxic-ischemic, toxic, and post-therapeutic processes as well. Although the clinical differential in the patient with metabolic disease may be limited to a few nonmetabolic processes, MRS may be helpful in the evaluation and differentiation of these diseases. A number of processes have been evaluated by proton MRS, including hypoxic-ischemic disease in the neonate, as well as multiple sclerosis.

Hereditary galactosemia results from deficiency in the enzyme galactose-L-phosphate uridyltransferase. This results in galactitol accumulation, which is theorized to cause subsequent brain edema. Studies of ^{1}H MRS of patients with acute galactose intoxication have demonstrated elevation of a doublet peak at 3.7 ppm, found to correspond in vitro to galactitol *(146,147)*. In an infant with galactosemia and encephalopathy, Berry et al. *(146)* found elevation of this peak, as well as decreased myo-inositol and choline. A study by Moller et al. of adult patients under dietary treatment revealed normal metabolite spectra *(148)*; similarly, Wang et al. found galactitol peak elevation only in galactosemia patients with encephalopathy. MRS provides a means to obtain in vivo assessment of brain galactitol, and the effectiveness of dietary therapy.

In pediatric multiple sclerosis, note is made of localized decrease in NAA, and elevations in choline and myo-inositol *(149)*. In acute lesions, note was made of decreased NAA without derangement of choline and MI. Although decreased NAA was noted in gray matter adjacent to multiple sclerosis plaques, no metabolic derangement was noted in normal-appearing white matter. In hypoxic-ischemic disease, studies have demonstrated a correlation between glutamine/glutamate elevation and severity of encephalopathy, with lactate elevation portending a poor outcome *(150,151)*. Results regarding NAA decreased have been more variable.

Evaluation of MR spectra of patients with toxin and drug exposure has been studied extensively. NAA was shown to be mildly decreased in chronic solvent abusers *(152)*. Evaluation of a group of patients exposed to carbon monoxide demonstrated NAA loss, initial choline elevation then loss, and persistent elevated lactate in patients who did poorly *(153,154)*. Children exposed to methamphetamines and cocaine in utero were noted to have elevations in total creatine without additional metabolic abnormalities, perhaps a reflection of altered energy metabolism *(155,156)*.

A study evaluating the differences between HIV encephalopathy and progressive multifocal leukoencephalopathy (PML) found a more pronounced NAA loss in PML lesion and increased lactate, as opposed to HIV encephalopathy, which demonstrated a more consistent choline elevation *(157)*. An evaluation of pediatric AIDS encephalopathy demonstrated NAA loss and a lactate peak, both of which improved following treatment *(158)*. Studies of high-dose chemotherapy and postirradiation changes of the brain demonstrated no significant abnormalities in metabolite ratios *(159,160)*. A series studying patients with developmental delay 2 yr or older without known medical history and with normal appearing MRI demonstrated decreased NAA/creatine and increased choline/creatine ratios *(161)*.

5. FUTURE STRATEGIES AND DEVELOPMENTS IN MR-BASED IN VIVO METABOLIC IMAGING

In theory, higher field systems should provide improved spectral resolution and increased SNR *(162)* However, issues of shimming, coil design, and alterations in metabolite T_1/T_2 may degrade these advantages. Gonen et al. *(163)* found in a comparison of 3D-CSI of the brain with 1.5 T vs 3.0 T systems assuming similar sequence, shim, and anatomy, that SNR was improved 23–46%. This improvement in SNR could be used to increase sensitivity to changes in metabolites detected on MRS or be exchanged for shorter acquisition time. They also found that spectral resolution was improved, with improved peak discrimination, most notably between choline and creatine. At a TE of 135 or 144 ms, a generally flatter baseline was also noted, easing peak area estimation. The application of phased array coil technology has allowed for prominent gain in SNR. The application of a GRASE sequence has been discussed with dual advantages of significant improvement in acquisition time as well as effective homonuclear decoupling *(164)*.

Carbon-based spectroscopy is of substantial interest in general, and particularly in metabolic imaging. However, the natural abundance of ^{13}C, the isotope that can be evaluated by MRS (^{12}C has no net nuclear spin), is extremely low, with a consequent decrease in expected available signal. Evaluation of ^{13}C spectra is rendered more difficult by extensive spin-spin coupling and is more technically demanding.

Although alteration in metabolites such as NAA, choline, creatine, glutamate/glutamine is possible, discrimination of other metabolites can be difficult because of the severe overlap of these metabolites. Some strategies to over come this include certain metabolite nulling sequences and two-dimensional MRS.

Modification of spectra with metabolite-nulling sequences allow for improved evaluation of lipids and other macromolecules, with improved resolution of peaks at 0.9 ppm, and 1.3 ppm, and 2.1 ppm. These correlated to spectra of methyl groups, methylene, and terminal methylene groups, respectively. Findings have been evaluated in the setting of acute stroke and multiple sclerosis. Mader et al. *(165)* found significant evaluation in macromolecular peaks in the setting of acute MS plaques in contrast to normal-appearing white matter and chronic MS lesions, perhaps representing the residues from cytosolic protein that may become elevated during the course of disruption.

2D spectroscopy allows for improved peak resolution because of the added second spectral dimension *(166)*. Well-resolved cross peaks are present that can aid in metabolite measurement. It has already been used in the past for in vitro spectroscopic evaluation of metabolic disorders; Iles et al. described the use of 2D MRS in evaluating the components of urine in patients with HMG CoA lyase deficiency *(72)*. In vivo two-dimensional proton MRS allows for evaluation of other metabolites, including aspartate, phosphocreatine, γ-aminobutyric acid, threonine, phosphoethanolamine and ethanolamine. It also allows for distinct separation of glutamine/glutamate from NAA and myo-inositol. Already, accounts of in vivo two-dimensional proton spectra have been published in healthy human subjects with good reproducibility *(167)*, as well as patients with brain tumors *(168)*.

In certain pathologic metabolic processes, abnormalities of MR spectra are well correlated directly with the abnormal process, e.g.,for example, markedly elevated NAA in Canavan's, anomalous peaks in PKU, MSUD, and HMG-CoA lyase deficiency. In other cases, MR spectra metabolite abnormalities correlate to pathologic processes and, in many cases, in advance of abnormality seen on routine MRI. As our experience with functional genomics improves to the level of a generalized understanding of gene, protein, and metabolite expression in various disorders and the interplay between them, these features may be better exploited in terms of diagnostic and imaging strategies. Increased sophistication of technologies of mass spectrometry and chromatography has allowed for improvements in the diagnosis and screening for inherited metabolic disease *(169)*. In parallel development, metabolite peaks that have not been routinely visualized in the past may be better visualized as ^1H and alternative spectra MRS technology improves. Another means by which imaging will evolve in the context of functional genomics, is with alternative MR-contrast agents. The recent completion of sequencing of the human genome is indeed a landmark event, which promises incredible tools in the field of biomedical research *(170)*. A variety of novel gadolinium-based and iron oxide-based agents have been synthesized, and it is now possible to target these agents through antibodies and genetic targeting. It may be possible in the future to monitor specific enzyme expression in vivo with these agents.

In conclusion, advanced MR techniques, with proton MRS playing a prominent role, provide an important adjunctive means of evaluated the patient with suspected leukodystrophy or metabolic white matter disease. Spectroscopic imaging allows for in vivo metabolic interrogation. Future developments in MRI and nonimaging technologies will likely allow us to advance from the metabolic to the metabolomic evaluation of patients. These exciting possibilities may allow for not only for reliable and efficient diagnosis of these diseases, but as a means to directly evaluate the metabolic effects of treatment and diet alteration, and the temporal evaluation of metabolic profiles as a whole.

REFERENCES

1. Warmuth C, Gunther M, Zimmer C. Quantification of blood flow in brain tumors: comparison of arterial spin labeling and dynamic susceptibility-weighted contrast-enhanced MR imaging. Radiology 2003;228:523–532.
2. Moore DF, Altarescu G, Ling GSF, et al. Elevated cerebral blood flow velocities in Fabry disease with reversal after enzyme replacement. Stroke 2002;33:525–531.
3. Fukutani Y, Noriki Y, Sasaki K, et al. Adult-type metachromatic leukodystrophy with a compound heterozygote mutation showing character change and dementia. Psychiatry Clin Neurosci 1999;53:425–428.
4. Tamagaki C, Murata A, Saito A, Kinoshita T. [Two siblings with adult-type metachromatic leukodystrophy: correlation between clinical symptoms and neuroimaging.] [Japanese]. Seishin Shinkeigaku Zasshi 2000;102:399–409.
5. Salmon E, Can der Linden M, Maerfens Noordhout A, Brucher JM. Early thalamic and cortical hypometabolism in adult-onset dementia due to metachromatic leukodystrophy. Acta Neurologica Belg 1999;1999:3.

6. Sawaishi Y, Hatazawa J, Ochi N, et al. Positron emission tomography in juvenile Alexander disease. J Neurol Sci 1999;165:116–120.
7. Al-Essa MA, Bakheet SM, Patay ZJ, Powe JE, Ozand PT. Clinical and cerebral fdg pet scan in a patient with krabbe's disease. Pediatr Neurol 2000;22:44–47.
8. Van Zijl PC, Barker PB. Magnetic resonance spectroscopy and spectroscopic imaging for the study of brain metabolism. Ann NY Acad Sci 1997;820:75–96.
9. Murphy-Boesch J, Stoyanova R, Srinivasan R, et al. Proton-decoupled 31P chemical shift imaging of the human brain in normal volunteeers. NMR Biomed 1993;6:173.
10. Gruetter R, Novotny EJ, Boulware SD. Localized 13C NMR spectroscopy in the human brain of amino acid labeling from 13C glucose. J Neurochem 1994;63:1377–1385
11. Birken DL, Oldendorf WH. N-Acetylaspartic acid: a literature review of a compound prominent in 1H-NMR spectroscopic studies of brain. Neurosci Biobehav Rev 1989;13:23–31.
12. Bjartmar C, Battistuta J, Terada N, Dupree E, Trapp BD. N-acetylaspartate is an axon-specific marker of mature white matter in vivo: a biochemical and immunohistochemical study on the rat optic nerve. Ann Neurol 2002;51:51–58.
13. Bitsch A, Bruhn H, Vougioukas V, et al. Inflammatory CNS demyelination: histopathologic correlation with in vivo quantitative proton MR spectroscopy. AJNR Am J Neuroradiol 1999;20:1619–1627.
14. Gonen O, Catalaa I, Babb JS. Total brain N-acetylasparatate: A new measure of disease load in MS. Neurology 2000;54:15–19.
15. Urenjak J WS, Gadian D, et al. Specific expression of N-acetylaspartate in neurons, oligodendrocute-type-2 astrocyte progenitors and immature oligodendrocytes in vitro. J Neurochem 1992;59:55–61.
16. Burri R SC, Herschkowitz N. N-Acetyl-aspartate is a major source of acetyl groups for lipid synthesis during rat brain development. J Neurosci 1991;13:403–411.
17. Castillo M KL, Scatliff J, Mukherji SK. Proton MR Spectroscopy in Neoplastic and Non-neoplastic Brain Disorders. MRI Clin North Am 1998;6:1–20.
18. Kreis R HL, Kuhlmann B, Boesch C, Bossi E, Huppi PS. Brain metabolite composition during early human brain development as measured by quantitative in vivo 1H magnetic resonance spectroscopy. Magn Reson Med 2002;48:949–958.
19. Griffin JL, Mann CJ, Scott J, Shoulders CC, Nicholson JK. Choline containing metabolites during cell transfection: an insight into magnetic resonance spectroscopy detectable changes. FEBS Lett 2001;509:263–266.
20. Brand A R-LC, Leibfritz D. Multinuclear NMR studies on the energy metabolism of glial and neuronal cells. Dev Neurosci 1993;15:189–198.
21. Lee JD, Arcinue E, Ross B. Organic osmolytes in the brain of an infant with hypernatremia. N Engl J Med 1994;331:439–442.
22. Kang PB, Hunter JV, Kaye EM. Lactic acid elevation in extramitochondrial childhood neurodegenerative diseases. J Child Neurol 2001;16:657–660.
23. Shen J, Novotny EJ, Rothman DL. In vivo lactate and beta-hydroxybutyrate editing using a pure-phase refocusing pulse train. Magn Reson Med 1998;40:783–788.
24. Seeger U, Mader I, Nagele T, Grodd W, Lutz O, Klose U. Reliable detection of macromolecules in single-volume 1H NMR spectra of the human brain. Magn Reson Med 2001;45:948–954.
25. Behar KL, Rothman DL, Spencer DD, Petroff OAC. Analysis of macromolecule resonances in 1H NMR spectra of human brain. Magn Reson Med 1994;32:294–302.
26. Salibi N, Brown MA. Clinical MR Spectroscopy. New York: Wiley-Liss; 1998:220.
27. Gonen O, Hu J, Stoyanova R, Leigh JS, Goelman G, Brown TR. Hybrid three dimensional (1D-Hadamard, 2D-chemical shift imaging) phosphorus localized spectroscopy of phantom and human brain. Magn Res Med 1995;33:300–308.
28. Gonen O, Arias-Mendoza F, Goelman G. 3D localized in vivo 1H spectroscopy of human brain by using a hybrid of 1D-Hadamard with 2D-chemical shift imaging. Magn Res Med 1997;37:644–650.
29. Gonen O, Murdoch JB, Stoyanova R, Goelman G. 3D multi-voxel proton spectroscopy of human brain using a hybrid of 8th order Hadamard encoding with 2D-chemical shift imaging. Magn Res Med 1998;39:34–40.
30. Spielman DM, Adalsteinsson E, Lim KO. Quantitative assessment of improved homogeneity using higher-order shims for spectroscopic imaging of the brain. Magn Reson Med 1998;40:376–382.
31. van Der Veen JW, Weinberger DR, Tedeschi G, Frank JA, Duyn JH. Proton MR spectroscopic imaging without water suppression. Radiology 2000;217:296–300.
32. Sijens PE, van den Bent MJ, Nowak PJ, van Dijk P, Oudkerk M. 1H chemical shift imaging reveals loss of brain tumor choline signal after administration of Gd-contrast. Magn Reson Med 1997;37:222–225.
33. Lin AP, Ross BD. Short-echo time proton MR spectroscopy in the presence of gadolinium. J Comput Assist Tomogr 2001;25:705–12.
34. Smith JK, Kwock L, Castillo M. Effects of contrast material on single-volume proton MR spectroscopy. AJNR Am J Neuroradiol 2000;21:1084–1089.
35. Murphy PS, Dzik-Jurasz AS, Leach MO, Rowland IJ. The effect of Gd-DTPA on T(1)-weighted choline signal in human brain tumours. Magn Reson Imaging 2002;20:127–130.
36. Hunter JV, Wang ZJ. MR spectroscopy in pediatric neuroradiology. Magn Reson Imaging Clin North Am 2001;9:165–89, ix.
37. Ross B, Bluml S. Magnetic resonance spectroscopy of the human brain. Anat Record 2001;265:54–84.
38. Hetherington H, Mason G, Pan J, et al. Evaluation of cerebral gray and white matter metabolite differences by spectroscopic imaging at 4.1T. Magn Reson Med 1994;32:565–571.
39. Pouwels PJ, Brockmann K, Kruse B, et al. Regional age dependence of human brain metabolites from infancy to adulthood as detected by quantitative localized proton MRS. Pediatr Res. 1999;46:474–485.
40. Huppi PS. MR imaging and spectroscopy of brain development. Magn Reson Imaging Clin North Am 2001;9:1–17, vii.
41. Inder TE, Huppi PS. In vivo studies of brain development by magnetic resonance techniques. Mental Retard Dev Disabil Res Rev 2000;6:59–67.
42. Grodd W, Krageloh-Mann I, Petersen D, Trefz FK, Harzer K. In vivo assessment of N-acetylaspartate in brain in spongy degeneration (Canavan's disease) by proton spectroscopy. Lancet 1990;336:437–438.
43. Grodd W, Krageloh-Mann I, Klose U, Sauter R. Metabolic and destructive brain disorders in children: findings with localized proton MR spectroscopy. Radiology 1991;181:173–181.
44. Tzika AA, Ball WS, Jr., Vigneron DB, Dunn RS, Kirks DR. Clinical proton MR spectroscopy of neurodegenerative disease in childhood. AJNR Am J Neuroradiol 1993;14:1267–1281; discussion 1282–1284.
45. Bruhn H, Kruse B, Korenke GC, et al. Proton NMR spectroscopy of cerebral metabolic alterations in infantile peroxisomal disorders. J Comput Assist Tomogr 1992;16:335–344.
46. Wang ZJ ZR. Proton MR Spectroscopy of pediatric brain metabolic disorders. Neuroimaging Clin North Am 1998;8:781–807.
47. Zimmerman RA, Wang, Zhiyue J. The value of proton MR spectroscopy in pediatric metabolic brain disease. Am J Neuroradiol 1997;18:1872–1879.
48. Kolodny EH. Genetic and metabolic aspects of leukodystrophies. In: Dangond F (ed.) Disorders of Myelin in the Central and Peripheral Nervous Systems. Woburn, MA: Butterworths, 2002.
49. Li BS, Wang H, Gonen O. Metabolite ratios to assumed stable creatine level may confound the quantification of proton brain MR spectroscopy. Magn Reson Imaging 2003;21:923–928.
50. Suzuki K, Armao D, Stone JA, Mukherji SK. Demyelinating diseases, leukodystrophies, and other myelin disorders. Neuroimaging Clin North Am 2001;11:vii, 15–35.
51. Nezu A, Kimura S, Takeshita S, Osaka H, Kimura K, Inoue K. An MRI and MRS study of Pelizaeus–Merzbacher disease. Pediatr Neurol 1998;18:334–337.
52. Takanashi J, Sugita K, Osaka H, Ishii M, Niimi H. Proton MR spectroscopy in Pelizaeus-Merzbacher disease. AJNR Am J Neuroradiol. 1997;18:533–535.
53. Plecko B, Stockler-Ipsiroglu S, Gruber S, et al. Degree of hypomyelination and magnetic resonance spectroscopy findings in

patients with Pelizaeus Merzbacher phenotype. Neuropediatrics 2003;34:127–136.
54. Novotny EJ, Jr., Avison MJ, Herschkowitz N, et al. In vivo measurement of phenylalanine in human brain by proton nuclear magnetic resonance spectroscopy. Pediatr Res 1995;37:244–249.
55. Kreis R, Pietz J, Penzien J, Herschkowitz N, Boesch C. Identification and quantitation of phenylalanine in the brain of patients with phenylketonuria by means of localized in vivo 1H magnetic-resonance spectroscopy. J Magn Reson Series B. 1995;107:242–251.
56. Koch R, Burton B, Hoganson G, et al. Phenylketonuria in adulthood: a collaborative study. J Inherited Metabolic Dis 2002;25: 333–346.
57. Weglage J, Moller HE, Wiedermann D, Cipcic-Schmidt S, Zschocke J, Ullrich K. In vivo NMR spectroscopy in patients with phenylketonuria: clinical significance of interindividual differences in brain phenylalanine concentrations. J Inherited Metabolic Dis 1998;21:81–82.
58. Moller HE, Ullrich K, Weglage J. In vivo proton magnetic resonance spectroscopy in phenylketonuria. Eur J Pediatr 2000;159:S121–S125.
59. Leuzzi V, Bianchi MC, Tosetti M, Carducci CL, Carducci CA, Antonozzi I. Clinical significance of brain phenylalanine concentration assessed by in vivo proton magnetic resonance spectroscopy in phenylketonuria. J Inherited Metabolic Dis 2000;23:563–570.
60. Pietz J, Kreis R, Schmidt H, Meyding-Lamade UK, Rupp A, Boesch C. Phenylketonuria: findings at MR imaging and localized in vivo H-1 MR spectroscopy of the brain in patients with early treatment. Radiology. 1996;201:413–420.
61. Thompson GN, Francis DE, Halliday D. Acute illness in maple syrup urine disease: dynamics of protein metabolism and implications for management. J Pediatr 1991;119:35–41.
62. Felber SR, Sperl W, Chemelli A, Murr C, Wendel U. Maple syrup urine disease: metabolic decompensation monitored by proton magnetic resonance imaging and spectroscopy. Ann Neurol 1993;33:396–401.
63. Heindel W, Kugel H, Wendel U, Roth B, Benz-Bohm G. Proton magnetic resonance spectroscopy reflects metabolic decompensation in maple syrup urine disease. Pediatr Radiol 1995;25:296–299.
64. Wajner M, Coelho JC. Neurological dysfunction in methymalonic acidemia is probably related to the inhibitory effect of methylmalonate on brain energy metabolism. J Inherited Metabolic Dis 1997;20:761–768.
65. Takeuchi M, Harada M, Matsuzaki K, Hisaoka S, Nishitani H, Mori K. Magnetic resonance imaging and spectroscopy in a patient with treated methylmalonic acidemia. J Comput Assist Tomogr 2003;27:547–551.
66. Trinh BC, Melhem ER, Barker PB. Multi-slice proton MR spectroscopy and diffusion-weighted imaging in methylmalonic acidemia: report of two cases and review of the literature. AJNR Am J Neuroradiol 2001;22:831–833.
67. Choi CG, Lee HK, Yoon JH. Localized proton MR spectroscopic detection of nonketotic hyperglycinemia in an infant. Korean J Radiol 2001;2:239–242.
68. Viola A, Chabrol B, Nicoli F, Confort-Gouny S, Viout P, Cozzone PJ. Magnetic resonance spectroscopy study of glycine pathways in nonketotic hyperglycinemia. Pediatr Res 2002;52:292–300.
69. Heindel W, Kugel H, Roth B. Noninvasive detection of increased glycine content by proton MR spectroscopy in the brains of two infants with nonketotic hyperglycinemia. AJNR Am J Neuroradiol 1993;14:629–635.
70. van der Knaap MS, Bakker HD, Valk J. MR imaging and proton spectroscopy in 3-hydroxy-3-methylglutaryl coenzyme A lyase deficiency. AJNR Am J Neuroradiol 1998;19:378–382.
71. Yalcinkaya C, Dincer A, Gunduz E, Ficicioglu C, Kocer N, Aydin A. MRI and MRS in HMG-CoA Lyase Deficiency. Pediatr Neurol 1999;20:375–380.
72. Iles R, Jago JR, Williams SR, Chalmers RA. 3-hydroxy-3-methylglutaryl-coenzyme A lyase deficiency studied using 2-dimensional proton nuclear magnetic resonance spectroscopy. FEBS Lett 1986;203:49–53.
73. Hanefeld F, Kruse B, Bruhn H, Frahm J. In vivo proton magnetic resonance spectroscopy of the brain in a patient with L-2-hydroxyglutaric acidemia. Pediatr Res 1994;35:614–616.
74. Salvan AM, Chabrol B, Lamoureux S, Confort-Gouny S, Cozzone PJ, Vion-Dury J. In vivo brain proton MR spectroscopy in a case of molybdenum cofactor deficiency. Pediatr Radiol 1999;29:846–848.
75. Engelbrecht V, Rassek M, Huismann J, Wendel U. MR and proton MR spectroscopy of the brain in hyperhomocysteinemia caused by methylenetetrahydrofolate reductase deficiency. AJNR Am J Neuroradiol 1997;18:536–539.
76. Horstman M, Neumaier-Probst E, Lukacs Z, Steinfeld R, Ullrich K, Kohlschutter A. Infantile cobalamin deficinecy with cerrebral lactate accumulation and sustained choline depletion. Neuropediatrics 2003;34:261–264.
77. Frerman FE, Goodman SI. Deficiency of electron transfer flavoprotein or electron transfer flavoprotein:ubiquinone oxidoreductase in glutaric acidemia type II fibroblasts. Proc Natl Acad Sci USA 1985; 82:4517–4520.
78. Takanashi J, Fujii K, Sugita K, Kohno Y. Neuroradiologic findings in glutaric aciduria type II. Pediatr Neurol 1999;20:142–145.
79. Shevell MI, Didomenicantonio G, Sylvain M, Arnold DL, O'Gorman AM, Scriver CR. Glutaric acidemia type II: neuroimaging and spectroscopy evidence for developmental encephalomyopathy. Pediatr Neurol 1995;12:350–353.
80. Lehnert W, Sperl W, Suormala TM, Baumgartner ER. Proprionic acidaemia: clinical, biochemical and therapeutic aspects. Experience in 30 patients. Eur J Pediatr. 1994;153:68–80.
81. Surtees RA, Matthews EE, Leonard JV. Neurologic outcome of proprionic acidemia. Pediatr Neurol 1992;8:333–337.
82. Bergman AJ, Van der Knaap MS, Smeitink JA, et al. Magnetic resonance imaging and spectroscopy of the brain in propionic acidemia: clinical and biochemical considerations. Pediatr Res 1996;40:404–409.
83. Chemelli AP, Schocke M, Sperl W, Trieb T, Aichner F, Felber S. Magnetic resonance spectroscopy (MRS) in five patients with treated propionic acidemia. J Magn Reson Imaging 2000;11:596–600.
84. Duncan DB, Herholz K, Kugel H, et al. Positron emission tomography and magnetic resonance spectroscopy of cerebral glycolysis in children with congenital lactic acidosis. Ann Neurol 1995;37:351–358.
85. Kim HS, Kim DI, Lee BI, et al. Diffusion-weighted image and MR spectroscopic analysis of a case of MELAS with repeated attacks. Yonsei Med J 2001;42:128–133.
86. Cross JH, Gadian DG, Connelly A, Leonard JV. Proton magnetic resonance spectroscopy studies in lactic acidosis and mitochondrial disorders. J Inherited Metabolic Dis 1993;16:800–811.
87. Krageloh-Mann I, Grodd W, Niemann G, Haas G, Ruitenbeek W. Assessment and therapy monitoring of Leigh disease by MRI and proton spectroscopy. Pediatr Neurol 1992;8:60–64.
88. Matthews PM, Andermann F, SIlver K, Karpati G, Arnold DL. Proton MR spectroscopic characterization of differences in regional brain metabolic abnormalities in mitochondrial encephalomyopathies. Neurology. 1993;43:2484–2490.
89. Barkovich AJ, Good WV, Koch TK, Berg BO. Mitochondrial disorders: analysis of their clinical and imaging characteristics. Am J Neuroradiol 1993;14:1119–1137.
90. Krageloh-Mann I, Grodd W, Schoning M, Marquard K, Nagele T, Ruitenbeek W. Proton spectroscopy in five patients with Leigh's disease and mitochondrial enzyme deficiency.[comment]. Dev Med Child Neurol 1993;35:769–776.
91. Moller HE, Wiedermann D, Kurlemann G, Hilbich T, Schuierer G. Application of NMR spectroscopy to monitoring MELAS treatment: a case report. Muscle Nerve. 2002;25:593–600.
92. Kugel H, Heindel W, Roth B, Ernst S, Lackner K. Proton MR spectroscopy in infants with cerebral energy deficiency due to hypoxia and metabolic disorders. Acta Radiol 1998;39:701–710.
93. Takahashi S, Oki J, Miyamoto A, Okuno A. Proton magnetic resonance spectroscopy to study the metabolic changes in the brain of a patient with Leigh syndrome. Brain Dev 1999;21:200–204.
94. Lin DDM, Crawford TO, Barker PB. Proton MR Spectroscopy in the Diagnostic evaluation of suspected mitochondrial disease. Am J Neuroradiol 2003;24:33–41.
95. Bianchi MC, Bianchi MC, Tosetti M, , Battini R, Manca ML, Mancuso M, et al. Proton MR spectroscopy of mitochondrial diseases: analysis of brain metabolic abnormalities and their possible diagnostic relevance. Am J Neuroradiol 2003;24:1958–1966.

96. Brockmann K, Bjornstad A, Dechent P, et al. Succinate in dystrophic white matter: a proton magnetic resonance spectroscopy finding characteristic for complex II deficiency. Ann Neurol. 2002;52: 38–46.
97. Rubio-Gozalbo ME, Heerschap A, Trijbels JM, De Meirleir L, Thijssen HO, Smeitink JA. Proton MR spectroscopy in a child with pyruvate dehydrogenase complex deficiency. Magn Reson Imaging. 1999;17:939–44.
98. Robinson JN, Norwitz ER, Mulkern R, Brown SA, Rybicki F, Tempany CM. Prenatal diagnosis of pyruvate dehydrogenase deficiency using magnetic resonance imaging. Prenatal Diagnosis 2001;21:1053–1056.
99. Bates TE, Strangward M, Keelan J, Davey GP, Munro PMG, Clark JB. Inhibition of N-acetylaspartate production: implications for 1H MRS studies in vivo. Neuroreport 1996;7:1397–1400.
100. Loes DJ PC, Krivit W. Globoid Cell Leukodystrophy: Distinguishing Early-Onset from Late-Onset Disease Using a Brain MR Imaging Scoring Method. Am J Neuroradiol 1999;20:316–323.
101. Zarifi MK, Tzika AA, Astrakas LG, Poussaint TY, Anthony DC, Darras BT. Magnetic resonance spectroscopy and magnetic resonance imaging findings in Krabbe's disease. J Child Neurol 2001;16:522–526.
102. Farina L, Bizzi A, Finocchiaro G, et al. MR imaging and proton MR spectroscopy in adult Krabbe disease. Ajnr: Am J Neuroradiol. 2000;21:1478–1482.
103. Faerber EN MJ, Smergel EM. MRI appearances of metachromatic leukodystrophy. Pediatr Radiol1999;29:669–672.
104. Kruse B, Hanefeld F, Christen HJ, et al. Alterations of brain metabolites in metachromatic leukodystrophy as detected by localized proton magnetic resonance spectroscopy in vivo. J Neurol 1993;241: 68–74.
105. Confort-Gouny S, Vion-Dury J, Chabrol B, Nicoli F, Cozzone PJ. Localised proton magnetic spectroscopy in X-linked adrenoleukodystropy. Neuroradiology 1995;37:568–575.
106. Izquierdo M, Adamsbaum C, Benosman A, Aubourg P, Bittoun J. MR spectroscopic imaging of normal-appearing white matter in adrenoleukodystrophy. Pediatr Radiol. 2000;30:621–9.
107. Pouwels PJ, Kruse B, Korenke GC, Mao X, Hanefeld FA, Frahm J. Quantitative proton magnetic resonance spectroscopy of childhood adrenoleukodystrophy. Neuropediatrics 1998;29:254–264.
108. Kruse B, Barker PB, Van Zijl PC, Duyn JH, Moonen CT, Moser HW. Multislice proton magnetic reosnance spectroscopic imaging in X-linked adrenoleukodystrophy. Ann Neurol. 1994;36:595–608.
109. Tzika AA, Ball WS, Jr., Vigneron DB, Dunn RS, Nelson SJ, Kirks DR. Childhood adrenoleukodystrophy: assessment with proton MR spectroscopy. Radiology 1993;189:467–480.
110. Engelbrecht V, Rassek M, Gartner J, Kahn T, Modder U. The value of new MRI techniques in adrenoleukodystrophy. Pediatr Radiol. 1997;27:207–215.
111. Sener RN. Atypical X-linked adrenoleukodystrophy: new MRI observations with FLAIR, magnetization transfer contrast, diffusion MRI, and proton spectroscopy. Magn Reson Imaging. 2002;20:215–219.
112. Korenke GC, Pouwels PJ, Frahm J, et al. Arrested cerebral adrenoleukodystrophy: a clinical and proton magnetic resonance spectroscopy study in three patients. Pediatr Neurol. 1996;15:103–7.
113. Rajanayagam V, Grad J, Krivit W, et al. Proton MR spectroscopy of childhood adrenoleukodystrophy. Am J Neuroradiol 1996;17:1013–1024.
114. Groenendaal F, Bianchi MC, Battini R, et al. Proton magnetic resonance spectroscopy (1H-MRS) of the cerebrum in two young infants with Zellweger syndrome. Neuropediatrics 2001;32:23–27.
115. Brockmann K, Dechent P, Meins M, et al. Cerebral proton magnetic resonance spectroscopy in infantile Alexander disease. J Neurol 2003;250:300–306.
116. Imamura A, Orii KE, Mizuno S, Hoshi H, Kondo T. MR imaging and 1H-MR spectroscopy in a case of juvenile Alexander disease. Brain Development. 2002;24:723–726.
117. Takanashi J, Sugita K, Tanabe Y, Niimi H. Adolescent case of Alexander disease: MR imaging and MR spectroscopy. Pediatr Neurol. 1998;18:67–70.
118. Matalon R, Michals-Matalon K. Biochemistry and molecular biology of Canavan disease. Neurochemistry Research 1999;24:507–513.
119. Marks HG, Caro PA, Wang Z, et al. Use of computed tomography, magnetic resonance imaging, and localized 1H magnetic resonance spectroscopy in Canavan's disease: a case report. Ann Neurol. 1991;30:106–110.
120. Austin SJ, Connelly A, Gadian DG, Benton JS, Brett EM. Localized 1H NMR spectroscopy in Canavan's disease: a report of two cases. Magn Reson Med 1991;19:439–445.
121. Wittsack HJ, Kugel H, Roth B, Heindel W. Quantitative measurements with localized 1H MR spectroscopy in children with Canavan's disease. J Magn Reson Imaging. 1996;6:889–893.
122. Leone P, Janson CG, Bilianuk L, et al. Aspartoacylase gene transfer to the mammalian central nervous system with therapeutic implications for Canavan disease. Ann Neurol. 2000;48:27–38.
123. van der Knaap MS, Barth PG, Stroink H. Leukoencephalopathy with swelling and a discrepantly mild clinical course in eight children. Ann Neurol. 1995;37:324–334.
124. Sener RN. van der Knaap syndrome: MR imaging findings including FLAIR, diffusion imaging, and proton MR spectroscopy. Eur Radiol 2000;10:1452–1455.
125. Mejaski-Bosnjak V, Besenski N, Brockmann K, Pouwels PJ, Frahm J, Hanefeld FA. Cystic leukoencephalopathy in a megalencephalic child: clinical and magnetic resonance imaging/magnetic resonance spectroscopy findings. Pediatr Neurol. 1997;16:347–350.
126. Brockmann K, Finsterbusch J, Terwey B, Frahm J, Hanefeld F. Megalencephalic leukoencephalopathy with subcortical cysts in an adult: quantitative proton MR spectroscopy and diffusion tensor MRI. Neuroradiology. 2003;45:137–142.
127. Topcu M, Gartioux C, Ribierre F. Vacuolating megalencephalic leukoencephalopathy with subcortical cysts, mapped to chromosome 22qtel. Am J Human Genet. 2000;66:733–739.
128. van der Knaap MS, Kamphorst W, Barth PG, Kraaijeveld CL, Gut E, Valk J. Phenotypic variation in leukoencephalopathy with vanishing white matter. Neurology 1998;51:540–547.
129. Tedeschi G, Bonavita S, Barton NW, et al. Proton magnetic resonance spectroscopic imaging in the clinical evaluation of patients with Niemann-Pick type C disease. J Neurol Neurosurg Psychiatry 1998;65:72–79.
130. Seitz D, Grodd W, Schwab A, Seeger U, Klose U, Nagele T. MR imaging and localized proton MR spectroscopy in late infantile neuronal ceroid lipofuscinosis. AJNR Am J Neuroradiol. 1998;19:1373–1377.
131. Brockmann K, Pouwels PJ, Christen HJ, Frahm J, Hanefeld F. Localized proton magnetic resonance spectroscopy of cerebral metabolic disturbances in children with neuronal ceroid lipofuscinosis. Neuropediatrics. 1996;27:242–248.
132. Confort-Gouny S, Chabrol B, Vion-Dury J, Mancini J, Cozzone PJ. MRI and localized proton MRS in early infantile form of neuronal ceroid lipofuscinosis. Pediatr Neurol 1993;9:57–60.
133. Breunig F, Weidemann F, Beer M, et al. Fabry disease: diagnosis and treatment. Kidney IntSuppl 2003:S181–S185.
134. Neufeld EF, Meunzer J. The mucopolysaccharidoses. In: Scriver CR, Beaudet AL, Sly WS, Valle D, eds. The Metabolic and Molecular Bases of Inherited Disease. New York: McGraw-Hill; 1995:2465–2494.
135. Bonavita S, Virta A, Jeffries N, Goldin E, Tedeschi G, Schiffmann R. Diffuse neuroaxonal involvement in mucolipidosis IV as assessed by proton magnetic resonance spectroscopic imaging. J Child Neurol 2003;18:443–449.
136. Takahashi Y, Sukegawa K, Aoki M, et al. Evaluation of accumulated mucopolysaccharides in the brain of patients with mucopolysaccharidoses by (1)H-magnetic resonance spectroscopy before and after bone marrow transplantation. Pediatr Res 2001;49:349–355.
137. Savage AV, Applegarth DA. Diagnosis of mucopolysaccharidoses using 1H-n.m.r. spectroscopy of glycosaminoglycans. Carbohydrate Res 1986;149:471–474.
138. Seto T, Kono K, Morimoto K, et al. Brain magnetic resonance imaging in 23 patients with mucopolysaccharidoses and the effect of bone marrow transplantation. Ann Neurol 2001;50:79–92.

139. De Laurenzi VD, Rogers GR, Hamrock DJ, et al. Sjogren-Larsson syndrome is caused by mutatuions in the fatty aldehyde dehydrogenase gnee. Nat Genet 1996;12:52–57.
140. Mano T, Ono J, Kaminaga T, et al. Proton MR spectroscopy of Sjogren-Larsson's syndrome. AJNR Am J Neuroradiol 1999;20: 1671–1673.
141. Miyanomae Y, Ochi M, Yoshioka H, et al. Cerebral MRI and spectroscopy in Sjogren-Larsson syndrome: case report. Neuroradiology 1995;37:225–228.
142. De Stefano N, Dotti MT, Mortilla M, Federico A. Magnetic resonance imaging and spectroscopic changes in brains of patients with cerebrotendinous xanthomatosis. Brain 2001;124:121–131.
143. Joutel A, Corpechot C, Ducros A, et al. Notch3 mutations in CADASIL, a hereditary adult-onset condition causing stroke and dementia. Nature 1996;383:707–710.
144. Joutel A, Andreux F, Gaulis S, et al. The ectodomain of the Notch3 receptor accumulates within the cerebrovasculature of CADASIL patients. J Clin Invest 2000;105:597–605.
145. Auer DP, Schirmer T, Heidenreich JO, Herzog J, Putz B, Dichgans M. Altered white and gray matter metabolism in CADASIL: a proton MR spectroscopy and 1H-MRSI study. Neurology 2001;56: 635–642.
146. Berry GT, Hunter JV, Wang Z, et al. In vivo evidence of brain galactitol accumulation in an infant with galactosemia and encephalopathy. J Pediatr 2001;138:260–262.
147. Wang ZJ, Berry GT, Dreha SF, Zhao H, Segal S, Zimmerman RA. Proton magnetic resonance spectroscopy of brain metabolities in galactosemia. Ann Neurol. 2001;50:266–269.
148. Moller HE, Ullrich K, Vermathen P, Schuierer G, Koch HG. In vivo study of brain metabolism in galactosemia by 1H and 31P magnetic resonance spectroscopy. Eur J Pediatr. 1995;154:S8–S13.
149. Bruhn H, Frahm J, Merboldt KD, et al. Multiple sclerosis in children: cerebral metabolic alterations monitored by localized proton magnetic resonance spectroscopy in vivo. Ann Neurol. 1992;32: 140–150.
150. Malik GK, Pandey M, Kumar R, Chawla S, Rathi B, Gupta RK. MR imaging and in vivo proton spectroscopy of the brain in neonates with hypoxic ischemic encephalopathy. Eur J Radiol 2002;43:6–13.
151. Amess PN, Penrice J, Wylezinska M, et al. Early brain proton magnetic resonance spectroscopy and neonatal neurology related to neurodevelopmental outcome at 1 year in term infants after presumed hypoxic-ischaemic brain injury. Dev Med Child Neurol 1999;41:436–445.
152. Noda S, Yamanouchi N, Okada S, et al. Proton MR spectroscopy in solvent abusers. Ann NY Acad Sci 1996;801:441–444.
153. Sakamoto K, Murata T, Omori M, et al. Clinical studies on three cases of the interval form of carbon monoxide poisoning: serial proton magnetic resonance spectroscopy as a prognostic predictor. Psychiatry Res 1998;83:179–192.
154. Murata T, Kimura H, Kado H, et al. Neuronal damage in the interval form of CO poisoning determined by serial diffusion weighted magnetic resonance imaging plus 1H-magnetic resonance spectroscopy. J Neurol Neurosurg Psychiatry 2001;71:250–253.
155. Smith LM, Chang L, Yonekura ML, et al. Brain proton magnetic resonance spectroscopy and imaging in children exposed to cocaine in utero. Pediatrics 2001;107:227–231.
156. Smith LM, Chang L, Yonekura ML, Grob C, Osborn D, Ernst T. Brain proton magnetic resonance spectroscopy in children exposed to methamphetamine in utero. Neurology 2001;57:255–260.
157. Simone IL, Federico F, Tortorella C, et al. Localised 1H-MR spectroscopy for metabolic characterisation of diffuse and focal brain lesions in patients infected with HIV. J Neurol Neurosurg Psychiatry 1998;64:516–523.
158. Pavlakis SG, Lu D, Frank Y, et al. Brain lactate and N-acetylaspartate in pediatric AIDS encephalopathy. Ajnr: Am J Neuroradiol. 1998;19: 383–385.
159. Brown MS, Stemmer SM, Simon JH, et al. White matter disease induced by high-dose chemotherapy: longitudinal study with MR imaging and proton spectroscopy [comment]. Ajnr: Am J Neuroradiol 1998;19:217–221.
160. Davidson A, Tait DM, Payne GS, et al. Magnetic resonance spectroscopy in the evaluation of neurotoxicity following cranial irradiation for childhood cancer. Br J Radiol 2000;73:421–424.
161. Filippi CG, Ulug AM, Deck MD, Zimmerman RD, Heier LA. Developmental delay in children: assessment with proton MR spectroscopy. AJNR Am J Neuroradiol. 2002;23:882–888.
162. Gruetter R, Weisdorf SA, Rajanayagan V, et al. Resolution improvements in in vivo 1H NMR spectra with increased magnetic field strength. J Magn Reson. 1998;135:260–264.
163. Gonen O, Gruber S, Li BS, Mlynarik V, Moser E. Multivoxel 3D proton spectroscopy in the brain at 1.5 versus 3.0 T: signal-to-noise ratio and resolution comparison. AJNR Am J Neuroradiol 2001;22: 1727–1731.
164. Dreher W, Leibfritz D. A new method for fast proton spectroscopic imaging: spectroscopic GRASE. Magn Reson Med 2000;44:668–672.
165. Mader I, Seeger U, Weissert R, et al. Proton MR spectroscopy with metabolite-nulling reveals elevated macromolecules in acute multiple sclerosis. Brain 2001;124:953–961.
166. Thomas MA, Yue K, Binesh N, et al. Localized Two-Dimensional Shift Correlated MR Spectroscopy of Human Brain. Magn Reson Med 2001;46:58–67.
167. Binesh N, Yue K, Fairbanks L, Thomas MA. Reproducibility of localized 2D correlated MR spectroscopy. Magn Reson Med 2002;48:942–948.
168. Thomas MA, Ryner LN, Mehta M, Turski P, Sorenson JA. Localized J-resolved 1H MR spectroscopy of human brain tumors in vivo. J Magn Reson Imaging. 1996;6:453–459.
169. Rashed MS. Clinical applications of tandem mass spectrometry: ten years of diagnosis and screening for inherited metabolic diseases. J Chromatogr B 2001;758:27–48.
170. Collins FS, Green ED, Guttmacher AE, Guyer MS. A vision for the future of genomics research. Nature 2003;422:835–847.

22 Childhood Mitochondrial Disorders and Other Inborn Errors of Metabolism Presenting With White Matter Disease

Adeline Vanderver, MD and Andrea L. Gropman, MD, FAAP, FACMG

SUMMARY

Advances in neuroimaging, particularly magnetic resonance imaging (MRI), have made possible the study of normal and abnormal brain myelination. Although many disorders present with nonspecific MRI findings, more detailed analysis with newer imaging modalities such as diffusion tensor imaging (DTI), magnetic resonance spectroscopy (MRS), and magnetization transfer imaging (MTI), allow detailed investigation of white matter microstructure, abnormalities that may precede and evade detection on conventional MRI. The purpose of this chapter is to discuss the utility of neuroimaging in the study and differentiation of mitochondrial disorders and other inborn errors of metabolism. The first section will discuss imaging modalities used and their application in the diagnosis and study of metabolic disease. The second section will focus on some characteristic disorders in which abnormal white matter is visualized on neuroimaging. Although not meant to be an exhaustive summary, several key disorders and their clinical and neuroimaging features will be presented. Finally, a table as well as some examples will serve to solidify the differentiation between the disorders and illustrate the best neuroimaging modalities with which to investigate etiology and, in some cases, follow disease progression and response to therapies.

Key Words: Neuroimaging; inborn errors of metabolism; white matter; leukodystrophy; mitochondrial cytopathy; magnetic resonance imaging; magnetic resonance spectroscopy; diffusion tensor imaging.

1. INTRODUCTION

1.1. INBORN ERRORS OF METABOLISM/ MITOCHONDRIAL DISORDERS

Inborn errors of metabolism are genetic disorders inherited in an autosomal-recessive or X-linked fashion, characteristically presenting in infancy or early childhood with acute, but often nonspecific features, and typically involving severe devastation to the central nervous system and other organs. Although individually rare, collectively these disorders can account for a prevalence of 1:5–6000 live births. Several groups of metabolic disorders, notably mitochondrial cytopathies, peroxisomal disorders, amino acid disorders, urea cycle defects, organic acidemias, and lysosomal disorders, can cause damage to the cerebral (and in some cases peripheral) myelin, which can be visualized by magnetic resonance imaging (MRI).

The pattern of white matter involvement may be nonspecific or, alternatively, suggestive, selective, or specific, allowing clinical differentiation. A number of imaging modalities may be used to assist in the diagnosis and follow up of such disorders.

1.2. APPLICATION OF MRI TO STUDY METABOLIC DISORDERS

MRI is based on the principles of nuclear magnetic resonance, a spectroscopic technique used by scientists to obtain microscopic chemical and physical information about molecules *(1)*. MRI became widely available in the mid 1980s and has become the preferred imaging modality for clinical and research anatomical imaging *(2)*. Unlike computed tomography (CT) scanning, MRI can be obtained in multiple planes and can produce highly resolute images. By using different pulse sequences for image acquisition, tissue contrast can be manipulated to yield images that are suited for specific purposes. The lack of ionizing radiation makes it safe for repeated scanning of an individual and particularly for the scanning of infants and young children.

Neuroimaging technologies can be applied to neurogenetic research problems and provide important information regarding brain function under pathophysiologic conditions and allow for correlation with the disease states. This information can help clarify unanswered questions about specific genetic influences in children. The neurological development of children with neurogenetic disorders enables the investigator to identify the anatomical and functional differences caused by each genetic syndrome. This may result in the development of biological therapies targeted to benefit selected brain regions that may be uniquely affected by each disorder.

From: *Bioimaging in Neurodegeneration*
Edited by P. A. Broderick, D. N. Rahni, and E. H. Kolodny

1.3. MRI PHYSICS

The property of MRI that allows visualization relies on the property of protons in a magnetic field. MRI physics are based on the concept of spin, which is a property of protons and neutrons found in atomic nuclei. A spinning proton defines a magnetic moment. However, no proton spins exactly on its longitudinal axis. Instead, a proton's magnetic moment precesses about it. Nuclei of a particular element in a given magnetic field strength have a specific precessional frequency, the Larmor frequency. Although many protons have spin, the most commonly evaluated element is hydrogen, which is the most physiologically abundant because approx 60–70% of the human body is composed of water. Each water molecule in turn consists of two hydrogen protons. The most common magnetically active protons studied include hydrogen (^{1}H) and phosphorus (^{31}P). When an individual is placed in a superconducting magnet (i.e., an MRI scanner) a percentage of the protons will align with the externally applied magnetic field, B_0, along an axis, Z. The application of a radiofrequency pulse is sufficient in duration and amplitude to perturb the protons from the *z*-axis into an XY plane. When the radiofrequency pulse is turned off, the protons are allowed to relax back to their original orientation. This relaxation depends upon the magnitude and chemical environment of the proton (i.e., tissue characteristics) and allows tissue contrast because gray matter, white matter, and cerebrospinal fluid (CSF) have differing numbers of protons. Two types of relaxation are seen, longitudinal and transverse, and contribute to the T_1 and T_2 properties of a tissue.

1.4. DIFFUSION TENSOR IMAGING

Diffusion tensor imaging (DTI) allows study of white matter microstructure by enabling the investigation of the orientation of brain pathways in vivo. In axons, water diffusion is impeded by cell walls and myelin sheaths. As a result, water movement along the axis of an axon is much larger than water movement perpendicular to it. DTI allows visualization of this movement by characterizing water diffusion in three-dimensional space *(3)*. The apparent diffusion coefficient (ADC) is a measure of the diffusion of a molecule. Small molecules such as free water have a high diffusion coefficient, whereas water that is bound to a large protein in cytosol will have a low diffusion coefficient. Temperature, shape, and size of the molecule, as well as integrity of the myelin sheaths, help to determine the ADC. Tissues with random microstructure or unrestricted media will have diffusion that is equal in all directions, or isotropic diffusion. Tissues with ordered microstructure will exhibit diffusion that is greater in some directions than in others, or anisotropic diffusion. Such images provide useful information regarding white matter myelination and its integrity.

1.5. APPLICATIONS OF DTI TO NEUROGENETICS

DTI has been applied to the study of white matter abnormalities most extensively in X-linked adrenal leukodystrophy, an X-linked neurodegenerative disorder involving predominantly white matter tracts *(4)*. Eichler et al. have used DTI in X-linked adrenal leukodystrophy to show fractional anisotropy (FA) decreases and isotropic ADC (IADC) increased over the zones toward the center of the lesion *(5)*. Abnormalities in diffusion could be observed in white matter that appeared to be unaffected on routine anatomic MRI. DTI has been used in pilot studies to investigate the behavior of water diffusion in cerebral structural abnormalities associated with mitochondrial cytopathies and cerebral dysgenesis, demonstrating its ability to observe abnormal white matter microscopic integrity before anatomic abnormalities on routine MRI. It is expected that DTI will continue to find applications in the study of neurometabolic disorders involving white matter structures.

1.6. MAGNETIC RESONANCE SPECTROSCOPY (MRS)

MRS is a method whereby the chemical composition of magnetically active nuclei, such as ^{1}H or ^{31}P, within a tissue can be determined. The technique requires a large main magnetic field to orient the nuclear magnetic spins parallel or antiparallel with it. No magnetic field gradients are applied. The frequencies of the nuclei will differ depending upon what chemical group they are in. For example, protons on a methyl group, such as in lactate, will give an MRS signal at a slightly lower frequency than protons on an aromatic ring such as phenylalanine. The differences in frequency are small in the order of parts per million (ppm). ^{1}H MRS is capable of producing information on a large number of brain chemicals. The most common chemicals studied via the identification and integration of the spectral peaks found in the "proton" (^{1}H) spectra include *N*-acetyl-λ–aspartate (NAA), creatine, phosphocreatine, choline, myo-inositol, lactate, glutamate, and glutamine. In addition, amino acids, lipids, and γ-aminobutyric acid (GABA) may be detected. Multiple techniques are available, including single voxel spectroscopy, which allows sampling from one brain region, and spectroscopic imaging, which enables one to study the distribution of chemicals in the brain.

Spectroscopy investigations are performed to study brain chemistry and ascertain from the individual ^{1}H visible chemicals some of the following information with regard to disease: NAA for its role in mitochondrial oxidative metabolism and as a putative marker for neuronal viability as well as its role in lipid synthesis (source of acetyl groups); creatine and phosphocreatine as creatine to phosphocreatine energy conversion mediated is by creatine kinase; choline, a precursor for neurotransmitter acetylcholine and membrane phospholipids, phosphatidylcholine, and sphingomyelin; myo-inositol, neuronal signaling of the phosphoinositide pathway, osmoregulation, cell nutrition, and detoxification; lactate, which is a byproduct of anaerobic metabolism, elevated concentrations resulting from glycolytic metabolism as happens in brain ischemia; and glutamate and glutamine, which are major excitatory and inhibitory neurotransmitters in the central nervous system (CNS; refs. *6* and *7*).

Proton MRS may complement MRI in diagnostic assessment and therapeutic monitoring of neurodegenerative disorders. It may enable detection of lesions or injury before abnormalities on routine MRI or DTI. MRS has low sensitivity, and metabolite concentrations are low, generative signals that are typically 10,000 times less intense than a water signal. One of the difficulties in interpretation of such spectra is the poor spectral resolution (overlap) of peaks from these compounds with those of underlying protons from macromolecules. By changing the imaging parameters (i.e., in the case of lactate) or adding editing software (i.e., as with GABA) one might be able better differentiate these compounds. Applying ^{1}H MRS to patients with neurogenetic/neurometabolic disorders requires normative data, which will need to take into account age and brain region. The relative amount of metabolites, often expressed relative to Cr, is affected by age. In addition to comparing ratios, quantitative MRS is useful in these settings.

1.7. APPLICATIONS OF MRS IN THE STUDY OF NEUROMETABOLIC DISORDERS

Despite its potential, MRS has had limited use in detecting pediatric neurogenetic and neurometabolic disorders. Spectra obtained are rarely specific to a single disorder; however, the analysis of metabolite peaks may allow separation into distinct groups and, when used in combination with clinical, biochemical, and molecular data, provide a predictor for disease severity, response to therapy, and outcome. MRS, however, may provide specific diagnostic information for creatine synthesis disorders, Canavan's disease, and the amino acidopathies phenylketonuria, maple syrup urine disease, and urea cycle disorders *(8–12)*. Increases in NAA can be demonstrated on proton MRS, offering an additional noninvasive diagnostic test for establishing the diagnosis of Canavan disease.

1.8. OXIDATIVE METABOLISM DISORDERS

Disorders of oxidative metabolism, which result in a shift to anaerobic glycolysis with lactate formation, can be studied by MRS and separated from overlapping lipids by recognition of the doublet nature and ability to invert it with long TE in spin echo sequences. Correlations between CSF lactate and lactate measured by ^{1}H MRS in mitochondrial cytopathies have been demonstrated *(13)*; thus, MRS can be used as a noninvasive alternative to CSF lactate sampling. Regional metabolite variations may suggest the metabolic basis for observed neurocognitive phenotypic features. Derangements in oxidative metabolism have been demonstrated by MRS to correlate with degree of clinical decompensation and response to therapy in Leigh syndrome *(14,15)*. Elevated lactate also may be demonstrated in other disorders of electron transport chain function, such as glutaric academia type II *(16)*.

Detection of CNS lactate has been shown to be useful in the diagnosis of mitochondrial disorders. Using multisection spectroscopy, Lin et al. studied 29 patients believed to have a mitochondrial disorder based upon biochemical, molecular, or pathologic criteria *(17)*. A high level of lactate on MRS correlated well with other markers of mitochondrial disease. However, it was found that abnormal CNS lactate concentrations could be missed by MRS as a result of differences in the type of mitochondrial cytopathy, timing of the study relative to disease exacerbation, or location of the region of interest relative to most affected tissue.

^{1}H MRS has been used to study patients with urea cycle disorders *(18,19)*. Findings in patients with late onset ornithine transcarbamylase deficiency demonstrate myoinositol depletion in association with glutamine accumulation, followed by choline depletion, the reverse of which is seen in hepatic encephalopathy and thus may be used to study responses to gene dosage and therapy.

1.9. NEUROTRANSMITTER DISORDERS

Recent improvements in MRI have allowed methods to measure synthesis rates and turnover of neurotransmitters. Glutamate, an excitatory amino acid/neurotransmitter, which plays a major role in many nervous system pathways, can be resolved from glutamine at field strengths of 2.0 Tesla and greater. There have been limited studies addressing glutamate concentrations in pediatric inborn errors of metabolism and neurogenetic disorders. Glycine, an inhibitory neurotransmitter, can be measured in the brain of children with defects in the glycine cleavage system *(20,21)*. In addition to obtaining measurements of brain glycine concentrations, the response to treatment can be assessed by obtaining serial measurements. GABA, another inhibitory neurotransmitter in the brain, is important in brain development and is elevated in succinic semialdehyde dehydrogenase deficiency, an inborn error of GABA degradation *(22)*.

1.10. ^{31}P MRS

This modality has been used to investigate altered energy status of the CNS and muscle tissue as may accompany mitochondrial cytopathies, leukodystrophies *(23)*, familial hemiplegic migraine, and Glutaric aciduria type 1. Proton decoupled ^{31}P MRS separates and quantifies the phosphomonoesters phosphorylcholine and phosphorylethanolamine and the phosphodiesters glycerophosphorylethanolamine and glycerophosphorylcholine. These metabolites, which are involved in myelin biosynthesis, are components of important biologic compounds, including lecithin, plasmalogen, and sphingomyelin *(24)*. Another application of ^{31}P MRS is in defining brain energy states. Total creatine peak on ^{1}H MRS is incompletely defined. ^{31}P MRS allows quantification of adenosine triphosphate, inorganic phosphate, phosphocreatine, and determination of pH.

2. SPECIFIC DISORDERS (*SEE ALSO* TABLES 1 AND 2, PP. 284–286)

2.1. LYSOSOMAL DISORDERS

Lysosomal disorders may present as a classic leukodystrophy or a leukoencephalopathy associated with systemic manifestations. The disorders discussed are not exhaustive of lysosomal disease with white matter disorders and are meant to be representative.

2.1.1. Metachromatic Leukodystrophy

Metachromatic leukodystrophy is also known as MLD; metachromatic leukoencephalopathy; cerebral sclerosis, diffuse, metachromatic form; sulfatide lipidosis; arylsulfatase A deficiency; arylsulfatase A (ARSA) deficiency; and cerebroside sulfatase deficiency.

2.1.1.1. General

Metachromatic leukodystrophy is a lysosomal storage disorder caused by the deficiency of arylsulfatase A, or cerebroside sulfatase activator (saposin B), causing the storage of the sphingolipid sulfatide. The disease is characterized by a progressive demyelination, which results in progressive neurologic dysfunction, including pyramidal signs, polyneuropathy, bulbar symptoms, and ataxia. The disease is inherited in an autosomal-recessive manner with identified genes (ARSA and SAP B). The pathogenesis of the disease is poorly understood.

2.1.1.2. Clinical Features

There are six allelic forms of MLD: congenital; late infantile, juvenile, and adult forms; partial cerebroside sulfate deficiency; and pseudo-arylsulfatase A deficiency and nonallelic forms, including cerebroside sulfatase activator (saposin B) deficiency.

Congenital forms are reported but poorly described. The late infantile form is the most frequent presentation. Onset is before 2 yr of age, and death occurs in early childhood. Early development is normal but onset of gait disturbance and hypotonia rapidly become manifest. Ataxia may develop. Speech disturbance and dysarthria are present. A progressive, often-painful polyneuropathy can occur. Flaccid paresis caused by the polyneuropathy is gradually superseded by a spastic quadriplegia. The child may have episodes of hypertonic extension and crying. Bulbar dysfunction, mutism, and feeding difficulties develop. Seizures may rarely occur. A vegetative state ensues. The CSF contains elevated protein.

The juvenile form is similar to the infantile form, but occurs in a 5- to 18-yr-old previously healthy child.

In the adult form of metachromatic leukodystrophy, initial symptoms are often psychiatric, leading to a misdiagnosis of schizophrenia, or behavioral, leading to a diagnosis of dementia *(25,26)*. Neurologic symptoms, such as chorea or dystonia, appear late and may be missed *(27,28)*. Adult patients with MLD may also present with recurrent neuropathies *(29,30)* or new seizure disorders *(31)*. The diagnosis of adult MLD often is suspected only because of white matter abnormalities detected by CT and/or MRI.

2.1.1.3. Biochemical Features

Metachromatic leukodystrophy is biochemically characterized by an accumulation of sulfatides (sulfogalactosylceramides) mainly in oligodendrocytes and macrophages/microglia and less significantly in neurons. The deficient enzyme is a lysosomal hydrolase, cerebroside sulfate sulfatase (arylsulfatase A; ref. *32*). ARSA activity can be measured by enzyme assay in leukocytes or fibroblasts.

Pseudoarylsulfatase A deficiency refers to a condition of apparent arylsulfatase A enzyme deficiency in persons without neurologic abnormalities. Specific alleles have been linked to this state. Low arylsulfatase A is not necessarily indicative of disease, which should be taken into consideration when screening for the disease. However, certain studies have suggested that slightly low ARSA levels are seen more often in patients with psychiatric disease or pervasive developmental disorders.

2.1.1.4. Genetic Features

Gene Map Locus: *22q13.31-qter*. There is suggestion of genotype/phenotype correlation in metachromatic leukodystrophy. Metachromatic leukodystrophy occurs with an estimated frequency of 1/40,000, with clusters in Arabs living in Israel and the western Navajo.

2.1.1.5. Imaging Features

MRI is not characteristic in metachromatic leukodystrophy. There is diffuse high-intensity signal in the cerebral white matter on T_2-weighted images, initially of periventricular white matter and the centrum semiovale white matter. Demyelination is more prominent in the occipital region. Other common manifestations include involvement of the corpus callosum, the internal capsule, and the corticospinal tract *(33)*. Although initially sparing other structures, the disease progresses to involvement of the arcuate fibers and the cerebellar white matter *(34)*. The signal abnormality is sometimes described as a "tigroid" or "leopard-skin" appearance of the deep white matter similar to that seen in Pelizaeus–Merzbacher disease *(33,35)*.

Functional studies, such as ^{18}F-deoxyglucose positron emission tomography, reveal a very peculiar pattern of metabolic impairment in thalamic areas, medial and fronto-polar regions, and in occipital lobes *(36)*.

^{1}H MRS reveals a marked reduction of the neuronal marker, NAA, in white and gray matter, reduction in choline *(37)*, and elevated lactate in demyelinated areas. Metachromatic leukodystrophy patients also show a generalized increase of myo-inositol in the brain *(38)*. Diffusion MRI may reveal a cytotoxic edema-like pattern *(39)*.

2.1.1.6. Pathology

The most characteristic findings on pathology is the accumulated substrate of the affected ARSA enzyme. Galactosphingosulfatides that are strongly metachromatic, doubly refractile in polarized light, and pink with periodic acid-Schiff (PAS) staining are found in excess in the white matter of the central nervous system, in the kidney, and in the urinary sediment.

2.1.1.7. Treatment

Some benefit has been reported in adult-onset metachromatic leukodystrophy after bone marrow transplantation *(40)*. Bone marrow transplant has been less efficacious in infantile or juvenile metachromatic leukodystrophy caused by ARSA or saposin-B deficiency *(41,42)*, although it still may be helpful in presymptomatic or early symptomatic cases *(43)*.

2.1.2. KRABBE'S DISEASE

Krabbe disease is also known as globoid cell leukodystrophy (GLD, GCL); globoid cell leukoencephalopathy; galactosylceramide β-galactosidase deficiency; galactocerebrosidase deficiency; and GALC deficiency.

2.1.2.1. General

The first descriptions of Krabbe's disease were in 1916 by Krabbe. Globoid cell leukodystrophy (Krabbe's disease) is an autosomal disease caused by genetic defects in a lysosomal enzyme, galactosylceramidase (galactosyl ceramide β-galactoside deficiency). Typically, the disease occurs among infants and takes a rapidly fatal course, but rarer late-onset forms also exist (late infantile, juvenile, adult). Clinical manifestations are a result of central and peripheral white matter destruction. On pathologic specimens, there is infiltration of the disappearing myelinated tissue with multinuclear macrophages laden with periodic acid-Schiff material (globoid cells). The galactosylceramidase gene (GALC) has been cloned, and molecular diagnosis is available. However, the diagnosis is usually made by testing for a deficiency in enzyme activity. Deficiency of one of the sphingolipid activator proteins, saposin A, may also be responsible for a late-onset, slowly progressive globoid cell leukodystrophy. Efforts to treat this disorder have focused on bone marrow transplantation.

2.1.2.2. Clinical Features

Krabbe's disease has an extremely variable phenotype that is somewhat dependent on the age of presentation.

Early infantile Krabbe's disease is characterized by spasticity, motor regression, and seizures. Irritability is a significant feature described as typical of presentation at this age. Peripheral neuropathy with areflexia may be a prominent initial symptom *(44)* or develop as the disease progresses. A rapid comprehensive decline of neurologic function with optic atrophy, spastic quadriparesis, and a vegetative state occurs during the course of 1 to 2 yr and usually results in an early death. Cherry red spots on the retina are seen but may be subtle early in the course of the disease.

Late infantile Krabbe's disease has onset of initial symptoms after 2 yr of age. Early symptoms may include gait disturbance ("toppling") and visual failure. These are progressively complicated by regression of motor and cognitive development as described in the early infantile form. Most reported cases died within their first decade.

Late-onset Krabbe's disease has onset of symptoms from the middle of the second decade. The course is much more slowly progressive. Peripheral neuropathy is more commonly a presenting feature *(45)*. It is often associated with a slowly progressive spastic paraparesis or quadriparesis that may be asymmetric.

2.1.2.3. Biochemical Features

Globoid cell leukodystrophy (Krabbe's disease) is caused by mutations in galactosylceramidase, a lysosomal enzyme that acts to digest galactosylceramide, a glycolipid concentrated in myelin, and psychosine (galactosylsphingosine). The diagnosis is most often made by leukocyte assay for activity of galactosylceramidase (galactosyl ceramide β-galactoside deficiency or GALC). There has been no clear association of levels of GALC activity and age at presentation.

Pseudodeficiency of GALC on enzyme assay may be seen and is a diagnostic pitfall. It is defined as the measurement of low activity (usually less than 15% of the normal mean for controls) of an enzyme in a healthy person. This may complicate the differential diagnosis of presymptomatic people who will present with adult-onset clinical disease and also affects carrier detection. Confirmation of the enzymatic deficiency usually requires additional enzyme studies on fibroblast in vivo or deoxyribonucleic acid (DNA) mutation analysis. Prenatal testing may be performed on cultured chorionic villi.

2.1.2.4. Genetic Features

Gene Map Locus: *14q31*. The gene coding for galactocerebrosidase (GALC) was localized to human chromosome 14 *(46)* and then mapped to chromosome *14q31 (47)*. Animal data suggest that genetic saposin A deficiency could also cause late-onset chronic leukodystrophy without GALC deficiency *(48)*.

There has been some evidence of correlation between a difference in mutation sites and the clinical features of globoid cell leukodystrophy *(49)* but most authors describe significant limitations in genotype/ phenotype correlation *(50,51)*. There is a higher-than-usual incidence in a Druze isolate in Israel *(52)*.

2.1.2.5. Imaging Features

Krabbe's disease was described before the advent of MRI, and there is a body of literature describing CT findings. Characteristic CT scan imaging for the early manifestations of the disease consist of symmetric hyperdensities involving the cerebellum, thalami, caudate, corona radiata, and brain stem. In the later stages of the disease, brain atrophy, low density in the white matter, and calcification-like, symmetric, punctate high-density areas in the corona radiata are seen (Fig. 1A,B). There have been described cases where CT findings preceded MRI abnormality or were a more sensitive indicator of white matter abnormality *(53–55)*. It is important to keep in mind that there have been reports of infantile- and adult-onset globoid cell leukodystrophy in which clinical signs preceded any imaging abnormality *(56)*.

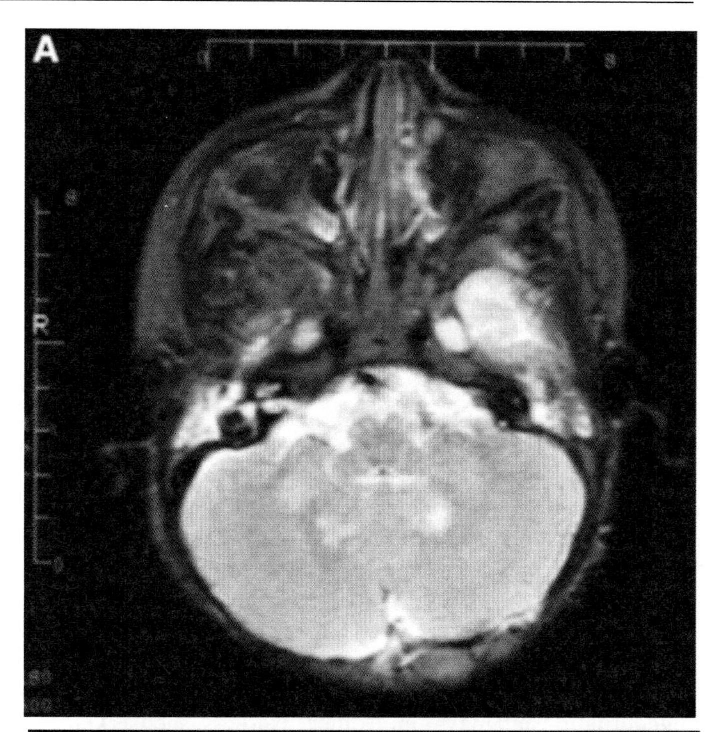

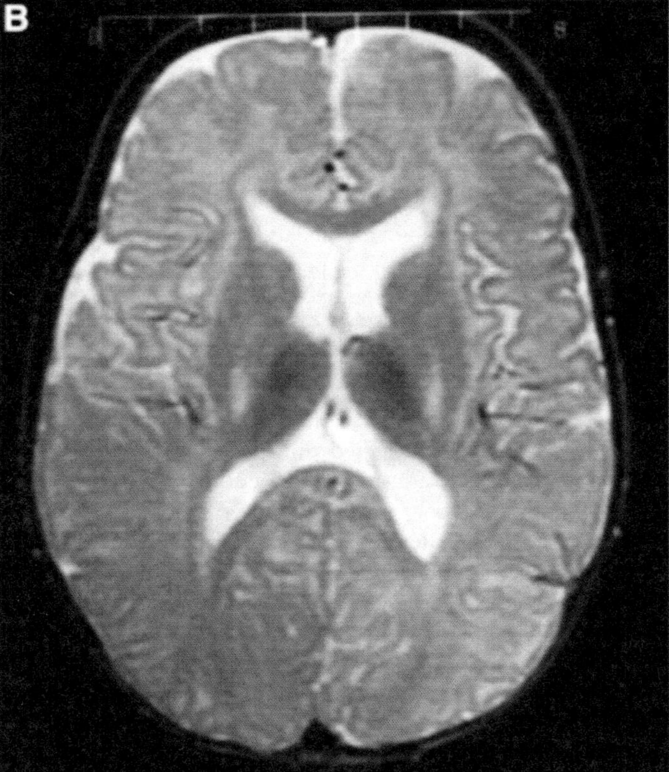

Fig. 1. (**A**) Krabbe's disease in an infant. Cerebellar abnormalities may be seen in infantile Krabbe's disease. (**B**) Krabbe's disease in an infant. The posterior portion of the internal capsule is more affected in this case.

The description of imaging abnormalities has increased since the use of MRI. The classic intracranial findings in early infantile globoid cell leukodystrophy on MRI are of T_2 hyperintensity without contrast enhancement of the posterior periventricular white matter and corticospinal tracts, sparing U fibers. Deep white matter abnormality around the atria and posterior limbs of the internal capsules is usually first seen followed by a significant decrease of white matter volume, generalized atrophy, and abnormal high signal in all white matter areas except the anterior limbs of the internal capsules. Posterior periventricular white matter, parietal lobes, corona radiata, centrum semiovale, and splenium of the corpus callosum appear predominantly affected. Characteristic linear patterns have been observed in the centrum semiovale on MRI *(57)*. The volume of the central gray nuclei may demonstrate abnormal high signal *(58)*, and cerebellar white matter lesions may be seen *(59)*. However, cerebellar white matter and deep gray matter involvement may be present only in early infantile-onset globoid cell leukodystrophy *(60)*.

The clinical symptomatology of a combined central and peripheral nervous system disorder is reflected in the imaging findings. There are a few cases where optic nerve enlargement has been noted to be a prominent feature *(61,62)*. Spinal nerve and cranial nerve enhancement may be a prominent, even isolated, finding and may mimic other disorders, such as Guillain Barré syndrome *(63)*. Spinal involvement has been shown on MRI as abnormal contrast enhancement of the lumbosacral nerve roots and cauda equina with no area of abnormal signal intensity or enhancement within the substance of the spinal cord *(64)*.

In adults, Krabbe's disease may present on MRI with selective corticospinal tract involvement (bilateral frontoparietal white matter, the centrum semiovale, and the posterior limb of the internal capsule with sparing of the periventricular white matter; ref. 65). These findings can sometimes be asymmetric *(66)*.

The application of MRS to Krabbe's disease has had interesting ramifications on the description of the various age related phenotypes. In infantile globoid cell leukodystrophy, pronounced elevation of both myo-inositol and choline-containing compounds in affected white matter reflect demyelination and glial proliferation. The accompanying decrease of NAA points to neuroaxonal loss. Gray matter shows similar, albeit much milder, alterations. Elevated lactate levels have been rarely reported *(67)*. In juvenile globoid cell leukodystrophy, MRS also shows a similar but milder pattern. In adult globoid cell leukodystrophy, results of MRS of affected white matter have been close to normal in some studies *(68)* and abnormal selectively in the pyramidal tract in other studies *(66)*.

DTI has been used in a limited fashion in Krabbe's disease. Early data suggest that diffusion tensor-derived anisotropy maps can provide a quantitative measure of abnormal white matter in patients with Krabbe's disease. DTI scans appear more sensitive than T_2-weighted images for detecting white matter abnormality and therefore may be useful as a marker of treatment response *(69)*.

2.1.2.6. Pathology

The pathology of Krabbe's disease is characteristic, consisting of a rapid and nearly complete disappearance of myelin structures and oligodendrocytes in the CNS. Damaged white matter is infiltrated with often-multinucleated macrophages ("globoid cells") that contain strongly PAS-positive materials. In peripheral nerves, uniformly thin myelin sheaths without evidence of demyelination, and the disappearance of Schwann cells is observed. Globoid cells may be seen in affected peripheral nerves *(70)*.

Pathophysiologic studies suggest that immunologic mechanisms may have a role in cellular apoptosis *(71)*. Psychosine may also play a significant role in apoptotic cell loss *(72)*.

2.1.2.7. Treatment

The only serious attempt at treating human patients is bone marrow transplantation, which can provide significant alleviation of symptoms and even reversal of CNS deterioration *(73)*, particularly in those patients with later-onset, more slowly progressive globoid cell leukodystrophy.

2.1.3. Niemann Pick Disease Type CI

Niemann Pick disease is also known as NPC, Niemann-Pick disease with cholesterol esterification block, neurovisceral storage disease with vertical supranuclear, and ophthalmoplegia

2.1.3.1 General

Niemann Pick disease type C is an autosomal-recessive disorder attributable to mutations in the NPC1 or 2 genes. It results from defective cholesterol esterification with accumulation of unesterified cholesterol and gangliosides in most cells and affects the CNS and those viscera with high endogenous rates of lipid turnover. The fatal damage in Niemann Pick disease is caused by neurodegeneration, which begins early in life.

2.1.3.2. Clinical Features

The presentation of Niemann Pick disease is that of a systemic storage disorder, with hepatosplenomegaly, jaundice and, on occasion, liver failure in infancy. It may also cause infiltration of the bone marrow and visceral organs with "sea blue" histiocytes. It has severe neurologic manifestations, including vertical supranuclear gaze palsy, hypotonia, dysarthria or loss of speech, dementia, spasticity, seizures, cerebellar ataxia, and dystonia. It has a highly variable phenotype and age of onset, but four major groups may be identified: early infantile, late infantile, juvenile, and adult presentation. An earlier onset is associated with faster progression and shorter lifespan.

2.1.3.3. Biochemical Features

Diagnosis may be confirmed by demonstrating impaired ability of cultured fibroblasts to esterify exogenously supplied cholesterol *(74)*. In addition, characteristic ultrastructural changes in dermal fibroblasts, such as osmiophilic pleomorphic lamellar inclusions and perivascular histiocytes, may help confirm diagnosis *(75)*.

2.1.3.4 Genetic Features

Gene locus: *18q11-q12*. In several patients with Niemann-Pick disease type C, mutations in the NPC1 gene have been identified *(76)*.

2.1.3.5. Imaging Features

White matter abnormalities on MRI are recognized, particularly with water suppressed proton magnetic resonance spectra documenting fat in the cerebral tissue *(77)*. Symmetric posterior white matter signal abnormalities have been described *(78)*. However, they are often associated with more diffuse abnormalities, and this disorder is not classically regarded as a leukodystrophy. ^{1}H MRS shows diffuse involvement of frontal and parietal cortices, centrum semiovale white matter, and the caudate nucleus *(79)*. The disorder, may, however, mimics white matter diseases, such as multiple sclerosis *(80)*.

^{1}H MRS has been used to study small series of patients with Niemann Pick disease . In patients with Niemann Pick disease, NAA/creatine was significantly decreased in the frontal and parietal cortices, centrum semiovale, and caudate nucleus; whereas choline/creatine was significantly increased in the frontal cortex and centrum semiovale *(79)*. In addition, the abnormal lipid signal seen on ^{1}H MRS in this disorder can be followed in patients who are treated with cholesterol-lowering agents *(77)*.

2.1.3.6. Pathology

The brain of patients with Niemann Pick disease is atrophied, and microscopic examination demonstrates ballooning of neurons caused by the accumulation of lipid material. In the cortex, normal or mildly ballooning cells may occur interspersed with very distended neurons. Neuronal loss is marked. There is variable degeneration and lipidosis present in the cerebellar cortex. Cortical white matter changes occur with severe neuronal damage.

2.1.3.7. Treatment

No definitive therapy exists for this disorder.

2.1.4. GM1 Gangliosidosis

GMI gangliosidosis is also known as β-galactosidase-1 deficiency and GLB1 deficiency.

2.1.4.1. General

GMI gangliosidosis is autosomal-recessive disorder resulting in neurovisceral storage of gangliosides. Two types are differentiated: type 1 GM1, which is more generalized, and type 2 GM1,which is the cerebral form.

2.1.4.2. Clinical Features

GM1 gangliosidosis has a variable age-related presentation. Neonates present with severe symptoms of rapid onset. Infants present with hepatosplenomegaly, macroglossia, Hurler-like bone changes, seizures, spasticity, and bulbar dysfunction. Juvenile and adult types present with milder features, including movement disorders and ataxia.

Children with type 2 GM1 may appear normal initially, and then develop progressive cognitive and motor deterioration, beginning variably from 6 mo to 5 yr of age. Patients develop hyperacusis, epilepsy, spastic tetraplegia, ataxia, and extrapyramidal features. The average life expectancy is between 3 and 10 yr of age. The disorder is caused by a deficiency of β galactosidase.

2.1.4.3. Biochemical Features

Type 1 GM1 features severe brain accumulation of monosialoganglioside GM1 with minor accumulation of the asialo derivative. In Type 2 GM1, there are marked increases of both GM1 and asialo GM1 in the brain. In the viscera, oligosaccharide accumulates. The deficiency of β–galactosidase can be assayed in blood lymphocytes or fibroblasts.

2.1.4.4. Genetic Features

Gene locus: *3p21.33*. Molecular genetic testing is available.

2.1.4.5. Imaging Features

Neuroimaging findings in patients with type 1 GM1 have rarely been reported. MRI in the infantile form is characteristic, with lack of normal myelination in the thalami, brainstem, and cerebellum. Delayed myelination also has been described. MRI of adult-onset patients shows signal abnormalities of the basal ganglia *(81,82)*. Brain gray matter also is affected and, in fact, gray matter abnormalities may predominate the imaging picture. Thalamic hypointensity may be seen on MRI *(83)*. Chen et al 1998 reported the imaging features in a single patient with late-onset disease.

2.1.4.6. Pathology

Diffuse neuronal storage with ballooning of neuronal cytoplasm in both cerebral and cerebellar cortices is seen, as well in the basal ganglia, brain stem, spinal cord. and dorsal root ganglion. Cerebral white matter gliosis and myelin loss also are observed.

2.1.4.7. Treatment

There is no treatment for this condition.

2.1.5 GM 2 Gangliosidosis

GM 2 gangliosidosis is also known as Tay Sachs, Sandhoff disease, and hexosaminidase deficiency

2.1.5.1. General

GM2 gangliosidoses are autosomal-recessive disorders of sphingolipid storage.

2.1.5.2. Clinical Features

GM2 gangliosidosis has several biochemical variants, including Tay Sachs and Sandhoff disease. Their clinical manifestations are similar. Infantile presentations result in progressive neurologic deterioration before 1 yr of age with spasticity, blindness, seizures, and acquired macrocephaly. A cherry-red spot is seen on funduscopic examination of the retina. Juvenile or adult-onset types patients have prominent ataxia and dementia.

2.1.5.3. Biochemical Features

Tay Sachs disease is caused by a deficiency of Hexosaminidase A. Sandhoff disease is caused by a deficiency of Hexosaminidase A and B. Enzyme analysis can be achieved in blood lymphocytes or skin fibroblasts.

2.1.5.4. Genetic Features

The gene locus is *5q31.3-q33.1* for Tay Sachs; the gene locus is *5q13* for Sandhoff disease.

2.1.5.5. Imaging Features

Bilateral homogeneous thalamic hyperdensity are visible on CT early in the course of the disorder. MRI findings include hyperintensity in both basal ganglia and thalamus followed by diffuse white matter lesions on T_2-weighted images *(84)*. Abnormal signal intensities are seen in the caudate nucleus, globus pallidum, putamen, cerebellum, and brainstem. Progressive atrophy develops in the last stages of disease, and thalamic lesions can become hypointense. Optic nerve involvement may be seen. There is no contrast enhancement. Unusual features have been described, such as an isolated brainstem lesion mimicking a mass lesion *(85)*. In adult and juvenile cases, findings may be limited to cerebellar atrophy without cerebral involvement.

^{1}H MRS has been used in a limited way in these disorders, mainly documenting findings indicating widespread demyelination as well as neuroaxonal loss *(86)*.

2.1.5.6. Pathology

The gross brain changes in this disorder vary with duration of disease but generally involves massive increase in weight and volume. Neurons appear large and distended owing to storage of lipid. Cell nuclei are often displaced to the periphery. There is diffuse axonal loss and symmetric myelination disturbances. Demyelination affects the centrum semiovale and may involve or spare subcortical U fibers.

2.1.5.7. Treatment

There is no treatment for this condition.

2.2. PEROXISOMAL DISORDERS

2.2.1. Adrenoleukodystrophy

Adrenoleukodystrophy is also known as Addison disease and cerebral sclerosis; adrenomyeloneuropathy (AMN); Siemerling–Creutzfeldt disease; Bronze Schilder disease; and melanodermic leukodystrophy.

2.2.1.1 General

Adrenoleukodystrophy is a rare, X-linked recessive, inherited metabolic disorder affecting cerebral white matter and adrenal cortex, leading to progressive neurological disability and death.

2.2.1.2. Clinical Features

The first manifestations of ALD may be psychiatric or behavioral; hence, diagnosis may be delayed. In the classic forms, progressive dementia ensues with disturbances of gait manifest by spastic paraplegia. Other features include cerebellar signs of clumsiness and incoordination, dysarthria, dysphagia, and neurosensory loss. Visual loss occurs because of optic atrophy or bilateral occipital white matter lesions. Seizures, when present, are multifocal in origin. Features may appear asymmetric in onset. Pace of deterioration is variable. Adrenomyeloneuropathy or AMN is a phenotypic variant and may occur in families with boys with classic adrenoleukodystrophy. This variant involves additionally, distal polyneuropathy. Females, because of their allelic heterogeneity and variant X-inactivation, may appear unaffected or display variable symptoms from mild to more significant. Features may mimic multiple sclerosis.

2.2.1.3. Biochemical Features

Patients with adrenoleukodystrophy accumulate high levels of saturated, very long chain fatty acids in their brain and adrenal cortex because the fatty acids are not broken down in the normal manner.

2.2.1.4. Genetic Features

Gene map locus *Xq28*. The *ALD* gene was discovered in 1993 and the corresponding protein found to be a member of a family of transporter proteins, not an enzyme. It is still a unknown as to how the transporter affects the function of the fatty acid enzyme and, how high levels of very long chain fatty acids result in the loss of myelin in nerve fibers.

2.2.1.5. Imaging Features

Cerebral X-linked adrenoleukodystrophy is typified by white matter demyelination that often starts in the parieto-occipital regions bilaterally and then extends across the corpus callosum (Fig. 2A,B). The disease then progresses anteriorly and laterally as a confluent lesion to involve white matter of the temporal, parietal, and frontal lobes with relative sparing of the subcortical arcuate fibers (Fig. 1; ref. *87*). In the minority of cases, the disease may start frontally and then progresses posteriorly. Pathologically, the affected areas can be divided into zones *(88)*. Zone A is a central zone of scarring consisting of gliosis as well as scattered astrocytes. It is notable for absence of oligodendroglia, axons, myelin, and inflammatory cells. The adjacent peripheral zone B shows numerous perivascular inflammatory cells and demyelination, with axonal preservation. Lastly, zone C, the outermost zone, harbors active destruction of myelin sheaths with lack of perivascular inflammatory cells (Fig. 2C).

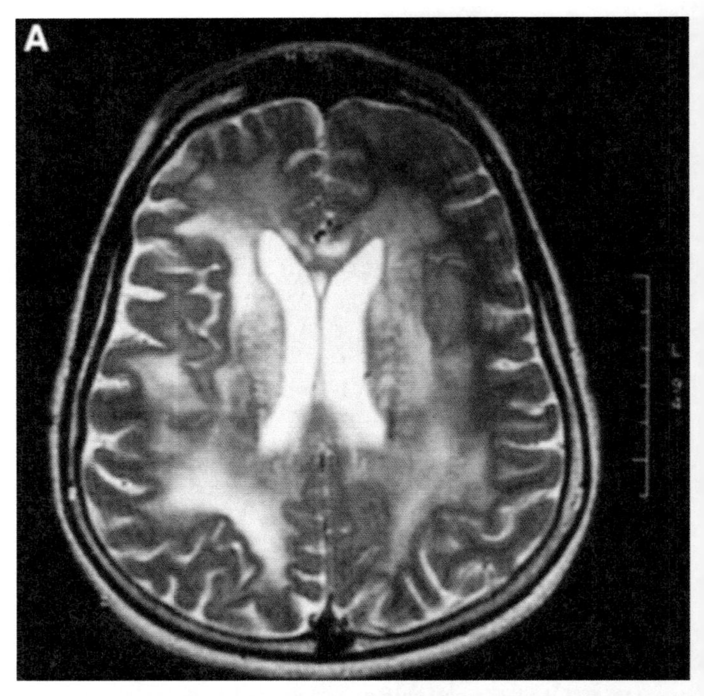

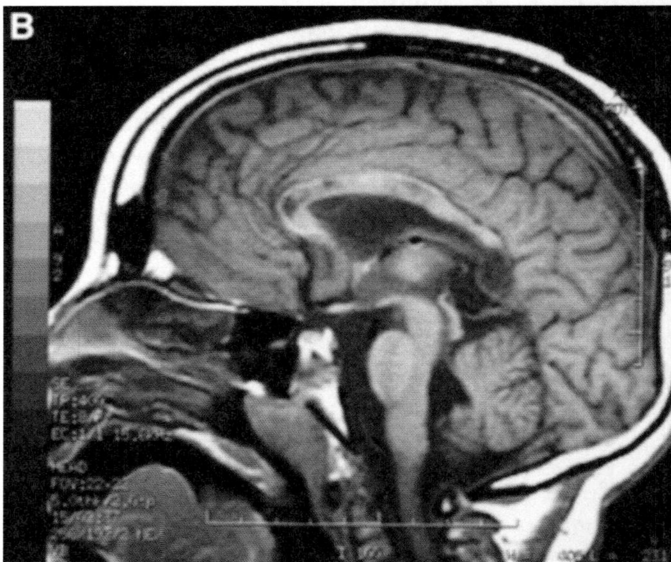

Fig. 2. (**A**) Adrenoleukodystrophy. Note the confluent white matter changes in this advanced case of adrenoleukodystrophy in a 10-yr-old child with a 2-yr history of progressive neurologic deterioration. (**B**) Adrenoleukodystrophy. There are significant abnormalities in the white matter of the corpus callosum, particularly the splenium. (**C**) Adrenoleukodystrophy. There is characteristic enhancement at the advancing border of demyelination, separating the white matter into distinct zones.

Various MR techniques have been used to evaluate and further characterize the white matter lesions in X-linked adrenoleukodystrophy. All three zones are shown to be visible on conventional MRI; however, the most pronounced signal

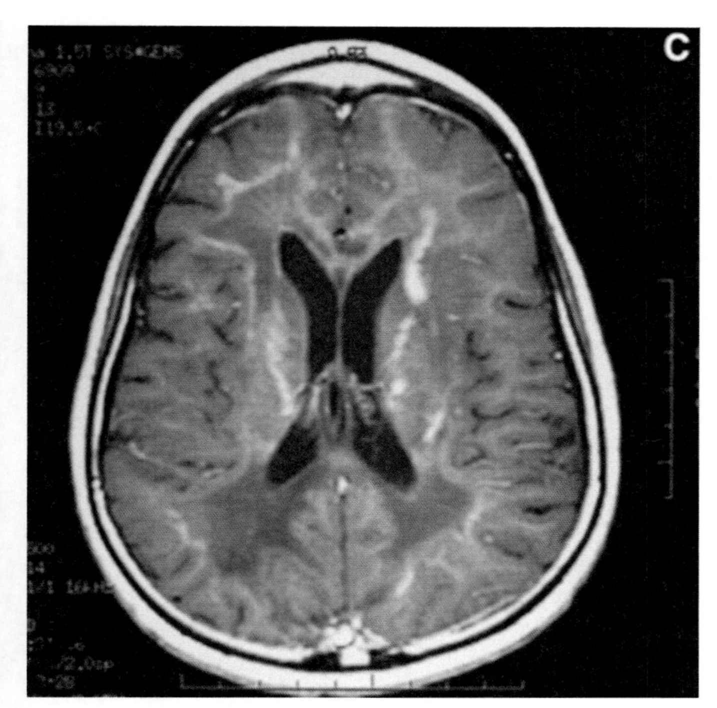

Fig. 2C.

change occurs in the central area (zone A). The adjacent zone of inflammation (zone B) shows contrast enhancement. The most peripheral zone of active but still partial demyelination (zone C) demonstrates pronounced signal changes. Magnetization transfer imaging and diffusion tensor MRI have also shown gradations within regions of T_2 hyperintensity, which have been studied to differentiate between the histopathologic zones *(89,90)*.

Multislice proton MRS imaging has demonstrated the presence of more extensive brain abnormalities in X-linked adrenoleukodystrophy than can be seen on conventional MRI, and thus may serve as a marker to predict subsequent areas destined for demyelination *(91)*.

Brain involvement evidenced by MRI is rare in female subjects heterozygous for X-linked adrenoleukodystrophy, including those who have clinical evidence of spinal cord involvement, although in some cases, these females can have white matter abnormalities resembling homozygotes. ^{1}H MRS in female heterozygotes have shown the NAA/choline ratio to be significantly reduced in the parieto-occipital white matter, but NAA/creatine and Cho/creatine ratios are not significantly different from the normal range. In the frontal white matter the choline/creatine ratio are significantly elevated, but other ratios did not demonstrate significant differences. NAA/choline and NAA/creatine ratios were significantly reduced along the white matter of the internal capsule and corticospinal projection fibers, suggesting that NAA was decreased. There were no significant differences in the calcarine or parietal gray matter *(92)*.

2.2.1.6. Pathology

Autopsy findings include a shrunken brain with decreased volume. The central gray matter appears cystic, cavitated, or indurated. Cortical thickness is preserved as the atrophy is caused by the loss of white matter volume. With extensive myelin loss, there may be secondary ventriculomegaly. Widespread demyelination is the rule; however, the process starts in the occipital regions bilaterally and extends across the splenium of the corpus callosum and extends outward and in a forward manner. Subcortical U fibers and cerebellar white matter are typically spared *(93, 94)*.

2.2.1.7. Treatment

Recent evidence suggests that a mixture of oleic acid and euric acid, known as "Lorenzo's Oil," administered to boys with X-linked adrenoleukodystrophy can reduce or delay the appearance of symptoms *(95)*; however, this remedy does not appear helpful in patients with clinically advanced disease. Bone marrow transplants can provide long-term benefit to boys who have early evidence of X-linked adrenoleukodystrophy, but the procedure carries the risk of mortality and morbidity and is not recommended for those whose symptoms are already severe or who have the adult-onset or neonatal forms. It remains to be proven whether asymptomatic adrenoleukodystrophy individuals will develop a less-severe course if they are placed on the Lorenzo's oils *(96)*. The one therapy demonstrated to be effective is bone marrow transplantation.

2.2.2. Peroxisomal Single Enzyme Defect

2.2.2.1. General

Defects in many of the enzymatic processes of the peroxisomes have been demonstrated to cause disease. Diseases caused by the deficiency of a single enzyme within the peroxisome include Refsum's disease (phytanoyl-CoA oxidase), X-linked adrenoleukodystrophy (fatty acid acyl-CoA synthetase/adrenal leukodystrophy protein [ALDP]), and β-oxidation disorders (acyl-CoA oxidase, bifunctional protein, and thiolase), among others. These last three disorders are characterized by the finding of white matter abnormalities. Defects in peroxisomal β-oxidation have been found to result from mutations in any of three genes: acyl-CoA oxidase (pseudo neonatal adrenoleukodystrophy), D-bifunctional enzyme, or 3-oxoacyl-CoA thiolase (peroxisomal thiolase or pseudo Zellweger).

2.2.2.2. Clinical Features

Clinically, disorders resulting from a defect in peroxisomal β-oxidation resemble the peroxisome biogenesis disorders, especially D-bifunctional protein. The clinical characteristics may include hypotonia, craniofacial dysmorphism, severe retardation, sensory defects, muscle weakness, hypomyelination in cerebral white matter, and hepatomegaly. Patients with D-bifunctional protein also may show disordered neuronal migration. In addition, most patients identified with these disorders die within the first two years. It is estimated that 10–15% of all patients with a phenotype of the peroxisome biogenesis disorders actually have defects in one of the peroxisomal fatty acid oxidation enzymes.

2.2.2.3. Biochemical Features

Depending upon the specific enzyme deficiency or class of disorder, peroxisomal disorders may show accumulation of very long chain fatty acids, and/or abnormalities in plasmalogen, phytanic acid, and pipecolic acid.

2.2.2.4. Genetic Features

The genetics of these groups of disorders is complex and beyond the scope of this chapter.

2.2.2.5 Imaging Features

MRI is remarkable for diffuse white matter changes accompanied by malformations of cerebral development, such as polymicrogyria (Fig. 3A,B).

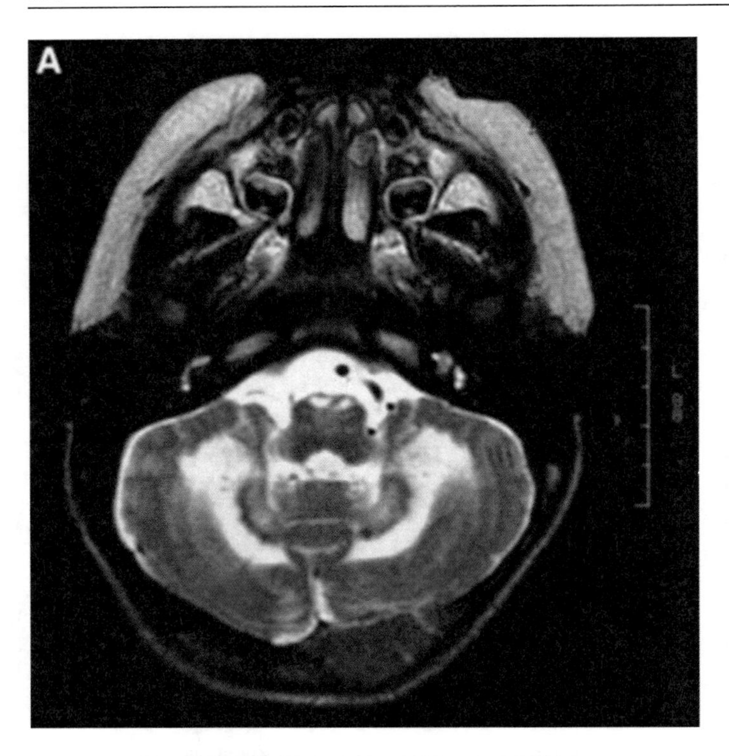

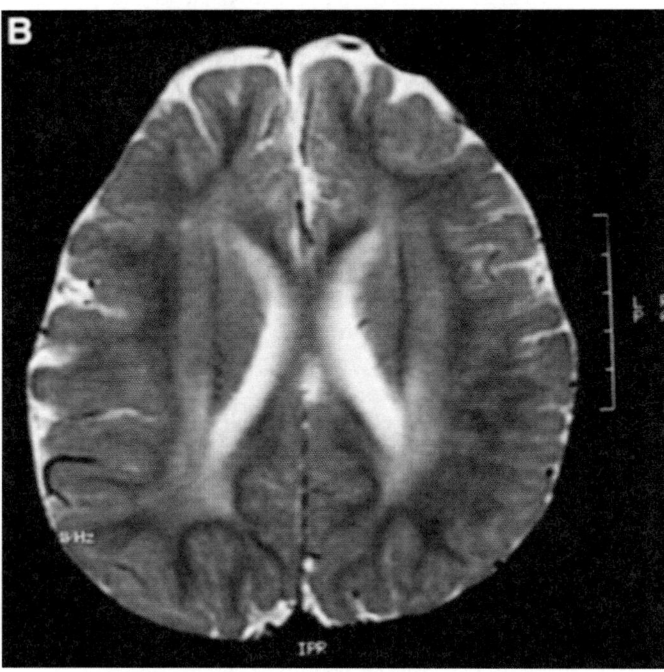

Fig. 3. (**A**) Single enzyme peroxisomal disorder. White matter abnormalities are seen characteristically in the brainstem in this case of D-bifunctional enzyme deficiency in a 3-yr-old child. (**B**) Single enzyme peroxisomal disorder. White matter abnormalities are less apparent in the cerebral myelin in this patient with D-bifunctional enzyme deficiency.

2.2.2.6. Pathology

Neuropathology is similar in other peroxisomal disorders (i.e. Zellweger syndrome).

2.2.2.7. Treatment

There is no definitive treatment for these disorders.

2.2.3. Zellweger Spectrum: Peroxisomal Biogenesis Disorders

Peroxisomal biogenesis disorders include cerebro-hepatorenal syndrome; CHR syndrome; and ZWS.

2.2.3.1. General

There also is a group of disorders resulting from a deficiency in the biogenesis of the peroxisome. These disorders are known as the peroxisome biogenesis disorders. Peroxisome biogenesis disorders are caused by defects in any of at least 14 genes whose products (peroxins) are required for the proper assembly of the peroxisome.

2.2.3.2. Clinical Features

This group includes the Zellweger spectrum, which is comprised of three disorders: Zellweger syndrome (or cerebro-hepatorenal syndrome), neonatal adrenoleukodystrophy, and infantile Refsum disease. These patients have multiorgan disease and prominent neurologic dysfunction.

Symptoms include severe hypotonia; poor sucking ability (in an infant), often requiring gavage feeding; epilepsy; hyporeflexia/areflexia; psychomotor retardation; nystagmus; and impaired hearing. Physical characteristics include a high forehead, large fontanel and wide sutures, external ear deformity, epicanthus, low broad nasal bridge, and shallow orbital ridges, resulting in a typical facies. Hepatomegaly and renal cysts, with progressive hepatorenal dysfunction, as well as impaired adrenocortical function, also are common features. Skeletal abnormalities such as calcific stippling (punctuate calcifications of long bones) are seen. Cardiac malformations are frequent. Global developmental delay is a hallmark of the Zellweger spectrum, and the cause may be related to malformations of cerebral development.

2.2.3.3. Biochemical Features

Patients with Zellweger syndrome have absence of peroxisomes and manifest biochemical consequences of impaired peroxisomal beta oxidation, including increased accumulation of very long chain fatty acids.

2.2.3.4. Genetic Features

Gene map locus *2p15, Chr.1, 1q22, 12p13.3, 7q21-q22, 6q23-q24.*

2.2.3.5. Imaging Features

MRI is remarkable for cerebral demyelination with sparing of subcortical fibers and pronounced central cerebellar demyelination as well as features of disordered cerebral migration such as polymicrogyria *(97).*

2.2.3.6. Pathology

Neuropathologic lesions in Zellweger syndrome comprise three major categories: (1) abnormalities in neuronal migration or differentiation, (2) defects in the formation or maintenance of central white matter, and (3) postdevelopmental neuronal degenerations. The central white matter lesions are typified by: (1) inflammatory demyelination, (2) noninflammatory dysmyelination, and (3) nonspecific reductions in myelin volume or staining with or without reactive astrocytosis *(98).*

2.2.3.7. Treatment

There is no definitive treatment in this disorder.

2.3. MITOCHONDRIAL CYTOPATHIES

2.3.1. General

Mitochondrial disorders are caused by mutations of nuclear or mitochondrial DNA encoded genes involved in oxidative phosphorylation. Mutations in these critical genes are associated with specific clinical syndromes with diverse presentations *(99,100)*. Because mitochondria are present in many of our organs and play a key role in energy metabolism, mitochondrial encephalomyopathies often present as multisystem disorders that may manifest with neurologic, cardiac, endocrine, gastrointestinal, hepatic, renal, and/or hematologic involvement.

2.3.2. Clinical Features

The clinical presentation in any one patient is influenced by factors such as specific gene mutation, heteroplasmy, secondary mutations, or polymorphisms in both nuclear and mitochondrial genomes, as well as environmental and immunological factors *(101,102)*. Neurologic features may reflect dysfunction of any part of the nervous system and encompass a broad range of common neurologic symptoms, including dementia, developmental delay, stroke, epilepsy, psychiatric illness, neuropathy, and myopathy. Children with mitochondrial disease may present with life-threatening illness in the newborn period; however, the majority of children come to clinical attention for nonspecific problems, including failure to thrive, developmental delay, seizures, hypotonia, and loss of developmental milestones *(103)*. Adult-onset disease may be typified by neuromuscular complaints such as ptosis, muscle weakness, and/or neuropathy with or without associated multisystem disease *(104–107)*.

2.3.3. Biochemical Features

Several laboratory studies may be useful to screen for impaired energy metabolism such as serum lactate, pyruvate, plasma amino acids, complete blood count, electrolytes, carnitine, acylcarnitine profile, ammonia, and creatine phosphokinase. Renal tubular acidosis as part of a Fanconi syndrome may be seen, especially in patients with complex IV defects. Renal disease is more common in pediatric presentations. Elevated lactate is suggestive but not specific for mitochondrial disorders. Lactate levels are not entirely reliable markers of mitochondrial dysfunction. Many children do not have elevated serum lactate or may have elevations only under certain provocations, such as after glucose loading, illness, or exercise. Blood lactate values may be spuriously elevated when a tourniquet is used, or as a result of a child struggling with the venipuncture. In these cases, arterial lactate level may be more reliable. In infants and young children with encephalopathy, CSF lactate may be elevated. Elevations of lactate also may be seen with organic acidemias, pyruvate carboxylase deficiency, fatty acid oxidation defects, biotinidase deficiency, and pyruvate dehydrogenase complex deficiency. Lactate/ pyruvate ratios are a marker of cytoplasmic redox status and are typically elevated in oxidative phosphorylation disorders. The ratio of β-hydroxybutyrate to acetoacetate (ketone body molar ratio) is another measure of redox state. Other abnormal studies may include elevations of pyruvate, and elevated alanine.

Serum CPK values are usually normal in mitochondrial disorders except in mitochondrial depletion and in those disorders associated with rhabdomyolysis. In both congenital and infantile forms of mitochondrial (mt)DNA depletion, the creatine kinase concentration may be greater than 1000 IU and should alert the physician to a possible diagnosis.

Defects in fatty acid metabolism may be associated with elevated plasma free fatty acids, hypoketonemia, hypocarnitinemia, and dicarboxylic aciduria. Intermediates of the Krebs cycle may suggest mitochondrial fatty acid oxidation disorder. Values may be abnormal only during a concurrent stressor. Many of these studies are more informative if performed after a brief fast.

2.3.4. Genetic Features

Mitochondrial cytopathies are caused by mutations in either mtDNA (typically resulting from point mutations in transfer RNAs or structural genes) or nuclear (n)DNA. Dozens of mutations have been identified. The specific genetics of these disorders is beyond the scope of this chapter.

2.3.5. Imaging Features

Mitochondrial disease is classically associated with deep gray-matter lesions (Fig. 4A). When white matter is involved, the lesions are typically subcortical and overshadowed by more significant disease in the gray matter. There is a spectrum of abnormalities seen that will vary based on the metabolic brain defect, stage of the disease, and age of the patient. It should be noted that although several of the mitochondrial cytopathies have characteristic imaging features, these are not considered diagnostic in exclusion of clinical, neurophysiological, biochemical, histological, and molecular findings. Common MRI findings in children may include one of several patterns *(108,109)*.

Gray matter nuclei involvement may be a predominant finding. These tend to be symmetric but may appear partial or patchy. In acute phases, they may appear swollen with a high signal appearance on T_2-weighted MRI. Chronically, they become shrunken. In the brainstem, the periaqueductal gray matter, pons, and mesencephalon are common sites of involvement (Fig. 4B). The cerebellum, particularly the dentate nuclei, may be affected.

MR lesions in corresponding locations therefore strongly suggest the presence of a defect in the energy-producing pathway. Putaminal involvement is reported to be a consistent feature in Leigh syndrome.

A frequent finding on pediatric MRI in patients with mitochondrial cytopathies is abnormal myelination. Abnormalities of myelin, including delayed myelination, leukodystrophic pattern, and demyelination, are common. Extensive areas of demyelination may be demonstrated in the cerebral hemispheres, near the corpus callosum and adjacent white matter (Fig. 4C). This type of finding may mimic a leukodystrophy and has recently been recognized as a finding consistent with a mitochondrial presentation. Therefore, infants with leukoencephalopathies, especially leukodystrophies, who do not have one of the more common causes of white-matter disease should be evaluated for a possible mitochondrial cytopathy.

Infarct-like, often transient lesions not confined to the vascular territories are the imaging hallmark of mitochondrial myopathy, encephalopathy, lactic acidosis, and stroke-like episodes (MELAS). Focal necrosis and laminar cortical

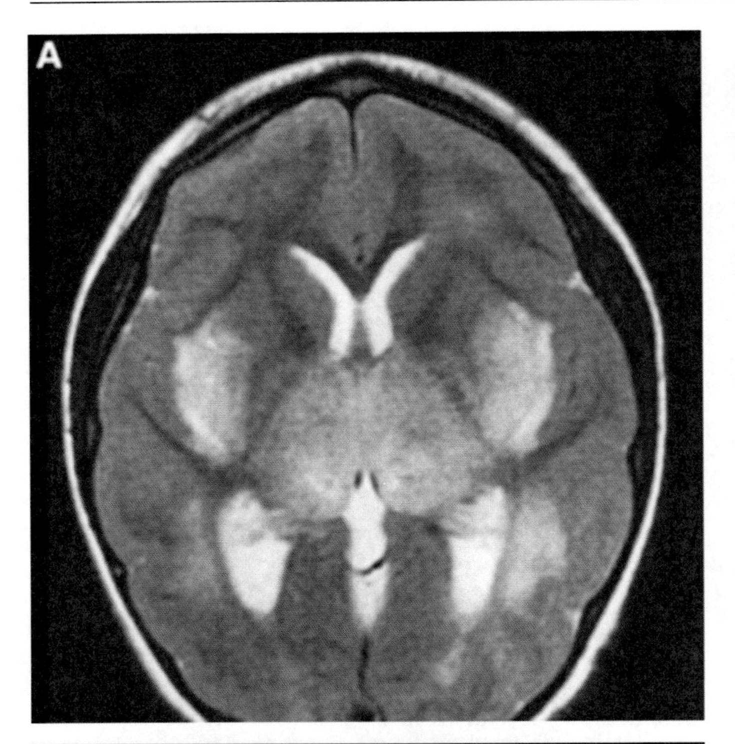

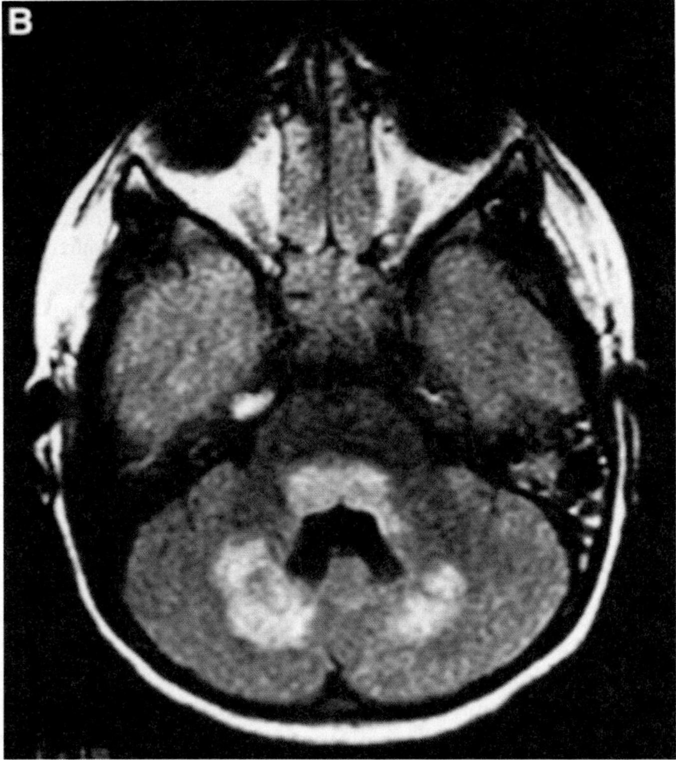

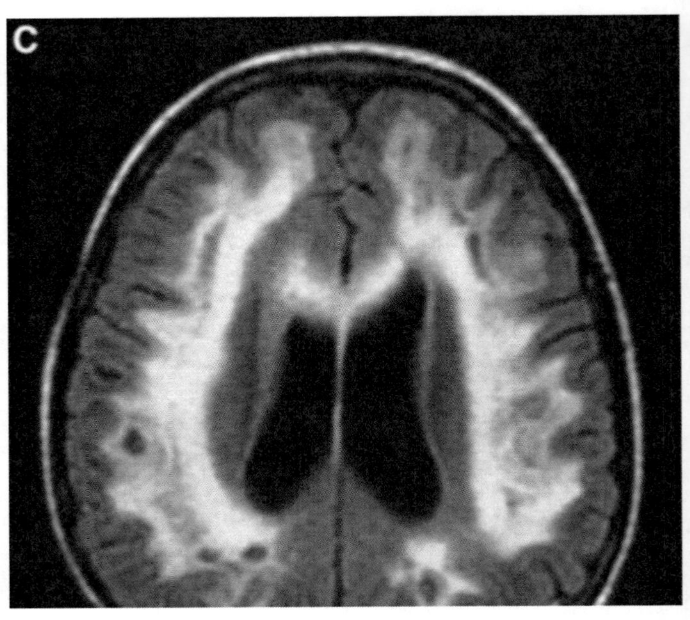

Fig. 4. (**A**) Leigh syndrome. Significant basal ganglia involvement is seen which is typical for this disorder. (**B**) Leigh syndrome. Typical brainstem and cerebellar white matter abnormalities are seen. (**C**) Leigh syndrome showing confluent white matter abnormalities.

necrotic changes are the histopathologic correlates of this disease, together with neuronal degeneration and mineral deposits within the basal ganglia.

Patients with MELAS have elevated resting blood lactate levels and elevated CSF lactate. CT scans reveal areas of low density affecting both gray and white matter *(110)*. Posterior infarcts are most common with an occipital, temporal, frontal ratio of 4 to 2 to 1 *(111)*. MRI scans in patients with MELAS show posteriorly located strokes that do not conform to a vascular territory as well as basal ganglia lucency. The etiology of the stroke like episodes remains undetermined. Electron and light microscopy studies have shown mitochondrial angiopathy associated with enlarged, succinate dehydrogenase staining mitochondria in the intracerebral vessel endothelium *(112)*. Diffusion MRI shows vasogenic edema. Other radiologic findings include basal ganglia calcification and cerebellar and cerebral atrophy (typically after long standing disease; ref. *113*). MRS may demonstrate elevations of lactate within cortical lesions during an acute episode, with resultant normalization after clinical resolution *(108,114)*. Using MRS, researchers have found elevations of brain lactate in patients with MELAS due to A3243G, as well as their oligosymptomatic family members, in whom the degree of lactate elevation was directly related to heteroplasmic levels of mutation *(115)*.

2.3.6. Pathology

Brain pathologic findings depend upon the subtype of mitochondrial disorder. The diagnosis of Leigh syndrome, which earlier could be made only by postmortem examination, is characterized by vascular proliferation and demyelination, which leads to necrosis and cavitation in the basal ganglia, midbrain, pons, and posterior column of the spinal cord. In muscle biopsy specimens, mitochondrial proliferation is a frequent result of mtDNA mutations and may be demonstrated histologically in Gomori trichrome stained muscle specimens. Fibers exhibiting excessive amounts of mitochondria may appear as purple clumps underneath the sarcolemma. This is so called "ragged red fiber" (RRF) which is the hallmark of mitochondrial dysfunction. These fibers also stain strongly for succinate dehydrogenase (complex II, ragged blue fibers), an electron transport chain enzyme encoded by the nDNA, and may be seen in patients with both mtDNA and nDNA mutations. The cytochrome *c* oxidase (COX) reaction plays an important role in the diagnosis of mitochondrial cytopathies. Three of the catalytic subunits of COX, namely COX I, II, and III, are encoded by mtDNA. The finding of COX-negative fibers suggests impairments of mitochondrial protein synthesis and may be seen in mutations of transfer RNA genes as well as mtDNA deletions. Often, these COX-negative fibers are associated with RRFs. Patients who harbor defects in mtDNA protein coding genes will have COX-negative RRFs on biopsy, and patients with complex I or II defects present with biopsies displaying COX-positive RRFs. These fibers represent the accumulation of mitochondria in response to a defect in oxidative phosphorylation. The COX staining reaction can show foci of scattered COX negative fibers that may correspond to ragged red fibers. This is suggestive of impaired mitochondrial protein synthesis. Severely decreased COX staining may be consistent with fatal infantile myopathies.

Electron microscopy is used to some degree in the diagnosis of mitochondrial myopathies. Ultrastructural analysis with electron microscopy may reveal intramitochondrial paracrystalline inclusions, or disrupted cristae.

2.3.7. Treatment

Although there is no definitive treatment of mitochondrial cytopathies, most physicians treating such patients will use various combinations of Coenzyme Q 10, B vitamins, and other factors that may act as cofactors of oxidative phosphorylation.

2.4. ORGANIC ACIDURIA

2.4.1. Canavan Disease

Canvan disease is also known as Canavan-Van Bogaert-Bertrand disease; spongy degeneration of CNS; aspartoacylase deficiency; ASP deficiency; aminoacylase 2 deficiency; and ACY2 deficiency.

2.4.1.1. General

Canavan's disease is an autosomal-recessive disorder caused by aspartoacylase deficiency. The deficiency of aspartoacylase leads to increased concentration of NAA acid in the brain and body fluids. This causes disruption of myelin, resulting in spongy degeneration of the white matter of the brain. The clinical features of the disease are hypotonia in early life, which evolves to spasticity, macrocephaly, head lag, and progressive severe mental retardation. Like Tay-Sachs and Nieman Pick disease, it is most prevalent in the Ashkenazi Jewish population. Research at the molecular level led to the cloning of the gene for aspartoacylase and the possibility of genetic diagnosis, the creation of animal models and continued research into the pathophysiology of this still poorly understood disorder.

2.4.1.2. Clinical Features

There are at least three different phenotypes of Canavan disease. A congenital form has been described, with marked hypotonia, feeding difficulties, lethargy and a rapid decline with death within months.

The infantile form of Canavan disease is the most common. Patients have an unremarkable gestation and appear normal at birth. Hypotonia becomes evident by 6 mo of life. Affected infants have poor head control, decerebrate posturing, irritability, poor feeding, and limited spontaneous motor activity. By the second year of life, hypotonia has given way to spasticity with tonic extensor spasms. Cortical blindness and optic atrophy develop. Sensorineural hearing loss may be present with developmental malformations of the organ of Corti *(116)*. Macrocephaly is a prominent finding at this stage. The patient may have myoclonic or other seizures and choreoathetosis. Autonomic crises may occur with vomiting, temperature instability and vasomotor disturbances. A vegetative state develops and affected infants generally die in early childhood.

The juvenile form is less common. Symptoms develop often after 5 yr of life. Visual loss, progressive cerebellar ataxia, regression of cognitive function, and spasticity are present. Macrocephaly is not an obligate feature. Some of these patients have had long-term survival, even into adulthood *(117)*.

2.4.1.3. Biochemical Features

The first reported cases of NAA aciduria in patients with leukodystrophy occurred nearly 40 yr ago *(118–120)*; however, it is not until much later that the association with Canavan disease was made *(121)*. Since that time, the measurement of NAA in urine, plasma, and amniotic fluid has proved to be a valuable diagnostic tool. Aspartoacylase can be assayed in cultured skin fibroblasts and cultured amniotic cells and chorionic villi.

2.4.1.4. Genetic Features

Gene Locus: 17pter-p13. The gene for aspartoacylase has been cloned *(122)*, and more than 40 mutations have been described, with two founder mutations among Ashkenazi Jewish patients (E285A and Y231X). Programs for carrier testing are currently in practice for the screening of Ashkenazi Jews for these two common mutations. Mutations in the aspartoacylase gene among non-Jewish patients are different and more diverse.

2.4.1.5. Imaging Features

Since diagnosis is available biochemically, there are limited recent studies on the MRI findings of Canavan disease. Conventional magnetic resonance imaging classically reveals early involvement of the arcuate fibers (U fibers). In most cases, the entirety of the cerebral white matter is then progressively affected, with relative sparing of the putamen *(123)*. There also may be early involvement of the lentiform nuclei and the heads of the caudate nuclei apparent on T_2-weighted images *(124)*.

Magnetic resonance proton spectroscopy reveals an increase of the NAA/choline ratio and an overall increase of the NAA concentration *(125)*.

2.4.1.6. Pathology

Classical descriptions of the pathology of Canavan disease include spongy degeneration (vacuolization) of cortical and subcortical tissues and extensive demyelination, with preservation of axons and oligodendroglia. Demyelination is accompanied by a moderate astrocytosis. Optic pathways manifest significant demyelination

2.4.1.7. Treatment

No effective treatment currently exists for Canavan disease. Virus-based gene transfer by recombinant adeno-associated virus containing the aspartoacylase gene is a possible therapeutic option in the future *(126)*.

2.4.2. Glutaric Aciduria Type 1

Glutaric aciduria type 1 is also known as Glutaricaciduria I; GA I; and glutaryl-CoA dehydrogenase deficiency.

2.4.2.1. General

Glutaric aciduria is an inherited biochemical disorder caused by a deficiency of the enzyme glutaryl CoA dehydrogenase, which is in the mitochondrial biochemical pathway for degradation of the lysine and tryptophan. It is characterized by progressive dystonia and athetosis as the result of gliosis and neuronal loss in the basal ganglia. If not treated, episodes of biochemical intoxication cause degeneration of the basal ganglia of the brain and progressive extrapyramidal features. Children who have glutaric aciduria and have suffered significant brain damage often mistakenly are given a diagnosis cerebral palsy.

2.4.2.2. Clinical Features

Glutaric aciduria type 1 is manifested in infancy by macrocephaly, hypotonia, choreoathetosis, dystonia, and encephalopathic crisis associated with an intercurrent illness or surgery. Rare adult onset cases are described.

2.4.2.3. Biochemical Features

GA1 is caused by a deficiency of glutaryl-CoA dehydrogenase.

2.4.2.4. Genetic Features

Gene map locus: *19p13.2*. The GCDH gene contains 11 exons and spans approx 7 kb *(127)*. A single mutation was found as the cause of glutaric acidemia in the Old Order Amish of Lancaster County, Pennsylvania. Mutation analysis is more heterogeneous in nonAmish patients.

2.4.2.5. Imaging Features

Marked widening of the operculae is found on MRI in patients with glutaric acidemia type 1. White matter changes have been reported in approx half of patients with organic acidurias *(128)*. Abnormal high signal intensity on T_2-weighted images is seen in the basal ganglia and periventricular white matter in approx two thirds of children. Abnormal high signal on T_2-weighted images may be seen in the dentate nucleus, substantia nigra, and the pontine medial lemniscus *(129)*. Features seen on MRI in adult-onset cases have been described as a diffuse leukoencephalopathy.

Strauss et al. *(130)* suggests that micrencephalic macrocephaly is a distinctive radiological feature of GA I. He studied a group of Old Order Amish, in which the disease has increased prevalence. In most neonates, an enlarged head circumference is the only presenting sign of the disorder. The authors pointed to radiologic signs of large fluid collections in the middle cranial fossae. Veins can be seen stretching tenuously across this space, where they are subject to distortion and rupture. Acute subdural hemorrhage can occur after minor head trauma and in some instances is accompanied by retinal hemorrhages. Investigation of child abuse preceded a correct metabolic diagnosis in some non-Amish children.

2.4.2.6. Pathology

Neuropathology in Glutaric aciduria type 1 may demonstrate temporal and frontal lobe hypoplasia, degeneration and necrosis of the putamen and the globus pallidus, status spongiosus of the cerebral white matter and, occasionally, heterotopic neurons in the cerebellum. There may be hypoplasia of the cerebral white matter *(130)*.

2.4.2.7. Treatment

Specific management includes pharmacological doses of L-carnitine and dietary protein restriction. Metabolic decompensation must be treated aggressively to avoid permanent brain damage.

2.5. AMINO ACIDURIA

2.5.1. Phenylketonuria

Phenylketonuria is also known as PKU; phenylalanine hydroxylase deficiency; PAH deficiency; oligophrenia phenylpyruvica; and Folling disease

2.5.1.1. General

Phenylketonuria is an inborn error of phenylalanine metabolism occurring in 1 of every 12,000 births in North America. In the absence of phenylalanine hydroxylase, and without protein restriction beginning in infancy, most persons with this disorder develop severe, irreversible mental retardation, as well as neurobehavioral symptoms such as seizures, tremors, gait disorders, athetoid movements, autism, and psychotic episodes These clinical manifestations of phenylketonuria rarely develop in children born after the mid-1960s, when routine screening was legislated and early treatment for phenylketonuria became commonplace.

2.5.1.2. Clinical Features

Infants with phenylketonuria appear normal at birth. Many of these infants have blue eyes and may show fairer hair and skin coloration than family members. Other clinical features include early vomiting, irritability, and eczema-like rash, and mousy odor of the urine. Some infants manifest signs of CNS dysfunction such as hypertonicity and exaggerated deep tendon stretch reflexes. If left untreated, children will develop mental retardation, seizures, and behavioral problems. Also observed are microcephaly, widely spaced teeth, poor somatic growth, and poor development of tooth enamel.

2.5.1.3. Biochemical Features

Patients with phenylketonuria lack activity in the phenylalanine hydroxylase enzyme. This enzyme normally converts the essential amino acid, phenylalanine, to tyrosine. Failure of the conversion to take place results in a buildup of phenylalanine, which through a mechanism that is not well understood, is toxic to the CNS and causes severe clinical symptoms. Biochemically, patients manifest elevations of phenylalanine and often-decreased tyrosine in the blood and CSF.

2.5.1.4. Genetic Features

Gene map locus: 12q24.1. Classical phenylketonuria is inherited in an autosomal-recessive manner and is the result of mutations in the phenylalanine hydroxylase gene. Many mutations have been described. Most patients are compound heterozygotes rather than homozygotes for one particular mutant allele.

2.5.1.5. Imaging Features

MRI studies in patients with phenylketonuria reveals white matter alterations that correlate with blood phenylalanine concentrations as well as brain phenylalanine concentrations measured by magnetic resonance spectroscopy *(131)*.

With elevated phenylalanine levels (>600 μmol/L), patients with phenylketonuria generally demonstrate symmetric patchy and/or band-like areas of enhanced signal intensity on T_2-weighted images, which involve the posterior/periventricular white matter, which is the last area to myelinate in humans. In those more severely affected patients, the lesions may extend to the frontal and subcortical white matter, and include the corpus callosum *(132–134)*.

In addition, signal changes within the corticospinal tract, extending from the internal capsule to the cerebral peduncles, have been described in some adolescent and adults with phenylketonuria *(135)*.

In one series, a high percentage of infratentorial white matter lesions, was reported (approx 25%). The focal subcortical white matter changes were associated with a decrease of signal intensity in T_1-weighted sequences *(136,137)*.

2.5.1.6. Pathology

Neuropathological abnormalities in patients with phenylketonuria consist of spongy degeneration of the myelin as well as numerous foci of demyelination. The severity of lesions increases with age. Myelin of the cerebral hemispheres, optic tracts, and cerebellum is most affected as compared with other brain regions.

2.5.1.7. Treatment

Early diagnosis of phenylketonuria is important because it is treatable by dietary means. Basically, this relies on the restriction of phenylalanine from the diet by using specific formulas and food charts. In some patients, tyrosine supplementation is advocated. The diet should be continued throughout life for maximal benefit.

2.5.2. Maple Syrup Urine Disease

Maple syrup urine disease is also known as MSUD; branched-chain ketoaciduria; branched-chain α-keto acid; dehydrogenase deficiency; BCKD deficiency; and keto acid decarboxylase deficiency

2.5.2.1. General

Maple syrup urine disease is caused by impaired metabolism of the branched chain amino acids, leucine, isoleucine, and valine. This arises as the result of a defect in oxidative decarboxylation, leading to buildup of ketoacids. Untreated, this condition leads to significant CNS impairment, including mental retardation and seizures.

2.5.2.2. Clinical Features

Infants with untreated maple syrup urine disease manifest symptoms in the first week of life. In classical cases, poor feeding, vomiting, opisthotonus, seizures, and respiratory abnormalities are observed. Infants may subsequently become comatose, with signs of increased intracranial pressure, including bulging anterior fontanel, seizures, signs of bulbar involvement, and ocular motility abnormalities. In milder cases of maple syrup urine disease, the clinical picture may comprise mild-to-moderate mental retardation with a clinical history of periods of coma, acidosis, lethargy, and hypoglycemia.

2.5.2.3. Biochemical Features

Increased levels of plasma branched-chain amino acids and urinary branched-chain keno acids are seen.

2.5.2.4. Genetic Features

Gene map loci: *19q13.1-q13.2, 6p22-p21, 1p31, 7q31-q32*. MSUD can be caused by mutation in at least four genes: branched-chain keto acid dehydrogenase, branched-chain keto acid dehydrogenase e1, β-polypeptide, dihydrolipoamide branched-chain transacylase, and dihydrolipoamide dehydrogenase. These genes encode the catalytic components of the branched-chain α-keto acid dehydrogenase complex, which catalyzes the catabolism of the branched-chain amino acids, leucine, isoleucine, and valine. A high frequency of maple syrup urine disease is seen among the Old Order Mennonites of Pennsylvania *(138–140)*.

2.5.2.5. Imaging Features

Imaging studies in patients with maple syrup urine disease have shown reversible brain edema during acute metabolic decompensation. Jan et al, using DWI, demonstrated marked restriction of proton diffusion consistent with a mechanism of cytotoxic or intramyelinic sheath edema *(141)*. The edema involved preferentially the brainstem, basal ganglia, thalami, cerebellar and periventricular white matter, and the cerebral cortex. ^{1}H MRS in this same population demonstrated the presence of an abnormal branched-chain amino acids and branched-chain α-keto acids peak at 0.9 ppm and elevations of lactate *(141)*. The imaging changes were noted to be reversible after treatment. The duration of lactate elevation correlates with the presence of brain edema and coma *(142)*.

Application of diffusion tensor imaging in maple syrup urine disease has demonstrated symmetric high signal in the globus pallidus, mesencephalon, dorsal pons, and nucleus dentatus, with observed restriction of the water mobility. ADC maps reveal low ADC values in these regions compared with those of apparently unaffected regions in the brain parenchyma *(143)*. It has been suggested that areas of increased signal and low ADC values are due to dysmyelination as a mechanism.

2.5.2.6. Pathology

Untreated maple syrup urine disease leads to arrest of normal myelination in the CNS.

2.5.2.7. Treatment

Treatment of maple syrup urine disease relies upon dietary restriction of the branched chain amino acids, which can be achieved via food lists as well, as special metabolic formulas. In addition, some variants of maple syrup urine disease show a responsiveness to thiamine; thus, thiamine may be used in some clinical settings.

2.5.3. Urea cycle Disorders (Ornithine Transcarbamylase Deficiency)

This disorder is also known as OTC deficiency.

2.5.3.1. General

The urea cycle disorders represent one of the most common groups of inborn errors of metabolism. Epidemiologic studies suggest that the incidence of these disorders is 1 in 30,000 *(144,145)*. Syndromes involving deficiencies of each of the five urea cycle enzymes as well as three related cofactors and transporters have been described *(146)*. The most common of these, ornithine transcarbamylase deficiency, is inherited in an X-linked manner, with an estimated incidence of 1 in 70,000 *(147)*. More than 240 mutations have been identified in the OTCD gene *(148)*. Approximately 60% of hemizygous males harbor a mutation around the enzyme active site and present with hyperammonemic coma in the newborn period. The remainder have more peripheral mutations in other parts of the gene which impart a less severe phenotype, typically with later onset of symptoms *(149)*.

2.5.3.2. Clinical Features

The majority of children with ornithine transcarbamylase deficiency, the most common and only X-linked disorder of ureagenesis, have substantial cognitive and motor deficits resulting from hyperammonemic episodes *(150,151)*. Neonatal onset disease mortality rate is high; survivors sustain brain injury, leading to mental retardation, cerebral palsy, and seizures *(152)*. Neonatal survivors have a mean IQ of 43. In males with partial deficiencies, disease onset is later and outcome is better; however, it is still associated with high mortality and morbidity with many manifesting cognitive, motor, and psychiatric sequelae *(153,154)*. Treatment of ornithine transcarbamylase deficiency depends upon protein and, hence, nitrogen restriction in combination with an alternative pathway of waste nitrogen excretion *(155)*.

In heterozygous females with ornithine transcarbamylase deficiency, the phenotype is broad because of allelic heterogeneity and differential X-inactivation patterns. Approx 85% of heterozygous females are considered to be asymptomatic, with the remainder showing symptoms ranging from behavioral and learning disabilities, protein intolerance, cyclical vomiting, and episodes of hyperammonemic coma *(156–158)*. Symptomatic women have been found to have mutations seen in neonatal onset disorder in hemizygous males *(149)* and develop hyperammonemia due to skewed X-inactivation. There is therefore a range of residual enzyme capacity, and thus urea synthetic capacity with associated clinical symptomatology *(159)*.

2.5.3.3. Biochemical Features

Patients with ornithine transcarbamylase deficiency manifest elevations of ammonia in blood and CNS, as well as elevations of plasma and CSF glutamine. Urinary excretion of orotic acid is increased. Heterozygotes may manifest metabolic abnormalities as result of skewed X-inactivation, after protein load or after allopurinol loading test.

2.5.3.4. Genetic Features

Gene map locus: Xp21.1. Tuchman reported that approx 10% to 15% of all molecular alterations associated with ornithine transcarbamylase deficiency were large deletions involving all or part of the OTC gene. Tuchman et al. *(160)* have reported approximately 90 different mutations associated with ornithine transcarbamylase deficiency.

2.5.3.5. Imaging Features

Several small series of patients with ornithine transcarbamylase deficiency who were studied with CT or MRI have been described *(161–165)*. The neuroimaging findings remain inconclusive because the staging may reflect different phases of the disease process. In a majority of cases, CT revealed areas of low density in the white matter which may appear both as symmetric as well as asymmetric in character, and were found to be partially reversible with treatment. MRI scanning subsequently has allowed better delineation of white matter injury in ornithine transcarbamylase deficiency.

The noninvasive detection of elevated brain glutamine by ^{1}H MRS has been documented in experimentally induced hyperammonemia *(166,167)*. Glutamine has been implicated in hyperammonemic encephalopathy. It has been shown that a rise in plasma glutamine levels precedes hyperammonemia *(168)*. The importance of glutamine in this process is further strengthened by the relationship between hyperammonemia, neurologic dysfunction, and cerebral spinal glutamine concentration observed in patients with hepatic encephalopathy. Additional evidence for this hypothesis was demonstrated by a report of elevated brain glutamine by MRS in patients with ornithine transcarbamylase deficiency with hyperammonemic encephalopathy *(169)*. They believe their results support the view that the encephalopathy associated with hyperammonemia may be related to the concentration of brain glutamine.

2.5.3.6. Pathology

Neuropathological findings in patients with urea cycle disorders share similar features with hepatic encephalopathy and hypoxic ischemic encephalopathy, that is, they depend upon both the duration of hyperammonemic coma as well as the interval between coma and death. Evidence from autopsy and imaging studies suggests that ornithine transcarbamylase deficiency results in white matter injury. However, these studies suffer in part from small series of patients being reported, weighting towards patients presenting with hyperammonemic coma, studying patients with different stages of disease, or predominance of CT reports, which are not ideal (compared with MRI) in delineating white matter pathology. In survivors of prolonged neonatal coma who live several months and then ultimately die, neuropathologic findings consist of cortical atrophy, ventriculomegaly, gliosis with Alzheimer's type II astrocytes, spongiform changes at the gray/white junction, ulegyria, and spongiform changes in the deep gray nuclei–basal ganglia and thalamus. Neuroimaging studies performed months later in neonatal coma survivors are consistent with these pathological findings, correlating with hypomyelination of white matter, myelination delay, cystic changes of the white matter, and gliosis of the deep gray matter nuclei *(170–173)*.

2.5.3.7. Treatment

Treatment of patients with ornithine transcarbamylase deficiency involves protein restriction, along with administration of sodium phenylbutyrate to activate alternative pathways of waste nitrogen synthesis and excretion. Citrulline may also be used.

2.6.1. Cockayne/ Xeroderma Pigmentosa

Cockayne/ Xeroderma Pigmentosa includes Cockayne syndrome, type A; CSA; excision-repair cross-complementing, group 8; and ERCC8.

2.6.1.1. General

Xeroderma pigmentosum, Cockayne syndrome, and trichothiodystrophy are rare autosomal-recessive inherited human disorders that are associated with impaired nucleotide excision repair activity *(174,175)*. Xeroderma pigmentosum is characterized clinically by severe hypersensitivity to sunlight, abnormal skin pigmentation, and a marked predisposition to skin cancer *(176,177)*.

2.6.1.2. Clinical Features

The major neurological complications of xeroderma pigmentosum and Cockayne syndrome include the following: microcephaly, motor neuron signs or segmental demyelination manifested by spasticity, hyporeflexia/areflexia, ataxia, chorea. Other findings include sensorineural deafness, supranuclear ophthalmoplegia, progressive neurological degeneration, neuronal loss, mental retardation, and dementia. The neurologic complications might dominate the clinical picture (i.e., overshadow cutaneous manifestations).

Xeroderma pigmentosum and Cockayne syndrome present as distinct clinical entities *(175)*, although both disorders are characterized by extreme sun sensitivity. Approximately 20% of patients with xeroderma pigmentosum have microcephaly, deafness, and progressive neurologic degeneration, features that are also present in Cockayne syndrome *(178)*. These deficits worsen slowly and may present later in the course of the disease, after the cutaneous features are manifest. Audiometry may reveal early high-tone hearing loss, whereas Electromyography and nerve conduction velocities may show an axonal (or mixed) neuropathy *(179,180)*.

2.6.1.3. Biochemical Features

The white matter shows loss of myelin lipids with increased levels of cholesterol esters

2.6.1.4. Genetic Features

Cockayne syndrome Gene map locus Chr.5; Xeroderma pigmentosa gene map loci *9q22.3, 2q21, 3p25, 19q13.2-q13.3, 11p12-p11, 16p13.3-p13.13, 13q33, 6p21.1-p12.*

2.6.1.5. Imaging Features

In patients with xeroderma pigmentosum neurologic disease, CT and MRI of the brain may reveal enlarged ventricles and thinning of the cortex *(181,182)*. Symmetric calcifications in the cerebellum and basal ganglia with a hyperintense rim is characteristic of Cockayne syndrome *(123)*. There may be areas of demyelination that vary in their extent. MRI with T_2-weighted sequences reveals periventricular hyperintensity and white matter hyperintensity *(183,184)*.

2.6.1.6. Pathology

The neurologic abnormalities in xeroderma pigmentosum result from a primary degeneration of normally developed neurons. The end result is axonopathy and gliosis *(185–187)*. The brain is atrophied and small with thickened leptomeninges.

Death of corticospinal neurons is responsible for the spasticity and extensor plantar responses seen on physical examination. Loss of neurons in the basal ganglia and substantia nigra pars compacta may lead to movement disorders *(188)*. Neuronal depletion also affects the locus ceruleus and the cerebellum with resultant ataxia, dysarthria, and abnormal eye movements. Death of neurons in the cord and spinal root ganglia accounts for a mixed sensorimotor distal axonopathy *(187)*.

2.6.1.7. Treatment

There is no treatment for this disorder.

2.6.2. Alexander Disease

2.6.2.1. General

Alexander's disease is a progressive disorder of white matter whose onset is usually in infancy or early childhood, with features of macrocephaly, frontal white matter changes on MRI with a caudal progression, and histologic accumulation of Rosenthal fibers in astrocytes. It is associated with mutations of the glial fibrillary acidic protein gene and thus appears to be a primary disorder of astrocytes.

2.6.2.2. Clinical Features

Alexander's disease is usually classified according to the age of onset. An infantile form with onset during the first 2 yr of life, a juvenile form with onset in childhood, mainly school age, and an adult phenotype have all been described. The clinical course can be very variable within these groups. Thus, this clinical classification is not a useful predictor of severity and progression of the disease.

The infantile phenotype may have even neonatal onset and may be very rapidly progressive *(189)*, often leading to death early in life. It is characterized by poor feeding in the infant, macrocephaly, and often intractable, and generalized seizures. Hypotonia and paraparesis are present with lack of developmental progression. There is usually no spasticity or ataxia. Hydrocephalus may be seen, with raised intracranial pressure caused by aqueductal stenosis as a result of pathologic astroglia proliferation *(190)*; indeed, seizures and signs of raised intracranial pressure may mislead the diagnosis.

Juvenile cases often do not have megalencephaly and tend to have predominant pseudobulbar and bulbar signs. They often have ataxia and spasticity with a more slowly progressive course. They may have intact cognition.

In adult cases, the signs are variable and very heterogeneous. The finding of a mutation in glial fibrillary acidic protein in an adult neuropathologically proven case of Alexander's disease defined this phenotype as part of the Alexander's disease spectrum *(191)*. One described patient presented with progressive imbalance and ocular motility abnormalities suggestive of brainstem or cerebellar dysfunction *(192)*. In addition, predominant bulbar dysfunction with associated sleep disturbance (sleep apnea), symptoms of dysautonomia, and dysmorphism were found in one kindred affected by the disease *(193)*. Additional described symptoms have included palatal myoclonus, progressive spastic paraparesis, and ataxia, hemiplegia *(194,195)*. Among adults the condition can also appear to fluctuate and so mimic multiple sclerosis and should be thought of in demyelinating disease without inflammatory features *(196)*. Some adult cases are familial in nature *(197)*.

2.6.2.3. Biochemical Features

The CSF can show an elevation of B-crystallin and heat shock protein

2.6.2.4. Genetic Features

Gene Map Locus: *17q21, 11q13*. Previously the diagnosis of Alexander's disease often rested on brain biopsy or autopsy samples. Sequence analysis of DNA samples from patients representing different Alexander's disease phenotypes revealed that most cases are associated with nonconservative mutations in the coding region of glial fibrillary acidic protein (GFAP) *(198)*. Multiple subsequent studies confirmed the finding of heterozygous mutations in the GFAP gene in cases where Alexander's disease was suggested by imaging or neuropathology. These mutations have been found in the 1A, 2A, and 2B segments of the conserved central rod domain of glial fibrillary acidic protein and also in the variable tail region. The first report of identification of the causative mutation of the GFAP gene for neuropathologically proven hereditary adult-onset Alexander's disease occurred a year after the description of the mutation in younger patients *(199)*, tying together the broad spectrum of presentations.

Alexander's disease may therefore now be diagnosed through analysis of patient DNA samples for mutations in the GFAP gene. Multiple studies of affected patients and their families confirm that the mutation is heterozygous, arising most often de novo, with no other affected family members. However, in adult onset cases, familial autosomal-dominant transmission has been described. In addition, in one study, the presence of an identical mutation in monozygotic twins with infantile Alexander's disease points to possibility of glial fibrillary acidic protein mutations in germ cells or very early postzygotic stages *(200)*. These last two important features need to be taken into consideration when offering genetic counseling for this usually sporadic disorder. Fetal testing is now an option for parents who have had a child that has Alexander's disease with an identified mutation and who wish to have additional children.

Phenotype/genotype correlation has been found in a severe infantile variant, but has otherwise not been described in the other variants. Indeed, mutations in the same coding region of the human glial fibrillary acidic protein have been reported to present as infantile, juvenile, and adult cases. This suggests that clinical severities of Alexander's disease are caused not only by the different sites and nature of mutations in glial fibrillary acidic protein but also by other modifying factor(s) *(195)*.

2.6.2.5 Imaging Features

Imaging features have been found to be variable depending on age of presentation. The classic presentation in infancy or childhood is associated with such typical MRI findings that before identification of mutations in glial fibrillary acidic protein, MRI findings were considered to be diagnostic. Five MRI criteria were defined: extensive cerebral white matter changes with frontal predominance, a periventricular rim with high signal on T_1-weighted images and low signal on T_2-weighted images, abnormalities of basal ganglia and thalami, brain stem abnormalities, and periventricular gray and white matter enhancement *(201)*.

In the adult type, MRI again has been shown to be much more heterogeneous. Diffuse, symmetric hyperintense abnormalities in brainstem and cerebellar white matter have been described on MRI. Prominent atrophy of the medulla oblongata and upper spinal cord also may be seen on MRI. In some cases no demyelination of the cerebral white matter was seen *(202)*. Other cases have described disseminated patchy white matter changes involving the corpus callosum, basal ganglia, and brainstem *(194)*.

MRI findings consistent with the diagnosis have been seen in two asymptomatic individuals whose white matter abnormality was identified incidentally and where mutation studies have confirmed the diagnosis *(203)*. Another case reports a child with isolated macrocephaly and mutation in the GFAP gene *(204)*. It is unclear whether these patients are presymptomatic or represent mild onset cases.

Localized proton MRS of the grey and white matter, basal ganglia, and cerebellum in patients with Alexander disease shows strongly elevated concentrations of myo-inositol in conjunction with normal or increased choline-containing compounds in all regions, a reduction of NAA, most pronounced in cerebral and cerebellar white matter, and accumulation of lactate in affected white matter *(194)*. The myo-inositol/creatine ratio is abnormal in both the demyelinated white matter and normal area in the MRS *(205)*. Serial MRS shows a progressively increasing lactate peak in voxels performed over the basal ganglia *(189)*.

Positron emission tomography with ^{18}F-fluorodeoxyglucose demonstrated, in a single case, hypometabolism in the frontal white matter corresponding to the areas with leukodystrophy. However, the overlying gray matter preserved normal glucose metabolism *(206)*.

2.6.2.6. Pathology

The pathological hallmark of all forms of Alexander disease is the presence of Rosenthal fibers, which are cytoplasmic inclusions in astrocytes that contain the intermediate filament protein GFAP in association with small heat-shock proteins. Astrocytes are dystrophic and associated with myelin abnormalities.

Because the description of the mutation in glial fibrillary acidic protein is associated with Alexander disease, much speculation has occurred on the pathophysiologic mechanism. Alexander's disease is likely a primary disorder of astrocytes and not a primary myelin dysfunction. It has been compared with other disorders of intermediary filament formation. Alexander's disease likely results from a dominant gain of function of the GFAP gene product. It has been suggested that the gain of function is caused by a partial block of filament assembly that leads to accumulation of an intermediate that participates in toxic interactions *(207)*. This toxic product would then result in demyelination. This hypothesis is supported by the phenotype of mice overexpressing human glial fibrillary acidic protein *(208)*. Nevertheless, glial fibrillary acidic protein-null mice display myelin abnormalities and blood–brain barrier dysfunction that are present in Alexander's disease, and the final pathophysiology remains to be determined *(209)*.

2.6.3.Vanishing White Matter Disease

Vanishing white matter disease is also known as myelinopathia centralis diffusa; childhood ataxia with central nervous system hypomyelination (CACH); Cree leukoencephalopathy; vanishing white matter leukodystrophy with ovarian failure; and ovarioleukodystrophy

2.6.3.1. General

Vanishing white matter disease, or CACH, is an autosomal-recessive neurodegenerative disease, most often beginning in early childhood with a chronic progressive course punctuated by episodic deterioration in the context of febrile illness or mild head trauma. The patients demonstrate progressive ataxia, spastic quadriplegia, and relatively preserved mental capacities with white matter rarefaction and a typical MRI picture.

First described in 1993 by Hanefeld et al. *(210)*, the disease as an entity was confirmed by several groups in the following years. The term "vanishing white matter disease" was coined *(211)* and is used as a synonym with the name "CACH," an acronym for the descriptive name childhood ataxia with diffuse central nervous system hypomyelination *(212)*. Initially the diagnosis was based on clinical and radiologic findings with no specific biochemical marker. Discovery of mutations in the eIF2B genes have led to diagnostic confirmation by mutation analysis.

2.6.3.2. Clinical Features

Clinical symptoms include a slowly progressive cerebellar ataxia, spasticity, variable optic atrophy, mild or no epilepsy, and relatively preserved mental capacities. In addition, there are episodes of rapid and major deterioration after febrile illnesses and minor head trauma. These episodes can be accompanied by emesis and lethargy or even progress to unexplained coma. Death may occur after coma. Recovery after these episodes is slow and incomplete.

An early childhood-onset type was the first to be described *(210–213)*. Initial psychomotor development appears normal or only mildly delayed. A chronic progressive ataxia and spasticity develop in the toddler years with relative preservation of cognitive function and variable optic atrophy. Additionally, episodes of severe deterioration, sometimes provoked by a minor fall or an infection, lead to loss of motor function over a few days with irritability, vomiting, lethargy, or coma. Some cases have reported an association with seizures *(214,215)* or chorea *(214)*, although these are not constant features. Head growth deceleration has been described in some cases *(216)*. Children remain severely disabled after episodes of deterioration and may succumb to a subsequent episode of deterioration with coma.

A more severe infantile phenotype, with onset before age 1 and severe evolution without relapsing episodes (death within 1 yr of disease duration) has been described *(217)*. Cree leukoencephalopathy is likely an allelic variant of the severe infantile phenotype of CACH/vanishing white matter disease. Cree leukoencephalopathy is a phenotype described in 1988 in the Cree and Chippewayan indigenous populations in Northern Quebec and Manitoba. Children appeared normal except for hypotonia until acute onset between 3 and 9 mo of age of seizures, vomiting, hyperventilation, diarrhea, and spasticity in the context of a febrile illness. Death ensued by 21 mo in the described cases after a course of developmental regression, blindness, and cessation of head growth with leukoencephalopathy on autopsy. Cree leukoencephalopathy has also been found to be caused by mutations in the eIF2B genes, reinforcing the clinical heterogeneity of this disorder *(218)*.

A severe neonatal phenotype has been described, including delayed fetal movements, intrauterine growth retardation, oligohydramnios, microcephaly, congenital cataracts, mild dysmorphic features, and joint contractures at birth. In addition, a systemic picture presents, with growth retardation, pancreatic abnormalities, hypoplastic kidneys, hepatosplenomegaly, cataracts, and ovarian dysgenesis in addition to the leukoencephalopathy *(219,220)*.

Adolescent or early adult cases described in the literature appear to have a milder course, with less frequent comas and spasticity dominating the ataxia. Indeed, there are sporadic, although much older reports in the literature of an adult onset orthochromatic cavitating leukodystrophy that may have represented the first described cases of vanishing white matter disease *(221)*. An adult-onset dementia with typical MRI features *(222)* also has been described. Another genetically confirmed patient presented with progressive speech difficulties and gait disturbance attributable to spastic paraparesis.

Also recently described is a phenotype that presents with premature ovarian dysfunction and white matter abnormalities on magnetic resonance imaging. This ovarioleukodystrophy is a condition in which progressive neurologic decline in adulthood, white matter abnormalities on MRI, and ovarian failure has been found to be associated with mutations in the eIF2B genes eIF2B2, -4, and -5 *(223)*. Previous reports of "ovarian dysgenesis" *(219)*, "bilateral streak ovaries" *(211)*, and "ovarian failure" *(224)* in patients with the earlier onset CACH/vanishing white matter phenotypes exist.

2.6.3.3. Biochemical Features

The only known biochemical marker described is the elevation of glycine in CSF and urine in patients with vanishing white matter/CACH *(223,225)*. The original description included two patients in whom the CSF glycine level reached the level considered diagnostic for nonketotic hyperglycinemia. The activity of the glycine cleavage system was found to be normal in lymphoblasts in these patients *(225)*. The cause of elevation of CSF glycine in the disease of vanishing white matter is unknown, and it is unclear whether it is caused by a primary disturbance of glycine metabolism or is secondary to excitotoxic brain damage.

2.6.3.4. Genetic Features

Gene Map Locus: *12 q24.1* (eIF2B1), *14q24 .3* (eIF2B2), *1 p34.1* (eIF2B3), *2p23.3* (eIF2B4), *3q27 .3* (eIF2B5). Initial studies demonstrated an autosomal-recessive inheritance pattern and linkage to chromosome *3q27* observed in a data set of 19 patients *(226)*. The results of genealogical studies suggested that seven parents in four Dutch families with vanishing white matter may have inherited an allele for the disease from a common ancestor who lived at least eight generations previously *(226)*. Further linkage analysis in siblings appeared to confirm a gene locus at *3q27 (227)*.

Subsequently, mutations in EIF2B5, encoding the epsilon subunits of the translation initiation factor eIF2B and located on chromosomes *3q27* were determined to cause vanishing white matter *(228)*. Patients without mutations in the EIF2B5 gene were found to be mutated in one of the other genes that encode eIF2B subunits: EIF2B1 to EIF2B4 *(228)*. The most common mutation remains a homozygous missense mutation in the EIF2B5 gene, although other homozygous or even compound. Vanishing white matter leukodystrophy with ovarian failure, or ovarioleukodystrophy, is caused by mutations in the EIF2B2, EIF2B4, and EIF2B5 genes. Patients with Cree leukoencephalopathy were found to have a founder homozygous mutation at R195H in the eIF2B-5 gene *(223)*.

Childhood ataxia with central hypomyelination/vanishing white matter leukoencephalopathy is related to mutations in all five genes of the eukaryotic translation initiation factor (eIF2B). Vanishing white matter is the first human disease related to mutations in any of the five genes encoding subunits of eukaryotic initiation factor eIF2B or any translation factor at all. EIF2B is a complex that is essential to the regulation of translation. Eukaryotic initiation factors (eIFs) are involved in translation initiation for the translation of messenger RNA into peptides. Among them, the guanine-nucleotide exchange factor eIF2B plays a key regulatory role by converting the protein synthesis initiation factor 2 (eIF2) from an inactive GDP-bound form to an active eIF2-GTP complex. The exchange of GDP for GTP is required for each round of translation initiation, and regulation of this step controls global rates of protein synthesis under diverse conditions. A malfunction of eIF2B under stress conditions (physical, chemical, oxidative, and thermal trauma) results in an abnormal "heat shock response" and an abnormal control of protein production during this period of stress contributing to cell dysfunction and death *(220)*. This may explain the rapid deterioration of people with vanishing white matter under stress such as a febrile illness or even minor head trauma.

2.6.3.5. Imaging Features

CT is rarely described in the literature because of nonspecific findings. When described, it is described as bilateral and symmetric hypoattenuation of white matter with sparing of the caudate nucleus and putamen. MRI characteristics initially formed the basis for diagnosis of this disorder however, and are felt to be typical.

Hemispheric involvement of cerebral white matter that may include U fibers *(229)* with gradual rarefaction resulting in a signal change isointense with CSF/ventricles on protein density (PD)/T_1-weighted and T_2-weighted images is prominent. Gradual cavitation results in the appearance of threads of residual tissue left in a space in which the white matter has otherwise "vanished" *(221)*. White matter space appears mildly swollen with thickening of the gyri. Basal ganglia and the internal capsule are generally preserved, although individual cases of thalamic and globus pallidus involvement are described. Pontine tegmentum, basis pontis and pyramidal tract abnormalities have been described in both the classically described patients and the later onset patients *(221,229)*. Cerebellar tissue can be atrophic, predominantly in the vermis.

A similar picture has been described inpatients with Cree leukoencephalopathy with severe cerebral and cerebellar white matter changes with additional involvement of the internal capsule and globus pallidus, but sparing of the putamen and caudate. Bilateral midbrain and pontine changes also are seen in these patients. Rarefaction and cavitation as described in the classically described patients is present.

Involvement of the white matter has been found in asymptomatic patients who later develop disease symptoms *(211,230)*. Indeed, significant white matter changes may already be present early in the course and evolve minimally despite clinical deterioration. In patients with ovarioleukodystrophy, white matter changes have been on occasion found in an incidental fashion in a patient with ovarian failure and no neurologic disturbance other than nonspecific symptoms such as headache *(231)*.

Proton MRS studies show a global decrease of metabolite signals (NAA, choline, creatine, myo-inositol) over the abnormal white matter with a less prominent decrease over the cortex *(232,233)*. Some increase in lactate and glucose over affected white matter are also described *(210,211,232)*.

In studies currently available as research tools, such as proton-decoupled phosphorus MRS, other specific findings have been described. In patients with vanishing white matter, cerebral concentrations of ethanolamine metabolites were abnormal. Glycerophosphorylethanolamine was reduced, and phosphorylethanolamine was increased, whereas the choline-containing phosphorylated metabolites were unchanged. Ethanolamine metabolites constitute the plasmalogens, metabolites that are involved in biosynthesis and catabolism of membrane phospholipids, and abnormalities may result in impairment of myelin membrane synthesis or myelin membrane transport in the brain of the patient with vanishing white matter in vivo *(234)*. These techniques may lend insight into the metabolic abnormalities of vanishing white matter disease.

A multiparametric MR study of a patient with adult-onset genetically confirmed vanishing white matter disease shows that the cortical adaptive capacities were relatively preserved (functional MRI; ref. *235*).

2.6.3.6. Pathology

Macroscopic pathologic changes include severe extensive changes in hemispheric cerebral and cerebellar white matter, with marked rarefaction or cavitation. The pontine central tegmental tracts may also be affected. The subcortical white matter (U fibers), internal capsule and corpus callosum appear preserved.

Microscopic analysis indicates that there is axonal and myelin sheath loss in the white matter *(213)*. Neuronal loss has been seen in the cerebellar cortex *(236)*. Macrophage infiltration and areas of reactive gliosis have been described, but the findings are minimal relative to the degree of white matter disease *(237)*. A relative increase in the number of oligodendrocytes has been described *(237,238)* with typical foamy oligodendrocytes. In one such case, in an active demyelinating lesion in the brainstem, oligodendrocytes showed typical signs of apoptosis, leading to the hypothesis of oligodendroglial death as a primary event *(238)*. In addition, characteristic shrinking and perivascular clustering of astrocytes with an overall reduced number of astrocytes has been described *(237)*.

Ultrastructural descriptions found white matter axonal swelling with complete loss of myelin sheaths at the periphery of cavitating lesions in the cerebrum *(237)*.

2.6.3.7. Treatment

There is currently no available treatment for Vanishing White Matter disease or CACH.

2.6.4. Pelizaeus–Merzbacher Disease

2.6.4.1 General

Pelizaeus–Merzbacher disease is an X-linked recessive leukodystrophy that is caused by a mutation in the proteolipid protein (PLP) gene on chromosome Xq22. The most common mutation is gene duplication followed in frequency by missense mutations, insertions, and deletions. The clinical spectrum is extremely variable and ranges from severe neonatal cases to relatively benign adult forms and X-linked recessive spastic paraplegia type 2. Classic findings on physical exam include transitory nystagmoid eye movements with rotatory movements of the head, spastic quadriparesis, ataxia, Parkinsonism, and dementia. Imaging and gross pathology reveal a characteristic "tigroid" appearance of affected white matter.

2.6.4.2. Clinical Features

Several phenotypes exist in Pelizaeus–Merzbacher disease. The classic type, with onset in infancy and death in late adolescence or young adulthood, is characterized by initial signs of abnormal head and eye movements. The head movements are described as rotary, "tremor," or "nodding." Eye movements are described as pendular, nystagmoid, or "wagging." The nystagmus disappears as the patient matures. Other initial symptoms include hypotonia and choreoathetosis. During the first two decades of life, ataxia, spasticity of the legs and then the arms, and involuntary movements or Parkinsonism become manifest, as well as optic atrophy, microcephaly, and dementia. Patients have been reported to live into their sixth decade.

The connatal type shows rapid progression and is fatal in infancy or childhood. These patients may show profound neonatal hypotonia and hyporeflexia and electromyographic features suggestive of neonatal spinal muscular atrophy *(239)*. They may have all the other features of classic Pelizaeus–Merzbacher disease , as well as respiratory distress and stridor.

The transitional form is intermediate. Some heterozygous females have manifestations of the disorder. The connatal form has also been described on rare occasions in females *(240)*.

2.6.4.3. *Biochemical Features*

There are no known biochemical features specific to this disorder.

2.6.4.4. *Genetic Features*

Gene Map Locus: *Xq22*. Pelizaeus–Merzbacher disease is most frequently caused by duplication of the PLP gene, but also may be caused by deletion as well as by point mutations. Animal models support PLP duplications as a molecular basis for the disease because transgenic mice with extra copies of the wild-type PLP gene exhibit a similar phenotype as humans with Pelizaeus–Merzbacher disease. Neurologic symptoms and severity of the disease in transgenic mice correlates with the level of over expression of the PLP gene. Mutations may be de novo or transmitted. Clinical severity has been found to be correlated with the nature of the mutation.

There remain a number of patients with the Pelizaeus–Merzbacher disease phenotype who do not have documented mutations on the PLP gene *(241,242)*. Some of these patients may have mutations of extra exon sequences of the PLP gene *(243,244)*.

Spastic paraplegia type 2 (SPG2) is allelic to the PLP gene and may represent part of the disease spectrum of this disorder. Prenatal diagnosis is available by analysis of the PLP gene.

2.6.4.5. *Imaging Features*

The CT scan of patients with classical Pelizaeus–Merzbacher disease may be normal initially and generally is not helpful in confirming the diagnosis of Pelizaeus–Merzbacher disease at an early stage. Over time, it may show marked cerebellar atrophy and focal areas of demyelination of cerebral white matter *(245)*.

On MRI, severe hypomyelination is a hallmark Pelizaeus–Merzbacher disease . There is correlation between the degree of hypomyelination and the severity of clinical handicap *(246)*. Affected white matter may have a tigroid or leopard skin appearance caused by patchy areas of myelin deposition. Cerebral white matter hypomyelination on MR images may be associated with brainstem lesions. Hypoplasia of the cerebellum and brainstem may be present, as well as diffuse brain atrophy. Optic atrophy may be seen *(247)*. In the connatal form, a complete absence of myelin in the brain is demonstrated and is felt to be pathognomic of this disorder *(123)*. Diffusion tensor imaging has demonstrated the existence of diffusional anisotropy in the corpus callosum, internal capsule, and white matter of the frontal lobe *(248)*.

Results of proton MRS imaging in Pelizaeus–Merzbacher disease have been heterogeneous. Some studies show diffuse or focal reductions in NAA in the affected white matter. Mild increases in choline and creatine levels have been observed *(249)*. In another study, localized spectra in the posterior portion of the centrum semiovale showed increased absolute concentrations of NAA, creatine, and myo-inositol *(250)*. In the connatal form of Pelizaeus–Merzbacher disease, a significant choline peak reduction is seen *(251)*.

2.6.4.6. *Pathology*

A cardinal finding in Pelizaeus–Merzbacher disease has been a lack of myelin in large parts of white matter with the preservation of islands of intact myelin, resulting in a "tigroid" appearance *(251a)*.

2.6.5. Sjögren–Larsson Syndrome

Sjögren–Larsson Syndrome is also known as SLS; ichthyosis, spastic neurological disorder, and oligophrenia; fatty alcohol:NAD+ oxidoreductase deficiency; FAO deficiency; fatty aldehyde dehydrogenase deficiency; and FALDH deficiency.

2.6.5.1. *General*

In 1988, Sjögren–Larsson Syndrome was shown to be an inborn error of lipid metabolism caused by deficient activity of fatty alcohol:NAD oxidoreductase *(252)*. Later research identified a defect in fatty aldehyde dehydrogenase, which is a component of the fatty alcohol:NAD oxidoreductase enzyme complex *(253)*. Sjögren–Larsson Syndrome is now the most widely recognized form of neuroichthyosis. Life expectancy is dependent upon the severity of neurologic symptoms and is comparable to that in patients with other nonprogressive or slowly progressive neurologic disease. Morbidity is associated with chronic neurologic disease and lifelong ichthyosis.

2.6.5.2. *Clinical Features*

The features of Sjögren–Larsson syndrome are appreciated even in utero. Fetuses affected by Sjögren–Larsson Syndrome exhibit histologic evidence of ichthyosis as early as the end of the second trimester *(254,255)*. Infants with Sjögren–Larsson Syndrome are often born several weeks prematurely. Most infants with Sjögren–Larsson Syndrome demonstrate cutaneous signs at birth, whereas the neurological symptoms usually develop within the first 2 yr of life. Symptoms thereafter persist throughout life.

Clinical diagnosis rests on identification of the following key findings: ichthyosis, which is generally the first symptom to prompt medical evaluation. Ichthyosis is most often apparent at birth. A collodion membrane (a parchment-like membrane covering the skin) is not commonly present at birth but has been observed in as many as 15% of patients with this disorder. Approximately 30% of patients with Sjögren–Larsson Syndrome first develop ichthyosis after the neonatal period, but some patients do not show cutaneous disease until later in life. Neurological features include mental retardation, spastic diplegia or tetraplegia often present by 2 yr of age, speech delay, and short stature is common, owing to a combination of growth delay and leg contractures. Ophthalmologic findings include retinal glistening white dots and retinal pigmentary changes. Photophobia is common in individuals with Sjögren–Larsson Syndrome; decreased visual acuity often occurs.

2.6.5.3. *Biochemical Features*

Diagnosis is made by demonstrating defective fatty alcohol oxidation directly in a skin biopsy using a histochemical staining method. However, this approach is not quantitative.

2.6.5.4. *Genetic Features*

Sjögren–Larsson Syndrome is caused by mutations in the FALDH gene, recently cloned (also known as ALDH10) located on chromosome *17p11.2*. *(256)*. Enzymatic and genetic testing provide a reliable means for diagnosing Sjögren–Larsson Syndrome and determining carrier status.

2.6.5.5. *Imaging Features*

The MRI and clinical findings are consistent with a leukoencephalopathy associated with a lipid abnormality. Brain MRI may be useful for detecting white matter disease, which is observed in approx 60–70% of patients with SLS *(257)*. MRI and ^{1}H MRS findings suggest an accumulation of long-chain fatty alcohol intermediates that result in delayed myelination and dysmyelination *(257)*. The myelin abnormality usually is present in periventricular regions, the centrum semiovale, the corpus callosum, and frontal and parietal lobes. There is sparing of subcortical U fibers , and the location of white matter abnormalities correlate with clinical symptoms and signs *(258)*. In infants with Sjögren–Larsson Syndrome, initial MRI scans may appear normal, with evidence of white matter disease emerging later as the patient ages *(259)*. Proton MRS of the brain in two subjects using single-voxel proton MRS (^{1}H-MRS) at short TE revealed two abnormal peaks at 1.3 ppm and 0.9 ppm. These two abnormal spectral peaks were seen in high-intensity areas on T_2-weighted images and also in basal ganglia of normal intensities and believed to represent abnormal lipids *(260–263)*. The specificity of this lipid peak in those with Sjögren–Larsson Syndrome is yet to be determined.

2.6.5.6. Pathology

Ophthalmological abnormalities of the retina have been reported in Sjögren–Larsson syndrome. The most consistent finding is the presence of glistening white dots on the fundus, usually present in the foveal and perifoveal areas.

Autopsy report of a patient with Sjögren–Larsson Syndrome has shown accumulation of lipoid substances that were stained lightly with PAS in the subpial, subependymal, and perivascular glial layers, the subpial and perivascular spaces, and the white matter of the cerebrum and brainstem. Also noted was proliferation of perivascular macrophages containing lipofuscin-like pigments, as well as a dense distribution of round bodies staining strongly with PAS in the subpial, subependymal, and perivascular glial layers and the white matter. Myelinated nerve fibers were reduced in the cerebral and cerebellar white matter. Neuronal cytoarchitecture was preserved in the cerebral neocortex, except for the insula, where there was random organization of the pyramidal neurons *(264)*.

2.6.5.7. Treatment

There is no definitive treatment for this disorder.

2.6.6. Merosin Deficiency

2.6.6.1. General

One important etiology of a leukoencephalopathy on MRI is the merosin-deficient form of congenital muscular dystrophy due to a laminin-α2 deficiency. These children often have quite dramatic and unexpected white matter findings with lack of upper motor neuron signs and minimal cognitive findings. The clinical picture consists of hypotonia and gross motor delays caused by the nonprogressive muscle disease or a peripheral neuropathy. The disease may be severe, with the child never gaining the ability to walk, or mild, with the child gaining the ability to walk between 2 and 3 yr of age. The MRI reveals a diffuse T_2-signal hyperintensity.

2.6.6.2. Clinical Features

The original clinical phenotype first described in the laminin α2 negative patients was homogenous, consisting of hypotonia and weakness at birth or in the first 6 mo of life, followed by severe muscle weakness, and contractures. These patients rarely were independent ambulators and often had creatine kinase levels >1000 U/L. Characteristic white matter hypodensity was seen on brain MRI with high T_2 signal in the periventricular and subcortical white matter. In the majority of cases there was no clinical evidence of CNS involvement *(265–267)*.

During the past few years, however, there have been a number of case reports demonstrating some variability in the clinical picture associated with laminin α2 deficiency. In contrast to the phenotype of laminin α2 deficiency in association with severe classical muscular dystrophy without clinical CNS involvement, more case reports have suggested a later onset of symptoms, with some patients remaining asymptomatic at the time of diagnosis *(268–270)*. Still others presented with mild, nonprogressive proximal weakness with achievement of ambulation and creatine kinase levels less than 1000 U/L *(271–275)*. Patients with classical congenital muscular dystrophy with laminin α2 deficiency may present with additional disease features including altered visual and somatosensory evoked potentials with minor neurological and perceptual-motor deficits *(276,277)*, evidence of peripheral demyelinating neuropathy *(278,279)*, as well as subclinical cardiac involvement *(280,281)*.

2.6.6.3. Biochemical Features

Creatine phosphokinase is elevated to some extent in these patients.

2.6.6.4. Genetic Features

Gene map locus: *6q2*. The disease is associated with mutations in the laminin-α2 gene. Laminin-α2 is associated with α-dystroglycan in muscle, but the protein is also found in CNS and Schwann's cells within the basal lamina. Laminin α2 is a muscle specific isoform of laminin that is localized to the basal lamina of muscle fibers and is thought to interact with myofiber membrane receptors, such as the integrins, and possibly the dystrophin-associated glycoprotein *(282)*.

2.6.6.5. Imaging Features

Classical imaging findings include striking white matter changes on T_2-weighted brain MRI. There is T_1 and T_2 prolongation in the central cerebrum *(283)*.

However, the significance of the white matter changes is not well understood.

In older children the white matter changes do not progress with age. They typically are more marked in older age groups and may involve the U fibers *(284)*.

Although merosin-deficient congenital muscular dystrophy with white matter abnormalities on neuroimaging is well documented in the literature, the association with cortical dysplasia has also been reported, especially involving the occipital cortex *(285–289)*.

Abnormalities of the cerebellum are common in other forms of CMD as well.

There is limited literature regarding the use of newer imaging modalities such as DTI and ^{1}HMRS in this disorder

2.6.6.6. Pathology

Muscle biopsy shows evidence of a dystrophic process. Immunofluorescence and immunoblotting analyses of muscle biopsies using antibodies directed against both the 80-kDa and 320-kDa fragments of laminin α2 may showed a barely detectable immunoreactivity in some of the patient's muscle fibers, although it is possible for some scattered myofibers to show patchy, discrete areas of highly positive immunostaining.

2.6.6.7. Treatment

There is no specific treatment for this disorder.

2.6.7. Smith-Lemli-Opitz Syndrome

Smith-Lemli-Opitz syndrome is also known as SLO syndrome, type I and RSH syndrome.

2.6.7.1. General

Smith-Lemli-Opitz syndrome is a congenital multiple anomaly syndrome caused by an abnormality in cholesterol metabolism due to deficiency of the enzyme 7-dehydrocholesterol reductase. It is characterized by prenatal and postnatal growth retardation, microcephaly, moderate-to-severe mental retardation, and multiple major and minor malformations. The malformations include distinctive facial features, cleft palate, cardiac defects, underdeveloped external genitalia in males, postaxial polydactyly, and 2-3 syndactyly of the toes. The clinical spectrum is wide and individuals have been described with normal development and only minor malformations *(290)*.

2.6.7.2. Clinical Features

Clinical diagnostic criteria have not been established for Smith-Lemli-Opitz syndrome. A pattern of congenital anomalies suggests the diagnosis. The features most commonly observed are microcephaly, postaxial polydactyly, 2-3 syndactyly of the toes, growth and mental retardation, cleft palate, and hypospadias in males. Neurologically, there is ptosis, hypotonia, and mental retardation.

2.6.7.3. Biochemical Features

Although serum concentration of cholesterol is usually low, it may be in the normal range in approx 10% of patients, making it an unreliable test for screening and diagnosis. Demonstration of the characteristic biochemical profile of elevated serum concentration of 7-dehydrocholesterol or an elevated 7-dehydrocholesterol:cholesterol ratio is diagnostic. Although most affected individuals have hypocholesterolemia, serum concentration of cholesterol values in normal and affected individuals can overlap *(291)*.

2.6.7.4. Genetic Features

Gene map locus: *11q12-q13*. Smith-Lemli-Opitz syndrome is inherited in an autosomal recessive manner. Molecular mutation analysis is available in affected individuals. Carrier detection is possible if the disease-causing mutations have been identified in an affected family member. Prenatal testing is available using (1) biochemical testing or (2) molecular genetic testing if the disease-causing mutations in the family are known.

2.6.7.5. Imaging

Very few neuroimaging studies have been conducted in this population. A single patient imaged in 1997 showed MRI findings of frontal lobe hypoplasia, cortical migration defect, and abnormalities of median line structures *(292)*. A recent study evaluating 18 patients by the use of MRI and MRS demonstrated rare structural abnormalities with abnormal CNS findings noted in five patients, including callosal abnormalities Dandy-Walker variant and arachnoid cyst. Holoprosencephaly was noted in one patient. MRS showed that the choline:NAA, lipid:NAA, and lipid:choline ratios were correlated with the clinical degree of disease severity and serum total sterol ratios (cholesterol/cholesterol + 7-dehydrocholesterol + 8-dehydrocholesterol). Choline:NAA was elevated in seven patients. There was a statistically significant positive correlation between the lipid:choline ratio and the serum cholesterol precursor, 8-dehydrocholesterol. In two patients ^{1}H MRS demonstrated abnormally elevated lipids prior to cholesterol therapy, which improved on therapy *(293)*.

2.6.7.6. Pathology

In severe cases, holoprosencephaly has been found. Other specific pathology has not been delineated in large series.

2.6.7.7. Treatment

Preliminary research on dietary cholesterol supplementation has yielded some promising results for treatment of Smith-Lemli-Opitz syndrome. Because Smith-Lemli-Opitz syndrome is a cholesterol deficiency syndrome, research trials have begun and include an increase in total caloric intake as well as an increase in cholesterol intake .

2.6.8. Megaloencephalic Leukoencephalopathy With Subcortical Cysts

Megaloencephalic leukoencephalopathy is also known as MLC; Van der Knaap disease; and vacuolating megaloencephalic leukoencephalopathy with subcortical cysts.

2.6.8.1. General

Vacuolating megaloencephalic leukoencephalopathy with subcortical cysts is an autosomal-recessive disorder characterized by acquired macrocephaly, developmental motor delay of varying degrees, slowly progressive cerebellar and pyramidal signs, sometimes intractable epilepsy and initially preserved intellectual function. It is related to different mutations in the MLC1 gene, encoding a putative membrane protein of still-unknown function.

2.6.8.2. Clinical Features

Megalencephalic leukoencephalopathy is a recently described *(294)* syndrome of cerebral leukoencephalopathy, megaloencephaly, slow progressive ataxia and spasticity, seizures, and preserved intellectual functioning for years after onset of the disorder. Onset of symptoms varies from 2 mo to 10 yr. Macrocephaly is acquired over the first years of life, and then appears to stabilize. Some patients had delay of independent walking, presumably because of early onset ataxia.

Progression often continues into adulthood. Most patients have a mild course, but there are more severe phenotypes with rapid progression of gait disturbance and spastic tetraparesis. An isolated case of unexplained coma after minor head trauma exists *(295)*.

Epilepsy was present in six of the eight initially described patients, with time of first seizure ranging from 1.5 to 12 yr. More than 60% of the published cases since then have been reported to have seizures *(296)*.

2.6.8.3. Biochemical Features

There are no known specific biochemical features in this disorder.

2.6.8.4. Genetic Features

The disease locus was mapped to the telomeric region of 22q (ref. *297* then the MLC gene was identified and cloned; ref. *298*). Although many cases have been confirmed to have mutations in MLC1, genetic heterogeneity exists, and the existence of at least a second MLC locus is postulated *(299,300)*.

In India, megalencephalic leukoencephalopathy with subcortical cysts occurs predominantly in the Agarwal community. A common mutation in the MLC1 gene has been seen in 31 Agarwal patients, which suggests a founder effect *(301)*.

Table1
Imaging Characteristics of the Leukodystrophies

Type of leukodystrophy	*VWM*	*Krabbe's disease*	*Canavan disease*	*MLC*
Computed tomography				
Findings		Hyperdensity basal ganglia, thalami		Swelling, diffuse hypodensity
Useful adjunctive tool?		Yes, may predate MRI		
Magnetic resonance imaging				
Hemispheric involvement	**Yes, swollen-appearing WM, broadened gyri**	**Yes, characteristic linear pattern in centrum semiovale**	Yes	Yes, swollen-appearing WM
Lobar involvement	No	Posterior predominant initially	No	No
Arcuate fibers	Involved	Spared	Involved early	
Cystic changes	No	No	No, not visible on imaging	**Yes, anterior temporal and frontal**
Brainstem changes	Pontine tegmentum	Corticospinal tracts	Yes	Spared
Spinal cord involvement	Rare reports	With involvement of spinal roots	Posterolateral demyelination	Not reported
Cerebellar findings	Variable atrophy, predominantly vermis	In the early infantile cases	Yes	Yes
Internal capsule	Spared	Spares anterior portion	Late	Spared
Basal ganglia	Spared	In the early infantile cases	**Sparing the putamen**, early involvement of the globus pallidus and caudate	Spared
Contrast enhancement	No	**May be seen in cranial nerves and lumbosacral spinal roots**	No	No
Magnetic resonance spectroscopy				
MRS of white matter				
NAA	Absent	↓infantile	↑↑	↓
Choline	Absent	↑ infantile		↓
Creatine	Absent			↓
Myo-inositol	Absent	↑infantile		nl
Lacate	↑	Rarely ↑	NR	NR
MRS of cortex	Less profound than in white matter	Same changes less profound	Not reported	Normal
MRS of basal ganglia			Not reported	Not reported
DT imaging	No studies available	Limited studies	No studies available	Limited studies

NR, not reported; VWM, vanishing white matter disease (also known as CACH); MLC, magalencephalic leukodystrophy with subcortical cyts.

The most pathognomonic manifestations for each disorder are written in bold.

Aicardi Goutieres	*Alexander's disease*	*Metachromatic leukodystrophy*	*Pelizaeus Merzbacher*	*Adrenoleukodystrophy*
Basal ganglia calcifications			Normal initially	
Yes			No	
Yes	**Frontal predominance, periventricular rim**	**Initially periventricular and centrum semiovale, may be tigroid**	**Severe hypomyelination, Tigroid appearance**	
No		More prominent occipital area		
		Spared until late		
No	Late	Not reported		
Yes	Yes	Not reported	Yes, hypoplasia	
Not reported	Yes, adult type	Not reported		
Yes	Yes, adult type	Spared until late	Yes, hypoplasia	
Not reported		Yes, early		
Yes, with calcifications	**Basal ganglia and thalami**	Not reported		
No	Yes, periventricular	No		**Yes, at border of advancing demyelination**
NR	↓	↓	Variable	
NR		↓	↑, except connatal	
NR			↑	
NR	↑	↑	↑	
NR	↑	↑	NR	
NR	More mildly affected	Same		
NR	Lactate peak			
Not reported	Not reported	Not reported	Abnormalities of corpus callosum, internal capsule, frontal lobe	

Table 2
Imaging Features of Selected Imaging Systems

Leukoencephalopathies	*Phenylketonuria*	*Maple syrup urine disease*	*Ornithine transcarbamylase deficiency (OTCD)*	*Mitochondrial cytopathies*
Computed tomography				
Findings	Malignant PKU may be associated with calcifications of basal ganglia	Cytotoxic edema	Low-density lesions in white matter, symmetric or asymmetric	Variable findings include gray and white matter pathologies
Useful adjunctive tool?	No	No	No	No/yes for evaluation of acute stroke-like events
Magnetic resonance imaging				
Hemispheric involvement	Foci of demyelination-spongy degeneration	Weak diffuse T_2 hyper-intensity in the cerebellar white matter and in the dorsal brainstem	T_1 and T_2 prolongations with round lesions in the deep white matter or small foci of T_2 and T_1 prolongation in the subcortical white matter. MRI might be normal in the early stage of the disease, and progress in proportion to the clinical stage of OTCD. Focal and patchy may look like hypomyelination or myelination delay	Predominantly gray matter, but also white matter involvement, may be calcifications. Lesions may be isolated or confluent
Lobar involvement	Diffuse	Diffuse edema	Diffuse	Frontal, parietal, occipital, temporal, corpus callosum
Arcuate fibers	No	No	No	No
Cystic changes	May be observed with severe disease		Yes, in neonatal onset disease	May be necrotic/cavitary lesions in basal ganglia and white matter in Leigh's disease
Brainstem changes	Less marked	Yes		Characteristic pattern in Leigh's disease with involvement of deep gray nuclei
Spinal cord involvement				May be involved
Cerebellar findings	Foci of demyelination-spongy degeneration	White matter edema		May see hypoplasia, or affectation of cerebellar peduncles
Interal capsule		Posteriori limb edema		
Basal ganglia	Lesions may be seen		Can be involved with spongiform changes	T_2 hyperintensities

Magnetic resonance spectroscopy				
MRS of white matter	Elevations of phenylalanine	Branched-chain amino acids (BCAA) and branched-chain alpha-keto acids (BCKA) peak at 0.9 ppm	Elevations of glutamine and glutamate (Glx)	
NAA	Unchanged		Unchanged	Decreased
Choline	Decreased		Decreased	
Creatine	Unchanged		Unchanged	Unchanged
Myo-inositol	Decreased		Depletion	Decreased
Lactate		Elevated		Elevations of lactate and/or decreased NAA may be seen
MRS of cortex	Elevations of phenylalanine		Elevations of glutamine reported	Elevations of lactate and/or decreased NAA may be seen
MRS of basal ganglia				Elevations of lactate and/or NAA may be seen
DT imaging	Low ADC reported in the literature	Hyperintensity of the cerebellar white matter, the brainstem, the cerebral peduncles, the thalami, the dorsal limb of the internal capsule and the centrum semiovale, (ADC) of these regions was markedly (>80%) decreased.	NR	

NR, not reported.

2.6.8.5. Imaging Features

The CT scan findings include supratentorial diffuse hypodensities in the white matter and swelling *(302,303)*. The MRI findings are characteristic for a discrepancy between the mild-appearing clinical picture and severe lesions on MRI. Cerebral hemispheric white matter appears diffusely swollen, obliterating the subarachnoid spaces. Cysts develop in the tips of the temporal lobes and frontoparietal subcortical area. Cerebellar hemispheres are mildly involved and not swollen. Central white matter structures, including the corpus callosum, internal capsule, and brainstem, are relatively spared *(294,304)*.

Proton MRS of white matter revealed marked reduction of NAA, creatine, and choline with normal values for myo-inositol, consistent with axonal loss and astrocytic proliferation *(305)*. Metabolic abnormalities are milder in the frontal white matter and more severe in the posterior white matter. The ^{1}H-MRSI pattern of the gray matter is normal *(306)*.

Diffusion tensor imaging showed an increased apparent diffusion coefficient and reduced anisotropy in affected white matter pointing to reduced cell density with an increased extracellular space *(307)*.

2.6.8.6. Pathology

Autopsy cases describe a spongiform leukoencephalopathy, without cortical involvement. Most vacuoles appear covered by single myelin lamellae or by oligodendroglial extensions *(308)*.

2.6.8.7. Treatment

Only symptomatic treatment of epilepsy is available at this time. Seizures appear to be well controlled with typical anticonvulsants.

2.6.9. Aicardi Goutieres

Aicardi Goutieres is also known as familial infantile encephalopathy with calcification of basal ganglia and chronic cerebrospinal fluid lymphocytosis.

2.6.9.1. General

The first reports of this syndrome were in 1984, by Aicardi and Goutieres, although retrospective analysis of the literature suggests that a group of siblings reported in the past as having had Fahr disease may have actually been affected with this disorder *(309)*. The first eight cases reported had a progressive familial encephalopathy in infancy, with calcification of the basal ganglia and chronic CSF lymphocytosis, leading rapidly to a vegetative state and early death. Multiple similar reports followed *(310–315)*. A raised level of CSF interferon-α was noted subsequently in many of these patients. It is thought to be autosomal recessive. Recent genomic studies have linked the entity in a number of cases to gene locus 3q21.

2.6.9.2. Clinical Features

Patients often appear normal initially. While still in infancy, they develop a progressive encephalopathy with microcephaly, spastic quadriplegia, dystonia, refractory seizures, visual loss, abnormal eye movements and profound retardation. CSF studies show a chronic lymphocytosis and/or raised levels of interferon-α. Serologic studies for TORCH infections, which are a group of congenital infections consisting of toxoplasmosis, other (syphilis, hepatitis B, varicella zoster, Epstein barr virus, parvovirus, and human immunodeficiency virys), rubella, cytomegalovirus, and herpes simplex virus, prompted by the finding of intracranial calcifications, show no evidence of infection. There is a rapid course toward a behavioral vegetative state and death occurs in early childhood. Some patients, however, stabilize after this early fulminant course *(316)* or have a much milder course *(317)*.

2.6.9.3. Biochemical Features

A raised level of interferon-α in the CSF constitutes a marker of the syndrome: this level, which falls with age, is higher in the CSF than in the serum, suggesting intrathecal synthesis *(316)*. A moderate CSF pleocytosis is also present. CSF examination (including interferon-α) should be performed early as typical abnormalities decrease or disappear with age *(318)*. Other abnormal CSF findings include extremely high neopterin and biopterin combined with lowered 5-methyltetrahydrofolate concentrations in two patients *(319)*.

2.6.9.4. Genetic Features

Gene Map Locus: *3p21*: Recent genomic studies have linked the entity in a number of cases to gene locus 3q21.

2.6.9.5. Imaging Features

CT is very important in the diagnosis of Aicardi Goutieres, demonstrating clearly the presence of calcifications at basal ganglia, specifically the lenticular nuclei. These are often bilateral and symmetrical. Calcifications may extend to the cerebral cortex, cerebellum, and Sylvian fissure. Hypodensity may also be seen in the white matter. Some patients have shown a generalized cortical atrophy, dilatation of the lateral, third, and fourth ventricles, widening of the surface CSF spaces, and hypoplasia of the posterior fossa structures.

MRI reveals diffuse leukodystrophy and progressive cerebral atrophy *(320)*. Dysmyelination may be present in the brainstem white matter similar to that in the cerebral white matter *(321)*.

2.6.9.6. Pathology

Calcifications are both present as concretions and as perivascular cuffs of calcium surrounding small vessels. Small vessel involvement (microangiopathy) is present. Microangiopathy may bean important pathogenic mechanism in Aicardi-Goutieres syndrome *(322)*.

2.6.9.7. Treatment

Long-term substitution with folinic acid (2–4 mg/kg/d) resulted in substantial clinical recovery with normalization of CSF folates and pterins in one patient and clinical improvement in another.

REFERENCES

1. Hashemi RJ, Bradley WG Jr. MRI the Basics. Baltimore: Lippincott, Williams and Wilkins; 1997.
2. Basser PJ. Inferring micro structural features and the physiological state of tissues from diffusion-weighted images. NMR Biomed 1995;8:333–344.
3. Wieshmann UC, Clark CA, Symms MR, et al. Reduced anisotropy of water diffusion in structural cerebral abnormalities demonstrated with diffusion tensor imaging. Magn Reson Imaging 1999;17:1269–1274.
4. Moser HW, Loes DJ, Melhem ER, et al: X-Linked adrenoleukodystrophy: overview and prognosis as a function of age and brain magnetic resonance imaging abnormality. A study involving 372 patients. Neuropediatrics 2000;31:227–239.
5. Eichler FS, Itoh R, Barker PB, et al. Proton MR spectroscopic and diffusion tensor brain mr imaging in X-linked adrenoleukodystrophy: initial experience. Radiology 2002;225:245–252.
6. Pfeuffer J, Tkac I, Provencher SW, Gruetter R. Toward an in vivo neurochemical profile: quantification of 18 metabolites in short-

echo-time (1)H NMR spectra of the rat brain. J Magn Reson 1999;14:104–120.
7. Tkac I, Andersen P, Adriany G, et al. In vivo 1H NMR Spectroscopy of the human brain at 7T. Magn Reson Med 2001;46:451–456.
8. deGrauw TJ, Cecil KM, Byars AW, Salomons GS. The clinical syndrome of creatine transporter deficiency. Mol Cell Biochem 2003; 244:45–48.
9. Weglage J, Wiedermann D, Denecke J, et al. Individual blood-brain barrier phenylalanine transport determines clinical outcome in phenylketonuria. J Inherit Metab Dis 2002;25:431–436.
10. Jan W, Zimmerman RA, Wang ZJ, et al. MR diffusion imaging and MR spectroscopy of maple syrup urine disease during acute metabolic decompensation. Neuroradiology 2003;45:393–399.
11. Choi CG, Yoo HW. Localized proton MR spectroscopy in infants with urea cycle defect. AJNR 2001;22:834–837.
12. Takanashi J, Kurihara A, Tomita M, et al. Distinctly abnormal brain metabolism in late onset ornithine transcarbamylase deficiency. Neurology 2002;59:210–214.
13. Cross JH, Gadian DG, Connelly A, Leonard JV. Proton magnetic resonance spectroscopy studies in lactic acidosis and mitochondrial disorders. J Inherit Metab Dis 1993;16:800–811.
14. Sylvain M, Mitchel GA, Shevell MI, et al. Muscle and brain magnetic resonance spectroscopy (MRS) and imaging (MRI) in children with Leigh's syndrome associated with cytochrome C oxidase deficiency: dependence on findings on clinical status. Ann Neurol 1993; 54:464.
15. Harada M, Tanouchi M, Arai K, et al. Therapeutic efficacy of a case of pyruvate dehydrogenase complex deficiency monitored by localized proton magnetic resonance spectroscopy. Magn Reson Imaging 1996;14:129–133.
16. Shevell MI, Didomenicantonio G, Sylvain M, et al. Glutaric academia type II: neuroimaging and spectroscopy evidence for developmental encephalomyopathy. Pediatr Neurol 1995;12:350–353.
17. Lin DD, Crawford TO, Barker PB. Proton MR spectroscopy in the diagnostic evaluation of suspected mitochondrial disease. AJNR Am J Neuroradiol 2003;24:33–41.
18. Choi CG, Yoo HW. Localized proton MR spectroscopy in infants with urea cycle defect. AJNR Am J Neuroradiol 2001;22:834–837.
19. Takanashi J, Kurihara A, Tomita M, et al: Distinctly abnormal brain metabolism in late onset ornithine transcarbamylase deficiency. Neurology 2002;59:210–214.
20. Gabis L, Parton P, Roche P, et al. In vivo 1H magnetic resonance spectroscopic measurement of brain glycine levels in nonketotic hyperglycinemia. J Neuroimaging 2001;11:209–211.
21. Viola A, Chabrol B, Nicoli F, et al . Magnetic resonance spectroscopy study of glycine pathways in nonketotic hyperglycinemia. Pediatr Res 2002;52:292–300.
22. Jakobs C, Jaeken J, Gibson KM, et al. Inherited disorders of GABA metabolism. 1993; J Inherit Metab Dis 16:704–715.
23. Bluml S, Philippart M, Schiffman R, et al: Membrane phospholipids and high energy metabolites in childhood ataxia and CNS hypomyelination. Neurology 2003;61:648–654.
24. Gillies RJ, Barry JA, Ross BD: In vitro and in vivo 13C and 31P NMR analyses of phosphocholine metabolism in rat glioma cells. Magn Reson Med. 1994;32:310–318.
25. Shapiro EG, Lockman LA, Knopman D, Krivit W. Characteristics of the dementia in late-onset metachromatic leukodystrophy. Neurology 1994;44:662–665.
26. Cengiz N, Ozbenli T, Onar M, Yildiz L, Ertas B. Adult metachromatic leukodystrophy: three cases with normal nerve conduction velocities in a family. Acta Neurol Scand 2002;105:454–457.
27. Kihara H. Genetic heterogeneity in metachromatic leukodystrophy. Am J Hum Genet 1982;34:171–181.
28. Waltz G, Harik SI, Kaufman B. Adult metachromatic leukodystrophy: value of computed tomographic scanning and magnetic resonance imaging of the brain. Arch Neurol 1987;44:225–227.
29. Felice KJ, Gomez- Lira M, Natowicz M, Grunnet ML, Tsongalis GJ, Sima AA, Kaplan RF. Adult-onset MLD: a gene mutation with isolated polyneuropathy. Neurology 2000;55:1036–1039.
30. Coulter-Mackie MB, Applegarth DA, Toone JR, Gagnier L, Anzarut AR, Hendson G. Isolated peripheral neuropathy in atypical metachromatic leukodystrophy: a recurrent mutation. Can J Neurol Sci 2002; 29:159–163.
31. Bostantjopoulou S, Katsarou Z, Michelakaki H, Kazis A. Seizures as a presenting feature of late onset metachromatic leukodystrophy. Acta Neurol Scand 2000;102:192–195.
32. Austin J, Armstrong D, Fouch S, Mitchell C, Stumpf DA, Shearer L, Briner O. Metachromatic leukodystrophy (MLD). VIII. MLD in adults: diagnosis and pathogenesis. Arch Neurol 1968;18:225–240.
33. Kim TS, Kim IO, Kim WS, Choi YS, Lee JY, Kim OW, Yeon KM, Kim KJ, Hwang YS. MR of childhood metachromatic leukodystrophy. AJNR Am J Neuroradiol 1997;18:733–738.
34. Zafeiriou DI, Kontopoulos EE, Michelakakis HM, Anastasiou AL, Gombakis NP. Neurophysiology and MRI in late-infantile metachromatic leukodystrophy. Pediatr Neurol 1999;21:843–846.
35. Faerber EN, Melvin J, Smergel EM MRI appearances of metachromatic leukodystrophy. Pediatr Radiol 1999;29:669–672.
36. Salmon E, Van der Linden M, Maerfens Noordhout A, Brucher JM, Mouchette R, Waltregny A, Degueldre C, Franck G. Early thalamic and cortical hypometabolism in adult-onset dementia due to metachromatic leukodystrophy. Acta Neurol Belg. 1999;99:185–188.
37. Sener RN. Metachromatic leukodystrophy. Diffusion MR imaging and proton MR spectroscopy. Acta Radiol 2003; 44:440–443.
38. Kruse B, Hanefeld F, Christen HJ, Bruhn H, Michaelis T, Hanicke W, Frahm J. Alterations of brain metabolites in metachromatic leukodystrophy as detected by localized proton magnetic resonance spectroscopy in vivo. J Neurol 1993;241:68–74.
39. Sener RN. Metachromatic leukodystrophy: diffusion MR imaging findings. AJNR Am J Neuroradiol 2002;23:1424–1426.
40. Solders G, Celsing G, Hagenfeldt L, Ljungman P, Isberg B, Ringden O. Improved peripheral nerve conduction, EEG and verbal IQ after bone marrow transplantation for adult metachromatic leukodystrophy. Bone Marrow Transplant 1998;22:1119–1122.
41. Landrieu P, Blanche S, Vanier MT, Metral S, Husson B, Sandhoff K, Fischer A.Bone marrow transplantation in metachromatic leukodystrophy caused by saposin-B deficiency: a case report with a 3-year follow-up period. J Pediatr 1998;133:129–32.
42. Malm G, Ringden O, Winiarski J, Grondahl E, Uyebrant P, Eriksson U, Hakansson H, Skjeldal O, Mansson JE. Clinical outcome in four children with metachromatic leukodystrophy treated by bone marrow transplantation. Bone Marrow Transplant 1996;17:1003–1008.
43. Stillman AE, Krivit W, Shapiro E, Lockman L, Latchaw RE . Serial MR after bone marrow transplantation in two patients with metachromatic leukodystrophy. AJNR Am J Neuroradiol 1994;15:1929–1932.
44. Korn-Lubetzki I, Dor-Wollman T, Soffer D, Raas-Rothschild A, Hurvitz H, Nevo Y. Early peripheral nervous system manifestations of infantile Krabbe disease. Pediatr Neurol 2003;28:115–118.
45. Sabatelli M, Quaranta L, Madia F, Lippi G, Conte A, Lo Monaco M, Di Trapani G, Rafi MA, Wenger DA, Vaccaro AM, Tonali P. Peripheral neuropathy with hypomyelinating features in adult-onset Krabbe's disease. Neuromusc Disord 2002;12:386–391.
46. Zlotogora J, Chakraborty S, Knowlton RG, Wenger DA. Krabbe disease locus mapped to chromosome 14 by genetic linkage. Am J Hum Genet 1990;47:37–44.
47. Cannizzaro LA, Chen YQ, Rafi MA, Wenger DA. Regional mapping of the human galactocerebrosidase gene (GALC) to 14q31 by in situ hybridization. Cytogenet. Cell Genet 1994;66:244–245.
48. Matsuda J, Vanier MT, Saito Y, Tohyama J, Suzuki K, Suzuki K. A mutation in the saposin A domain of the sphingolipid activator protein (prosaposin) gene results in a late-onset, chronic form of globoid cell leukodystrophy in the mouse. Hum Mol Genet 2001;10:1191–1199.
49. Furuya H, Kukita Y, Nagano S, Sakai Y, Yamashita Y, Fukuyama H, et al. Adult onset globoid cell leukodystrophy (Krabbe disease): analysis of galactosylceramidase cDNA from four Japanese patients. Hum Genet 1997;100:450–456.
50. De Gasperi R, Sosa MAG, Sartorato EL, Battistini S, MacFarlane H, Gusella, JF, Krivit W, Kolodny EH. Molecular heterogeneity of

late-onset forms of globoid-cell leukodystrophy. Am J Hum Genet 1996;59:1233–1242.
51. Wenger DA, Rafi MA, Luzi P. Molecular genetics of Krabbe disease (globoid cell leukodystrophy): diagnostic and clinical implications. Hum Mutat 1997;10:268–279.
52. Zlotogora, J Regev, R.; Zeigler, M.; Iancu, T. C.; Bach, G. Krabbe disease: increased incidence in a highly inbred community. Am J Med Genet 1985;21:765–770.
53. Finelli DA, Tarr RW, Sawyer RN, Horwitz SJ. Deceptively normal MR in early infantile Krabbe disease. AJNR Am J Neuroradiol 1994;;15:167–171.
54. Choi S, Enzmann DR.Infantile Krabbe disease: complementary CT and MR findings. AJNR Am J Neuroradiol 1993;14:1164–1166.
55. Kwan E, Drace J, Enzmann D. Specific CT findings in Krabbe disease. AJR Am J Roentgenol 1984;143:665–670.
56. Zafeiriou DI, Anastasiou AL, Michelakaki EM, Augoustidou-Savvopoulou PA, Katzos GS, Kontopoulos EE. Early infantile Krabbe disease: deceptively normal magnetic resonance imaging and serial neurophysiological studies. Brain Dev 1997;19:488–491.
57. Sasaki M, Sakuragawa N, Takashima S, Hanaoka S, Arima M. MRI and CT findings in Krabbe disease. Pediatr Neurol 1991;7:283–288.
58. Farley TJ, Ketonen LM, Bodensteiner JB, Wang DD. Serial MRI and CT findings in infantile Krabbe disease. Pediatr Neurol 1992;8:455–458.
59. Barone R, Bruhl K, Stoeter P, Fiumara A, Pavone L, Beck M. Clinical and neuroradiological findings in classic infantile and late-onset globoid-cell leukodystrophy (Krabbe disease). Am J Med Genet 1996;63:209–217.
60. Loes DJ, Peters C, Krivit W. Globoid cell leukodystrophy: distinguishing early-onset from late-onset disease using a brain MR imaging scoring method. AJNR Am J Neuroradiol 1999;20:316–323.
61. Jones BV, Barron TF, Towfighi J. Optic nerve enlargement in Krabbe's disease. AJNR Am J Neuroradiol 1999;20:1228–1231.
62. Hittmair K, Wimberger D, Wiesbauer P, Zehetmayer M, Budka H. Early infantile form of Krabbe disease with optic hypertrophy: serial MR examinations and autopsy correlation. AJNR Am J Neuroradiol 1994;15:1454–1458.
63. Vasconcellos E, Smith M. MRI nerve root enhancement in Krabbe disease. Pediatr Neurol 1998;19:151–152.
64. Given CA , Santos CC, Durden DD. Intracranial and spinal MR imaging findings associated with Krabbe's disease: case report. AJNR Am J Neuroradiol 2001;22:1782–1785.
65. Satoh JI, Tokumoto H, Kurohara K, Yukitake M, Matsui M, Kuroda Y, et al. Adult-onset Krabbe disease with homozygous T1853C mutation in the galactocerebrosidase gene. Unusual MRI findings of corticospinal tract demyelination. Neurology 1997;49:1392–1399.
66. Farina L, Bizzi A, Finocchiaro G, Pareyson D, Sghirlanzoni A, Bertagnolio B, et al. MR imaging and proton MR spectroscopy in adult Krabbe disease. AJNR Am J Neuroradiol 2000;21:1478–1482.
67. Zarifi MK, Tzika AA, Astrakas LG, Poussaint TY, Anthony DC, Darras BT. Magnetic resonance spectroscopy and magnetic resonance imaging findings in Krabbe's disease. J Child Neurol 2001;16:522–526.
68. Brockmann K, Dechent P, Wilken B, Rusch O, Frahm J, Hanefeld F.Proton MRS profile of cerebral metabolic abnormalities in Krabbe disease. Neurology 2003;60:819–825.
69. Guo AC, Petrella JR, Kurtzberg J, Provenzale JM. Evaluation of white matter anisotropy in Krabbe disease with diffusion tensor MR imaging: initial experience. Radiology 2001;218:809–815.
70. Sabatelli M, Quaranta L, Madia F, Lippi G, Conte A, Lo Monaco M, et al. Peripheral neuropathy with hypomyelinating features in adult-onset Krabbe's disease. Neuromusc Disord 2002;12:386–391.
71. Itoh M, Hayashi M, Fujioka Y, Nagashima K, Morimatsu Y, Matsuyama H. Immunohistological study of globoid cell leukodystrophy. Brain Dev 2002;24:284–290.
72. Jatana M, Giri S, Singh AK. Apoptotic positive cells in Krabbe brain and induction of apoptosis in rat C6 glial cells by psychosine. Neurosci Lett 2002;330:183–187.
73. Krivit W, Shapiro EG, Peters C, Wagner JE, Cornu G, Kurtzberg J , et al. Hematopoietic stem-cell transplantation in globoid-cell leukodystrophy. N Engl J Med 1998;338:1119–1126.
74. Argoff CE, Kaneski CR, Blanchette-Mackie EJ, Comly M, Dwyer NK, Brown A, Brady RO, Pentchev PG. Type C Niemann-Pick disease: documentation of abnormal LDL processing in lymphocytes. Biochem Biophys Res Commun 1990;171:38–45.
75. Boustany RN, Kaye E, Alroy J. Ultrastructural findings in skin from patients withNiemann-Pick disease, type C. Pediatr Neurol 1990;6:177–183.
76. Carstea ED, Morris JA, Coleman KG, Loftus SK, Zhang D, Cummings C, et al. Niemann-Pick C1 disease gene: homology to mediators of cholesterol homeostasis. Science 1997;277:228–231.
77. Sylvain M, Arnold DL, Scriver CR, Schreiber R, Shevell MI. Magnetic resonance spectroscopy in Niemann-Pick disease type C: correlation with diagnosis and clinical response to cholestyramine and lovastatin. Pediatr Neurol 1994; 10:228–232.
78. Palmeri S, Battisti C, Federico A, Guazzi GC. Hypoplasia of the corpus callosum in Niemann-Pick type C disease. Neuroradiology 1994;36:20–22.
79. Tedeschi G, Bonavita S, Barton NW, Betolino A, Frank JA, Patronas NJ, Alger JR, Schiffmann R. Proton magnetic resonance spectroscopic imaging in the clinical evaluation of patients with Niemann-Pick type C disease. J Neurol Neurosurg Psychiatry 1998;65:72–79.
80. Grau AJ, Brandt T, Weisbrod M, Niethammer R, Forsting M, Cantz M, Vanier MT, Harzer K.Adult Niemann-Pick disease type C mimicking features of multiple sclerosis. J Neurol Neurosurg Psychiatry 1997; 63:552.
81. Chen CY, Zimmerman RA, Lee CC, Chen FH, Yuh YS, Hsiao HS. Neuroimaging findings in late infantile GM1 gangliosidosis. AJNR Am J Neuroradiol 1998;19:1628–1630.
82. Campdelacreu J, Munoz E, Gomez B, Pujol T, Chabas A, Tolosa E. Generalised dystonia with an abnormal magnetic resonance imaging signal in the basal ganglia: a case of adult-onset GM1 gangliosidosis. Mov Disord 2002;17:1095–1097.
83. Kobayashi O, Takashima S. Thalamic hyperdensity on CT in infantile GM1-gangliosidosis. Brain Dev 1994;16:472–474.
84. Mugikura S, Takahashi S, Higano S, Kurihara N, Kon K, Sakamoto K. MR findings in Tay-Sachs disease. J Comput Assist Tomogr 1996;20:551–555.
85. Nassogne MC, Commare MC, Lellouch-Tubiana A, Emond S, Zerah M, Caillaud C, et al. Unusual presentation of GM2 gangliosidosis mimicking a brain stem tumor in a 3-year-old girl. AJNR Am J Neuroradiol 2003;24:840–842.
86. Alkan A, Kutlu R, Yakinci C, Sigirci A, Aslan M, Sarac K. Infantile Sandhoff's disease: multivoxel magnetic resonance spectroscopy findings. J Child Neurol 2003;18:425–428.
87. Loes DJ, Hite S, Moser H, Stillman AE, Shapiro E, Lockman L, Latchaw RE, Krivit W. Adrenoleukodystrophy: a scoring method for brain MR observations. AJNR Am J Neuroradiol 1994;15:1761–1766.
88. Schaumburg HH, Powers JM, Raine CS, Suzuki K, Richardson EP Jr. Adrenoleukodystrophy: a clinical and pathological study of 17 cases. Arch Neurol 1975;32:577–591.
89. Melhem ER, Breiter SN, Ulug AM, Raymond GV, Moser HW. Improved tissue characterization in adrenoleukodystrophy using magnetization transfer imaging. AJR Am J Roentgenol 1996;166: 689–695.
90. Ito R, Melhem ER, Mori S, Eichler FS, Raymond GV, Moser HW. Diffusion tensor brain MR imaging in X-linked cerebral adrenoleukodystrophy. Neurology 2001;56:544–547.
91. Eichler FS, Barker PB, Cox C, Edwin D, Ulug AM, Moser HW, Raymond GV. Proton MR spectroscopic imaging predicts lesion progression on MRI in X-linked adrenoleukodystrophy. Neurology 2002;58:901–907.
92. Fatemi A, Barker PB, Ulug AM, Nagae-Poetscher LM, Beauchamp NJ, Moser AB, Raymond GV, Moser HW, Naidu S. Neurology 2003;60:1301–1307.
93. Moser HW. Adrenoleukodystrophy: phenotype, genetics, pathogenesis and therapy. Brain 1997;120:1485–1508.
94. Powers JM, Moser HW. Peroxisomal disorders: genotype, phenotype, major neuropathologic lesions, and pathogenesis. Brain Pathol 1998;8:101–120.

95. Senior K. Lorenzo's oil may help to prevent ALD symptoms. Lancet Neurol 2002;1:468.
96. Moser HW, Kok F, Neumann S, et al. Adrenoleukodystrophy update: genetics and effect of Lorenzo's oil therapy in asymptomatic patients. Int Pediatr 9 1994;196–204.
97. Barth PG, Gootjes J, Bode H, Vreken P, Majoie CB, Wanders RJ.Late onset white matter disease in peroxisome biogenesis disorder. Neurology. 2001;57:1949–1955.
98. Powers JM, Moser HW. Peroxisomal disorders: genotype, phenotype, major neuropathologic lesions, and pathogenesis. Brain Pathol 1998;8:101–120.
99. DiMauro S, Schon EA. Mitochondrial respiratory chain diseases. N Engl J Med 2003;348;2656–2668.
100. Hart PE, DeVivo DC, Schapira AHV. Clinical features of the mitochondrial encephalomyopathies. In: Schapira AHV, DiMauro S, eds. Mitochondrial Disorders in Neurology 2 . Boston: Butterworth Heinemann; 2002:35–68.
101. Schon EA, Bonilla E, DiMauro S. Mitochondrial DNA mutations and pathogenesis. J Bioenerg Biomembr 1997;29:131–149.
102. Wallace DC. Mitochondrial diseases in mouse and man. Science 1999;283:1482–1488.
103. Darin N, Oldfors A, Moslemi A-R Holme E, Tulinius M. The incidence of mitochondrial encephalomyopathies in childhood: clinical features and morphological, biochemical, and DNA abnormalities. Ann Neurol 2001;49:377–383.
104. Servidei S. Mitochondrial encephalomyopathies: gene mutation. Neuromuscular Disord 1998;8:18–19.
105. Andreu AL, Hanna MG, Reichmann H, Bruno C, Penn AS, Tanji K, et al. Exercise intolerance due to mutations in the cytochrome b gene of mitochondrial DNA. N Engl J Med 1999;341:1037–1044.
106. Vissing J, Ravn K, Danielsen ER, Duno M, Wibrand F, Wevers RA, Schwartz M. Multiple mtDNA deletions with features of MNGIE. Neurology 2002;59:926–929.
107. Rowland LP, Blake DM, Hirano M, DiMauro S, Schon EA, Hays AP, et al. Clinical syndromes associated with ragged red fibers. Rev Neurol 1991;147:467–473.
108. Barkovich AJ, Good WV, Koch TK, Berg BO. Mitochondrial disorders: analysis of their clinical and imaging characteristics. Am J Neuroradiol 1993;14:1119–1137.
109. Valanne L, Ketonen L, Majander A, Suomalainen A, Pihko H. Neuroradiologic findings in children with mitochondrial disorders. AJNR Am J Neuroradiol 1998;19:369–367.
110. Pavlakis SG, Phillips PC, DiMauro S, De Vivo DC, Rowland LP. Mitochondrial myopathy, encephalopathy, lactic acidosis, and stroke-like episodes: a distinctive clinical syndrome. Ann Neurol 1984;16:481–488.
111. Allard JC, Tilak S, Carter AP. CT and MR of MELAS syndrome. Am J Neuroradiol 1988;9:1234–1238.
112. Ohama E, Ohara S, Ikuta F, Tanaka K, Nishizawa M, Miyatake T. Mitochondrial angiopathy in cerebral blood vessels of mitochondrial encephalomyopathy. Acta Neuoropathol 1987;74:226–233.
113. Sue CM, Crimmins DS, Soo YS, Pamphlett R, Presgrave CM, Kotsimbos N, Jean-Francois MJ, Byrne E, Morris JG. Neuroradiological features of six kindreds with MELAS tRNALeu A3243G point mutation: implications for pathogenesis. J Neurol Neurosurg Psychiatry 1998;65, 233–240.
114. Matthews PM, Andermann F, Silver K, Karpati G, Arnold DL. Proton MR spectroscopic characterization of differences in regional brain metabolic abnormalities in mitochondrial cytopathies. Neurology 1993;43:2484–2490.
115. Dubeau F, De Stefano N, Zifkin BG, Arnold DL, Shoubridge EA. Oxidative phosphorylation defect in the brains of carriers of the tRNAleu (UUR) A3242G mutation in a MELAS pedigree. Ann Neurol 2000;47:179–185.
116. Ishiyama G, Lopez I, Baloh RW, Ishiyama A. Canavan's leukodystrophy is associated with defects in cochlear neurodevelopment and deafness. Neurology. 2003;60:1702–1704.
117. Feigelman, T, Shih VE, Buyse ML. Prolonged survival in Canavan disease. Dysmorph Clin Genet 1991;5:107–110.
118. Hagenfeldt L, Bollgren I, Venizelos N. N-acetylaspartic aciduria due to aspartoacylase deficiency—a new etiology of childhood leukodystrophy. J Inherit Metab Dis 1967;10:135–141.
119. Kvittingen EA; Guldal G, Borsting S, Skalpe IO, Stokke O, Jellum E. N-acetylaspartic aciduria in a child with a progressive cerebral atrophy. Clin Chim Acta 1986;158:217–227.
120. Divry P, Vianey-Liaud C, Gay C, Macabeo V, Rapin F, Echenne B. N-acetylaspartic aciduria: report of three new cases in children with a neurological syndrome associating macrocephaly and leucodystrophy. J Inherit Metab Dis 1988;11:307–308..
121. Matalon R, Michals K, Sebesta D, Deanching M, Gashkoff P, Casanova J. Aspartoacylase deficiency and N-acetylaspartic aciduria in patients with Canavan disease. Am J Med Genet 1988;29:463–471.
122. Kaul R, Gao GP, Balamurugan K, Matalon R. Cloning of the human aspartoacylase cDNA and a common missense mutation in Canavan disease. Nature Genet. 1993;5:118–123.
123. Valk J and Van der Knaap MS. Magnetic Resonance of Myelin, Myelination, and Myelin Disorders. New York: Springer Verlag; 1989.
124. Toft PB, Geiss-Holtorff R, Rolland MO, Pryds O, Muller-Forell W, Christensen E, et al. Magnetic resonance imaging in juvenile Canavan disease. Eur J Pediatr. 1993;152:750–753.
125. Wittsack HJ, Kugel H, Roth B, Heindel W. Quantitative measurements with localized 1H MR spectroscopy in children with Canavan's disease. J Magn Reson Imaging 1996;6:889–893.
126. Janson C, McPhee S, Bilaniuk L, Haselgrove J, Testaiuti M, Freese A, et al. Clinical protocol. Gene therapy of Canavan disease: AAV-2 vector for neurosurgical delivery of aspartoacylase gene (ASPA) to the human brain. Hum Gene Ther 2002;13:1391–1412.
127. Biery BJ, Stein DE, Morton DH, Goodman SI. Gene structure and mutations of glutaryl-coenzyme A dehydrogenase: impaired association of enzyme subunits that is due to an A421V substitution causes glutaric acidemia type I in the Amish. Am J Hum Genet 1996;59:1006–1011.
128. Brismar J, Ozand PT. CT and MR of the brain in glutaric acidemia type I: a review of 59 published cases and a report of 5 new patients. AJNR Am J Neuroradiol 1995;16:675–683.
129. Twomey EL, Naughten ER, Donoghue VB, Ryan S. Neuroimaging findings in glutaric aciduria type 1. Pediatr Radiol 2003;33:823–830.
130. Strauss KA, Morton DH. Type I glutaric aciduria, part 2: a model of acute striatal necrosis. Am J Med Genet 2003;121:53–50.
131. Moller HE, Weglage J, Bick U, Wiedermann D, Feldmann R, Ullrich K. Brain imaging and proton magnetic resonance spectroscopy in patients with phenylketonuria. Pediatrics 2003;112:1580–1583.
132. Bick U, Fahrendorf G, Ludolph AC, Vassallo P, Weglage J, Ullrich K. Disturbed myelination in patients with treated hyperphenylalaninemia: evaluation with magnetic resonance imaging. Eur J Pediatr 1991;150:185–195.
133. Bick U, Ullrich K, Stöber U, et al. White matter abnormalities in patients with treated hyperphenylalaninemia: magnetic resonance relaxometry and proton spectroscopy findings. Eur J Pediatr 1993;152:1012–1020.
134. Thomson AJ, Tillotson S, Smith I, Kendall B, Moore SG, Brenton DP. Brain MRI changes in phenylketonuria. Brain 1993;116:811–821.
135. Leuzzi V, Trasimeni G, Gualdi GF, Antonozzi I. Biochemical, clinical and neuroradiological (MRI) correlations in late-detected PKU patients. J Inherit Metab Dis 1995;18:624–634.
136. Möller H, Pietz J, Kreis R, Schmidt H, Meyding-Lamadé UK, Rupp A, Boesch C. Phenylketonuria: findings at MR imaging and localized in vivo H-1 MR spectroscopy of the brain in patients with early treatment. Radiology 1996;201:413–420.
137. Weglage J, Bick U, Wiedermann D, Feldmann R, and Ullrich K. Brain imaging and proton magnetic resonance spectroscopy in patients with phenylketonuria. Pediatrics 2003;112:1150–1583.
138. Auerbach VH, DiGeorge A. Maple syrup urine disease. In: Hommes FA, Van den Berg CJ, eds. Inborn Errors of Metabolism. London: Academic Press; 1973;337.

139. Naylor EW. Newborn screening for maple syrup urine disease (branched chain ketoaciduria). In: Bickel H, Guthrie R, Hammersen G, eds. Neonatal Screening for Inborn Errors of Metabolism. Berlin: Springer-Verlag; 1980:19.
140. Marshall L, DiGeorge A. Maple syrup urine disease in the Old Order Mennonites (abstract). Am J Hum Genet 1981;33:139A.
141. Jan W, Zimmerman RA, Wang ZJ, Berry GT, Kaplan PB, Kaye EM. MR diffusion imaging and MR spectroscopy of maple syrup urine disease during acute metabolic decompensation. Neuroradiology 2003;45:393–399.
142. Felber SR, Sperl W, Chemelli A, Murr C, Wendel U. Maple syrup urine disease: metabolic decompensation monitored by proton magnetic resonance imaging and spectroscopy. Ann Neurol 1993;33:396–401.
143. Sener RN. Diffusion magnetic resonance imaging in intermediate form of maple syrup urine disease. J Neuroimaging 2002;12:368–370.
144. Nagata N, Matsuda I, Oyanagi K. Estimated frequency of urea cycle enzymopathies in Japan. Am J Med Genet 1991;39:228–229.
145. Applegarth DA, Toone JR, Lowry RB. Incidence of inborn errors of metabolism in British Columbia, 1969–1996. Pediatrics 2000; 105:e10.
146. Brusilow SW, Horwich AL. Urea cycle enzymes. In: Scriver CR, Beaudet AL, Valle D, Sly WS, Childs B, Kinzler KW, Vogelstein B, eds. The Metabolic and Molecular Bases of Inherited Disease. 8th ed. New York: McGraw-Hill; 2001:1909–1964.
147. Dionisi-Vici C, Rizzo C, Burlina AB, Caruso U, Sabetta G, Uziel G, Abeni D. Inborn errors of metabolism in the Italian pediatric population: a national retrospective survey. J Pediatr 2002;140:321–327.
148. Tuchman M, Jaleel N, Morizono H, Sheehy L, Lynch MG. Mutations and polymorphisms in the human ornithine transcarbamylase gene. Hum Mutat 2002;19:93–107.
149. McCullough BA, Yudkoff M, Batshaw ML, Wilson JM, Wilson JM, Raper SE, et al. Genotype spectrum of ornithine transcarbamylase deficiency: correlation with the clinical and biochemical phenotype. Am J Med Genet 2000;93:313–319.
150. Campbell AGM, Rosenberg LE, Snodgrass PJ, Nuzum CT. Ornithine transcarbamylase deficiency: a cause of lethal neonatal hyperammonemia in males. N Engl J Med 1973;288:1–6.
151. Kang ES, Snodgrass PJ, Gerald PS. Ornithine transcarbamylase deficiency in the newborn infant. J Pediatr 1973;82:642–649.
152. Batshaw ML, Brusilow S, Waber L, Blom W, Brubakk AM, Burton BK, et al. Treatment of inborn errors of urea synthesis: activation of alternative pathways of waste nitrogen synthesis and excretion. N Engl J Med 1982;306:1387–1392.
152. Rowe PC, Newman SL, Brusilow SW. Natural history of symptomatic partial ornithine transcarbamylase deficiency. N Engl J Med 1986;314:541–547.
154. Di Magno EP, Lowe JE, Snodgrass PJ, Jones JD. Ornithine transcarbamylase deficiency-a cause of bizarre behavior in a man. N Engl J Med 1986;315:744–747.
155. Batshaw ML, MacArthur RB, Tuchman M. Alternative pathway therapy for urea cycle disorders: twenty years later. J Pediatr 2001;138: S46–S54; discussion S54–S55.
156. Batshaw ML, Hyman SL, Mellits ED, Thomas GH, DeMuro R, Coyle JT. Behavioral and Neurotransmitter Changes in the Urease-Infused Rat: A model of congenital hyperammonemia. Pediatr Res 1986;20:1310–1315.
157. Maestri NE, Clissold DMA, Brusilow S. Neonatal onset ornithine transcarbamylase deficiency. J Pediatr 1999;134:268–272.
158. Maestri NE, Lord C, Glynn M, Bale A, Brusilow SW. The phenotype of ostensibly healthy women who are carriers for ornithine transcarbamylase deficiency. Medicine 1998;77:389–397.
159. Yudkoff M, Daikhin Y, Nissim I, Jawad A, Wilson J, Batshaw M. In vivo nitrogen metabolism in ornithine transcarbamylase deficiency. J Clini Invest 1996;89:2167–2173.
160. Tuchman M, Plante RJ, Garcia-Perez MA, Rubio V. Relative frequency of mutations causing ornithine transcarbamylase deficiency in 78 families. Hum Genet 1996;97:274–276.
161. Kendall BE, Kingsley DPE, Leonard JV, Lingam S, Oberholzer VG. Neurological features and computed tomography of the brain in children with ornithine carbamoyl transferase deficiency. J Neurol, Neurosurg Psychiatry 1983;46:28–34.
162. Msall M, Batshaw ML, Suss R, Brusilow SW, Mellits ED. Neurologic outcome in children with inborn errors of urea synthesis. Outcome of urea-cycle enzymopathies. N Engl J Med 1984;310:1500–1505.
163. Takayanagi M, Ohtake A, Ogura N, Nakajima H, Hoshino M. A female case of ornithine transcarbamylase deficiency with marked computed tomographic abnormalities of the brain. Brain Dev 1984;6:58–60.
164. Christodoulou J, Qureshi A, McInnes RR, Clarke JTR. Ornithine transcarbamylase deficiency presenting with strokelike episodes. J Pediatr 1993;122:423–425.
165. Kurihara K. Takanashi J, Tomita M, Kobayashi K, et al. (2003) Magnetic resonance imagine in late onset ornithine transcarbamylase deficiency. Brain Dev 25:40–44.
166. Fitzgerald SM, Hetherington HP, Behar KL, Shulman RG. Effects of acute hyperammonemia on cerebral amino acid metabolism and pH, in vivo, measured by 1H and 31P nuclear magnetic resonance. J Neurochem 1989;52:741–749.
167. Bates TE, Williams SR, Kauppinen RA, Gadian DG. Observations of cerebral metabolites in an animal model of acute liver failure. J Neurochem 1989;53:102–110.
168. Batshaw ML, Roan Y, Jung AL, Rosenberg LA, Brusilow SW. Cerebral dysfunction in asymptomatic carriers of ornithine transcarbamylase deficiency. N Engl J Med 1980;302:482–485.
169. Connelly A, Cross JH, Gadian DG, Hunter JV, Kirkham FJ, Leonard JV. Magnetic Resonance Spectroscopy Shows increased Brain glutamine in ornithine carbamoyl transferase deficiency. Pediatr Res 1993;33:77–81.
170. Krieger I, Snodgrass PJ, Roskamp J. Atypical clinical case of ornithine transcarbamylase deficiency due to a new (comparison with Reye's disease). J Clin Endrocrinol Med 1979;388–392.
171. Kornfeld M, Woodfin BM, Papile L, Davis LE, Bernard LR. Neuropathy of Ornithine Carbamoyl Transferase Deficiency. Acta Neuropathol (Berl) 1985;65:261–264.
172. Dolman CL, Clasen RA, Dorovini-Zis K. Severe Cerebral damage in ornithine transcarbamylase deficiency. Clin Neuropathol 1988;7:10–15.
173. Yamanouchi H, Yokoo H, Yuharea Y, Marayama K, Sasaki A, Hirato J, Nakazato Y. An Autopsy case of ornithine transcarbamylase deficiency. Brain Dev 2002;24:91–94.
174. Sugasawa K, Ng JM, Masutani C, Iwai S, van der Spek PJ, Eker AP, et al. Xeroderma pigmentosum group C protein complex is the initiator of global genome nucleotide excision repair. Mol Cell 1998; 2:223–232.
175. Bootsma D, et al. Nucleotide excision repair syndromes: xeroderma pigmentosum, Cockayne syndrome, and trichothiodystrophy. In: Vogelstein B, Kinzler KW, eds. The Genetic Basis of Human Cancer. 2nd ed. New York: McGraw-Hill; 2002.
176. Kraemer KH, Slor H Xeroderma pigmentosum. Clin Dermatol 1985;3:33–69.
177. Kraemer KH. Lee MM, Andrews AD, Lambert WC. The role of sunlight and DNA repair in melanoma and nonmelanoma skin cancer. The xeroderma pigmentosum paradigm. Arch Dermatol 1994; 130:1018–1021.
178. Rapin I, Lindenbaum Y, Dickson DW, Kraemer KH, Robbins JH. Cockayne syndrome and xeroderma pigmentosum. Neurology 2000;55:1442–1449.
179. Roytta M, Anttinen A. Xeroderma pigmentosum with neurological abnormalities. A clinical and neuropathological study. Acta Neurol Scand 1986;73:191–199.
180. Kanda T, Oda M, Yonezawa M, Tamagawa K, Isa F, Hanakago R, Tsukagoshi H. Peripheral neuropathy in xeroderma pigmentosum. Brain 1990;113:1025–1044.
181. Mimaki T, Itoh N, Abe J, Tagawa T, Sato K, Yabuuchi H, Takebe H. Neurological manifestations in xeroderma pigmentosum. Ann Neurol 1986;20:70–75.
182. Mimaki T, Tagawa T, Tanaka J, Sato K, Yabuuchi H. EEG and CT abnormalities in xeroderma pigmentosum. Acta Neurol Scand 1989;80:136–141.

183. Sugita K, Takanashi J, Ishii M, Niimi H. Comparison of MRI white matter changes with neuropsychologic impairment in Cockayne syndrome. Pediatr Neurol 1992;8:295–298.
184. Battistella PA, Peserico A. Central nervous system dysmyelination in PIBI(D)S syndrome: a further case. Childs Nerv Syst 1996;12:110–113.
185. Hakamada S, Watanabe K, Sobue G, Hara K, Miyazaki S. Xeroderma pigmentosum: neurological, neurophysiological and morphological studies. Eur Neurol 1982;21:69–76.
186. Origuchi Y, Eda I, Matsumoto S, Furuse A. Quantitative histologic study of sural nerves in xeroderma pigmentosum. Pediatr Neurol 1987;3:356–359.
187. Lindenbaum Y, Dickson D, Rosenbaum P, Kraemer K, Robbins I, Rapin I. Xeroderma pigmentosum/cockayne syndrome complex: first neuropathological study and review of eight other cases. Eur J Paediatr Neurol 2001;5:225–242.
188. Stojkovic T, Defebvre L, Quilliet X, Eveno E, Sarasin A, Mezzina M, Destee A. Neurological manifestations in two related xeroderma pigmentosum group D patients: complications of the late-onset type of the juvenile form. Mov Disord 1997;12:616–619.
189. Bassuk AG, Joshi A, Burton BK, Larsen MB, Burrowes DM, Stack C. Alexander disease with serial MRS and a new mutation in the glial fibrillary acidic protein gene. Neurology 2003 61:1014–1015.
190. Springer S, Erlewein R, Naegele T, Becker I, Auer D, Grodd W, Krageloh-Mann I. Alexander disease—classification revisited and isolation of a neonatal form. Neuropediatrics 2000;31:86–92.
191. Namekawa M, Takiyama Y, Aoki Y, Takayashiki N, Sakoe K, Shimazaki H, et al. Identification of GFAP gene mutation in hereditary adult-onset Alexander's disease. Ann Neurol 2002;52:779–785.
192. Martidis A, Yee RD, Azzarelli B, Biller J.Neuro-ophthalmic, radiographic, and pathologic manifestations of adult-onset Alexander disease. Arch Ophthalmol 1999;117:265–267.
193. Stumpf E, Masson H, Duquette A, Berthelet F, McNabb J, Lortie A, et al. Adult Alexander disease with autosomal dominant transmission: a distinct entity caused by mutation in the glial fibrillary acid protein gene. Arch Neurol 2003;60:1307–1312.
194. Brockmann K, Meins M, Taubert A, Trappe R, Grond M, Hanefeld F. A novel GFAP mutation and disseminated white matter lesions: adult Alexander disease? Eur Neurol 2003;50:100–105.
195. Kinoshita T, Imaizumi T, Miura Y, Fujimoto H, Ayabe M, Shoji H, et al. A case of adult-onset Alexander disease with Arg416Trp human glial fibrillary acidic protein gene mutation. Neurosci Lett 2003;350:169–172.
196. Gordon N. Alexander disease. Eur J Paediatr Neurol 2003;7:395–399.
197. Johnson AB, Brenner M. Alexander's disease: clinical, pathologic, and genetic features. J Child Neurol. 2003;18:625–632.
198. Brenner M, Johnson AB, Boespflug-Tanguy O, Rodriguez D, Goldman JE, Messing A. Mutations in GFAP, encoding glial fibrillary acidic protein, are associated with Alexander disease. Nat Genet 2001;27:117–120.
199. Namekawa M, Takiyama Y, Aoki Y, Takayashiki N, Sakoe K, Shimazaki H, et al. Identification of GFAP gene mutation in hereditary adult-onset Alexander's disease. Ann Neurol 2002;52:779–785.
200. Meins M, Brockmann K, Yadav S, Haupt M, Sperner J, Stephani U, Hanefeld F.Infantile Alexander disease: a GFAP mutation in monozygotic twins and novel mutations in two other patients. Neuropediatrics 2002;33:194–198.
201. van der Knaap MS, Naidu S, Breiter SN, Blaser S, Stroink H, Springer S, et al. Alexander disease: diagnosis with MR imaging. AJNR Am J Neuroradiol 2001;22:541–552.
202. Stumpf E, Masson H, Duquette A, Berthelet F, McNabb J, Lortie A, et al. Adult Alexander disease with autosomal dominant transmission: a distinct entity caused by mutation in the glial fibrillary acid protein gene. Arch Neurol 2003;60:1307–1312.
203. Gorospe JR, Naidu S, Johnson AB, Puri V, Raymond GV, Jenkins SD, et al. Molecular findings in symptomatic and pre-symptomatic Alexander disease patients. Neurology 2002;58:1494–1500.
204. Guthrie SO, Burton EM, Knowles P, Marshall R. Alexander's disease in a neurologically normal child: a case report. Pediatr Radiol 2003;33:47–49.
205. Imamura A, Orii KE, Mizuno S, Hoshi H, Kondo T. MR imaging and 1H-MR spectroscopy in a case of juvenile Alexander disease. Brain Dev 2002;24:723–726.
206. Sawaishi Y, Hatazawa J, Ochi N, Hirono H, Yano T, Watanabe Y, Okudera T, Takada G. Positron emission tomography in juvenile Alexander disease. J Neurol Sci 1999;165:116–120.
207. Li R, Messing A, Goldman JE, Brenner M. GFAP mutations in Alexander disease. Int J Dev Neurosci 2002;20:259–268.
208. Messing A, Goldman JE, Johnson AB, Brenner M. Alexander disease: new insights from genetics. J Neuropathol Exp Neurol 2001;60:563–573.
209. Mignot C, Boespflug-Tanguy O, Gelot A, Dautigny A, Pham-Dinh D, Rodriguez D.Alexander disease: putative mechanisms of an astrocytic encephalopathy. Cell Mol Life Sci 2004;61:369–385.
210. Hanefeld F, Holzbach U, Kruse B, Wilichowski E, Christen HJ, Frahm J. Diffuse white matter disease in three children: an encephalopathy with unique features on magnetic resonance imaging and proton magnetic resonance spectroscopy. Neuropediatrics 1993;24: 244–248.
211. van der Knaap MS, Barth PG, Gabreels FJ, Franzoni E, Begeer JH, Stroink H, Rotteveel JJ, Valk J. A new leukoencephalopathy with vanishing white matter. Neurology 1997;48:845–855.
212. Schiffmann R, Moller JR, Trapp BD, Shih HHL, Farrer RG, Katz DA, et al. Childhood ataxia with diffuse central nervous system hypomyelination. Ann. Neurol 1994;35:331–340..
213. van der Knaap MS, Kamphorst W, Barth PG, Kraaijeveld CL, Gut E, Valk J. Phenotypic variation in leukoencephalopathy with vanishing white matter. Neurology 1998;51:540–547.
214. Sugiura C, Miyata H, Oka A, Takashima S, Ohama E, Takeshita K. A Japanese girl with leukoencephalopathy with vanishing white matter. Brain Dev 2001;23:58–61.
215. Topcu M, Saatci I, Apak RA, Soylemezoglu F. A case of leukoencephalopathy with vanishing white matter. Neuropediatrics 2000;31:100–103.
216. Rosemberg S, Leite Cda C, Arita FN, Kliemann SE, Lacerda MT. Leukoencephalopathy with vanishing white matter: report of four cases from three unrelated Brazilian families. Brain Dev 2002 ;24:250–256.
217. Francalanci P, Eymard-Pierre E, Dionisi-Vici C, Boldrini R, Piemonte F, Virgili R, et al. Fatal infantile leukodystrophy: a severe variant of CACH/VWM syndrome, allelic to chromosome 3q27. Neurology 2001;57:265–270.
218. Fogli A, Dionisi-Vici C, Deodato F, Bartuli A, Boespflug-Tanguy O, Bertini E. A severe variant of childhood ataxia with central hypomyelination/vanishing white matter leukoencephalopathy related to EIF21B5 mutation. Neurology 2002;59:1966–1968.
219. Boltshauser E, Barth PG, Troost D, Martin E, Stallmach T. Vanishing white matter" and ovarian dysgenesis in an infant with cerebro-oculo-facio-skeletal phenotype. Neuropediatrics 2002;33:57–62.
220. van der Knaap MS, Van Berkel CG, Herms J, Van Coster R, Baethmann M, Naidu S, et al. eIF2B-related disorders: antenatal onset and involvement of multiple organs. Am J Hum Genet 2003;73:1199–1207.
221. van der Knaap MS, Kamphorst W, Barth PG, Kraaijeveld CL, Gut E, Valk J. Phenotypic variation in leukoencephalopathy with vanishing white matter. Neurology 1998;51:540–547.
222. Prass K, Bruck W, Schroder NW, Bender A, Prass M, Wolf T, et al. Adult-onset Leukoencephalopathy with vanishing white matter presenting with dementia. Ann Neurol 2001;50:665–668.
223. Fogli A, Rodriguez D, Eymard-Pierre E, Bouhour F, Labauge P, Meaney BF, et al. Ovarian failure related to eukaryotic initiation factor 2B mutations. Am J Hum Genet 2003;72:1544–1550.
224. Verghese J, Weidenheim K, Malik S, Rapin I. Adult onset pigmentary orthochromatic leukodystrophy with ovarian dysgenesis. Eur J Neurol 2002;9:663–670.
225. van der Knaap MS, Wevers RA, Kure S, Gabreels FJ, Verhoeven NM, van Raaij-Selten B, et al. Increased cerebrospinal fluid glycine: a biochemical marker for a leukoencephalopathy with vanishing white matter. Ann Neurol 1999; 14:728–731.

226. Leegwater PA, Konst AA, Kuyt B, Sandkuijl LA, Naidu S, Oudejans CB, et al. The gene for leukoencephalopathy with vanishing white matter is located on chromosome 3q27. J Hum Genet 1999;65:728–734.
227. Francalanci P, Eymard-Pierre E, Dionisi-Vici C, Boldrini R, Piemonte F, Virgili R, et al. Fatal infantile leukodystrophy: a severe variant of CACH/VWM syndrome, allelic to chromosome 3q27. Neurology 2001;57:265–270.
228. van der Knaap MS, Leegwater PA, Konst AA, Visser A, Naidu S, Oudejans CB, et al. Mutations in each of the five subunits of translation initiation factor eIF2B can cause leukoencephalopathy with vanishing white matter. Ann Neurol 2002;51:264–270.
229. Topcu M, Saatci I, Apak RA, Soylemezoglu F. A case of leukoencephalopathy with vanishing white matter. Neuropediatrics 2000;31:100–103.
230. Rosemberg S, Leite Cda C, Arita FN, Kliemann SE, Lacerda MT. Leukoencephalopathy with vanishing white matter: report of four cases from three unrelated Brazilian families. Brain Dev 2002;24:250–256.
231. Fogli A, Rodriguez D, Eymard-Pierre E, Bouhour F, Labauge P, Meaney BF, et al. Ovarian failure related to eukaryotic initiation factor 2B mutations. Am J Hum Genet 2003;72:1544–1550.
232. Tedeschi, G, Schiffmann R, Barton NW, Shih HHL, Gospe SM Jr, Brady RO, et al Proton magnetic resonance spectroscopic imaging in childhood ataxia with diffuse central nervous system hypomyelination. Neurology 1995;45:1526–1532.
233. Schiffmann R, Moller JR, Trapp BD, Shih HHL, Farrer RG, Katz DA , et al. Childhood ataxia with diffuse central nervous system hypomyelination. Ann Neurol 1994;35:331–340.
234. Bluml S, Philippart M, Schiffmann R, Seymour K, Ross BD. Membrane phospholipids and high-energy metabolites in childhood ataxia with CNS hypomyelination. Neurology 2003;61:648–654.
235. Gallo A, Rocca MA, Falini A, Scaglione C, Salvi F, Gambini A, et al. Multiparametric MRI in a patient with adult-onset leukoencephalopathy with vanishing white matter. Neurology 2004;62:323–326.
236. Sugiura C, Miyata H, Oka A, Takashima S, Ohama E, Takeshita K. A Japanese girl with leukoencephalopathy with vanishing white matter. Brain Dev 2001;23:58–61.
237. Francalanci P, Eymard-Pierre E, Dionisi-Vici C, Boldrini R, Piemonte F, Virgili R, et al. Fatal infantile leukodystrophy: a severe variant of CACH/VWM syndrome, allelic to chromosome 3q27. Neurology 2001;57:265–270.
238. Bruck W, Herms J, Brockmann K, Schulz-Schaeffer W, Hanefeld F. Myelinopathia centralis diffusa (vanishing white matter disease): evidence of apoptotic oligodendrocyte degeneration in early lesion development. Ann Neurol 2001;50:532–536.
239. Kaye EM, Doll RF, Natowicz MR, Smith FI. Pelizaeus-Merzbacher disease presenting as spinal muscular atrophy: clinical and molecular studies. Ann Neurol 1994;36:916–919.
240. Ziereisen F, Dan B, Christiaens F, Deltenre P, Boutemy R, Christophe C. Connatal Pelizaeus-Merzbacher disease in two girls. Pediatr Radiol 2000;30:435–438.
241. Mimault C, Giraud G, Courtois V, Cailloux F, Boire JY, Dastugue B, et al. Proteolipoprotein gene analysis in 82 patients with sporadic Pelizaeus-Merzbacher Disease: duplications, the major cause of the disease, originate more frequently in male germ cells, but point mutations do not. The Clinical European Network on Brain Dysmyelinating Disease. Am J Hum Genet. 1999;65:360–369.
242. Cailloux F, Gauthier-Barichard F, Mimault C, Isabelle V, Courtois V, Giraud G, et al. Genotype-phenotype correlation in inherited brain myelination defects due to proteolipid protein gene mutations. Clinical European Network on Brain Dysmyelinating Disease. Eur J Hum Genet. 2000;8:837–845.
243. Boespflug-Tanguy O, Mimault C, Melki J, Cavagna A, Giraud G, Pham Dinh D, et al. Genetic homogeneity of Pelizaeus-Merzbacher disease: tight linkage to the proteolipoprotein locus in 16 affected families. PMD Clinical Group. Am J Hum Genet. 1994;55:461–467.
244. Hobson GM, Davis AP, Stowell NC, Kolodny EH, Sistermans EA, de Coo IF, et al. Mutations in noncoding regions of the proteolipid protein gene in Pelizaeus-Merzbacher disease. Neurolog. 2000;55: 1089–1096.
245. Statz A, Boltshauser E, Schinzel A, Spiess H. Computed tomography in Pelizaeus–Merzbacher disease. Neuroradiology. 1981;22: 103–105
246. Plecko B, Stockler-Ipsiroglu S, Gruber S, Mlynarik V, Moser E, Simbrunner J, et al. Degree of hypomyelination and magnetic resonance spectroscopy findings in patients with Pelizaeus Merzbacher phenotype. Neuropediatrics 2003;34:127–136.
247. Wang PJ, Young C, Liu HM, Chang YC, Shen YZ. Neurophysiologic studies and MRI in Pelizaeus-Merzbacher disease: comparison of classic and connatal forms. Pediatr Neurol 1995;12:47–53.
248. Ono J, Harada K, Sakurai K, Kodaka R, Shimidzu N, Tanaka J, et al. MR diffusion imaging in Pelizaeus-Merzbacher disease. Brain Dev. 1994;16:219–223.
249. Pizzini F, Fatemi AS, Barker PB, Nagae-Poetscher LM, Horska A, Zimmerman AW, et al. Proton MR spectroscopic imaging in Pelizaeus-Merzbacher disease. AJNR Am J Neuroradiol. 2003;24: 1683–1689.
250. Takanashi J. Inoue K, Tomita M, Kurihara A, Morita F, Ikehira H, et al. Brain N-acetylaspartate is elevated in Pelizaeus-Merzbacher disease with PLP1 duplication. Neurology 2002;58:237–241.
251. Spalice A, Popolizio T, Parisi P, Scarabino T, Iannetti P. Proton MR spectroscopy in connatal Pelizaeus-Merzbacher disease. Pediatr Radiol 2000;30:171–175.
251a. Sasaki A, Miyanaga M, Ototsuji M, et al. Two autopsy cases with Pelizaeus-Merzbacher disease phenotype of adult onset, without mutation of proteolipid protein gene. Acta Neuropathol (Berl) 2000;99:7–13.
252. Rizzo WB, Dammann AL, Craft DA. Sjögren-Larsson syndrome: impaired fatty alcohol oxidation in cultured fibroblasts due to deficient fatty alcohol:nicotinamide adenine dinucleotide oxidoreductase activity J Clin Invest 1988;81:738–744.
253. Rizzo WB, Dammann AL, Craft DA, Black SH, Henderson Tilton A, Africk D, et al. Sjögren-Larsson syndrome: inherited defect in the fatty alcohol cycle. J. Pediat. 1989;115:228–234.
254. Tabsh K, Rizzo WB, Holbrook K, Theroux N. Sjögren-Larsson syndrome: technique and timing of prenatal diagnosis. Obstet Gynecol 1993;82:700–703.
255. Rizzo WB, Craft DA, Kelson TL, Bonnefont JP, Saudubray JM, Schulman JD, et al. Prenatal diagnosis of Sjögren-Larsson syndrome using enzymatic methods. Prenat Diagn 1994;14:577–581.
256. De Laurenzi V, Rogers GR, Hamrock DJ, Marekov LN, Steinert PM, Compton JG, et al. Sjögren-Larsson syndrome is caused by mutations in the fatty aldehyde dehydrogenase gene. Nat Genet 1996; 12:52–57.
257. van Domburg PH, Willemsen MA, Rotteveel JJ, de Jong JG, Thijssen HO, Heerschap A, et al. Sjögren-Larsson syndrome: clinical and MRI/MRS findings in FALDH-deficient patients. Neurology 1999;52:1345–1352.
258. Van Mieghem F, Van Goethem JW, Parizel PM, van den Hauwe L, Cras P, De Meirleire J, et al. MR of the brain in Sjögren-Larsson syndrome. AJNR Am J Neuroradiol 1997;18:1561–1563.
259. Willemsen MA, Lutt MA, Steijlen PM, Cruysberg JR, van der Graaf M, Nijhuis-van der Sanden MW, et al. Clinical and biochemical effects of zileuton in patients with the Sjögren-Larsson syndrome. Eur J Pediatr 2001a;160:711–717.
260. Mano T, Ono J, Kaminaga T, Imai K, Sakurai K, Harada K, et al. Proton MR spectroscopy of Sjögren-Larsson's syndrome. AJNR Am J Neuroradiol 1999;20:1671–1673.
261. Miyanomae Y, Ochi M, Yoshioka H, Takaya K, Kizaki Z, Inoue F, et al. Cerebral MRI and spectroscopy in Sjögren-Larsson syndrome: case report. Neuroradiology 1995;37:225–228.
262. Kaminaga T, Mano T, Ono J, Kusuoka H, Nakamura H, Nishimura T. Proton magnetic resonance spectroscopy of Sjögren-Larsson syndrome heterozygotes. Magn Reson Med 2001;45:1112–1115.
263. Willemsen MA, IJlst L, Steijlen PM, Rotteveel JJ, de Jong JG, van Domburg PH, et al. Clinical, biochemical and molecular genetic characteristics of 19 patients with the Sjögren-Larsson syndrome. Brain 2001b;124:1426–1437.
264. Yamaguchi K, Handa T. Sjögren-Larsson syndrome: postmortem brain abnormalities. Pediatr Neurol 1998;18:338–341.

265. Philpot J, Sewry C, Pennock J, Dubowitz V. Clinical phenotype in congenital muscular dystrophy: correlation with expression of merosin in skeletal muscle. Neuromusc Disord 1995;5:301–305.
266. Hayashi YK, Koga R, Tsukahara T, Ishii H, Matsuishi T, Yamashita Y, Nonaka I, Arahata K. Deficiency of laminin alpha 2-chain mRNA in muscle in a patient with merosin negative congenital muscular dystrophy. Muscle Nerve 1995;18:1027–1030.
267. North KN, Specht K, Sethi RK, Shapiro F, Beggs AH. Congenital muscular dystrophy associated with merosin deficiency. J Child Neurol 1996;11:291–295.
268. Naom I, D'Alessandro M, Topaloglu H, Sewry C, Ferlini A, Helbling Leclerc A, et al. Refinement of the laminin alpha 2 locus to human chromosome 6q2 in severe and mild merosin deficient congenital muscular dystrophy. J Med Genet 1997;34:99–104.
269. Farina L, Morandi L, Milanesi I, Ciceri E, Mora M, Moroni I, et al. Congenital muscular dystrophy with merosin deficiency: MRI findings in five patients. Neuroradiology 1998;40:807–811.
270. Morandi L, Di Blasi C, Farina L, Sorokin L, Uziel G, Azan G, et al. Clinical correlations in 16 patients with total or partial laminin alpha2 deficiency characterized using antibodies against 2 fragments of the protein. Arch Neurol 1999;56:209–215.
271. Tan E, Topaloglu H, Sewry C, Zorlu Y, Naom I, Erdem S, et al. Late onset muscular dystrophy with cerebral white matter changes due to partial merosin deficiency. Neuromusc Disord 1997;7:85–89.
272. Naom I, D'Alessandro M, Sewry CA, Philpot J, Manzur AY, Dubowitz V, Muntoni F. Laminin alpha 2-chain gene mutations in two siblings presenting with limb-girdle muscular dystrophy. Neuromusc Disord 1998;8:495–501.
273. Dubowitz V. 50th ENMC International Workshop: Congenital Muscular Dystrophy, 28 February to 2 March 1997, Naarden, The Netherlands. Neuromusc Disord 1997;7:539–547.
274. Di Blasi C, Mora M, Pareyson D, Farina L, Sghirlanzoni A, Vignier N, et al. Partial laminin alpha 2 deficiency in a patient with a myopathy resembling inclusion body myositis. Ann Neurol 2000;47:811–816.
275. Cohn RD, Herrmann R, Sorokin L, Wewer UM, Voit T. Laminin alpha2 chain-deficient congenital muscular dystrophy: variable epitope expression in severe and mild cases. Neurology 1998;51:94–100.
276. Mercuri E, Muntoni F, Berardinelli A, Pennock J, Sewry C, Philpot J, Dubowitz V. Somatosensory and visual evoked potentials in congenital muscular dystrophy: correlation with MRI changes and muscle merosin status. Neuropediatrics 1995;26:3–7.
277. Mercuri E, Dubowitz L, Berardinelli A, Pennock J, Jongmans M, Henderson S, et al. Minor neurological and perceptuo-motor deficits in children with congenital muscular dystrophy: correlation with brain MRI changes. Neuropediatrics 1995;26:156–162.
278. Shorer Z, Philpot J, Muntoni F, Sewry C, Dubowitz V. Demyelinating peripheral neuropathy in merosin-deficient congenital muscular dystrophy. J Child Neurol 1995;10:472–475.
279. Mecuri E, Pennock J, Goodwin F, Sewry C, Cowan F, Dubowitz L, Dubowitz V, Muntoni F. Sequential study of central and peripheral nervous system involvement in an infant with merosin deficient congenital muscular dystrophy. Neuromusc Disord 1996;6:425–429.
280. Cil E, Topaloglu H, Caglar M, Ozme S. Left ventricular structure and function by echocardiography in congenital muscular dystrophy. Brain Dev 1994;16:301–303.
281. Spyrou N, Philpot J, Foale R, Camici PG, Muntoni F. Evidence of left ventricular dysfunction in children with merosin-deficient congenital muscular dystrophy. Am Heart J 1998;136:474–476.
282. Helbling-Leclerc A, Zhang X, Topaloglu H, Cruaud C, Tesson F, Weissenbach J, et al. Mutations in the laminin alpha 2-chain gene (LAMA2) cause merosin-deficient congenital muscular dystrophy. Nat Genet 1995;11:216–218.
283. Helbling-Leclerc A, Zhang X, Topaloglu H, Cruaud C, Tesson F, Weissenbach J, et al. Mutations in the laminin alpha 2-chain gene (LAMA2) cause merosin-deficient congenital muscular dystrophy. Nat Genet 1995;11:216–218.
284. Helbling-Leclerc A, Zhang X, Topaloglu H, Cruaud C, Tesson F, Weissenbach J, et al. Mutations in the laminin alpha 2-chain gene (LAMA2) cause merosin-deficient congenital muscular dystrophy. Nat Genet 1995;11:216–218.
285. Brett FM, Costigan D, Farrell MA, Heaphy P, Thornton J, King MD. Merosin-deficient congenital muscular dystrophy and cortical dysplasia. Eur J Paediatr Neurol 1998;2:77–82.
286. Mackay MT, Kornberg AJ, Shield L, Phelan E, Kean MJ, Coleman LT, Dennett X. Congenital muscular dystrophy, white-matter abnormalities, and neuronal migration disorders: the expanding concept. J Child Neurol 1998;13:481–487.
287. Pini L, Merlini FMS, Tome M, Chevallay, Gobbi G. Merosin-negative congenital muscular dystrophy, occipital epilepsy with periodic spasms and focal cortical dysplasia. Report of three Italian cases in two families. Brain Dev 1996;18:316–322.
288. Sunada Y, Edgar TS, Lotz BP, Rust RS, Campbell KP. Merosin-negative congenital muscular dystrophy associated with extensive brain abnormalities. Neurology 1995;45:2084–2089.
289. Van der Knaap MS, Smit LME, Barth PG Catsman-Berrevoets CE, Brouwer OF, Begeer JH, de Coo IF, Valk J. Magnetic resonance imaging in classification of congenital muscular dystrophies with brain abnormalities. Ann Neurol 1997;42:50–59.
290. Porter FD. Human malformation syndromes due to inborn errors of cholesterol synthesis. Curr Opin Pediatr 2003;15:607–613.
291. Kelley RI. Diagnosis of Smith-Lemli-Opitz syndrome by gas chromatography/mass spectrometry of 7-dehydrocholesterol in plasma, amniotic fluid and cultured skin fibroblasts. Clin Chim Acta 1995;236:45–58.
292. Trasimeni G, Di Biasi C, Iannilli M, Orlandi L, Boscherini B, Balducci R, Gualdi GF. MRI in Smith-Lemli-Opitz syndrome type I. Childs Nerv Syst 1997;13:47–49.
293. Caruso PA, Poussaint TY, Tzika AA, Zurakowski D, Astrakas LG, Elias ER, et al. MRI and (1)H MRS findings in Smith-Lemli-Opitz syndrome. Neuroradiology 2004;46:3–14.
294. van der Knaap MS, Barth PG, Stroink H, van Nieuwenhuizen O, Arts WF, Hoogenraad F, et al. Leukoencephalopathy with swelling and a discrepantly mild clinical course in eight children. Ann Neurol 1995;37:324–334.
295. Bugiani M, Moroni I, Bizzi A, Nardocci N, Bettecken T, Gartner J, et al. Consciousness disturbances in megalencephalic leukoencephalopathy with subcortical cysts. Neuropediatrics 2003;34:211–214.
296. Higuchi Y, Hattori H, Tsuji M, Asato R, Nakahata T. Partial seizures in leukoencephalopathy with swelling and a discrepantly mild clinical course. Brain Dev. 2000;22:387–389.
297. Topcu M, Gartioux C, Ribierre F, Yalcinkaya C, Tokus E, Oztekin N, et al. Vacuoliting megalencephalic leukoencephalopathy with subcortical cysts, mapped to chromosome 22qtel. Am J Hum Genet 2000;66:733–739.
298. Leegwater PA, Yuan BQ, van der Steen J, Mulders J, Konst AA, Boor PK, et al. Mutations of MLC1 (KIAA0027), encoding a putative membrane protein, cause megalencephalic leukoencephalopathy with subcortical cysts. Am J Hum Genet 2001;68:831–838.
299. Patrono C, Di Giacinto G, Eymard-Pierre E, Santorelli FM, Rodriguez D, De Stefano N, et al. Genetic heterogeneity of megalencephalic leukoencephalopathy and subcortical cysts. Neurology 2003;61:534–537.
300. Blattner R, Von Moers A, Leegwater PA, Hanefeld FA, Van Der Knaap MS, Kohler W. Clinical and genetic heterogeneity in megalencephalic leukoencephalopathy with subcortical cysts (MLC). Neuropediatrics 2003;34:215–218.
301. Singhal BS, Gorospe JR, Naidu S. Megalencephalic leukoencephalopathy with subcortical cysts. J Child Neurol 2003;18:646–652.
302. Gulati S, Kabra M, Gera S, Ghosh M, Menon PS, Kalra V.Infantile-onset leukoencephalopathy with discrepant mild clinical course. Indian J Pediatr 2000;67:769–773.
303. Koeda T, Takeshita K. Slowly progressive cystic leukoencephalopathy with megalencephaly in a Japanese boy. Brain Dev 1998;20:245–249.
304. Brockmann K, Finsterbusch J, Terwey B, Frahm J, Hanefeld F.Megalencephalic leukoencephalopathy with subcortical cysts in an adult: quantitative proton MR spectroscopy and diffusion tensor MRI. Neuroradiology 2003;45:137–142.

305. De Stefano N, Balestri P, Dotti MT, Grosso S, Mortilla M, Morgese G, Federico A. Severe metabolic abnormalities in the white matter of patients with vacuolating megalencephalic leukoencephalopathy with subcortical cysts. A proton MR spectroscopic imaging study. J Neurol 2001;248:403–409.
306. Brockmann K, Finsterbusch J, Terwey B, Frahm J, Hanefeld F. Megalencephalic leukoencephalopathy with subcortical cysts in an adult: quantitative proton MR spectroscopy and diffusion tensor MRI. Neuroradiology 2003;45:137–142.
307. van der Knaap MS, Barth PG, Vrensen GF, Valk J. Histopathology of an infantile-onset spongiform leukoencephalopathy with a discrepantly mild clinical course. Acta Neuropathol (Berl) 1996;92:206–212.
308. Babbitt DP, Tang T, Dobbs J, Berk R. Idiopathic familial cerebrovascular ferrocalcinosis (Fahr's disease) and review of differential diagnosis of intracranial calcification in children. Am J Roentgenol 1969;105:352–358.
309. Giroud M, Gouyon JB, Chaumet F, Cinquin AM, Chevalier-Nivelon A, Alison M, et al. A case of progressive familial encephalopathy in infancy with calcification of the basal ganglia and chronic cerebrospinal fluid lymphocytosis. Childs Nerv Syst 1986;2:47–48.
310. Mehta L, Trounce JQ, Moore JR, Young ID. Familial calcification of the basal ganglia with cerebrospinal fluid pleocytosis. J Med Genet 1986;23:157–160.
311. Tolmie JL, Shillito P, Hughes-Benzie R, Stephenson JBP. The Aicardi-Goutieres syndrome (familial, early onset encephalopathy with calcifications of the basal ganglia and chronic cerebrospinal fluid lymphocytosis). J Med Genet 1995;32:881–884.
312. Kumar D, Rittey C, Cameron AH, Variend S. Recognizable inherited syndrome of progressive central nervous system degeneration and generalized intracranial calcification with overlapping phenotype of the syndrome of Aicardi and Goutieres. Am J Med Genet 1998;75:508–515.
313. Crow YJ, Jackson AP, Roberts E, van Beusekom E, Barth P, Corry P, et al. icardi-Goutieres syndrome displays genetic heterogeneity with one locus (AGS1) on chromosome 3p21. Am J Hum Genet 2000;67:213–221.
314. Bonnemann CG, Meinecke P. Encephalopathy of infancy with intracerebral calcification and chronic spinal fluid lymphocytosis—another case of the Aicardi-Goutieres syndrome. Neuropediatrics 1992;23:157–161.
315. Lanzi G, Fazzi E, D'Arrigo S. Aicardi-Goutieres syndrome: a description of 21 new cases and a comparison with the literature. Eur J Paediatr Neurol 2002;6:A9–22.
316. McEntagart M, Kamel H, Lebon P, King MD. Aicardi-Goutieres syndrome: an expanding phenotype. Neuropediatrics 1998;29:163–167.
317. Lebon P, Badoual J, Ponsot G, Goutieres F, Hemeury-Cukier F, Aicardi J. Intrathecal synthesis of interferon-alpha in infants with progressive familial encephalopathy. J Neurol Sci 1988;84:201–208.
318. Blau N, Bonafe L, Krageloh-Mann I, Thony B, Kierat L, Hausler M, Ramaekers V. Cerebrospinal fluid pterins and folates in Aicardi-Goutieres syndrome: a new phenotype. Neurology 2003;61:642–647.
319. Polizzi A, Pavone P, Parano E, Incorpora G, Ruggieri M. Lack of progression of brain atrophy in Aicardi-Goutieres syndrome. Pediatr Neurol 2001;24:300–302.
320. Kato M, Ishii R, Honma A, Ikeda H, Hayasaka K. Brainstem lesion in Aicardi-Goutieres syndrome. Pediatr Neurol 1998;19:145–147.
321. Barth PG. The neuropathology of Aicardi-Goutieres syndrome. Eur J Paediatr Neurol 2002;6:A27–31.

23 Mitochondrial Disease

Brain Oxidative Metabolism Studied by ^{31}P, ^{1}H, and ^{13}C Magnetic Resonance Spectroscopy, Functional Magnetic Resonance Imaging, and Positron Emission Tomography

GRAHAM J. KEMP, MA, DM, FRCPath, ILTM

SUMMARY

This chapter summarizes how noninvasive methods can be used to study brain energy metabolism in vivo and what abnormalities they detect in mitochondrial disease. ^{31}P magnetic resonance spectroscopy (MRS) can measure transient changes in nonoxidative adenosine triphosphate (ATP) synthesis and at steady state reflects the balance of ATP demand and mitochondrial ATP supply. ^{1}H MRS can be used to measure lactate, which accumulates when glycolysis exceeds pyruvate oxidation: the place of this in normal brain is a matter of debate. ^{13}C MRS can measure a variety of carbon fluxes and can be complemented by ^{15}N MRS studies of nitrogen fluxes. The response to functional activation depends heavily on mitochondrial function and control. Positron emission tomography studies showed that blood flow and glucose consumption are increased in activation, with a smaller increase in oxygen consumption. In contrast, ^{13}C MRS studies suggested large increases in oxidative metabolism. Blood oxygen level-dependent functional magnetic resonance imaging, which measures the hyperoxygenation that results during functional activation from increased blood flow in excess of oxygen use, shows that glucose oxidation is coupled to flow, although increases less, because of the need to maintain oxygen extraction while maintaining a PO_2 gradient to drive oxygen into the cell and a cellular PO_2 sufficient for cytochrome oxidase. This has implications for mitochondrial control. ^{31}P MRS studies of activation have given contradictory results, but it now appears that in young people at least there is no decrease in phosphocreatine (as is seen in exercising muscle, which is some respects an analog of activated brain), and a paradoxical increase during the immediate postactivation phase. The implication is that mitochondrial control is dominated by "open loop" influences, not directly a consequence of increased ATP turnover. Mitochondrial disease has been relatively little studied by these methods: low "resting" phosphocreatine and elevated lactate concentrations have been reported but are probably a feature of severe brain disease. Phosphocreatine reportedly decreases during recovery from activation, suggesting a failure of mitochondrial function rather different to that in affected muscle. There are a number of unresolved questions about the cerebral metabolic pathophysiology in these patients, and application of the methods reviewed here may help to answer them, possibly to useful clinical effect.

Key Words: Brain metabolism; cerebral blood flow; cerebral glucose consumption; cerebral oxygen consumption; electron transport chain; functional MRI metabolic regulation; magnetic resonance spectroscopy; mitochondria; oxidative phosphorylation; tricarboxylic acid cycle.

1. INTRODUCTION

Much of brain metabolism is oxidative *(1)*; therefore, mitochondria are of crucial importance to brain function. Brain metabolism is not readily accessible to study, especially in the human, but some noninvasive techniques are available *(2)*. Positron emission tomography (PET) has been used to measure the uptake of glucose and (more problematically) oxygen at rest and during activation. ^{31}P magnetic resonance spectroscopy (^{31}P MRS) measures cell pH and phosphorus metabolite concentrations, from which inferences can be made about mitochondrial regulation and disease. Interactions between oxygen supply and oxidative metabolism underlie blood oxygenation level-dependent (BOLD) functional magnetic resonance imaging (fMRI), which has been used to study the mitochondrial contribution to the brain activation response. Finally, ^{13}C magnetic resonance spectroscopy (^{13}C MRS) is a powerful technique with a number of applications and is used to measure energy metabolism during brain activation; clinical applications in the diagnosis or monitoring of mitochondrial disease may come from this. This chapter reviews briefly the metabolic background (Section 2), the information available from these techniques (Section 3) and its relationship to metabolic regulation (Section 4) at rest and during activation (Sec-

From: *Bioimaging in Neurodegeneration*
Edited by P. A. Broderick, D. N. Rahni, and E. H. Kolodny

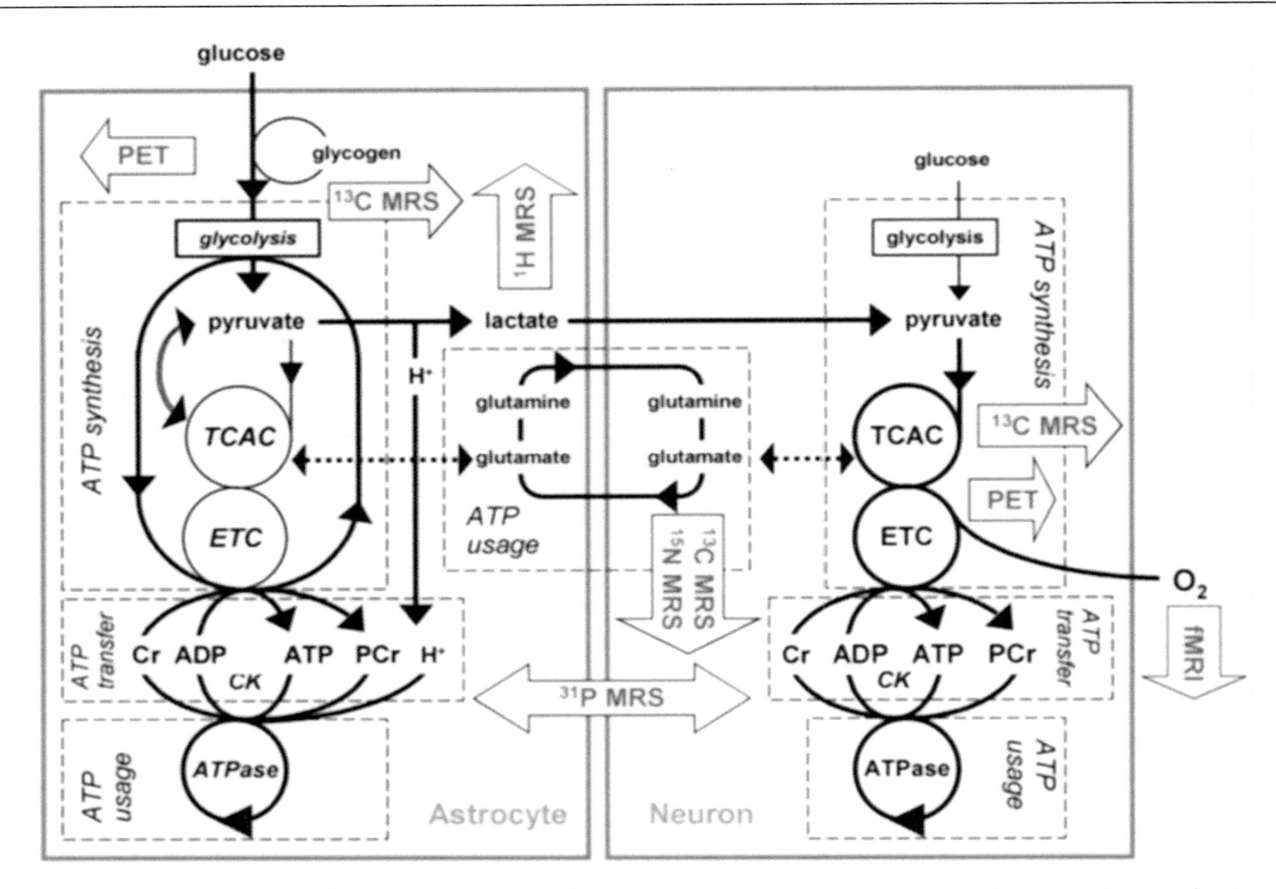

Fig. 1. Brain energy metabolism studied noninvasively. The figure outlines the relevant features of brain metabolism: *see* Section 2 for a description. The block arrows indicate some of the ways in which it can be studied noninvasively (*see* Section 3). The figure omits the entry of acetate into the astrocyte TCAC *(3)*, ketone bodies into both cell types (both of which can be important in pathological states; ref. *4*), and GABA cycling *(5)*. The figure also omits mitochondrial CK: although this has been much studied *(6,7)*, its functional significance for the measurements discussed here is not clear *(8,9)*.

tion 5), and what abnormalities are seen in mitochondrial disease (Section 6).

2. BRAIN METABOLISM

2.1. ENERGY METABOLISM IN THE BRAIN: ADENOSINE TRIPHOSPHATE (ATP) PRODUCTION

Figure 1 summarizes the general layout of brain cell energy metabolism *(1)*. A natural approach is in terms of ATP supply and demand *(10)*, but an immediate difficulty in brain is the need to consider at least two cell types: neurons and glia (notably astrocytes). Each has the same basic apparatus of energy metabolism, although there are quantitative differences. The ATP production block in Fig. 1 comprises anaerobic glycolysis, the tricarboxylic acid cycle (TCAC), and the electron transport chain. ATP is produced by substrate-level phosphorylation in glycolysis and the TCAC (not shown) and by oxidative phosphorylation. Glycolysis to lactate can proceed without oxygen (generating 2 ATP), and the TCAC and electron transport chain can be considered together as oxidative ATP synthesis (generating a further 34 ATP; ref. *11*). In whole brain, glucose is the principal fuel. However, glia have metabolically active glycogen stores *(12–14)*, which probably become important when ATP supply transiently falls below demand, for example, prolonged focal activation and mild hypoxia *(15)*. Also, lactate is a potential fuel for neurons, possibly the preferred fuel (refs. *16* and *17*; although this view has been criticized; ref. *18*). The whole brain fully oxidizes glucose *(19)*. Figure 1 reflects an influential view that the astrocyte is the main site of glucose uptake and its glycolysis to lactate *(3)*, this being released and then taken up for oxidation by neurons *(20,21)*. The extent of direct glucose oxidation in the neuron in vivo is unknown. However the concept of neuronal lactate usage has been criticized *(18,22)* and remains controversial *(23)*. There also may be some neuronal lactate production *(23)*, and anaerobic glycolysis may be important at certain cellular sites, for example, dendritic spines (which rarely contain mitochondria; ref. *22*) and the Na^+,K^+-ATPase (which seems to prefer glycolytically derived ATP; ref. *24*).

2.2. NEUROTRANSMITTER CYCLING

The metabolism of glutamate has been studied using ^{13}C MRS (Section 3.3.). Glutamate, which is synthesized in the neuron from glutamine (by glutaminase), is released as a neurotransmitter, then taken up by the astrocyte and reconverted to glutamine (by glutamine synthase), which passes back to the neuron as a substrate for glutamate synthesis (Fig. 1; refs. *25* and *26*. Important in ^{13}C MRS studies is the exchange between the mitochondrial TCAC intermediate 2-oxoglutarate and the cytosolic glutamate pool, which is mediated by aminotransferases *(22,28)* and glutamate dehydrogenase *(29,30)*. Glutamine synthase flux reflects largely this glutamate/glutamine shuttle *(26,28,31)*. Glutamine synthase is also a means of ammonia detoxification *(31–33)*.

2.3. ENERGY METABOLISM IN BRAIN: ATP TRANSFER AND ATP BUFFERING

In what one might think of as the standard model of energy transfer, developed in the context of skeletal muscle, ATP passes between the sites of its production to the sites of its usage, whereas adenosine diphosphate (ADP) and inorganic phosphate (Pi) return *(34,35)*. The brain is less straightforward: as we discuss in Section 2.4., ATP is used at many cellular sites. Nevertheless, the same principles apply: the presence of cytosolic creatine kinase (CK) at high activity (and therefore always close to equilibrium) assures that most "ATP" travels from sites of production to sites of usage as (phosphocreatine) PCr and that most "ADP" returning travels as free Cr, which facilitates diffusion (the "spatial buffering" role of CK; ref. *34*). Cytosolic CK has another important function: it stabilizes ATP concentration ("temporal buffering") because temporary mismatch between ATP synthesis and use is matched by changes in PCr concentration, so the free concentration of ADP remains low (although its relative rise may be large), and ATP falls very little unless there is net loss of adenine nucleotides *(36)*. If ATP production can match the changed ATP usage rate, PCr and ADP reach a new steady state.

PCr splitting is accompanied by net "consumption" of protons *(37,38)*, tending to alkalinize the cell, whereas lactate accumulation tends to acidify it. As in exercising muscle *(38,39)*, the resulting change in cell pH depends on cytosolic buffering capacity *(40,41)* as well as various membrane ion transport processes *(42)*. Cell pH recovery from acidification *(42,43)* also will depend on ion transport processes, as in muscle *(44,45)*. These principles are important in understanding ^{31}P MRS measurements of the activation response (Section 5.6.) and in modelling the relationship between brain electrical activity and metabolism *(46)*.

2.4. ENERGY METABOLISM IN THE BRAIN: ATP USAGE

The human brain, although accounting only for 2% of body weight, accounts for 20% of the body's oxygen consumption *(47)*. In muscle, most ATP is used by the force-generating actomyosin ATPase, with a lesser amount by the sarcoplasmic Ca^{2+}-ATPase *(48)*. The brain is more complicated. A recent analysis *(49)* concluded that in primate brain, approx 75% of the energy used for signaling is used to reverse the ion movements generating postsynaptic currents, 7% for presynaptic responses, 10% to support action potentials, 2% to maintain neuronal resting potential, and 6% in glia for the resting potential and glutamate recycling *(49,50)*.

This is at variance *(50)* with the claim, based on ^{13}C-labeling studies (Section 5.2.), that a large fraction of energy is used for glutamate recycling *(25,26)*. In this model, the uptake of glutamate by the astrocyte stimulates glucose uptake and lactate efflux by activating the Na^+/K^+-ATPase *(51)*, closing the "neurotransmitter cycle" *(52)*; this costs 2 ATP (1 for the Na^+/K^+-ATPase to extrude the 3 Na^+ ions used for glutamate uptake and 1 for glutamate synthetase), which could be supplied by anaerobic glycolysis of 1 glucose *(26)*. This analysis has been criticized on technical grounds *(22)*; it not clear, for example, that astrocyte ATP production is only anaerobic *(22)*. Nevertheless, a correlation between transmitter recycling and energy production *(25,26)* is to be expected *(50)*. γ-Aminobutyric acid (GABA) cycling also costs ATP *(26)*, and overall a lesser amount of energy is used for inhibitory processes *(50)*.

In Fig. 1, the double arrow in the astrocyte represents two additional processes: anaplerotic replenishment of TCAC intermediates from pyruvate (via pyruvate carboxylase; ref.*53*), and the reverse process, pyruvate recycling from TCAC metabolites (via malic enzyme or phosphoenolpyruvate carboxykinase; refs. *4* and *54*); both of these cost ATP.

3. METHODS FOR STUDYING BRAIN METABOLISM IN VIVO

3.1. ^{31}P Magnetic Resonance Spectroscopy (MRS)

^{31}P MRS can measure PCr, Pi, and ATP directly (although quantitation is not trivial; ref. *55*), and cell pH is obtainable from the frequency ("chemical shift") of the Pi peak. In muscle, free ADP concentration can be obtained indirectly using the CK equilibrium, usually by measuring PCr/ATP and pH, and by assuming concentrations of ATP and total creatine (= PCr + free creatine; ref. *56*), which has obvious limitations. Such a calculation can be performed for brain *(57,58)*, but uncertainties about the cellular location of the metabolites limit its interpretation.

The related technique of magnetization transfer can be used to measures the exchange flux through CK, which alters as the concentrations of CK reactants change in response to activation and other stimuli *(59–61)*. New high-field ^{31}P MRS methods offer improved signal-to-noise and sensitivity for human studies, with better measurements of metabolite concentrations and CK flux *(62)*. A recent report describes the use of magnetization transfer at high field to measure Pi-to-ATP flux, which (as in muscle; ref. *63*) can be made to estimate oxidative ATP synthesis rate *(62)*.

3.2. ^{1}H MRS

^{1}H MRS in vivo can detect total creatine (with sufficient resolution, PCr and Cr separately; ref.*64*), lactate and glucose *(65, 66)*, and some compounds that are not directly concerned with energy metabolism, such as *N*-acetyl aspartate and myo-inositol *(67)*, and (recently) reduced glutathione *(68)*. We discuss below lactate measurements in the context of the activation response and mitochondrial disease (*see* Section 5.6.).

3.3. ^{13}C MRS

^{13}C can to a certain extent differentiate between glial and neuronal responses. It can be used in vivo, although studies ex vivo on brain extracts offer the most information. ^{13}C MRS can obtain signal in vivo from glial glutamine and glycogen, from glucose and lactate (probably in both cell types), and from neuronal glutamate, aspartate, and GABA *(69)*. The power of ^{13}C labeling derives from detection of isotopomeric labeling patterns *(28,70,71)*. Ex vivo analysis distinguishes two compartments: one (= glia) with glutamine synthetase and a TCAC mainly mediated by pyruvate carboxylase, and the other (= neurons) with a faster TCAC mainly mediated by pyruvate dehydrogenase *(3,4,72)*. Such methods also reveal heterogeneity within both cell types *(73)*. Using [1-^{13}C]glucose infusion in vivo, measurements of glucose transport *(74–76)*, glycogen turnover *(12)*, and glycolytic rate *(28,74)* are reasonably straightforward. Other rate measurements depend on modeling. Ignoring acetyl CoA production from ketone bodies and net lactate production, the glucose consumption rate (CMR_{glc}) is half, and the oxygen consumption rate (CMR_{O2}) three times the rate of TCAC flux *(77–79)*. Previous in vivo analyses used single-compartment glutamate C3/C4 kinetics to measure TCAC rate and mitochondrial exchange *(78,79)*, later adding a glial compartment to model glutamate/glutamine exchange *(31)* and distinguishing between glial and neuronal TCAC rates *(80)*. Pyruvate carboxylase can be measured from C3/C2 labeling *(28,72,5,80)*. Estimates of these interconnected fluxes tend to correlate, but their covariance can be reduced by making additional measurements *(28)*.

Because [1-^{13}C]glucose incorporates in both cell types *(69)*, [2-^{13}C]glucose can be used to distinguish astrocyte and neuronal TCAC rate, glutamate cycling, and anaplerosis *(53)*. Labeled acetate can be used to study glial metabolism specifically *(69)*. These have been recent advances in measurement of CMR_{glc} *(81)* and glycogen turnover *(82)*, and the improved spectral resolution and sensitivity offered by very high field ^{13}C MRS opens up new possibilities *(83)*. We discuss ^{13}C MRS measurements in the context of the activation response (*see* Section 5.2.).

3.4. ^{15}N MRS

^{15}N MRS can also differentiate between glial and neuronal responses but has so far not been used in humans. Infused $^{15}NH_4$ can be used to estimate glial-specific *(69)* flux through glutamate dehydrogenase *(29,30)*, glutamine synthase *(30,32,33)*, and Pi-activated glutaminase *(84)*. Branched-chain aminotransferase can be measured using [^{15}N]leucine *(85)*. GABA shunt flux can be measured ex vivo from the labeling kinetics of GABA and glutamate *(86)*.

3.5. fMRI

fMRI measures changes in the BOLD signal, which reflects brain oxygenation via the effects of deoxyhemoglobin on the relaxation properties of water *(87,88)*, and has been widely exploited in neuropsychology. There are many MRI-based methods to measure regional cerebral blood flow (CBF) and cerebral blood volume (CBV; refs. *87,89–93*). The nature of the signal and the implications for metabolic regulation will be discussed below 5.4., along with its use to estimate relative changes in CMR_{O2} (Section 5.4.).

3.6. Other Noninvasive Techniques: PET and Near Infrared Spectroscopy (NIRS)

PET can be used, with appropriate tracers, to measure regional CMR_{O2}, oxygen extraction fraction, CBF, and glucose uptake CMR_{glc} *(94,95)*. Some applications to the study of the activation response will be discussed (Section 5.1.). Brain oxygenation and the redox state of cytochrome oxidase also can be studied noninvasively by NIRS *(96)*. Neither of these techniques can distinguish between neurons and glia.

4. REGULATION OF CEREBRAL METABOLISM AND BLOOD FLOW

In understanding metabolic responses, the first steps are to measure metabolite concentrations and metabolic fluxes (these either by labeling methods such as ^{13}C MRS, or dynamically from changing concentrations, e.g., using ^{31}P and ^{1}H MRS). The next step is to analyze metabolic regulation: the central concern here is the relationship between fluxes and concentrations, in relation to appropriate kinetic models.

4.1. Open-Loop and Closed-Loop Control in Metabolic Regulation

^{31}P MRS fits well with a simple model of metabolic regulation, usually presented in the context of aerobically exercising skeletal muscle *(8)*. Some principles are set out in the upper half of Fig. 2. Changes in PCr measure the mismatch between ATP usage and supply, and a PCr fall usually means a rise in ADP. ADP has been posited as the error signal in a closed-loop feedback mechanism in muscle, where many studies show a roughly hyperbolic (Michaelis–Menten) relationship between oxidative ATP synthesis and free ADP concentration resembling that seen in isolated mitochondria *(8,97,107)*. Other candidates for the error signal are difficult to distinguish experimentally *(108)*. In low-power aerobic exercise, muscle shows exponential PCr kinetics *(109)* consistent with closed-loop control *(8)*: similar activation-related changes in brain PCr and ADP are discussed below (Section 5.6.). In general, if changes in potential regulators are not sufficient to explain alterations in oxidation rate in vivo, we must posit open-loop mechanisms ("feed-forward" or "parallel activation") by which the stimulus directly activates oxidation (e.g., via Ca^{2+}-stimulation of mitochondrial dehydrogenases; Fig. 2; ref. *98*). Some argue for this in muscle *(98)*, and others argue that it is unnecessary because the ADP dependence is actually cooperative *(107)*, but in general the greater the open-loop component, the smaller the metabolite changes during activation *(10,97)*. In muscle, glycolytic ATP synthesis increases at high ATP turnover, and the degree to which this is explained by open- and closed-loop mechanisms is a matter of debate *(38,99,100,110)*: the analogy to brain activation will be discussed in Section 5.6.

The importance of this distinction lies in the flow of information. Activation of the neuron stimulates ATP usage; therefore, neuronal PCr should tend to decrease (closed-loop feedback; Fig. 2). However, the small or absent PCr falls seen in some studies (Section 5.6.) suggest that closed-loop mechanisms predominate. In the model assumed in Fig. 1, the ATP-consuming uptake of transmitter glutamate released from the neuron is the stimulus to astrocyte glucose uptake and glycolysis *(26,52)*; this appears to be a closed-loop mechanism tending to decrease astrocyte PCr (although it is the neuron that benefits from most of the oxidative ATP yield; ref. *26*). Astrocyte lactate must be high enough to sustain a gradient into the neuron: this is an open-loop signal stimulus to neuronal lactate oxidation, deriving ultimately from transmitter cycling, tending to reduce any fall in neuronal PCr. Neuronal glutamine uptake is presumably driven by its release from the astrocyte, but neuronal glutaminase activity *(84)* would be stimulated by any rise in Pi accompanying a PCr fall. A decrease in PCr tends to raise pH, but this is opposed by lactate production (largely in the astrocyte?) or lactate uptake (in the neuron): there is little information on which effect dominates in vivo (Section 5.6.).

4.2. Regulation of CBF

Like ATP synthesis, CBF is subject to open- and closed-loop influences in activation, and there are many possible

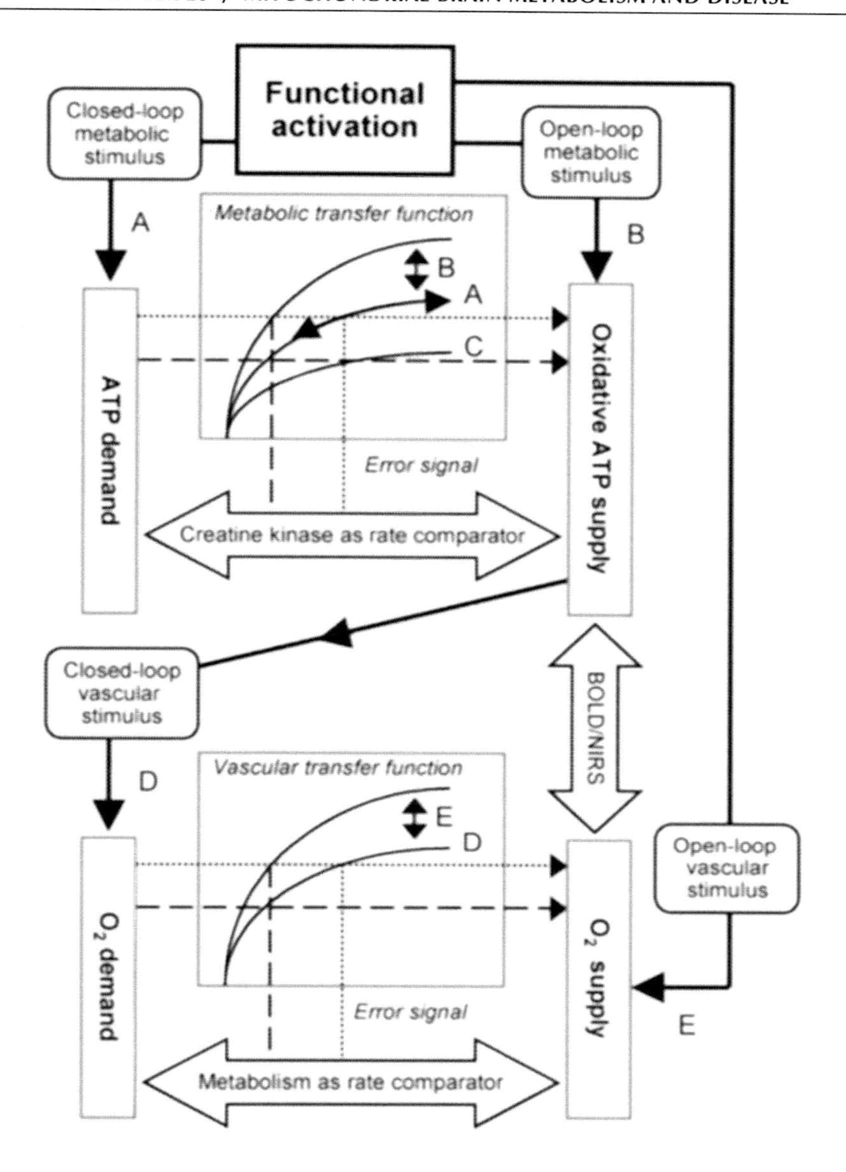

Fig. 2. Regulation of cerebral metabolism and cerebral blood flow. The figure sets out some principles useful in understanding the metabolic and physiological regulation discussed in this chapter. The basic points are open-loop and closed-loop control (Section 4.1.; refs. *2,8,10,97,99,100,101*). The upper half of the figure describes the regulation of mitochondrial ATP synthesis. The ATP demand is set by a stimulus related to functional activation: in muscle this would be, roughly, the ATPase rate required to generate the desired force *(48)*; in the brain it is the energy required for signaling (Section 2.4.). Mitochondrial ATP synthesis is controlled by an error signal such as ADP concentration *(8)*: the relationship between these is called here the "metabolic transfer function" (Section 4.1.). This works, in engineering terms, as closed-loop feedback control (specifically, integral control; ref. *101*), because the size of the error signal is related to the time-integrated mismatch between the ATP demand (input) and ATP supply (output). Thus, the CK system acts as an integrating rate comparator: the more different the supply is from the demand, and the longer it is so, the larger the error signal and consequently also the restoring force (i.e., the rate of ATP synthesis). The system tracks alterations in demand by moving up and down the "resting" transfer function (the transfer function and the target rate are both labeled *A*). By contrast, an open-loop signal acts in parallel (sometimes called "feed-forward" regulation) to increase oxidative ATP synthesis directly, without having to let any error signal adjust itself to a mismatch; this in effect scales up the transfer function (the open loop signal and the "stimulated" transfer function are both labeled *B*). Thus the response to activation (the transition from the dashed to the dotted horizontal line) could be entirely closed-loop, with an increase in the error signal (the transition from the dashed to the dotted vertical line, moving along the unstimulated transfer function), or entirely open-loop (no change in error signal, moving up the dashed vertical line), or anywhere in between; the closed-loop error signal could even fall (Section 5.6.; ref. *98*). This analysis describes the situation at steady state: if ATP demand exceeds supply then PCr will fall and ADP rise until the new steady state is reached; open-loop mechanisms can in principle respond faster during transients *(98)*. This ignores the contribution of glycolytic ATP synthesis, which is complicated in muscle *(100,110)* and much more so in brain (Section 2.1.).The upper panel also shows the effect of mitochondrial dysfunction, which lowers the maximum (refs. *102* and *103*; and perhaps changes the shape; ref. *104*) of the transfer function (now labeled *C*). There is some scope to compensate for the resulting reduction in output rate by letting the feedback signal increase *(105)*; it also might be possible to compensate by increasing an open-loop signal *(10)*. The lower half of the figure describes the control of blood flow (Section 4.2.). It is similar in principle to the control of metabolism in the upper panel (Section 4.1.). The transfer function describes how flow responds to signals released in response to the mismatch between oxygen supply and demand; in this sense, cellular metabolism acts as a rate comparator between these two. In this

mediators *(1)*. Some principles are set out in Fig. 2. A vasoactive signal responding to the mismatch between ATP demand and oxidative ATP supply would represent closed-loop feedback, a signal parallel to stimulation of ATP turnover would be open-loop. Examples of directly closed loop signals would by CO_2, H^+, lactate, or adenosine (insofar at it reflected cellular ATP depletion). It is evident *(50)* that open-loop influences dominate because during activation, oxygen supply exceeds the demands of neuronal activity in size *(111,112)* and area in *(113)*. There is evidence that glutamate causes Ca^{2+}-dependent release of vasodilator factors by postsynaptic cells *(50)* and astrocytes *(114)*, and it has been suggested that in addition to the spatially restricted vasodilatation caused by fast neurotransmitters, there might more widespread regulation by dopaminergic fibers *(50)*. Thus, the BOLD response may reflect open-loop responses to neuronal signaling rather than energy consumption *(50)*. Figure 2 also attempts to show the interactions between the control of flow and oxygen consumption (see legend).

5. THE RESPONSE TO FUNCTIONAL ACTIVATION

5.1. Glucose Oxidation, Oxygen Usage, and Blood Flow During Activation

The brain is almost entirely oxidative in the unstimulated state *(13,115)*, and responds to functional activation (e.g., somatosensory or visual stimulation, cognitive tasks) with an increase in ATP turnover fueled by glucose. Invasive techniques *(115,116)* and PET studies *(94,95,117,118)* have demonstrated increased glucose uptake (CMR_{glc}) in proportion to increased CBF. Consistent with this, ^1H MRS shows a transient fall in brain glucose during activation *(66,119)*.

The contribution of oxidation has been the subject of debate. Some invasive *(13,115,116)* and PET studies *(94)* found increases in oxygen uptake (CMR_{O2}) were small compared with the increases in CMR_{glc} and CBF. This excess CBF response is responsible for the hyperoxygenation during activation observed by NIRS *(96)* and BOLD fMRI *(66,120)*. This does not mean that flow and oxygen consumption are uncoupled *(121)*, as analysis of both BOLD fMRI response *(87,89,111,122–124)* and ^{13}C MRS (Section 5.4.; ref. *112*) show that CMR_{O2} is closely correlated with CBF over a wide range, although it changes less than CBF. Thus, fMRI reflects metabolic events despite detecting mainly a vascular response.

The disproportionate change in CBF can be explained in terms of an oxygen limitation to oxidative metabolism *(112,125)*: (1) if cell PO_2 is near zero, then capillary PO_2 must rise to sustain increased O_2 flux into the tissue. (2) To maintain the normally high O_2 extraction ratio (70%) and therefore sustain increased O_2 delivery, CBF must increase disproportionately *(121,124)*. Capillary bed oxygen diffusivity has been modeled as constant *(125)* or flow-dependent *(112,126)*, but neither accounts particularly well for dynamic responses *(127)*. To accommodate real, nonzero, cell PO_2, the model assumes that this is maintained at a fixed value that acts as an oxygen buffer or reserve *(121)*, so that although small changes in O_2 usage can be accommodated with small changes in CBF, larger changes require much bigger CBF increases *(128)*. There has been some debate about the size of this reserve *(129,130)*.

In addition to the technical implications for spatial localization of neuronal function fMRI *(121)*, this finding has implications for control of flow: although it does not rule out a closed-loop mechanism, it would require a very high gain, and an open-loop parallel activation mechanism would be simpler and more robust (cf. the argument for parallel activation in muscle mitochondrial regulation; ref. *10*).

A different aspect of "uncoupling" is the apparent decrease in glucose/oxygen stoichiometry, which suggests predominant nonoxidative ATP production: suggestions for its cause include net lactate efflux into the blood *(131)* and cycling between glucose and glycogen ("glycogen shunt" ref. *132*). However, invasive studies show aerobic metabolism must eventually become important *(116)*, and indeed it would seem perverse for it not to, given its greater efficiency. We return to this in Sections 5.3. and 5.4. after discussing ^{13}C measurements of TCAC rates (Section 5.2.).

5.2. TCAC Flux During Activation

^{13}C MRS also can study the activation response, albeit with relatively poor time resolution. In stimulated brain slice in vitro, ^{13}C MRS suggests increased TCAC flux, roughly in accordance with an approx 50% increase in CMR_{O2} *(133)*. In rat somatosensory cortex in vivo, ^{13}C MRS suggests that TCAC flux roughly trebles during forepaw stimulation, with similar increases in CMR_{glc} and CMR_{O2} *(77)*, so that most energy required for activation is supplied by glucose oxidation. Although within the capacity of brain glucose transport *(75)*, this increase is rather greater than fMRI and PET suggest (Sections 5.1. and 5.4.; refs. *77* and *89*, perhaps because of differences in the baseline *(134)*. Further analysis of the activation experiments, necessarily simplifying some complex kinetics *(28)*, suggested that CMR_{glc} is proportional to glutamate cycling over a range of activity (Section 2.4.; ref. *26*). More recently, ^{13}C MRS studies of human visual cortex found only 30–60% increase in CMR_{O2} inferred from TCAC flux *(134,135)*, which is similar to PET-measured increases in CMR_{glc} *(94)*. Furthermore, the absence of rapid lactate labeling implies that if lactate is an important fuel for neurons (Section 2.1.), then its produc-

Fig. 2. *(continued)* context, the stimulus to oxygen demand is the rate of ATP synthesis, which is in turn the output of the metabolic control system in the upper half of the figure: both this target rate of oxygen supply and the "unstimulated" transfer function are labeled *D*. As discussed in Section 4.2., flow is dominated by open-loop influences (labeled *E*), working in parallel with metabolic signals and mechanisms; at steady state (which is what the figure shows) this would match oxygen supply and demand, but in the short term the increase in flow exceeds that in demand, and results in hyperoxygenation: thus the tissue oxygenation state which is measured by the BOLD fMRI response (and also the NIRS response) is a comparator of oxidative oxygen supply and oxidative ATP supply (=oxygen use). This analysis neglects the influence of cellular oxygen on mitochondrial ATP synthesis *(103,106)*, whereby mitochondrial ATP synthesis is inhibited (i.e., the metabolic transfer function scaled downwards) when the size or duration of the mismatch between oxygen supply and demand results in sufficiently severe deoxygenation; in the brain, the dominant open-loop control of blood flow ensures that deoxygenation occurs only transiently, before hyperoxygenation supervenes (Section 5.4.).

tion and use, which are postulated to be in different cells, must be very tightly coupled *(135)*.

5.3. Invasive Measurements of Oxygen/Glucose Consumption

In a recent invasive rat brain study *(13)*, cerebral oxygen/glucose ratio was low during sensory stimulation and high during recovery: excess glucose consumption during stimulation was not accounted for by lactate efflux or accumulation. Brain lactate levels nearly doubled during stimulation and then normalized, whereas brain glycogen decreased during activation and then decreased further during recovery. It seems that the disproportionately greater increases in CMR_{glc} compared with $CMRO_2$ reflects only a temporal displacement: during activation, glycolysis increases more than oxidative metabolism, leading to accumulation of intermediary products that are subsequently oxidized during recovery; this is confounded by astrocyte glycogenolysis and perhaps functional segregation of glucose and glycogen usage *(13,136)*. It is clear that brain metabolism is complex and not fully characterized by global fuel uptake or changes in local concentrations of glucose and lactate *(136)*.

5.4. Metabolic Implications of the Standard BOLD fMRI Response

BOLD also is a complicated phenomenon, depending on several processes that change on different timescales *(121)*. Although CBF changes dominates the fMRI response, regional CMR_{O2} nevertheless increases, tending to reduce the BOLD effect *(87,89,111)*. This can be used to estimate CMR_{O2}. "BOLD calibration" methods *(89,111,137,138)* do this by comparing the BOLD and CBF responses to activation and to a purely vasodilator stimulus such as hypercapnia *(11)*, although there are some technical criticisms to be made *(134)*. CO_2 calibration is not necessary in newer approaches such as "multimodal MRI" (where the signal change is directly related to changes in CBF, CBV, and CMR_{O2}; refs. *11,126,139*) or similar methods *(140)*. Agreement between this sort of analysis and ^{13}C MRS is now good *(11,126,141,142)*, although not all aspects are well understood *(143)*. An interesting development is the addition of metabolism to BOLD modeling: a recent report finds that the best metabolic control model involved parallel activation (Section 4.1.) as well as closed-loop control by ADP concentration *(46)*.

Whether oxygen consumption and flow are coupled during dynamic changes is still open *(121)*. In invasive animal work, the first event after sensory stimulation is an increase in O_2 consumption, with accompanying deoxygenation *(113,144–146)*. It is a matter of debate *(121,127,147)* whether this is what is seen as the initial dip in the BOLD signal, succeeded by the familiar increase in BOLD secondary to a decrease in deoxyhemoglobin. Because oxygen consumption and neuronal activity are co-localized but the delayed blood flow is not, an fMRI that is focused on this initial phase should yield much higher spatial resolution *(144,145)*. There also are implications for mitochondrial control (e.g., how limited is it by the initial deoxygenation) as well as control of flow (is this locally closed-loop, but regionally open-loop?).

5.5. What is the BOLD Signal Reporting?

BOLD fMRI uses alterations in brain hemodynamics to infer changes in neural activity. But what activity does it reflect? Attenuation of BOLD response with an inhibitor of neuronal glutamate release is consistent with the proposed link between neurotransmitter cycling and energy metabolism *(126)*. A number of studies have correlated BOLD with electrical activity *(148,149)*, and recent work suggests that BOLD reflects input and intracortical processing *(50,149,150)* rather than spiking output *(151)* directly. If most brain energy is used to power postsynaptic currents and action potentials (Section 2.4.) and, furthermore, if hemodynamic responses are driven by neurotransmitter-related signaling and not directly by the local energy needs of the brain, then understanding the BOLD response should focus on the neural signaling mechanisms controlling blood flow rather than the sites of energy usage *(50)*. In any case, neuropsychologic interpretation of BOLD data must reckon with the large baseline rate of glucose oxidation *(11,47)*.

5.6. Phosphorus Metabolites, pH, and Lactate During Activation

The use of ^{31}P MRS to study energy metabolism is well established in skeletal muscle *(38,100,152–154)*. The standard pattern in exercising muscle is a fall in PCr, perhaps to a steady state, accompanied by an initial rise in pH (due to proton consumption by net PCr depletion) succeeded by a fall in pH if and when lactate begins to accumulate. The relationship between the work rate and the PCr change measures (in various ways of which the details are debated) the size of the feedback signals necessary to drive glycolytic *(38,100,110)* and oxidative ATP synthesis *(98,104,107,108)*. After the end of exercise, PCr is resynthesized by oxidative ATP synthesis *(108,155)*, and pH recovers as protons are expelled from the cell *(44,45)*. The initial rate of PCr depletion measures ATP turnover *(156)*, the initial rate of PCr resynthesis measures end-exercise rate of oxidative ATP synthesis *(108)*, and a measure of effective mitochondrial capacity can be obtained in various ways from the recovery kinetics of PCr *(8,102,104, 109,153)*.

There is some analogy between muscle exercise and brain activation, but in brain the ^{31}P MRS data are much harder to acquire. The earliest reports were of a PCr decrease during visual activation (refs. *157–160*; which resembles the muscle response) and also afterwards (ref. *157*; which does not). Magnetization transfer studies also detected an increase in CK exchange flux *(60,160)*, which probably reflect a rise in ADP concentration *(161)*, which could be the result of a fall in PCr concentration. Other studies found no decrease in the concentration of PCr *(60)* or a decrease only among older subjects *(162)*. The implication of this is that no appreciable perturbation of the CK equilibrium is required to stimulate glycolytic or oxidative metabolism in the young brain, so that open-loop mechanisms dominate (Section 4.1.); in the brains of older patients, it might be that vascular insufficiency made this no longer enough.

In the most recent report, there was no change in PCr during prolonged visual activation but PCr rose significantly during recovery, and pH increased during activation with a slow return to rest values *(163)*. Although this may correspond temporally to increased postactivation glucose oxidation *(115)*, which is assumed to restore gradients, the rise in PCr means that energy is supplied in excess of requirement *(163)*, implying a dominant open-loop mechanism.

It remains unclear whether there is much glycolytic ATP synthesis during activation. ^{1}H MRS shows transient lactate increase in some *(66,158,164)* but not all *(165)* studies, and invasive studies also have shown increased lactate production *(115,116)*. Cell pH, which depends on the balance between PCr splitting and lactate production, is reported as possibly increasing *(158,163)* or constant *(159)*, or possibly increased after stimulation in older subjects only *(162)*. The metabolic architecture of brain is more complicated than muscle and so the transient lactate change reflects transfer of astrocytic lactate to neurons rather than simple accumulation *(166)*, although little is known about how the MRS measurements reflect this, nor of how pH, lactate, PCr and pH changes are reconciled in the two cells.

6. THE RESPONSE TO MITOCHONDRIAL DISEASE

Mitochondria are important in brain metabolism, and mitochondrial disease often has significant brain effects. Here we describe the relatively limited work on brain metabolism in vivo in mitochondrial disease. In a steady-state ^{31}P MRS study, the simplest prediction is that in the presence of mitochondrial dysfunction of whatever cause *(104)*, decreased PCr and increased ADP will reflect an increased mitochondrial error signal (Fig. 2; refs. *105,167,168*). In the brain, “”resting” ^{31}P MRS studies have supplied researchers with variable results: early reports of reduced PCr *(58,169)* have been criticized on technical grounds *(170)*. In recent work, as in some earlier studies *(171)*, patients with no clinical brain involvement had normal PCr/ATP levels *(163)*. By analogy with muscle, one might predict a steady-state increase in anaerobic glycolysis (in astrocytes, or also in neurons?) if such closed-loop adaptations are insufficient to normalize oxidative ATP synthesis, and there have been case reports in mitochondrial encephalopathy of increased cerebral lactate measured by ^{1}H MRS *(172,173)* in addition to evidence uncovered by PET of increased glycolysis *(172,174,175)* and decreased CMR_{O2} *(175)*. Whether this is detectable in a particular case is a quantitative matter: where oxidative impairment is severe enough, lactate will surely be increased at rest *(163)*.

In the muscle during exercise, the standard pattern in mitochondrial disease is an increased decline in PCr levels and rise in ADP during exercise, with slow postexercise recovery of PCr and ADP *(105,167)*. In brain there has been little work on pathophysiologic changes in the activation response: in two patients with mitochondrial myopathy, encephalopathy, lactic acidosis, and stroke-like episodes (i.e., MELAS), the PCr fall in response to visual stimulation was increased *(176)*, as expected if oxidative ATP production is important in this response. In a recent study *(163)*, PCr was unchanged from resting values in controls (Section 5.6.) and in patients; but in recovery from activation, when PCr rose in controls (Section 5.6.), PCr decreased in patients. This is consistent with the idea that measurements remain normal until oxidative metabolism is stressed during the postactivation phase *(163)*. There are clearly a number of outstanding questions, and improved signal-to-noise and spectral resolution now becoming available at high field should help *(62)*. Because ^{31}P MRS can measure fundamental cellular bioenergetics (albeit with limitations), one can imagine that MRS studies of brain activation will prove an important diagnostic tool in such patients *(163)*. There do not seem to have been any PET or fMRI studies of brain activation in mitochondrial disease.

REFERENCES

1. Zauner A, Daugherty WP, Bullock MR, Warner DS. Brain oxygenation and energy metabolism: part I-biological function and pathophysiology. Neurosurgery 2002;51:289–301.
2. Kemp GJ. Non-invasive methods for studying brain energy metabolism: what they show and what it means. Dev Neurosci 2000;22:418–428.
3. Hassel B, Bachelard H, Jones P, Fonnum F, Sonnewald U. Trafficking of amino acids between neurons and glia in vivo. Effects of inhibition of glial metabolism by fluoroacetate. J Cereb Blood Flow Metab 1997;17:1230–1238.
4. Künnecke B, Cerdan S, Seelig J. Cerebral metabolism of [1,2–$^{13}C_2$]glucose and [U-$^{13}C_4$]3-hydroxybutyrate in rat brain as detected by ^{13}C NMR spectroscopy. NMR Biomed 1993;6:264–277.
5. Shen F, Petersen KF, Behar KL, et al. Determination of the rate of the glutamate/glutamine cycle in the human brain by in vivo ^{13}C NMR. Proc Natl Acad Sci USA 1999;96:8235–8240.
6. Chen L, Roberts R, Friedman DL. Expression of brain-type creatine kinase and ubiquitous mitochondrial creatine kinase in the fetal rat brain: evidence for a nuclear energy shuttle. J Comp Neurol 1995;363:389–401.
7. Friedman DL, Roberts R. Compartmentation of brain-type creatine kinase and ubiquitous mitochondrial creatine kinase in neurons: evidence for a creatine phosphate energy shuttle in adult rat brain. J Comp Neurol 1994;343:500–511.
8. Kemp GJ, Manners DN, Bastin ME, Clark JF, Radda GK. A theoretical model of the mitochondrion/creatine kinase/myofibril system in muscle. Mol Cell Biochem 1998;184:249–289.
9. Saks VA, Ventura Clapier R, Aliev MK. Metabolic control and metabolic capacity: two aspects of creatine kinase functioning in the cells. 1996;1274:81–88.
10. Korzeniewski B. Parallel activation in the ATP supply-demand system lessens the impact of inborn enzyme deficiencies, inhibitors, poisons or substrate shortage on oxidative phosphorylation in vivo. Biophys Chem 2002;96:21–31.
11. Rothman DL, Behar KL, Hyder F, Shulman RG. In vivo NMR studies of the glutamate neurotransmitter flux and neuroenergetics: implications for brain function. Annu Rev Physiol 2003;65:401–427.
12. Choi I-Y, Tkac I, Ugurbil K, Gruetter R. Noninvasive measurements of [1–^{13}C]glycogen concentrations and metabolism in rat brain in vivo. J Neurochem 1999;73:1300–1308.
13. Madsen PL, Cruz NF, Sokoloff L, Dienel GA. Cerebral oxygen/glucose ratio is low during sensory stimulation and rises above normal during recovery: excess glucose consumption during stimulation is not accounted for by lactate efflux from or accumulation in brain tissue. J Cereb Blood Flow Metab 1999;19:393–400.
14. Griffin JL, Rae C, Radda GK, Matthews PM. Delayed labelling of brain glutamate after an intra-arterial [^{13}C]glucose bolus: evidence for aerobic metabolism of guinea pig brain glycogen store. Biochim Biophys Acta 1999;1450:297–307.
15. Gruetter R. Glycogen: the forgotten cerebral energy store. J Neurosci Res 2003;74:179–183.
16. Smith D, Pernet A, Hallett WA, Bingham E, Marsden PK, Amiel SA. Lactate: a preferred fuel for human brain metabolism in vivo. J Cereb Blood Flow Metab 2003;23:658–664.
17. Bouzier Sore AK, Voisin P, Canioni P, Magistretti PJ, Pellerin L. Lactate is a preferential oxidative energy substrate over glucose for neurons in culture. J Cereb Blood Flow Metab 2003;23:1298–1306.
18. Chih CP, Lipton P, Roberts EL. Do active cerebral neurons really use lactate rather than glucose? Trends Neurosci 2001;24:573–578.
19. Pellerin L, Magistretti PJ. How to balance the brain energy budget while spending glucose differently. J Physiol 2003;546:325.
20. Pellerin L, Pellegri G, Martin JL, Magistretti PJ. Expression of monocarboxylate transporter mRNAs in mouse brain: support for a

distinct role of lactate as an energy substrate for the neonatal vs. adult brain. Proc Natl Acad Sci USA 1998;95:3990–3995.
21. Schurr A, Miller JJ, Payne RS, Rigor BM. An increase in lactate output by brain tissue serves to meet the energy needs of glutamate-activated neurons. J Neurosci 1999;19:34–39.
22. Chih CP, Roberts Jr EL. Energy substrates for neurons during neural activity: a critical review of the astrocyte-neuron lactate shuttle hypothesis. J Cereb Blood Flow Metab 2003;23:1263–1281.
23. Pellerin L, Magistretti PJ. Food for thought: challenging the dogmas. J Cereb Blood Flow Metab 2003;23:1282–1286.
24. Silver IA, Deas J, Erecinska M. Ion homeostasis in brain cells: differences in intracellular ion responses to energy limitation between cultured neurons and glial cells. Neuroscience 1997;78:589–601.
25. Sibson NR, Shen J, Mason GF, Rothman DL, Behar KL, Shulman RG. Functional energy metabolism: in vivo ^{13}C-NMR spectroscopy evidence for coupling of cerebral glucose consumption and glutamatergic neuronal activity. Dev Neurosci 1998;20:321–330.
26. Sibson NR, Dhankhar A, Mason GF, Rothman DL, Behar KL, Shulman RG. Stoichiometric coupling of brain glucose metabolism and glutamatergic neuronal activity. Proc Natl Acad Sci USA 1998;95:316–321.
27. Sonnewald U, White LR, Odegård E, et al. MRS study of glutamate metabolism in cultured neurons/glia. Neurochem Res 1996; 21:987–993.
28. Gruetter R, Seaquist ER, Kim S, Ugurbil K. Localized in vivo ^{13}C-NMR of glutamate metabolism in the human brain: initial results at 4 tesla. Dev Neurosci 1998;20:380–388.
29. Kanamori K, Ross BD. Steady-state in vivo glutamate dehydrogenase activity in rat brain measured by ^{15}N NMR. J Biol Chem 1995;270:24805–4809.
30. Lapidot A, Gopher A. Quantitation of metabolic compartmentation in hyperammonemic brain by natural abundance ^{13}C-NMR detection of ^{13}C-^{15}N coupling patterns and isotopic shifts. Eur J Biochem 1997;243:597–604.
31. Sibson NR, Dhankhar A, Mason GF, Behar KL, Rothman DL, Shulman RG. In vivo ^{13}C NMR measurements of cerebral glutamine synthesis as evidence for glutamate-glutamine cycling. Proc Natl Acad Sci USA 1997;94:2699–2704.
32. Kanamori K, Ross BD, Kuo EL. Dependence of in vivo glutamine synthetase activity on ammonia concentration in rat brain studied by ^{1}H - ^{15}N heteronuclear multiple-quantum coherence-transfer NMR. Biochem J 1995;311:681–688.
33. Shen J, Sibson NR, Cline G, Behar KL, Rothman DL, Shulman RG. ^{15}N-NMR spectroscopy studies of ammonia transport and glutamine synthesis in the hyperammonemic rat brain. Dev Neurosci 1998;20:434–443.
34. Meyer RA, Sweeney HL, Kushmerick MJ. A simple analysis of the "phosphocreatine shuttle". Am J Physiol 1984;246:C365–C377.
35. Kushmerick MJ. Energy balance in muscle activity: simulations of ATPase coupled to oxidative phosphorylation and to creatine kinase. Comp Biochem Physiol B Biochem Mol Biol 1998;120:109–123.
36. Connett R. Analysis of metabolic control: new insights using a scaled creatine model. Am J Physiol 1988;254:R949–R959.
37. Kushmerick MJ. Multiple equilibria of cations with metabolites in muscle bioenergetics. Am J Physiol 1997;272:C1739–C1747.
38. Kemp GJ, Roussel M, Bendahan D, Le Fur Y, Cozzone PJ. Interrelations of ATP synthesis and proton handling in ischaemic exercise studied by ^{31}P magnetic resonance spectroscopy. J Physiol 2001;535: 901–928.
39. Bendahan D, Kemp GJ, Roussel M, Le Fur Y, Cozzone PJ. ATP synthesis and proton handling in short periods of exercise and subsequent recovery. J Appl Physiol 2003;93:2391–2397.
40. Roberts EL, Chih CP. The pH buffering capacity of hippocampal slices from young adult and aged rats. Brain Res 1998;779:271–275.
41. Corbett RJ, Laptook AR, Garcia D, Ruley JI. Cerebral acid buffering capacity at different ages measured in vivo by ^{31}P and ^{1}H nuclear magnetic resonance spectroscopy. J Neurochem 1992;59:216–226.
42. Nishimura M, Johnson DC, Hitzig BM, Okunieff P, Kazemi H. Effects of hypercapnia on brain pHi and phosphate metabolite regulation by ^{31}P-NMR. J Appl Physiol 1989;66:2181–2188.
43. Adler S, Simplaceanu V, Ho C. Brain pH in acute isocapnic metabolic acidosis and hypoxia: a ^{31}P-nuclear magnetic resonance study. Am J Physiol 1990;258:F34–F40.
44. Kemp GJ, Thompson CH, Sanderson AL, Radda GK. pH control in rat skeletal muscle during exercise, recovery from exercise and acute respiratory acidosis. Magn Reson Med 1994;31:103–109.
45. Kemp GJ, Thompson CH, Taylor DJ, Radda GK. Proton efflux in human skeletal muscle during recovery from exercise. Eur J Appl Physiol 1997;5:462–471.
46. Aubert A, Costalat R, Valabregue R. Modelling of the coupling between brain electrical activity and metabolism. Acta Biotheor 2001;49:301–326.
47. Raichle ME, Gusnard DA. Appraising the brain's energy budget. Proc Natl Acad Sci USA 2002;99:10237–10239.
48. Jeneson JA, Westerhoff HV, Kushmerick MJ. A metabolic control analysis of kinetic controls in ATP free energy metabolism in contracting skeletal muscle. Am J Physiol 2000;279:C813–C832.
49. Attwell D, Laughlin SB. An energy budget for signaling in the grey matter of the brain. J Cereb Blood Flow Metab 2001;21: 1133–1145.
50. Attwell D, Iadecola C. The neural basis of functional brain imaging signals. Trends Neurosci 2002;25:621–625.
51. Pellerin L, Magistretti PJ. Excitatory amino acids stimulate aerobic glycolysis in astrocytes via an activation of the Na^{+}/K^{+} ATPase. Dev Neurosci 1996;18:336–342.
52. Magistretti PJ, Pellerin L, Rothman DL, Shulman RG. Energy on demand. Science 1999;283:496–497.
53. Sibson NR, Mason GF, Shen J, et al. In vivo ^{13}C NMR measurement of neurotransmitter glutamate cycling, anaplerosis and TCA cycle flux in rat brain during [2–^{13}C]glucose infusion. J Neurochem 2001;76:975–989.
54. Hassel B, Sonnewald U. Glial formation of pyruvate and lactate from TCA cycle intermediates: implications for the inactivation of transmitter amino acids? J Neurochem 1995;65:2227–2234.
55. Buchli R, Martin E, Boesiger P. Comparison of calibration strategies for the in vivo determination of absolute metabolite concentrations in the human brain by ^{31}P MRS. NMR Biomed 1994;7: 225–230.
56. Arnold DL, Matthews PM, Radda GK. Metabolic recovery after exercise and the assessment of mitochondrial function in vivo in human skeletal muscle by means of P-31 NMR. Magn Reson Med 1984;1:307–315.
57. Nioka S, Chance B, Lockard SB, Dobson GP. Quantitation of high energy phosphate compounds and metabolic significance in the developing dog brain. Neurol Res 1991;13:33–38.
58. Eleff SM, Barker PB, Blackband SJ, et al. Phosphorus magnetic resonance spectroscopy of patients with mitochondrial cytopathies demonstrates decreased levels of brain phosphocreatine. Ann Neurol 1990;27:626–630.
59. Sauter A, Rudin M. Determination of creatine kinase kinetic parameters in rat brain by NMR magnetization transfer. Correlation with brain function. J Biol Chem 1993;268:13166–13171.
60. Chen W, Zhu XH, Adriany G, Ugurbil K. Increase of creatine kinase activity in the visual cortex of human brain during visual stimulation: a ^{31}P NMR magnetization transfer study. Magn Reson Med 1999;38:551–557.
61. Cadoux-Hudson TA, Blackledge MJ, Radda GK. Imaging of human brain creatine kinase activity in vivo [published erratum appears in FASEB J 1990 Mar;4(5):1525]. FASEB J 1989;3:2660–2666.
62. Lei H, Zhu XH, Zhang XL, Ugurbil K, Chen W. In vivo ^{31}P magnetic resonance spectroscopy of human brain at 7 T: an initial experience. Magn Reson Med 2003;49:199–205.
63. Brindle KM, Blackledge MJ, Challis RAJ, Radda GK. ^{31}P NMR magnetization-transfer measurements of ATP turnover during steady-state isometric muscle contraction in the rat hind limb in vivo. Biochemistry 1989;28:4887–4893.
64. Tkác I, Starcuk Z, Choi IY, Gruetter R. In vivo ^{1}H NMR spectroscopy of rat brain at 1 ms echo time. Magn Reson Med 1999;41:649–656.
65. Gruetter R, Novotny EJ, Boulware SD, Rothman DL, Shulman RG. ^{1}H NMR studies of glucose transport in the human brain. J Cereb Blood Flow Metab 1996;16:427–438.

66. Frahm J, Krüger G, Merboldt KD, Kleinschmidt A. Dynamic uncoupling and recoupling of perfusion and oxidative metabolism during focal brain activation in man. Magn Reson Med 1996;35:143–148.
67. Brooks J, Roberts N, Kemp G, Martin P, Whitehouse G. MRI based compartmentation and its application to measuring metabolite concentrations in the frontal lobe. Magn Reson Med 1999;41:883–888.
68. Terpstra M, Henry PG, Gruetter R. Measurement of reduced glutathione (GSH) in human brain using LCModel analysis of difference-edited spectra. Magn Reson Med 2003;50:19–23.
69. Rothman DL. Studies of metabolic compartmentation and glucose transport using in vivo MRS. NMR Biomed 2001;14:149–160.
70. Mason GF, Behar KL, Lai JC. The ^{13}C isotope and nuclear magnetic resonance: unique tools for the study of brain metabolism. Metab Brain Dis 1996;11:283–313.
71. Taylor A, McLean M, Morris P, Bachelard H. Approaches to studies on neuronal/glial relationships by ^{13}C-MRS analysis. Dev Neurosci 1996;18:434–442.
72. Lapidot A, Gopher A. Cerebral metabolic compartmentation. Estimation of glucose flux via pyruvate carboxylase/pyruvate dehydrogenase by ^{13}C NMR isotopomer analysis of D-[U-^{13}C]glucose metabolites. J Biol Chem 1994;269:27198–27208.
73. Sonnewald U, Hertz L, Schousboe A. Mitochondrial heterogeneity in the brain at the cellular level. J Cereb Blood Flow Metab 1998;18:231–237.
74. Mason GF, Behar KL, Rothman DL, Shulman RG. NMR determination of intracerebral glucose concentration and transport kinetics in rat brain. J Cereb Blood Flow Metab 1992;12:448–455.
75. Gruetter R, Ugurbil K, Seaquist ER. Steady-state cerebral glucose concentrations and transport in the human brain. J Neurochem 1998;70:397–408.
76. Van Zijl PC, Davis D, Eleff SM, Moonen CT, Parker RJ, Strong JM. Determination of cerebral glucose transport and metabolic kinetics by dynamic MR spectroscopy. Am J Physiol 1997;273:E1216–E1227.
77. Hyder F, Rothman DL, Mason GF, Rangarajan A, Behar KL, Shulman RG. Oxidative glucose metabolism in rat brain during single forepaw stimulation: a spatially localized $^{1}H[^{13}C]$ nuclear magnetic resonance study. J Cereb Blood Flow Metab 1997;17:1040–1047.
78. Mason GF, Rothman DL, Behar KL, Shulman RG. NMR determination of the TCA cycle rate and alpha-ketoglutarate/glutamate exchange rate in rat brain. J Cereb Blood Flow Metab 1992;12:434–447.
79. Mason GF, Gruetter R, Rothman DL, Behar KL, Shulman RG, Novotny EJ. Simultaneous determination of the rates of the TCA cycle, glucose utilization, alpha-ketoglutarate/glutamate exchange, and glutamine synthesis in human brain by NMR. 1995;15:12–25.
80. Hassel B, Sonnewald U, Fonnum F. Glial-neuronal interactions as studied by cerebral metabolism of [2–^{13}C]acetate and [1–^{13}C]glucose: an ex vivo ^{13}C NMR spectroscopic study. J Neurochem 1995;64:2773–2782.
81. Cohen DM, Wei J, O'Brian Smith E, Gao X, Quast MJ, Sokoloff L. A method for measuring cerebral glucose metabolism in vivo by 13C-NMR spectroscopy. Magn Reson Med 2002;48:1063–1067.
82. Oz G, Henry PG, Seaquist ER, Gruetter R. Direct, noninvasive measurement of brain glycogen metabolism in humans. Neurochem Int 2003;43:323–329.
83. Henry PG, Tkac I, Gruetter R. ^{1}H-localized broadband ^{13}C NMR spectroscopy of the rat brain in vivo at 9.4 T. Magn Reson Med 2003;50:684–692.
84. Kanamori K, Ross BD. In vivo activity of glutaminase in the brain of hyperammonaemic rats measured by ^{15}N nuclear magnetic resonance. Biochem J 1995;305:329–336.
85. Kanamori K, Ross BD, Kondrat RW. Rate of glutamate synthesis from leucine in rat brain measured in vivo by ^{15}N NMR. J Neurochem 1998;70:1304–1315.
86. Hassel B, Johannessen CU, Sonnewald U, Fonnum F. Quantification of the GABA shunt and the importance of the GABA shunt versus the 2-oxoglutarate dehydrogenase pathway in GABAergic neurons. J Neurochem 1998;71:1511–1518.
87. Kim S-G, Rostrup E, Larsson HBW, Ogawa S, Paulson OB. Determination of relative CMR_{O2} from CBF and BOLD changes: significant increase of oxygen consumption rate during visual stimulation. Magn Reson Med 1999;41:1152–1161.
88. Chen W, Zhu XH, Kato T, Andersen P, Ugurbil K. Spatial and temporal differentiation of fMRI BOLD response in primary visual cortex of human brain during sustained visual simulation. Magn Reson Med 1998;39:520–52.
89. Davis TL, Kwong KK, Weisskoff RM, Rosen BR. Calibrated functional MRI: mapping the dynamics of oxidative metabolism. Proc Natl Acad Sci USA 1998;95:1834–1839.
90. Schwarzbauer C, Heinke W. Investigating the dependence of BOLD contrast on oxidative metabolism. Magn Reson Med 1999;41:537–543.
91. Kida I, Kennan RP, Rothman DL, Behar KL, Hyder F. High-resolution CMR_{O2} mapping in rat cortex: a multiparametric approach to calibration of BOLD image contrast at 7 Tesla. J Cereb Blood Flow Metab 2000;20:847–860.
92. Mandeville JB, Jenkins BG, Kosofsky BE, Moskowitz MA, Rosen BR, Marota JJ. Regional sensitivity and coupling of BOLD and CBV changes during stimulation of rat brain. Magn Reson Med 2001;45:443–447.
93. Payen JF, Väth A, Koenigsberg B, Bourlier V, Decorps M. Regional cerebral plasma volume response to carbon dioxide using magnetic resonance imaging. Anesthesiology 1998;88:984–992.
94. Fox PT, Raichle ME, Mintun MA, Dence C. Nonoxidative glucose consumption during focal physiologic neural activity. Science 1988;241:452–464.
95. Roland PE, Eriksson L, Stone-Elander S, Widen L. Does mental activity change the oxidative metabolism of the brain? J Neurosci 1987;7:2373–2389.
96. Kida I, Yamamoto T, Tamura M. Interpretation of BOLD MRI signals in rat brain using simultaneously measured near-infrared spectrophotometric information. NMR Biomed 1996;9:333–338.
97. Kemp GJ. Studying metabolic regulation in human muscle. Biochem Soc Trans 2000;28:100–103.
98. Korzeniewski B. Regulation of ATP supply during muscle contraction: theoretical studies. Biochem J 1998;330:1189–1195.
99. Conley KE. Blei ML, Richards TL, Kushmerick MJ, Jubrias SA. Activation of glycolysis in human muscle in vivo. Am J Physiol 1997;273:C306–C315.
100. Conley KE, Kushmerick MJ, Jubrias SA. Glycolysis is independent of oxygenation state in stimulated human skeletal muscle in vivo. J Physiol 1998;511:935–945.
101. Kemp GJ. Interactions of mitochondrial ATP synthesis and the creatine kinase equilibrium in skeletal muscle. J Theor Biol 1994;170:239–246.
102. Kemp GJ, Taylor DJ, Radda GK. Control of phosphocreatine resynthesis during recovery from exercise in human skeletal muscle. NMR Biomed 1993;6:66–72.
103. Kemp GJ, Roberts N, Bimson WE, Bakran A, Frostick SP. Muscle oxygenation and ATP turnover when blood flow is impaired by vascular disease. Spectroscopy. Int J 2002;16:317–334.
104. Paganini AT, Foley JM, Meyer RA. Linear dependence of muscle phosphocreatine kinetics on oxidative capacity. Am J Physiol 1997;272:C501–C510.
105. Taylor DJ, Kemp GJ, Radda GK. Bioenergetics of skeletal muscle in mitochondrial myopathy. J Neurol Sci 1994;127:198–206.
106. Richardson RS, Leigh JS, Wagner PD, Noyszewski EA. Cellular PO_2 as a determinant of maximal mitochondrial O_2 consumption in trained human skeletal muscle. J Appl Physiol 1999;87:325–331.
107. Jeneson JAL, Wiseman RW, Westerhoff HV, Kushmerick MJ. The signal transduction function for oxidative phosphorylation is at least second order in ADP. J Biol Chem 1996;271:27995–27998.
108. Harkema SJ, Meyer RA. Effect of acidosis on control of respiration in skeletal muscle. Am J Physiol 1997;272:C491–C500.
109. Meyer RA. A linear model of muscle respiration explains monoexponential phosphocreatine changes. Am J Physiol 1988;254:C548–C553.
110. Lambeth MJ, Kushmerick MJ. A computational model for glycogenolysis in skeletal muscle. Ann Biomed Eng 2002;30:808–827.

111. Hoge RD, Atkinson J, Gill B, Crelier GR, Marrett S, Pike GB. Linear coupling between cerebral blood flow and oxygen consumption in activated human cortex. Proc Natl Acad Sci USA 1999;96:9403–9408.
112. Hyder F, Shulman RG, Rothman DL. A model for the regulation of cerebral oxygen delivery. J Appl Physiol 1998;85:554–564.
113. Malonek D, Grinvald A. Interactions between electrical activity and cortical microcirculation revealed by imaging spectroscopy: implications for functional brain mapping. Science 1996;272:551–554.
114. Zonta M, Angulo MC, Gobbo S, et al. Neuron-to-astrocyte signaling is central to the dynamic control of brain microcirculation. Nat Neurosci 2003;6:43–50.
115. Madsen PL, Linde R, Hasselbalch SG, Paulson OB, Lassen NA. Activation-induced resetting of cerebral oxygen and glucose uptake in the rat. J Cereb Blood Flow Metab 1998;18:742–748.
116. Ueki M, Linn F, Hossmann KA. Functional activation of cerebral blood flow and metabolism before and after global ischemia of rat brain. J Cereb Blood Flow Metab 1988;8:486–494.
117. Seitz RJ, Roland PE. Vibratory stimulation increases and decreases the regional cerebral blood flow and oxidative metabolism: a positron emission tomography (PET) study. Acta Neurol Scand 1992;86:60–67.
118. Katayama Y, Tsubokawa T, Hirayama T, Kido G, Tsukiyama T, Iio M. Response of regional cerebral blood flow and oxygen metabolism to thalamic stimulation in humans as revealed by positron emission tomography. J Cereb Blood Flow Metab 1986;6:637–641.
119. Chen W, Novotny EJ, Zhu XH, Rothman DL, Shulman RG. Localized ^{1}H NMR measurement of glucose consumption in the human brain during visual stimulation. Proc Natl Acad Sci USA 1993;90:9896–9900.
120. Krüger G, Kleinschmidt A, Frahm J. Dynamic MRI sensitized to cerebral blood oxygenation and flow during sustained activation of human visual cortex. Magn Reson Med 1996;35:797–800.
121. Buxton RB. The elusive initial dip. Neuroimage 2001;13:953–958.
122. Hoge RD, Atkinson J, Gill B, Crelier GR, Marrett S, Pike GB. Investigation of BOLD signal dependence on cerebral blood flow and oxygen consumption: the deoxyhemoglobin dilution model. Magn Reson Med 1999;42:849–863.
123. Kim SG, Ugurbil K. Comparison of blood oxygenation and cerebral blood flow effects in fMRI: estimation of relative oxygen consumption change. Magn Reson Med 1997;38:59–65.
124. Hoge RD, Pike GB. Oxidative metabolism and the detection of neuronal activation via imaging. J Chem Neuroanat 2001;22:43–52.
125. Buxton RB, Frank LR. A model for the coupling between cerebral blood flow and oxygen metabolism during neural stimulation. J Cereb Blood Flow Metab 1997;17:64–72.
126. Hyder F, Kida I, Behar KL, Kennan RP, Maciejewski PK, Rothman DL. Quantitative functional imaging of the brain: towards mapping neuronal activity by BOLD fMRI. NMR Biomed 2001;14:413–431.
127. Mayhew J, Johnston D, Martindale J, Jones M, Berwick J, Zheng Y. Increased oxygen consumption following activation of brain: theoretical footnotes using spectroscopic data from barrel cortex. Neuroimage 2001;13:975–987.
128. Hudetz AG. Mathematical model of oxygen transport in the cerebral cortex. Brain Res 1999;817:75–83.
129. Mintun MA, Lundstrom BN, Snyder AZ, Vlassenko AG, Shulman GL, Raichle ME. Blood flow and oxygen delivery to human brain during functional activity: theoretical modeling and experimental data. Proc Natl Acad Sci USA 2001;98:6859–6864.
130. Gjedde A. Cerebral blood flow change in arterial hypoxemia is consistent with negligible oxygen tension in brain mitochondria. Neuroimage 2002;17:1876–1881.
131. Shulman RG, Hyder F, Rothman DL. Lactate efflux and the neuroenergetic basis of brain function. NMR Biomed 2001;14: 389–396.
132. Shulman RG, Hyder F, Rothman DL. Cerebral energetics and the glycogen shunt: neurochemical basis of functional imaging. Proc Natl Acad Sci USA 2001;98:6417–6422.
133. Lukkarinen J, Oja JM, Turunen M, Kauppinen RA. Quantitative determination of glutamate turnover by ^{1}H-observed, ^{13}C-edited nuclear magnetic resonance spectroscopy in the cerebral cortex ex vivo: interrelationships with oxygen consumption. Neurochem Int 1997;31:95–104.
134. Chen W, Zhu XH, Gruetter R, Seaquist ER, Adriany G, Ugurbil K. Study of tricarboxylic acid cycle flux changes in human visual cortex during hemifield visual stimulation using ^{1}H-{^{13}C} MRS and fMRI. Magn Reson Med 2001;45:349–355.
135. Chhina N, Kuestermann E, Halliday J, et al. Measurement of human tricarboxylic acid cycle rates during visual activation by ^{13}C magnetic resonance spectroscopy. J Neurosci Res 2001;66:737–746.
136. Dienel GA, Wang RY, Cruz NF. Generalized sensory stimulation of conscious rats increases labeling of oxidative pathways of glucose metabolism when the brain glucose-oxygen uptake ratio rises. J Cereb Blood Flow Metab 2002;22:1490–1502.
137. van Zijl PC, Eleff SM, Ulatowski JA, et al. Quantitative assessment of blood flow, blood volume and blood oxygenation effects in functional magnetic resonance imaging. Nat Med 1998;4:159–167.
138. Mandeville JB, Marota JJ, Ayata C, Moskowitz MA, Weisskoff RM, Rosen BR. MRI measurement of the temporal evolution of relative $CMRO_2$ during rat forepaw stimulation. Magn Reson Med 1999;42:944–951.
139. Hyder F, Renken R, Kennan RP, Rothman DL. Quantitative multimodal functional MRI with blood oxygenation level dependent exponential decays adjusted for flow attenuated inversion recovery (BOLDED AFFAIR). Magn Reson Imaging 2000;18:227–235.
140. An H, Lin W, Celik A, Lee YZ. Quantitative measurements of cerebral metabolic rate of oxygen utilization using MRI: a volunteer study. NMR Biomed 2001;14:441–447.
141. Hayashi T, Watabe H, Kudomi N, et al. A theoretical model of oxygen delivery and metabolism for physiologic interpretation of quantitative cerebral blood flow and metabolic rate of oxygen. J Cereb Blood Flow Metab 2003;23:1314–1323.
142. Raichle ME. Cognitive neuroscience. Bold insights. Nature 2001;412:128–130.
143. Gsell W, De Sadeleer C, Marchalant Y, MacKenzie ET, Schumann P, Dauphin F. The use of cerebral blood flow as an index of neuronal activity in functional neuroimaging: experimental and pathophysiological considerations. J Chem Neuroanat 2000;20:215–224.
144. Thompson JK, Peterson MR, Freeman RD. Single-neuron activity and tissue oxygenation in the cerebral cortex. Science 2003;299:1070–1072.
145. Vanzetta I, Grinvald A. Increased cortical oxidative metabolism due to sensory stimulation: implications for functional brain imaging. Science 1999;286:1555–1558.
146. Devor A, Dunn AK, Andermann ML, Ulbert I, Boas DA, Dale AM. Coupling of total hemoglobin concentration, oxygenation, and neural activity in rat somatosensory cortex. Neuron 2003;39:353–359.
147. Lindauer U, Royl G, Leithner C, et al. No evidence for early decrease in blood oxygenation in rat whisker cortex in response to functional activation. Neuroimage 2001;13:988–1001.
148. Arthurs OJ, Boniface SJ. What aspect of the fMRI BOLD signal best reflects the underlying electrophysiology in human somatosensory cortex? Clin Neurophysiol 2003;114:1203–1209.
149. Logothetis NK, Pauls J, Augath M, Trinath T, Oeltermann A. Neurophysiological investigation of the basis of the fMRI signal. Nature 2001;412:150–157.
150. Logothetis NK. The neural basis of the blood-oxygen-level-dependent functional magnetic resonance imaging signal. Philos Trans R Soc Lond B Biol Sci 2002;357:1003–1037.
151. Smith AJ, Blumenfeld H, Behar KL, Rothman DL, Shulman RG, Hyder F. Cerebral energetics and spiking frequency: the neurophysiological basis of fMRI. Proc Natl Acad Sci USA 2002;99:10765–10770.
152. Kemp GJ, Roberts N, Bimson WE, et al. Mitochondrial function and oxygen supply in normal and in chronically ischaemic muscle: a combined ^{31}P magnetic resonance spectroscopy and near infra-red spectroscopy study in vivo. J Vasc Surg 2001;34:1103–1110.
153. Kemp GJ, Radda GK. Quantitative interpretation of bioenergetic data from ^{31}P and ^{1}H magnetic resonance spectroscopic studies of skeletal muscle: an analytical review. Magn Reson Quart 1994;10:43–63.

154. Kushmerick MJ, Conley KE. Energetics of muscle contraction: the whole is less than the sum of its parts. Biochem Soc Trans 2002;30:227–231.
155. Quistorff B, Johansen L, Sahlin K. Absence of phosphocreatine resynthesis in human calf muscle during ischaemic recovery. Biochem J 1993;291:681–686.
156. Foley JM, Meyer RA. Energy cost of twitch and tetanic contractions of rat muscle estimated in situ by gated ^{31}P NMR. NMR Biomed 1993;6:32–38.
157. Kato T, Murashita J, Shioiri T, Hamakawa H, Inubushi T. Effect of photic stimulation on energy metabolism in the human brain measured by ^{31}P-MR spectroscopy. J Neuropsych Clin Neurosci 1996;8:417–422.
158. Sappey Marinier D, Calabrese G, Fein G, Hugg JW, Biggins C, Weiner MW. Effect of photic stimulation on human visual cortex lactate and phosphates using ^{1}H and ^{31}P magnetic resonance spectroscopy. J Cereb Blood Flow Metab 1992;12:584–592.
159. Rango M, Castelli A, Scarlato G. Energetics of 3.5 s neural activation in humans: a ^{31}P MR spectroscopy study. Magn Reson Med 1997;38:878–883.
160. Mora B, Narasimhan PT, Ross BD, Allman J, Barker PB. ^{31}P saturation transfer and phosphocreatine imaging in the monkey brain. Proc Natl Acad Sci USA 1991;88:8372–8376.
161. Bittl JA, DeLayre J, Ingwall JS. Rate equation for creatine kinase predicts the in vivo reaction velocity: ^{31}P NMR surface coil studies in brain, heart, and skeletal muscle of the living rat. Biochemistry 1987;26:6083–6090.
162. Murashita J, Kato T, Shioiri T, Inubushi T, Kato N. Age-dependent alteration of metabolic response to photic stimulation in the human brain measured by ^{31}P MR-spectroscopy. Brain Res 1999;818:72–76.
163. Rango M, Bozzali M, Prelle A, Scarlato G, Bresolin N. Brain activation in normal subjects and in patients affected by mitochondrial disease without clinical central nervous system involvement: a phosphorus magnetic resonance spectroscopy study. J Cereb Blood Flow Metab 2001;21:85–91.
164. Prichard J, Rothman D, Novotny E, et al. Lactate rise detected by ^{1}H NMR in human visual cortex during physiologic stimulation. Proc Natl Acad Sci USA 1991;88:5829–5831.
165. Kauppinen RA, Eleff SM, Ulatowski JA, Kraut M, Soher B, van Zijl PC. Visual activation in alpha-chloralose-anaesthetized cats does not cause lactate accumulation in the visual cortex as detected by [^{1}H]NMR difference spectroscopy. Eur J Neurosci 1997;9:654–661.
166. Frahm J, Krueger G, Merboldt KD, Kleinschmidt A. Dynamic NMR studies of perfusion and oxidative metabolism during focal brain activation. Adv Exp Med Biol 1997;413:195–203.
167. Arnold DL, Taylor DJ, Radda GK. Investigations of human mitochondrial myopathies by phosphorus magnetic resonance spectroscopy. Annals of Neurology 1985;18:189–196.
168. Argov Z, Bank WJ, Maris J, Peterson P, Chance B. Bioenergetic heterogeneity of human mitochondrial myopathies: phosphorus magnetic resonance study. Neurology 1987;37:257–262.
169. Barbiroli B, Montagna P, Martinelli P, et al. Defective brain energy metabolism shown by in vivo ^{31}P MR spectroscopy in 28 patients with mitochondrial cytopathies. J Cereb Blood Flow Metab 1993;13:469–474.
170. Matthews PM, Arnold DL. Phosphorus magnetic resonance spectroscopy of brain in mitochondrial cytopathies. Ann Neurol 1990;28:839–840.
171. Matthews PM, Berkovic SF, Shoubridge EA, et al. In vivo magnetic resonance spectroscopy of brain and muscle in a type of mitochondrial encephalomyopathy (MERRF). Ann Neurol 1991;29:435–438.
172. Duncan DB, Herholz K, Kugel H, et al. Positron emission tomography and magnetic resonance spectroscopy of cerebral glycolysis in children with congenital lactic acidosis. Ann Neurol 1995;37:351–358.
173. Pavlakis SG, Kingsley PB, Kaplan GP, Stacpoole PW, O'Shea M, Lustbader D. Magnetic resonance spectroscopy: use in monitoring MELAS treatment. Arch Neurol 1998;55:849–852.
174. Molnar MJ, Valikovics A, Molnar S, et al. Cerebral blood flow and glucose metabolism in mitochondrial disorders. Neurology 2000;55:544–548.
175. Sano M, Ishii K, Momose Y, Uchigata M, Senda M. Cerebral metabolism of oxygen and glucose in a patient with MELAS syndrome. Acta Neurol Scand 1995;92:497–502.
176. Kato T, Murashita J, Shioiri T, Terada M, Inubushi T, Kato N. Photic stimulation-induced alteration of brain energy metabolism measured by ^{31}P-MR spectroscopy in patients with MELAS. J Neurol Sci 1998;155:182–185.

Index

Acetylcholinesterase (AChE), senile plaque association, 64
AD, *see* Alzheimer's disease
Adrenoleukodystrophy, *see* Leukodystrophy
Aicardi Goutieres, 285, 288
Alexander's disease,
  biochemical features, 277
  clinical features, 234, 277
  genetics, 277, 278
  magnetic resonance imaging findings, 235, 278
  magnetic resonance spectroscopy findings, 251, 278
  pathologic findings, 235
  pathology, 278
Altropane tracer, Parkinson's disease imaging, 38, 39
Alzheimer's disease (AD),
  β–amyloid,
    aggregation,
      fibrillization process, 62–64
      fluorescence assays, 65, 66
      inhibitors, 68, 69
      peptide fragment studies, 69
      protofibrils, 69, 70
      regulation, 64, 65
    neurotoxicity, 62, 66–70, 85
    senile plaque composition, 61
  dementia,
    clinical features, 75, 76, 85
    imaging of inflammation,
      cobalt tracers as calcium probes, 78
      computed tomography, 77
      magnetic resonance imaging, 77, 78
      microglia, 79, 80
      positron emission tomography, 78–80
      radiopharmaceutical imaging of receptors, 79
      serotonergic system, 80
      single-photon emission tomography, 78–80
    inflammation pathophysiology, 76, 77
  epidemiology, 95
  history of study, 61
  neurodegeneration regional specificity, 86
  oxidative stress,
    mechanisms, 85, 86
    neurotoxicity, 85–87
    serotonin oxidation, *see* Tryptamine-4,5-dione
  positron emission tomography functional staging,
    activation studies in diagnosed patients, 109–111
    high-risk patient imaging, 111, 112
    overview, 107
    resting state brain metabolism correlations with cognition and behavior, 107–109
  postmortem studies,
    oxidative phosphorylation downregulation, 112, 113, 116
    synapse loss, 112
  progression prediction with magnetic resonance imaging/spectroscopy,
    cross-sectional studies,
      diagnosed patients, 95, 96
      high-risk patients, 96–99
      pathologic correlation studies, 99
    longitudinal studies,
      pathologic correlation studies, 100–102
      prediction in normal and high-risk groups, 100
    treatment, 102, 103
  serotonergic system defects, 86
β–Amyloid, *see* Alzheimer's disease
α1-Antichymotrypsin, β–amyloid aggregation regulation, 65
Atomic force microscopy, β–amyloid aggregation, 62, 63
ATP, 298, 299
Brain boundary shift integral (BBSI), Alzheimer's disease progression, 101, 102
BRODERICK PROBE®,
  detectable substances, 200
  nitrous oxide imaging in epilepsy patients, xi–xiii
  size, 200
  L-tryptophan bioimaging in temporal lobe epilepsy subtyping, 142–146
  white matter electrochemical signal detection in temporal lobe epilepsy,
    findings, 201, 202
    magnetic resonance imaging comparison, 204
    overview, 200, 201
    presurgical evaluation and classification of patients, 201
    recording, 200
    sensor placement, 201
CADASIL, magnetic resonance spectroscopy findings, 255
Canavan's disease,
  biochemical features, 273
  clinical features, 233, 234, 273
  genetics, 273
  magnetic resonance imaging findings, 234, 273
  magnetic resonance spectroscopy findings, 251, 273
  pathology, 234, 273
  treatment, 273

β–CBT tracer, Parkinson's disease imaging, 42
Central nervous system hypomyelination, *see* Vanishing white matter disease
Cerebral metabolic rate,
invasive measurements, 301
measurement in activation studies, 300
mitochondrial disease effects, 302
regulation, 300
Cerebrotendinous xanthomatosis (CTX), magnetic resonance spectroscopy findings, 255
β–CFT tracer, Parkinson's disease imaging, 42, 46, 50
β–CIT tracers, Parkinson's disease imaging, 16, 18, 41, 42, 46
Cockayne syndrome, 276, 277
Computed tomography (CT),
Aicardi Goutieres, 285, 288
Alzheimer's dementia imaging, 77
Krabbe's disease findings, 265, 266
ornithine transcarbamylase deficiency findings, 276
vanishing white matter disease, 280
Corpus callosotomy, *see* Epilepsy
Cortical dysplasia,
classification, 131, 132
positron emission tomography,
flumazenil positron emission tomography,
childhood epilepsy syndromes, 135
dual pathology, 135
electroencephalography correlations, 135–137
epileptic foci detection, 135
fluorodeoxyglucose positron emission tomography,
electroencephalography importance, 132
Lennox–Gastaut syndrome, 134
magnetic resonance imaging comparison, 134, 135
microscopic cortical dysplasia, 132, 133
West syndrome, 133, 134
miscellaneous tracers, 138
rationale, 131, 132
serotonergic system, 137
Cortical spreading depression (CSD), intrinsic optical signal imaging, 150, 151
CT, *see* Computed tomography
CTX, *see* Cerebrotendinous xanthomatosis
β–Cyclodextrin, β–amyloid inhibition, 68
Dopaminergic system,
dopamine synthesis, 25
imaging, *see* Parkinson's disease
pathways, 25, 26
receptors, 26
tracers for positron emission tomography and single-photon emission tomography, 14–16
Electroencephalography (EEG),
flumazenil positron emission tomography correlation in cortical dysplasia,
importance with fluorodeoxyglucose positron emission tomography, 132
α-methyl-L-tryptophan positron emission tomography correlations in epilepsy, 126, 127
periodic lateralized epileptiform discharges, 193–196
Epilepsy,
inhibitory processes in prevention, 151
interictal spikes, 151
intrinsic optical signal imaging, *see* Intrinsic optical signal imaging
nitrous oxide cerebral imaging with BRODERICK PROBE®, xi–xiii
palliative surgery, 180
positron emission tomography, *see* Cortical dysplasia; α–Methyl-L-tryptophan positron emission tomography
serotonergic system, 123, 124, 141, 142
single-photon emission tomography imaging of cerebral blood flow in periodic lateralized epileptiform discharges, 193–196
surgery,
indications, 179, 180
intraoperative magnetic resonance imaging,
corpus callosotomy, 186, 188
cost-benefit analysis, 190
depth electrode placement guidance, 188, 189
electrocorticography considerations, 189
functional magnetic resonance imaging, 189, 190
hemispherectomy, 188
historical perspective, 180, 181
hypothalamic hamartoma, 188
neocortical epilepsy, 186
rationale, 177, 178
technology, 178, 179
temporal lobe epilepsy, 181, 186
temporal lobe epilepsy subtyping with L-tryptophan bicimaging,
BRODERICK PROBE® measurements, 142, 143
mesial versus neocortical temporal lobe epilepsy, 142, 146, 147
overview, 141, 142
serotonin level correlations, 143–146
statistical analysis, 143
tissue preparation, 142
validation, 143
white matter electrochemical signal detection in temporal lobe epilepsy,
BRODERICK PROBE®,
findings, 201, 202
magnetic resonance imaging comparison, 204
overview, 200, 201
presurgical evaluation and classification of patients, 201
recording, 200
sensor placement, 201
neuromolecular imaging, 200, 204
faradaic electrochemistry, 204
gray matter comparison, 204
overview, 199, 200
Fabry's disease, 229, 253
Fetal tissue transplantation, *see* Parkinson's disease
Flumazenil positron emission tomography, *see* Cortical dysplasia
Fluorescence resonance energy transfer (FRET), β–amyloid aggregation assay, 65, 66
Fluorodopa tracer, Parkinson's disease imaging, 14, 28–31, 39, 46, 49, 50
fMRI, *see* Functional magnetic resonance imaging
FRET, *see* Fluorescence resonance energy transfer
Functional magnetic resonance imaging (fMRI),
activation studies, 300, 301
BOLD signal, 297
cerebral blood flow regulation studies, 300, 302
intraoperative magnetic resonance imaging in epilepsy, 189, 190
Parkinson's disease, 9

Gaucher's disease, 230
Glutaric acidemia type I,
clinical features, 274
genetics, 274
magnetic resonance imaging, 274
pathology, 274
treatment, 274
Glutaric acidemia type II, features and magnetic resonance spectroscopy findings, 247, 248
Hemispherectomy, *see* Epilepsy
HMG-CoA lyase deficiency, features and magnetic resonance spectroscopy findings, 247
5-Hydroxytryptophan, cerebrospinal fluid levels of oxidation products in Alzheimer's disease, 89, 90
IBF tracer, Parkinson's disease imaging, 14, 16
IBZM tracer, Parkinson's disease imaging, 14, 16
Intrinsic optical signal (IOS) imaging,
advantages, 149
cortical spreading depression, 150, 151
epilepsy,
animal acute epilepsy models,
electrophysiology measurements, 153
ferret single wavelength imaging, 154, 155, 157, 158
ictal onset versus offset, 161, 162
mirror focus, 156
optical imaging of interictal spikes, 153, 154
rat multiwavelength imaging, 155, 156, 158–160
rodent models, 152, 153
surgery, 153
variability in signals, 160, 161
animal chronic epilepsy models, 162–165
intraoperative use,
epileptiform events, 169
historical perspective, 165, 166
multiple wavelength imaging of cortical stimulation, 167, 169
somatosensory architecture, 167
technical challenges, 166
technique, 167
neocortical epilepsy, 151
prospects for study,
analysis, 170, 171
inverted signal, 170
wavelength findings, 170
relationship between epileptiform events and functional architecture, 162
treatment guidance, 151, 152
functional architecture imaging, 150
origin of signals, 149, 150
IOS imaging, *see* Intrinsic optical signal imaging
Iron, magnetic resonance imaging for estimation in brain, 6, 7
KD, *see* Krabbe's disease
Krabbe's disease (KD),
age of onset, 215
biochemical features, 265
clinical features, 215, 216, 229, 264, 265, 266
gene defects, 215, 216
genetics, 265
magnetic resonance imaging,
differential diagnosis, 219–221
findings, 216–219, 229, 230, 266
magnetic resonance spectroscopy findings, 221, 222, 249, 266
pathologic findings, 229, 266
treatment, 266
Laminin, β–amyloid aggregation regulation, 65
Lennox–Gastaut syndrome, fluorodeoxyglucose positron emission tomography, 134
Leukodystrophy, *see also* Krabbe's disease,
adrenoleukodystrophy/adrenomyeloneuropathy,
biochemical features, 268
clinical features, 231, 232, 268
genetics, 268
magnetic resonance imaging findings, 232, 233, 268–269
magnetic resonance spectroscopy findings, 250, 251, 268–269
pathology, 232, 269
treatment, 269
Alexander's disease, *see* Alexander's disease
Canavan's disease, *see* Canavan's disease
clinical presentation, 211
genetic diseases and clinical features, 211, 212
laboratory tests, 211
lysosomal disorders, 225, 226
magnetic resonance imaging,
globoid cell leukodystrophy, *see* Krabbe's disease
overview, 209, 210, 225
patterns, 210
treatment monitoring, 210, 211
magnetic resonance spectroscopy, *see also* Magnetic resonance spectroscopy,
localization techniques, 243
overview, 240, 241
prospects, 255, 256
proton signal neurochemistry, 241–243
technical factors, 243, 244
mucopolysaccharidoses,
clinical features, 230
magnetic resonance imaging findings, 230
magnetic resonance spectroscopy findings, 253, 255
sphingolipidoses,
Fabry's disease, 229, 253
Gaucher's disease, 230
GM1/GM2 gangliosidoses, 226, 227, 228, 266, 267
metachromatic leukodystrophy, *see* Metachromatic leukodystrophy
sudanophilic leukodystrophies, 235, 236, 237
Magnetic resonance imaging (MRI), *see also* Functional magnetic resonance imaging,
Aicardi Goutieres, 285, 285, 288
Alzheimer's dementia imaging, 77, 78
Alzheimer's disease, *see* Alzheimer's disease
BRODERICK PROBE® neuromolecular imaging, 204
fluorodeoxyglucose positron emission tomography comparison in cortical dysplasia, 134, 135
intraoperative imaging in epilepsy surgery, *see* Epilepsy
Parkinson's disease, *see* Parkinson's disease
perfusion imaging, 239, 240
principles, 261, 262
relaxation times, 3, 209, 210
Smith-Lemli-Opitz syndrome, 283
vanishing white matter disease, 280
white matter,
diffusion tensor imaging, 262
epilepsy findings, 204

leukodystrophy findings,
adrenoleukodystrophy/adrenomyeloneuropathy, 232, 233, 268–269
Alexander's disease, 235, 278
Canavan's disease, 234
Fabry's disease, 229
Gaucher's disease, 230
globoid cell leukodystrophy, *see* Krabbe's disease
GM1/GM2 gangliosidoses, 227, 228
metachromatic leukodystrophy, 226, 264
mucopolysaccharidoses, 230
overview, 209, 210, 225
patterns, 210
perfusion imaging, 239, 240
sudanophilic leukodystrophies, 237
treatment monitoring, 210, 211
myelin composition, 209
neurogenetic applications, 262
Magnetic resonance spectroscopy (MRS),
activation studies,
mitochondrial disease, 302
phosphorous metabolites, pH, and lactate, 301, 302
tricarboxylic acid cycle flux during activation, 300, 301
adrenoleukodystrophy, 250, 251, 268–269
Alexander's disease, 251, 278
Alzheimer's disease, *see* Alzheimer's disease
brain energetics,
carbon spectroscopy, 299, 300
nitrogen spectroscopy, 300
phosphorous spectroscopy, 299, 300
proton spectroscopy, 299
CADASIL, 255
Canavan's disease, 251
cerebral metabolism studies, 300
cerebrotendinous xanthomatosis, 255
developing brain studies, 244
findings in specific diseases, 245–255
glutamate recycling studies, 298, 299
glutaric acidemia type II, 247, 248
HMG-CoA lyase deficiency, 247
Krabbe's disease, 221, 222, 249, 265, 266
localization techniques, 243
lysosomal storage diseases, 253
maple syrup urine disease, 246
metachromatic leukodystrophy, 249, 250
methylmalonic acidemia, 246
mitochondrial disorders, 248, 249
mucopolysaccharidosis, 253
multiple sclerosis, 255
neurotransmitter disorders, 263
nonketotic hyperglycinemia, 246, 247
oxidative metabolism disorders, 263
Parkinson's disease, *see* Parkinson's disease
Pelizaeus–Merzbacher disease, 244–246, 281
phenylketonuria, 246
phosphorous spectroscopy applications, 263, 299
principles, 240, 241, 262, 263
propionic acidemia, , 248
proton signal neurochemistry, 241, 242, 243
resolution enhancement, 255, 256
Sjögren–Larsson syndrome, 255, 282
Smith-Lemli-Opitz syndrome, 283
technical factors, 243, 244
vacuolating megaloencephalic leukoencephalopathy, 251, 253, 283, 288
vanishing white matter disease, 280
voxel size and processing, 244
Zellweger syndrome, 251
Malformations of cortical development, *see* Cortical dysplasia
Maple syrup urine disease (MSUD), 246, 275
MCI, *see* Mild cognitive impairment
Megaloencephalic leukoencephalopathy (MLC), 251, 253, 283, 283, 288
Merosin deficiency, 282, 283
Metachromatic leukodystrophy (MLD), 226, 249, 250, 263, 264
Methylenetetrahydrofolate reductase deficiency, magnetic resonance spectroscopy findings, 247
Methylmalonic acidemia (MMA), features and magnetic resonance spectroscopy findings, 246
α–Methyl-L-tryptophan positron emission tomography,
cortical dysplasia findings, 137
mathematical modeling, 124
neocortical epilepsy studies,, 124, 125, 126, 127
serotonergic system evaluation, 124, 137
tracer uptake via kynurenine pathway, 127
Mild cognitive impairment (MCI), progression prediction with magnetic resonance imaging/spectroscopy,
cross-sectional studies,
diagnosed patients, 95, 96
high-risk patients, 96–99
pathologic correlation studies, 99
longitudinal studies, 100–102
Mitochondrial disorders,
brain energetics, 297–299
cytopathies, 271
magnetic resonance imaging findings, 271, 272
pathology, 272, 273
functional activation studies, 302
magnetic resonance spectroscopy findings, 248, 249
prevalence, 261
MLC, *see* Megaloencephalic leukoencephalopathy
MLD, *see* Metachromatic leukodystrophy
MMA, *see* Methylmalonic acidemia
MPS, *see* Mucopolysaccharidoses
MRI, *see* Magnetic resonance imaging
MRS, *see* Magnetic resonance spectroscopy
MS, *see* Multiple sclerosis
MSA, *see* Multiple system atrophy
MSUD, *see* Maple syrup urine disease
Mucopolysaccharidoses (MPS), 230, 253, 255
Multiple sclerosis (MS), magnetic resonance spectroscopy findings, 255
Multiple system atrophy (MSA), 4–6, 16
Neuromolecular imaging, 199–201, *see also* BRODERICK PROBE®
Niemann-Pick disease, 253, 266, 267, 268
Nitrous oxide,
anesthesia, xi
cerebral imaging with BRODERICK PROBE® in epilepsy, xi–xiii
Nonketotic hyperglycinemia (NKH), features and magnetic resonance spectroscopy findings, 246, 247
Ornithine transcarbamylase (OTC) deficiency, 275, 276
Oxidative phosphorylation,
brain energetics, 299–300
downregulation in Alzheimer's disease, 112, 113, 116

Parkinson's disease (PD),
challenges in early disease imaging, 37, 38
diagnostic criteria, 13
fetal tissue transplantation,
neurotrophic factor treatment of grafts, 54
positron emission tomography imaging,
complication monitoring, 53, 54
dopamine receptor imaging, 52, 53
dopamine transporter imaging, 50, 51
fluorodopa imaging, 49, 50
functional brain activation imaging, 53
metabolic marker imaging, 53
multiple tracer imaging, 55
surgery, 54, 55
tissue sources, 54
functional magnetic resonance imaging, 9
magnetic resonance imaging,
brain iron quantification, 6, 7
differential diagnosis, 4–6
pulse sequences, 3–4
voxel-based morphometry, 6
magnetic resonance spectroscopy, 8, 9
positron emission tomography,
activation and regional cerebral blood flow patterns, 26–28, 47, 48
familial Parkinson's disease, 31
fluorodopa imaging, 14–16, 28–31, 39, 49, 50
levodopa complication monitoring, 31, 32
multimodality and multitracer studies, 16
placebo effect studies, 32
postsynaptic dopaminergic function, 46, 47
presynaptic dopaminergic function, 46
tracers, 13–16, 28, 42
single-photon emission tomography,
altropane tracer, 38, 39
positron emission tomography comparison in early disease, 39–41
multimodality and multitracer studies, 16
serotonergic system imaging, 18
tracers,
blood flow and metabolism, 13, 14
dopaminergic system, 14–16, 41, 42
treatment, 48, 49
PD, *see* Parkinson's disease
Pelizaeus–Merzbacher disease (PMD),
clinical features, 244–246, 280, 281
genetics, 281
magnetic resonance imaging findings, 281
magnetic resonance spectroscopy findings, 244–246, 281
pathology, 281
Periodic lateralized epileptiform discharges (PLEDs), single-photon emission tomography cerebral blood flow imaging, 193–196
Peroxisomal disorders, 269, 270, *see also* Zellweger syndrome
PET, *see* Positron emission tomography
Phenylketonuria (PKU), 246, 274, 275
PK11195, microglia imaging in Alzheimer's dementia, 79, 80
PLEDs, *see* Periodic lateralized epileptiform discharges
PMD, *see* Pelizaeus–Merzbacher disease
Positron emission tomography (PET),
Alzheimer's dementia imaging, 78–80
brain energetics studies,
activation studies, 300
parameters, 300
tricarboxylic acid cycle flux during activation, 300, 301
epilepsy, *see* Cortical dysplasia; α–Methyl-L-tryptophan positron emission tomography
mapping, 45
Parkinson's disease, *see* Parkinson's disease
resolution, 45
tracers,
dopaminergic system tracers, 14–16
glucose metabolism, 13, 14
Progressive supranuclear palsy (PSP), 4–6, 16
Propionic acidemia, magnetic resonance spectroscopy findings, 248
Rat, epilepsy models and intrinsic optical signal imaging, 152, 153, 155, 156, 158–160, 162–165
Serotonergic system,
Alzheimer's dementia imaging, 80
Alzheimer's disease effects, 86
epilepsy findings, 123, 124
serotonin oxidation, *see* Tryptamine-4,5-dione
single-photon emission tomography imaging in Parkinson's disease, 18
Single-photon emission tomography (SPECT),
Alzheimer's dementia imaging, 78–80
cerebral blood flow imaging in periodic lateralized epileptiform discharges, 193–196
Parkinson's disease, *see* Parkinson's disease
tracers,
dopaminergic system tracers, 14–16, 37–42
glucose metabolism, 13, 14
serotonergic system tracers, 18
Sjögren–Larsson syndrome (SLS), 255, 281, 282
Smith Lemli Opitz syndrome, 283
TRODAT-1 tracer, Parkinson's disease imaging, 42
Tryptamine-4,5-dione,
cerebrospinal fluid levels in Alzheimer's disease, 89, 90
neurotoxicity, 87–89
serotonin oxidation product, 87
stability and decomposition, 87, 88
synthesis and purification, 87
Tuberous sclerosis complex (TSC), α–methyl-L-tryptophan positron emission tomography findings, 124, 125
Vacuolating megaloencephalic leukoencephalopathy, see Megaloencephalic leukoencephalopathy
Vanishing white matter disease, 278, 279, 280
VBM, *see* Voxel-based morphometry
Voltammetry, conventional in vivo, 200, *see also* Neuromolecular imaging; BRODERICK PROBE®
Voxel-based morphometry (VBM), magnetic resonance imaging in Parkinson's disease, 6
West syndrome,
flumazenil positron emission tomography, 135
fluorodeoxyglucose positron emission tomography, 133, 134
White matter, *see* Leukodystrophy; Magnetic resonance imaging; Magnetic resonance spectroscopy; *specific diseases*
Xeroderma pigmentosa, 276, 277
Zellweger syndrome,
magnetic resonance imaging findings, 269, 270
magnetic resonance spectroscopy findings, 251
pathology, 270